Basic
Microbiology

Basic
Microbiology

SEVENTH EDITION

WESLEY A. VOLK
School of Medicine, University of Virginia

HarperCollins*Publishers*

Sponsoring Editor: Bonnie Roesch
Project Coordination, Text and Cover Design: York Production Services
Photo Researcher: Karen Koblick
Production Manager: Michael Weinstein
Compositor: York Production Services
Printer and Binder: Von Hoffmann Press, Inc.
Cover Printer: The Lehigh Press, Inc.

Cover photos: Top left, radiolarian skeleton, © 1974 Peter Arnold, Inc.; top right, radiolarian, Indian Ocean,
© Peter Arnold, Inc.; bottom right, AIDS virus, HIV, © Peter Arnold, Inc.; bottom left, E-coli, © Peter Arnold, Inc.;
back cover, perineum condylomata accum. venereal wart (viral), © Peter Arnold, Inc.

For permission to use copyrighted material, grateful acknowledgment is made to the copyright holders on
pp. 587–588, which are hereby made part of this copyright page.

BASIC MICROBIOLOGY, Seventh Edition

Copyright © 1992 by HarperCollins Publishers Inc.

Library of Congress Cataloging-in-Publication Data

Volk, Wesley A.
    Basic microbiology / Wesley A. Volk. — 7th ed.
        p. cm.
    Includes bibliographical references and index.
    ISBN 0-06-046849-1
    1. Microbiology. [1. Microbiology.] I. Title.
QR41.2.V64 1992
576—dc20

91-20886
CIP

92   93   94   9   8   7   6   5   4   3   2

To Joan R. Volk

Contents in Brief

Contents
in Detail

Preface

Basic Microbiology, Seventh Edition, is designed for students who are entering careers in the health science fields as well as for those who merely want to learn about the world around them and the role of microbiology in their everyday lives. Like its predecessors, the 7th edition presents the fundamentals of microbiology in a manner that is comprehensible to students with a minimal background in biology and chemistry. However, because it is not possible to understand what microorganisms are and how they are involved in our daily lives without knowing something of their molecular biology, the reader will find quite a bit of chemistry and genetics in this text.

APPROACH AND ORGANIZATION

As its title suggests *Basic Microbiology* describes the importance of microorganisms in all aspects of our lives, but it is primarily devoted to their function as disease producers and the ways in which humans can control such organisms. Thus, approximately half the text is devoted to immunology and medical microbiology, with the remainder concerned with introductory material on bacteriology, virology, and eucaryotic microorganisms.

The organization of this edition is essentially the same as in previous editions. All infectious disease organisms are grouped together according to their major portal of entry.

Unit I, consisting of 15 chapters, offers a broad overview of what microorganisms are, what they do, and how we are able to control them. This unit also includes a discussion of the properties of bacterial, plant, and animal viruses, as well as an enhanced coverage of eucaryotic microorganisms.

Unit II is comprised of 6 chapters dealing with host responses to infection and the use of antisera and vaccines to prevent infections.

The 8 chapters in Unit III provide descriptions of human diseases caused by microorganisms. Insofar as possible, virulence factors are described for each pathogen, permitting the reader to understand how a particular organism is able to cause an infection. Discussions of epidemiology and current or potential vaccines also are included for each infectious agent.

Unit IV provides a brief, 3 chapter, introduction to the microbiology of water and sewage, the microbiology of food and milk, and agricultural and industrial microbiology.

NEW TO THIS EDITION

Each chapter of the text has been thoroughly reviewed and revised to incorporate the most up-to-date information as it relates to basic microbiological concepts and their relationship to health and medical applications.

As a result of feedback from users of the previous edition, the parasitic worms are back again in this edition. Coverage is found in Chapters 15, 26, and 28.

Chapter 14 has expanded coverage of the role of viruses in human cancer and the function of antioncogenes, while Chapter 18 expands on the coverage of antibody synthesis. In Chapters 26, 27, and 29 the latest material on AIDS, hepatitis, and Lyme disease are included respectively. Coverage of food preservation has been expanded in Chapter 31.

A totally new feature has been added for this edition—**Microbiology Milestones**. These boxed features are now found in each chapter of the text and include interesting historical and epidemiological information. Written in an easy-to-read manner, these brief essays will lend interest and relevance to the basic textual coverage and help students appreciate the study of microbiology.

All of the line art in this edition has been redrawn for greater clarity and pedagogical effectiveness. In addition, you will notice the new trim size and design of the text, developed to increase the ease of use by the students.

ANCILLARY SUPPORT

A complete package of ancillary materials for both the instructor and student has been developed for this edition.

Instructor's Manual Written by George Wistreich of East Los Angeles College, this thorough manual that reviews the content of the text and offers teaching tips and resources is available upon adoption.

Testbank Written by Howard Lorson and Donna Hoel of Cayahoga Community College, this complete testbank is available in both printed and computerized formats. Testmaster, the computerized format, is available in both IBM and Macintosh versions.

Transparencies A set of 75 acetate transparencies of art from the textbook is available upon adoption.

Student Study Guide Prepared by R. Wilson Gorham of Northern Virginia Community College, this totally new learning guide will include chapter summaries, objectives, and numerous learning activities for students to use as a review. A mastery test for each chapter is included to help students prepare for examinations.

Microbiology Coloring Book A new text by Edward Alcamo and Lawrence Elson is now available to support the study of microbiology. Using the proven effectiveness of interactive participation by coloring on the part of the student, the understanding of the structure and function of microorganisms is greatly enhanced.

Lab Manual Microbiology in Practice 5th edition by Lois Beishir of Antelope Valley College was newly revised in 1991. It is proven as a practical guide for laboratory investigations for the allied health microbiology laboratory. It is accompanied by an instructor's manual of its own.

ACKNOWLEDGMENTS

I would like to thank the numerous users and nonusers of the previous editions of this text who have shared their comments with me about the text and their suggestions for improvements. In particular I would like to thank the following reviewers of the 7th edition manuscript for their thoughtful contributions. They include, David Jenkins, University of Alabama at Birmingham; Carl E. Knight, Eastfield College; Billie S. Lane, Chattanooga State Technical Community College; Michael L. Lockhart, Northeast Missouri State University; John C. Makemson, Florida International University; Francis Maxin, Community College of Allegheny County; Roger Nichols, Weber State University; and Clarence Wolfe, Northern Virginia Community College.

Wesley A. Volk

Basic
Microbiology

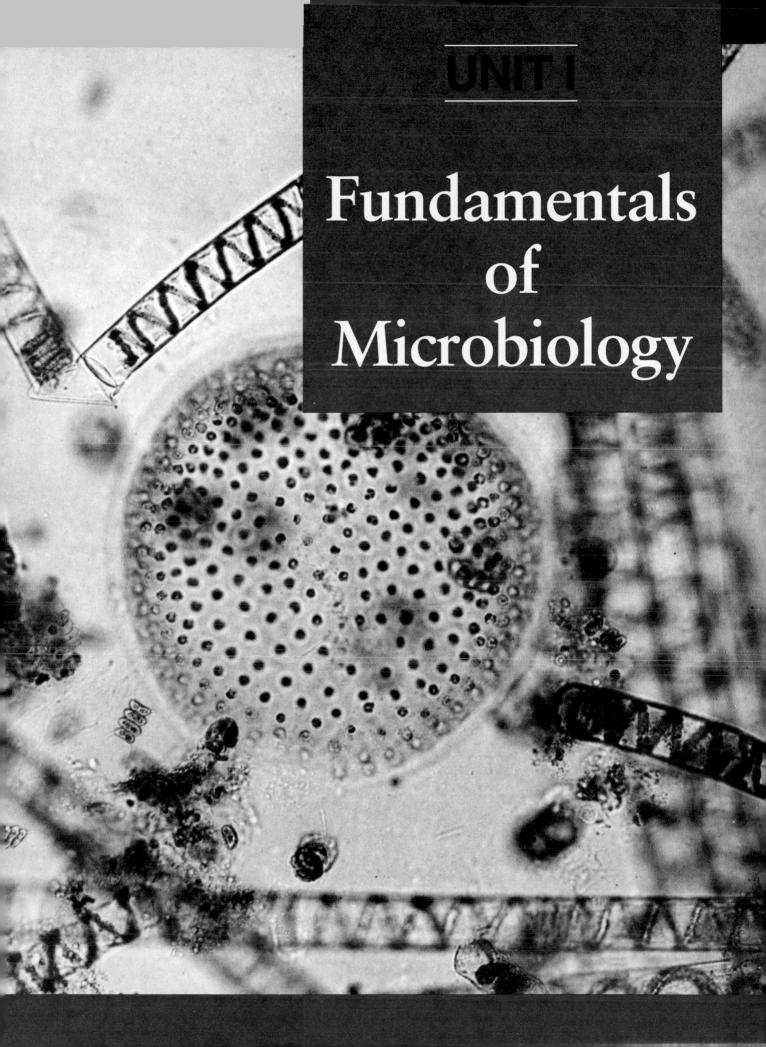

UNIT I

Fundamentals of Microbiology

Chapter 1

Introduction to Microorganisms

OBJECTIVES

After reading this chapter, you should be able to

1 Define bacteriology, mycology, protozoology, and virology

2 Discuss the theory of spontaneous generation and the types of experiments that disproved the generation of microorganisms from inert matter.

3 List several benefits to society that have resulted from a knowledge of microorganisms.

4 Explain the purpose of Koch's postulates.

5 Differentiate between the structure and gene organization occurring in procaryotic and eucaryotic cells.

6 Describe some proposed events that resulted in the origin of the first primitive procaryotic cell.

7 Outline the principal divisions of microbiology and the major interests of scientists in each discipline.

WHAT IS MICROBIOLOGY?

The word **microbiology** is a broad term meaning the study of living organisms that are individually too small to be seen with the naked eye. It includes the study of bacteria (**bacteriology**), viruses (**virology**), yeasts and molds (**mycology**), protozoa (**protozoology**), some algae, and some forms of life that do not fit well into any of these groups. Such forms of life are given the name **microorganisms**. Sometimes they are called microbes or, in the vernacular, germs.

It is difficult to imagine that just a little over a century ago Louis Pasteur and a small number of his colleagues were trying to convince the medical profession that these little organisms cause disease and that one kind of microorganism is responsible for the production of wine while a different organism causes wine to spoil. Once these ideas were proved and accepted, the study of microorganisms and their metabolic processes became an important science.

The information derived from studies in the field of microbiology has made possible great advances in our ability to control many infectious diseases. One example of this is seen in our modern-day methods of sewage disposal and water sanitation, which have virtually eliminated epidemics of typhoid fever and cholera from the Western world. Other examples include the complete elimination of smallpox and the major curtailment of paralytic polio through the use of mass immunizations. The availability of antibiotics represents yet another advance in our control of infectious diseases. In addition, microorganisms have been used to study many normal biochemical processes that subsequently have been shown to occur in higher forms of life. Thus, much that we now know about human metabolism was first observed in microorganisms. The field of molecular genetics—which explains how genes control the activity of a cell—had its origin in the study of microorganisms. This has given rise to the field of recombinant DNA genetics which is concerned with the techniques involved in transferring specific genes from any cell into a microorganism in such a manner that the newly acquired genetic material will be expressed, resulting in the synthesis of a new molecule such as human growth factor, human insulin, and many other biologically important substances.

It is clear, therefore, that the field of microbiology includes more than just the study of disease-producing microorganisms; it also comprises subdivisions such as agricultural microbiology, industrial microbiology, and sanitary engineering, as well as a study of our immune responses following microbial infections.

PRACTICAL APPLICATIONS OF MICROBIOLOGY

A practical knowledge of microbiology is particularly important in medicine and allied health fields. For example, one of the primary responsibilities of hospital personnel is the safeguarding of patients, and a large part of this responsibility consists of protecting the patient from hospital-acquired infections. Under normal hospital conditions, no place exists where the patient is not in some peril of microbial invasion.

One purpose of this text is to educate the student as to the existence of microorganisms, to describe their anatomy and physiology, and to show something of what they can do and how they can be controlled. Equipped with this background, the student will have no difficulty in understanding the various measures that are prescribed to control microbes. A cardinal rule of warfare is: Know the enemy. The individual who knows something of the peculiar attributes of each medically important species of microorganism will be able to take advantage of its specific vulnerabilities.

In the course of this study, the student will also learn something of the relationships between microbes and humans quite apart from the hospital situation—for example, the measures that the community takes to safeguard the health of its members and also the growing concept of preventive medicine.

It should be emphasized, however, that by no means are all microorganisms normally **pathogenic** (capable of causing disease) or **parasitic** (growing in or on a host such as a plant or animal). With the vast majority of microorganisms, humanity in general enjoys a state of peaceful coexistence. In fact, many microbes are involved in cycles of nature in which various materials are degraded so that they can be used again by higher forms of life. Such microbes are essential to the continuance of our species. Others are used to synthesize chemicals and antibiotics, while still other microorganisms are necessary for the manufacture of a broad array of food products and beverages (cheeses, pickles, sauerkraut, wine, and beer, to name a few). This book will describe organisms that are essential for our continued well-being as well as those that are responsible for human misery through their ability to cause disease.

EVOLUTION OF THE STUDY OF MICROORGANISMS

Although the existence of microorganisms had long been suspected, their presence was verified by micro-

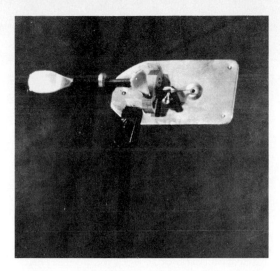

Figure 1.1 Leeuwenhoek's microscope utilized a single bi-convex lens to view bacteria suspended in a drop of liquid placed on a movable pin.

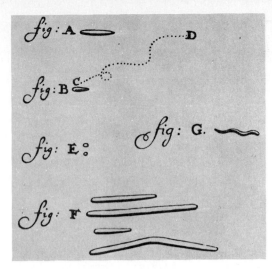

Figure 1.2 Although his microscope was capable of only 200- to 300-fold magnification, Leeuwenhoek was able to achieve these remarkable drawings submitted to the Royal Society of London.

scopic observation comparatively recently. About 1674, the invisible world of microorganisms was discovered by the Dutch merchant Anton van Leeuwenhoek (1632–1723).

Leeuwenhoek was an amateur scientist who devoted a great deal of time to his hobby of grinding lenses (see Figure 1.1). With these, he observed microorganisms in rainwater, seawater, scrapings from between the teeth, fermenting mixtures, and many other materials. Many of the minute organisms—which included protozoa, yeast, and bacteria—were seen in motion, and he referred to them as "animalcules." The accuracy of his drawings is surprising when one realizes that the highest magnification possible with his lenses was about ×300 (see Figure 1.2). This, contrasted with today's compound microscope, which provides a magnification of ×1000, makes Leeuwenhoek's observations even more remarkable. Judging from his drawings, Leeuwenhoek's lenses were the best of his time and, although he selfishly kept his lens-grinding techniques secret, he wrote voluminous letters to the Royal Society of London explaining his observations in great detail.

THE THEORY OF SPONTANEOUS GENERATION

Leeuwenhoek's observations, though their importance was not fully recognized until nearly 200 years later, came at a time when the theory of spontaneous generation was being challenged. This theory had been advanced to explain the occurrence of flies on putrefying meat, mice in decomposing fodder, and snakes in stagnant water. To disprove these ideas, it was necessary to carry out carefully controlled experiments. Some, such as the experiment of Francesco Redi, were quite simple. By placing meat in a jar and covering the jar with gauze to prevent direct access by flies, Redi proved that maggots were not spontaneously generated on decaying meat. As the meat decomposed, flies attracted by the odor laid their eggs only on the gauze covering. The absence of developing maggots on the putrefying meat provided crucial evidence against their spontaneous development, and by the middle of the eighteenth century, the theory of spontaneous generation of most forms of life had largely been laid to rest. But it was still widely believed that microorganisms arose spontaneously. John Needham (1713–1781), in a paper published in 1749, claimed that microorganisms arose in his infusions whether he boiled them, covered them, or took any other precautions he could devise. A great controversy developed during this period when Lazzaro Spallanzani (1729–1799), a priest, claimed that Needham had not taken sufficient precautions to prevent microorganisms in the air from entering his heated infusions. Like many people today, Spallanzani's contemporaries found it difficult to accept totally new concepts, so his arguments were widely ignored and the controversy concerning the spontaneous generation of microorganisms continued for another hundred years, until the middle of the nineteenth century.

The experiments of Louis Pasteur and John Tyndall provided the final disproof of spontaneous gen-

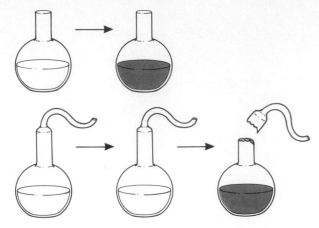

Figure 1.3 Pasteur's swan-necked flasks remained sterile because the bend in the neck excluded dust particles.

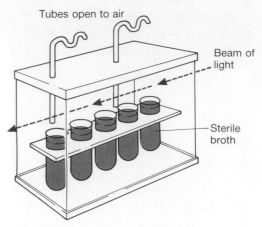

Figure 1.4 Similarly, broth remained sterile in Tyndall's dust-free incubation chamber. In both cases the broth was exposed to air, but dust was excluded.

eration. Pasteur poured meat infusions into flasks and then drew the top of each flask into a long, curved neck that would admit air but not dust (Figure 1.3). He found that after the infusions were heated, they would remain sterile indefinitely unless he broke the neck of the flask, which allowed dust to enter the infusion. He further demonstrated that if he placed a series of these flasks along a busy road, opened them, and then resealed them a few minutes later, microorganisms would grow in nearly all the flasks. On the other hand, if he performed the same experiment on the top of a mountain, where there was little activity, practically none of his flasks became contaminated.

Although Pasteur's experiments appeared to end forever the controversy of spontaneous generation, this idea was not laid to rest until the work of the English physicist John Tyndall became known. Tyndall had observed that when one shines a bright beam of light through air, the pathway of the light can be seen because it is refracted by the dust particles in the air. If, however, the air were 100 percent free of dust, one could not see the beam. Tyndall constructed a specially designed box (Figure 1.4). After the dust in the box had settled so that he could no longer see a beam of light pass through the air, he carefully placed tubes of sterile infusions in the box. As long as the air was not disturbed, the infusions remained sterile even though they were open to the air—again demonstrating that microorganisms existed on the dust particles in the air and that they were not spontaneously generated in hay or in meat infusions.

Tyndall's work resulted in a second interesting observation. He noted that on some occasions, only a few minutes of boiling would sterilize his hay infusions, whereas at other times even $5\frac{1}{2}$ h of boiling was not sufficient. He eventually recognized that there were heat-stable forms of bacteria—what we now know as **endospores.** He found that if an infusion was

MICROBIOLOGY MILESTONES

Pasteur's swan-necked flasks, some of which had remained sterile for four years, were displayed during a public lecture at the Sorbonne on April 7, 1864. Pasteur ended his lecture on that date with the following statements:

And, therefore, gentlemen, I could point to that liquid and say to you, I have taken my drop of water from the immensity of creation, and I have taken it full of the elements appropriated to the development of inferior beings. And I wait, I watch, I question it, begging it to recommence for me the spectacle of the first creation. But it is dumb, dumb since these experiments were begun serveral years ago; it is dumb because I have kept it from the only thing that man

cannot produce, from the germs that float in the air, from life, for life is a germ and a germ is life. Never will the generation of spontaneous generation recover from the mortal blow of this simple experiment.

Pasteur went on to add:

There is now no circumstance known in which it can be affirmed that microscopic beings came into this world without germs, without parents similar to themselves. Those who affirm it have been duped by illusions, by ill-conducted experiments, spoilt by errors that they either did not perceive or did not know how to avoid.

heated to boiling for 10 min, the heat-labile forms, or vegetative cells, were destroyed. If this infusion was then allowed to remain at room temperature overnight, most of the endospores would germinate to form vegetative cells that are easy to kill.

Fermentation

Pasteur's early studies and work were in chemistry and crystallography. However, as a result of his efforts to disprove the theory of spontaneous generation, he became deeply interested in the biological activities of microorganisms and devoted most of his time to this field. One of his first tasks as a microbiological troubleshooter was to find out why the production of alcohol from sugar beets would occasionally result in the formation of lactic acid instead of ethanol. He soon observed that there was more than one type of microorganism involved. The undesirable lactic acid fermentation resulted from contamination with rod-shaped bacteria; the ethanol production resulted from the activity of yeast cells. This observation was followed by others in which Pasteur proved that each type of bacterium is able to carry out the conversion of sugar to particular end products. Thus, one type of bacterium causes the formation of lactic acid from sugar, another forms butyric acid, and so on. Pasteur observed that these fermentation processes took place in the absence of air. He was the first to use the terms **aerobic** and **anaerobic** to describe, respectively, those requiring air and those unable to grow in the presence of air. Pasteur's discoveries were soon utilized in industrial fermentation, but the idea that microorganisms could also cause disease in humans and animals was less readily accepted.

Yet another new concept of the role of microorganisms came to the fore during the 1880s, when Serge Winogradsky isolated a number of bacteria that utilized only inorganic compounds for their growth. These organisms obtain their energy from the oxidation of compounds such as ammonia, nitrites, iron, and reduced sulfur; their carbon source is derived entirely from gaseous carbon dioxide. It is now known that such organisms, termed **chemoautotrophs,** are essential to convert inorganic compounds into forms usable by green plants.

The Germ Theory of Disease

The contagious nature of certain diseases, such as leprosy, has been recognized since biblical times. One of the first specific cases of the association of a microorganism with disease was made in 1834, when Agostino Bassi proved that a disease in silkworms was the result of a fungus infection. In a similar situation, Pasteur was called on in 1865 to study the silkworm disease that was destroying the silk industry in France. He established criteria whereby infected silkworm moths could be recognized microscopically. When females free of infection were used, the silkworms remained free of disease. Joseph Lister (1827–1912), an English physician, soon put to practical use the emerging concept that disease and infection were the result of invading microorganisms. He is credited with the first attempts to prevent postsurgical infection by the use of antiseptic technique. He used dilute phenol both in wound dressings and as an aerosol during surgical procedures. Crude as this method may appear, it marked the beginning of our efforts to control infectious microorganisms.

Once the microbial etiology of infectious diseases was accepted, scientists began to isolate and identify the causative agents of the many severe diseases of the day. Thus, the last quarter of the nineteenth century was a time of great activity that resulted in an impressive number of exciting discoveries. Robert Koch (1843–1910), introduced the scientific approach to the field of medical microbiology. He established rules (now known as Koch's postulates) that were necessary to establish a cause-and-effect relationship between a microorganism and a disease. In summary these rules stated:

1. The same organism must be found in all cases of a given disease.
2. The organism must be isolated and grown in pure culture.
3. The organisms from the pure culture must reproduce the disease when inoculated into a susceptible animal.
4. The organism must then again be isolated from the experimentally infected animal.

Working from these postulates, Koch was able to establish the etiology of such diseases as anthrax, cholera, and tuberculosis. It should be borne in mind, however, that Koch's postulates cannot always be used in their entirety to prove the etiology of a disease. For example, one cannot grow the organisms that cause leprosy in pure culture (although they can be grown in experimental animals).

Another important contribution from Koch's laboratory was the use of agar to solidify culture media. In a paper published in 1881, Koch stressed the fact that a solid surface is essential for separating mixtures of bacteria in order to obtain pure cultures. Koch observed that if a boiled potato was cut in half and exposed to air, small, discrete colonies of bacteria would soon appear. Each colony arose from the multiplication of a single bacterium and thus represented a **pure culture.** He reasoned that if a nutrient medium could be made solid, the isolation of pure cultures of

TABLE 1.1 A Few of the Pioneers and Their Major Contributions to Microbiology

Anton van Leeuwenhoek	1674	First person to observe microscopic life
Francesco Redi	1650	
Lazzaro Spallanzani	1776	All carried out experiments that refuted spontaneous generation
Louis Pasteur	1858	
John Tyndall	1877	
Edward Jenner	1798	Developed vaccine for smallpox
Agostino Bassi	1835	Associated silkworm disease with a fungus infection
Ignaz Semmelweis	1850	Provided evidence that childbed fever was transmitted to patients by physicians
Louis Pasteur	1857	Described microbial fermentations
	1866	Preserved wine by gentle heating
	1885	Developed vaccine for rabies
Joseph Lister	1867	Introduced use of dilute phenol as a disinfectant
Robert Koch	1876–1882	Isolated causative agents of anthrax, tuberculosis, and cholera
G. T. A. Gaffky	1880	Isolated causative agent of typhoid fever
Hans Christian Gram	1884	Developed Gram stain for bacteria
Ellie Metchnikoff	1884	Described phagocytosis and cellular immunity
Emil von Behring	1890	Developed an antitoxin for diphtheria
S. Kitasato	1894	Isolated plague bacillus
Walter Reed	1990	Described epidemiology of yellow fever

bacteria would become a simple procedure. In working toward this end, Frau Hesse, a worker in Koch's laboratory, apparently became the first to use agar to solidify a nutrient medium. The unique properties of agar making it so valuable to the microbiologist are that it will melt only at about the temperature of boiling water and then not resolidify until it is cooled to approximately 43°C. Thus, if 2 percent agar is added to a liquid medium, it can be dispensed while hot (and sterile) into tubes or petri dishes, where it will solidify when cooled. Koch's use of agar for this purpose not only proved to be a major advance in bacteriologic technique but once and for all silenced those critics who strongly proclaimed that there was only one species of bacteria that might assume different shapes and forms. Koch's pure cultures bred true, ending all resistance to the concept that the microbial world contained many different species of bacteria.

There have been many other scientists who earned recognition in the history of medical microbiology—in fact, some died after being infected by the organisms they were studying. Even a brief list would include such names as Ignaz Semmelweis, who was reviled and persecuted by his colleagues because of his claim that childbed fever (a frequently fatal infection caused by organisms of the genus *Streptococcus*) was transmitted to the patient by the physicians themselves; G. T. A. Gaffky, who isolated the typhoid bacillus; F. A. J. Loeffler, who identified the diphtheria bacillus; and S. Kitasato, who cultured the plague bacillus. Early in the twentieth century, Howard Taylor Ricketts and Stanislaus von Prowazek accidentally contracted typhus during the course of their research and died of it. There are many other scientific giants on whose shoulders we stand today, and the reader is encouraged to refer to Appendix A of this text for a more complete review of the history of microbiology. Table 1.1 provides a summary of the accomplishments of some early leaders in medical microbiology.

Today, over a hundred years after Pasteur's momentous investigations, our research in microbiology aims at understanding the etiology and control of cancer, genetic control of biochemical synthesis, and a host of other facts about microorganisms. To begin, the student must know the structure of microorganisms and how this structure may influence the properties of a cell.

PROCARYOTIC AND EUCARYOTIC CELLS

If one were to compare the biochemical activities of cells derived from such diverse sources as bacteria, spinach, and rat liver, amazing similarities would be seen. All the cells would be found to have their heritable characteristics coded in deoxyribonucleic acid (DNA). All would utilize one mechanism for the formation and storage of energy. All would have essentially identical methods of protein synthesis, nucleic acid synthesis, and polysaccharide synthesis. This incredible biochemical unity throughout the living world has only minor variations on the major theme. Yet when we examine cells morphologically, we find some important variations. These variations can be resolved into cells of two principal types: eucaryotic and procaryotic. Let us compare the cellular organization, gene structure, and chemical composition of these two very different cell types.

Cellular Organization

The most obvious difference between eucaryotic and procaryotic cells is the structural simplicity of the procaryotic cell. Procaryotic cells are able to divide into two independent cells in as little as 15 min, as compared with an 18- to 24-h dividing time for eucaryotic cells. The following sections examine some of the major properties that differentiate these two cell types.

Eucaryotic Cellular Organization

The presence of many intracellular membranes that compartmentalize the cell into a number of discrete organelles is a major characteristic of eucaryotic cells. Each organelle has specific functions necessary for cell maintenance. **Smooth** and **rough endoplasmic reticula** are membranous structures that act as sites of biosynthesis of certain proteins and hormones. The **Golgi apparatus** consists of stacks of smooth membranes involved in the processing (by adding sugar residues) of secretory proteins. **Lysosomes** are membrane-bound structures containing a variety of hydrolytic enzymes that take part in the breakdown of complex materials, including the killing of ingested microorganisms (see Chapter 17). **Mitochondria,** double-membraned struc-

tures, generate **adenosine triphosphate** (ATP), a compound in which energy is stored in the form of high-energy phosphate bonds (see Chapter 4). In photosynthetic plant cells, the light-trapping pigment, chlorophyll, is contained in a membrane-surrounded organelle called a **chloroplast,** which converts light energy into chemical bond energy (ATP). In addition, eucaryotic cells possess long, hollow cylinders called **microtubules,** which function in determining cell shape. All these organelles are surrounded by a cellular plasma membrane, which acts to contain the intracellular organelles within the eucaryotic cell. Some of the eucaryotic organelles may be seen in the schematic drawing in Figure 1.5 and the electron micrographs in Figure 1.6.

Procaryotic Cellular Organization

The type of cell characteristic of the bacteria and the cyanobacteria is called procaryotic. It is bounded by a plasma membrane but has no other separate membrane-bound organelles. Unlike eucaryotic cells, procaryotic cells do not possess mitochondria or, if photosynthetic, chloroplasts.

Cytoplasmic streaming—a flow of contents often seen in eucaryotic cells—is not seen in procaryotic cells, but it is probably not necessary in a cell sufficiently small for simple diffusion to move material around inside the cytoplasm. Most procaryotic cells possess a cell wall external to the cytoplasmic mem-

Figure 1.5 Diagram of the main structural components of a eucaryotic cell.

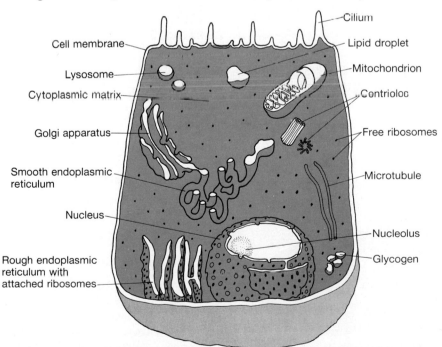

Cell membrane

Lysosome

Cytoplasmic matrix

Golgi apparatus

Smooth endoplasmic reticulum

Nucleus

Rough endoplasmic reticulum with attached ribosomes

Cilium

Lipid droplet

Mitochondrion

Centrioles

Free ribosomes

Microtubule

Nucleolus

Glycogen

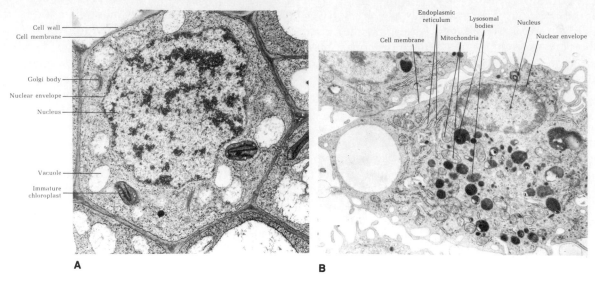

Figure 1.6 (a) Section of cell from stem of a young pea plant, *Pisum sativum* (× 9945). (b) Section of an animal cell, in this case a macrophage from a mouse (× 6240).

brane that contains muramic acid, a compound not found in eucaryotic cells (see Chapter 3). Also, procaryotic cells possess ribosomes smaller than those found in the cytoplasm of eucaryotic cells. And although many eucaryotic cells can engulf particulate matter by a process called phagocytosis, procaryotic cells are unable to take in any material unless it is first made soluble. Figure 1.7 shows a drawing of a procaryotic cell; prominent features are labeled. Figure 1.8 illustrates the simplicity of the procaryotic cell as compared with eucaryotic cell types.

Gene Structure and Organization

Eucaryotic Genes

Eucaryotic cells possess a true nucleus that is separated from the cytoplasm of the cell by a well-defined two-

Figure 1.7 Diagram of the major procaryotic cell components.

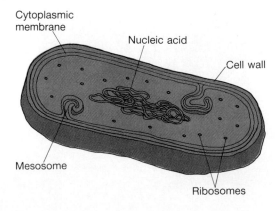

layer nuclear membrane, the nuclear envelope. Within this nucleus, the DNA and several kinds of proteins (histones) are organized into linear strands called chromosomes. The number of chromosomes in a eucaryotic nucleus is fixed for a given species. Certain fungi may have 1 or 2, humans have 46, and other species of plants and animals have other numbers (see Figure 1.9). When a eucaryotic cell divides, it goes through a rather elaborate process to provide each daughter cell with a full set of chromosomes. During this process, called mitosis, the strands of DNA replicate, providing each chromosome with two identical sets of genetic information visible as two sets of arms, or chromatids, on the chromosome. After the nuclear envelope disintegrates, a spindle forms, with the chromosomes on a plane halfway between the poles of the spindle. Microtubule fibers of the spindle attach to the chromosomes and the chromatids are pulled apart, each chromatid now becoming a new chromosome and being pulled to one pole or the other of the spindle, where the new nucleus will form. This nuclear division, called karyokinesis, is usually followed by cytokinesis, a division of the material outside the nucleus (the cytoplasm) somewhere between the two new nuclei. The two daughter cells produced through mitosis contain genetic information identical to that of the parent cell.

Procaryotic Genes

The procaryotic cell does not possess a true nucleus because its DNA is not separated from the cytoplasm by a nuclear envelope. Also, the DNA of the procaryotic cell does not exist in multiple distinct chromosomes, as in eucaryotic cells, but in a single continuous

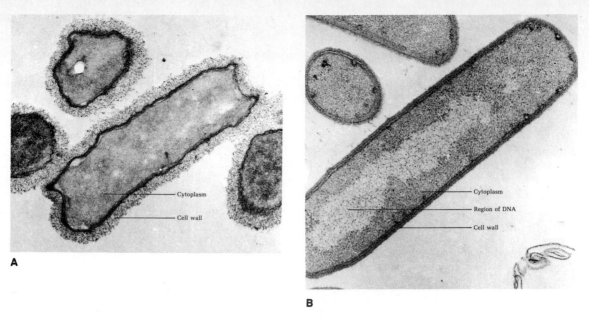

Figure 1.8 (a) Bacterium *Bacillus mucroides* (× 26,000). (b) Section of bacterium *Klebsiella aerogenes*. Original magnification × 20,730.

thread; however, many procaryotic cells also possess small pieces of extrachromosomal DNA, called plasmids, that confer certain auxiliary metabolic capabilities to the cells.

It is interesting to note that plasmids do not appear to be necessary for the normal replication of a bacterium, but they do encode for ancillary functions such as antibiotic resistance, toxins, sex pili, bacteriophage receptors, and cell surface components. Any one of these properties may be essential for survival in certain environments but none appear necessary for

growth. Some bacteria contain a number of different plasmids, and many of these may occur as multiple copies within a cell. As described in Chapter 8, plasmids may be readily transferred among related species, resulting in the spread of such characteristics as antibiotic resistance among large populations of bacteria.

The procaryotic chromosome exists as a covalently closed circular DNA structure completely lacking the histones that are characteristic of eucaryotic chromosomes. Moreover, as is discussed in Chapter 7, a procaryotic gene is always transcribed directly

Figure 1.9 Karyotype of a cell from a human male. (a) A chromosome spread during metaphase. (b) The 23 paired chromosomes. Note that for the male, pair 23 does not match.

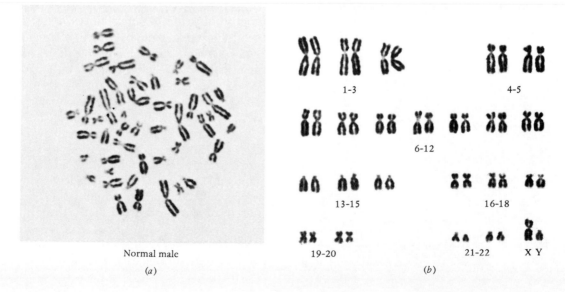

Normal male

(a)

1-3   4-5

6-12

13-15   16-18

19-20   21-22   X Y

(b)

into messenger ribonucleic acid (RNA), whereas many eucaryotic genes have intervening sequences (introns) that are first transcribed into a precursor RNA. These sequences must then be enzymatically removed and the correct gene sequences spliced together to make a functional messenger RNA.

As one might surmise, the procaryotic cell does not require an elaborate mitotic process to ensure the distribution of its DNA to daughter cells. The DNA of a procaryotic cell is attached to the plasma membrane at a limited number of sites. After chromosome separation has occurred, the cytoplasmic membrane grows inward between the sites of DNA attachment, dividing the cell into two identical daughter cells. This process is called **binary fission** (Figure 1.10).

Chemical Composition

Although there are many similarities between these two cell types, each possesses unique structural components. For example, eucaryotic membranes routinely contain cholesterol (as well as other sterols), whereas no procaryotic organism can synthesize sterol compounds even though members of one bacterial genus, *Mycoplasma,* incorporate pre-formed sterols from the growth medium into their cytoplasmic membranes. The complex lipids (sphingolipids, cerebrosides) found

---

Figure 1.10 An electron micrograph ($\times$ 120,000) showing binary fission. Note that one wall is completed between two cells while another is just beginning to lay down a new wall to divide the two cells.

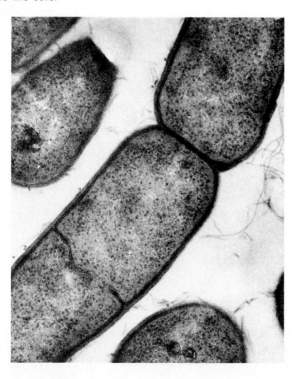

in eucaryotic cell membranes are not seen in procaryotes. In addition, most procaryotic cells are unique in that their cell walls contain muramic acid, a compound not found in eucaryotic cells.

Table 1.2 summarizes the major properties of eucaryotic and procaryotic cells.

THE ORIGIN OF LIFE

After discussing in considerable detail how the theory of spontaneous generation was disproved, this section will describe some of the current concepts of how microorganisms were originally spontaneously generated.

A study of fossil remains reveals that procaryotic organisms originated about 3.5 billion years ago. It has been hypothesized that over a period of millions of years the high temperatures, ultraviolet light from the sun, and the discharges from electrical storms catalyzed the formation of organic amino acids from the inorganic gases in the primordial earth. The fact that this could occur was demonstrated by Stanley Miller and Harold Urey when they mimicked the temperature and electrical discharges within a closed system containing water, hydrogen, methane, and ammonia. These and other experiments showed that most of the known amino acids, as well as other organic molecules, were produced. During additional millions of years, such amino acids combined to form peptides, polypeptides, and, eventually, proteins. Other organic compounds also were formed until finally there came into being a mixture capable of directing the synthesis of simple organic molecules. Once this stage was reached, it was only a matter of time until such molecules were able to associate into a self-duplicating unit that eventually became the original procaryotic cell.

A more recent theory concerning the origin of life proposes that life began as an RNA world. This concept followed the discoveries of Thomas Cech and Sidney Altman (corecipients of the Nobel Prize in chemistry in 1989) who demonstrated that RNA does possess considerable enzymatic ability in that it can catalyze some reactions in much the same way as protein enzymes do. Thus, according to RNA-world proponents, about 4 billion years ago all available genetic information was encoded in RNA and all metabolic reactions were catalyzed by RNA, and it was not until many millions of years later that RNA was translated into protein and transcribed into DNA (with no good concept of which occurred first). The problem, however, appears to be unsolvable because the existence of an RNA-world requires that a functional RNA capable of reproducing itself would have to be formed

TABLE 1.2   Comparison of Eucaryotic and Procaryotic Cells

	Eucaryotes	Procaryotes
Structure		
Organelles	Many types always present	Never
Nucleus	Nuclear membrane and spindle apparatus	DNA in contact with cytoplasm
Composition		
Lipids		
Complex phospholipids and sphingolipids	Common	Very infrequent
Sterols	Always	Only in mycoplasma
Ribosomes	80S = 60S + 40S subunits	70S = 50S + 30S subunits
Cell wall	Not present or of cellulose	Peptidoglycan with muramic acid
Genetic organization		
Chromosomes	Many	One circular (+ plasmids)
Diploidy	Usually diploid	Haploid
Histones and nucleosomes	Present	Absent
Coupling of transcription and translation	Separated in time and space; long-lived messages	Coupled Short-lived messages
Colinearity of gene and product	Intervening sequences often present	Colinear

by spontaneous chemical reactions, and the probability of this occurring seems infinitesimally small. However, as stated by Gerald F. Joyce from the Scripps Clinic in La Jolla, Calif., "There is no theory of the origin of life that is without problems."

In all likelihood, these first cells had extremely limited synthetic abilities and were able to exist only because they were immersed in a rich milieu of organic compounds. But once this "cell" was able to reproduce itself, the stage was set for the evolutionary process to begin, which through eons of trial and error evolved into a myriad of different cells, each fitting into a specific niche or environment. We can also suppose that because of the absence of oxygen, early procaryotes were anaerobic. During their subsequent growth, the cyanobacteria (photosynthetic procaryotes) evolved, with their ability to capture energy from the sun and to use that energy for the photolysis of water, resulting in the formation of gaseous oxygen. Then, as oxygen accumulated in our atmosphere, conditions were established for the emergence of the eucaryotic cell type.

Did this evolutionary process occurring over billions of years result in the selection of just two types of cells—procaryotic and eucaryotic? As for many questions, the answer to this is both "Yes" and "No." You will learn from reading this text that there are many variations of procaryotic cells, all of which possess the major characteristics listed in Table 1.2 for procaryotic cells. They vary, however, in shape, size, cell wall structure, and metabolic characteristics. In addition, one group of procaryotes possesses a number of biochemical molecules that are characteristic of eucaryotic cells. It has been proposed that these microorganisms be placed in a separate kingdom called the Archaebacteria. Similarly, there are many variations of eucaryotic cells as exemplified by the presence or absence of a cell wall (plant and animal cells) and by profuse differences in cell functions. Keep in mind, however, that the evolution of all of this heterogeneity took billions of years, as depicted schematically in Figure 1.11.

The Microbial World

In this introductory chapter, we have thus far listed some of the early events that established the field of microbiology and have described the major characteristics of procaryotic and eucaryotic cells. It seems now to be an appropriate time to delineate just what the study of microbiology entails.

Microbiology encompasses the study of all procaryotic organisms that have been placed in a separate kingdom termed the **Monera**. In addition, the microbial world contains the unicellular eucaryotes included in the algae and protozoa. Such organisms have been placed in a separate kingdom designated as the **Protista**. There is also, however, a very large group of multicellular eucaryotic worms, termed the **Helminths** that are usually included in a study of the microbial world. And, finally, there are the filamentous and unicellular eucaryotic cells, which are placed in the king-

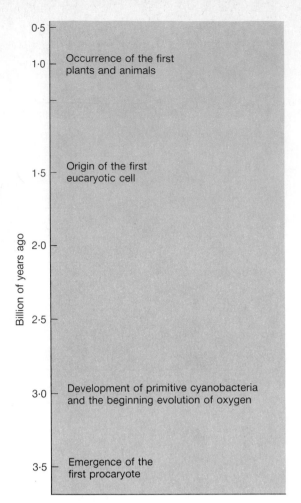

Billion of years ago

0.5

1.0 — Occurrence of the first
plants and animals

1.5 — Origin of the first
eucaryotic cell

2.0

2.5

3.0 — Development of primitive cyanobacteria
and the beginning evolution of oxygen

3.5 — Emergence of the
first procaryote

Figure 1.11 Evolutionary chart depicting the sequential biological events evolving since the "big bang" occurred at least 4.6 billion years ago.

dom **Fungi.** Microbiology also involves the study of viruses, which do not really fit into any of the above kingdoms. The following sections will provide a brief description of each of these various facets of microbiology.

Eucaryotic Organisms

Fungi

The fungi include organisms known as yeasts and molds. Some grow exclusively as single-celled yeasts, others solely as multicellular molds, and a few (particularly some fungi that produce human disease) exist as either yeasts or molds depending on their conditions of growth. The exceedingly complex classification of the fungi is based in part on the appearance and method of formation of their sexual spores, the type and appearance of their asexual spores, and the overall appearance of the entire organism.

Yeasts The microscopic one-celled fungi known as yeasts characteristically reproduce by forming buds on the mother cell that, when mature, pinch off to become new single yeast cells (see Figure 1.12). The majority of yeasts also produce sexual spores following the fusion of two separate cells. Many yeasts convert carbohydrates to ethyl alcohol and, by virtue of this characteristic, have contributed to the amenities of civilization through their use in the manufacture of alcoholic beverages (*Saccharomyces cerevisiae* and *Saccharomyces carlsbergensis*). Some strains of *S. cerevisiae* are also used to raise (leaven) bread because of their ability to produce large amounts of gaseous carbon dioxide in the bread dough. Only a few yeasts produce disease in humans.

Molds The multicellular fungi known as molds are considerably more complex than yeasts. Although individually they are microscopic organisms, many molds become readily visible as "mildew" on clothes, food, and leather in damp weather. Molds develop characteristically as branching, hairlike growths, and most form both sexual and asexual spores. Figure 1.13 illustrates typical cell branching as it occurs in molds. Some molds are responsible for the flavor of fine cheeses (Roquefort, Camembert, and Brie), and one, *Penicillium chrysogenum*, is the source of the antibiotic penicillin. On the other hand, a few molds (and yeasts) cause serious disease in humans.

Protozoa

Morphologically, protozoa exhibit a wide variety of shapes and sizes. Some are oval or spherical, others elongated, and some may change shape as they move along a surface. Some species may be as small as 5 or 10 μm in diameter, whereas others will reach diameters of 1 to 2 mm and thus be visible to the unaided eye. Although these are definitely eucaryotic organisms, they do not fit well into a classification with either plant or animal cells and are classified as Protista. Most protozoa are free-living organisms that are involved in the recycling of organic matter. Some, however, cause serious diseases in humans and animals, such as malaria, dysentery, trypanosomiasis, and leishmaniasis. These diseases are discussed in Unit III.

Most protozoa can reproduce both asexually as well as sexually and, although they do not form spores as do the fungi, many can secrete a thick coating around themselves to protect them from an adverse environment. We shall describe these organisms more fully in Chapter 15.

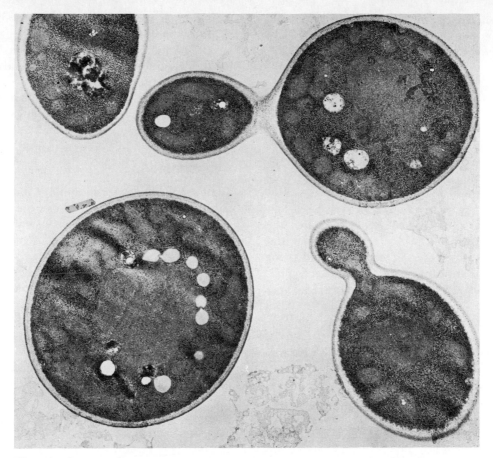

Figure 1.12 An electron micrograph of the beer yeast *S. carlsbergensis*. The round cell at the left is not budding. The cell on the bottom right has started a bud, whereas the cell on the top has completed budding and has laid down a cell wall that separates the bud from the mother cell.

Figure 1.13 Hyphal strands of *Microsporum canis* (× 250).

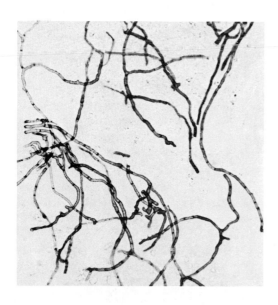

Multicellular Animal Parasites

Since some of these parasites may be several meters in length, they certainly do not qualify as microorganisms. They do, however, infect hundreds of millions of humans as well as a large percentage of animals, causing unimaginable misery in many parts of the world. They are referred to collectively as the helminths and may be subdivided into two large groups known as the flatworms and the roundworms. All have very complex life cycles, frequently requiring two separate species of animal hosts to complete their reproductive cycle. It is also noteworthy that the life cycle of even the largest of these organisms has a microscopic stage, and it is usually this entity that serves as the infectious agent in the dissemination of the disease.

Many of the human diseases caused by these worms have rather exotic names such as filariasis, loa loa, schistosomiasis, while others are more generally known in the Western world as tapeworms, hookworms, pinworms, and trichinosis. The life cycles and epidemiology of many of these multicellular parasites are described in Unit III.

Procaryotic Organisms

Bacteria

Bacteria are all unicellular, even though in some cases aggregates may appear to be multicellular. Although all are microscopic, bacteria are much smaller than the protozoa or the true fungi.

How do we measure a microorganism? It obviously cannot be measured in feet and inches any more than the height of a person can be measured in miles. The yardstick of the microscopic world is the micrometer (symbol: μm). One micrometer equals one-thousandth of a millimeter. There are 25.4 mm in 1 in. Therefore, 1 μm equals 1/25,400 in. A pinhead is about 2 mm in diameter. A typical protozoan, *Paramecium*, is about 200 μm long. Therefore, about 10 of these organisms, lined up end to end, would span the diameter of the pinhead.

If these creatures seem to be small, they are gigantic when compared with bacteria. Some common bacterial cells are about 1 μm long: 2000 of them would reach across the pinhead at its widest point. At best, the usual laboratory microscope magnifies up to about 1000 times; it gives a good overall view of most bacterial cells but few details.

Of the range of microorganisms that this book covers, the bacteria receive the lion's share of our attention for a number of reasons. First of all, bacteria are a major cause of human diseases. Moreover, the more we learn about the vital processes of bacteria, the easier it is to see the logic of the techniques employed in their control and to understand the methods used to manage the other microorganisms. Finally, there is a fringe benefit in the detailed study of bacteria. The life processes of the bacterial cell are remarkably similar to those of every other living cell, including the cells of the human body. Therefore, some knowledge of these primitive organisms greatly enhances the student's understanding of human physiology.

Cyanobacteria

The cyanobacteria, formerly known as the blue-green algae, are distributed worldwide in both fresh and marine waters. They obtain their energy by a type of photosynthesis that is essentially identical to that carried out by higher plants. Cyanobacteria are, however, procaryotes and therefore have all the structural characteristics described for procaryotic cells. Thus, their chlorophyll is not found in chloroplasts, as in eucaryotic photosynthetic organisms, but rather in specialized photosynthetic lamellae called thylakoids (see Figures 1.14 and 1.15). Some filamentous cyanobacteria produce occasional large, thickened cells within the filament that are called heterocysts. The primary function of the heterocyst is to convert atmospheric nitrogen into ammonia (nitrogen fixation), thereby making the gaseous nitrogen available for cellular metabolism.

Viruses

Viruses are the smallest known agents that can direct their own replication. Viruses are ultramicroscopic, too small to be seen with the conventional microscope, although they can be visualized using the electron microscope.

As you will learn in Chapter 14, viruses require living cells for growth, and they reproduce within these

Figure 1.14 *Nostoc punctiforme* ($\times$ 80,000).

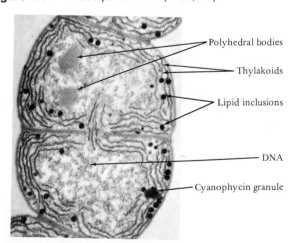

Figure 1.15 *Synechococcus leopoldiensis* ($\times$ 54,000).

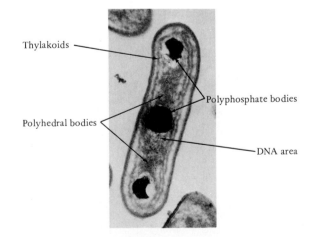

host cells by mechanisms different from bacterial binary fission. In essence, the minimal components of a virion (a single virus particle) are a molecule of RNA or DNA (but not both) enclosed in a protein coat. Some may also contain one or more enzymes, and some may be surrounded by a lipid membrane acquired when the virion budded from the host cell. All, however, lack an ability to synthesize ATP and none possess functional ribosomes. Thus, it is obvious that such agents must use components of the host cell in order to replicate. As a result, they cannot be classified with any other organisms and are separated into families on the basis of size, shape, presence or absence of a viral envelope, type of nucleic acid within the virion (DNA or RNA), the form of the viral nucleic acid (double-stranded, single-stranded, linear, or circular), and the existence of single or multiple pieces of nucleic acid within the virion.

Most cells, if not all, are susceptible to viral infections. Thus, in addition to animal viruses, there are numerous plant viruses as well as viruses that infect procaryotic cells. Those infecting procaryotic cells are called **bacteriophages** to differentiate them from the plant and animal viruses that infect eucaryotic cells. The structure and replication of all of these types of viruses is described in Chapter 14.

HOW BACTERIA ARE NAMED

Chapter 12 in this text is concerned with the classification of bacteria; but even at this early stage of study the student will be exposed to a number of bacterial names.

Our present system of naming bacteria is derived from the classification of plants devised in 1735 by Carolus von Linnaeus, a Swedish naturalist. In this scheme, plants are grouped into various categories designated as orders, families, genera, and species. The final name has two parts, consisting of both a genus and a species name. Microbiologists have, in general, adopted this method for the naming of bacteria, and bacteria are specifically referred to by a genus and a species name. All bacteria within a given genus possess many similarities, and, based on minor variations, a genus may contain any number of species. For example, all members of the genus *Streptococcus* are morphologically and physiologically similar, but *Streptococcus pyogenes* and *Streptococcus mutans* can be differentiated from each other based on habitat and minor biochemical properties. Using this system, bacteria are referred to by a binomial name comprising the genus and the species to which they belong.

THE DIVISIONS OF MICROBIOLOGY

From the brief description of microbiology in this chapter, it is apparent that one can subdivide microbiology into a number of branches such as bacteriology, virology, mycology, and parasitology. One can also arbitrarily partition this broad subject into the functional subdivisions listed below.

Medical microbiology is the study of microorganisms that cause disease in humans and animals. In medical microbiology we are concerned with epidemiology, that is, how the organism is spread from one infected host to another; we are interested in knowing specifically how the microorganism causes disease after gaining entry to the host; and we want to know how to prevent the disease, or if that is not possible, how to treat the infection. Medical microbiology is discussed in Unit III.

Immunology is the study of host reactions to a foreign substance in the body. Until recently, the major foreign substance considered capable of entering a host was an invading microorganism. As a result, this term means the study of immunity, and immunologists (those who study immunology) have been grouped with microbiologists since the beginning of modern microbiology. We shall see in Unit II that immunology is no longer confined to the study of immunity to infectious diseases but includes allergic reactions to many substances, adverse reactions to organ transplants that result in their rejection, and—the newest aspect—immunity to malignant cells, or cancer cells.

Industrial microbiology includes a large variety of microbial activities. Many commercial products—such as alcoholic beverages, organic solvents, and antibiotics—are products of microbial activities. The industrial use of microorganisms is a tremendously large field; we shall discuss a number of these applications in Unit IV.

Agricultural microbiology encompasses the study of microorganisms that are either harmful or beneficial to the production of agricultural products. Workers in this profession are interested in diseases of plants and livestock, as well as in the role of bacteria in soil fertility. Thus, the agricultural microbiologist should be knowledgeable in plant pathology, plant physiology, and veterinary diseases.

Other specialties covered by microbiology include food microbiology, dairy microbiology, soil microbiology, and the very important field of water sanitation and sewage disposal. Table 1.3 summarizes the major divisions of microbiology.

The reader will note that there is a short list of supplementary reading at the end of each chapter.

TABLE 1.3 Divisions of Microbiology

Mecical microbiology	Industrial microbiology
Agricultural microbiology	Sewage disposal
Dairy microbiology	Water sanitation
Food microbiology	Immunology
Soil microbiology	Virology

Whenever possible, readily available books or review articles are listed to supplement the information covered in this text. Other textbooks of microbiology available in your library may add to your understanding of the role of microorganisms in our ecology and as causes of human disease. The supplementary reading list at the end of Chapter 1 lists a number of such textbooks that the reader may wish to consult.

Microbial genetics is not really a separate division of microbiology because recombinant DNA technology is being used to produce new strains of bacteria, yeast, and mammalian cells that will benefit all mankind. We shall have occasion to describe a number of such examples later in this text, but a few cogent examples would include the following: (1) the production of human insulin by an engineered strain of *Escherichia coli;* (2) the synthesis of a vaccine for hepatitis B by the yeast, *Saccharomyces cerevisiae;* and (3) the construction of a strain of *Pseudomonas* that will lower the temperature at which frost will kill fruit trees. There are many more examples, but to complete our introduction, let us look at just one incredibly exciting aspect of genetic manipulation—human gene therapy. This involves the introduction of genes into an individual's cells in an attempt to correct a genetic defect. The first of such attempts was to introduce a gene that encodes for the enzyme adenosine deaminase into bone marrow stem cells of patients who are born without a functioning gene for this enzyme. Such individuals routinely die from the associated immune disorder, and it appears that the successful introduction of this gene has lead to a complete reversal of this defect.

There are many other human gene disorders that may lend themselves to such techniques, and it appears probable that we shall hear a great deal more about such procedures in the future.

An appropriate conclusion to our introduction of microbiology is shown in Table 1.4 where twentieth-century Nobel prize winners are listed. A brief review of their accomplishments illustrates the tremendously important biological advances that have resulted from research in microbiology.

TABLE 1.4 Nobel Laureates Whose Contributions Were Primarily in the Fields of Microbiology and Immunology

Year	Name	Contribution
1901	Emil von Behring	For his work on serum therapy, especially its application against diphtheria
1902	Ronald Ross	For his research on malaria
1905	Robert Koch	For his investigations and discoveries in relation to tuberculosis
1907	Alphonse Laveran	In recognition of his work on the role played by protozoa in causing disease
1908	Elie Metchnikoff Paul Ehrlich	In recognition of their work on immunity
1919	Jules Bordet	For his discoveries relating to immunity, especially concerning complement
1928	Charles Nicolle	For his work on typhus
1930	Karl Landsteiner	For his discovery of blood groups
1931	Otto Warburg	For his discovery of the nature and mode of action of respiratory enzymes
1939	Gerhard Donagk	For his reserach explaining the antibacterial effects of prontosil (leading to the discovery of the sulfa drugs
1945	Alexander Fleming Ernst Chain Howard Florey	For the discovery of penicillin and its curative effect in various infectious diseases and for the development of the procedures used in the commercial production of penicillin
1951	Max Theiler	For his discoveries and for a vaccine against yellow fever
1952	Selman Waksman	For his discovery of streptomycin, the first antibiotic against tuberculosis
1953	Hans Krebs	Discovery of the citric acid cycle
	Fritz Lipmann	Discovery of coenzyme A
1954	John Enders Thomas Walker Frederick Robbins	For discovery of the ability of poliovirus to grow in cultures of various types of tissue
1958	George Beadle Edward Tatum	For their discovery that genes act by regulating chemical events
	Joshua Lederberg	For his discovery concerning genetic recombination and the organization of genetic material of bacteria
1959	Severo Ochoa Arthur Kornberg	For discoveries in the synthesis of RNA and DNA

TABLE 1.4 *(Continued)*

1960	Sir MacFarlane Burnet Sir Peter Medawar	For the discovery of acquired immunological tolerance
1962	Francis Crick James Watson Maurice Wilkins	For discoveries concerning the molecular structure of nucleic acids
1965	Francois Jacob Andre Lwoff Jacque Monod	For their discoveries concerning genetic control of enzyme and virus synthesis
1966	Peyton Rous	For the discovery of tumor-inducing viruses
1958	Robert Holley H. G. Khorana Marshall Nirenberg	For their interpretation of the genetic code and its function in protein synthesis
1969	Max Delbruck Alfred Hershey Salvadore Luria	For their discoveries concerning the replication mechanism and the genetic structure of viruses
1972	Gerald Edelman Rodney Porter	For their work on the structure of antibodies
1975	Renato Dulbecco	For research on oncogenic DNA viruses and cancer
	David Baltimore Howard Temin	For their work on retroviruses and the discovery of reverse transcriptase
1978	Daniel Nathans Hamilton Smith Werner Arber	For their work on restriction enzymes (which are essential for genetic engineering)
1984	Cesar Milstein Georges J. Kohler	For developing the procedures to produce monoclonal antibodies
1987	Susumu Tonegawa	For his description of the genetics of antibody synthesis
1989	Michael Bishop Harold Varmus	For their discovery that virus-encoded cancer-causing oncogenes originate in eucaryotic cells

SUMMARY

Microbiology is the study of organisms that are individually too small to be seen with the naked eye. The earliest observations of microorganisms were made in the seventeenth century by Leeuwenhock. The next 200 years eventually saw the death of the theory of spontaneous generation and the birth of new concepts concerning the origin, industrial applications, and disease potential of microorganisms. Pasteur, Tyndall, Spallanzani, Koch, Semmelweis, Eberth, Kitasato, Loeffler, Winogradsky, Prowazek, and Ricketts stand out among the pioneers in microbiology.

Eucaryotic cells possess many intracellular membranes: smooth and rough endoplasmic reticula, Golgi apparatuses, lysosomes, mitochondria, chloroplasts, microtubules, and others; these compartmentalize the cell into numerous organelles. Procaryotic cells have no separate membrane-bound organelles, and they have smaller ribosomes than those found in the cytoplasm of eucaryotic cells. Also, procaryotic cells cannot ingest particulate matter, as can many eucaryotic cells. In addition, eucaryotic cells possess a true nucleus composed of a number of strands of DNA (chromosomes), all enclosed in a double-membrane structure (nuclear envelope) that separates the nucleus from the cytoplasm. Eucaryotic chromosomes also contain proteins (called histones), and the genetic information within a single gene is frequently separated by noncoding intervening sequences called introns. Procaryotes, on the other hand, contain a single circular chromosome that is not enclosed within a membrane. Moreover, the genetic material within a procaryotic gene is always contiguous, so that transcription leads directly to the formation of a messenger RNA without the necessity of processing out the intervening sequences. Procaryotic cells may also contain small pieces of extrachromosomal DNA, known as plasmids, which encode for many ancillary properties of the cell. Chemically, the eucaryotic cell membrane routinely contains sterols (such as cholesterol) and very complex lipids known as sphingolipids and cerebrosides. None of these are made in procaryotic cells. Most procaryotic cells, however, possess a cell wall external to their plasma membrane that contains muramic acid, a substance not found in eucaryotic cells.

The first primitive microorganism is believed to have originated about 3.5 billion years ago. Some theories propose that proteins were the first organic molecules that comprised the beginning of life, whereas others believe that it is possible that life began as an RNA-world. Subsequent evolution has resulted in a myriad of both procaryotic and eucaryotic cell types. Eucaryotic microorganisms comprise the fungi (yeasts and molds), algae, the protozoa, and the multicellular worms. Procaryotic cell types can be broadly categorized as bacteria and cyanobacteria. Viruses are not cells but consist primarily of nucleic acid surrounded by a protective coat. Viruses can replicate only when they are within a susceptible procaryotic or eucaryotic cell.

There are a number of different disciplines within the broad field of microbiology, of which the major ones are medical microbiology, immunology, industrial microbiology, and agricultural microbiology. Other subdivisions include food microbiology, dairy microbiology, soil microbiology, and water sanitation and sewage disposal.

QUESTIONS FOR REVIEW

1. What is a microorganism?
2. How did the following contribute to the development of microbiology: Leeuwenhoek, Redi, Spallanzani, Pasteur, Tyndall, Lister, Koch, Semmelweis, Winogradsky?
3. Differentiate between procaryotic and eucaryotic cells. What cell type is characteristic of bacteria? List the organelles found in eucaryotic cells and give the function of each.
4. What were the specific experiments carried out by Pasteur and Tyndall that disproved once and for all the theory of spontaneous generation?
5. List three benefits to our standard of living that have resulted from our study and knowledge of microorganisms.
6. What is meant by aerobic and anaerobic?
7. What are Koch's postulates and how are they used to establish the cause of a disease?
8. What unit is used to measure the size of microorganisms?
9. It is proposed that the universe came into existence 4.6 billion years ago. What time frames are assigned to the origin of the first primitive cell and its evolution to the first plants and animals?
10. List the major divisions of microbiology and the areas of interest within each division.
11. What are helminths?
12. What is a plasmid and what are some of the functions encoded in plasmids?

SUPPLEMENTARY READING

Brock TD, Smith DW, Madigan MT: *Biology of Microorganisms*, 4th ed. Englewood Cliffs, NJ, Prentice-Hall, 1984.

Brock TD: *Milestones in Microbiology*, Englewood Cliffs, NJ, Prentice-Hall, 1961.

Cairns-Smith AG: The first organisms. *Sci Am* 252(6):90, 1985.

Clark G: *Staining Procedures*, 3rd ed. Baltimore, Williams & Wilkins, 1973.

Dobell C: *Anton van Leeuwenhoek and His Little Animals*, New York, Harcourt Brace, 1960.

Fuerst R: *Microbiology in Health and Disease*, 15th ed. Philadelphia, Saunders, 1983.

Horgan J: In the beginning. *Sci Am* 264(2):116, 1991.

Jawetz E, et al: *Review of Medical Microbiology*, 18th ed. Los Altos, CA, Lange Medical Publishers, 1989.

Joklik WK, Willet HP, Amos DB, Wilfert CM: *Zinsser Microbiology*, 19th ed. Norwalk, CT, Appleton-Century-Crofts, 1988.

Lechavalier HA, Solotorovsky M: *Three Centuries of Microbiology*, New York, McGraw-Hill, 1965.

Sherris JC, et al: *Medical Microbiology: An Introduction to Infectious Disease*, 2nd ed. New York, Elsevier, 1990.

Volk WA, Benjamin DC, Kadner RJ, Parsons JT: *Essentials of Medical Microbiology*, 4th ed. Philadelphia, Lippincott, 1990.

Waldrop MM: Did life really start out in an RNA world? *Science* 246:1248, 1989.

Wilson AC: The molecular basis of evolution. *Sci Am* 253(4):164, 1985.

Chapter 2

Laboratory Equipment and Procedures

OBJECTIVES

A study of this chapter should result in a clear understanding of

1 The different types of microscopes used to study microorganisms and the resolving power of each type.

2 Techniques for studying live bacteria.

3 Types of stains used to visualize bacteria and the details of the Gram stain.

4 Techniques for the isolation of a pure culture and several types of media used to grow bacteria.

5 Terms such as pH, aerobic, anaerobic, facultative, and sterile.

There are certain basic techniques that the student of microbiology must learn to use in the laboratory. These techniques are used in growing bacteria, isolating them in pure culture (containing only one kind of bacterium), observing them, and, finally, identifying the organisms.

THE MICROSCOPE

Light Microscopes

Accounts in seventeenth-century scientific literature convey the excitement with which the Royal Society of London awaited letters and descriptions from Leeuwenhoek. We have advanced a long way in the past 350 years with respect to the equipment available for the microscopic study of microorganisms. The microscope you will use in the laboratory is no longer the "simple" (single lens) microscope of the type made and used by Leeuwenhoek but rather what we call a **compound microscope** because it has two sets of lenses (see Figure 2.1). One set is next to the object to be studied and, therefore, is called the objective. The other set, the ocular, is (as the name implies) the one next to your eye.

Both objectives and oculars are designed for different magnifications. The objectives usually are mounted in a rotating wheel known as a turret or revolving nosepiece; any one objective may be rotated into place depending on the magnification desired.

Oculars are made in different magnifications— that is, × 5, × 10, and × 15 (× meaning power to times actual size)—but most frequently the × 10 is used. How do these two lenses work together to give a higher magnification of the specimen than is possible with one lens alone?

Suppose that the objective is × 10, which means that it makes an image 10 times as big as the object under study. Here you have, in effect, a simple magnifying glass. Now, suppose that you view this magnified image through a second lens (the ocular), which

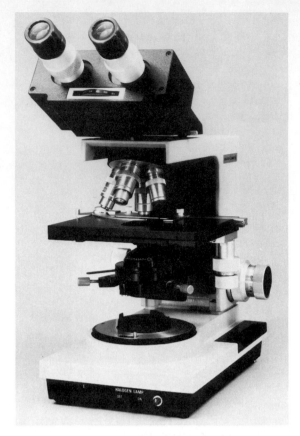

Figure 2.1 Modern binocular (two eyepieces) microscope. Note the mechanical stage to facilitate the movement of the slide and the built-in halogen lamp to illuminate the slide.

will magnify it five more times. How much larger than the object is the final image that you see? The answer is 50. The total power of any microscope, then, can be determined simply by multiplying the magnifying power of the objective lens by the magnifying power of the ocular lens.

The usual laboratory microscope has three objective lenses: low-power, high-power, and oil-immersion. The latter is the highest-powered of them all and is used almost exclusively to study bacteria. A drop of clear oil is placed directly on the cover glass

MICROBIOLOGY MILESTONES

Leeuwenhoek wrote more than 375 letters to the Royal Society of London describing his microscopic observations. In one such letter he wrote:

> I have had several gentlewomen in my house, who were keen on seeing the little eels in vinegar; but some of 'em were so disgusted at the spectacle, that they vowed they'd never use vinegar again. But what if one should tell such people in the future that there are more animals living in the skum on the teeth in man's mouth, than there are men in the whole kingdom? Especially in those who don't ever clean their teeth. For my part, I judge that all the people living in our United Netherlands are not as many as the living animals I carry in my mouth this very day.

———
From translation by Clifford Dobell: *Antony van Leeuwenhoek and His Little Animals.* London, Staples Press, 1932.

over the specimen, and the oil-immersion objective is lowered into the oil. The light rays traveling from the specimen through glass and oil do not bend as much as they would if they were to pass through glass and air; therefore, the image is clearer. The so-called high-power objective is actually intermediate in the power range; because it does not require oil, it is usually referred to as high-dry.

The oil-immersion objective magnifies approximately 97 times; hence, with a × 10 ocular, the total magnification of the objective would be × 970. It is possible to obtain somewhat higher power with different lens combinations, but there is a definite limit to magnification by this means. The high-dry usually has a × 44 objective; thus, with a × 10 ocular the overall magnification is 440. The low-power is a 10 × 10, or a hundredfold, magnification.

The **resolving power** of any microscope is a measure of its ability to discriminate between two adjacent objects. The absolute limit of the resolving power is roughly half the wavelength of the light used to illuminate the specimen. The wavelength of visible light ranges from 400 to 800 nm (nanometers), and thus, the smallest objects that can be observed in a light microscope must have a diameter of at least 200 nm (0.2 μm). Because most bacterial cells are only 0.3 to 1 μm in diameter, light microscopes are unable to provide information on their internal structure. Therefore, they are used mainly to visualize the shape of the cell and its reaction to different staining procedures.

Bacterial sizes are always expressed in metric units, as shown in Table 2.1.

Ultraviolet Microscope

A variation on the ordinary light microscope is the ultraviolet microscope. Because ultraviolet light has a shorter wavelength than visible light (200 to 300 nm), the use of ultraviolet light for illumination can increase the resolving power to twice that of the light microscope. The limit of resolution then becomes 0.1 μm. Because ultraviolet light is invisible to the human eye, the image must be recorded on a photographic plate or fluorescent screen. These microscopes use quartz

lenses, and they are too intricate and expensive for routine use.

Fluorescence Microscope

Fluorescent dyes are characterized by their ability to absorb short wavelengths of light and to emit the absorbed light at a longer wavelength. As a result, a bacterium that is stained with a fluorescent dye will become visible when viewed with ultraviolet light (the wavelength of which is too short to be seen with the human eye). Staining with auramine, a yellow fluorescent dye, is routinely done when looking at specimens that might contain *Mycobacterium tuberculosis,* the causative agent of tuberculosis. Because most bacteria do not stain with this dye, the mycobacteria are easily seen as brightly colored organisms against a dark background.

Fluorescence microscopy is also frequently used to detect or identify microorganisms by coupling a fluorescent dye to a specific antibody that will bind only to the target bacterium. The specificity of antigen-antibody reactions is thereby coupled to a sensitive and dramatic assay to determine the presence of a specific organism, because cell-bound fluorescence will occur only if that organism is present. This powerful technique can reveal the presence of any antigen on any cell as long as specific antibody is available. Application of this technique with larger eucaryotic cells can provide information about the cellular location of that specific antigen. Examples of fluorescent antibody techniques will be found throughout this text.

Darkfield Microscope

The darkfield microscope is used for viewing living bacteria, particularly those so thin that they approach the limit of resolution of the compound microscope. Many pathogenic spirochetes fall into this category, particularly the syphilis spirochete, *Treponema pallidum.*

The darkfield microscope differs from the ordinary compound light microscope only in having a special condenser that produces a hollow cone of visible light. The rays of light from this hollow cone do not

TABLE 2.1   Relationship of Metric Sizes

Size	Abbreviation	mm	μm	nm
Millimeter	mm	1	1000	1,000,000
Micrometer	μm	0.001	1	1,000
Nanometer	nm	0.000001	0.001	1

Note: 1 in. equals 25.4 mm and thus 25,400.

go directly up into the objective lens. Instead, they are reflected away at a slight angle from the top of the slide (see Figure 2.2). However, any light rays touching the specimen will be reflected directly into the objective, with the result that the specimen looks completely white against a black background (Figure 2.3). With no specimen in the field, the field would look dark, since there would be nothing to cause the light to be reflected up into the objective.

Phase-Contrast Microscope

The ideal way to observe living matter is in its natural state: unstained and alive. As a rule, however, a microscopic fragment of living matter (such as animal tissue or bacteria) is practically transparent, and individual details do not stand out. This difficulty can be overcome with the use of the phase-contrast microscope.

The principle of this device is complicated. For example, when an ordinary microscope is used, the nucleus of an unstained living cell is invisible. Since the nucleus is nevertheless present in the cell, its presence will alter very slightly the relationship of the light that passes through the nucleus with the light passing

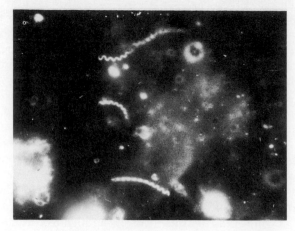

Figure 2.3 Darkfield view of exudate containing *Treponema pallidum,* the causative agent of syphilis. Because of its very small diameter, this organism would be exceedingly difficult to see using light microscopy and routine staining procedures.

through the material around the nucleus. This relationship, imperceptible to the human eye, is called a **phase difference,** and an arrangement of filters and diaphragms on the phase-contrast microscope will translate this phase difference into a difference in brightness—that is, into areas of light and shade that can be discerned by the eye—rendering the hitherto invisible nucleus (or any other structure) visible.

Electron Microscope

There are two types of electron microscopes, and each uses a beam of electrons rather than visible or ultraviolet light to delineate the specimen being examined. The transmission electron microscope (TEM) is used to view subcellular components (even nucleic acid molecules), and the scanning electron microscope (SEM) is employed to study the surface structure of a specimen.

Transmission Electron Microscope

The TEM (Figure 2.4) has been used by microbiologists for over 40 years to discover the structural details of microorganisms, including viruses. Because the wavelength of the electron beam used in this microscope is very short (approximately 0.005 nm), its resolving power is greater than 1 nm, and magnifications of as much as 1 million diameters are possible.

In general, the electron beam in a TEM passes through the specimen and then enters an electromagnetic objective, where the image is magnified before being projected onto a fluorescent screen. The image one sees is a result of the selective absorption of electrons by different parts of the specimen being viewed.

Figure 2.2 Light path through a darkfield microscope. Note that an opaque disk eliminates the light in the center of the beam, forming an intense hollow cone of light. The light reaching the object is at such an angle that it cannot enter the objective unless a particle (i.e., bacterium) in the specimen reflects the light upward. Thus, an empty field will appear dark whereas objects in the field will appear bright.

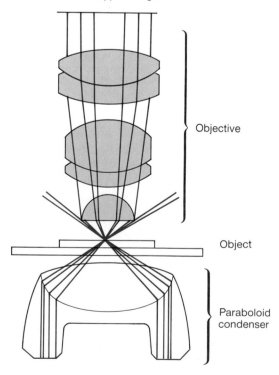

Objective

Object

Paraboloid condenser

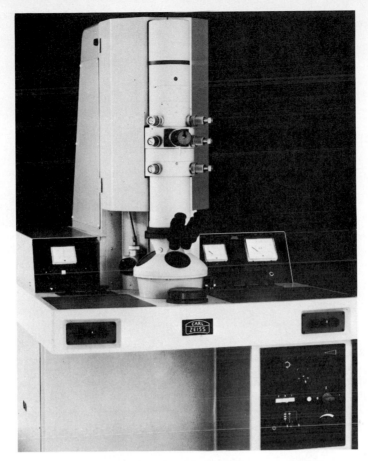

Figure 2.4 The operation of an electron microscope requires a great deal of experience and skill to obtain clear results. However, the magnification obtainable is many times that possible with the light microscope.

There are several major problems involved in using a transmission electron microscope: (1) it requires a skilled technician for operation; (2) it is a very expensive piece of equipment; (3) it requires the use of very thin specimens (which may be easily distorted); (4) the specimen being examined must be contained in a very high vacuum in order that the electrons may move effectively; (5) live specimens, therefore, cannot be examined; and (6) objects show no color.

This means that preparations must be dried and fixed with chemicals prior to study. When a specimen is dried under high vacuum directly from the frozen state, many of the distortions caused by conventional drying procedures can be avoided. Thin sections (0.1 μm or less) yield precise pictures, and examination of serial thin sections permits a detailed reconstruction of organelles or even whole cells in three dimensions.

An important development in electron microscopy is the use of uranyl acetate as a negative stain. Uranyl acetate is electron-dense, allowing structural

details of the surface of a cell to be observed while the background and empty areas remain opaque (Figure 2.5).

Shadow casting is another important technique for revealing surface details. An electron-dense metal such as platinum or chromium is vaporized under high vacuum and deposited at an angle on the preparation. The uncoated area on the opposite side acquires a shadow, and the resulting electron micrograph shows a three-dimensional effect (see Figure 2.6).

Freeze etching is a technique in which the unfixed sample is frozen at a very low temperature and then fractured by a sharp blow with a knife. The newly exposed surfaces are replicated by deposition of carbon metal. The fracture planes often extend along cell surfaces, yielding detailed views of surface structures, especially the intramembrane distribution of proteins (Figure 2.7). A novel and related technique is ion bombardment in which viruses or cells are eroded in an ion beam, allowing the examination of internal structures that were otherwise inaccessible (Figure 2.8).

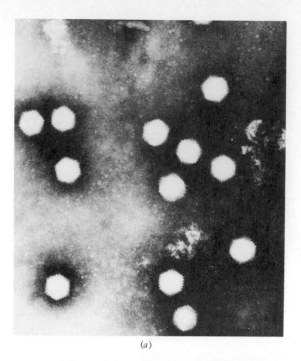

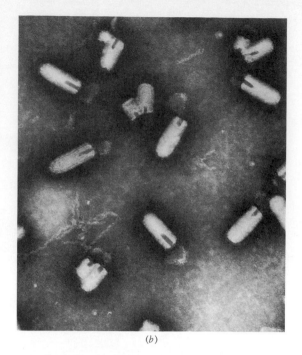

(a)

(b)

Figure 2.5 (a) Adenovirus negatively stained with uranyl acetate. Adenovirus exists as an icosahedron (20 faces), which appears in this micrograph as a six-sided structure. (b) Negative stain of vesicular stomatitis virus, which occurs as a bullet-shaped helical virus. Note that the flat end of many of these virions has opened, releasing part of the enclosed nucleic acid.

Figure 2.6 Electron micrograph of the spirochete *Spirochaeta stenostrepta* shadowed with platinum (× 12,100).

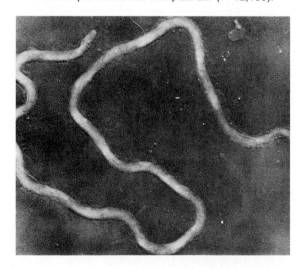

Figure 2.7 Freeze-etched micrograph of a large, gram-negative spirillum. Note how the outer membrane is removed in one area, exposing part of the cytoplasmic membrane.

Regularly structured protein overlying outer membrane

Outer membrane   Cytoplasmic membrane

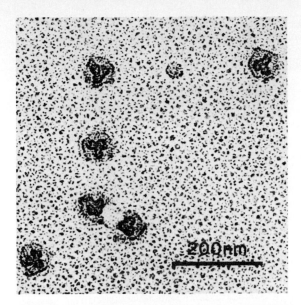

Figure 2.8 Ion bombardment of adenoviruses. Note how the argon etching has removed part of the external protein coat, revealing small internal spherical particles that contain the viral DNA and at least one viral protein.

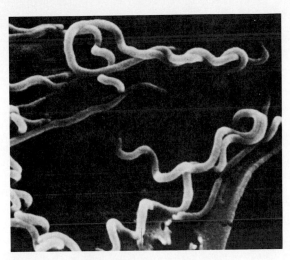

Figure 2.9 The three-dimensional qualities of scanning electron microscopy clearly reveal the corkscrew shape of cells of the syphilis-causing spirochete *T. pallidum,* attached here to rabbit testicular cells grown in culture (× 8000).

Scanning Electron Microscope

The scanning electron microscope employs a beam of electrons, but instead of being simultaneously transmitted through the entire field, the electrons are focused as a very fine probe or spot that is moved back and forth over the specimen. As the probe electrons strike the surface of the specimen, secondary electrons are emitted and then collected by a cathode ray tube.

The strength of the signal will be seen as dark or light areas on the collector, providing an image of the specimen's surface. Photographs taken of the cathode ray tube appear as spectacular three-dimensional micrographs, as is shown in Figure 2.9.

Figure 2.10 depicts the relationship between specimen size and the type of microscope that could be used to view it.

Figure 2.10 Relationship between size and type of microscope that could be used to visualize the specimen. Note that the transmission electron microscope (TEM) would be used on eucaryotic cells only to study fine intracellular details. SEM, scanning electron microscope; UV, ultraviolet microscope; mm, millimeter; μm, micrometer; nm, nanometer.

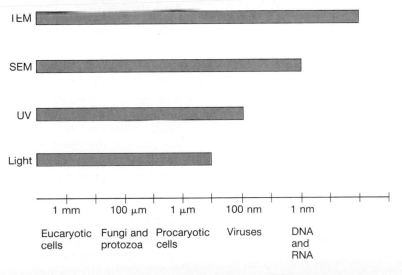

TABLE 2.2 Summary of Microscope Types

Type	Characteristics
Light microscope	Uses of visible light; normal magnification, × 1000; limit of resolution, 0.2 μm
Darkfield microscope	Uses a hollow cone of visible light so that only light refracted by the specimen enters the objective lens; magnification and resolution are the same as for the light microscope
Phase-contrast microscope	Detects phase changes in visible light as it passes through a specimen, which is more refractile (dense) than the surrounding medium; magnification and resolution are the same as for the light microscope
Fluorescence microscope	Uses ultraviolet light to view specimens that have been stained with a fluorescent dye that emits the absorbed ultraviolet light at a visible wavelength; magnification and resolution are the same as for the light microscope
Transmission electron microscope	Ultrathin sections of a specimen are prepared and negatively stained with an electron opaque heavy metal. The sections are placed in chamber from which the air is then evacuated. An electron gun produces a beam of electrons that travels through the vacuum to the specimen. Electromagnetic coils are used to focus the electrons from the specimen to form a magnified image that is viewed on a fluorescent screen or is recorded on a photographic plate. Magnification, × 10,000,000; resolution, 0.005 μm
Scanning electron microscope	A beam of electrons is directed onto the surface of a specimen and the emitted secondary electrons are collected to produce a three-dimensional image of the specimen on a cathode ray tube or photographic paper; magnification × 10,000; resolution, 0.02 μm

Table 2.2 summarizes the characteristics of commonly used microscopes.

TECHNIQUES FOR MICROSCOPIC STUDY OF BACTERIA

Living Bacteria

Living bacteria are difficult to see with the average light microscope because they appear almost colorless when viewed individually, even though the culture as a whole may be highly colored. However, it is sometimes necessary and desirable to look at living bacteria under the microscope, particularly to determine whether they are motile.

Laboratory Equipment and Procedures

One satisfactory means of observing living bacteria is in a hanging-drop preparation. A drop of liquid culture (or organisms suspended in water) is placed in the center of a coverslip. A special slide with a hollow depression in the center is used. The depression is ringed with a thin film of petrolatum and then turned upside down over the cover glass so that the drop of culture is in the center of the depression. The entire slide with the coverslip is then quickly turned over so that the drop of culture actually hangs from the coverslip down into the depression. The slide is placed on the stage of a microscope and the organisms are observed using the high-dry or the oil-immersion objective (see Figure 2.11). One can also use a normal flat slide to determine motility (i.e., the ability to move);

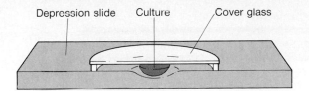

Figure 2.11 This side view of a hanging-drop preparation shows the drop of culture hanging from the center of the cover glass above the depression slide.

however, a thin ridge of petrolatum must be applied under the edge of the coverslip to prevent convection currents caused by evaporation from the wet mount.

Keep in mind that a bacterium is considered motile only if it seems to be going in a definite direction. Even nonmotile bacteria will bounce back and forth rapidly (**brownian movement**) due to bombardment from molecules of water.

Stained Bacteria

Bacteria are far more frequently observed in stained smears than in the living state. By **stained bacteria** we mean organisms that have been colored with a chemical dye so as to make them easier to see and study. In general, stained smears of bacteria reveal size, shape, arrangement, and the presence of some internal structures such as granules and spores. Special stains are used for observing capsules or flagella as well as certain internal cellular details. Stains that reveal chemical differences in bacterial structure may also be employed. These are called differential stains.

To prepare bacteria for staining, a small amount of culture is spread in a drop of water on a glass slide. This is called a **smear.** The smear is dried at room temperature, and the bacteria may be firmly bound (fixed) by passing the slide quickly (smear slide up) two or three times through the flame of a Bunsen burner. When cool, the smear is ready to be stained. Alternatively, many laboratories use methanol to fix the bacteria to the slide because heat may cause distortions in the appearance of the stained bacteria. This is done by placing a few drops of methanol on the air-dried smear and again allowing the smear to air-dry. The following sections outline a few of the more common procedures used to stain bacteria.

Staining Reactions

Stains are salts composed of a positive and a negative ion, one of which is colored (i.e., the chromophore). In basic dyes, the chromophore is in the positive ion (i.e., $dye^+ \ Cl^-$), whereas in acid dyes it is in the negative ion (i.e., $Na^+ \ dye^-$).

The marked affinity of bacteria for basic dyes is due primarily to the large amount of nucleic acid in the cell's cytoplasm. Thus, when the bacterium is stained, the negative charges in the nucleic acid of the bacterium react with the positive ion of a basic dye. Crystal violet, safranin, and methylene blue are a few of the basic dyes commonly used.

In contrast, acidic dyes are repelled by the overall negative charge of bacterium. Thus, staining a bacterial smear with an acidic dye has the effect of coloring only the background area. Since the bacterial cell is colorless against a colored background, this technique is very valuable for observing the overall shape of extremely small cells. This process is referred to as **negative staining.**

Another form of negative staining is used for the detection of capsules around cells. Because capsules are not usually stained by normal staining procedures, such cells are suspended in a solution of India ink. The India ink particles are too large to penetrate into the capsule and thus provide a black background in which capsules are seen as a clear area surrounding the cells.

Staining Techniques

Simple Stain This, as the name implies, is the simplest type of staining. One merely covers the fixed smear with any one of the following dyes: gentian violet, crystal violet, safranin, methylene blue, basic fuchsin, or other basic dyes. After 30 to 60 s the slide is washed off under the water tap and the smear gently blotted dry. It is now ready to be looked at under the microscope. After locating the stained specimen with the high-dry objective, one can observe it under oil immersion by placing a drop of oil directly on the stained smear and lowering the oil-immersion objective into the oil.

Gram Stain In 1884 the Danish physician Christian Gram devised a special stain that is probably the most important one used in bacteriology. It is a differential stain, so called because it divides all the true bacteria into two physiologic groups, thereby greatly facilitating the identification of a species. The staining procedure has four steps: (1) the smear is flooded with gentian or crystal violet; (2) after 60 s, the violet dye is washed off and the smear is flooded with a solution of iodine; (3) 60 s later, the iodine is washed off and the slide is washed with 95 percent ethyl alcohol for 15 to 30 s; and (4) the slide is counterstained for 30 s with either safranin (a red dye) or Bismarck brown. (The Bismarck brown usually is used by people who are color blind to red.) The length of time that each dye is left on the smear is not critical,

TABLE 2.3 Appearance of Cells during the Gram Stain

	Gram-positive	Gram-negative
After crystal violet	Purple	Purple
After iodine	Purple	Purple
After 95 percent alcohol wash	Purple	Colorless
After safranin	Purple	Red

and many laboratories have modified this procedure to allow each dye to remain in contact with the fixed organisms for only a few seconds. The time of decolorization, however, is critical because overdecolorization may lead to erroneous results. Also, many people prefer to use a 50:50 mixture of acetone and ethyl alcohol for the decolorization step because this solution acts faster than the 95 percent alcohol alone. Either decolorizing agent should give the same result.

The violet dye and the iodine form a complex aggregate. Some genera of bacteria are readily decolorized by the 95 percent alcohol wash, and these organisms are called **gram-negative** organisms; those that retain the complex are called **gram-positive** organisms. Because gram-negative bacteria are colorless after the alcohol wash, one always counterstains with a different color dye before looking at the smear under the microscope. The usual counterstain is the red dye safranin; hence gram-negative bacteria are red. Because alcohol does not wash out the blue dye complex from gram-positive cells, safranin counterstain has no effect, and gram-positive cells appear blue or bluish purple. Table 2.3 summarizes the appearance of both types of cells after each step of the Gram staining procedure. Table 2.4 lists the Gram reactions of some selected common bacteria.

It is now known that this staining difference results from the fact that the cell walls of gram-positive bacteria are different from those of gram-negative bacteria; the dye-iodine complex is trapped between the thick cell wall and the cytoplasmic membrane of the gram-positive organisms, whereas the removal of lipids from the gram-negative cell wall by the alcohol wash permits the dye-iodine complex to be washed out of the cell. One should be aware that only relatively young cultures should be used for Gram stains because older gram-positive cells frequently lose the ability to retain the crystal violet-iodine complex and, therefore, can appear as gram-negative organisms.

It is essential to know the Gram reaction (positive or negative) as well as the overall appearance of the bacterium in order to identify it. In addition, other general characteristics of an organism are associated with the Gram reaction. For example, most gram-positive bacteria are easily killed by low concentrations of penicillin, gramicidin, or gentian violet, whereas gram-negative bacteria are more resistant to these compounds but are considerably more sensitive to streptomycin. We discuss the reasons for these differences in more detail in Chapter 11.

Acid-Fast Stain The acid-fast stain (also called the Ziehl-Neelsen stain) is used primarily to help identify organisms in the genus *Mycobacterium*. This genus contains many disease-producing organisms, including the causative agents of tuberculosis and leprosy. The mycobacteria are called acid-fast because once they are stained with carbolfuchsin (a red dye), their unique chemical properties retain the dye even though the smear is washed with acid alcohol (95 percent ethanol containing 3 percent hydrochloric acid), a procedure that removes the stain from essentially all other organisms in a smear. One then counterstains with methylene blue and on examining the stained smear, all

TABLE 2.4 Gram Reaction of Selected Common Bacteria

Gram-positive	Gram-negative	
Bacillus	*Bacteroids*	*Neisseria*
Clostridium	*Bordetella*	*Pasteurella*
Corynebacterium	*Brucella*	*Proteus*
Lactobacillus	*Enterobacter*	*Pseudomonas*
Listeria	*Escherichia*	*Salmonella*
Micrococcus	*Franciscella*	*Shigella*
Staphylococcus	*Haemophilus*	*Vibrio*
Streptococcus	*Klebsiella*	*Yersinia*

acid-fast organisms will be stained red (from the carbofuchsin) and all non-acid-fast cells will appear blue.

Using this stain, the presence of only a few cells of *Mycobacterium tuberculosis* can be discerned in smears of sputum that contain other organisms normally present in the mouth and upper respiratory tract.

Other Stains A number of additional specialized staining procedures are used to stain parts of the bacterial cell. They include techniques for staining capsules, cell walls, nucleic acid, flagella, endospores, and other structures. All these staining procedures involve the use of two or more special dyes, but none is used routinely in the identification of a bacterium.

Table 2.5 summarizes the characteristics of the commonly used bacterial stains.

Preparation of a Pure Culture

Suppose that you have a specimen of material (e.g., saliva) that contains many different types of bacteria, and you wish to take from it one or more kinds of organisms for study. Since it is impractical to try to pick out individual bacterial cells from the material, a less direct method must be resorted to.

The first thing to do is to prepare a solid or semisolid plate composed of substances that the bacteria can use as nutrients. If you spread out the specimen of saliva on this nutrient plate, each bacterial cell, theoretically, will grow and multiply. Within a day, more or less, the **streak plate** will be covered with clumps, or **colonies,** of microorganisms visible to the naked eye, each colony composed of millions of organisms arising from the same cell.

Now, if you take a sterile inoculating needle, touch it to the colony of the particular organism to be studied, and then transfer these cells to a previously sterilized quantity of appropriate culture media (this is called **inoculation**), these bacteria will continue to multiply. Because this growth takes place apart from any other type of organism, it produces a **pure culture.** The properties of organisms in a pure culture can be studied by very rigid methods with accurate results because the influence of other living cells on the organisms is excluded.

Pure-Culture Techniques

To obtain the bacterial colonies from which pure cultures are made, two techniques are employed: the streak-plate and pour-plate methods.

The plate, on which the bacteria is spread, contains a mixture of nutrient substances and agar. Agar is a derivative of seaweed and has the property of dissolving in boiling water and then solidifying on cooling. The mixture of agar and nutrients is known as the **medium.**

In the **streak-plate method,** the sterile agar medium is melted, cooled to about 45°C, poured into a sterile petri dish (a glass dish about 3 in. in diameter), and allowed to solidify. Then, with the wire inoculating loop full of the mixed culture (i.e., the specimen of saliva or other material), streaks are made across the surface of the agar. There are several different methods for streaking, but in all methods the object is to deposit most of the organisms in the first few streaks. Thus, as one continues to streak the loop back and forth from one section to another of the petri dish, fewer bacteria will remain on the loop. If done properly, the last streaks should leave individual bacteria separated sufficiently from each other so that, after growth, colonies that have developed from individual bacteria will be well separated from each other (see Figure 2.12). A single colony can then be transferred to a sterile medium, resulting in a pure culture.

TABLE 2.5 Summary of Common Bacterial Stains

Stain	Description
Simple stain	Uses a single basic dye to color the bacterium
Gram stain	Differential stain using crystal violet, iodine, an alcohol wash, and safranin (a red dye). Gram-positive bacteria retain the crystal violet and appear blue. Gram-negative bacteria lose the crystal violet and appear red from the counterstain.
Acid-fast stain	Used primarily to stain the very lipid-rich mycobacteria. Acid-fast organisms retain the carbolfuchsin after washing with acid alcohol and appear red. Non-acid-fast organisms take up the methylene blue counterstain and appear blue.
Negative stain	Uses a single acidic dye in which the negatively charged dye is repelled by the negatively charged bacterium, leaving the organisms colorless against a stained background.

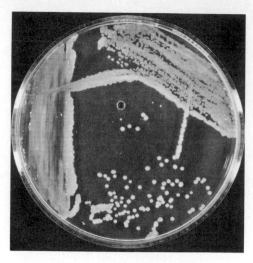

Figure 2.12 The streak-plate technique is used to isolate bacterial colonies for pure cultures. Each individual colony represents the progeny of a single cell. Note how the number of cells decreases as one goes from the heavy part of the plate to an area of isolated single colonies.

The **pour-plate method** consists of inoculating a dilution of the mixed culture into a test tube containing melted agar that has been cooled to 45°C. The contents are mixed to disperse the bacteria throughout the medium, and the mixture then is poured into a sterile petri dish and allowed to solidify. Alternatively, the inoculum is placed in the empty petri dish and the melted medium poured over it. The dish is swirled to mix before the medium solidifies. Colony growth takes place within the medium as well as on the surface (see Figure 2.13). The object in both procedures is to sep-

Figure 2.13 Pour plate prepared from dilution of a bacterial culture. Plates are prepared by inoculating tubes of melted and cooled nutrient agar with various dilutions of bacterial culture. The melted inoculated agar tubes are then poured into petri dishes and allowed to solidify. As with the streak plate, each colony represents the progeny of a single bacterial cell.

arate individual bacterial cells from each other so that they will grow into isolated colonies in a solid medium. One can take cells from one colony and obtain a pure culture. A second plate is usually restreaked with organisms from an isolated colony to ensure a pure culture.

Culture Media

The medium on which bacteria are grown will vary in its composition according to the requirements of the particular species. Some bacteria grow well on a very simple medium containing only inorganic salts plus an organic carbon source, such as glucose. Others may require a very complex medium to which blood or other complex materials are added. Almost all routine media can be purchased commercially as dry powders. Thus, to prepare a medium, one need only weigh out the desired amount of powder, add water, dispense, and sterilize before use. However, the student should have some knowledge of how media are prepared and which medium to use to grow a specific organism.

Infusion Media One of the most common media used for the routine cultivation of bacteria is called nutrient broth. This medium is prepared by boiling ground meat with water and filtering off the solid material to yield a clear liquid called an **infusion**. Peptone, which is partially degraded protein, and, frequently, 0.5 percent sodium chloride, are added to the liquid to provide carbon, nitrogen, and certain inorganic substances for bacterial growth. After the pH is adjusted so that the material is neither too acid nor too alkaline, the medium is ready for sterilization and, subsequently, for use as a culture medium. Usually the medium is dispensed in screw-cap tubes before it is sterilized. Many different species of bacteria can grow in this medium, resulting in a uniform cloudiness (turbidity). This turbidity is evidence of growth.

To grow bacteria on a solid surface rather than suspended in a liquid broth, 1.5 to 2 percent agar is added to any medium that one chooses to make solid. Once the agar has been melted in the medium, it can be dispensed in tubes that are then closed and sterilized. After sterilization and while the mixture is still fluid, the tubes are sometimes slanted so that, after solidification, there will be a large surface area to use for bacterial growth (Figure 2.14). This is referred to as a slant, in contrast to a somewhat fuller tube that has been allowed to solidify in an upright position, called a deep. The latter is inoculated by stabbing (an inoculating needle is pushed into semisolid medium in an upright tube, resulting in a growth of organisms along the stab).

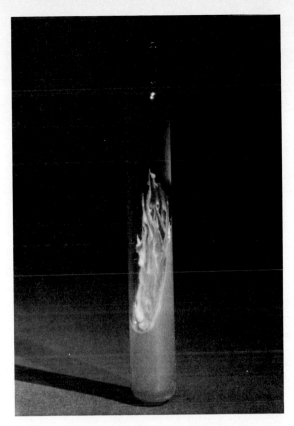

Figure 2.14 Agar slants are used to culture bacteria. Because the screw cap prevents desiccation of the agar medium, many bacteria can be kept for weeks or months on slants before it is necessary to transfer them to fresh medium.

A medium that contains complex substances such as beef extract, yeast extract, tryptones, and blood also may be called an **artificial** or **complex medium**. In contrast, we refer to one for which we can write the chemical formula of each ingredient in the medium as a **synthetic** or **defined medium**. Synthetic media may be very complicated and will vary widely according to the particular organism that one wishes to grow. For the most part, synthetic media are used only in growing microorganisms in research laboratories.

There are many other infusion media similar to nutrient broth or nutrient agar. Some of these are veal infusion, trypticase soy, and brain-heart infusion.

Carbohydrate Media Various sugars (carbohydrates) may be added to a nutrient broth base. This type of medium is used to determine if the bacterial species being identified is able to use a particular sugar for growth. Commonly employed in media of this type are such sugars as glucose, mannose, galactose, sucrose, maltose, and lactose. In addition, some sugar alcohols such as mannitol, glycerol, and dulcitol are used.

When a carbohydrate is used, many bacteria will form both acid and gas. Since we cannot see either the acid being produced or the sugar disappearing, a dye (called an indicator) that changes color in the presence of acid is added to the medium. Thus, to determine if the sugar in question was used, one need only observe the color of the medium after 24 to 48 h of growth. To determine if gas is produced, we can trap the gas so that it can be seen. This can be accomplished by placing a small test tube (upside down) in the liquid carbohydrate medium. If gas is produced, it will be trapped inside the tube (displacing the liquid medium), and we shall be able to see gas bubbles.

Selective and Differential Media Dyes frequently are added to a medium to inhibit the growth of certain bacteria while not interfering with the growth of others. This type of medium is called a **selective medium**, since it will select certain organisms. The addition of bile salts, or dyes such as brilliant green to a medium can make it selective for pathogenic gram-negative organisms because these additions will inhibit the growth of gram-positive bacteria.

An acid indicator may also be added to a solid medium so that a colony of bacteria that forms acid will be colored and can be differentiated from one that does not. This type of medium is called a **differential medium**. Both selective and differential media are particularly useful in the identification and isolation of enteric pathogens. For example, *Escherichia coli*, a normal inhabitant of the human intestinal tract, ferments lactose to form acids. On the other hand, many disease-producing organisms of the intestinal tract (such as the typhoid bacillus) do not ferment lactose to form acid. Thus, colonies of *Salmonella typhi* can be differentiated from *E. coli* on a differential medium such as Hektoen's agar, which contains an indicator that will be changed in color by the lactose users.

Enrichment Cultures Bacteria present in very small numbers in some natural environments are frequently difficult to isolate from the mixed population. If a suitable substrate and other conditions are provided that favor the growth of these organisms but are unsuitable for others, they can become dominant. Liquid media are used for enrichment cultures that provide nutrients and environmental conditions that favor the growth of the particular organism being sought but are not suitable for the growth of other types. After repeated transfers in enrichment media, differential plating methods can be used for pure culture isolation. For example, if N_2 is the only nitrogen source provided while other requirements are met, this would allow only bacteria that could use this form of nitrogen to grow. Other examples include selenite broth to enrich *Salmonella* and the presence of basic dyes to enrich gram-negative organisms.

Control of pH and Temperature

The pH of a medium is a measure of its acidity or alkalinity. By definition it is a measure of the activity of the hydrogen ion concentration. If we look at 100 percent pure water, the water is ionized so that it contains a concentration of 10^{-7} mole (mol) of hydrogen ions and 10^{-7} mol of hydroxyl ions per mole of water. Because they are exactly equal, the solution is neutral and the pH is expressed as the negative exponent of the hydrogen ion concentration, which in this case is 7. The thing to keep in mind is that for any aqueous solution, the product of the hydrogen ion concentration and the hydroxyl ion concentration must always be 10^{-14}. Thus, if we increase the hydrogen ion concentration 10-fold, so that it is now 10^{-6}, then the hydroxyl concentration will change to 10^{-8} mol per mole of water. (Note that 10^{-8} is 100 times smaller than 10^{-6}.) In this case we say that the pH is 6; since the hydrogen ion concentration is greater than the hydroxyl ion concentration, we refer to the solution as being acidic. Similarly, if we decrease the hydrogen ion concentration of a neutral solution by adding a 10-fold excess of hydroxyl ions, the hydrogen ion concentration will be 10^{-8} and the hydroxyl ion concentration 10^{-6}. This solution has a pH of 8 and is alkaline. As you can see, a pH below 7 is acidic, whereas a pH above 7 is alkaline. Most bacteria grow best at or near neutrality; therefore, before use, most media are adjusted to a pH near 7. A few bacteria grow at the extremes of the pH range (e.g., 8.5 or 2.0), so the pH must be adjusted according to the species being cultivated.

Temperature, too, must be controlled for the growth of bacteria. Some bacteria will grow at temperatures below 10°C, whereas others will grow at temperatures as high as 70°C (hot enough to cause a severe burn). However, the organisms that cause disease usually grow best at or near normal human body temperature, 37°C. Actually, most disease-producing organisms are killed when exposed to a temperature of 45 to 50°C for more than a few minutes. (See Table 2.6 for a comparison of Celsius and Fahrenheit temperatures.)

Oxygen Requirements

Bacteria can be divided into several general groups on the basis of their oxygen requirements. The first group, the **strict aerobes,** require free oxygen in order to grow. Since our atmosphere is approximately 20 percent oxygen, growing them is no problem as long as the bacteria are exposed to air.

TABLE 2.6 **Comparison of Celsius and Fahrenheit Temperatures**

Celsius	Fahrenheit
0°	32°
20°	68°
37°	98.6°
50°	122°
70°	158°
100°	212°
121°	249.8°

The **strict (obligate) anaerobes** will not grow in the presence of free oxygen and may actually be killed by its presence. The mechanism of this toxicity is discussed in Chapter 5. A number of techniques have been devised to grow anaerobic bacteria. One common method is to add a reducing agent, such as sodium thioglycolate, that will react with the free oxygen in the medium. In other cases, special cultural equipment is employed to remove the oxygen mechanically and replace the atmosphere with hydrogen and carbon dioxide. This may be done in a small, anaerobic jar (Figure 2.15) or, for laboratories involved in a large amount of anaerobic bacteriology, a large glove box may be used in which the atmosphere is maintained free of oxygen (Figure 2.16).

Figure 2.15 Petri dishes are placed inside the GasPak jar and water is added to the GasPak envelope. This produces hydrogen and carbon dioxide. The jar also contains palladium, which catalyzes the reaction of oxygen with hydrogen to form water. The carbon dioxide generated is to support the growth of carbon dioxide–requiring organisms.

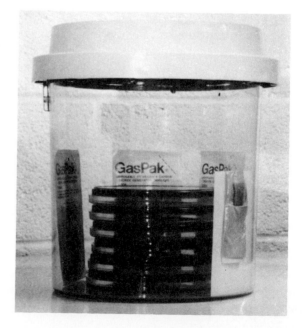

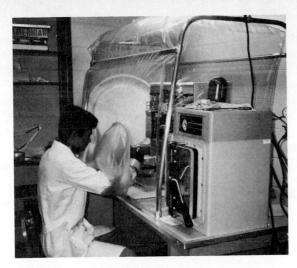

Figure 2.16 Anaerobic glove box containing an atmosphere of 85 percent N_2, 10 percent H_2, and 5 percent CO_2.

A subgroup of the obligate anaerobic bacteria comprises organisms that grow best with reduced oxygen tension but not necessarily under completely anaerobic conditions. Such organisms are designated as **microaerophilic,** and their preference for the presence of low levels of oxygen probably is associated with the involvement of oxygen in certain steps of intermediary metabolism, such as the biosynthesis of pyrimidines and unsaturated fatty acids.

A large group of organisms is designated as **facultative aerobes** or **facultative anaerobes.** These organisms can grow anaerobically and under such conditions will ferment carbohydrates to form stable fermentation products such as lactic acid, acetic acid, and so forth. When they are grown in the presence of air, however, the facultative organisms will change their metabolism to an aerobic one in which carbohydrates are oxidized to water and carbon dioxide.

Finally, there is a group of bacteria referred to as **aerotolerant** because these organisms will grow in the presence of air but do not possess an oxidative metabolism. They do not use oxygen in their metabolism but carry out a fermentative degradation of carbohydrates even in the presence of oxygen.

Sterilization Methods

In order that pure cultures may be produced, a medium must be sterile before inoculation; that is, we must be sure that there are no living organisms in the medium when it is inoculated. The method most commonly used to sterilize media is to place them in an autoclave (really a large and elaborate form of pressure cooker). The autoclave uses steam under pressure to raise the temperature of the material being sterilized to a level at which all forms of life will be killed. For routine sterilization, the autoclave is usually operated at 15 lb/in^2 of steam pressure. At this pressure the temperature will be 121°C. The usual exposure period at this temperature is 15 to 20 min. If large flasks of media are sterilized, a longer period will be required, because it takes time for the heat to penetrate the material.

Certain media (such as many carbohydrate media) that cannot stand high temperatures may be sterilized at a lower temperature by continuing the process for a longer period.

Enriching substances, such as blood serum, are frequently added to plain media, but these substances are unable to withstand the heat of sterilization. Material of this type can be sterilized by filtering the liquid through specially designed membrane filters, which have very small pores. Once the enrichment substance has been sterilized, it can be added to the medium when the latter is sufficiently cool but still liquid (45–50°C).

Glassware, test tubes, fermentation tubes, bottles, pipettes, flasks, and petri dishes are usually sterilized in an autoclave or a hot-air oven. The hot-air oven requires a higher temperature and a longer time to sterilize an object than does the autoclave, but it is usually more convenient, since glassware has to be dried after sterilization by the latter method (see Chapter 9). Hot-air sterilization usually is carried out at 170°C for 2 to 3 h; petri dishes usually are placed in cans or wrapped in paper, and test tubes and bottles are plugged with cotton or other enclosure to prevent later contamination by microorganisms in the atmosphere.

How Bacteria Are Identified

In order to begin the process of identification, a certain minimal amount of information is required. This minimal information includes

1. Size, shape, and arrangement of the organism;
2. Gram-staining reaction;
3. If motile;
4. The overall size and appearance of the bacterial colony.

With these minimal observations it is sometimes possible to decide to which group or family the unknown organism belongs, and sometimes even the correct genus can be chosen.

Further identification of genus and species requires biochemical information. The specific biochemical information required may vary from family to family. It may require that you determine which sugars are metabolized, whether the unknown organism can break down gelatin or urea, or even whether the organism can grow on a medium that contains ammonium salts as the sole source of nitrogen. It is not possible to know how much of this specific information will be required for a complete identification until one has decided to which family or hierarchy the unknown organism belongs. In some cases it is a very simple process—for instance, when a family contains only one or two genera and each genus is made up of only one or two species. In other cases it can be a very difficult task to decide to which of possibly 100 species your particular unknown belongs.

Immunologic tests are also used in the final identification of certain bacteria. One test involves mixing known antiserum with the unknown organism to see if the bacteria will clump (agglutinate). As an example, consider the disease typhoid fever. The blood of a person who has recovered from typhoid fever (or has been immunized against it) contains antibodies that causes *Salmonella typhi* to clump. These particular antibodies are specific and would have little or no effect on other bacteria. It is easy to see how blood samples containing specific antibodies can be used to identify an unknown microbe.

SUMMARY

Certain special equipment is required in working with bacteria, whose minute size is a determining factor in many of the methods that have been discussed.

Various types of microscopes have been introduced. The compound microscope, which provides a magnification of about 1000 times, is used most commonly. Darkfield, ultraviolet, phase-contrast, and electron microscopes can resolve microbes not visible under normal light. The resolution with the electron microscope is much greater than that with light microscopes, since it uses a beam of electrons instead of visible light to illuminate a specimen. Living bacteria are examined for motility in hanging-drop or sealed wet-mount preparations.

Staining procedures reveal various details concerning morphology and chemical structure. The Gram stain is the differential stain used most commonly. It permits the division of bacteria into two groups: (1) gram-positive, or those that retain the violet stain after washing with a decolorizing agent, and (2) gram-negative, those that lose the stain in contact with a decolorizer and then take the counterstain. The Ziehl-Neelsen, or acid-fast, stain is also a differential stain. It is most useful in identifying the genus *Mycobacterium*, and particularly *M. tuberculosis*. Various simple stains such as methylene blue, safranin, and malachite green may be used, depending on the information desired. Special stains are used to visualize other structural details including cell walls, flagella, nuclear material, cell membranes, inclusion granules, and capsules.

A culture is a medium containing living organisms. The medium provides the nutrients for bacterial growth, and various recipes for media have been devised to favor the growth of particular species. Selective and differential media are useful in isolating and distinguishing between some species. Other conditions that affect the growth of bacteria are pH, temperature, and atmosphere.

Cultivation of bacteria for practical purposes requires pure cultures. A pure culture contains only one species. For isolating bacteria in pure culture, two procedures are commonly employed: the streak-plate and pour-plate methods.

To ensure purity of cultures, all media and equipment used in the handling of cultures must be sterilized. The autoclave, which uses steam under pressure, is used most commonly. Other methods include hot-air sterilization for certain equipment and filtration for heat-labile media.

Species identification makes use of many characteristics. These include morphology, motility, biochemical characteristics, oxygen requirements, Gram-staining reaction, and, in some, immunologic characteristics.

QUESTIONS FOR REVIEW

1. What is meant by the resolving power of a microscope?
2. Compare the different microscopes in use. What are the advantages of each?
3. What magnifications are possible with the low-power, high-dry, and oil-immersion objectives of the compound microscope?
4. Explain the usefulness of shadow casting in electron microscopy.
5. What is a hanging-drop preparation? For what purpose is it usually used?
6. Define true motility. What is brownian movement?
7. What is the purpose of staining bacteria?
8. What are stains? Why are basic stains usually used in staining bacteria?
9. Explain the Gram-stain procedure.
10. Explain the acid-fast stain. What is the importance of this stain?
11. What is a culture? A culture medium? A colony?

12. Explain the methods commonly employed for obtaining pure cultures.
13. What is a synthetic medium? A selective medium? A differential medium? An enrichment culture?
14. Define pH.
15. How are chemical changes such as acid or gas production determined in the various media?
16. What methods are employed to eliminate atmospheric oxygen for the growth of anaerobes?
17. Why must media be sterilized? What methods are used?
18. What features are considered in the identification of a species?

SUPPLEMENTARY READING

Bennig G, Rohrer H: The scanning tunneling microscope. *Sci Am* 253(2):50, 1985.

Bretscher MS: The molecules of the cell membrane. *Sci Am* 253(4):100, 1985.

Everhart T, Hayes T: The scanning electron microscope. *Sci Am* 226(1):54, 1972.

Balows A, et al (eds): *Manual of Clinical Microbiology*, 5th ed. Washington, DC, American Society for Microbiology, 1991.

Spencer M: *Fundamentals of Light Microscopy*. New York, Cambridge University Press, 1982.

Chapter 3

Bacterial Morphology

OBJECTIVES

A study of this chapter should provide an understanding of

1. Major groupings and shapes of bacteria.

2. Flagella, axial filaments, fimbriae and pili, capsules, the cytoplasmic membrane, and cytoplasmic inclusions such as storage granules, chromatophores, and endospores.

3. The differences between the cell wall structure for gram-positive and gram-negative cells.

The word **morphology** means the study of form and structure; therefore, we shall begin our study of the bacteria by having a look at the various shapes of bacterial cells. Three distinct forms are generally recognized; as a result, bacteria are divided into three main groups on the basis of their shape:

1. Cocci—spherical
2. Bacilli—cylindrical or rod-shaped
3. Spiral forms—curved rods or spirals

In addition there are some very rare bacteria that are either star-shaped or rectangular. These shapes have been described only recently and it is unlikely that most microbiologists will see either of them.

COCCI

The cocci (*sing.* coccus, meaning berry) look like miniature spheres or round berries under the microscope. They exist in several different arrangements, and because the specific grouping of the cells may be characteristic for a specific genus, this knowledge is a help in identifying an unknown organism. Some cocci characteristically exist singly; others are found in pairs, cubes, or long chains, depending on the manner in which they divide and then adhere to each other after division. For example, the species *Streptococcus pneumoniae* is characterized by cells that divide into two cocci, which in subsequent divisions break apart, usually leaving two cocci adhering to each other (see Figure 3.1). A coccus that continues to divide in one plane but does not break apart very often forms chains of cocci; this is characteristic of the genus *Streptococcus*. Cocci that divide into three planes at right angles to one another form cubical packets; this method of division occurs in the genus *Sarcina*. Cocci that divide into two planes to form tetrads of cells occur in the genus *Pediococcus*. Cocci that divide into two planes to form irregular clusters are classified in either the genus *Staphylococcus* or the genus *Micrococcus*.

BACILLI

Bacilli (*sing.* bacillus, meaning little staff) are bacteria shaped like rods or cylinders. They vary a great deal in their proportions. Some resemble cigarettes while another form (the fusiform bacillus) with tapered ends is more like a cigar. Some bacilli are almost as broad as they are long and, because they resemble cocci, they are called coccobacilli. Unlike the cocci, bacilli divide in only one plane and may be observed as single cells, in pairs, or in short or long chains. They are unlike the cocci in that the length of the chains is not an identifying characteristic.

Vibrios are curved rods resembling commas. Sometimes they grow together in serpentine or S-shaped strands.

SPIRAL FORMS

This group comprises a large variety of cylindrical bacteria that, instead of being straight like the bacilli,

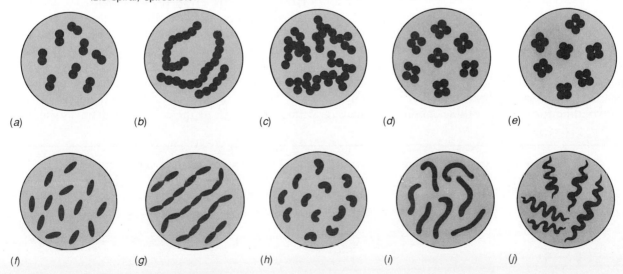

Figure 3.1 Morphologic types of bacteria: (*a*) diplococci; (*b*) streptococci; (*c*) staphylococci; (*d*) sarcinae; (*e*) *Pediococcus*; (*f*) bacilli; (*h*) spiral, *Vibrio*; (*i*) rigid spiral, *Spirillum*; and (*j*) flexible spiral, spirochete.

(*a*) (*b*) (*c*) (*d*) (*e*)

(*f*) (*g*) (*h*) (*i*) (*j*)

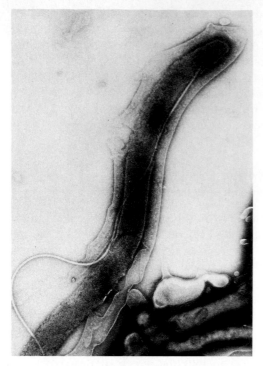

Figure 3.2 One end of a free-living spirochete. The protoplasmic cylinder, axial filaments, and external sheath are clearly visible in this negatively stained preparation (× 25,920).

be very difficult to identify the basic shape of the bacteria.

REPRODUCTION

Although the growth of bacteria is discussed in detail in Chapter 5, a brief word on how bacteria multiply is necessary at this point so that the student can better comprehend subsequent parts of this chapter. Bacteria are for the most part asexual; they multiply by an elongation of the cell followed by a division of the enlarged cell into two cells. Most often the cell splits in half; we call this binary fission (Figure 3.3).

BACTERIAL CELL STRUCTURE

If you keep in mind what you already know about a typical animal cell, you will find that you know something about bacterial cells; also, the features that are peculiar to bacteria will stand out all the more clearly by contrast.

Let us consider each principal part of a bacterium.

- Appendages—flagella, fimbriae, and pili
- Surface layers—capsule, cell wall, plasma membrane, and mesosomes
- Cytoplasm—nuclear material, plasmids, ribosomes, inclusions, and chromatophores
- Special structures

Flagella

Many bacteria are able to swim about by themselves through a liquid, often at remarkable speeds for their size. The ability of an organism to move by itself is called **motility.** Almost all spiral bacteria and about half the bacilli are motile, whereas essentially all cocci are nonmotile. The propulsive mechanism of most bacteria with which we shall be concerned is a helical threadlike appendage called a flagellum (*pl.* flagella).

Flagella are long and slender, generally several times the length of the cell. Their diameters may vary

are convoluted in varying degrees. The spiral bacteria are divided as follows:

1. Spirilla (*sing.* **spirillum**) are actual spirals or helices, like corkscrews. Their cell bodies are relatively rigid and essentially all are motile by means of polar flagella.

2. Spirochetes also are spiral bacteria, but they differ from the spirilla in that they are able to flex and wriggle their bodies while moving about. Their movement results from the contraction of the axial filaments, which spiral around the organism between the plasma membrane and the cell wall (see Figure 3.2).

Although the shape of a bacterium is determined by its heredity, the student should remember that a variety of outside conditions can also radically alter the shape of a bacterial cell. As bacteria age, they may swell or show rudimentary branching. Such forms are called **involution forms.** Under these conditions it may

Figure 3.3 Reproduction of bacteria by binary fission. The time required for one cell to become two cells will vary according to the growth conditions, but it may be as short as 10 min for some bacteria growing under ideal conditions.

from 12 to 30 nm, but, since this size is below the limit of resolution of the light microscope, flagella are not visible in routine stained smears of bacteria. Flagella can be seen with the light microscope only if a special substance called a **mordant** is used to build up the diameter of the flagellum. A stain is then applied to the mordant to color the thickened flagellum. Electron micrographs of flagella may be taken without staining, and these very highly magnified pictures reveal the structure of flagella to be amazingly similar to lengths of rope, even to the coiled strands.

The positions at which flagella are inserted into the bacterial cell are characteristic for a genus. Figure 3.4 presents a schematic depiction of the various types of bacterial flagellation and an electron micrograph of a cell showing a number of flagella.

If one removes the cell wall enzymatically from a bacterium, the flagella can be seen to originate in the cytoplasmic membrane. As is shown schematically and by electron microscopy in Figure 3.5, the flagellum is attached to the bacterium by means of a hook and a series of plates or rings that appear to anchor the flagellum to each layer of cell membrane and cell wall. Each flagellum has a wavy structure, and actual motility appears to result from the rotation of the flagellum in a manner similar to that of a boat propeller.

The driving force for rotation appears to result from the passage of protons from the exterior to the cytoplasm through the basal body; the entry of each proton causes the rotation of the rotor by a fixed degree. The fact that bacteria exhibit chemotaxis—that is, they move toward a gradient of various attractants (usually nutrient substrates)—demonstrates that movement is not entirely random. High-speed moving pictures have shown that when the flagellum is rotating counterclockwise, the bacterium will travel in a more or less straight line; but if the direction of flagellar movement is reversed, the organism will tumble aimlessly. When the bacterium is moving up a gradient of a chemotactic attractant, the straight-line movements continue longer and the tumbling occurs only briefly. However, if the bacterium is moving away from an attractant or accidentally toward a repellent, the tumbling occurs frequently until the net movement is properly redirected.

The mechanism whereby bacteria sense and adapt to their environment is a fascinating example of sensory control, involving transmembrane signaling and protein methylation. A major component in the process of chemotaxis is a family of transducer proteins that span the cytoplasmic membrane and have binding sites for specific chemoattractants and repellents on their exterior face. Binding of chemoeffectors trans-

Figure 3.4 Types of bacterial flagellation; (a) monotrichous (polar), (b) amphitrichous, (c) lophotrichous, (d) peritrichous, (e) cell of *Salmonella* sp. showing peritrichous flagellation.

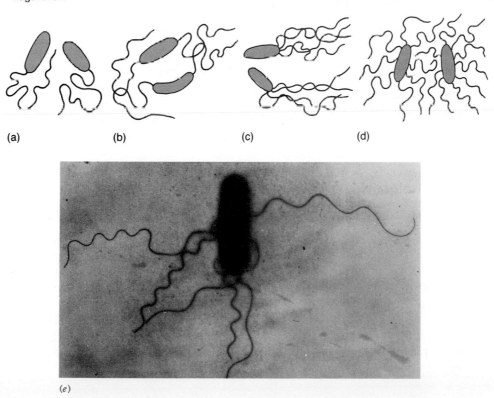

(a)   (b)   (c)   (d)

(e)

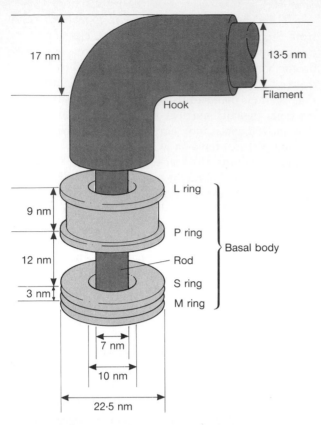

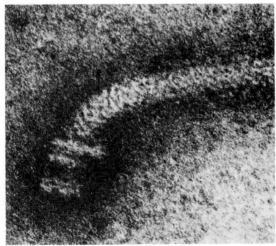

Figure 3.5 The electron micrograph and accompanying drawing illustrate the detailed structure of a flagellum and suggest how it is anchored to the bacterial cell. It is proposed that in *Escherichia coli* the L and P rings are anchored to the outer membrane and the S and M rings are anchored to the inner membrane. Flagella from *Bacillus subtilis* (which is gram-positive and has no outer membrane) lack the L and P rings.

mits a signal to the cytoplasmic portion of the protein, resulting in a conformational change that reflects the concentration of specific chemoeffectors in the medium. This signal is then transmitted to the flagellum's basal body to control the frequency of switching of the direction of flagellar rotation.

If you isolate flagella (by vigorously shaking a bacterial suspension followed by differential centrifugation) and make the solution acidic (pH 3), the protein of the flagella will dissociate into a large number of identical subunits. These units, called flagellin, will rapidly reaggregate at a neutral pH to form a structure indistinguishable from the original flagellum. One might assume, then, that during growth, the bacterium synthesizes the smaller subunits of flagellin and that these smaller subunits are then capable of spontaneously aggregating to form the flagella. It should be noted that the flagella on procaryotic organisms and their mechanism of imparting motility are very different from flagella existing on eucaryotic cells. Eucaryotic flagella are complex fibrils that move in a wavelike pattern and are composed of microtubules containing the protein tubulin.

Flagella are not the only means by which bacteria move about. Some types exhibit a gliding motility—they crawl over surfaces by waves of contraction produced within the cytoplasm. These organisms are extremely important in the decaying of dead organic material but are of no medical importance because they are never associated with human disease.

Motility can be observed most satisfactorily in young cultures (18 to 24 h old, or less) because bacteria tend to become nonmotile in older cultures. An old culture may become so crowded with inert living and dead bacteria that it is difficult to find a motile cell. In addition, the production of acid and other toxic products may result in the loss of bacterial motility in older cultures.

Axial Filaments

An axial filament, found only in the spirochetes, consists of protein fibrils wound spirally around the organism and attached to the two poles of the cell (Figure 3.6). They are located just beneath the outer membrane, where they function as a flagellalike structure to impart a rapid motility to the spirochete.

Fimbriae and Pili

Many gram-negative bacteria possess filamentous appendages that are straighter and considerably thinner and shorter than flagella (see Figure 3.7). These appendages are referred to as either fimbriae or pili by many persons, but because they provide two different functions, others no longer use these terms synonymously.

Fimbriae belong to a class of proteins called lectins, which recognize and bind to specific sugar residues in cell surface polysaccharides. They are also frequently called adhesins. As a result, bacteria pos-

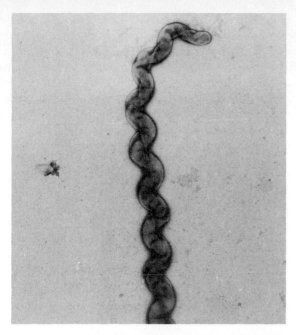

Figure 3.6 Electron micrograph of a spirochete in the genus *Leptospira*. Note the helical shape of the cell and the axial filament that runs beneath the outer membrane.

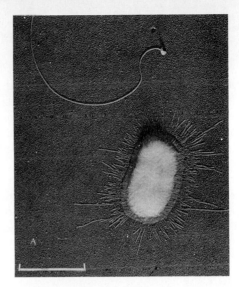

Figure 3.7 *Proteus vulgaris*. This cell is covered with fimbriae. An unattached flagellum is in the background to illustrate size differences between flagella and fimbriae. Scale = 1 μm.

sessing fimbriae have a tendency to adhere to each other as well as to animal cells. Interestingly, the ability of certain organisms (such as *Neisseria gonorrhoeae*, *Bordetella pertussis*, and the enterotoxigenic *Escherichia coli*) to cause disease is associated with the possession of fimbriae, and mutations resulting in the loss of fimbriae are accompanied by a loss of virulence.

Pili are morphologically and chemically similar to fimbriae, but we shall restrict this designation to appendages specifically involved in the transfer of DNA during bacterial conjugation. They are present in much smaller numbers per cell (fewer than 10) and are usually longer than the adhesion fimbriae described above. The function of the pilus will be discussed in Chapter 8.

Capsules

Many bacteria secrete material that adheres to the exterior of the bacterial cell. The general term for such material is the **glycocalyx**. In some cases the glycocalyx exists as a discrete, organized, thickened material around each cell or pair of cells. It is then referred to as a **capsule** (see Figure 3.8). Capsules may vary considerably in thickness. In some species they may be so thin that they can be detected only by chemical analysis. In other instances, the glycocalyx may exist as an unorganized, loosely attached, polysaccharide mass and, in such cases, it is usually called a **slime layer**.

The composition of the capsule is constant in a particular bacterial strain, but it varies widely even between organisms classified in the same genus and species. For example, capsules from various types of *S. pneumoniae* are all composed of very large molecules with molecular weights approaching 1 million. However, if one isolates the capsular material and hydrolyzes the polysaccharides to their component monosaccharides, one obtains different sugars from different types. Capsules tend to have a low affinity for dyes and as a result are not usually seen in routinely stained smears. They may, however, be visible in wet mounts of the organisms, particularly if the organisms are suspended in a dilution of India ink.

The actual value of the capsule to the bacterium is not always evident, although it appears to prevent

Figure 3.8 Encapsulated cells of *S. pneumoniae*. The capsules around the pairs of cells are swollen by use of type-specific antibody.

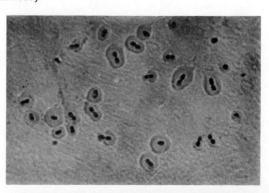

desiccation of the organism under adverse conditions. However, the presence of a capsule can be of great importance in medical microbiology because it may act as a protective layer that resists ingestion by the host's phagocytic cells (cells that engulf and digest foreign material). Furthermore, because the resistance provided by the capsule can be negated in the presence of specific antibodies, immunity to some infectious agents is directed against the organism's capsule.

It is noteworthy that certain oral bacteria secrete a glycocalyx that is responsible for their adhesion to the tooth surface. This is called a **dental plaque,** and the bacterial activity within such plaques is ultimately responsible for both tooth decay and periodontal disease.

Cell Wall

The bacterial cell wall has received intense scrutiny because it is essential to the bacterial cell and is chemically unlike any structure present in animal tissues. It, therefore, is an obvious target for drugs that can attack and kill bacteria without harm to the host. **It is precisely at the point of synthesis of the bacterial cell wall that the action of many antibiotics used in the treatment of disease is directed.**

The primary, and undoubtedly most necessary, function of the cell wall is to provide a strong, rigid structural component that can withstand the osmotic pressures caused by the high chemical concentrations of inorganic ions in the cell. Without the cell wall, a bacterium under normal environmental conditions would take up water and burst, much as would an inflated inner tube or football bladder without a tire or leather covering to support it. Most bacterial cell walls have in common a structural component called **peptidoglycan.** (Exceptions include the mycoplasma, the extreme halophils, and the archaebacteria.) This component provides the rigidity necessary to maintain the integrity of the cell. Peptidoglycan is a very large molecule—one that covers the entire cell—made up of N-acetylglucosamine and N-acetylmuramic acid, a structure that is unique to procaryotic cells. To each molecule of N-acetylmuramic acid is attached a tetrapeptide consisting of four amino acids; to provide necessary additional strength to this molecule, bridges of amino acids may cross-connect the tetrapeptides attached to the N-acetylmuramic acid. Most components of the cell wall structure are cross-linked by covalent bonds, and any substance that prevents either the formation or transport of the individual components to the cell wall weakens the structure and kills the cell. Figure 3.9 illustrates the structure of a small cross section of the peptidoglycan of a cell wall.

As noted earlier, bacteria can be divided into two large groups—gram-positive and gram-negative—on the basis of a differential staining technique called the Gram stain. These two groups differ mainly in their cell walls and, because of these differences, each will be discussed separately.

Figure 3.9 (a) Peptidoglycan structure showing cross-linking of linear polymers of N-acetylglucosamine and N-acetylmuramic acid as it exists in the gram-positive staphylococci. Note that the peptides are linked by a pentaglycine bridge. (b) The peptidoglycan structure of most gram-negative bacteria. Note here that the tetra peptides may be occasionally free and, if not, are joined directly to each other. Note also that a lipoprotein is linked to occasional molecules of N-acetylmuramic acid. The other end of the lipoprotein anchors the peptidoglycan to the outer membrane.

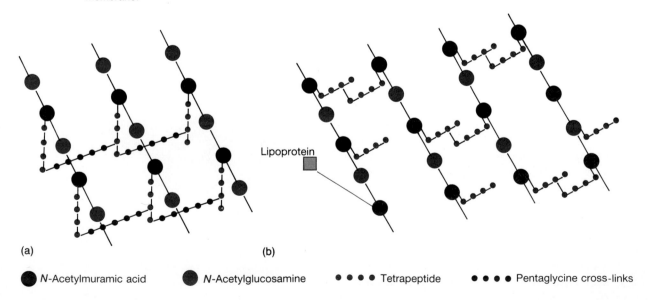

(a)

(b)

Lipoprotein

● N-Acetylmuramic acid ● N-Acetylglucosamine ●●●● Tetrapeptide ●●●● Pentaglycine cross-links

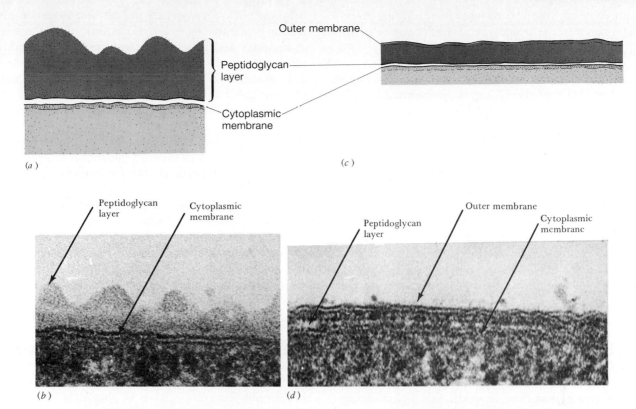

Figure 3.10 (a) Drawing of a gram-positive cell wall showing the very thick peptidoglycan exterior to the cytoplasmic membrane. (b) Electron micrograph of a gram-positive cell wall. (c) Drawing of a gram-negative cell wall showing outer membrane. Note also that the peptidoglycan is only one molecule thick and that it lies between the outer membrane and the cytoplasmic membrane. (d) Electron micrograph of a gram-negative cell wall. Note that the O antigen is not visible.

Gram-positive Bacteria

The cell walls of gram-positive organisms are quite thick (20 to 80 nm) and consist of from 60 to 80 percent peptidoglycan, which is extensively cross-linked in three dimensions to form a thick polymeric mesh (see Figure 3.10). All gram-positive cells possess a linear polymer of N-acetylmuramic acid and N-acetylglucosamine, but there is variation in the length and composition of the peptide bridge that cross-links the tetrapeptide from one N-acetylmuramic acid to the adjoining polymer.

Some gram-positive organisms also contain cell wall substances called teichoic acids that are linked to the muramic acid of the peptidoglycan layer. The teichoic acids occur in two major forms, ribitol teichoic acid and glycerol teichoic acid (see Figure 3.11). In general, they are long polymers of either ribitol (a five-carbon sugar alcohol) or glycerol (a three-carbon sugar alcohol) linked to each other through phosphodiester bridges. However, all of the hydroxyl groups of the ribitol or glycerol subunits are linked to various sugars or amino acids. Because these teichoic acids are highly antigenic (i.e., they will induce a host to make specific antibodies that will react with the teichoic acids), it seems that they extend through the peptido-

Figure 3.11 Teichoic acids are usually long chains of glycerol or ribitol joined together with phosphodiester bridges. The R group for glycerol teichoic acid is frequently D-alanine, whereas for the ribitol-type teichoic acid it may be glucose, succinate, N-acetylglucosamine, D-alanine, or short oligosaccharides. Substitutions may vary considerably even within a species.

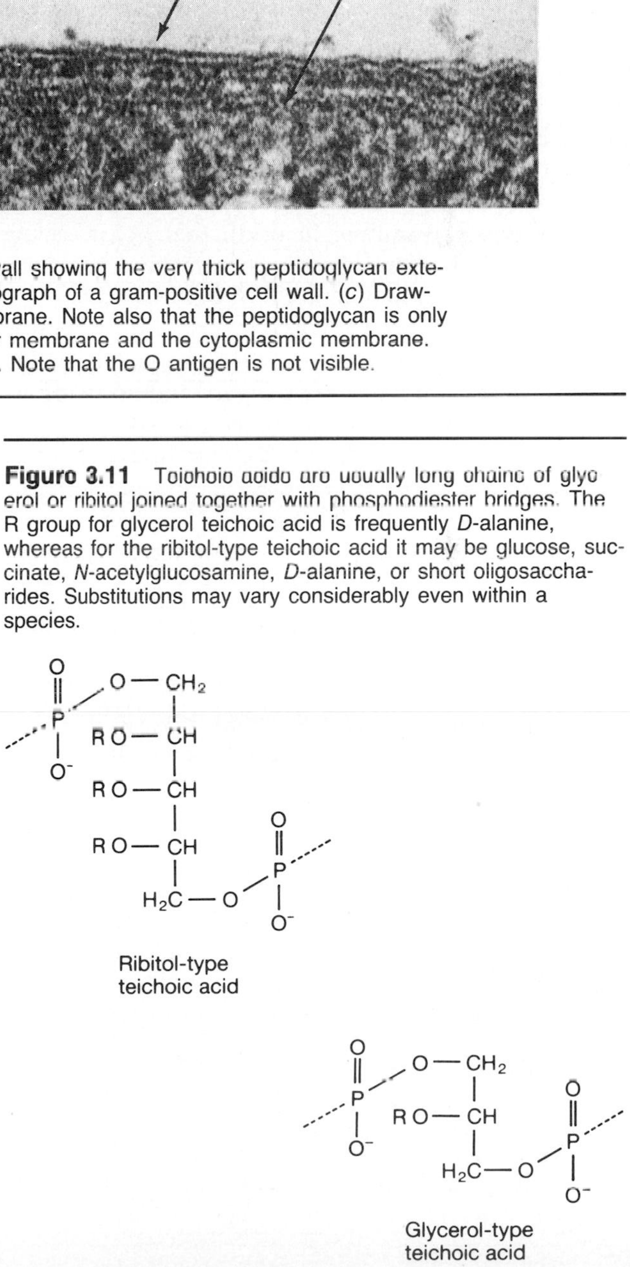

Ribitol-type teichoic acid

Glycerol-type teichoic acid

glycan layer and thereby provide the antigenic determinants used in the serological identification of many groups and species of gram-positive bacteria.

For the most part, protein is not found as a constituent of the gram-positive cell wall. An exception is the M protein of the Group A streptococci.

Gram-negative Cell Walls

The cell walls of gram-negative bacteria are chemically more complex than those of the gram-positive bacteria. For example, the gram-negative cell walls contain less peptidoglycan (10 to 20 percent dry weight of cell wall), but exterior to their peptidoglycan layer they possess a second "membrane" structure, composed of proteins, phospholipids, and lipopolysaccharide (fatty acids linked to polysaccharides; see Figure 3.10). This lipopolysaccharide component of the gram-negative bacterial cell wall is extremely important because of its toxicity to animals. Because of this toxicity and because the material is an integral part of the bacterial cell, it has been given the name **endotoxin.** It is this material that is responsible for the high fevers and, in many cases, irreversible shock (drop in blood pressure) resulting in death that occur during infections with gram-negative organisms. Structural studies have disclosed that the lipopolysaccharide is composed of two major components: a disaccharide of glucosamine to which are bound fatty acids (called lipid A) and a long chain of sugars and sugar phosphates linked to the lipid A moiety. It is known that the toxicity of the molecule resides in the lipid A portion.

Interestingly, all of the cell wall lipopolysaccharide (LPS) exists solely in the outer leaflet (outer layer) of the outer membrane, making this membrane considerably more rigid than normal cell membranes.

Moreover, the gram-negative outer membrane is not readily penetrated by hydrophobic (insoluble in water) compounds and is, therefore, resistant to dissolution by detergents. This property enables many gram-negative bacteria to grow in the presence of detergents that are lethal for gram-positive cells.

Gram-negative cell walls also contain a lipoprotein that is covalently linked to the peptidoglycan layer. This lipoprotein is also linked to three fatty acids that are embedded in the outer membrane, serving to anchor this membrane to the peptidoglycan part of the cell wall.

When an individual is infected (or artificially immunized) with gram-negative bacteria, the body responds by making antibodies to the polysaccharide chain of the lipopolysaccharide. The polysaccharide is frequently called the O antigen or somatic antigen. In many cases, such antibodies protect against a subsequent infection by the same strain of bacteria. They are also used for the serological classification of many gram-negative organisms. Figure 3.12 shows the structure of a typical lipopolysaccharide. This general type of structure is probably similar for all gram-negative bacteria, but the type and arrangement of the sugars in the polysaccharide will vary from species to species.

Neither in gram-positive nor in gram-negative organisms will the cell wall take up enough of the ordinary basic laboratory dyes to become readily visible. As a result, when you observe a stained bacterium under the microscope, you are seeing the part of the bacterium that lies inside the cell wall. A special stain can be used that stains only the cell wall, but the stain is not routinely used.

It is possible to dissolve the peptidoglycan portion of the cell wall with enzymes. The enzyme most commonly used for this purpose is called **lysozyme;** it hydrolyzes the bond between N-acetylglucosamine

---

MICROBIOLOGY MILESTONES

During the 1890s and early 1900s, William B. Coley, a surgeon at Memorial Hospital in New York, achieved considerable notoriety (and success) treating cancer patients with injections of killed bacteria. These preparations, which became known as "Coley's toxins," usually contained a gram-positive organism (*Streptococcus pyogenes*) and a gram-negative organism (*Serratia marcescens*). Success was, however, sporadic in that some tumors diminished or disappeared altogether, but many were unaffected. When radiation therapy and chemotherapy became available, Coley's toxins were no longer used.

Subsequent research demonstrated that the active component of Coley's toxins was the lipopolysaccha-ride (LPS) present in the outer membrane of gram-negative bacteria (also called endotoxin). Moreover, it is now known that the antitumor effect was due to the release of two cytokines by macrophages after interacting with LPS. These cytokines, designated as tumor necrosis factor (also called cachectin) and interleukin-1, not only kill certain tumor cells but are also responsible for the many other biological effects of LPS, namely, fever, tissue injury, shock, and death. Studies have also shown that anitbodies to tumor necrosis factor will protect animals from shock and death following an injection with gram-negative organisms. It seems highly probable that a similar procedure would be effective for the treatment of gram-negative sepsis in humans.

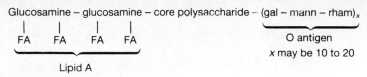

Glucosamine – glucosamine – core polysaccharide – (gal – mann – rham)$_x$
| | | |
FA   FA   FA   FA

Lipid A

O antigen
x may be 10 to 20

Figure 3.12 Structure of a typical lipopolysaccharide from a gram-negative bacterium. FA, fatty acid; gal, galactose; mann, mannose; rham, rhamnose. Lipid A forms the outer leaflet of the outer membrane. The polysaccharide designated as the O antigen projects outward from the outer membrane. The sugars making up the O antigen may vary from one organism to another. This structure is also referred to as "endotoxin" because of its toxicity in animals.

and *N*-acetylmuramic acid, causing the cell wall to break into a number of small pieces. Under normal environmental conditions, this results in the complete rupture of the cell. If, however, the bacterium is placed in a solution of sugar having the same osmotic pressure as the inside of the cell, the removal of the cell wall results in the liberation of the cell contents surrounded by a thin, fragile membrane (see Figure 3.13). This thin membrane located just inside the cell wall and containing the fluid contents of the cell is called the cytoplasmic membrane.

When one removes the peptidoglycan layer from a gram-positive bacterium in the presence of 10 to 20 percent sucrose, the remaining cell, surrounded by only the cytoplasmic membrane, is called a **protoplast.** However, when the same procedure is carried out on a gram-negative cell, the protein-lipopolysaccharide outer membrane remains attached to the cytoplasmic membrane. As a result, the gram-negative bacterium devoid of its peptidoglycan layer is referred to as a **spheroplast.**

Figure 3.13 Protoplast of *Clostridium botulinum* escaping through the ruptured cell wall 30 to 60 min after penicillin and lysozyme were added. Scale = 1 μm.

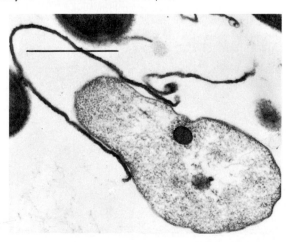

Basis for the Gram Reaction

The response of an organism in the Gram stain is a consequence of its cell wall structure rather than the presence of any particular chemical. Treatment with crystal violet and Gram's iodine results in the formation of large, insoluble blue dye complexes inside of the cell. During the ethanol-washing step, cell membranes are dissolved. The blue dye complex, however, is retained in gram-positive cells by the thick peptidoglycan mesh, whereas it is readily washed out through the very thin peptidoglycan layer remaining in gram-negative cells after both membranes have been dissolved. Gram-positive cells from old cultures may stain gram-negative because their peptidoglycan layer is thinner or partly disrupted.

Figure 3.14 provides a comparison of the cell wall structures of gram-positive and gram-negative bacteria. Note that the lipopolysaccharide (endotoxin) is located exclusively in the outer leaflet of the outer membrane.

Cytoplasmic Membrane

The fragile cytoplasmic membrane, located just inside the rigid cell wall, has a number of functions: (1) in aerobic organisms it transports electrons and protons released during the oxidation of bacterial "foodstuffs" to oxygen (to form water), and it converts the energy liberated by such oxidations into chemical energy that can be used by the cell; (2) it contains some of the enzymes necessary for the synthesis and transport of peptidoglycan, teichoic acids, and outer membrane components; (3) it secretes extracellular hydrolytic enzymes; (4) it ensures the segregation of nuclear material (DNA) to daughter cells during cell division; and (5) it controls the transport of most compounds entering and leaving the cell.

The cytoplasmic membrane accounts for 8 to 10 percent of the dry weight of the cell. Chemically, it is composed of phospholipids (which contain glycerol, fatty acids, and phosphate) and proteins, which are

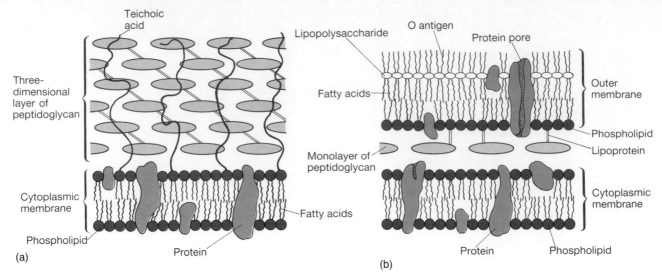

Figure 3.14 (*a*) Gram-positive cell wall. (*b*) Gram-negative cell wall.

embedded in it. Figure 3.15 shows a schematic "fluid-mosaic" model of the phospholipid orientation, in which the circular areas represent the hydrophilic ("water-loving") ends of the phospholipid molecule and the wavy lines the internal hydrophobic ("water-hating") fatty acids. Interspersed throughout this phospholipid membrane are proteins that perform the various transport and enzyme functions associated with the membrane.

Mesosomes

Invaginations of the cytoplasmic membrane, usually irregular in shape, are referred to as **mesosomes** (Figure 3.16). These structures have been observed primarily in thin sections of gram-positive bacteria, and

several functions have been postulated for them. They provide for an increase in membrane surface, which may be useful for respiration and transport. It has been proposed that mesosomes that form near the septum of gram-positive cells participate in cell replication by serving as organs of attachment for the bacterial chromosomes (Figure 3.17). Mesosomes that form elsewhere along the membrane may be involved in the secretion of extracellular proteins or in the metabolism of hydrocarbons.

Cell Sap or Cytoplasm

The cell sap (which contains the soluble constituents of the cytoplasm) and the structural components of the cytoplasm are surrounded by the cytoplasmic

Figure 3.15 Fluid-mosaic model for membrane structure. The solid bodies represent the globular integral proteins, which at long range are randomly distributed in the plane of the membrane. The small circles represent the hydrophilic ends of the membrane phospholipids, and the wavy lines represent the component fatty acids of the phospholipids.

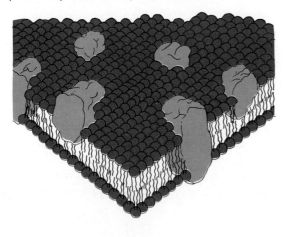

Figure 3.16 A number of mesosomes have invaginated from the cytoplasmic membrane in this section of a *Bacillus fastidiosus* cell. A larger mesosome not continuous with the membrane in the plane of this section is located to the right.

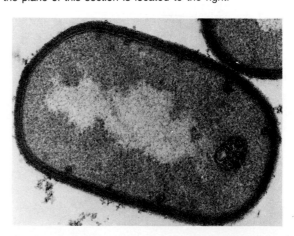

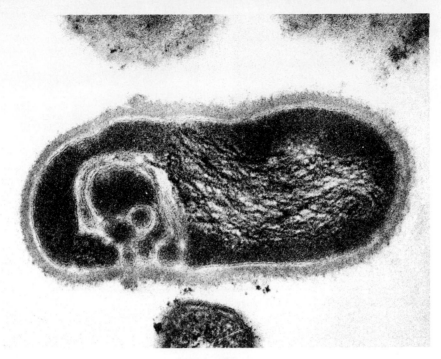

Figure 3.17 In this cell of *Corynebacterium parvum,* the chromatin making up the bacterial nucleoid is the fibrous skein filling most of the cell. It is attached to the membrane of the large mesosome (× 113,000).

membrane. The cytoplasm, including the nuclear area, consists of about 80 percent water. In addition to water there are nucleic acids, proteins, carbohydrates, lipids, inorganic ions, and a variety of compounds of low molecular weight. Ribosomes and various inclusions are also found within the cytoplasm. In certain photosynthetic species, pigments will be found in chromatophores. A special structure, the endospore, is also developed within the cytoplasm in some species. These various structures will be discussed separately in the following pages.

Bacterial Chromosome

The bacterial chromosome exists as a single, circular molecule of double-stranded DNA. It is complexed with small amounts of proteins and RNA, but it is not associated with histones as is eucaryotic DNA. Moreover, bacterial DNA does not contain introns (intervening sequences that are not expressed as proteins) and, as a result, it is transcribed directly into messenger RNA without requiring RNA splicing as occurs in eucaryotic cells.

Bacterial chromosomes do possess some degree of higher organization, but this is not detectable in electron micrographs. Rather, one sees an amorphous fibrous area in immediate contact with the cytoplasm (Figure 3.17).

Plasmids

Plasmids are relatively small, circular pieces of double-stranded DNA (2 million to 200 million daltons) that exist separately from the bacterial chromosome (Figure 3.18). They are capable of autonomous replication and encode for many auxiliary functions (such as antibiotic resistance), which are usually not necessary for bacterial growth. Many plasmids can be transferred from one bacterium to another by a process called **conjugation** or through laboratory manipulations. It is this latter technique, in which one inserts a piece of foreign DNA into a plasmid and then puts the plasmid into a susceptible bacterium, that provides the basis for **recombinant DNA technology.** Using these techniques, plasmids in *Escherichia coli* and yeast are able to encode for the synthesis of many foreign proteins, such as human insulin and human growth factor.

Ribosomes

Observation with the electron microscope shows that the cytoplasm is densely packed with ribosomes. Ribosomes are composed of RNA and proteins and comprise the structural units required for the synthesis of proteins. When not involved in the synthesis of protein, they exist within the bacterium in two subunits

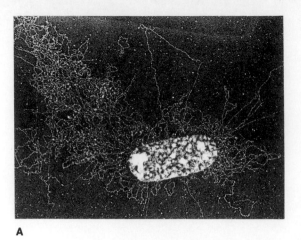

A

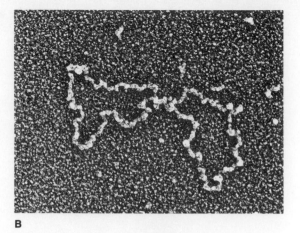

B

Figure 3.18 (a) Disrupted cell of *E. coli;* the DNA has spilled out and a plasmid can be found slightly to the left of top center. (b) Enlargement of a plasmid (about 1 μm from side to side). This plasmid provides the *E. coli* cell bearing it with resistance to the substance colicin.

designated 50S and 30S. The S represents the sedimentation velocity as measured in Svedberg units, which is a measure of size and density. As soon as the 30S ribosomal subunit attaches to a molecule of messenger RNA to initiate protein synthesis, the 50S subunit joins it to form an intact 70S ribosome.

Inclusions

In addition to nuclear material, bacterial cytoplasm may contain cell inclusions—bits of material that are not integral parts of the cell structure. These may be granules of various kinds: glycogen, polyhydroxybutyric acid droplets, inorganic phosphate, sulfur, or compounds containing nitrogen (see Figure 3.19).

One common inclusion is composed of a high-molecular-weight polymer of polyphosphate. These particular granules, which appear to serve as a reserve source of phosphate and energy, are called **metachromatic granules** because they take on a reddish color when stained with methylene blue. Metachromatic granules are also given the collective name **volutin**. They are commonly found in many microorganisms, including bacteria, fungi, algae, and protozoa.

For certain bacteria known as the sulfur bacteria, the oxidation of sulfur granules produces energy. In the case of members of the genus *Thiobacillus,* the granules of sulfur may be taken directly into the cell from the outside environment. However, some bacterial types also oxidize compounds containing sulfur, such as thiosulfate, and deposit the resulting elemental sulfur granules inside the cell. These granules can then serve as a reserve source of energy for the cell.

Chromatophores

Chromatophores represent special membrane systems found in certain photosynthetic bacteria and cyanobacteria. These procaryotic cells lack chloroplasts (in photosynthetic eucaryotic cells, the chloroplasts contain chlorophyll). In some species, the pigments are found in lamellae located beneath the cell membrane; in others, discrete vesicles or chromatophores are found throughout the cytoplasm. In electron micro-

Figure 3.19 Bacillus showing large inclusion of polyhydroxybutyric acid (PHB). When cells like these are inoculated into a fresh medium, the PHB is used by the cell to provide energy for synthesis of new enzymes, etc.

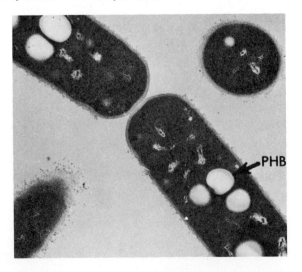

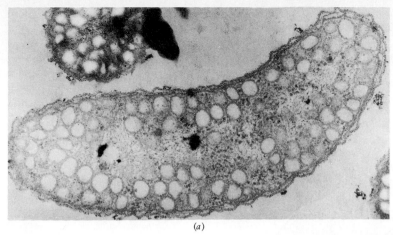

(a)

Figure 3.20 (a) *Rhodospirillum rubrum,* a purple, nonsulfur, photosynthetic bacterium possessing multiple vesicular chromatophores. (b) *Ectothiorhodospira mobilis,* a purple, sulfur, photosynthetic bacterium in which the photosynthetic pigments occur in lamellae located just below the cytoplasmic membrane.

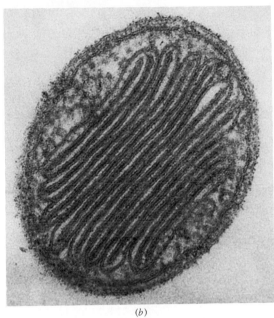

(b)

graphs the membrane surrounding the vesicles appears to be similar to the cytoplasmic membrane. Chromatophores vary in size from one species to another, from 15 to 100 nm. Figure 3.20 illustrates the major membrane chromatophores found in photosynthetic procaryotes.

Endospores

Only two genera of medical importance, *Bacillus* and *Clostridium,* have the ability to develop a specialized structure called an **endospore.** These genera, however, contain the causative agents of anthrax, tetanus, botulism, and gas gangrene, and later we shall see what is the role of the endospore in the epidemiology of these diseases. Other endospore-producing procaryotes are found in several genera of soil bacteria: *Desulfotomaculum, Sporolactobacillus,* and *Sporosarcina.* An endospore is a minute, highly durable body formed within the cell and capable of developing into a new vegetative organism. The process of endospore formation is generally as follows: after the active growth period—that is, approaching the stationary growth phase (see Chapter 5)—a structure called a **forespore** develops within the cell. Later it becomes refractile and forms a thick wall (see Figure 3.21). When sectioned, the endospore can be seen to consist of an outer coat, a cortex, and the core containing the nuclear structure. The diameter of the endospore may be the same as, larger than, or smaller than that of the vegetative cell, and it may be located in the center (as shown in Figure 3.22), subterminally, or at

the end of the cell, depending on the species. As the endospore matures, the vegetative cell wall dissolves, freeing the mature endospore.

Because of this very thick wall, endospores cannot be stained by the usual staining procedures. Thus, in a Gram-stained preparation, the endospore appears as a clear or unstained body within the cell. However, special spore stains have been devised in which the spore is heated with the stain, or the stain is mixed with a surface-active detergent that helps the stain to penetrate the cell wall.

When a vegetative cell forms an endospore, it makes some new enzymes, produces a totally different cell wall, and changes shape. In other words, sporulation is a simple form of cell differentiation; for this reason, the process is being studied extensively to learn

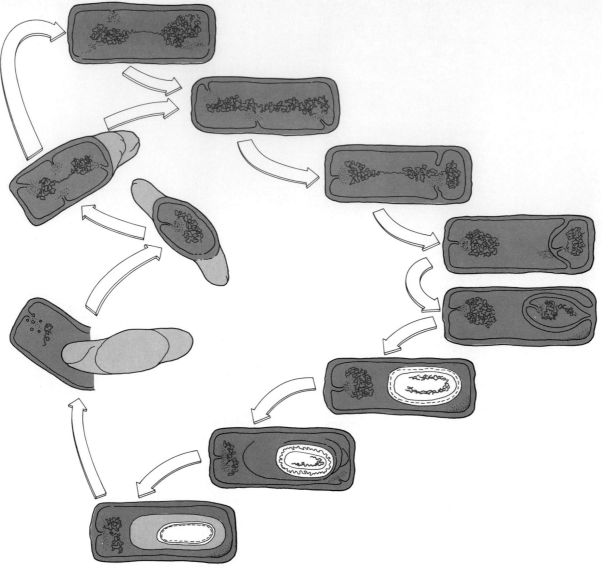

Figure 3.21 Diagrammatic summary of sporulation in a *Bacillus* species. Stages 1 through 3 represent steps in forespore development. Note the inversion of the cytoplasmic membrane resulting in the outer leaflet of this membrane becoming the inner leaflet of the spore membrane. At stage 4, cortex development commences (dotted line) and continues through stage 5, when the coat protein is deposited. Stage 6 is characterized by a dehydration of the spore protoplast and an accumulation of dipicolinic acid and calcium in the spore. At stage 7 the spore is completely refractile and a lytic enzyme acts to release the spore. Also shown are germination A; outgrowth to a primary cell B; from which the cell may, under special conditions, enter sporulation by a shortcut C, "the microcycle," but normally undergoes logarithmic growth (spiral arrow).

what events trigger these enzymatic and morphological changes.

The most obvious property possessed by an endospore, one not characteristic of a vegetative cell, is its imperviousness to heat and chemicals. For example, an endospore may remain viable for many years—perhaps centuries—under normal soil conditions. Furthermore, while most vegetative cells are killed by temperatures beyond 60 to 70°C, endospores may survive in boiling water for an hour or more. There is no simple explanation for this resistance, but the fact that endospores contain very little free water undoubtedly is a major reason for their thermal stability. Endospores also contain large amounts of calcium dipicolinate, a substance not found in the vegetative cells. It has been postulated that this compound may

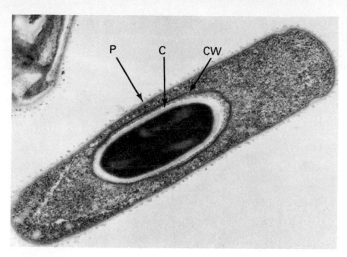

Figure 3.22 Electron micrograph of a mature endospore prior to lysis of the vegetative cell. P, peptidoglycan layer of the vegetative cell; CW, cell wall of the spore; C, very thick cortex of the spore. ($\times$ 24,500).

play a role in the heat resistance of the endospore, but any mechanism by which this could occur remains unknown.

It has been shown that endospores contain approximately five times more of the amino acid cysteine per milligram of total nitrogen than do their corresponding vegetative cells. It is postulated that, as a result, this high content of disulfide groups may account for the marked resistance of the endospore to ultraviolet irradiation. Endospores' resistance to chemicals that destroy vegetative cells can quite likely be explained by the impermeability of the thick spore coat (see Figure 3.23).

As long as environmental conditions are adverse to the growth of the bacterium, the endospore will remain a spore. However, if conditions become favorable for growth, several changes may take place. The refractility and heat resistance of the endospore disappear and the spore can then be stained by the usual procedures. If adequate nutrients are available, the endospore "sprouts" or germinates and once again becomes a vegetative bacterial cell reproducing in the normal manner.

Sporulation (the formation of endospores) is not part of the reproductive process of endospore-forming bacteria (although molds do multiply by means of spores). However, in the case of bacteria, one cell forms one endospore, which after germination is again only one cell. The formation of endospores is not necessary for the organism to continue to grow; thus, it is not necessarily part of the normal growth cycle.

Most endospore-forming bacteria are inhabitants of the soil, but bacterial endospores exist almost everywhere, including the atmosphere, where they ride on invisible particles of dust. The fact that endospores are so hard to destroy is the principal reason for the lengthy and elaborate sterilization procedures employed in hospitals, canneries, and other places where sterilization is required.

Table 3.1 provides a summary of the bacterial structures described in this chapter.

Figure 3.23 Spore of *Bacillus fastidiosus* ($\times$ 53,590).

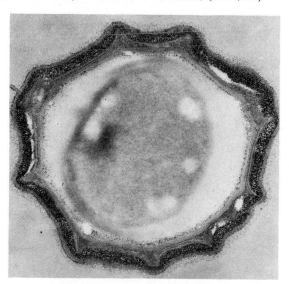

TABLE 3.1 Structures of the Bacterial Cell

	Structure	Chemical Composition	Importance
Appendages	Flagella	Protein (flagellin)	Provides for locomotion
	Fimbriae	Protein	Confers adhesive properties on certain bacteria
	Pili	Protein	Involved in the sexual transfer of DNA between cells
	Axial filaments	Protein (bundles of elastic fibers)	Characteristic of spirochetes; alternate expansion and contraction of the filament may account for movement
Surface layer	Capsule	Polysaccharide usually	Aids in protecting cell from hostile environment
		Polypeptide in some species	Contributes to virulence of certain pathogens
			Aids in preventing phagocytosis
			May be antigenic
			May be food reservoir in some
			May be waste products from cell
			May aid in preventing soil erosion by pasting soil particles together
	Cell wall (generalized)	Peptidoglycan	Protects cell from osmotic lysis
		Network of polysaccharide chains containing N-acetyl-glucosamine and N-acetyl-muramic acid connected by polypeptide cross-links	Peptidoglycan layer makes cell wall rigid
	Gram-positive bacteria	Contain teichoic acids	
		Some streptococci have an external protein antigen (M protein)	
	Gram-negative bacteria	Lipopolysaccharide and lipoprotein external to the peptidoglycan layer	
	Cytoplasmic membrane	Lipoprotein: lipid about 40%; protein about 60%; small amounts of carbohydrate	Osmotic regulator; contains many transport systems for various mineral ions, sugars, amino acids
			Site of several enzyme systems; responsible for functions performed by mitochondria in eucaryotic cells
			Transports waste outward as well as transporting "building blocks" of cell wall to the peptidogylcan layer; damage to membrane may cause cell death.

TABLE 3.1 *(Continued)*

	Structure	Chemical Composition	Importance
Cytoplasm and its structures	Cell sap or cytoplasm	Nucleic acids, proteins, carbohydrates, lipids, inorganic ions	Primary site of synthetic processes
		Low-molecular-weight compounds, such as amino acids	Center of functional activities
		Water about 80%	
	Chromosome	DNA fibrils (single chromosome)	Carrier of genetic information—information center for the direction of specific cellular synthesis
	Plasmids	Circular DNA	Encodes for ancillary functions such as antibiotic resistance
	Ribosomes (in a cell-synthesizing protein system, mostly found in chains called polysomes)	RNA 60%, protein 40%	Involved in a protein synthesis
	Mesosomes (cytoplasmic membrane—large, irregular, convoluted invaginations)	Protein and lipid	Provides increased membrane surface for supporting enzymes and, possibly, transport systems
			May have a role in wall synthesis and the separation of chromosomes during cell division
	Inclusions		
	Volutin	Polymetaphosphate (inorganic phosphate)	Reserve materials
	Granulose	Starchlike polymer of glucose	
	Other granules	Sulfur, lipids, glycogen, polyhydroxybutyrate	
	Chromatophores (special membrane systems)	Protein and lipid	Contain pigments concerned with photosynthesis in certain species of bacteria
Special structures	Endospores	Lack many enzymes; have small amounts of others	Extremely resistant to heat
		Small water content, high calcium content; protein and polysaccharide antigens	

SUMMARY

Bacteria occur in three distinct forms or shapes: cocci (round or oval), bacilli (rod-shaped), and spirilla (curved rods). The cocci may retain certain characteristic groupings as a result of the planes in which division occurred. The bacilli may occur singly or in chains, but the spirilla always occur as single cells.

Bacteria are measured in micrometers. The average diameter of a coccus may be 1 μm or less, whereas a bacillus or spirillum may be 2 to 5 μm long and 0.5 to 1 μm in diameter.

Bacteria reproduce by an elongation of the cell, followed by the division of the cell into two cells. This process is called binary fission.

The bacterial cell may or may not be motile, but if motile, it possesses long threadlike appendages called flagella, which are responsible for its movement. An exception to this is found in certain soil bacteria, which lack flagella and exhibit a gliding motility.

The cell wall of a bacterium is a rigid structure that provides support for the fragile protoplast it encloses. The major structural component of the wall is peptidoglycan. In addition, gram-positive cells may contain polymers of ribitol phosphate and/or glycerol phosphate, which are collectively termed teichoic acids. The cell walls of gram-negative bacteria contain a protein-lipopolysaccharide complex that, because of its toxicity to animals and its close association with the bacterial cell, is called endotoxin.

The protoplast is surrounded by a semipermeable cytoplasmic membrane that controls the entrance and exit of compounds into and from the cell. The bacterial cell contains a chromosome, not surrounded by a nuclear envelope, and ribosomes; it may or may not contain extrachromosomal DNA, known as plasmids, or reserve materials composed of polymetaphosphate that include metachromatic granules, granules of glycogen, polyhydroxybutyric acid, sulfur, or compounds containing nitrogen. Certain bacteria contain mesosomes, chromatophores, or an axial filament with special functions. Protoplasts and spheroplasts are formed by removing the peptidoglycan of gram-positive and gram-negative cells, respectively. Certain bacterial species can form a resting body referred to as an endospore.

QUESTIONS FOR REVIEW

1. What are the major shapes of bacteria and what is each called?
2. What are some typical groupings of cocci and how are they designated?
3. What arrangements of flagella exist on bacteria? Can you see flagella under the light microscope?
4. Why are bacterial capsules important in causing disease?
5. How do the cell walls of gram-positive species differ from those of gram-negative species? What do they have in common? Which contain endotoxin? Which contain teichoic acids?
6. What structure of bacteria may be affected by lysozyme? What is the result?
7. Describe how flagella function to move a bacterium toward an attractant.
8. Define chemotaxis.
9. What functions are served by fimbriae?
10. What functions may be served by mesosomes?
11. What is volutin?
12. What would be a distinguishing characteristic of bacteria that contain chromatophores?
13. Where are the axial filaments located in spirochetes and what function do they serve?
14. What is a plasmid? How does it differ from the bacterial chromosome?
15. What is an endospore? What genera of bacteria form endospores?
16. Differentiate between protoplasts and spheroplasts.
17. Why do gram-positive bacteria retain the blue dye and gram-negative organisms lose it during the Gram staining technique?

SUPPLEMENTARY READING

Aldrich HC, Todd WJ: *Ultrastructural Techniques for Microorganisms.* New York, Plenum, 1986.

Berg HC: How bacteria swim. *Sci Am* 223(2):36, 1975.

Beutler B, Cerami A: Tumor necrosis, cachexia, shock and inflammation: A common mediator. *Annu Rev Biochem* 57:505, 1988.

Brock TD: The bacterial nucleus: A history. *Microbiol Rev* 52:397, 1988.

DePamphilis ML, Adler J: Fine structure and isolation of the hook-basel body complex of flagella from *Escherichia coli* and *Bacillus subtilis.* *J Bacteriol* 105:384, 1971.

Hancock I, Poxton I (eds): *Bacterial Cell Surface Techniques.* Chichester, England, Wiley Interscience, 1988.

Inouye M (ed): *Bacterial Outer Membranes as Model Systems.* New York, Wiley Interscience, 1987.

Macnab RM, Aizawa SI: Bacterial motility and the bacterial flagellar motor. *Annu Rev Biophys Bioeng* 13:51, 1984.

Nikaido H, Nakae T: The outer membrane of gram-negative bacteria. *Adv Microbiol Physiol* 20:164, 1979.

Ordal G: Bacterial chemotaxis: a primitive sensory system. *Bioscience* 30:408, 1980.

Rogers HJ: *Bacterial Cell Structure.* Aspects of Microbiology Series, Washington, DC, American Society for Microbiology, 1983.

Shaw PJ, et al: Three-dimensional architecture of the cell sheath and septa of *Methanospirillum hungatei.* *J Bacteriol* 161:750, 1985.

Chapter 4

Properties of Biological Molecules

OBJECTIVES

After studying this chapter, you should

1. Know the basic components of proteins, polysaccharides, and lipids.

2. Understand how phospholipids interact to form a cell membrane.

3. Be able to describe what is meant by the tertiary structure of proteins.

4. Know what enzymes do and be able to list the general types of enzymes.

5. Understand how energy is liberated and stored.

Only a little over 100 years ago, Pasteur suggested that microorganisms could cause chemical changes in their environment. Specifically, he proposed that the conversion of grape juice into wine was a direct result of the action of yeast present in the grape juice. Even Pasteur could not imagine the mechanism whereby this was accomplished, but he concluded that it was a property of the living yeast cell. Imagine the surprise of the scientific community when some years later Hans Buchner produced alcohol from sugar using the cell-free contents of the yeast. Buchner was able to break the yeast cells open and, using the soluble contents of the yeast cell, convert sugar to alcohol. This experiment demonstrated that it is possible to take cells apart and study the individual reactions involved in thousands of microbial conversions.

We have learned a great deal since the experiments of Pasteur and Buchner. We now know that the conversion of sugar into alcohol is not a simple splitting of the sugar molecule to yield alcohol; rather, it is a series of a dozen separate, successive reactions each of which is directed by a specific molecule within the cell. In this chapter we discuss what these directing molecules are, what they do, and how they work— not only for the production of alcohol from sugar but also for the many thousands of reactions carried out by all living cells. But first let us look at some of the major kinds of molecules that must be synthesized in order for a cell to replicate itself.

BIOLOGICAL MOLECULES

Thousands of different kinds of large molecules are found in all cells (including bacteria). If we examine these biologically important macromolecules, we find that they comprise various types designated as proteins, polysaccharides, lipids, and nucleic acids. We shall discuss the nucleic acids in Chapter 7, but here let us take a brief look at the structures of each of the other types of macromolecules. Before we can begin to discuss their chemical structure, however, we must establish a small vocabulary of chemical terms. There are a number of standard chemical groups that impart specific properties to any molecule in which they exist. Some of the more common of these groups are shown in Table 4.1; the reader must become familiar with their names and configurations in order to follow the descriptions of the molecules of life.

Chemical Bonds

A review of any beginning chemistry text will familiarize the reader with basic atomic structure, in which each atom consists of a positively charged nucleus surrounded by successive shells of orbiting electrons. Moreover, one sees that the outermost shell of atomic electrons almost always has a minimum of eight electrons for maximum stability (exceptions are hydrogen and helium). Thus, atoms with less than that number of electrons will exchange or share electrons with other atoms, resulting in the formation of new molecules joined by chemical bonds. Depending on whether the electrons are shared or transferred results in the formation of either a covalent or an ionic bond.

Covalent bonds are formed when two atoms share one or more electrons in their outer shell of electrons. This may be as simple as a molecule of hydrogen (H_2), oxygen (O_2), or nitrogen (N_2), or it may involve different atoms, such as a molecule of methane (CH_4) or a molecule of glucose ($C_6H_{12}O_6$). The basic structure of most organic molecules consists of atoms of carbon, hydrogen, oxygen, nitrogen, sulfur, and phosphorus joined together with covalent bonds. Such bonds vary in strength, but in general they are said to be strong because it takes rather drastic conditions (such as boiling in strong acids or alkali) to break

MICROBIOLOGY MILESTONES

A textbook written 40 years ago described the molecular biology of diphtheria as follows: "The diphtheria bacilli become established on the mucous surfaces of the nose and throat and secrete a powerful toxin which is chiefly responsible for the symptoms." That's all! There was no concept of the molecular role of diphtheria toxin, nor was it even known that many other bacteria, such as *Escherichia coli*, *Shigella dysenteriae*, *Bordetella pertussis*, *Bacillus anthracis*, *Vibrio cholerae*, and many, many more, also secrete powerful toxins that are primarily responsible for the symptoms of the diseases they cause.

Although the symptoms of diseases caused by these toxin-secreting bacteria vary widely, all such toxins have one thing in common. They act as enzymes. Thus, it is the product of an enzymatic reaction that causes the cell death in diphtheria, fluid secretion in cholera and other diarrheas, the cell death and bloody diarrhea in dysentery, and many of the symptoms of whooping cough and anthrax, resulting from the disruption of hormonally regulated reactions. We see, therefore, that although enzymes are essential for life, the introduction of uncontrolled enzymes into our bodies may result in illness and, frequently, death.

TABLE 4.1 Structures of Common Chemical Groups

Chemical Group	Name of Group	Biological Class in Which Molecule Is Found[a]
$-\overset{\overset{\displaystyle O}{\|\|}}{C}-OH$	Carboxyl	Acids
$-\overset{\overset{\displaystyle O}{\|\|}}{C}-H$	Aldehyde	Sugars
$-OH$	Hydroxyl	Many biological molecules, but characteristic of alchohols
$-\overset{\overset{\displaystyle H}{\|}}{\underset{\underset{\displaystyle H}{\|}}{C}}-H$	Methyl	Many organic molecules
$-\overset{\overset{\displaystyle O}{\|\|}}{C}-$	Keto	Many biological molecules, especially sugars
$-NH_2$	Amino	Amines and amino acids
$-SH$	Sulfhydryl	Amino acids

[a] Many of these chemical groups are found in a wide range of organic compounds. Listed here are only those types of biological molecules with which we shall be concerned in bacterial nutrition and metabolism. Note also that all carbon atoms have four bonds and can, therefore, react with four different molecules; oxygen atoms possess two bonds, nitrogen atoms three bonds, and hydrogen atoms only a single bond.

them. Figure 4.1 provides a schematic diagram of covalent bonds.

Ionic bonds are formed when an atom donates one or more electrons to another atom that requires such electrons to fill its outer shell. This process thus leaves the electron donor with a positive charge and the electron acceptor with a negative charge. The electrostatic attraction between these two atoms results in an ionic bond (see Figure 4.2).

Hydrogen bonds are formed when a covalently bound hydrogen atom is electrostatically attracted to another bound nitrogen or oxygen molecule. Such attraction occurs because the nucleus of a bound hydrogen atom has a slight positive charge, whereas the nucleus of a covalently bound oxygen or nitrogen atom has a slight negative charge. Such bonds are considerably weaker than covalent bonds, but hydrogen bonds are extremely important in establishing the final tertiary structure of very large molecules of protein or nucleic acids.

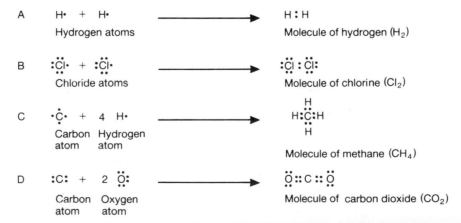

Figure 4.1 Schematic example of covalent bonds in which atoms bind together, sharing electrons (represented by a dot) in their outer atomic shell.

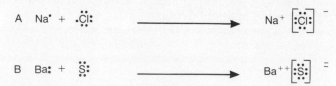

Figure 4.2 Schematic illustration of ionic bonds in which one electron, or more, is transferred from one atom (leaving it with a positive charge) and given to a second atom (giving it a negative charge). The ions are thus held together by electrostatic attraction.

Proteins

Proteins are composed of one or more chains of amino acids joined to each other through very strong covalent bonds. The specific bond that joins one amino acid to another is called a **peptide bond,** and, depending on the number of amino acids involved, the resulting molecule is referred to as a **dipeptide, tripeptide, tetrapeptide,** or **polypeptide** (meaning many amino acids). Some proteins may contain as few as 50 or 60 amino acids (e.g., the hormone insulin), whereas others may have many thousands of amino acids all joined together. Regardless of the size, if we were to break all of the peptide bonds in a protein and liberate the free amino acids, we would find that there are only 20 different amino acids. Thus, the hundreds of thousands of different proteins all contain the same 20 building blocks but differ in the sequence in which they exist in the protein.

Each of the 20 amino acids consists of a backbone of carbon atoms (varying from 2 to 11) to which are linked molecules of hydrogen (H), oxygen (O), nitrogen (N), and, in some cases, sulfur (S), and all possess a terminal carboxyl group and an amino group joined to the carbon adjacent to the carboxyl carbon. Figure 4.3 shows the structure of five amino acids and illustrates how they can be linked together through peptide bonds to form a protein. If a protein is boiled in strong acid, the peptide bonds are broken and free amino acids are liberated; however, we shall see later that these very stable bonds can be easily broken by certain cellular enzymes.

Note, however, that the formation of a peptide bond is really a dehydration synthesis, since water is released during its formation. Similarly, when such bonds are broken, the hydrogen and hydroxyl components of water are recombined with the newly liberated amino acids. Because of the insertion of water during the breaking of these bonds, such a reaction is referred to as **hydrolysis.**

However, this is not all that is involved in protein structure. The long chains of amino acids fold back and forth on each other in a very specific manner to form what is termed the protein's **tertiary** structure.

Figure 4.3 (a) The structural formulas of five common amino acids. (b) Two amino acids are joined by splitting out a molecule of water (HOH), resulting in the formation of a bond between the amino nitrogen from one amino acid and the carboxyl carbon of the other amino acid.

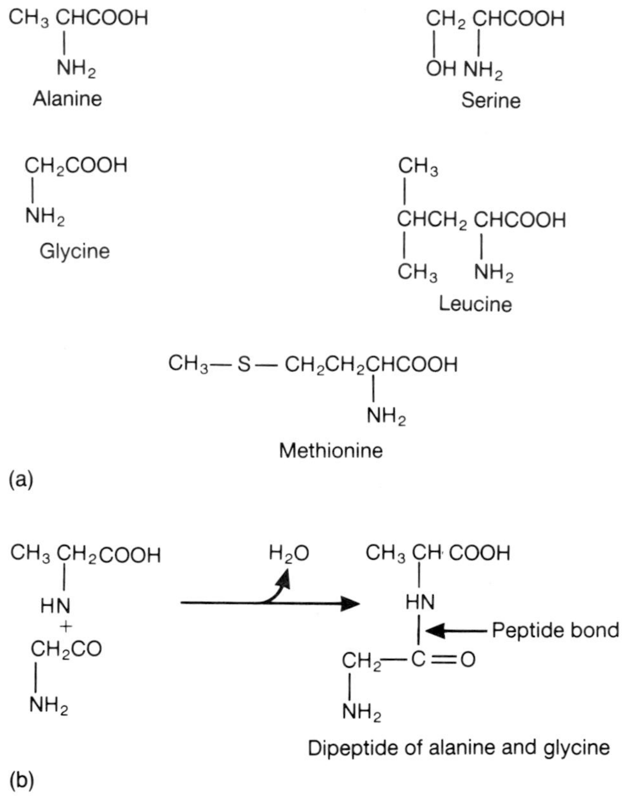

CH_3 CHCOOH
|
NH_2
Alanine

CH_2 CHCOOH
| |
OH NH_2
Serine

CH_2COOH
|
NH_2
Glycine

CH_3
|
CHCH$_2$ CHCOOH
| |
CH_3 NH_2
Leucine

CH_3— S — CH_2CH_2CHCOOH
|
NH_2
Methionine

(a)

CH_3 CH_2COOH H_2O CH_3 CH·COOH
| |
HN   HN   ← Peptide bond
+ |
CH_2CO CH_2— C=O
| |
NH_2 NH_2

Dipeptide of alanine and glycine

(b)

The tertiary structure is maintained by disulfide bonds (—S—S—) formed between two sulfur-containing amino acids and also by hydrogen bonds that are weak electrical interactions between covalent bonds carrying opposite charges (Figure 4.4). The specific function of virtually all proteins is a property of its tertiary structure, and any treatment that alters that structure (such as heating) will destroy the function of that protein. Such altered proteins are no longer functional and are referred to as **denatured.**

What are the functions of proteins? In general, they can be categorized as either structural or catalytic. As the name implies, structural proteins contribute to

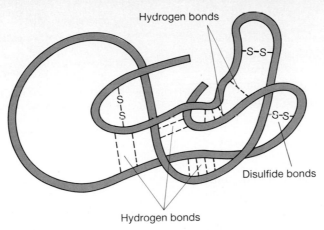

Figure 4.4 Simplified protein illustrating how disulfide bonds and hydrogen bonds determine its final tertiary structure.

the anatomy of certain cellular constituents such as ribosomes, membrane receptors, the porins (openings) in the outer membrane of gram-negative bacteria, and bridges that interconnect various constituents in the cell envelope of many bacteria. By and large, however, the majority of proteins are involved in directing the thousand-and-one chemical reactions required for cell growth and maintenance. Such proteins are called **enzymes,** and their only function is to direct specific chemical reactions necessary for the growth and maintenance of the cell. We shall discuss these proteins in greater detail later in this chapter.

Polysaccharides

A polysaccharide is a long chain of sugar molecules, each joined to the next by a **glycosidic bond.** Such bonds are formed by splitting out a molecule of water as shown in Figure 4.5. Many different sugars are found as components of polysaccharides. Some polysaccharides, such as glycogen or xylan, are chains of only a single type of sugar (i.e., glucose or xylose). Other polysaccharides, such as the lipopolysaccharide found in the outer membrane of gram-negative cells or the capsules of *Streptococcus pneumoniae,* may contain as many as nine different types of sugar residues. Figure 4.5 shows the structural formula of several common sugars and illustrates how two molecules of glucose can be joined together to form the disaccharide maltose. Simple sugars (monosaccharides) may have as many as nine carbon atoms, but the most common ones have five or six and are known as pentoses and hexoses, respectively. And, as with proteins, boiling a polysaccharide in an acid solution or treatment with the correct enzymes will break the glycosidic bonds by the insertion of water, liberating the free monosaccharides.

Polysaccharides are necessary for several functions, some of which vary from one cell type to another. For example, many plant cells have cell walls that are composed entirely of cellulose. The rigid cell

Figure 4.5 (a) Structural formulas of three common sugars. (b) When a molecule of water (H_2O) is split out, a glycosidic bond is formed between the two molecules of glucose with the formation of a disaccharide, maltose.

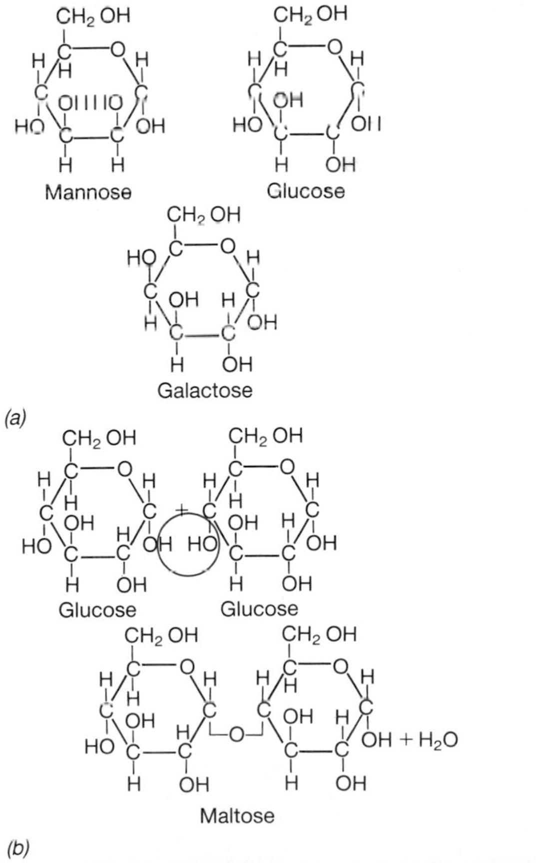

(a)

Mannose

Glucose

Galactose

(b)

Glucose   Glucose

Maltose

walls found in most bacteria are also composed primarily of polysaccharides. Other polysaccharides, such as glycogen and starch, are stored as granules within a cell as food reserves. In addition, many proteins have chains of polysaccharide attached to specific amino acids. Such **glycoproteins** are exceedingly common as components of cell surfaces.

Fatty Acids and Lipids

Simply fatty acids are molecules that terminate with a carboxyl group but otherwise usually contain only carbon and hydrogen, as shown in Figure 4.6. Fatty acids with more than eight carbon atoms are insoluble in water but are soluble in organic solvents, such as chloroform or methanol. When fatty acids are linked to glycerol through an **ester bond** (by splitting out a molecule of water), they form a lipid known as a **triglyceride** (see Figure 4.6). Heating the triglyceride in acid, or treatment with specific enzymes, breaks the ester linkage by inserting a molecule of water and

Phosphatidylethanolamine

Phosphatidylcholine

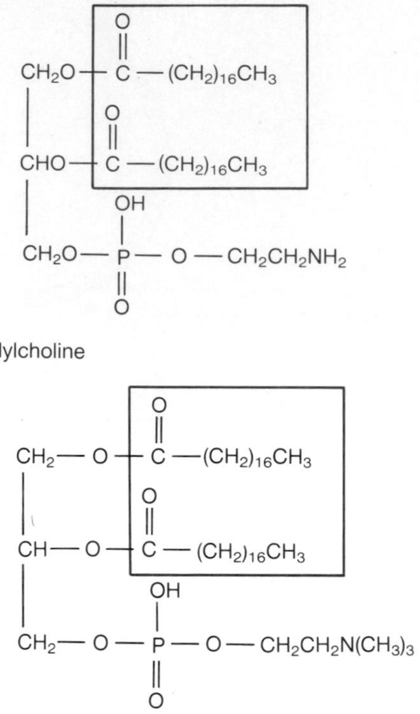

Figure 4.7 Structural formulas of two major phospholipids that make up a major portion of the lipids found in cell membranes. That portion of each molecule enclosed in the box is the hydrophobic part of the phospholipid and, as a result, remains embedded in the membrane. The remaining hydrophilic part of the molecule extends outward from the membranes.

Figure 4.6 (a) Structural formulas of four simple fatty acids. (b) Where a molecule of water (HOH) is split out from each hydroxyl group of glycerol and each carboxyl group of palmitic acid, an ester bond linking the fatty acid to the glycerol is formed.

$CH_3(CH_2)_{12}COOH$
Myristic acid

$CH_3(CH_2)_{14}COOH$
Palmitic acid

$CH_3(CH_2)_{16}COOH$
Stearic acid

$CH_3(CH_2)_{18}COOH$
Arachidonic acid

(a)

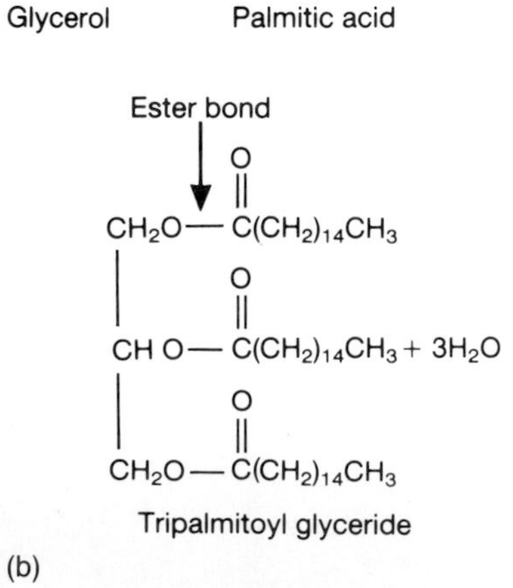

(b)

liberates the free fatty acids and glycerol. In many glycerides, the number 1 and 2 carbons of glycerol are esterified to fatty acids (Figure 4.7), but the number 3 carbon is linked through a phosphodiester linkage to choline or ethanolamine (Figure 4.7). Such structures, designated as phospholipids, possess the usual property in which part of the molecule is soluble in water (glycerol and phosphoethanolamine or phosphocholine) and is termed hydrophilic (water-loving), whereas the fatty acid components of the phospholipid are insoluble in water and are thus hydrophobic (water-fearing). When placed in an aqueous environment, phospholipids form a structure that is two molecules thick and oriented in such a manner that their hydrophilic head groups are facing outward into the aqueous environment while their hydrophobic groups face inward. This is the basic structure of all cell membranes (Figure 4.8). In addition, fatty acids are occasionally covalently bound to polysaccharides or proteins to form structures known as lipopolysaccharides and lipoproteins, respectively. As one could surmise, such a union imparts unusual properties, and these complex lipids are used primarily as structural components of the cell.

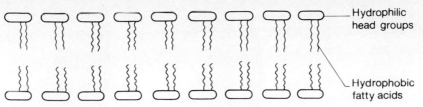

Hydrophilic
head groups

Hydrophobic
fatty acids

Figure 4.8 Basic membrane structure. Phosopholipids form the structural part of a membrane by aligning themselves into a bilayer in which their hydrophilic head groups face outward into the aqueous environment while their hydrophobic fatty acids face inward away from the aqueous environment. Although not shown here, biological membranes also contain proteins that carry out many reactions necessary to the cell.

ENZYME CATALYSIS

We have now briefly described the structural features of those macromolecules that are the major building blocks of the cell. We shall turn our attention now to molecules involved in directing chemical reactions, resulting in the growth and maintenance of the cell. Molecules whose biological function is to direct specific chemical reactions are called **enzymes.** Let us see what kinds of enzymes are found in a typical cell and discuss briefly how they function to direct the biosynthesis of cell components and provide the necessary energy for these reactions.

To begin our discussion, let us review a simple chemical reaction:

$$H_2 + \tfrac{1}{2}O_2 \xrightarrow{\text{Pt}} H_2O$$

The platinum (Pt) itself is not changed; rather, it speeds up the reaction so that it is many thousands of times faster than it would be otherwise. Such a substance is called a **catalyst.**

The molecules in the cell cytoplasm that convert sugar into alcohol are called enzymes. Enzymes are very large protein molecules. They can be thought of as organic catalysts in that they speed up biological reactions by millions of times. Enzymes are, in fact, very active. A **substrate** is a compound on which an enzyme exerts its catalytic effect. An average enzyme molecule will react with 300 to 400 molecules of its substrate each second; many are even more active. But note: like the inorganic catalysts, enzymes are not used up in the reaction. The student should bear in mind that there are many thousands of enzymes in the biological world. The purpose of this large array of enzymes is twofold: (1) to break down or oxidize food material to provide energy and (2) to use this energy for the synthesis of new cell material.

It is beyond the intent of this text to go into a detailed discussion of how an enzyme works to effect a change in its substrate. However, we can certainly list some general facts about enzyme reactions that will provide the student with a schematic or visual concept of these processes. First, we know that enzymes are moderately specific toward the substrates with which they will react. The words *moderately specific* are used to indicate that some enzymes will react with only one substrate, whereas others may react with several closely related substrates. The limitation resides in the fact that the substrate must physically fit into an active site on the enzyme. In some cases, closely related substrates can fit close enough structurally to undergo a reaction, whereas in others only one substrate will fit into the active center of the enzyme. This concept is schematically diagrammed in Figure 4.9. How the enzyme accomplishes the changes on its substrate can best be explained by stating that the effect of the enzyme is to lower the energy of activation for the specific reaction. Energy of activation can be compared to being in a valley and having to climb over a mountain to get to the next valley, which is lower than the one from which you started. The overall direction of the move is downhill, but the journey is partially uphill. Our figurative mountain is the energy of activation for a particular reaction as shown schematically in Figure 4.10. In other words, the reactants must acquire sufficient energy, probably manifested as a conformational change, to pass over an "energy barrier" before a reaction can take place. A simple chemical demonstration would be to mix hydrogen and oxygen together in a vessel. Even though the reaction of hydrogen and oxygen to form water will liberate considerable energy and should occur spontaneously (like going downhill), nothing will happen unless we put in a spark that will provide the necessary energy of activation or provide a catalyst (such as platinum) that will lower the energy of activation. Enzymes are protein molecules that catalyze biological reactions by lowering the energy of activation for a specific reaction. Thus, a reaction, such as the hydrolysis of the disaccharide, sucrose, would occur in the presence of the enzyme **invertase** at a rate many millions of times faster than in its absence.

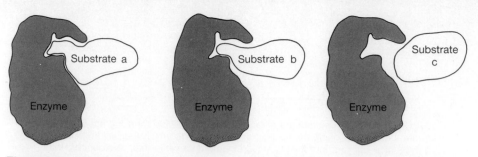

Figure 4.9 Schematic diagram of structural relationship between an enzyme and its substrate. (*a*) Active center of enzyme physically conforms very closely to structure of substrate to be modified. (*b*) Active center of enzyme fits moderately well to substrate conformation and in all likelihood will catalyze its enzymatic reaction, but at a slower rate than that for substrate a. (*c*) Substrate c is too large to conform to active center and, hence, the enzyme displays no activity with this substitute.

Figure 4.10 Schematic diagram depicting energy of activation.

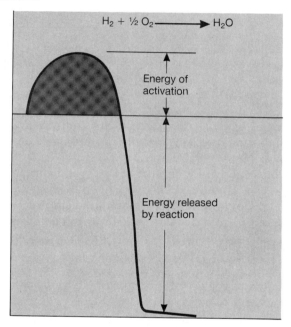

Classification of Enzymes

Because of the many types of reactions catalyzed by enzymes, it is essential to have a system for naming and classifying these complex proteins. Most enzymes are named by adding the suffix **ase** to the substrate or product of their reaction. Thus, an enzyme that removes phosphate groups would be termed a **phosphatase;** one that hydrolyzes a peptide bond, a **peptidase;** and one that hydrolyzes maltose to glucose, a **maltase.** Enzymes also have been classified into various categories according to the types of reactions they catalyze. One comprehensive classification is that of the Commission of Enzymes of the International Union of Biochemistry, in which they are classified

into six main groups, as follows:

  Oxidoreductases

  Transferases

  Hydrolases

  Lyases

  Isomerases

  Ligases (synthetases)

Oxidoreductases

Oxidoreductases carry out the specific energy-releasing reactions for the cell. This is most often accomplished through enzymes called **dehydrogenases,** which, by removing hydrogen, also remove electrons. As these electrons are passed to an electron (or hydrogen) acceptor, energy is released that the cell is able to trap and store as chemical energy. *Oxidation is the only major chemical source of energy for a cell,* and biological oxidation is most frequently accomplished by the removal of hydrogen. Over 200 different oxidoreductases have been described.

Transferases

Transferases are a very large group of enzymes that catalyze the transfer of chemical groups from one substrate to another. They do not cause the liberation of energy from a substrate but instead convert a substrate to a compound that may then be oxidized or used for the synthesis of cellular material. A number of special names are used to indicate reaction types (e.g., kinase) to indicate a phosphate transfer from adenosine triphosphate (ATP) (or other phosphate donor) to the named substrate (i.e., glucokinase transfers a phosphate group from ATP to glucose). A pyrophosphokinase is used for a similar transfer of a pyrophosphate, and a phosphomutase is used in catalyzing an intramolecular phosphate transfer. The following types of groups may also be transferred by enzymes of this class:

1. One-carbon groups, such as carboxyl, formyl, or methyl groups
2. Two-carbon aldehyde groups
3. Acyl groups, such as acetyl
4. Glycosyl groups (any enzyme transferring a simple sugar to form a larger sugar or starch is included here)
5. Nitrogenous groups, particularly the transfer of amino groups to keto acids to form new amino acids necessary for protein synthesis

Hydrolases

Hydrolases are so called because they hydrolyze large molecules into smaller usable components. Included in this group are enzymes that break down starch or cellulose to glucose or maltose, proteins to amino acids, and fats to glycerol and fatty acids. In bacteria these enzymes are excreted by the cell into its external environment (and are, therefore, called exoenzymes); thus, large insoluble compounds can be broken down in the presence of water into soluble molecules that can enter the bacterial cell and serve as nutrients. General examples of hydrolytic enzymes are:

- **Cellulases**—hydrolyze cellulose to glucose
- **Amylases**—hydrolyze starch to maltose
- **Proteases**—hydrolyze proteins to amino acids
- **Lipases**—hydrolyze fats to glycerol and fatty acids
- **Nucleases**—hydrolyze ribonucleic acid (RNA) and deoxyribonucleic acid (DNA) into smaller soluble molecules

Many of the exoenzymes or hydrolases of bacteria are therefore really digestive juices. In humans the digestive juices break food down into substances that are usable by the body. The bacterial exoenzymes do much the same thing. They are of particular interest to us because they are largely responsible for the ability of bacteria to "digest" a most extraordinary range of substances. Not only fence posts and other wooden objects but many other things of organic origin can serve ultimately as bacterial food—paper, asphalt paving, bituminous coatings, dead animal and plant bodies, excrement, and petroleum (including gasoline), among other things.

Some exoenzymes are potent toxins and contribute to the disease-producing potential of a bacterium by catalyzing reactions that are harmful to the host.

We shall have occasion to discuss a number of these in Unit III of this text.

Lyases

Lyases (1) remove groups from substrates nonhydrolytically, usually leaving double bonds, or (2) add groups to double bonds. This most commonly involves the removal of water (L-malate $\rightarrow$ L-fumarate + H_2O), the removal of ammonia (serine + H_2O $\rightarrow$ pyruvate + NH_3 + H_2O), or the removal of a carboxyl group (L-lysine $\rightarrow$ cadaverine + CO_2).

Isomerases

Isomerases catalyze the process of isomerization, which results in the transfer of two hydrogen atoms from an adjacent carbon to an aldehyde group. The result is the reduction of the aldehyde group to a primary alcohol group and the formation of a keto group on the carbon atom that donated its hydrogen atoms (D-mannose $\rightarrow$ D-fructose). Other enzymes in this group include **racemases,** which convert L-isomers to D-isomers and vice versa (i.e., L-glutamate $\rightarrow$ D-glutamate, or L-lysine $\rightarrow$ D-lysine), and epimerases, which intraconvert sugars by changing the position of specific hydroxyl groups.

Ligases

Ligases catalyze the linking together of two molecules. They have been known for many years as synthetases. They take part in many of the steps involved in the synthesis of macromolecules such as proteins and many other compounds used as intermediates in nucleic acid biosynthesis.

It should be stressed that the action of cellular enzymes is neither sporadic nor disorganized. All cells, including bacteria, have enzyme systems in which the enzymes act in an orderly sequence until a particular series of reactions has been completed. Many enzymic systems act in a kind of chain reaction; the product of one reaction becomes the substrate for the next reaction, and so on. You will see examples of this in the next chapter.

Coenzymes

A number of enzymes (particularly those that remove some portion of the substrate molecule, such as dehydrogenases, lyases, and transferases) require a second molecule to accept the portion removed and carry it to the next or final acceptor. These carrier molecules, called **coenzymes,** are not proteins, but are for the most part small organic compounds. It should be pointed out that the B vitamins we eat in our food (or take in a vitamin tablet) are used by our bodies as precursors to form coenzymes. Bacteria require the same B vitamins and the same coenzymes. Many bacteria are able to synthesize all of their own requirements, but others require an external source of one or more of these substances.

Factors That Influence Enzyme Activity

Since enzymes are proteins and exist in living cells, it is no surprise that even with the proper enzyme, coenzyme, and substrate, certain conditions must be met in order for the enzyme to be active.

pH

The hydrogen ion concentration (i.e., the acidity or alkalinity) of the solution markedly affects the activity of an enzyme. This is because the amino acids making up the active center of the enzyme must be in the proper ionization state in order to be active. Some enzymes are active at rather low acid pH values, pH 3 to 4, and others may be active at alkaline pH values as high as 11 or 12, but by far the majority of enzymes demonstrate maximum activity in the neutral range pH 6 to 8. One can ascertain for each enzyme the minimum, maximum, and optimum pH for activity (see Figure 4.11). In addition to requiring the proper pH for enzymatic activity, many enzymes require divalent cations such as Mg^{2+}, Mn^{2+}, and Ca^{2+}, presumably to bind the substrate to the active center of the enzyme.

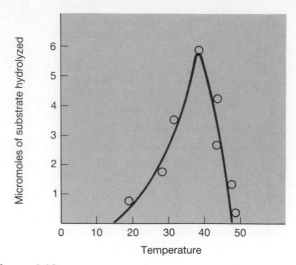

Figure 4.12 Effect of temperature on enzyme activity.

Temperature

The rate at which an enzyme will react increases up to an optimum temperature, and then decreases until the maximum temperature is reached (see Figure 4.12). The decrease in activity as the temperature increases above the optimum is usually due to destruction of the enzyme. Most enzymes have their optimum activity at a temperature between 30 and 40°C.

Substrate Concentration

As shown in Figure 4.13, the rate of enzyme activity will increase as the substrate concentration increases, until the enzyme is saturated with its substrate. The

Figure 4.11 Effect of pH on enzyme activity.

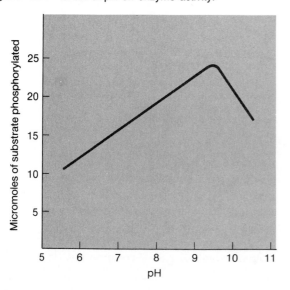

Figure 4.13 Effect of substrate concentration on enzyme activity.

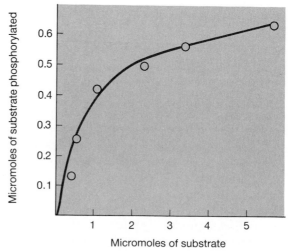

concentration of substrate that saturates an enzyme is a measure of the affinity an enzyme has for its substrate. Thus, an enzyme with a high affinity will display maximum activity at very low substrate concentrations, whereas one with a low substrate affinity will require much higher concentration before reaching its maximal rate.

ENERGY LIBERATION AND STORAGE

It is not within the scope of this text to go into a detailed discussion of the thermodynamics of energy storage in biological systems. The general concepts are not difficult, however, and the beginning student should not have difficulty in understanding the way in which a cell obtains and stores energy to be used for the synthesis of new cellular material. *These same principles are used by all living things, including bacteria and humans.*

When chemical reactions are carried out in a test tube, the energy liberated is usually lost as heat. However, a prerequisite of living cells is that at least part of the energy liberated by the oxidation of a substrate be trapped and stored so that it can be used for the many vital processes that require energy (e.g., the synthesis of new cellular material, moving a ribosome along messenger RNA during the synthesis of protein, dancing, or playing football).

Let us consider the oxidation of the sugar glucose to carbon dioxide and water, as shown in the following equation:

$$\underset{\text{Glucose}}{C_6H_{12}O_6} + \underset{\text{Oxygen}}{6O_2} \rightarrow \underset{\text{Carbon dioxide}}{6CO_2} + \underset{\text{Water}}{6H_2O}$$

Although this appears as a single reaction, the equation actually represents the end result of 20 to 30 individual reactions and the liberation of 675,000 calories of energy per mole of glucose (180 g of glucose). Only a few of the reactions involved are oxidations, and it is these oxidations that liberate the energy the cell can trap and store as chemical energy for later use. Chemical energy is trapped and used in the form of high-energy phosphate bonds in ATP.

ATP possesses three phosphate groups (see Figure 4.14). The two terminal phosphates are linked by high-energy phosphate bonds, so called because each is capable of liberating approximately 8000 calories of energy per mole; the phosphate bond joined to the ribose part of ATP is referred to as a low-energy phosphate bond because it possesses only about 3000 ca-

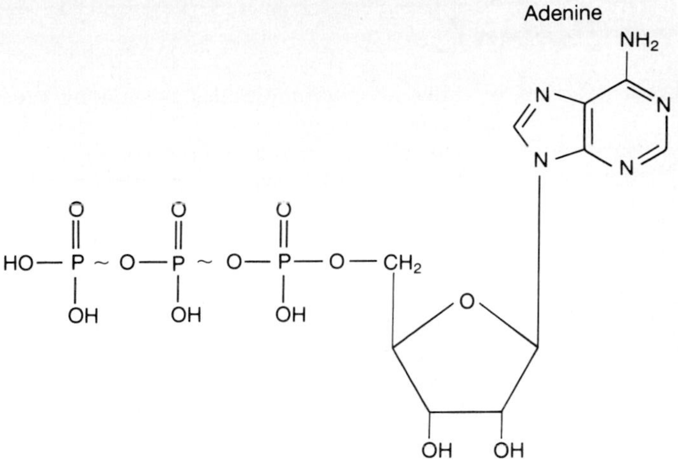

Figure 4.14 Structure of ATP. Note that terminal phosphate bonds designated as "~" are high-energy bonds.

lories of energy per mole. It is the high-energy phosphate bonds, formed during the oxidation of nutrients, that trap and store the liberated energy.

The precise mechanisms by which these bonds are formed are complex, and the details will not be discussed in this text. Suffice it to say that there are two ways to trap the energy liberated by the oxidation of a substrate: (1) substrate phosphorylation, which occurs during the degradation of a substrate (see Chapter 6) and (2) oxidative phosphorylation, which traps the energy liberated by the passage of electrons from the substrate through the cytochrome system. The latter process is complex but it appears to result from a chemical charge difference formed between the inner and outer surface of the cytoplasmic membrane during the stepwise passage of electrons to oxygen. It should be pointed out that the aerobic degradation of a substrate (resulting in oxidative phosphorylation) yields far more energy for the cell than does an anaerobic process that utilizes only substrate phosphorylation. For example, the anaerobic breakdown of glucose to form lactic acid (a major source of muscle energy) yields a net of only two energy-rich phosphate bonds for each molecule of glucose degraded. On the other hand, the complete oxidation of glucose to carbon dioxide and water results in the formation of 36 energy-rich bonds. Thus, it is apparent that aerobic metabolism is considerably more efficient than anaerobic metabolism.

Subsequent chapters of this text will provide some details concerning precisely how carbohydrates are degraded by bacteria and how the accumulated ATP is used in the synthesis of cell components.

SUMMARY

Thousands of different macromolecules are among the structural and catalytic components of a cell. In general, such molecules can be categorized as proteins, polysaccharides, and fats, or lipids. Proteins consist of long chains of amino acids that are joined to each other by peptide bonds. Similarly, polysaccharides are long chains of sugar molecules, each of which is joined to the next by a glycosidic bond. And fats are formed when three molecules of a fatty acid become linked to a molecule of glycerol via ester bonds. Phospholipids are a special type of fat in which one hydroxyl group of glycerol is linked to phosphorylcholine or phosphorylethanolamine while the other two hydroxyls of the glycerol molecule are ester-linked to fatty acids. Phospholipids are major components of cell membranes.

The original observation by Pasteur that the conversion of grape juice to wine was accomplished by living yeast cells has been extended during the past 100 years to the point that we can now separate many of the complex chemical changes carried out by cells of all kinds into individual reactions. These reactions are catalyzed by specific proteins called enzymes. The substrate specificity of any enzyme is a result of the substrate's physically fitting into the active center of the enzyme in order to undergo the specified reaction. There are many ways to classify enzymes. The official international classification divides them into six main classes based on the type of reaction catalyzed by the enzyme, namely, oxidoreductases, transferases, hydrolases, lyases, isomerases, and ligases.

Many enzymes that remove a group such as H, NH_2, or CO_2 require an immediate acceptor for the group removed. This acceptor is a small organic molecule called a coenzyme. After accepting the chemical group removed from the substrate by the enzyme, the coenzyme transfers it to another substrate and, in turn, is recycled to accept another group. Most of the B vitamins necessary for nutrition are precursor molecules for the biosynthesis of coenzymes by the cell.

The activity of an enzyme is affected by the external environment. Thus, it must have the correct pH and temperature for maximum velocity. Various metal cations such as Mg^{2+}, Mn^{2+}, and Ca^{2+} are frequently necessary for enzyme activity.

Energy resulting from oxidative reactions in the cell is trapped and stored as high-energy phosphate bonds in ATP. During the aerobic oxidation of a substrate to carbon dioxide and water, electrons are transferred to oxygen to form water. The energy released during this electron exchange is trapped by a process called oxidative phosphorylation, which ultimately results in the formation of energy-rich phosphate bonds in the form of ATP.

QUESTIONS FOR REVIEW

1. Define "covalent bond," and show how it differs from an ionic bond.
2. What new concept of cell metabolism was gained from the work of Hans Buchner?
3. List what you know concerning the structure of proteins, polysaccharides, and fats.
4. What property does a phospholipid possess that enables it to form a membrane structure spontaneously?
5. What is meant by the "tertiary" structure of proteins?
6. What are the properties of a catalyst?
7. Why are enzymes considered to be organic catalysts?
8. What is meant by the statement that enzymes are moderately specific?
9. What seems to be the major criterion that determines whether an enzyme will react with a substrate?
10. List the six classes of enzymes and give a brief example of each.
11. What is a coenzyme?
12. How is enzyme activity affected by pH? By temperature?
13. How is energy stored in a cell?
14. Are all phosphate bonds high-energy bonds? Explain.
15. What is oxidative phosphorylation?

SUPPLEMENTARY READING

Becker WM: *Energy and the Living Cell.* Philadelphia, Lippincott, 1977.

Blackburn S: *Enzyme Structure and Function.* New York, Dekker, 1976.

Ferdinand W: *The Enzyme Molecule.* New York, Wiley, 1976.

Hinkle P, McCarty R: How cells make ATP. *Sci Am* 238(3):104, 1978.

Kamp G: *Cell Biology,* 2nd ed. New York, McGraw-Hill, 1984.

Weinberg RA: The molecules of life. *Sci Am* 253(4):48, 1985.

Chapter 5

Bacterial Nutrition

BACTERIAL NUTRITION

All organisms, whether they be bacteria, humans, or trees, need a constant supply of food in order to continue to live. The kind of food used and the methods by which it is assimilated and utilized is called **nutrition**. Whatever the organism, nutrition has a twofold purpose: (1) to synthesize (build up) protoplasm and (2) to supply energy for all life processes.

Throughout the living world, protoplasm is largely a homogeneous substance—it is essentially the same whether we find it in ourselves or in bacteria. Therefore, it follows that most living things need more or less the same basic food. A better word to use when discussing the direct requirements of a cell is **nutrients.** Food is the raw material from which nutrients are derived, and it can assume an infinite variety of forms. Nutrients are the substances, inorganic and organic, that actually pass in solution through the cytoplasmic membrane. To obtain nutrients from food, the cell must be able to **digest** the food—that is, to convert the large, complex molecules of protein, carbohydrate, and lipid to small, simple molecules that go readily into solution and so can enter the cell.

Precisely what nutrients do cells need to synthesize protoplasm? Here is the basic list:

1. A carbon source (e.g., carbohydrate)
2. A nitrogen source (e.g., amino acids or ammonia)
3. Certain inorganic ions
4. Essential metabolites (vitamins; possibly amino acids)
5. Water

As noted, the cell also needs a source of **energy.** Without energy (the capacity to do work), the cell could not synthesize protoplasm and carry on its other life processes. Bacteria usually obtain energy from chemical oxidations.

Where does a bacterium obtain its carbon, nitrogen, inorganic ions, and so on? Most bacteria obtain these materials from organic matter—that is, they attack organic matter, decompose it, and use what they need. This process includes three steps:

1. Decomposition of matter containing protein, carbohydrate, or lipid
2. Absorption of the simple forms of these materials
3. Synthesis of protein, carbohydrate, and lipid within the cell

There is one great difference between the nutrition of animals and that of bacteria. Most animals, from humans to protozoa, are capable of ingesting solid food. Many amoebas, for instance, can ingest solid food particles into their vacuoles. These particles are broken down by the organism's enzymes and taken into the protoplasm, where they supply energy and build up new protoplasm. On the other hand, bacteria cannot ingest solid food. Bacteria and true fungi must receive their nutrients in a solution of water, which means that digestion must take place *outside* the organism. The following sections will describe the specific types of nutrients that are required for bacterial growth.

Bacterial Nutrients

We have already discussed the fact that nutrients must be water soluble to enter the bacterial cell. Let us now briefly concern ourselves with the scope of required nutrients.

Carbon Source

With the exception of some synthetic plastics and complex organics such as a few of the pesticides, there is probably no carbon-containing compound that cannot be used by some bacterium as a carbon source for the synthesis of its protoplasm. Carbon sources even include such diverse substances as wood, asphalt, gasoline, or the carbon dioxide in the air. Most disease-producing organisms, however, obtain their carbon by metabolizing simple carbohydrates and proteins.

Nitrogen Source

Because all proteins and nucleic acids contain nitrogen, it is obvious that substantial amounts of nitrogen are required for growth. Some organisms can obtain the required nitrogen by using nitrogen gas from the air (nitrogen fixation); others can utilize inorganic sources of nitrogen, such as ammonium salts; some may require organically bound nitrogen, such as glutamine, asparagine, or peptide digests.

Inorganic Ions

All organisms require phosphate, both for use as a component of cellular structures and for the storage of energy. Sulfur, usually as sulfate, can be used by most bacteria to synthesize their sulfur-containing amino acids (cysteine and methionine). Some ions, such as Mg^{2+}, K^+, and Ca^{2+}, act as cofactors for certain enzymes and must be added to a growth medium. In addition, a number of ions are required in such minute amounts that contaminants in the medium may provide adequate supplies.

One ion that merits special consideration is iron. Many bacteria secrete compounds (given the general name of siderophores) that will react with iron to solubilize it so that it can be brought into the bacterial cell. It is interesting that, because essentially all of the host's iron is tightly bound to iron-transporting proteins (i.e., transferrin and lactoferrin), the growth of a bacterium within a host might well depend on the ability of the infecting organism to produce siderophores that can successfully compete with the iron-binding protein of the host. It has been postulated that one major difference between virulent and avirulent tubercle bacilli may be the possession of a tightly bound iron-binding siderophore (mycobactin) that permits the virulent organisms to compete with the host for available iron by obtaining iron from the host's transferrin and/or lactoferrin.

Essential Metabolites

It is readily apparent that in order to reproduce all bacteria must be able to synthesize their proteins, carbohydrates, fats, and nucleic acids—the constituents of protoplasm. But the complexity of the raw materials that bacteria utilize to accomplish this synthesis varies considerably with the particular species. Some organisms have extraordinary powers of synthesis and are able to reproduce using only inorganic materials in their cellular nutrition (*Nitrosomonas, Nitrobacter, Thiobacillus*). On the other hand, some parasitic bacteria cannot be made to grow in the laboratory unless they are supplied with highly complex organic materials such as whole blood (*Haemophilus*). Such bac-

teria have a limited enzymatic endowment and hence an equally limited ability to synthesize. For example, the synthesis of nucleic acid requires at least four to six different purines and pyrimidines, as well as the sugars ribose and deoxyribose; the synthesis of protein requires at least 20 different amino acids. Thus, a bacterium that cannot synthesize any one of these essential building blocks must have them supplied or it cannot grow. Such requirements are collectively termed **essential metabolites**. Another category of essential metabolites is vitamins: nicotinic acid (niacin), thiamine, riboflavin, pantothenic acid, and so on. Many of these vitamins are needed to form the coenzymes necessary for enzymatic action (Table 5.1). And some pathogenic bacteria such as *T. pallidum*, the cause of syphilis, and *Mycobacterium leprae*, the cause of leprosy, have never been grown on an artificial medium. Table 5.2 contains a summary of the principal bacterial nutrients, their function, and a few examples of each type.

Eucaryotic Cell Nutrients

Many types of eucaryotic cells can be grown in cell cultures, but in all cases the nutritional requirements for the growth of these cells is considerably more complex than that of procaryotic cells. In essence, eucaryotic cells require all of the nutrients described for bacteria, but in addition, most (if not all) require protein

TABLE 5.1 Summary of Water Soluble B Vitamins and Their Functions in Biological Reactions

Vitamin	Function
B_1 (thiamine)	Phosphorylated to form the coenzyme cocarboxylase; serves as a cofactor in the oxidative decarboxylation of alpha keto acids, such as pyruvic acid
B_2 (riboflavin)	Phosphorylated and adenylated to form flavin mononucleotide and flavin adenine dinucleotide, respectively; both nucleotides are active in electron transport
B_6 (pyridoxine)	Functions as a coenzyme in the deamination, transamination, and decarboxylation of amino acids
B_{12} (cyanocobalamin)	Coenzyme required for transfer of methyl groups and in the formation of succinic acid from methylmalonyl coenzyme
Niacin (nicotinic acid)	Functions as a coenzyme in the electron transport by nicotinamide adenine dinucleotide and nicotinamide adenine dinucleotide phosphate
Pantothenic acid	Component of coenzyme A, which functions in the transfer of acyl groups
Biotin	Functions as a carrier of activated carbon dioxide in carboxylation reactions
Folic acid	Accepts one-carbon fragments, which are used in the synthesis of purines, pyrimidines, and methionine

TABLE 5.2   Summary of Bacterial Nutrients

Type	Function	Examples
Carbon source	Provides energy through oxidation and provides the structural component of the cell wall	May include virtually any carbon-containing compound; varies from CO_2 in the air to very complex organic substances
Nitrogen source	Provides nitrogen for the synthesis of amino acids, nucleic acids, and coenzymes	Some species use N_2 of air, others inorganic compounds such as NO_3^- or NH_4^+; others require organic sources of nitrogen such as glutamine or asparagine
Inorganic Ions	Necessary cofactors for enzymes; storage of energy; electron transport systems	Mg^{2+}, Mn^{2+}, Fe^{2+}, PO_4^{2-}, Na^+, K^+, and even Mo for organisms fixing gaseous nitrogen
Essential metabolites	To provide complex organic compounds that an organism is unable to synthesize	Vitamins, amino acids, purines, pyrimidines, coenzymes, heme

growth factors that bind to specific receptors on the eucaryotic cell membrane. These factors are normally secreted by various cells within the body, and a number of them have been named either according to their source or in reference to the type of cell on which they act. Thus, we see names such as *platelet-derived growth factor* (PDGF), *epidermal growth factor* (EGF), *endothelial growth factor* (EGF), *granulocyte colony-stimulating factor* (G-CSF), and many more. Others have been given names such as *interleukin, insulinlike growth factor, hemopoietin,* and *erythropoietin.* The genes for most of these growth factors have now been cloned in bacteria, and purified growth factor is thus available for many of them.

It is considerably beyond the scope of this text to provide any detail as to the function of such protein growth factors. Suffice it to say that they act on regulatory proteins through their binding to specific receptors on the eucaryotic cell membrane. Once bound, they initiate a transmembrane signal that serves to drive the cell to its next phase in the growth cycle. In general, this is accomplished in one of two ways: (1) a cytoplasmic or membrane-bound protein kinase is activated, which then phosphorylates several proteins by transferring the phosphate from adenosine triphosphate (ATP) to the amino acid tyrosine within such proteins, or (2) through a very complicated series of reactions beginning with the hydrolysis of a membrane-bound phosphatidylinositol and ending with the activation of a protein kinase C, which phosphorylates serine and threonine in several cytoplasmic proteins. All of these phosphorylations have a regulatory role, with the ultimate function of inducing cell division in eucaryotic cells.

Conditions Necessary for Bacterial Growth

pH

In addition to adequate nutrition, a number of other conditions must be met in order to grow bacteria. For example, the medium must have the correct pH. Most bacteria do not grow under highly alkaline conditions and, with the exception of the cholera bacillus (*Vibrio cholerae*), essentially none does well at a pH greater than 8. Most pathogens grow best at a neutral pH (pH 7) or one that is slightly more alkaline (pH 7.4). Some bacteria grow at a pH of 6, and, not infrequently, one finds organisms that grow well at a pH of 4 to 5. Very rarely does an organism thrive below a pH of 4, although certain autotrophic bacteria are exceptions.

Because many bacteria produce metabolic products that are acidic or basic, it is necessary to incorporate into the growth medium a **buffer** that will stabilize the pH. A number of compounds can be used, but the most commonly employed buffer used in bacteriological media is a combination of KH_2PO_4 (monobasic potassium phosphate) and K_2HPO_4 (dibasic potassium phosphate). This buffer acts in the following manner to control the pH:

$$HPO_4^{2-} + \underset{\substack{\text{Excess} \\ \text{acid}}}{H^+} \longrightarrow {}_2PO_4^-$$

$$H_2PO_4^- + \underset{\substack{\text{Excess} \\ \text{base}}}{OH^-} \longrightarrow HPO_4^{2-} + H_2O$$

Thus, in the presence of excess acid, the protons react with HPO_4^{2-} to form the $H_2PO_4^-$ (a weak acid); in the opposite situation, the excess hydroxyl ions react with $H_2PO_4^-$ to form HPO_4^{2-} and water.

Temperature

The temperature also must be controlled. Bacteria in general may be placed in one of three groups according to the temperature range in which they grow best, as shown in Figure 5.1. By far the largest group, and those with which we are concerned, are called **mesophiles** because their optimum growth temperature falls between 20 and 40°C, the range between normal room temperature and just slightly above normal body temperature. Another very important group of organisms is called the **psychrophiles,** with a growth range of 0 to 20°C. Organisms in this group have optimum growth temperatures below 15°C and may grow, though slowly, at temperatures as low as 0°C. This fact is extremely important to remember, because placing material in a refrigerator does not ensure that microbial growth will not take place. The third category of bacteria is made up of the **thermophiles.** These organisms are certainly not a health problem because their optimum growth usually takes place above 45 or 50°C. Some may even grow at temperatures above 95°C. Organisms in the thermophilic group are found in nature in hot springs or rotting compost piles that have generated considerable internal heat as a result of their metabolic activities.

Oxygen

The requirement for oxygen (or its absence) was discussed in Chapter 2, but we shall return to this subject

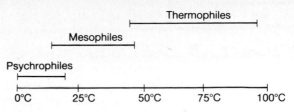

Figure 5.1 Temperature ranges for bacterial growth.

in defining its role as a final electron acceptor in the metabolism of aerobic organisms. It is not surprising to learn that the metabolism of aerobic bacteria (like our own metabolism) requires oxygen as a final electron acceptor. What is surprising is oxygen's lethal effect on obligate anaerobes, and only recently has a possible explanation for this phenomenon been described. In brief, it is now known that in the presence of oxygen all organisms normally produce minuscule amounts of a free radical of oxygen (O_2^-) called superoxide. Superoxide is very toxic, and aerobically respiring cells are believed to survive only because they possess the enzyme superoxide dismutase, which scavenges the superoxide radical as follows:

$$O_2^- + O_2^- + 2H^+ \longrightarrow H_2O_2 + O_2$$

Most obligate anaerobes lack superoxide dismutase and cannot, therefore, tolerate the presence of free oxygen.

You will note that hydrogen peroxide (H_2O_2) is an end product of superoxide breakdown by superoxide dismutase. This compound is also very toxic and will kill the cells if it is not destroyed. Aerobic bacteria accomplish this by producing the enzyme catalase, which carries out the following reaction:

MICROBIOLOGY MILESTONES

Humans have long been obsessed with both the origin of life on this world and whether life of any kind exists on other worlds. Within our solar system, only Mars seems to have an environment that could possibly support some form of life as we know it. Let us look at some of the extreme environments that support microbial life on this planet to see if analogous forms could have developed in an environment where temperatures only periodically rise above freezing and very dry conditions exist.

From the dry valleys of the Antarctic, bacteria have been isolated that were able to grow at temperatures ranging from −2 to 40°C. Organisms have also been found growing in hot springs where the temperatures approach 100°C. Some are found in springs in which the pH of the water may be as acidic as pH 1 or as al-

kaline as pH 11 and, in spite of the fact that salt has long been used as a food preservative, extreme halophilic microorganisms can be found growing in lakes such as the Great Salt Lake and the Dead Sea, which are devoid of all other forms of life. Organisms have also been found in mine effluents containing concentrations of heavy metals such as arsenic and antimony that are toxic for most forms of life. Others have been shown to be resistant to very high levels of radiation, apparently as a result of their ability to rapidly repair radiation-induced damage.

So, the ability of microorganisms to live in extreme environments goes well beyond conditions that most of us think of as "tolerable," so perhaps life on Mars is possible.

$$2H_2O_2 \xrightarrow{\text{catalase}} 2H_2O + O_2$$

Most obligately anaerobic bacteria (as well as the aerotolerant lactic acid bacteria) are unable to synthesize the enzyme catalase. They are in double jeopardy in the presence of oxygen because (1) they cannot destroy superoxide and (2) they cannot destroy hydrogen peroxide enzymatically.

BACTERIAL GROWTH

Supposing all conditions are favorable, let us see how bacteria grow and how we can quantitate the amount of growth at any specific time.

Reproduction of Bacteria

Essentially everyone working with bacteria is concerned with the growth of microorganisms. People working in a hospital must grow the organisms from an infected patient so that a diagnosis can be made. Bacteria also must be grown in the presence of drugs or antibiotics to determine if they are sensitive or resistant, so that the patient may be properly treated. A bacteriologist in a food processing plant must also grow the bacteria present on finished food products (such as frozen dinners) to evaluate the safety of such foods. Even the biochemist interested in microbial metabolism must grow organisms to study the inner workings of the bacterial cell.

What is growth? Growth of an animal or plant means increasing the number of constituent cells; this is also the case with microorganisms. Thus, if you place 10 bacterial cells in 1 mL of a favorable medium and 24 h later find 10 million bacteria per milliliter, you have had bacterial growth. In fact, you have had a millionfold bacterial growth. This increase in bacterial numbers takes place by a process called binary fission. By **binary fission** we mean that each bacterium replicates its constituents, forms a new cell wall across its short diameter (transversely), and then breaks apart into two new cells. Each of these then may divide into two more cells, and so on and on. The overall result of this type of growth is an exponential or logarithmic increase in bacterial numbers. Thus, the progeny of a single bacterium will double with each division, yielding 2, 4, 8, 16, 32 bacteria, and so on, as shown in Figure 5.2.

Measurement of Growth

How would you measure bacterial growth so that you would know that there are 10 million bacteria present in a medium? Quite obviously, one could not count 10 million bacteria in 1 mL of medium. However, this can be accomplished by diluting the culture or medium with sterile water to a point at which 1 mL of the dilution contains few enough bacteria that they can be counted (preferably between 30 and 300). One then mixes a measured quantity of the dilution with a melted nutrient agar medium. The mixture is poured into a petri dish, allowed to harden, and incubated for one or two days to allow each individual cell to multiply until it forms a colony (a mass of bacteria) visible to the naked eye. Now all that is necessary is to count the colonies to know how many living bacteria were present in your dilution. There are electronic instruments that will count the colonies for you, requiring only about 1 s to scan the entire petri dish.

Figure 5.2 Schematic diagram of binary fission illustrating logarithmic growth. Note that during the first 20 min, one new cell was produced; however, during the third 20-min generation time, four new cells were produced. By the twelfth generation (after 4 h) 2048 new cells would be produced during the 20-min generation time.

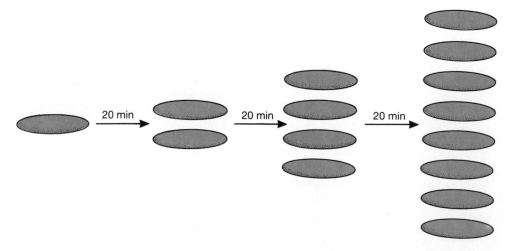

A simple equation to calculate the number of bacteria in the original sample is as follows:

B_o = number of bacteria in 1 mL of original sample

D = dilution factor (for a dilution of 1:10,000, D = 10,000)

C = number of colonies counted

Then

$$B_o = (D)(C)$$

Let us try an example. If one plated out a dilution of 1:100,000 (see Figure 5.3) and after growth counted 78 colonies, what would be the number of bacteria in the original culture?

$$B_o = (100,000)(78)$$

$$B_o = 7,800,000 \text{ per mL}$$

Similarly, one could blend 10 g of a frozen meat pie into 90 mL of sterile water and then dilute this for colony counts to determine the number of bacteria per gram of frozen pie.

A simple formula for calculating the dilution factor for each dilution is as follows:

$$\text{Dilution factor} = \frac{\text{Amount transferred} + \text{amount in bottle}}{(\text{Amount transferred})(\text{amount plated})}$$

For example, if you transfer 0.1 mL into 99.9 mL of water and plate out 0.1 mL, the dilution factor will be

$$\frac{0.1 + 99.9}{(0.1)(0.1)} = 10,000$$

Thus, if 111 colonies grew from 0.1 mL of this dilution, the original culture contained $(111)(10,000)$ = 1,110,000 organisms per milliliter. Viable bacteria can also be enumerated by filtering a sample through a membrane filter that will retain any bacteria present. The filter is then placed on top of an appropriate agar medium in a sterile petri dish. After 36 to 48 h, the colonies growing on the membrane can be counted.

It should be emphasized, however, that this technique counts only viable bacteria that are able to grow on the medium used and under the conditions that the plate was cultured. Thus, a specimen that contained 1 million viable cells of *Escherichia coli* and 10 million viable cells of *Bacteroides fragilis* (an obligate anaerobe) would show only the growth of *E. coli* if incubated aerobically. On the other hand, such dilution plates comprise the only method of estimating numbers of viable bacteria in any specimen.

A statistical method used to estimate the number of coliform bacteria in a sample of drinking water is

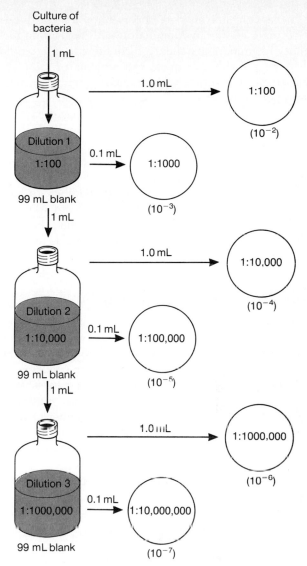

Figure 5.3 Serial dilutions for counting bacteria. Circles represent platings, and all dilutions are expressed as the dilution per milliliter of the original culture. Thus, when 0.1 mL of a 10^{-2} dilution is plated, this is expressed as a 10^{-3} dilution, since it represents one-tenth as many cells as would be used by plating 1.0 mL of the 10^{-2} dilution.

called the most probable number (MPN). This procedure is described in Chapter 30.

There are other techniques to determine total bacterial counts or cell mass, but these procedures include all cells, regardless of whether they are dead or alive. For example, one can centrifuge a suspension of bacteria and either determine the dry weight of the washed pellet or dissolve the pellet in strong acid and determine the total nitrogen present. The latter value would reflect the amount of amino acids and, to a lesser extent, the amount of nucleic acid present.

One very common way of evaluating bacterial growth is to measure the turbidity of a growing culture using a colorimeter or a spectrophotometer. If one used plate counts in conjunction with measuring turbidity to prepare a standard turbidity curve, it would then be possible to equate the amount of light absorbed by the culture with an actual number of bacteria per milliliter.

It is also possible to count the number of bacteria in a stained smear. This can be accomplished by spreading 0.01 mL of suspension over an area of 1 cm². After staining, the number of bacteria in each of several microscope fields is counted. Of course, if you wish to equate this value with actual numbers of bacteria, it is again necessary to determine from a known standard how many bacteria occur per field because different microscopes have different size fields for their oil-immersion lenses.

There is also available a commercially designed slide called a *Petroff-Hauser Counting Chamber*. It consists of a slide with a slight indentation of known volume, which is inscribed with squares of a known area. To use it, one merely lays a coverslip over the slide and allows the indented well to fill with the bacterial suspension to be counted (using capillary attraction). The number of bacteria in each of several squares is counted and the average count is multiplied by a factor of 1,250,000 to yield the number of total cells per milliliter in the original suspension. Keep in mind, however, that like all direct counts, both living and dead bacteria are included.

Finally, one can count bacteria electronically by passing a culture through a very small orifice in a Coulter particle counter. Counters of this type are used routinely to count blood cells, but they can be adapted to count bacteria also.

Most of the time, however, you will not be concerned with exactly how much growth occurred but rather with the problem of whether growth took place at all. To determine this, you have only to look at the culture. After reaching about 10 million bacteria per milliliter, a liquid medium will have become turbid (cloudy) and a solid medium will have visible growth either on the surface or down into the medium.

Phases of Growth

Under favorable conditions almost all bacteria are able to reproduce very rapidly. The time required for one organism to divide into two is referred to as the generation time. In some bacteria, such as *E. coli*, the average generation time may be as little as 20 min, whereas in others (e.g., *M. tuberculosis*) it is about 15 to 20 h. The generation time during active growth varies with each species of bacteria, although in the majority it will be less than 1 h. An illustration of the practical importance of appreciating the generation

time of bacteria is provided by the rapid handling of urine cultures in the doctor's office and in hospitals. Suppose that a urine sample contaminated with only 500 *E. coli* per milliliter is left in the hospital ward for a few hours before it is taken to the diagnostic clinic. Assuming a generation time of 20 min, those 500 *E. coli* would become 256,000 organisms in 3 h or over 2 million in 4 h. What may have started off as a small contamination could now be incorrectly interpreted as a serious infection in the patient.

Because bacterial growth occurs exponentially, the results can be plotted as the logarithm of the number of cells versus the time of growth. In this manner a growth curve of bacteria (Figure 5.4) is obtained that can be divided into four phases.

Lag Phase

When bacteria are inoculated into a new medium, reproduction usually does not begin immediately. The lag phase is a period of adaptation to the environment, and its duration may be from an hour to several days. The length of time depends on the kind of bacteria, the age of the culture, and the available nutrients in the medium provided.

The lag phase is a lag in multiplication only, for the cells are very active metabolically. Inclusions are used up, and there is active synthesis of enzymes and other essential constituents. In the latter part of the lag phase there is some increase in overall size; some bacilli are reported as increasing to two to three times their original length. By the end of this period, the

---

Figure 5.4 Typical growth curve illustrating the phases of growth occurring when bacteria are inoculated into a culture medium. (*a*) Lag phase: cells begin to synthesize inducible enzymes and use stored food reserves. (*b*) Logarithmic growth phase: the rate of multiplication is constant. (*c*) Stationary phase: death rate is equal to rate of increase. (*d*) Death phase: cells begin to die at a more rapid rate than that of reproduction. The slope of this phase varies from one genus to another and may last from less than one day to several months.

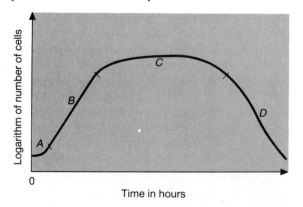

cells have usually lost all of their reserve storage granules.

Log Phase

This is the period of most rapid reproduction and the one in which the typical characteristics of the active cells are usually observed. Because the majority of antibiotics inhibit either cell wall synthesis or protein synthesis, it is during this phase of growth that the cell is most susceptible to the action of many antibiotics. During this phase the generation time is constant for a given environment for each species; if one plots the logarithm of the number of organisms against time, one obtains a straight line.

The generation time of an organism can be determined during this phase. As noted earlier, each generation results in a doubling of the cell number. With this information, the following can be used to calculate the generation time. (Most mathematics books will have log tables to the base 10, which can be used for the calculation of generation time.)

B_o = number of bacteria at beginning of time interval

B_t = number of bacteria at end of any interval of time (t)

g = generation time, usually expressed in minutes

t = time, usually expressed in minutes

n = number of generations

$B_t = B_o \times 2^n$

By taking the logarithms of both sides of the equation, and substituting t/g for n, we get the following equation:

$$g = \frac{t \log 2}{\log B_t - \log B_o}$$

Let us take an example to illustrate this:

B_o = 2.83×10^7 cells per milliliter, or log 7.45

B_t = 4.5×10^8 cells per milliliter, or log 8.65

t = 135 min

$\log 2 = 0.301$

Then

$$g = \frac{135 \times 0.301}{8.65 - 7.45} = 33.9$$

The generation time for this organism is 33.9 min.

The generation time varies with species of organism, concentration of available nutrients in the medium, and temperature of incubation. Other conditions, such as pH, and oxygen availability for aerobes, would also have an influence. As has already been described, some species multiply rapidly when provided with favorable conditions. Individual cells are slightly smaller during the log phase than during the lag period, and they do not usually accumulate granular inclusions.

One very important concept concerning the log growth of bacteria is that at any point during this period some cells are just beginning to divide, others are partially finished, and still others are finishing division. Researchers who wanted to learn at exactly what point in reproduction certain enzymes were formed had to have all cells at exactly the same stage of division simultaneously. This can be accomplished by several techniques, as by chilling a culture for a short period followed by a return to the normal growth temperature, or by limiting an essential metabolite during early growth and then supplying additional amounts of this metabolite. A far better technique (termed a "baby machine") is to adsorb rapidly dividing bacteria in the pores of a membrane filter and collect newly formed cells by gently backwashing the filter with a growth medium. The cells collected over a short period of time are in a physiologically similar state and will produce several generations of **synchronous growth** (all the cells of a culture are dividing at almost the same time) in a suitable medium. Also, one can centrifuge a culture of bacteria on a sucrose gradient, which separates the cells by size. Any one size represents cells at a similar point in the growth cycle that can be grown in synchrony for several generations.

Bacterial cells can be maintained in the logarithmic phase by continually transferring them to a fresh medium of the same constitution. The process can be continued automatically by using a special device called a chemostat. The **chemostat** consists of a growth chamber connected to a reservoir of fresh medium. The fresh medium is continually supplied to the bacteria in the growth chamber. One particular nutrient of the medium may be provided in limited concentration to control the rate of growth. As fresh medium enters the chamber, the cells are removed (see Figure 5.5). The chemostat has proved to be very useful in providing large amounts of bacteria to be used for physiological studies.

Stationary Phase

As the culture grows older and approaches the maximum population of bacteria that the medium can

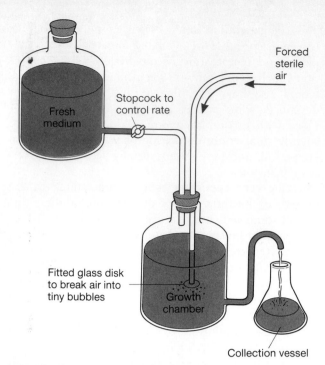

Figure 5.5 Chemostat used for continuous cultures. Rate of growth can be controlled either by controlling the rate at which new medium enters the growth chamber or by limiting a required growth factor in the medium.

Labels in figure:
Forced sterile air
Stopcock to control rate
Fresh medium
Fitted glass disk to break air into tiny bubbles
Growth chamber
Collection vessel

support, the rate of reproduction slows down and some cells die. When the rate of reproduction equals the rate of death, the overall number of bacteria remains constant. This is referred to as the **stationary phase.**

In addition to the limited supply of nutrients remaining in the medium, waste products of metabolism tend to accumulate and may be toxic to the organisms. For aerobic organisms, the oxygen supply may be inadequate for the large number of cells.

Endospores develop in spore-forming genera, and granular inclusions are usually observed. Involution (abnormal) forms may appear, particularly as the death phase is reached.

Death Phase

Once the rate of death exceeds the rate of reproduction, the actual number of bacteria declines. In this death phase, reproduction will usually have stopped. With certain species, it may take weeks, months, or even longer before the end of this phase is reached. In this period, involution forms (odd-shaped abnormal cells) may appear. It would be very difficult to identify an organism if you looked only at these old cultures. Dead cells may or may not dissolve, or lyse.

SUMMARY

Bacteria require nutrients both to synthesize protoplasm and to provide a source of energy. In general, nutrients can be categorized as a carbon source, a nitrogen source, inorganic ions, essential metabolites, and water. In addition, the pH, temperature, and presence or absence of oxygen may influence bacterial growth.

The number of bacteria present within a culture or sample may be measured by plating out appropriate dilutions and counting the number of colonies appearing after growth has occurred. Other methods for estimating bacterial growth include the determination of dry weight of cells, total nitrogen in cells, turbidity of a liquid culture, and counting the stained bacteria in a known volume. Bacteria can also be counted electronically using a Coulter particle counter.

The generation times for different species of bacteria vary from 20 min to 20 h. When bacteria are inoculated into a new medium, the resulting growth can be divided into four phases, which are designated as lag phase, log phase, stationary phase, and death phase.

QUESTIONS FOR REVIEW

1. What are the basic nutrients required for bacterial growth?
2. What additional factors are usually required for eucaryotic cell growth?
3. Define psychrophile, mesophile, and thermophile.
4. Summarize how a buffer acts to stabilize the pH of a medium.
5. What is superoxide and what is its effect on growing cells?
6. If you added 0.1 mL of a bacterial culture to 99.9 mL of sterile water, what is your dilution factor? If 73 colonies grew from 0.1 mL of this dilution, how many bacteria were there per milliliter of the original culture?
7. Draw a bacterial growth curve and label the phases of growth. What occurs during each phase of growth?

SUPPLEMENTARY READING

Becker WM: *Energy and the Living Cell.* Philadelphia, Lippincott, 1977.

Bullen JJ, Rogers HG, Griffiths E: Role of iron in bacterial infection. *Curr Topics Microbiol Immunol* 80:1, 1978.

Ingraham J, Maaloe O, Neidhardt F: *Growth of the Bacterial Cell.* Sunderland, MA, Sinauer Associates, 1983.

Koch AL: Turbidity measurements in microbiology. *ASM News* 50:473, 1984.

Krieg NR, Hoffman PS: Microaerophily and oxygen toxicity. *Annu Rev Microbiol* 40:107, 1986.

Mandelstam J, McQuillen K, Dawes I (eds): *Biochemistry of Bacterial Growth,* 3rd ed. New York, Wiley, 1982.

Morris JG: The physiology of obligate anaerobiosis. *Adv Microbiol Physiol* 12:169, 1975.

Chapter 6

Bacterial Metabolism

OBJECTIVES After studying this chapter, you should understand

1 The difference between respiration and fermentation.

2 How energy is captured during oxidative phosphorylation.

3 The function of each photosystem used in green plant and cyanobacterial photosynthesis.

4 How noncyanobacterial photosynthetic organisms form a reductant in the absence of photosystem II.

5 The difference between autotrophs and heterotrophs and how each obtains its energy from chemical oxidations.

6 How stable fermentation products are formed and what some of these products are.

*M*etabolism is the sum of all the chemical processes that occur within a cell. Although the work of bacterial exoenzymes is carried on outside the cell, the production of these enzymes is one aspect of metabolism. When the required nutrients have entered the cell, the endoenzymes reconvert the simple nutrients to the complex ingredients of protoplasm, within which energy is stored.

Think of metabolism as consisting of two opposite processes occurring simultaneously. The first aspect of metabolism is the synthesis of protoplasm and the use of energy. This is called **anabolism**. The second aspect of metabolism, an opposite process, is called **catabolism**; it consists of the oxidation of a substrate, accompanied by a capture of energy (Figure 6.1). As long as the bacterium lives and carries on its life processes (growing, moving, and reproducing), it will always need energy. It is apparent, therefore, that a very important part of the study of bacterial metabolism is concerned with learning just how a bacterium obtains its energy. Let us now review some general principles of energy capture that apply to all cells.

OXIDATION-REDUCTION

Oxidation is defined as a loss of electrons; **reduction** means the gain of electrons. Because electrons cannot exist free, there is always a corresponding reduction for each oxidation that occurs. Oxidation also results in a liberation of energy; it is the function of this chapter to describe how bacteria are able to trap this released energy for use in subsequent synthetic reactions.

Oxidative Phosphorylation

Most biological oxidations are catalyzed by enzymes called **dehydrogenases.** These dehydrogenases transfer the released electrons and protons (H^+) to an intermediate coenzyme electron acceptor, such as nicotinamide adenine dinucleotide (NAD^+) or nicotinamide adenine dinucleotide phosphate ($NADP^+$), to form NADH or NADPH, respectively. Oxidative phosphorylation occurs when these high-energy electrons are passed through an electron-transport chain to eventually reduce oxygen or an inorganic oxidant.

Current concepts of just how a cell is able to convert the energy in these electrons are best described by the chemiosmotic model of Mitchell. By referring to Figure 6.2, we can summarize this model as follows:

1. Electrons flowing to oxygen are passed through various carriers (flavoproteins, quinones, nonheme iron-containing proteins, and cytochromes).

2. This electron transport results in the passage of protons (H^+) from the cytoplasm to the external environment, that is, from the inside to the outside of the bacterial cell. Some electron carriers, such as flavins and quinones, acquire hydrogens when they are reduced and thus carry both electrons and protons across the cytoplasmic membrane. Other electron carriers, such as cytochromes, carry only electrons. When a carrier can accept only electrons, the protons will be released into the medium outside the cell, resulting in the net movement of protons across the membrane and the separation of charge. This separation of charge generates a substantial difference in the chemical concentration of protons (i.e., a pH gradient) across the membrane.

3. Most cells maintain an internal pH of 7.5, and thus, a pH gradient occurs when the external pH is

CATABOLISM: Complex foodstuffs $\xrightarrow{\text{Breakdown}}$ Energy + Simple molecules

ANABOLISM: Energy + Simple molecules $\xrightarrow{\text{Buildup}}$ Cellular constituents

Figure 6.1 Catabolism releases the energy and produces the intermediates used during anabolism to synthesize cellular constituents.

Figure 6.2 Electron transport and oxidative phosphorylation according to the chemiosmotic model of Mitchell. The proton liberated from the oxidation of NADH is passed through the membrane to the external medium. The accompanying electron is passed by a nonheme iron carrier to coenzyme Q · H, which picks up another proton and electron. As shown, the electrons terminate in the cytochrome system, where they are eventually accepted by O_2 to form H_2O. External protons reenter the cell through a proton channel where, using the protonmotive force generated by the pH difference and the electric potential difference, an ATPase converts ADP + inorganic phosphate to ATP.

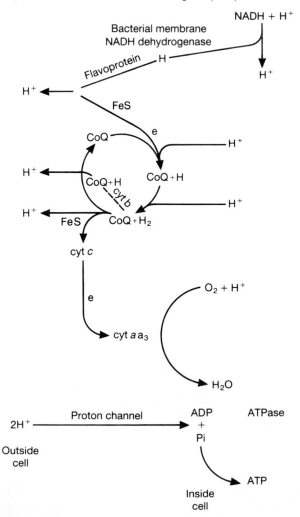

less than 7.5. The membrane potential and the pH gradient together comprise the electrochemical potential across the membrane commonly called the **protonmotive force.** This force attracts protons back into the cell and is coupled to the performance of a wide variety of metabolic and mechanical work.

4. Membranes possess specific proton channels containing an enzyme complex called adenosine triphosphatase (ATPase). When these extruded protons flow back through this channel from exterior to interior, energy is released which, in the presence of the ATPase, is converted to chemical energy by the phosphorylation of adenosine diphosphate (ADP) to ATP. This energy is stored in high-energy phosphate bonds (designated by wavy lines) where it remains available for the synthesis of cellular constituents.

$$\underset{\text{Adenosine diphosphate}}{\text{Adenosine} - P \sim P} + \text{inorganic P} \xrightarrow{\text{Energy}}$$

$$\underset{\text{Adenosine triphosphate}}{\text{Adenosine} - P \sim P \sim P}$$

NUTRIENT TRANSPORT

Only a few substances, such as water, O_2, N_2, fatty acids, and other lipid-soluble compounds, are able to enter a bacterial cell by passive diffusion. All other nutrients must first bind to a specific transport system present in the cytoplasmic membrane. Such transport systems can be divided into three general categories. **Active transport** systems, also called **permeases,** are proteins that use metabolic energy to move a solute across the membrane into the cytoplasm. The energy used is in some cases the protonmotive force and in others results from the hydrolysis of ATP. In either case, the result is the accumulation of a much higher concentration of solute within the cell than in the external environment. **Group translocation** occurs when a solute is modified during its transport in such a way that it cannot be transported out of the cell at any appreciable rate. The best characterized group translocation process is the phosphoenolpyruvate-sugar transferase system (PTS), which catalyzes the simultaneous transport and phosphorylation of a number of sugars and sugar alcohols. **Facilitated diffusion** also uses specific membrane-bound carrier proteins to move a substance into a cell but no metabolic energy is required, and the substance is not accumulated against a concentration gradient as in the case of active transport. The carrier merely transports the substance across the membrane, allowing its concentration to equalize that of the external environment.

There are two ultimate sources of energy for living cells. Organisms that utilize sunlight are called **photosynthetic** or phototrophic, and those that obtain their energy from chemical oxidations are described as **chemosynthetic** or chemotrophic.

PHOTOSYNTHESIS

All green plants containing the green pigment chlorophyll obtain their energy directly from light and their carbon source from the gaseous carbon dioxide (CO_2) in the air. Because essentially all life as we know it depends directly or indirectly on photosynthesis, it is a process of immense importance to biologists.

It has been known for many years that the overall photosynthetic reaction for plants can be written as follows:

$$CO_2 + 2H_2O \xrightarrow[\text{chlorophyll}]{\text{light}} H_2O + \underset{\text{Carbohydrate}}{CH_2O} + O_2$$

By multiplying each reactant by six, it can be seen that the incorporation of six molecules of CO_2 leads to the formation of one molecule of glucose:

$$6CO_2 + 12H_2O \xrightarrow[\text{chlorophyll}]{\text{light}} 6H_2O + \underset{\text{Glucose}}{C_6H_{12}O_6} + 6O_2$$

Because glucose is considerably more reduced than CO_2, it is apparent that the incorporation of CO_2 to form glucose requires both energy and a source of electrons and protons (H^+) to reduce the CO_2. In green plants (including the cyanobacteria), each of these requirements comes from a separate photosystem—one providing energy and the other providing the electrons and protons necessary to reduce the CO_2.

Photosystem I

Photosystem I provides much of the energy necessary for the reduction of CO_2 to carbohydrate as well as for any other reactions involved in the biosynthesis and maintenance of the cell. It does this by converting the energy in a photon of light into chemical energy that can be used by the cell. This process begins when a photon of light is absorbed by a molecule of chlorophyll; the chlorophyll becomes excited and emits an electron at a high state of energy. By losing an electron, the chlorophyll molecule becomes oxidized. The remaining function of photosystem I is to capture the excess energy in the excited electron as it returns to its former state in the chlorophyll molecule. This occurs as the electron is passed through a series of electron acceptors back to the oxidized chlorophyll molecule in a manner analogous to oxidative phosphorylation. The extruded protons then reenter the cell through proton channels, and ADP is phosphorylated to form ATP using the released energy. Because the emission of electrons and their return to photosystem I is a cyclic event, the resulting capture of energy is referred to as **cyclic phosphorylation** (see Figure 6.3).

Photosystem II

The availability of ATP alone, however, is not sufficient to reduce CO_2 to carbohydrate. The cell must also provide a reductant in the form of a reduced coenzyme, NADPH. Photosystem II also contains chlorophyll, and it uses the energy obtained from a photon of light to provide this reductant. It does so by using the light for the photolysis of water, as shown below:

$$H_2O \xrightarrow[\text{chlorophyll}]{\text{light}} 2e^- + 2H^+ + \tfrac{1}{2}O_2$$

Note that it is this reaction that generates oxygen in green-plant photosynthesis.

The protons from the photolysis of water and electrons from photosystem I are then used to reduce NADP as shown below:

$$NADP^+ + 2H^+ + 2 \text{ electrons} \longrightarrow NADPH + H^+$$

In the presence of ATP and the appropriate enzymes, the major biosynthetic reaction of photosynthesis occurs when CO_2 is reduced to form a molecule of glucose:

$$6CO_2 + 12NADPH + 12H^+ \xrightarrow{\text{ATP}} C_6H_{12}O_6 + 12NADP^+ + 6H_2O$$

It must be emphasized that photosystems I and II do not operate independently, and it is only through the cooperation of these two systems that NADP is reduced to NADPH. By referring to Figure 6.3, the reader can see that the electrons for the reduction of NADP come from photosystem I. Thus, not all elec-

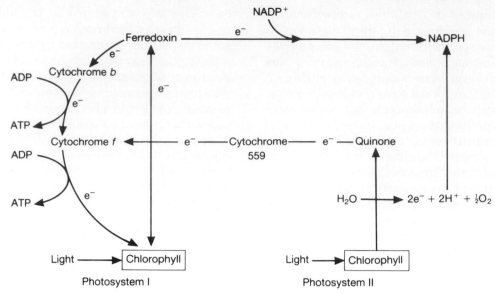

Figure 6.3 Green-plant photosynthesis. Note that photosystem I can carry out cyclic phosphorylation or can siphon off electrons from reduced ferredoxin to form NADPH. In the latter case, electrons from photosystem II are passed to cytochrome *f* of photosystem I to replace the emitted electron. Note also that the photolysis of water by photosystem II provides the protons for the formation of NADPH. Not all intermediate electron carriers are shown.

trons generated in photosystem I are used for cyclic photophosphorylation. Some are siphoned off to reduce NADP. But what happens to the electrons generated by photosystem II? As shown in Figure 6.3, they enter photosystem I, where they are used to reduce light-oxidized chlorophyll. Because these electrons move from photosystem II to photosystem I, the resulting capture of energy is termed **noncyclic phosphorylation.** The protons formed by the photolysis of water by photosystem II are transferred directly to NADP where, along with the electrons from photosystem I, they convert NADP to NADPH. The electrons released during the photolysis of water are captured by the oxidized chlorophyll in photosystem II.

Procaryotes designated as cyanobacteria carry out photosynthetic reactions identical to those described for green-plant photosynthesis. There is, however, a large group of photosynthetic bacteria that utilize an anaerobic photosynthetic system that, unlike the cyanobacteria, never results in the liberation of free oxygen.

Noncyanobacteria Photosynthesis

Noncyanobacteria photosynthesis occurs in a manner similar to that described for green-plant photosynthesis except that these bacteria lack photosystem II for the photolysis of water. Hence, *water is never a source of the reductant, and oxygen is never formed as a product.* These organisms supply the necessary

reductant by the chemical oxidation of a hydrogen donor, but the overall reaction is the same as that shown for green-plant photosynthesis except that water is replaced by a different reduced substrate, which we can write as H_2A. This equation for noncyanobacteria photosynthesis can be written as follows:

$$CO_2 + 2H_2A \xrightarrow[\text{chlorophyll}]{\text{light}} H_2O + CH_2O + 2A$$

Based on their pigments and the type of reductant used in photosynthesis (represented by H_2A), such bacteria have been classified into three families: Chlorobiaceae, Chromaticeae, and Rhodospirillaceae.

Chlorobiaceae—also called the green-sulfur bacteria—is a large family of which the best-studied genus is *Chlorobium*. These organisms can use several different sulfur-containing compounds as well as hydrogen gas as photosynthesis reductants. A specific formula for photosynthesis by the green-sulfur bacteria may be any one of the following, depending on the reductant available (CH_2O represents synthesized carbohydrate):

1. $CO_2 + 2H_2S \xrightarrow{\text{light}} (CH_2O) + H_2O + 2S$
2. $3CO_2 + 2S + 5H_2O \xrightarrow{\text{light}} 3(CH_2O) + 2H_2SO_4$
3. $2CO_2 + Na_2S_2O_3 + 3H_2O \xrightarrow{\text{light}}$
   $\qquad\qquad 2(CH_2O) + Na_2SO_4 + H_2SO_4$
4. $CO_2 + 2H_2 \xrightarrow{\text{light}} (CH_2O) + H_2O$

Chromaticeae—also called the purple-sulfur bacteria—differ from green-sulfur bacteria primarily because they contain a number of red and purple carotenoid pigments in their cells. The best-studied genus of this family is named *Chromatium*. These organisms, however, use the same photosynthetic reductants as do the green-sulfur bacteria; as a result, they can carry out the reactions listed above.

Rhodospirillaceae—also called the nonsulfur purple bacteria—are morphologically similar to the Chromaticeae, but they are unable to use sulfur compounds as photosynthetic reductants. The two genera of this family most extensively studied are *Rhodospirillum* and *Rhodopseudomonas*. These organisms are capable of using hydrogen or various organic compounds as reductants; the following equation is one example:

$$CO_2 + 2CH_3CHOHCOOH \xrightarrow{light} (CH_2O) + H_2O + 2CH_3COCOOH$$

Many strains of Rhodospirillaceae are also able to assimilate organic compounds such as acetate into their metabolic systems; most are also capable of using atmosphere nitrogen as their nitrogen source.

One important property of bacterial photosynthesis not seen in green-plant photosynthesis is that it can occur only in the complete absence of oxygen. However, some members of the family Rhodospirillaceae are capable of nonphotosynthetic growth in the presence of oxygen if given a sufficiently rich medium on which to grow.

Table 6.1 summarizes the various photosynthetic reactions.

CHEMOAUTOTROPHIC METABOLISM

Organisms not possessing chlorophyll must obtain their energy through chemical oxidations. The members of one group of bacteria have the amazing ability to transform inorganic matter into proteins, carbohydrates, fats, nucleic acids, vitamins, and so on—a feat that is utterly impossible for animals. These bacteria are called either **autotrophs** or **chemolithotrophs,** and their mode of nutrition is termed **autotrophic** or **chemolithotrophic.** Many autotrophic bacteria are unable to utilize organic compounds. Like the photosynthetic organisms, they use carbon dioxide as their sole source of carbon for cellular synthesis. However, as we already know, it requires energy and NADPH to convert carbon dioxide into cellular material. The photosynthetic bacteria obtain this necessary energy from light, but the autotrophs must obtain their energy through chemical oxidations. The unusual property of the autotrophs is that they obtain their energy by the oxidation of inorganic compounds. Even though this may seem odd, the process of energy capture is the same as described for oxidative phosphorylation. Thus, electrons released by the oxidation of sulfur, ammonia, and so on, are passed via an electron-transport chain, causing the extrusion of protons from the membrane. The pH potential thus formed is converted to high-energy phosphate bonds when these protons reenter the cell through the proton channels. Once ATP is formed, the biosynthetic pathways of the cell are analogous to those of the photosynthetic organisms.

The autotrophic bacteria have been placed in a number of different genera based on the type of inorganic substance oxidized as an energy source. Some examples of these bacteria and their energy-yielding reactions are

Thiobacillus (oxidation of sulfur)
$$2S + 3O_2 + 2H_2O \longrightarrow 2H_2SO_4$$

Nitrosomonas (oxidation of ammonia)
$$2NH_4Cl + 3O_2 \longrightarrow 2HNO_2 + 2HCl + 2H_2O$$

Nitrobacter (oxidation of nitrites)
$$2NaNO_2 + O_2 \longrightarrow 2NaNO_3$$

Various genera (oxidation of hydrogen)
$$2H_2 + O_2 \longrightarrow 2H_2O$$

Siderocapsa (oxidation of iron compounds)
$$4FeCO_3 + O_2 + 6H_2O \longrightarrow 4Fe(OH)_3 + 4CO_2$$

TABLE 6.1 Summary of Photosynthetic Reactions

	Eucaryotic	Cyanobacter	Noncyanobacter
Source of protons for reduction of CO_2	H_2O	H_2O	Sulfur compounds, H_2, and acetate
Oxidized end product	O_2	O_2	S, SO_4, or H_2O
Photosystems involved	I and II	I and II	I only
Atmospheric environment	Aerobic	Aerobic	Anaerobic

HETEROTROPHIC METABOLISM

Most bacteria with which we will be concerned have lost the power to synthesize protoplasm from inorganic sources and rely instead on ready-made organic material for food. These organisms are called **heterotrophs** ("nourished by others"); therefore, their nutrition is **heterotrophic.** It should be understood, however, that there is no abrupt division between the autotrophs and the heterotrophs. Some heterotrophs need many more organic substances than others do, whereas others are practically autotrophic except for their source of carbon and energy. They cannot extract these elements from the atmosphere, but if supplied with one organic substance, such as glucose (a simple sugar), for carbon and energy, they can utilize inorganic materials for their other needs. Table 6.2 summarizes the classification of bacteria according to their energy and carbon source.

METABOLIC PATHWAYS

It has been stressed throughout this chapter that biosynthetic processes require energy and that energy is derived from oxidative reactions. Because electrons cannot exist free, we know that for every oxidation there must be a corresponding reduction. The heterotrophic bacteria are, by and large, categorized according to the nature of their end products of metabolism. For the most part, *these end products represent the final electron acceptors of their metabolic pathways.*

What is the nature of these final electron acceptors? It varies according to the enzyme systems of the organism and may be either the free oxygen of the atmosphere or various organic or inorganic molecules. Bacteria that must use oxygen as a final hydrogen acceptor are known as **obligate aerobes.** Others are so constituted as to find the oxygen of the air literally poisonous; they can only live in the absence of air. These are the **obligate anaerobes.** Still other kinds of bacteria have enzyme systems so versatile that they can use either oxygen or some organic compound as a final hydrogen acceptor. Such organisms are **facultative.** Finally, there are bacteria that seem to grow best in the presence of small amounts of oxygen; they are referred to as **microaerophilic organisms.**

We may also categorize metabolic pathways as **fermentative** or **respiratory.** These terms are really descriptive of the final electron acceptors used by a bacterium, and it is important to be able to differentiate between them. Respiratory metabolism occurs when the electrons released by oxidation are passed through the electron transport chain, protons are extruded through the bacterial membrane, and energy is captured by oxidative phosphorylation. Fermentation, on the other hand, is an anaerobic process and does not use the membrane-bound electron transport chain and does not result in oxidative phosphorylation. Rather, electrons and protons are transferred directly from an oxidized substrate to another organic intermediate compound, eventually forming stable fermentation products. Such fermentation products accumulate because the fermenting organism is unable to oxidize them any further under the anaerobic conditions present during fermentation. Moreover, many fermentative organisms do not even possess an electron transport chain and are, therefore, unable to carry out the oxidative phosphorylation reactions under any conditions.

Microbial Fermentation

During fermentation, the intermediate products formed by the catabolism of an organic substrate (such as glucose) serve as final electron acceptors, resulting in the formation of stable fermentation products. For example, many organisms convert sugars to pyruvic acid. In such cases, however, NADH is also formed; it must pass its acquired electrons on to some acceptor if the organism is to continue to metabolize. This is accomplished by using the pyruvic acid, or some metabolic product formed from pyruvic acid, as a final electron acceptor.

It is also important to note that because no electron transport chain is used during fermentation, high-

TABLE 6.2 Classification Based on Energy and Carbon Source

Group	Energy Source	Carbon Source
Chemoheterotrophs	Oxidation or organic compounds	Organic
Chemoautotrophs	Oxidation of inorganic substances such as ammonia, sulfides, and ferrous compounds	CO_2
Photoheterotrophs	Light	Organic
Photoautotrophs	Light	CO_2

energy phosphate bonds are not formed through oxidative phosphorylation but rather by a process known as **substrate phosphorylation.** In this case, a phosphorylated intermediate is oxidized and the energy released through oxidation is converted directly into a high-energy bond. The resulting high-energy phosphate bond can then be transferred to ADP to form ATP, as shown in Figure 6.4.

Figure 6.4 Embden-Meyerhof pathway for the dissimilation of glucose. Note that one molecule of fructose-1,6-diphosphate is split into two three-carbon compounds that are in equilibrium with each other. Thus, as the glyceraldehyde-3-phosphate is oxidized, the dihydroxacetone phosphate is converted to more glyceraldehyde-3-phosphate. For each molecule of the six-carbon glucose metabolized, there are two of the three-carbon intermediates. It can also be seen that two molecules of ATP are required to initiate the pathway, but four molecules of ATP are formed by the dissimilation of one molecule of glucose to pyruvic acid. Also, as shown in this figure, there are two molecules of NADH that must be reoxidized by passing their electrons and protons to a final electron acceptor.

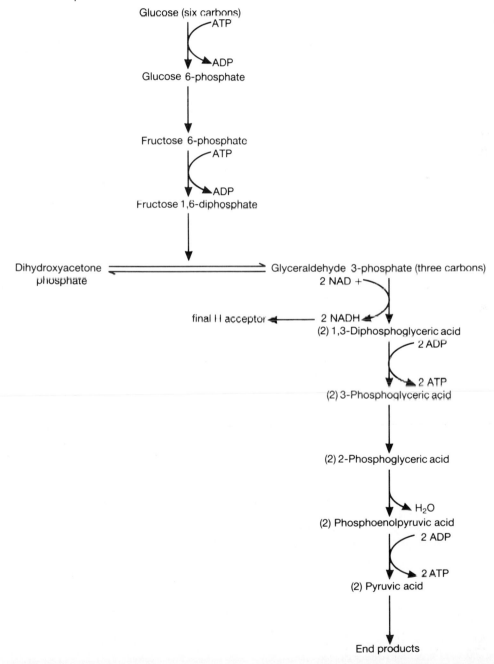

Pathways of Fermentation

The Embden-Meyerhof pathway is a common mechanism for the conversion of glucose to pyruvic acid. As can be seen in Figure 6.4, two molecules of ATP are required per molecule of glucose to initiate this pathway, but four molecules of ATP are obtained through substrate phosphorylation by the conversion of glucose to pyruvic acid, giving a net yield of 2 mol of ATP per mole of glucose metabolized. (Note that the six-carbon fructose-1,6-diphosphate is cleaved into two three-carbon molecules—dihydroxyacetone phosphate and glyceraldehyde-3-phosphate. Because dihydroxyacetone phosphate is constantly converted to glyceraldehyde-3-phosphate as the latter is further metabolized, every step following glyceraldehyde-3-phosphate occurs twice for each original glucose molecule.) Furthermore, although Figure 6.4 indicates that pyruvic acid is the final product of this pathway, *there are still two molecules of NADH that must be oxidized.* Thus, depending on the type of microorganism, pyruvic acid ($CH_3COCOOH$) will be additionally metabolized to yield final fermentation products, as shown in color in steps 1 through 6.

1. Homolactic acid fermentation (some streptococci and lactobacilli)

2. Alcoholic fermentation (yeast)

3. Mixed acid fermentation (*E. coli* and some other enteric bacteria)

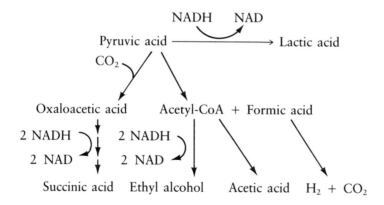

4. Butylene glycol fermentation (*Enterobacter, Bacillus, Pseudomonas*)

    This fermentation produces small amounts of the same end products as the mixed acid fermentation, but, in addition, a large part of the pyruvic acid is converted to 2,3-butylene glycol as shown below.

$$2CH_2COCOOH \longrightarrow CH_2COHCOOH + CO_2$$

Pyruvic acid

Acetolactic acid
$$
\begin{array}{c}
| \\
C{=}O \\
| \\
CH_3
\end{array}
$$

$$CH_3CHOHCHOHCH_3 \longleftarrow CH_3CHOHCOCH_3$$

2,3-Butylene glycol

Acetoin

NAD ← NADH

CO₂

Also, as will be discussed in Chapter 30, the presence of a small amount of acetoin in the growth medium is used as a criterion for differentiating *Enterobacter* and *Escherichia*.

5. Propionic acid fermentation (*Propionibacterium* and *Veillonella*)

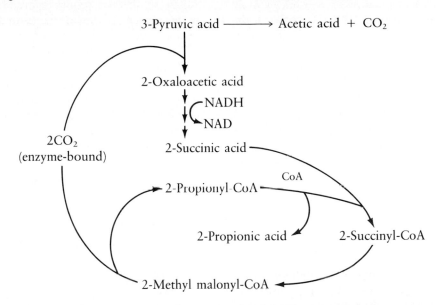

The energy incorporated into the propionyl-CoA bond is preserved by the reaction of propionyl-CoA with succinic acid to form succinyl-CoA and free propionic acid. Furthermore, the CO_2 arising from the decarboxylation of methyl malonyl-CoA remains bound to a biotin-containing enzyme that transfers the CO_2 directly to the pyruvic acid to form oxaloacetic acid. In addition to this mechanism for the formation of oxaloacetic acid, these organisms can also form oxaloacetic acid by a reaction of phosphoenolpyruvic acid with free CO_2.

6. Butyric acid, butanol, acetone fermentation (*Clostridium*)

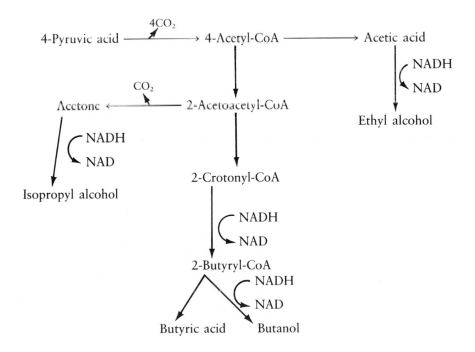

As you can see, there is great variation as to which compounds can be used as final electron acceptors. Hence, the final fermentation products are also variable. In some cases, as in the lactic acid or alcoholic fermentation, only a single acceptor is used, forming a single fermentation product. Others, such as the mixed acid or butyric acid fermentation, utilize multiple electron acceptors, forming a large number of fermentation products. We shall discuss the industrial importance of some of these products in Unit V.

Not all bacteria metabolize sugars via the Embden-Meyerhof pathway, but the alternate pathways of dissimilation result in the same type of fermentative end products already described.

Table 6.3 provides a summary of the major types of microbial fermentations.

Microbial Respiration

Respiration is defined as the use of an electron transport chain to pass electrons to a final inorganic electron acceptor. Energy is obtained through oxidative phosphorylation, but the process may use either oxygen as a final electron acceptor (aerobic respiration) or other inorganic compounds (anaerobic respiration).

Aerobic Respiration

Many organisms are able to use oxygen as a final hydrogen acceptor. In such cases, it is not necessary to reduce intermediates as in fermentation; as a result, such intermediates can be completely oxidized to CO_2 and H_2O. This is a tremendous advantage to the organism, because the amount of energy available from the complete oxidation of a molecule of glucose is many times greater than that obtained from the fermentation of glucose. This occurs because the stepwise passage of each pair of electrons from NADH to oxygen through a series of cytochrome carriers results in the formation of three molecules of ATP. This, along with the energy gained by the oxidation of pyruvate to acetate, results in an additional yield of 36 molecules of ATP being generated from the metabolism of each molecule of glucose to CO_2 and H_2O. When this is compared with only two molecules of ATP gained during the fermentation of one molecule of glucose to ethanol or lactic acid (which aerobes also capture), it is obvious that aerobic metabolism is considerably more efficient than is fermentation.

Let us now turn our attention to how pyruvic acid is converted to CO_2 and H_2O and how it results in the release of excess energy for the cell. This is accomplished through a degradative pathway called the tricarboxylic acid cycle or, by some, the Krebs cycle, as shown in Figure 6.5. First, note that it is a cycle—it goes 'round and 'round—and each time a molecule of oxaloacetate comes around, a molecule of acetate (derived from pyruvate) enters the cycle and is joined to the four-carbon oxaloacetate to form the six-carbon acid for which this cycle is named. Second, each complete turn of the cycle results in a series of oxidations (resulting in the reduction of NAD or flavin adenine dinucleotide—FAD) and the release of two molecules of CO_2. Thus, the six-carbon citric acid comes back to the beginning of the cycle as a four-carbon acid ready to accept another molecule of acetate. Finally, all those molecules of NADH and $FADH_2$ will result in oxidative phosphorylation by passing their electrons through a series of cytochromes to oxygen, yielding water and resulting in the formation of three molecules of ATP for each pair of electrons going through the transport system.

Net energy gains for fermentation and respiration are summarized in Table 6.4.

TABLE 6.3   Summary of Major Microbial Fermentations

Type	Organisms Involved	End Products
Alcoholic	*Saccharomyces (yeast)*	Ethanol, CO_2
Lactic	*Streptococcus, Bacillus, Lactobacillus*	Lactic acid
Mixed acid	*Escherichia coli* and other enterics	Lactic acid, acetic acid, succinic acid, ethanol, H_2, and CO_2
Butylene glycol	*Enterobacter and Klebsiella*	2,3-butylene glycol plus small amounts of mixed acid products
Propionic acid	*Propionibacterium Veillonella*	Propionic acid, acetic acid, and CO_2
Butyric acid—butanol	*Clostridium*	Butyric acid, butanol, isopropanol, ethanol, acetic acid, and CO_2

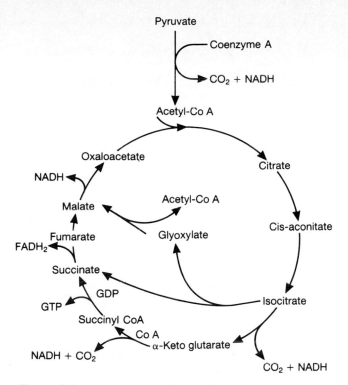

Figure 6.5 Tricarboxylic acid cycle. Note each turn of the cycle liberates eight protons, which results in ATP formation via oxidative phosphorylation. The shaded area illustrates the glyoxylate cycle, which provides for additional four-carbon intermediates to be formed to replenish normal intermediates lost in biosynthetic reactions.

Anaplerotic Sequences

Intermediates of the tricarboxylic acid (TCA) cycle are used for the biosynthesis of amino acids, nucleic acids, and other important constituents of the cell. Removal of these intermediates from the cycle for the purpose of making biosynthetic intermediates results in the de-

pletion of the four-carbon compounds required for continued operation of the cycle. Hence, a mechanism must be available to regenerate the metabolites lost to biosynthesis; these reactions are called **anaplerotic reactions.** As an example, many bacteria use the enzyme phosphoenolpyruvate carboxylase to form the four-carbon oxaloacetic acid from the triose phosphates of the Embden-Meyerhof pathway. Without this activity, cell growth using hexoses as a carbon source would be impossible.

Glyoxylate Cycle

Another anaplerotic reaction pathway that is especially important for cells using acetic acid or fatty acids as a carbon source is the glyoxylate cycle. This cycle consists of two reactions: (1) the splitting of glyoxylate from isocitric acid and (2) the addition of the two-carbon glyoxylic acid to acetyl-CoA to form the four-carbon malic acid. As shown in Figure 6.5, the glyoxylate pathway bypasses the two decarboxylation steps of the TCA cycle. Operation of this cycle is not necessary for the generation of energy from acetic acid, but it is required for the continued operation of the TCA cycle when compounds are removed from it for biosynthetic purposes.

Anaerobic Respiration

There is one final group of organisms that are set apart because they are neither aerobic nor fermentative. These bacteria are mostly anaerobes, but, rather than using metabolic intermediates, they use inorganic ions as their final electron acceptors. Such organisms can be subdivided into three types: **sulfate reducers, nitrate reducers,** and the **methane bacteria.** Keep in mind, however, that even though their metabolism is anaer-

TABLE 6.4   Net Energy Gain from the Dissimilation of One Molecule of Glucose

	Energy Gain per Molecule of Glucose
Anaerobic glycolysis (substrate phosphorylation)	2 ATP
Aerobic metabolism (oxidative phosphorylation)	
From glycolysis (oxidation of 2NADH)	6 ATP
Metabolism of acetyl-CoA (oxidation of 2NADH)	6 ATP
Tricarboxylic acid cycle	
Metabolism of succinyl-CoA	2 GTP
Oxidation of 6 NADH	18 ATP
Oxidation of 2 $FADH_2$	4 ATP
Total energy-rich bonds formed	38

obic, electrons released by oxidation are passed through the electron transport chain and energy is captured by oxidative phosphorylation. The only real difference between aerobic and anaerobic respiration is that the latter process uses an inorganic ion other than free oxygen as a final electron acceptor.

Sulfate Reducers *Desulfovibrio* and *Desulfotomaculum* (an endospore former) are obligately anaerobic organisms that use sulfate as a final electron acceptor, reducing it to the level of sulfide. The overall reaction can be written as follows:

$$SO_4{}^{2-} + 8e^- + 8H^+ \longrightarrow S^{2-} + 4H_2O$$

These organisms require an organic source of carbon and are, therefore, heterotrophs.

Nitrate Reducers Most of the organisms that are able to use nitrate as a final electron acceptor can be thought of as facultative anaerobes. Thus, under anaerobic conditions these organisms can use nitrate if it is available; if not, they will carry out either an aerobic or fermentative metabolism. A few of the organisms capable of replacing oxygen with nitrate include *Escherichia*, *Enterobacter*, *Bacillus*, *Pseudomonas*, *Micrococcus*, and *Rhizobium*. Many of these organisms are able to reduce nitrates all the way to nitrogen gas, as shown below:

$$2NO_3{}^- + 12e^- + 12H^+ \longrightarrow N_2 + 6H_2O$$

This process, called denitrification, is a serious agricultural problem because it results in the loss of nitrates from the soil. It can, however, be very useful when it removes nitrogen from sewage or other wastewaters.

Methane Bacteria There are several genera of methane bacteria that are able to use carbon dioxide as an electron acceptor and thereby reduce it to methane, as shown below:

$$CO_2 + 8e + 8H^+ \longrightarrow CH_4 + 2H_2O$$

These organisms are found in the rumen of cud-chewing animals, and it is estimated that in one cow these bacteria may produce as much as 60 L of methane per day. They are also found in the black mud of ponds and lakes where organic matter is undergoing decomposition, and they are a major part of the flora found in terminal fermentation tanks in sewage disposal plants. We shall return to this subject in the last unit of this text.

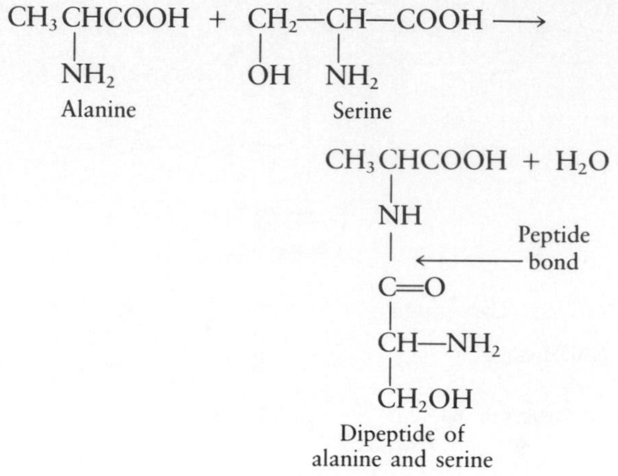

Figure 6.6 A peptide bond links the amino acids together to form large proteins. Note that in the formation of a peptide bond, water is split out from the amino group of one amino acid and the carboxyl group of another amino acid. Thus, when one hydrolyzes a protein, water is inserted into these peptide bonds, which results in a release of the amino acids. Many proteolytic enzymes can accomplish this hydrolytic step.

DISSIMILATION OF NONCARBOHYDRATE SUBSTRATES

Bacteria are remarkably versatile in the variety of compounds they can degrade; in fact, there is probably no biological substance that some organism cannot break down. Among the common organic compounds that can be degraded by bacteria are proteins, nucleic acids, and fats.

Protein Dissimilation

Proteins are very large molecules composed of amino acids linked by peptide bonds (see Figure 6.6). The first step in the use of proteins requires that the protein be degraded into individual amino acids. This is accomplished by the secretion of enzymes—called **proteases**—that hydrolyze the peptide bonds and thus release the individual amino acids. The amino acids can then be taken into the cell, where they can be used again for protein synthesis or further degraded to yield energy or building blocks for anabolic reactions.

Fat Dissimilation

The hydrolytic breakdown of fats occurs through the action of fat-splitting enzymes, or lipases. Simple lipids are degraded to glycerol and fatty acids, as shown in the following equation:

$$\begin{array}{l} CH_2-OOC_4H_7 \\ | \\ CH-OOC_4H_7 + 3H_2O \xrightarrow{\text{lipase}} \\ | \\ CH_2-OOC_4H_7 \\ \quad \text{Tributyrin} \end{array}$$

$$\begin{array}{l} CH_2OH \\ | \\ CHOH + 3CH_3(CH_2)_2COOH \\ | \qquad\qquad\qquad \text{Butyric} \\ CH_2OH \qquad\qquad \text{acid} \\ \text{Glycerol} \end{array}$$

The liberated glycerol can then be metabolized via the Embden-Meyerhof pathway, and the fatty acid can be degraded via acetate through the citric acid cycle.

SUMMARY

Bacteria must have nutrients from which they can synthesize their protoplasm and a source of energy to carry out the anabolic reactions. Energy is generated by the movement of protons, released during oxidation reactions, across the cell membrane. The resulting pH gradient is used either as a protonmotive force or to phosphorylate ADP to ATP. Nutrients are transported into the cell either through active transport, group translocation, or by facilitated diffusion. Some bacteria are photosynthetic, and their sole source of energy is light. Light-induced reactions occurring during photosynthesis cause the formation of high-energy phosphate bonds and the reduction of NADP to NADPH. Water provides the reductant for the latter reaction in the case of green plants and the cyanobacteria; noncyanobacterial photosynthetic bacteria never use water as a source of reductant, but they may use a variety of compounds containing sulfur, hydrogen, or even organic molecules. Nonphotosynthetic bacteria obtain their energy from chemical oxidations. The chemoautotrophs oxidize a variety of inorganic molecules as energy sources. These include NH_3, H_2S, S, NO_2, H_2, and Fe^{2+}. Heterotrophic bacteria, however, obtain their energy from the oxidation of organic molecules. Those using oxygen as a final electron acceptor are called aerobes, while those unable to use oxygen as a final electron acceptor are called anaerobes. The anaerobes may use inorganic ions such as nitrate, sulfate, and carbon dioxide as final acceptors, or they may use an organic compound formed during the breakdown of the substrate. This latter type of anaerobic metabolism is called fermentation. The fermentation of carbohydrates results in the formation of a large number of organic acids and alcohols. The pathways by which these compounds are formed vary somewhat from organism to organism, but the major one is the Embden-Meyerhof pathway. Most aerobic organisms that oxidize carbohydrates to carbon dioxide and water use the tricarboxylic acid cycle for the final oxidation of acetate. Bacteria are able to degrade a great diversity of compounds, including carbohydrates, proteins, nucleic acids, and fats.

QUESTIONS FOR REVIEW

1. Define *anabolism; catabolism.*
2. What are the ultimate sources of energy for living cells?
3. What is meant by "proton = motive force"?
4. Outline how ATP is formed during oxidative phosphorylation.
5. How does active transport differ from facilitated diffusion?
6. Why does a photosynthesizing cell require NADPH?
7. Write the overall chemical reaction for green-plant photosynthesis. How does this differ from noncyanobacterial photosynthesis?
8. Name three families of photosynthetic bacteria. How are they differentiated from one another?
9. Define *chemoautotroph*. List five different compounds that autotrophs can oxidize as sources of energy.
10. Define *heterotroph.*
11. List six different types of fermentation. Include the end products of each of these fermentations.
12. How does the final electron acceptor in anaerobic respiration differ from that in aerobic respiration?

SUPPLEMENTARY READING

Ames GFL: Bacterial periplasmic transport systems: Structure, mechanism, and evolution. *Annu Rev Biochem* 55:397, 1986.

Anraku Y: Bacterial electron transport chains. *Annu Rev Biochem* 57:101, 1988.

Chory J, et al: Structure, function and synthesis of the photosynthetic membranes of *Rhodopseudomonas sphaeroides. ASM News* 50:144, 1984.

Dills SS, et al: Carbohydrates transport in bacteria. *Microbiol Rev* 44:385, 1980.

Drews G: Structure and functional organization of light-harvesting complexes and photochemical reaction centers in membranes of photosynthetic bacteria. *Microbiol Rev* 49:59, 1985.

Gottschalk G: *Bacterial Metabolism*. New York, Springer-Verlag, 1979.

Haddock BA, Jones CW: Bacterial respiration. *Bacteriol Rev* 41:47, 1977.

Hinckle PC, McCarthy RE: How cells make ATP. *Sci Am* 238(3):104, 1978.

Ingledew WJ, Poole RK: The respiratory chains of *Escherichia coli. Microbiol Rev* 48:222, 1984.

Jones CW: *Bacterial Respiration and Photosynthesis*. Aspects of Microbiology Series. Washington, DC, American Society for Microbiology, 1982.

Lehninger AL: *Principles of Biochemistry*. New York, Worth, 1982.

Miller K: The photosynthetic membrane. *Sci Am* 241(4):102, 1979.

Mitchell PL: Keilin's respiratory chain concept and its chemiosmotic consequences. *Science* 206:1148, 1979.

Nicholls DG: *Bioenergetics: An Introduction to the Chemiosmotic Theory*. New York, Academic Press, 1983.

Shuman HA: The genetics of active transport in bacteria. *Annu Rev Genet* 21:155, 1987.

Smith EL, et al: *Principles of Biochemistry: General Aspects,* 7th ed. New York, McGraw-Hill, 1983.

Chapter 7

Bacterial Genetics: Gene Function

OBJECTIVES The study of this chapter should provide a clear understanding of the following points

1 The structure of DNA and the mechanism whereby
DNA replicates itself so as to preserve the message
encoded in the DNA.

2 The steps involved in the transcription of DNA into
mRNA.

3 The mechanism whereby an mRNA is translated into
protein.

4 The difference between the inducible expression and
the repression of an operon.

5 The meaning of feedback inhibition and how it can
control enzyme activity.

6 The various types of mutations that occur and how one
can increase the rate of such mutations.

7 The purpose of the Ames test.

8 Mechanisms used by the cell for DNA repair.

In this chapter we discuss deoxyribonucleic acid (DNA) control of cellular activities. Because this control is mediated through the proteins and enzymes synthesized by the cell, we must first see how the information present in the DNA is translated into specific proteins. We begin by discussing the structure of DNA itself.

THE NATURE OF DNA

Typically, a DNA molecule consists of two polynucleotide chains that are joined to each other by hydrogen bonding, resulting in a spiral (helical) structure.

Each individual nucleotide consists of a pentose sugar (2-deoxyribose), a phosphate group, and a nitrogenous base (purine or pyrimidine). Two types of purines, adenine (A) and guanidine (G) and two different pyrimidines, thymine (T) and cytosine (C), are found in the final DNA molecule (Figure 7.1). To each of these bases is attached the five-carbon sugar 2-deoxyribose. The purines and pyrimidine bases are all joined together into a very long molecule by phosphodiester bonds that connect the number 5 carbon of one molecule of 2-deoxyribose to the number 3 carbon of the adjacent 2-deoxyribose. In brief, DNA is a large molecule consisting of a chain of 2-deoxyribose molecules joined together by phosphodiester bridges; joined to the number 1 carbon of each 2-deoxyribose is one of the purine or pyrimidine bases. It is the specific arrangement or sequence of these nucleotide bases that carries the genetic information in the cell. In order to understand how it is able to copy itself as it is passed to progeny cells, one must know the physical structure of DNA.

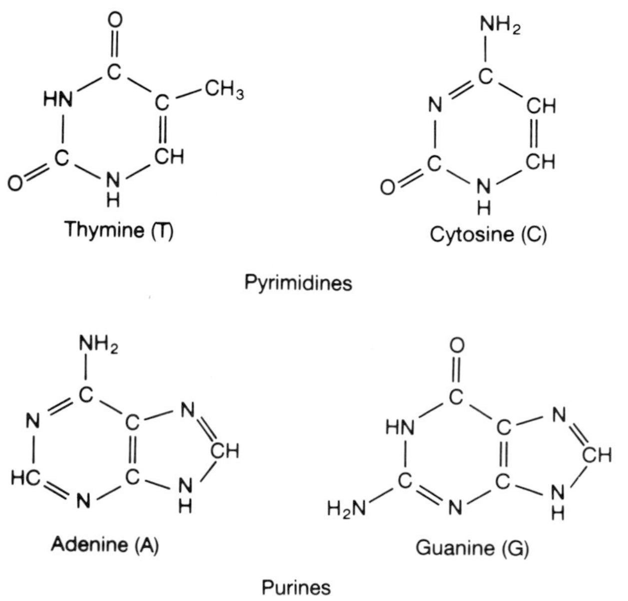

The DNA within the cell exists as a long double strand wound together in a configuration referred to as a double helix. The two strands are wound together in such a manner that each pyrimidine and purine base of strand 1 is joined through hydrogen bonds to a complementary base on strand 2. Obviously the complementary bases must be structurally compatible so that hydrogen bonding can occur. The compatible pairs are adenine (A) and thymine (T), and guanine (G) and cytosine (C). Thus, within the double helix, each A of strand 1 is across from and hydrogen-bonded to a molecule of T on strand 2; each G is across from and hydrogen-bonded to C (Figure 7.2). Synthesis of the DNA that will be passed to progeny cells provides the mechanism for making an exact copy through the use of complementary bases. To accomplish this an enzyme called DNA polymerase uses an existing strand of DNA as a template to line up the correct complementary bases, and the nucleotides are then joined together. When strand 1 is replicated, it yields a single strand identical to strand 2; when strand 2 is replicated, it yields a single strand identical to strand 1. The final result is two helices, each of which contains one original template strand and one new strand, as shown in Figure 7.3. Because each newly synthesized molecule of double-stranded DNA contains one of the original strands and one newly synthesized strand, this is referred to as **semiconservative replication.**

The DNA within a cell exists as an antiparallel structure, meaning that the two strands run in opposite directions so that each end of the DNA helix contains the 5' end of one strand and the 3' end of the other strand. Because DNA replication can occur only in the 5' to 3' direction, only one strand can be replicated continuously from the 5' end. Replication of the other strand takes place in a discontinuous fashion whereby short fragments are synthesized in the 5' to 3' direction, as shown in Figure 7.3. These DNA segments are then covalently linked together by an enzyme called DNA ligase.

During bacterial growth, chromosome DNA replication begins at a specific origin and proceeds in both directions around the circular chromosome. Two replication forks therefore function simultaneously, and each fork adds 750 to 1000 nucleotides per second. This requires that the unwinding machinery that separates the two strands during replication must operate at about 4500 revolutions per minute (Figure 7.4). A part of the DNA (which contains the origin of replication) remains attached to the bacterial membrane, and it is thought that the growth of the membrane

Figure 7.1   The purine and pyrimidine bases of DNA.

Thymine (T)

Cytosine (C)

Pyrimidines

Adenine (A)

Guanine (G)

Purines

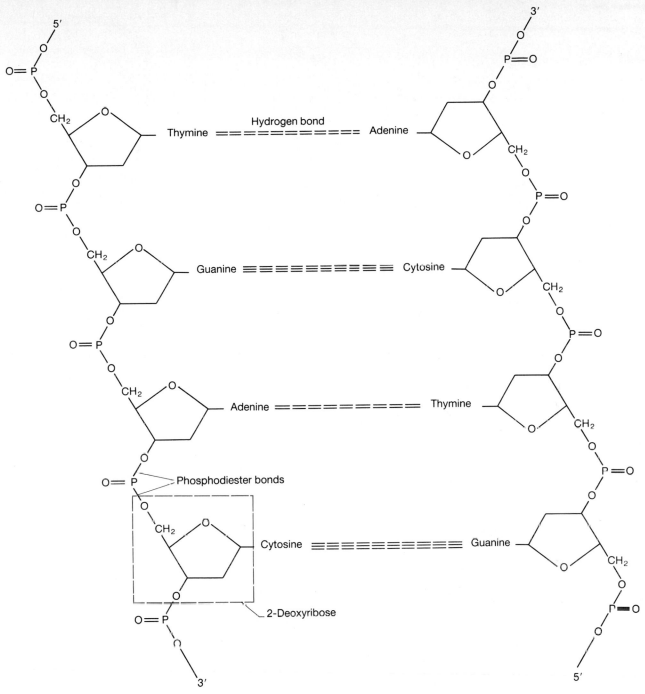

Figure 7.2 A sequence of four nucleotides showing how DNA is composed of a long polymer of 2-deoxyribose molecules each joined to each other through a phosphodiester bond. To the number 1 carbon of each 2-deoxyribose molecule is linked one of the purine or pyrimidine bases, shown in Figure 7.1. Each purine or pyrimidine is paired with its complementary nucleotide base on a second strand of DNA through hydrogen bonds (shown in color). Hydrogen bonds are very weak bonds and the two strands of DNA can be easily dissociated by merely heating the DNA.

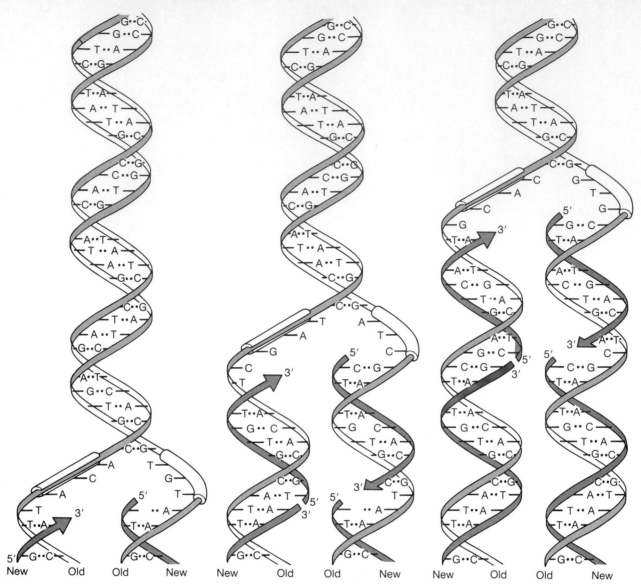

Figure 7.3 Highly schematic representation of the replication of DNA. As strands of the parent molecule unwind—most likely with the aid of unwinding proteins, which are more numerous and cover more bases than shown here—complementary nucleoside-5′-triphosphates are bound to the exposed bases. The triphosphates react with 3′-hydroxyl groups of the preceding nucleotide in the growing strands with the formation of a new 3′,5′-phosphodiester linkage and the loss of inorganic pyrophosphate. Thus, the new chains are formed in the 5′ → 3′ direction along both parental strands. As the new chains grow, sections of new double helices (each with many more bases than shown here) are formed on each of the parent strands. The daughter strands in each new double helix are stitched together at their ends with the enzyme DNA ligase.

between the two origin complexes accounts for the separation of the two progeny chromosomes.

Transcription of DNA to RNA

Although the above mechanism explains how DNA copies itself, it does not explain how DNA controls cellular activity through the synthesis of protein. This is done indirectly through the synthesis of a different type of nucleic acid called **ribonucleic acid** (RNA). RNA is very similar in structure to DNA but differs in that it contains the pyrimidine uracil (U) in place of thymine (T) and contains the five-carbon sugar ribose instead of 2-deoxyribose (see Figure 7.5). Furthermore, except in the case of double-stranded RNA viruses, RNA does not exist in the cell as a double helix like DNA. In RNA synthesis, DNA is used as the template; the enzyme **RNA polymerase** joins together the bases of RNA as each base reacts with its complementary base on the DNA template. It should be noted that only one strand of DNA (called the positive strand) is used as a template for RNA syn-

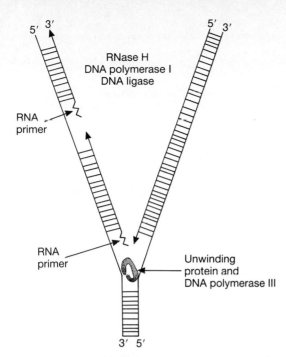

Figure 7.4 The origin of chromosome replication is termed a *replication fork* because the two strands of DNA must be separated before they can be replicated. In this highly schematic representation an unwinding protein separates the two strands of DNA and a second protein binds to each strand to keep them apart. Since replication can occur only from the 5' end toward the 3' end of each DNA strand, it is continuous in one strand (represented on the right-hand fork). On the other strand, however, the DNA polymerase must wait until a segment of about a thousand bases becomes unwound and then replicates back toward the replication fork in a 5' → 3' direction. Replication is initiated for each of these discontinuous fragments (called Okazaki fragments) by a primase that synthesizes a short chain of RNA, which acts as a primer for the DNA polymerase. As each segment is completed, RNase H removes the RNA primer, polymerase I fills in the gaps, and DNA ligase joins each segment together.

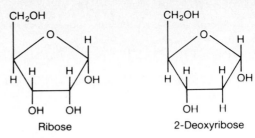

Figure 7.5 Ribose is a component of RNA, and 2-deoxyribose is a component of DNA.

thesis. Thus, a segment of DNA having a base sequence of AGTCTGACT would result in an RNA sequence of UCAGACUGA. This process is called **transcription,** because the message carried in the DNA is transcribed to (or written into) the RNA through complementary base pairing of the RNA nucleotides with the template DNA. Since the resulting RNA now carries the message to make the specific protein coded in the DNA segment, it is called messenger RNA—usually written mRNA. Figure 7.6 summarizes these steps.

In bacteria, the points on the DNA at which RNA polymerase binds and at which RNA synthesis starts are very specific sites called **promoters.** After binding, the polymerase opens the DNA double helix, forming a very stable open promoter complex. RNA synthesis then proceeds at a relatively constant rate until the polymerase reaches a specific termination site, at which point it either separates spontaneously or is released enzymatically from the DNA.

It is interesting to note that the mRNA obtained from the transcription of bacterial DNA is an exact

MICROBIOLOGY MILESTONES

Much water has flowed over the dam and a myriad of new techniques for manipulating DNA have evolved during the four decades since Watson and Crick published their results on the structure of DNA. It is now possible to locate many genes within the chromosome as well as start and stop sites of DNA transcription. Moreover, DNA sequencing has become a standard procedure and the detailed amino acid sequence of many proteins has now been proposed based on the nucleotide sequences occurring in the gene encoding for an enzyme, toxin, or cell membrane receptor.

All these techniques coupled with limited success in the actual isolation of specific human genes has now spawned the most colossal undertaking in the biological world. The goal of this project, officially termed as the **Human Genome Initiative,** is to identify and map all of the genes occurring in the 23 pairs of human chromosomes. Or, to state it another way, scientists want to locate where on each chromosome is located

each of up to 100,000 genes and, once found, to sequence the nucleotide bases that make up the gene.

This enormous project is being led by Dr. James Watson—the same Watson who received the Nobel Prize in 1963 for his work with Francis Crick on the structure of DNA. It is to be a worldwide effort, possibly with different chromosomes being allocated to each country involved in this work. The project is estimated to cost 3 billion dollars and require 15 years to complete.

How might this information benefit humans? There are about 4000 human diseases that are the result of genetic defects, some of which stem from a single base change in the gene. In addition, it is believed by many that alcoholism, schizophrenia, and Alzheimer's disease may have a genetic component. All in all, it seems highly probable that the information gained by this project will benefit mankind for hundreds of years.

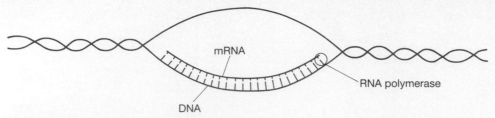

Figure 7.6 Transcription. DNA strands separate, permitting the enzyme RNA polymerase to use one strand as a template for the synthesis of mRNA. The polymerase begins at a start codon and continues making complementary RNA until it reaches a stop codon on the DNA. The mRNA is then released so that it can be translated into protein.

complementary copy of that DNA and, as such, functions directly as mRNA without any additional changes. Such is not the case during the transcription of eucaryotic DNA. The DNA in eucaryotic cells frequently contains many sequences that do not contribute to the final nucleotide sequence of the mRNA. Such sequences are called **intervening sequences,** or **introns,** and they must be removed from the initial transcript before yielding a functional mRNA. This processing of the initial eucaryotic RNA transcript cuts out the introns and then joins the remaining sequences (called **exons**), which will be expressed as mRNA. This process is referred to as RNA splicing. After splicing, which occurs in the eucaryotic cell nucleus, the mature mRNA enters the cytoplasm, where it serves as a code to direct protein synthesis. Figure 7.7 is a schematic illustration comparing procaryotic and eucaryotic transcription.

Translation of RNA to Protein

The process by which mRNA directs the synthesis of a specific polypeptide (protein) is called **translation,** because the information carried in the base sequences of the mRNA must now be translated into the amino acid sequence of the polypeptide. Each contiguous series of three nucleotides in the mRNA is called a **codon,** and each codon directs the insertion of one specific amino acid into the polypeptide. For example, the three nucleotides UUG constitute the codon for

Figure 7.7 Transcription. (*a*) Procaryotic. (*b*) Eucaryotic.

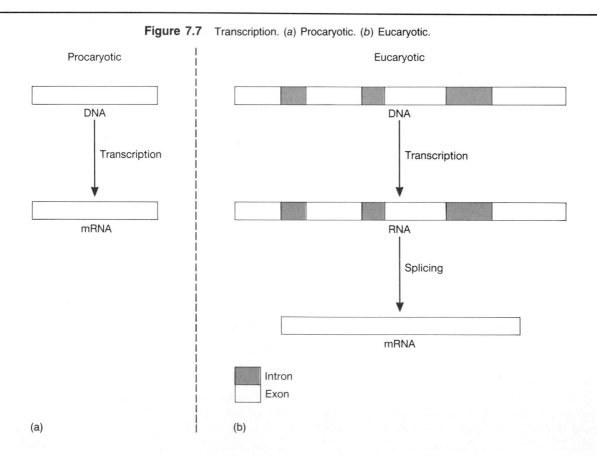

(a)　　　　　　　　　　(b)

the amino acid leucine. The translators that will read these codons are found in a class of small RNA molecules (70 to 80 nucleotides long) called transfer RNAs (tRNAs). Each tRNA possesses a specific region in its tightly folded molecule ensuring that the correct amino acid will be inserted into the growing polypeptide. This is accomplished in the following manner: (1) There are one or more different tRNAs for each of the 20 amino acids. (2) There is also a specific enzyme for each amino acid that joins the correct amino acid to its specific tRNA. As a result, the tRNA for leucine can accept only leucine; the tRNA for glycine can accept only glycine; and so forth for the entire 20 amino acids, as shown in Figure 7.8. (3) Each tRNA also possesses a specific **anticodon,** consisting of three nucleotides that are complementary to the three nucleotide codons on the mRNA. Thus, the anticodon on leucine tRNA is AAC; it will bind to the mRNA only with the codon calling for leucine (UUG).

Now that we have discussed how the mRNA is read by the amino acid–bound tRNAs, let us put it all together to see how a polypeptide is actually synthesized. The concepts are not difficult, and the reader is encouraged to follow the schematic illustration shown in Figure 7.9 along with the text.

Figure 7.8 Schematic structure of transfer RNA (tRNA). The tRNA for each amino acid differs in its anticodon and in the bases that make up the D loop and the TψC loop. It is the base composition (and hence structure) of these loops that determines which amino acid is added to the 3′ end of the tRNA. The three bases designated as "anticodon" base pair with their complementary bases on the mRNA, permitting the correct amino acid to be joined to the growing peptide.

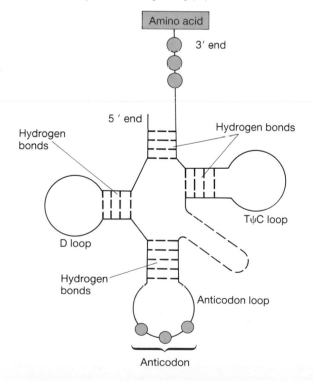

Role of the Ribosome in Protein Synthesis

For protein synthesis to occur, structural units are required that will present each codon to the tRNA, directing the sequential addition of each amino acid to the growing polypeptide. Such structural units, called ribosomes, are composed of proteins and RNA. Each ribosome exists as two independent subunits within the bacterial cell, one sedimenting at 30S and the other at 50S (S represents a Svedberg unit and is a measure of the rate a particle will sediment in a centrifugal field). To initiate protein synthesis, the 30S ribosomal subunit binds to a specific ribosome binding site on the mRNA. In the presence of a tRNA linked to its amino acid, the 50S subunit then joins the 30S subunit (already attached to the mRNA) to form an intact ribosome. The ribosome then moves along the mRNA, exposing the next codon. A peptide bond is then formed between the amino acids, and the ribosome moves again to expose the next codon and transfer the dipeptide to the peptide site. This process continues until the ribosome reaches a termination (stop) codon on the mRNA, at which time the completed polypeptide is released from its tRNA and the ribosome dissociates into its two subunits. Actually, as soon as a ribosome moves from the first to the next codon of the mRNA, another ribosome joins the first codon, and so on, until there are many ribosomes along the entire length of the mRNA, each directing the synthesis of a polypeptide. The occurrence of multiple ribosomes on a single molecule of mRNA is called a **polysome,** and it can be seen that the polysomes permit a single strand of mRNA to be simultaneously translated into many molecules of the same polypeptide (Figure 7.10).

Organization of Cellular DNA

The previous sections make it clear that heritable characteristics are encoded in a cell's DNA and that, in order to be expressed, this genetic language must first be transcribed into RNA. Let us now consider how this information is organized within the cell.

Bacterial cells contain a single, continuous molecule of double-stranded DNA, called a **chromosome.** Eucaryotic cells, however, may contain multiple chromosomes, which in most cases are present as pairs. In the case of sexual reproduction, one chromosome of each pair comes from the male and the other from the female. In humans, each cell contains 23 such pairs, for a total of 46 chromosomes. The cells carrying complementary pairs of chromosomes are called **diploid,** whereas cells that contain a single chromosome, such as bacteria, are termed **haploid.**

Each chromosome contains segments of DNA that can be ultimately transcribed and translated into

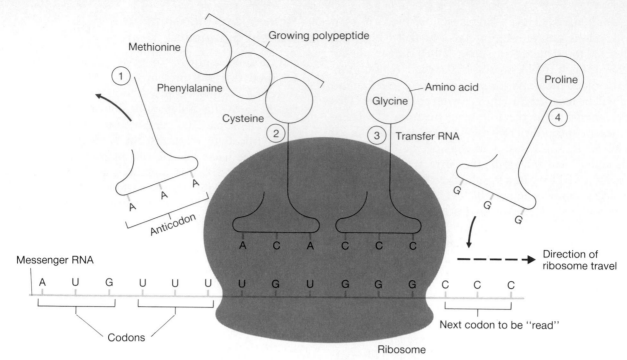

Figure 7.9 Summary of protein synthesis. Formation of a polypeptide takes place on the surface of a ribosome as it moves along a molecule of messenger RNA (mRNA) in the 5′ → 3′ direction. Using energy from ATP, specific activating enzymes attach particular amino acids to their appropriate tRNAs with high-energy bonds. As the ribosome shifts to a new three-base codon on the mRNA, it releases a tRNA from the vacated codon (tRNA 1). In the diagram this places the tRNA bearing the growing polypeptide chain (tRNA 2) in the left-hand binding site on the ribosome, and a tRNA with an anticodon complementary to the codon in the newly "read" right-hand position (tRNA 3) assumes its place by base pairing. The next events will be the formation of a peptide bond between the growing polypeptide and the amino acid on tRNA 3 (using the energy acquired during activation), the release of tRNA 2, the shifting of the mRNA one more codon to the right, and the attachment of tRNA 4 bearing its associated amino acid. These steps are repeated until a stop codon on the mRNA is reached; the completed polypeptide, which spontaneously folds into a precise three-dimensional configuration as it is synthesized, is then freed with the aid of a release factor.

a single polypeptide. Such discrete segments are called **genes.** When describing the genetic endowment of a cell, we speak of its **genotype.** If, however, we are describing the observed properties of a cell, we are referring to its **phenotype.** It will become apparent in the subsequent sections of this chapter that these two terms are not equivalent. Thus, a cell may have genes that are not expressed, thereby possessing a genotype different from its phenotype.

Regulation of Gene Expression

The continual transcription of all the cellular DNA would be wasteful to the cell because it would be producing enzymes to break down nutrients that are not present or forming biosynthetic intermediates that might be available from the environment. The expression of many genes, therefore, is subject to a form of regulation based in part on the requirement of the cell for the product of a specific gene. Thus, some genes

are continually expressed and, as such, are referred to as **constitutive genes:** others are synthesized only when induced by their specific substrate or a structurally similar analogue. Such enzymes are labeled **inducible enzymes.**

Inducible Expression of the Lactose Operon

Many enzymes involved in the catabolic degradation of their substrate are not synthesized unless the cells are in a medium containing the substrate for those enzymes. For example, when *E. coli* is grown using glycerol as a carbon source, the cells will synthesize only very minuscule amounts of the enzymes required for the degradation of lactose. If, however, lactose is added to the medium, the cells will rapidly start to synthesize carrier molecules (called permeases) to transport the lactose into the cell and enzymes to hydrolyze this disaccharide to glucose and galactose (β-galactosidase). An enzyme that can acetylate lactose

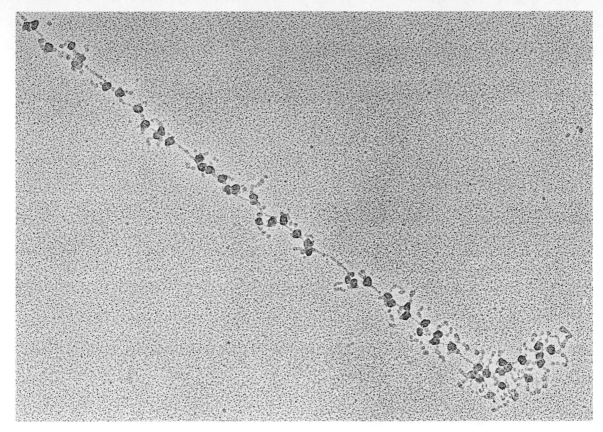

Figure 7.10 Electron micrograph of a polysome in which multiple ribosomes are translating a single molecule of mRNA. As the ribosomes move along the mRNA (from top to bottom), one can see the growing peptide chains increasing in length.

is also produced, but this enzyme is not necessary for the metabolism of the lactose.

The three genes induced by the presence of lactose make up an **operon,** which is defined as contiguous genes under the control of the same regulatory elements. Operons, or the clustering of related genes, are common in bacteria but very rare in eucaryotic cells. The lactose operon has a single promoter (attachment site for RNA polymerase), and the mRNA synthesized from this promoter codes for all three gene products. Because such mRNA results from the transcription of more than one gene, it is termed **polycistronic mRNA.** Polycistronic mRNA is not found in eucaryotes and is thus unique to procaryotes. Regulation of the lactose operon is mediated by a protein product encoded in the *lac I* regulatory gene. This protein is produced constitutively and binds to a site on the DNA (called the operator site) near the promoter (where the RNA polymerase binds) for the lactose operon; when bound, it prevents passage of the RNA polymerase and hence transcription. Because this protein prevents expression of the lactose operon, its effect is called **negative control,** and mutants that have lost the ability

to make this regulatory protein will transcribe the lactose operon constitutively. The regulatory gene for the lactose operon is located immediately adjacent to that operon, but regulatory genes for other operons may be considerably removed from the genes they control.

When lactose enters the cell, it binds to the regulatory gene product (called the **repressor**), changing its conformation so that it can no longer bind to the operator site. This, then, permits the RNA polymerase to transcribe the structural genes of the lactose operon. Figure 7.11 illustrates the regulation of the lactose operon.

Repression of Amino Acid Synthesis

Repressor proteins also control the genes involved in the biosynthesis of essential nutrients such as amino acids. Such pathways are regulated by repression, because the presence of an amino acid in the medium turns off production of the enzymes necessary for its biosynthesis. This mechanism is similar to the induc-

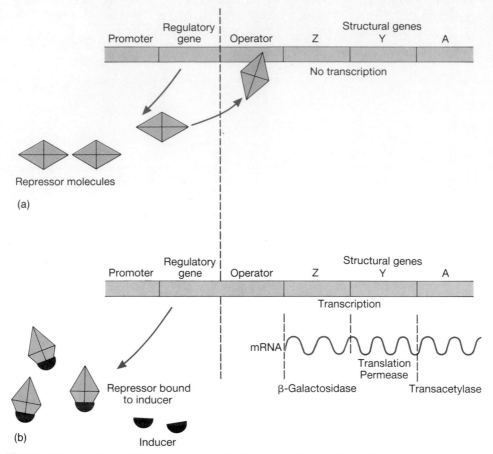

Figure 7.11 Jacob and Monod operon model for the regulation of inducible enzyme synthesis. (*a*) The regulatory gene produces a repressor that binds to the operator site, preventing transcription of the operon. (*b*) The inducer (an analogue of lactose) binds to the repressor and thus permits transcription to occur.

tion of the lactose operon in that a regulatory gene codes for an operon repressor. In this case, however, the regulatory gene encodes for a protein that will bind to the operator site only if its conformation is altered by first binding to the amino acid it is regulating. In other words, the repressor for tryptophan biosynthesis becomes effective only in the presence of an excess of tryptophan. In such a situation, the repressor binds to the tryptophan, forming a functional repressor that can then attach to the operator site for the tryptophan operon. This type of regulatory system, therefore, will prevent the synthesis of an amino acid when there is already an excess of that amino acid present. It should be apparent that repression and induction differ only in whether the effector stabilizes the repressor form so it can bind (tryptophan case) or not bind (lactose case), respectively, to the operator site. Figure 7.12 illustrates this type of regulation of the tryptophan operon.

Tryptophan synthesis is also regulated by structural changes that occur in the mRNA in the presence of tryptophanyl-tRNA. In this case, excess tryp-tRNA binds to the mRNA, resulting in a loop formation that prevents the passage of the ribosome.

Positive Regulatory Control

Some inducible operons are regulated in a manner fundamentally opposite from the lactose operon. Examples include the operons coding for enzymes used for the catabolism of the sugars maltose and arabinose. In these cases, the regulatory gene product forms a complex with the sugar, and it is only after this complex binds near the promoter site that RNA polymerase binding occurs and the operon is transcribed. This is called **positive control,** and it should be apparent that any mutant lacking the regulatory gene product would be unable to metabolize the corresponding sugar.

Regulation of Enzyme Activity

As might be surmised, enzyme repression is a somewhat coarse control mechanism; it responds only when there are rather large changes in substrate availability and, even then, the response is slow because enzymes already synthesized are usually fairly stable in bacteria. There is, however, a much finer enzyme control in which the end product of a pathway will bind to one

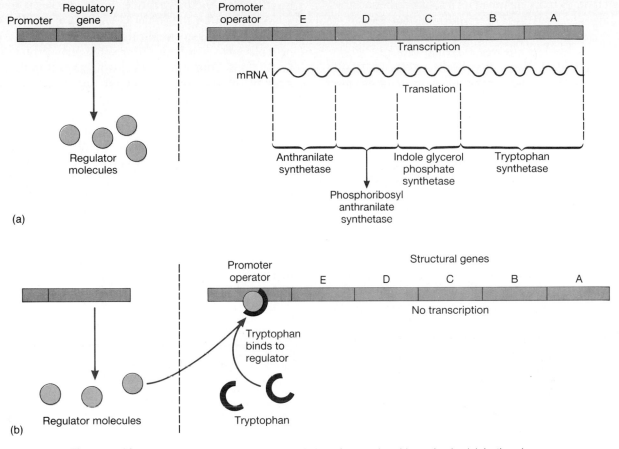

(a)

(b)

Figure 7.12 A simplified scheme for the regulation of tryptophan biosynthesis. (a) In the absence of tryptophan, regulator molecules are unable to bind to the operator, and transcription occurs. (b) Regulator molecules become activated after binding to tryptophan, blocking the operator and preventing transcription. Note: there is a second level of regulation of tryptophan biosynthesis, modulated by tryptophanyl tRNA, that is not shown here.

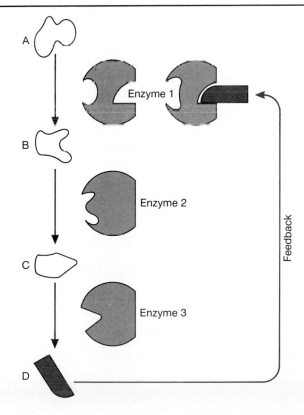

of the enzymes catalyzing an early reaction of that pathway (usually the first) and, while bound, will inhibit the action of the enzyme so that substrate cannot enter the pathway and more end product cannot be made (Figure 7.13). This mechanism, called **feedback inhibition,** does not prevent synthesis of an enzyme; it only inhibits its activity. When the available end product of the pathway is used by the bacterium, the inhibited enzyme is freed to catalyze its reaction, and more end product soon results.

An enzyme that is subject to feedback inhibition possesses two binding sites, one that binds to its substrate and a second that binds the inhibiting end product. Such an enzyme is called an **allosteric enzyme;**

Figure 7.13 Feedback inhibition. Product D, resulting from a series of enzymatic reactions beginning with substrate A, reacts with enzyme 1, changing the configuration of its substrate binding site and causing the enzyme to become inactive. This prevents the conversion of A to B. As the concentration of D falls, enzyme 1 and product D dissociate, allowing enzyme 1 to become active again. Feedback inhibition represents a much finer control than enzyme repression.

when it binds to its end product inhibitor, the substrate binding site undergoes a conformational change so that the substrate binds to the enzyme less effectively, or the enzyme has a lower maximal rate.

It should be noted that both enzyme repression at the gene level and inhibition at the enzyme level can occur simultaneously.

Mutations in Microorganisms

Mutation means heritable genetic change, and it is an essential part of evolution. It is generally believed that the myriad microorganisms existing today all evolved from a common ancestor that lived billions of years ago. This evolution would have resulted from chemical changes in the DNA that occurred occasionally as a result of cellular error during reproduction.

When these changes occur during normal growth, they are referred to as **spontaneous mutations.** The time scale for mutation rate is not expressed in terms of hours or days but rather of generations. Since many bacteria may reproduce themselves in as little as 30 min, the number of mutations occurring may be quite large even though the number of mutations per generation is no higher than that seen in higher organisms.

Molecular Types of Mutations

Because all information carried in the DNA is eventually converted to specific proteins, any change in the base sequence of the DNA can result in a change in the molecular composition of the resulting protein. In general, such changes are caused by a substitution, an insertion, or a deletion of one or more base pairs in the DNA molecule.

Base Pair Substitution Each amino acid in a polypeptide is encoded in a sequence of three nucleotides in the DNA, and any change in one of these nucleotides may cause a different amino acid to be inserted into the polypeptide. For example, if during DNA replication the triplet GAA (coding for leucine) should be mistakenly replicated as GTA, leucine would be replaced by histidine during subsequent protein synthesis. Mutations of this type, which result in the substitution of one amino acid for another, have been termed **missense mutations** (see Figure 7.14). Depending on the function of the resulting protein and the position of the new amino acid in the polypeptide, such a change may or may not result in a detectable change in the ability of the changed protein to function normally. It also should be noted that the four bases of DNA have the ability to code for 64 different triplet sequences. Many of the 20 amino acids, thus, possess more than one triplet code (Table 7.1). If in our example of DNA replication the base substitution from GAA had been to GAT, there would have been no detectable mutation, because both GAA and GAT code for leucine.

Insertion or Deletion of DNA Bases Insertion or deletion of only one base might appear not to cause a major change in the resulting protein composition. But, because each amino acid requires a triplet code, the loss or gain of a single base in the DNA results in the following type of **frame shift mutation:**

DNA: AAC GAA CGC TGA
RNA: UUG CUU GCG ACU

Deletion of the first A in the DNA shifts the "reading" frame as follows:

DNA: ACG AAC GCT GA
RNA: UGC UUG CGA CU

This would completely change the tetrapeptide encoded in this segment of DNA from leucine–histidine–alanine–threonine to cysteine–leucine–arginine–leucine. Unless such a loss is compensated by the insertion of a new base very close to the deleted base, it is

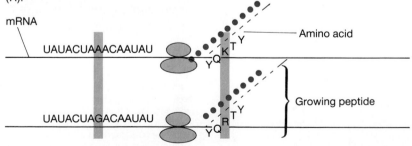

Figure 7.14 Missense mutation. The change of a single base in the mRNA may result in an amino acid substitution, causing changes in the function of the peptide. In the schematic shown, a change of adenine to guanine changes the codon from AAA to AGA and the resulting amino acid from lysine (K) to arginine (R).

TABLE 7.1 The Genetic Code Showing All Possibilities of Triplet Bases in the DNA and mRNA and the Amino Acids Encoded in Them

DNA	mRNA	A.A.	DNA	mRNA	A.A.	DNA	mRNA	A.A.	DNA	mRNA	A.A.
AAA	UUU	Phe	GAA	CUC	Leu	TAA	AUU	Ile	CAA	GUU	Val
AAG	UUC	Phe	GAG	CUC	Leu	TAG	AUC	Ile	CAG	GUC	Val
AAT	UUA	Leu	GAT	CUA	Leu	TAT	AUA	Ile	CAT	GUA	Val
AAC	UUG	Leu	GAC	CUG	Leu	TAC	AUG	Met	CAC	GUG	Val
AGA	UCU	Ser	GGA	CCU	Pro	TGA	ACU	Thr	CGA	GCU	Ala
AGG	UCC	Ser	GGG	CCC	Pro	TGG	ACC	Thr	CGG	GCC	Ala
AGT	UCA	Ser	GGT	CCA	Pro	TGT	ACA	Thr	CGT	GCA	Ala
AGC	UCG	Ser	GGC	CCG	Pro	TGC	ACG	Thr	CGC	GCG	Ala
ATA	UAU	Try	GTA	CAU	His	TTA	AAU	Asn	CTA	GAU	Asp
ATG	UAC	Try	GTG	CAC	His	TTG	AAC	Asn	CTG	GAC	Asp
ATT	UAA	non[a]	GTT	CAA	Gln	TTT	AAA	Lys	CTT	GAA	Glu
ATC	UAG	non[a]	GTC	CAG	Gln	TTC	AAG	Lys	CTC	GAG	Glu
ACA	UGU	Cys	GCA	CGU	Arg	TCA	AGU	Ser	CCA	GGU	Gly
ACG	UGC	Cys	GCG	CGC	Arg	TCG	AGC	Ser	CCG	GGC	Gly
ACT	UGA	non[a]	GCT	CGA	Arg	TCT	AGA	Arg	CCT	GGA	Gly
ACC	UGG	Trp	GCC	CGG	Arg	TCC	AGG	Arg	CCC	GGG	Gly

[a] Nonsense or termination codons.

obvious that a completely new polypeptide will be coded for by the segment of DNA following the deletion.

Nonsense Mutations There are three codons—UAG, UAA, and UGA—whose normal function is to cause the termination of synthesis of the polypeptide chain. Any mutation resulting in the formation of one of these termination codons will cause the release of an incomplete polypeptide. This has been termed a **nonsense mutation**. These codons have been given the trivial names of amber (UAG), ochre (UAA), and opal (UGA).

Conditional Lethal Mutations The general term **conditional lethal mutations** is assigned to mutations that are lethal under one set of circumstances but not lethal under "permissive" conditions. The most common type of conditional lethal mutation is called a **temperature-sensitive mutant** (ts). Such mutants may be able to function normally at 30°C but cannot grow at a higher temperature, such as 42°C. Temperature-sensitive mutants have provided important tools for the study of DNA replication, RNA synthesis, and protein synthesis.

Tests That Identify Mutants

As one might surmise, many mutants can be isolated by growing the culture on a medium that will support the growth of only the desired organism. Thus, if one wished to isolate a particular mutant that could synthesize all of its own amino acids, one would plate the culture on a medium that contained no amino acids, and any colonies that grew would represent the desired mutant.

Suppose, however, one wished to study the transport of the amino acid leucine across the cell membrane and wanted an organism that had an absolute requirement for leucine. If leucine is incorporated into the medium, both those requiring it and those not requiring it will grow. A technique to solve this dilemma, called **replica plating,** was devised by the Lederbergs. To do this, a number of organisms were streaked out on an agar plate, which in this case contained leucine. After 4 or 5 h, each organism had developed into a microcolony, and a sterile pad of velvet was pressed down gently over the plate. A few bacteria from each colony adhered to the velvet, so that, when the velvet was pressed down onto a second plate, it produced an exact copy of the original master plate. However, because the copy plate did not contain leucine, organisms requiring leucine were unable to grow. One can note the positions of those master plate colonies that failed to grow on the copy plate and select the leucine-requiring mutants. Figure 7.15 summarizes the steps involved in replica plating.

Methods of Inducing Mutations

So far we have been discussing spontaneous mutations, those that occur during the normal growth of an organism at a usual rate between one in a million and one in 100 million cells. There are techniques by which one can increase the overall rate of mutations in a bacterial culture. They are designed to increase the "mistakes" a cell makes while it copies its DNA during reproduction. Thus, they are not specific for any one

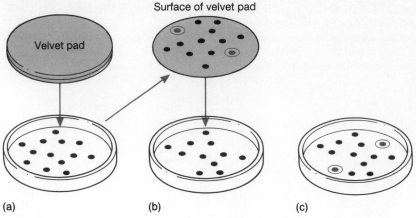

Figure 7.15 Replica plating. (*a*) A sterile velvet pad is pressed lightly down on the leucine-containing medium. (*b*) The pad is then pressed lightly on a sterile agar plate that does not contain leucine in the medium. (*c*) By noting the position of the colonies on the leucine-containing medium that failed to grow on the medium without leucine, one can subculture leucine-requiring colonies from the first plate.

gene locus; moreover, most of the resulting mutations kill the organism. However, since they do increase the overall incidence of mutants, they are called **mutagenic agents,** and the resulting mutant organisms are called **induced mutants.** Let us look at a few mutagenic agents and see how they work.

One can grow organisms in the presence of unusual pyrimidines that will be incorporated in their DNA. If 5-bromouracil—a molecule structurally similar to thymine, a normal constituent of DNA—is used, the organism incorporates the 5-bromouracil in its DNA where it should have put thymine. When this organism containing the 5-bromouracil reproduces, some of the 5-bromouracil bases will pair with guanine instead of the correct complementary purine, adenine (Figure 7.16). Thus, the new cell will have a number

of guanine residues where it should have had adenine and, as a result, the message in the DNA will be changed so that an incorrect or nonfunctional protein may result from the transcription and subsequent translation of that segment of DNA.

Another method of increasing mutation rates is to treat cells with alkylating agents such as nitrogen mustard or ethylene oxide. These agents react with the guanine in the DNA and, as a result, frequently cause the wrong complementary base to pair with guanine during replication of the DNA. That would, of course, change the message in that segment of DNA.

Treatment of cells with ultraviolet light also results in an increased mutation rate. This type of treatment causes the formation of thymine dimers, as shown schematically in Figure 7.17. In other words,

Figure 7.16 Effect of growth in the presence of a mutagen. (*a*) Normal DNA, mRNA, and resulting peptide. (*b*) DNA after substitution of 5-bromouracil (B) for some thymidine bases. Note how this substitution changes the resulting mRNA and the amino acid composition of the final peptide.

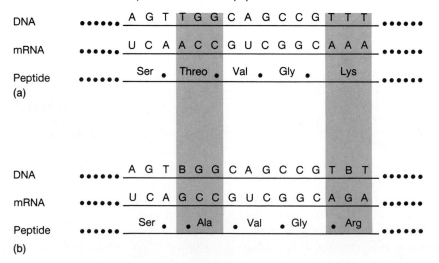

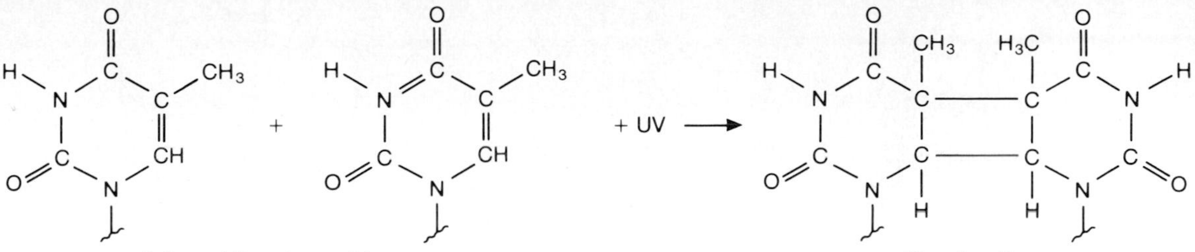

Figure 7.17 Formation of a thymine dimer by ultraviolet light. Repair takes place by one of several routes. In the light a photoenzyme may recognize and bind to the dimer in the strand of DNA. This complex then absorbs a photoreactivating wavelength of light, which results in the splitting of the dimer to restore the monomers. In the dark, excision repair or the less understood postreplication gap repair may occur. Excision repair consists of the removal from the strand of DNA of the dimer and several bases on either side by an excision enzyme followed by replacement of the excised bases by DNA polymerase, using the information on the complementary strand.

the ultraviolet light causes adjacent thymines in the DNA to coalesce in pairs. This causes distortion of the DNA and results in mistakes during the replication of the DNA and, hence, a change in the message carried by the DNA. It is interesting to note that visible light will frequently reverse this effect by activating a repair enzyme that hydrolyzes out the thymine dimers and allows a normal repair to take place. This overall repair process is referred to as **photoreactivation.**

DNA Repair

In addition to the photoactivated enzyme that excises thymine dimers, there are several other DNA repair systems in bacteria that function to improve cell viability and repair errors in DNA. Similar systems are probably present in all cells.

Excision–Repair System Three enzymatic activities are involved in the excision and repair of incorrect DNA. First, one of a number of endonucleases will recognize an error in the DNA, such as the lack of a base, a thymine dimer, an alkylated base, or the presence of uracil in place of thymine. The endonuclease will then cleave the damaged DNA strand just before the site of the error. Second, DNA polymerase I binds to that nick and proceeds to remove 50 to 100 bases of DNA, including the damaged part, while simultaneously resynthesizing new DNA using the intact complementary strand as a template. Third, a DNA ligase reseals the nick left where DNA polymerase I stops.

SOS Repair The excision–repair system is efficient, accurate, and effective as long as the chromosome replication fork has not come to a thymine dimer or an alkylated base, either of which will block passage of the replicating enzyme. If, however, the replication fork is stopped, inducible repair systems

are formed. Such systems are termed SOS repair; they include inhibition of cell division and respiration, increased genetic recombination activity, and the synthesis of an error-prone repair system. This system has the ability to replicate through damaged DNA, although it is much more likely to make mistakes in replication than do the normal DNA polymerases. In fact, all the mutations induced by ultraviolet irradiation are the result of mistakes caused by this enzyme. However, without this enzyme, DNA replication would be permanently blocked at the site of DNA damage, resulting in the death of the cell.

Mutations and Carcinogen Testing

Carcinogens are chemicals that induce an increased incidence of cancer; their recognition and the reduction of human exposure to them is a matter of national policy. Unfortunately, many carcinogens have been recognized only from epidemiological studies in humans (cigarette smoking), and then usually long after exposure has occurred. Rats and mice are now used to evaluate the carcinogenic potency of a chemical, but such studies are expensive and often subject to conflicting interpretation.

Several recently developed tests for the detection of carcinogens have employed microorganisms. The most widely used of these is the **Ames test,** developed by and named after Bruce Ames. It uses special strains of *Salmonella typhimurium* that carry several mutations. These strains synthesize a very short cell wall lipopolysaccharide, so they will take many compounds into their cells more readily than wild-type cells. Also, their DNA repair systems are defective and they synthesize constitutively an error-prone DNA polymerase as described under SOS repair. Most important, both strains of *Salmonella* used in this test are unable to synthesize the amino acid histidine. The *his* mutation

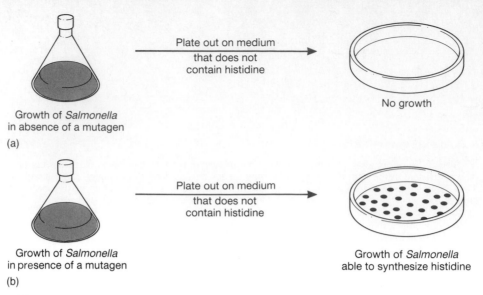

Figure 7.18 Ames test. (*a*) Strains of *Salmonella* used in this test cannot synthesize histidine. (*b*) Growth in the presence of a putative mutagen causes base substitutions and/or frame shifts, correcting the gene defect for histidine biosynthesis.

In the flask labeled (a): Growth of *Salmonella* in absence of a mutagen — Plate out on medium that does not contain histidine → No growth

In the flask labeled (b): Growth of *Salmonella* in presence of a mutagen — Plate out on medium that does not contain histidine → Growth of *Salmonella* able to synthesize histidine

in one strain is the result of a base substitution; in the other strain it is caused by a frame shift.

The Ames test is usually carried out by mixing each strain with the test compound along with an extract from rat liver containing the microsomal enzymes that are known to convert many chemicals to their carcinogenic form. This mixture is then plated on a medium lacking histidine to determine the frequency of mutation to histidine independence. Because each colony that grows represents a mutation to histidine synthesis, the number of colonies is a measure

of the effectiveness of the mutagenic agent used (see Figure 7.18).

The value of the Ames test is based on the fact that about 80 percent of carcinogens, as measured in animals, are also bacterial mutagens and will therefore increase the rate of mutation to histidine independence in this test. Although this does not mean that carcinogenesis is a result of a mutation, the Ames test does provide a fast, inexpensive, easy, and fairly reliable procedure for screening potential carcinogens.

SUMMARY

DNA carries within its molecule the messages that control the activities of the cell. These messages are coded for by the sequences of the four nucleotide bases present in the DNA molecule. The DNA exists in the cell in the form of a double helix, two strands of DNA nucleotides wound around each other in a helical structure; each base of strand 1 is directly across from its complementary base on strand 2. In the replication of the DNA, the message is preserved by the fact that the DNA polymerase uses an existing strand of DNA as a template to line up the correct complementary bases and join them together. Thus, when strand 1 is copied, it yields a single strand identical to strand 2; when strand 2 is copied, it yields a DNA strand identical to strand 1. The DNA uses its message for the synthesis of specific proteins by first transcribing its base sequence to RNA. The RNA formed as a result of complementary base pairing with the DNA is called messenger RNA because it translates its message directly into protein. Each series of three nucleotides in the mRNA codes for one amino acid. This codon reacts

with its specific anticodon, which is on the particular transfer RNA carrying the amino acid being coded for.

Eucaryotic DNA contains intervening sequences (introns) of nucleotides, which are transcribed to form RNA but must be spliced out to form a functional mRNA. Procaryotic DNA does not contain such sequences, and the transcribed RNA can be used directly as mRNA.

The translators that read the information encoded in the mRNA are found in a class of small RNA molecules called transfer RNAs. There is at least one specific tRNA for each of the 20 amino acids; each tRNA carries a special triple-base anticodon that will recognize the triplet codon on the mRNA. Protein synthesis requires that the mRNA bind to a particulate body called a ribosome. The ribosome moves down the mRNA to expose a new codon for recognition by the correct amino acid–charged tRNA, following which a peptide bond is formed between that amino acid and the growing polypeptide on the ribosome.

The control of enzyme synthesis at the genetic level of DNA occurs through repressors that can react with an operator site and thus prevent the transcription of that segment of DNA under control of the operator promoter. In the case of inducible enzymes, the substrate, or inducer, reacts with the repressor and thus allows the operon to be transcribed. A second mechanism of enzyme control is feedback inhibition, in which a product of a series of enzymatic reactions reacts with the enzyme itself, temporarily inactivating it.

Mutations occur in bacteria as a result of mistakes in the replication of DNA. They can be distinguished from adaptation by the fact that they occur randomly and spontaneously. The rate of mutation can be increased through the use of mutagenic agents that react with DNA in such a manner as to increase mistakes made during DNA replication. Molecular types of mutations include base pair substitutions, frame shift mutations, and nonsense mutations.

Bacteria have a number of systems by which they can repair altered DNA. Two of these are known as the excision–repair system and SOS repair. The Ames test is a technique by which one can postulate the carcinogenicity of a substance by determining the extent to which that substance will act as a mutagenic agent.

QUESTIONS FOR REVIEW

1. How does DNA replicate itself so as to preserve the message coded for in the DNA?
2. List all the steps involved in the transcription and translation of this message into a specific protein.
3. What is an intron? An exon?
4. Give an example of how a cell's genotype may differ from its phenotype.
5. What is a codon? An anticodon?
6. Describe the role of the ribosome in protein synthesis.
7. If a bacterium that is synthesizing its own amino acid tryptophan is transferred to a medium containing tryptophan, it will cease to synthesize tryptophan. Explain what happens to stop this synthesis.

8. How does the control of the tryptophan operon differ from that of the lactose operon?
9. What is meant by positive regulatory control? Feedback inhibition?
10. What is a mutagenic agent? Name two.
11. Describe replica plating.
12. Describe two ways in which a cell can repair its DNA.
13. What is the Ames test?

SUPPLEMENTARY READING

Cech TR: RNA as an enzyme. *Sci Am* 255(5):64, 1986.
Darnell JE Jr: RNA. *Sci Am* 253(4):68, 1985.
Doolittle RF: Proteins. *Sci Am* 253(4):88, 1985.
Dovoret R: Bacterial tests for potential carcinogens. *Sci Am* 241(2):40, 1979.
Drake JW, Balz RH: Biochemistry of mutagenesis. *Annu Rev Biochem* 45:11, 1976.
Eigen M, et al: The origin of genetic information. *Sci Am* 244(4):88, 1981.
Felsenfeld G: DNA. *Sci Am* 253(4):58, 1985.
Howard-Flanders P: Inducible repair of DNA. *Sci Am* 245(5):72, 1981.
Kazarian HH Jr: The nature of mutation. *Hospital Practice* 20(2):55, 1985.
Lake J: The ribosome. *Sci Am* 245(2):84, 1981.
Maniatis T, Ptashne M: A DNA operator–repressor system. *Sci Am* 234(1):64, 1976.
Nomura M: The control of ribosome synthesis. *Sci Am* 250(1):102, 1984.
Roth JR: Frameshift suppression. *Cell* 24:601, 1981.
Walker G: Mutagenesis and inducible responses to deoxyribonucleic acid damage in *Escherichia coli*. *Microbiol Rev* 48:60, 1984.
Williamson B: Gene therapy. *Nature* 298:416, 1982.
Witkin EM: Ultraviolet mutagenesis and inducible DNA repair in *Escherichia coli*. *Bacteriol Rev* 40:869, 1976.

Chapter 8

Bacterial Genetics: DNA Transfer

OBJECTIVES

After studying this chapter, you should

1 Understand the nature of transformation and be familiar with the importance of the techniques used in recombinant DNA technology.

2 Know the difference between restricted and general transduction.

3 Be able to list the major steps involved in both plasmid and chromosome transfer through conjugation.

4 Be able to list a variety of functions that are usually encoded in plasmids.

5 Comprehend what is a transposable element.

6 Know what restriction enzymes are and how these are used to clone segments of foreign DNA.

7 Understand why a cDNA must be used when cloning human genes in a bacterium.

8 Be able to list several human genes that have been cloned in *E. coli,* as well as to suggest a few new possibilities for cloned genes.

The nineteenth century was a stimulating and exciting time for biological scientists. We have already studied some of the notable advances resulting from the work of Pasteur: a biological explanation for fermentation, a germ theory for disease, and a disproof of spontaneous generation at all levels of life. During this same period Johann Gregor Mendel was carrying out his studies on the inheritance of various characteristics in the garden pea. It was these studies, first published in 1865, that formed the basis of what is now referred to as mendelian genetics. Mendel's work, and subsequent investigations generated by it, have led to the following general conclusions.

1. The genetic material in a eucaryotic cell appears, for the most part, in linear structures called chromosomes.

2. In the **somatic** eucaryotic cells (all body cells except sex cells) of higher plants and animals, there are normally two sets of chromosomes, one set from the female and one from the male; hence all chromosomes in body cells occur in pairs.

3. The characteristics expressed by the genes on the chromosomes can be categorized as either recessive or dominant. For a recessive trait to be expressed, both chromosomes of a pair must have the recessive gene; a dominant gene will be expressed even if only one chromosome has the dominant gene.

We could add to this list, but these three properties formed the major framework of genetics as it evolved through the first half of the twentieth century. The burning questions still unanswered during the 75 years after Mendel's original publication were (1) What was the true genetic material by which a species—human, tree, insect, bacterium, or virus—could pass on its morphological and biochemical characteristics to its descendants? (2) How was this information coded in the genetic material so that it could be read and used by the cell?

In this chapter we will describe how the genetic material in bacteria can be passed from cell to cell.

TRANSFORMATION

Bacterial transformation, originally described by Frederick Griffith in 1928, was undoubtedly the first major discovery in the field of modern genetics. It not only demonstrated that it is possible to transfer genetic material from one bacterial cell to another, but it subsequently led to the realization that this genetic material was DNA (deoxyribonucleic acid) and only DNA. Griffith studied the transformation of one type of *Streptococcus pneumoniae* into a different type. As

you will learn later, *S. pneumoniae* is divided into almost 100 different types based on chemical differences in their capsules (see Chapter 3). Thus, type I produces a capsule different from type II, and so on. These encapsulated organisms are very virulent for the mouse; the injection of only one or two such bacteria will almost invariably kill this animal. If, however, the organism mutates so as to lose the ability to make a capsule, it also loses its virulence; the injection of such "rough" variants will not kill a mouse.

Griffith carried out the following experiment:

1. Into one group of mice he injected living, nonencapsulated *S. pneumoniae* that had been type II before they lost the ability to make a capsule.

2. A second group of mice received heat-killed encapsulated type I organisms.

3. The third group of mice were injected with a mixture of living nonencapsulated type II organisms and heat-killed encapsulated type I bacteria.

After one or two days, Griffith observed the following results: The mice that had received the nonencapsulated type II cells were alive, as were those that had received the heat-killed encapsulated type I organisms. However, the mice that had received a mixture of these two cell types were dead. Moreover, when Griffith reisolated the *S. pneumoniae* from the dead mice, it was type I. Because the type I organisms had been killed prior to injection into the mouse, where did these virulent, encapsulated type I organisms come from? The only answer was that some substance from the killed type I cells was able to get inside the living, nonencapsulated type II organisms and transform these cells to living, encapsulated type I organisms. Figure 8.1 summarizes these experiments. It took 16 years more before the work of Oswald Avery, Colin MacLeod, and Maclyn McCarty demonstrated that the substance transferring the genetic capability to the pneumococci was DNA. The second question—how genetic information is coded into the DNA and how the cell reads this code—had to wait a few more years until the structure of DNA was elucidated by James D. Watson and Francis Crick. But let us now go back to discuss bacterial transformation in a little more detail.

Transformation has been shown to occur in quite a number of bacterial genera, of which *S. pneumoniae* (pneumococcus), *Haemophilus*, *Neisseria*, *Bacillus*, and *Rhizobium* are examples.

Nature of the DNA in Transformation

When DNA is extracted from a donor organism, the bacterial chromosome becomes fragmented into 200 to 500 pieces. A recipient bacterium is capable of

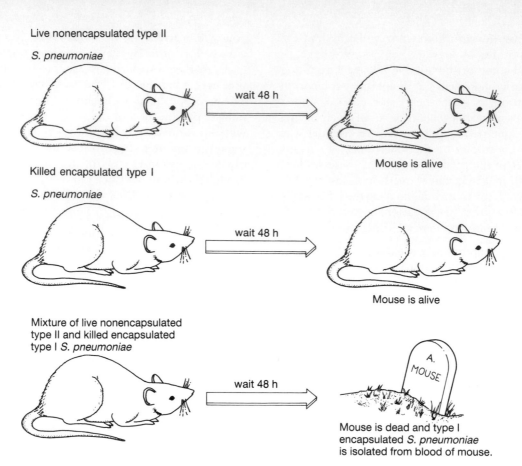

Live nonencapsulated type II

S. pneumoniae

wait 48 h

Mouse is alive

Killed encapsulated type I

S. pneumoniae

wait 48 h

Mouse is alive

Mixture of live nonencapsulated
type II and killed encapsulated
type I *S. pneumoniae*

wait 48 h

A. MOUSE

Mouse is dead and type I
encapsulated *S. pneumoniae*
is isolated from blood of mouse.

Figure 8.1   Griffith's original transformation experiment. The isolation of living encapsulated type I *S. pneumoniae* from the dead mouse could occur only if some genetic substance from the killed type I organisms were able to enter the living type II cells. This substance was subsequently shown to be DNA.

---

taking up a maximum of about 10 of these pieces of DNA (whose molecular weight ranges from 300,000 to 10,000,000 daltons) and, thus, the amount of DNA involved in any one transformation represents, at most, 2 to 5 percent of the total donor DNA. Only double-stranded DNA is effective in transformation, and because the process differs between gram-positive and gram-negative cells, we shall discuss each case separately.

Transformation in Gram-Positive Cells

In organisms such as *Streptococcus* and *Bacillus,* many of the cells will form surface structures that will bind and take up the transforming DNA. Such structures are usually formed during the early stationary phase of growth, and cells producing them are termed **competent.**

Double-stranded DNA binds to these proteins located on the surface of competent cells and is taken into the cell by an energy-dependent process that involves the breakdown of one of the DNA strands. The single-stranded molecule of DNA entering the cell must then integrate into the host cell chromosome by homologous recombination between two crossover points, resulting in the permanent incorporation of the newly acquired DNA into the recipient chromosome. As one might surmise, unless the DNA taken up is closely related to that of the recipient, it will not be able to integrate and will, therefore, not be expressed.

Transformation in Gram-Negative Cells

Gram-negative bacteria, such as *Haemophilus* and *Neisseria,* possess special structures on the surface of their outer membranes that bind both homologous and heterologous DNA. As a result, such organisms are always competent. The uptake of the DNA into the cell, however, is species specific, and heterologous DNA is excluded. The specificity of the uptake process is dependent on the recognition of a specific sequence of 14 base pairs on DNA from the same species, but is present much less frequently on DNA obtained from other organisms. After entry, the double-stranded

DNA can be expressed immediately, but the permanent maintenance of the newly acquired DNA fragment requires its integration into the host chromosomes by homologous recombination.

Importance of Transformation

In all likelihood, transformation occurs in nature following the lysis of an organism and the release of its DNA into the immediate environment. It is difficult to evaluate whether transformation plays an important role in the ability of an organism to cause disease, but it is certainly conceivable that recombinants arising from transformation between strains of low virulence could give rise to highly virulent organisms. Transformation is also used to map the bacterial chromosome, because the frequency of cotransformation of two characteristics is a measure of their distance apart on the chromosome.

For the research scientist, transformation provides a tool by which DNA can be purified and altered chemically in the laboratory and then reintroduced into a bacterium to ascertain the effect of the in vitro alteration. Moreover, as will be discussed later in this chapter, transformation has proved to be extremely valuable as a tool for the introduction of recombinant DNA molecules into bacterial cells. These procedures have been responsible for the production of innumerable genetic hybrids (cells possessing DNA that originated in different types of cells) and involve techniques for making cells artificially competent by treatment with solutions of $CaCl_2$.

TRANSDUCTION

Transduction, initially described in 1952 by Norton Zinder and Joshua Lederberg, is a second method whereby genetic material from one bacterial cell can be transferred to a second bacterium. In this case, however, the DNA is not passed naked, as in transformation, but is carried to the recipient bacterium inside a bacterial virus, called a **bacteriophage.**

The details of bacteriophage infection and replication are discussed in Chapter 14, and only the introductory details of these events will be provided here to enable the reader to understand the mechanism of transduction. When a bacteriophage (also called phage) infects a bacterium, the phage is first adsorbed to specific receptor sites on the surface of the bacterium, after which it injects its nucleic acid into the bacterial cytoplasm. At this point—depending on the particular phage—one of two events occurs. In the one case, the phage nucleic acid takes over the synthetic machinery of the bacterium by coding for enzymes that destroy the bacterial DNA, and it then directs the synthesis of more phage nucleic acid and phage protein coats. After the assemblage of perhaps 100 to 300 complete phage particles, the bacterium breaks open and releases the mature phage particles. This sequence of events terminates with the lysis and death of the bacterium and is, therefore, called the **lytic cycle,** and the infecting phage is referred to as a "virulent" phage.

In the second case, an entirely different sequence of events can occur following the injection of the phage nucleic acid into the bacterial cell. The phage DNA may code for a repressor that prevents the transcription of the phage DNA into messenger RNA. Thus, no enzymes are formed to destroy the bacterial DNA, no phage protein coats are synthesized, and the phage nucleic acid continues to be replicated along with the bacterial nucleic acid. Such phages are termed "temperate" phages, and the phage DNA is called a **prophage.** Bacteria carrying a prophage are referred to as **lysogenic,** because occasionally the repressor protein fails to prevent prophage transcription, permitting the prophage DNA to enter the lytic cycle, resulting in the formation of mature phage particles.

Depending on whether the original prophage was integrated into the chromosome of bacterium A or existed as extrachromosomal DNA (i.e., a plasmid), two types of transduction may occur: (1) restricted (or specialized) transduction and (2) general transduction.

Restricted Transduction

Restricted transduction occurs only in situations in which the prophage has been integrated into a specific site in the bacterial chromosome, as is the case with lambda phage (λ phage) in E. coli.

In the usual sequence of events, when a strain of E. coli lysogenic for λ phage enters the lytic cycle, the λ genome forms a loop and is excised from the bacterial chromosome. Subsequent replication and transcription of the excised λ DNA results in the final liberation of mature λ phage. About one time in a million, however, the λ genome loops out improperly and the excised portion of the chromosome includes an adjacent gene from the bacterial chromosome (see Figure 8.2). Because such phages can still integrate into an E. coli chromosome, a subsequent infection of an E. coli cell will result in lysogenization and integration of the partial phage genome along with the bacterial gene picked up from the previous bacterium. As you can see, however, this type of transduction is restricted to bacterial genes that are adjacent to the integrated prophage; it is, therefore, called restricted transduction.

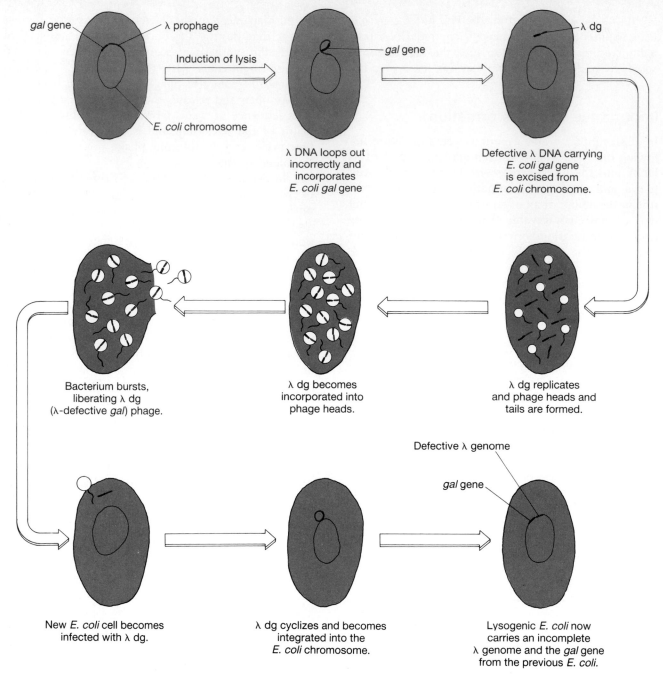

Figure 8.2 Specialized transduction. The λ prophage is incorrectly excised from the *E. coli* chromosome and as a result incorporates the adjacent *gal* gene from the *E. coli* into the λ genome. Subsequent infection with this defective λ phage results in the integration of the defective DNA as well as the *gal* gene into the chromosome of the newly infected *E. coli*.

Generalized Transduction

Generalized transduction is the phage-mediated transfer of bacterial DNA that occurs when a phage not integrated into the bacterial chromosome enters the lytic cycle (see Figure 8.3). As the phage begins the lytic cycle, phage enzymes hydrolyze the bacterial chromosome into a number of small pieces of DNA; as a result, any part of the bacterial chromosome may become incorporated into the phage capsid during the

final assembly of the phage. Such phages are released from the infected bacterium just as are normal phage particles; when they encounter another bacterium possessing the proper phage receptor, they adsorb and inject their DNA content into the recipient bacterium. Obviously, such DNA does not result in the production of more phage particles, but it can undergo homologous recombination with the corresponding re-

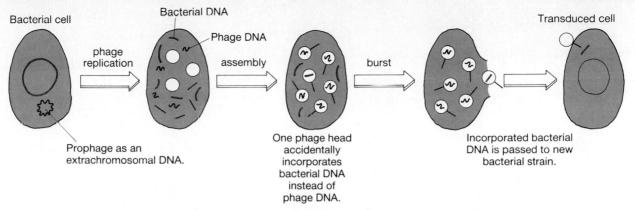

Bacterial cell

Bacterial DNA

Phage DNA

phage replication

assembly

burst

Transduced cell

Prophage as an extrachromosomal DNA.

One phage head accidentally incorporates bacterial DNA instead of phage DNA.

Incorporated bacterial DNA is passed to new bacterial strain.

Figure 8.3 Generalized transduction. When the extrachromosomally located prophage begins its replicative cycle, the bacterial DNA is hydrolyzed into small fragments that, during assembly, can be accidentally incorporated into a phage head. Subsequent infection of a bacterium with such a phage releases functional bacterial DNA into the newly infected cell.

gion of the recipient's chromosome, resulting in the expression of new characteristics by the cell. Because any portion of the DNA from an infected cell (chromosomal or extrachromosomal) may be packaged and carried to a recipient cell, this process is referred to as **generalized transduction**. The frequency with which such mispackaging occurs can be appreciated by the observation that about 0.3 percent of the progeny of phage P1 growing on *E. coli* will contain bacterial DNA instead of phage DNA. Because the amount of bacterial DNA that can be packaged into a phage head represents only about 2 percent of the bacterial chromosome, only closely linked genes can be cotransduced by a single phage particle. By determining the frequency of cotransduced genes, one can determine their order along the bacterial chromosome.

CONJUGATION

Conjugation is the transfer of DNA by direct contact between the cells. This mechanism is much more efficient than transformation or transduction; moreover, the donor cell is not destroyed in the process. Donor cells differ from recipient cells in that they contain a plasmid (extrachromosomal DNA) that encodes for a number of DNA transfer functions. The transfer of DNA by conjugation is rare among gram-positive bacteria (and probably fundamentally different) but very common among gram-negative bacteria. It is the major mechanism for the transfer of drug resistance and can occur between both closely related strains and unrelated genera.

Conjugation in *E. coli* was first described in 1946 by Joshua Lederberg and Edward Tatum, and most early work was concerned with the transfer of the bacterial chromosome from one cell to another. It is

now recognized that chromosome transfer is a relatively rare and specialized aspect of conjugation but that plasmid transfer occurs frequently and rapidly among gram-negative bacteria.

What are **plasmids**? They are extrachromosomal genetic elements consisting of circular double-stranded DNA, ranging from 0.1 to 5 percent of the size of the bacterial chromosome. They may encode any of a wide variety of auxiliary functions (including drug resistance), but all possess, as a minimum, the ability for self-replication, usually employing host cell enzymes. In general, plasmids are not necessary for cellular maintenance, but they often provide useful metabolic capabilities.

Plasmid Transfer

Bacterial conjugation is a one-way process and is not the result of cell fusion to form diploid strains, as occurs in eucaryotic cells. Donor cells contain a self-transmissible plasmid that possesses 15 to 25 genes, called the *tra* genes (for transfer), which encode for various structures and enzymes necessary for conjugation to occur. One essential structure encoded by one of the *tra* genes is the sex pilus, a filament extending out from the cell surface. The sex pilus can be distinguished from the many fimbriae a cell may possess because, in general, the sex pilus is longer and is present in only 3 to 10 copies per cell. When a donor cell comes into contact with a recipient cell, plasmid transfer involves the following steps:

1. The plasmid-coded sex pilus interacts with a receptor on the surface of the recipient cell (Figure 8.4).

2. Stimulated by the contact, the pilus withdraws into the donor cell, pulling the recipient into intimate contact. A conjugal bridge through which

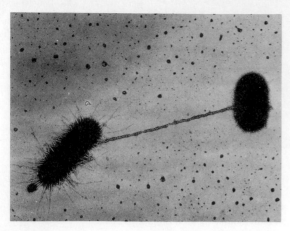

Figure 8.4 Conjugating *E. coli.* The male cell (with numerous short fimbriae) is connected to the female bacterium by an F pilus, as is described in the text (× 3000).

the DNA can be transferred is thus formed between the two cells.

3. Also triggered by this process is the synthesis of a *tra*-gene-encoded nuclease, which cleaves one strand of the plasmid DNA at a specific site called the origin of transfer.

4. The free 5′ end of the cleaved plasmid strand is then threaded through the conjugation bridge, presumably directed by other *tra*-encoded enzymes.

5. Although this step is not essential for the transfer to occur, both the transferred strand and the strand remaining in the donor are normally cop-

Figure 8.5 Simplified model of a sex pilus–dependent plasmid transfer. Following formation of a conjugation bridge, one strand of the plasmid DNA is nicked at a specific site (*oriT*) by a plasmid-encoded endonuclease. The 5′ end of the nicked DNA (possibly attached to a protein) is then transferred to the recipient. Both exposed single-stranded regions are simultaneously replicated during the transfer process.

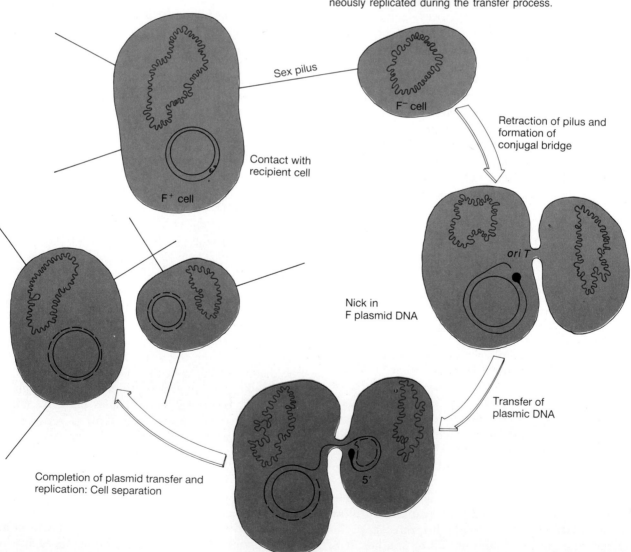

ied by host cell DNA polymerases, yielding double-stranded circular copies in both cells.

6. Following DNA transfer, the conjugation bridge breaks down and the two cells separate. The *tra* genes are expressed in the recipient cell so that it has now also become a donor cell. In this way, a plasmid can spread very rapidly through a bacterial population (Figure 8.5).

Unlike transformation and transduction, in which the transfer and expression of the donor DNA occurs only among closely related organisms, plasmid transfer by conjugation exhibits much less specificity. In fact, some host-range plasmids can transfer themselves into many different species of gram-negative bacteria.

Chromosome Transfer

One of the best-studied sex factors has been given the name of F (for fertility), and cells carrying an F plasmid are referred to as F^+ cells. Conjugation between an F^+ cell and a recipient (F^- cell) results in a rapid transfer of the F plasmid to the F^- cell, converting it to an F^+ cell. On rare occasions, however, the F plasmid becomes integrated into the bacterial chromosome, and such cells are able to transfer their chromosomes at frequencies approaching that at which an F^+ cell transfers the F plasmid. Such cells are called **Hfr cells** to indicate the *high frequency* of chromosomal recombinants.

The transfer of the bacterial chromosome by an Hfr cell begins at the site at which the F factor has integrated and proceeds around the chromosome in a fixed direction (Figure 8.6). The transfer of the entire chromosome requires about 120 min, but chromosome transfer is not usually completed, and the probability that a specific gene will be transferred depends on the distance of that gene from the origin of chromosome transfer. It is possible, therefore, to order the genes onto a genetic map based on the number of minutes required for a specific gene to be transferred to a recipient. Interestingly, the *tra* genes of the integrated F factor are the last to be transferred; thus, very few recipients become Hfr cells.

F Plasmid Integration

The circular F plasmid is integrated into the bacterial chromosome by a single recombination event; as described for restricted transduction, the reverse may also occur, in which the F factor is excised from the chromosome. Abnormal excision may give rise to F plasmids carrying bacterial genes from one or both sides of the site where the F factor had been integrated, as shown in Figure 8.7. Such plasmids, called F' plasmids, are rapidly transferred to recipient cells, providing another method whereby chromosomal genes may be dispersed throughout a cell population.

Plasmid Functions

Plasmids are widely distributed in nature, and most bacteria isolated from clinical materials will contain at least one and often as many as six different plasmids. They occur in both gram-positive and gram-negative bacteria and may be categorized according to the functions they encode.

Antibiotic resistance is frequently plasmid coded, as is resistance to mercury and silver salts. Plasmids may also be responsible for resistance to ultraviolet

Figure 8.6 Transfer of the bacterial chromosome by Hfr cells containing an integrated F factor. DNA transfer to a recipient begins at *oriT* and proceeds to transfer the bacterial chromosome, which, only if completed, will include the F factor.

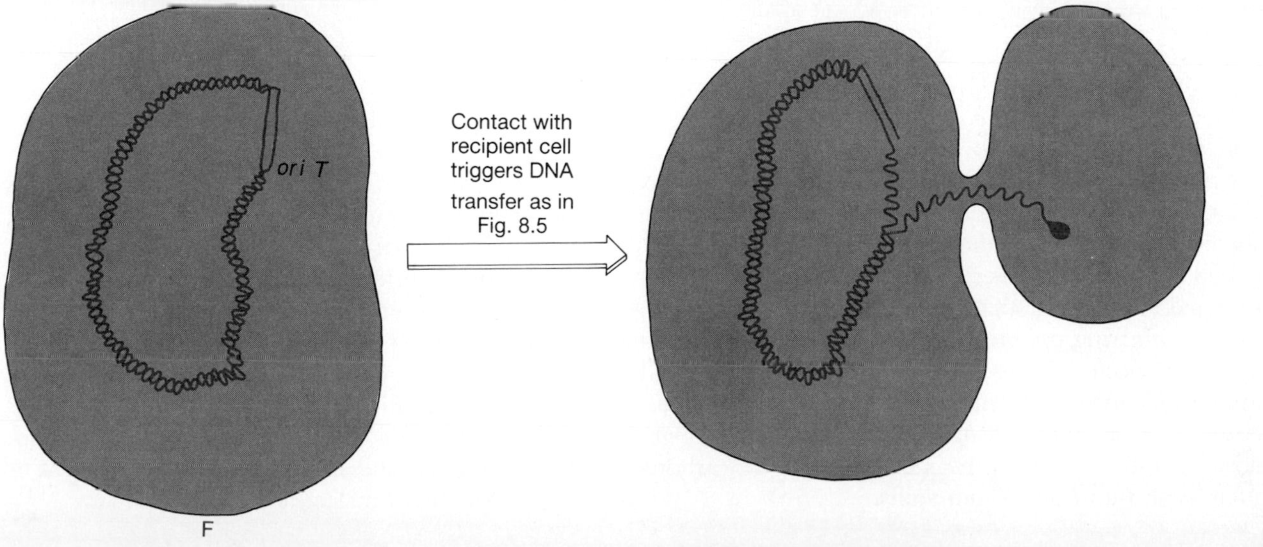

Contact with recipient cell triggers DNA transfer as in Fig. 8.5

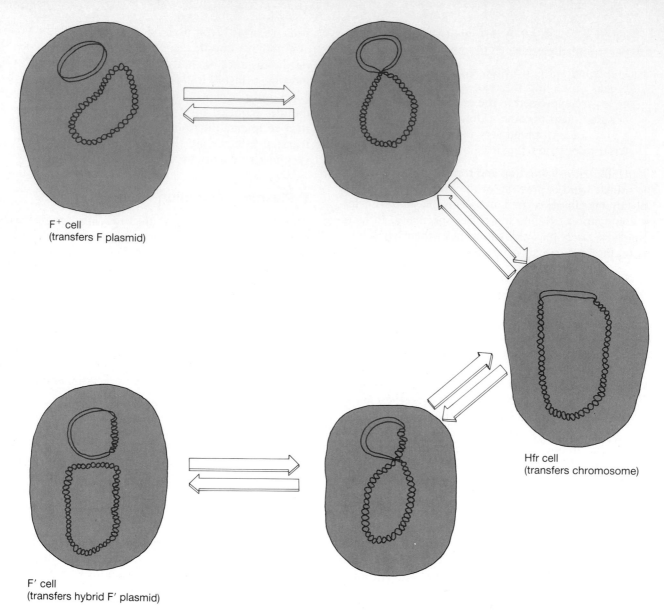

F⁺ cell
(transfers F plasmid)

Hfr cell
(transfers chromosome)

F′ cell
(transfers hybrid F′ plasmid)

Figure 8.7 Excision of the F plasmid. As described for phage λ, the integrated F plasmid can integrate and be excised from the chromosome in either a precise manner (reforming an F⁺ cell) or an unprecise manner, generating an F′ or hybrid plasmid.

light by encoding for DNA repair enzymes. Plasmids also encode for fimbriae, which are involved in the adherence of bacteria to other cells. Hemolysin production of *E. coli*, enterotoxins responsible for diarrhea, and many exotoxins—such as those responsible for botulism, tetanus, and scarlet fever—are also encoded in plasmids. Other plasmids code for the ability to metabolize new sugars and new carbon sources, thereby blurring the biochemical distinctions between different species. Some additional properties encoded in plasmids include the nitrogen-fixing enzymes in *Rhizobium*, the production of plant tumors by *Agrobacterium*, and the ability to metabolize hydrocarbons such as those in petroleum spills.

Conjugation in Gram-Positive Bacteria

Cell-to-cell transfer of DNA has also been shown to occur in some gram-positive bacteria. This, too, is a plasmid-coded process, but the basic mechanism appears to be different from the process occurring in gram-negative bacteria. There is no indication of the presence of sex pili. Instead, donor cells are stimulated to produce an adhesive substance after recognizing a pheromone produced by suitable recipient cells. The adhesive material causes the aggregation of donor and recipient cells, and DNA transfer occurs. The mechanism for this transfer, however, is unclear.

TRANSPOSABLE ELEMENTS

Early genetic studies of antibiotic resistance in bacteria revealed the puzzling finding that the DNA that encodes antibiotic resistance appeared to move from one plasmid to another, or to a phage, or to the bacterial chromosome. The term **transposon** is the general name used for such mobile genetic elements that can insert almost randomly into DNA by nonhomologous recombination. Transposons encode for a wide variety of characteristics, including drug resistance, toxins, or nothing.

Properties of Transposons

All transposons possess the following common features:

1. Transposons are incapable of autonomous replication; hence, they must be integrated into a replication unit (replicon) such as a phage, plasmid, or host chromosome.

2. Transposons have special sequences at the end of their DNA that are essential for their ability to integrate into another region of DNA. In most cases, these terminal sequences are inverted repeats of 15 to 40 base pairs in length, which means that the sequences at the two ends of the transposable element are identical. The inverted repeats at the end of the transposon must be recognized in a very specific manner by the enzymes that carry out transposition.

3. Integration of a transposon can occur at a large number of sites in the target DNA. Some transposons integrate completely at random; others prefer to integrate at sites that have sequences similar to those at the ends of the transposon.

4. Transposon insertion requires at least one enzyme, termed a **transposase**, which is encoded by and is specific for each transposon. Transposases probably recognize sequences at the ends of the transposon and catalyze the ligation of the transposon into the target site after cutting the recipient DNA in a staggered pattern.

Types of Transposons

Some transposons, also known as **insertion sequences,** are short (800 to 1500 base pairs) and encode only for their own transposase. Their presence, therefore, can be detected only by means of their ability to disrupt the function of the gene into which they insert. Interestingly, if a copy of the same insertion sequence should integrate on both ends of a gene, a transposon would be created that would permit the transfer of the surrounded gene into other sites on the chromosome or onto phages or plasmids harbored within the cell.

A major implication of the mobility of transposons is that the same toxin- or drug-inactivating enzyme can be present in many different species or encoded on different plasmids or phages. For example, the same penicillin-inactivating enzyme is produced by resistant strains of *E. coli*, *Neisseria gonorrhoeae*, and *Haemophilus influenzae*, although in each species the transposon resides on a different plasmid. Similarly, the *E. coli* ST enterotoxin is found also in related organisms such as *Yersinia enterocolitica* and is sometimes encoded on a plasmid and in other cases may be carried in a temperate phage.

Transposable elements have been described in gram-positive and gram-negative bacteria, archaebacteria, fungi, *Drosophila*, and plants. Many tumor viruses can integrate into random sites in the animal cell chromosome and, because their nucleic acid possesses terminal direct or inverted repeat sequences, they also have the general properties of a transposon.

MICROBIOLOGY MILESTONES

The history of transposable elements actually began in 1947 when Barbara McClintock published her first report of movable genes in maize. Her data showed that a number of mutations, causing color changes in the corn kernels, were associated with the movement (i.e., transposition) of one or more genes, designated *Ds* and *Ac,* from one location on the chromosome to another. At that time, however, neither the structure of DNA nor the techniques to sequence DNA were available, and this concept of "jumping genes" was impossible to comprehend at the molecular level. Her pioneering contributions were, however, eventually recognized when she was awarded the Nobel Prize in 1983.

Almost 20 years after McClintock's initial reports, mobile genetic elements were reported to occur in bacteria. These were first seen as antibiotic resistance genes that were "picked up" by plasmids as they traveled from one bacterium to another.

It is now known that transposable genes are widespread in all forms of life, and it has been proposed by Nina Fedoroff that they may have been extremely important in evolution. This concept is based on the fact that transposable elements can amplify a small genetic disturbance that could result, according to Dr. Fedoroff, in a "genetic earthquake." The observation that transposable elements can both turn genes off as well as amplify their expression supports their past and present role in organismic evolution.

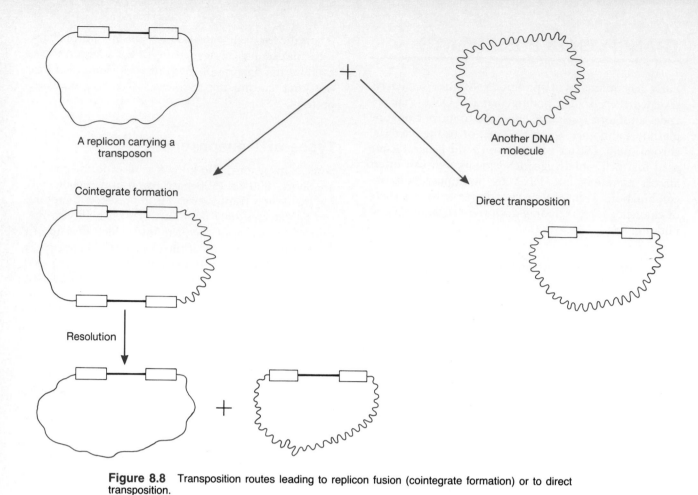

Figure 8.8 Transposition routes leading to replicon fusion (cointegrate formation) or to direct transposition.

Figure 8.9 Genetic structure of two transposons. Transposon Tn3 is approximately 5 kb in length and has 37-base-pair inverted repeat sequences at its ends. It carries three genes, *tnpA*, transposase, which catalyzes the formation of cointegrates; *tnpR*, resolvase, which separates cointegrates into two separate replicons; and *bla*, β-lactamase, which confers resistance to β-lactam antibiotics. Transposon Tn10 (9.5 kb) carries genes for tetracycline resistance flanked by two 1,5 kb insertion sequences (IS-10-L and -R), one of which encodes the transposase. *tetR* is a regulatory gene for *tetA*, the gene encoding for resistance to tetracycline.

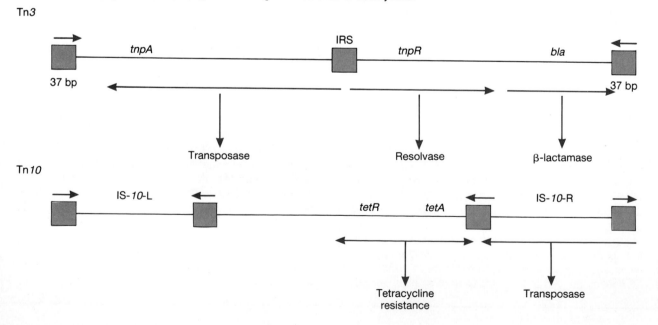

Figure 8.8 illustrates two mechanisms whereby a transposon may move from one DNA molecule to another. Note that despite the terminology of "jumping" genes, the original transposon does not actually move but rather is a replicated copy that becomes inserted into the new DNA. Schematic examples of two transposons, each encoding for resistance to an antibiotic, are depicted in Figure 8.9. It is particularly obvious that Tn10 was originally formed when two short insertion sequences became integrated on each end of the tetracycline resistance gene, creating a new transposon that included the antibiotic resistance gene.

Bacterial Recombination

We have used the word "recombination" since the beginning of this chapter, and it would seem that now is a good time to describe just what it means. Recombination describes a mechanism of genetic exchange between a gene in a recipient organism and an analogous segment of DNA from a donor organism. Most such recombinational events occur only between closely related DNAs, as shown in Figure 8.10. In such cases, the donor DNA positions itself (probably through base pairing between complementary nucleotides) alongside a homologous area of DNA in the recipient. Cellular enzymes nick that segment of the recipient DNA, and the donor DNA then replaces the excised fragment.

Restriction Enzymes

Because there are so many DNA transfer mechanisms operating among bacteria, one might expect that foreign DNA could enter a bacterium and easily exert a detrimental effect on the recipient cell. One of the prime defenses against such occurrences employs one or more of an almost universal group of endonucleases called restriction enzymes. Each organism has its own restriction enzyme that recognizes very specific DNA

sequences, four to eight bases long, and either cleaves both strands at the same site or makes staggered cuts on each strand (Figure 8.11). The cell's own DNA is not cleaved because it has been marked by the action of a modification enzyme that puts a methyl group on one of the bases in the sequence that is recognized by its own restriction enzyme. Since the majority of unrelated DNA is not so modified, it is cleaved by the cell's restriction enzyme.

Why, then, is it important to know about restriction enzymes? We shall see in the next section that the entire field of genetic engineering depends on the use of specific restriction enzymes for the construction of hybrid plasmids. The important concept for the reader to grasp is that a specific restriction enzyme will cleave DNA only if the sequences shown in Figure 8.11 are present and that each strand of the DNA is cleaved only where indicated by the arrows in this figure.

Recombinant DNA Technology

Knowledge of the properties of plasmids, transformation, and restriction enzymes has generated one of the most widely discussed and powerful techniques in modern molecular biology, termed **genetic engineering** or **recombinant DNA technology**. These methods permit the isolation, manipulation, and production in large quantity of segments of DNA from any type cell. The implications of these techniques are profound.

In brief, recombinant DNA technology is the construction of hybrid plasmids in vitro by inserting a fragment of foreign DNA into a plasmid that is able to replicate itself when introduced into E. coli or the yeast Saccharomyces cerevisiae. These plasmids are introduced into E. coli or the yeast by transformation. One then screens for the colonies carrying the desired hybrid plasmid.

Hybrid Plasmid Construction

If a plasmid is cleaved by a restriction enzyme that cleaves the DNA at a single site yielding staggered cuts (see Figure 8.12), the linear fragment formed will have complementary sequences at each end of the molecule. Since these sequences will base pair under proper conditions (low temperature), they are referred to as "sticky ends." Treatment with a DNA ligase will reform the original covalently closed circle. Foreign DNA from any source can be treated with the same restriction enzyme, cleaving it into fragments of various sizes *each possessing the same sticky ends as the cleaved plasmid DNA.* The foreign DNA fragments are then mixed with the cleaved plasmid DNA, and the entire mixture is annealed so that the sticky ends will pair at random. Treatment with a DNA ligase

---

Figure 8.10 Homologous recombination. The donor DNA base pairs with a similar area of the recipient chromosome and becomes integrated into the chromosome after excision of a fragment of the recipient DNA. In this case, the cell acquires the new characteristics encoded in dexyh.

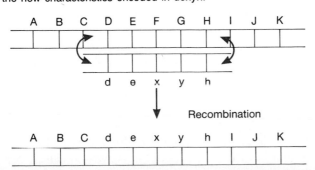

Restriction endonuclease	Cleavage points on base sequence recognized	Product of DNA cleavage
EcoRI	——— GAATTC ——— (↓) / ——— CTTAAG ——— (↑)	AATTC ——— / ——— CTTAA
HindIII	——— AAGCTT ——— (↓) / ——— TTCGAA ——— (↑)	AGCTT ——— / ——— TTCGA
Bg1II	——— AGATCT ——— (↓) / ——— TCTAGA ——— (↑)	GATCT ——— / ——— TCTAG
BamHI	——— GGATCC ——— (↓) / ——— CCTAGG ——— (↑)	GATCC ——— / ——— CCTAG
HaeIII	——— GGCC ——— (↓) / ——— CCGG ——— (↑)	CC ——— GG / GG ——— CC

Figure 8.11 Examples of base sequences occurring on double-stranded DNA that are recognized by specific restriction endonucleases. Arrows indicate points of cleavage. Note that all except *Hae*III give staggered cuts in which one strand of DNA is cleaved several bases away from where the second strand is cleaved. Because the nucleotide bases are complementary to each other, they are referred to as "sticky ends."

will then yield covalently closed circles of DNA. Figure 8.12 shows that, when a plasmid designated pBR322 is used, some of the re-formed circles yield only the original plasmid (see bottom left). Remember, however, that both the plasmid and the foreign DNA were cleaved with the same restriction enzyme and, therefore, the foreign DNA has the same sticky ends as the plasmid DNA. Thus, some of the foreign DNA fragments will base pair with the sticky ends of the cleaved plasmid and, after joining with a DNA ligase, will become covalently inserted into the plasmid, as shown in the bottom right of Figure 8.12. The DNA reaction mixture can then be used to transform *E. coli* or the yeast *S. cerevisiae*, and by using the appropriate techniques, one can select out only colonies possessing the desired fragment of foreign DNA.

DNA may also be cloned into a temperate bacteriophage, which then may be used to infect a bacterium to produce a lysogenic culture (see Chapter 14). Such cultures do not have the ability to undergo lysis because much of the phage DNA has been removed to permit the insertion of the recombinant DNA.

When transfecting a bacterium with a plasmid containing mammalian DNA (such as human growth hormone), one must be aware that procaryotic cells cannot process and splice transcribed RNA as do eucaryotic cells. Therefore, all intervening sequences (in-

trons) must be removed before the DNA is cloned into an appropriate vector. This is accomplished in the following manner. (1) Instead of isolating the eucaryotic DNA, one isolates the mRNA for the gene in question. (Note that the mRNA has already undergone RNA processing and splicing to remove any introns). (2) The mRNA is then transcribed by a reverse transcriptase (see Chapter 14) to obtain a DNA that is complementary to the mRNA (termed cDNA). (3) Because the cDNA does not contain any of the original introns, it can be used to construct a plasmid vector for infecting a procaryote such as *E. coli*.

Recombinant DNA technology has been also used to produce mammalian cell hybrids. In such cases, however, the vector carrying the recombinant nucleic acid must be a retrovirus that will integrate into the host cell chromosome. Retroviruses are RNA-containing viruses (see Chapter 14) that, in order to replicate, must first convert their RNA into a complementary DNA (cDNA) and then insert the cDNA into the host cell chromosome. Once integrated, such DNA becomes a permanent part of that cell and all of its progeny.

Why bother to insert the foreign DNA into a plasmid or a bacteriophage? Why not merely transform a culture of *E. coli* directly with the foreign DNA? The answer is simply that foreign DNA by itself will not be replicated within the bacterium. The DNA

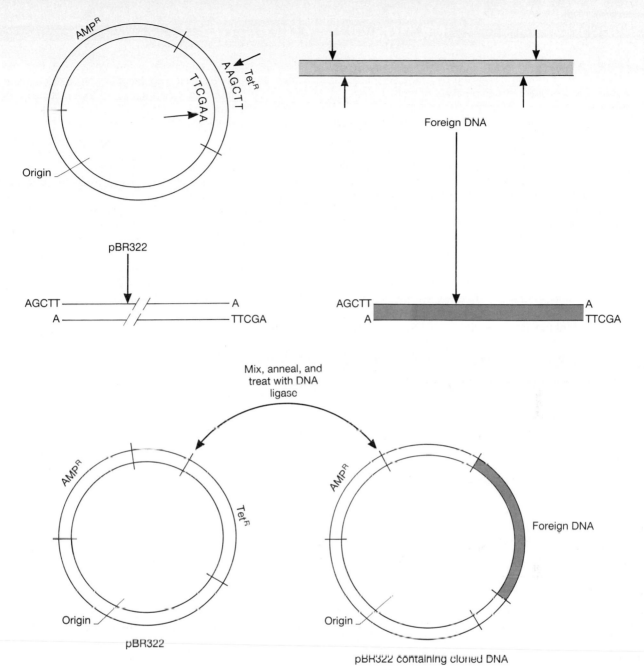

Figure 8.12 Cloning of foreign DNA into a plastic vector (pBR322). Both pBR322 and foreign DNA are cleaved with endonuclease, HindIII, yielding staggered cuts with sticky ends. Reannealing causes some of the foreign DNA to base pair with the sticky ends in pBR322. Because HindIII cuts in pBR322 are inside a gene for tetracycline resistance, foreign DNA reannealing with pBR322 inactivates the tet^R gene. Thus, bacteria transformed with pBR322 containing foreign DNA will be ampicillin-resistant and tetracycline-sensitive, whereas those receiving only the original pBR322 plasmid will be resistant to both antibiotics.

must have an origin of replication (promoter site) that is recognized by the cell's DNA polymerase; thus, when the foreign DNA is inserted into a normal plasmid, the bacterium will recognize its origin of replication and replicate the entire plasmid, including any inserted foreign DNA.

Genetic engineering has also been used to produce microorganisms that can be used to protect or augment plant growth, kill insects, control weeds, and degrade toxic chemicals. Because of the potential risks of introducing genetically altered microbes into the environment, the use of such organisms is very tightly

Human gene therapy is losing its science fiction status and is currently being discussed by many scientists as a promising solution to a number of human genetic diseases.

A few of the more common genetic defects include cystic fibrosis, sickle cell anemia, severe immunodeficiency disease, adenosine deaminase deficiency, purine nucleoside phosphorylase deficiency, and familial hypercholesterolemia. There are many more, but the specific genetic defect of most such diseases is not yet sufficiently characterized to be considered for gene therapy. Let us look at a few genetic diseases to see what might be done to correct them.

Individuals with familial hypercholesterolemia are missing a receptor on the surface of their liver cells that binds low-density lipoprotein (LDL) cholesterol. Those with two copies of the defective gene frequently die of a heart attack in their teens whereas those with only a single copy of the defective gene may have a first heart attack between 20 and 40 years of age.

Working with a strain of rabbits in which a similar disease occurs, scientists at the Whitehead Institute have succeeded in cloning the LDL cholesterol gene into a retrovirus that was then used to infect cell cultures of the defective rabbit liver. Once inside the cell, the new gene directed the synthesis of the LDL receptor and, in essence, the defect was cured in this in vitro situation. Techniques to put these cured liver cells back into a genetically defective rabbit are still "down the road," but the myriad of research projects on this and other genetic deficiencies gives strong promise that success may be just around the corner.

Adenosine deaminase (ADA) deficiency is also due to a single gene defect. Children with this defect fail to develop a functional immune system and routinely die in early childhood. As was described for familial hypercholesterolemia, a functional gene for ADA was cloned into a retrovirus, which was then used to infect lymphocytes from children with ADA deficiency, resulting in a "test tube" cure. It is noteworthy that these cured lymphocytes have been successfully used to treat patients with an ADA deficiency.

---

controlled by the Environmental Protection Agency (EPA) and thus far only a very few have been authorized for field testing.

One of the most publicized of these is the use of altered strains of *Pseudomonas syringae* and *P. fluorescens* known as "ice-minus" bacteria. Normal flora bacteria contribute to frost damage in plants, such as strawberries and potatoes, by the secretion of a protein that acts as a nucleating center for ice crystal formation on the leaves. Ice-minus bacteria have been engineered that lack the gene for the ice-nucleating protein, and it has been proposed that one can supplant the normal flora by spraying plants with the "engineered" strain. Preliminary trials suggest such bacteria do provide some protection against frost damage.

A few other genetically engineered microorganisms (most of which have not been tested under field conditions) include (1) amplification of the nitrogen-fixing genes in *Rhizobium meliloti*, a symbiont of alfalfa roots, (2) a genetically altered bacterium that is reported to interfere with the fungus that causes Dutch Elm disease, and (3) the development of strains of bacteria and viruses that kill insects.

Plant research has also resulted in the genesis of a variety of new genetically acquired characteristics. Among these are plants that are resistant to infectious agents or weed-control agents. Others have been engineered to produce more or better food.

It appears, therefore, that the limits of newly engineered species of microorganisms, animals, and plants is still beyond the realm of prediction, and it seems certain that future editions of this text will describe many more genetic alterations that are beneficial to humans.

Pros and Cons

It is generally agreed that the type of genetic manipulation just described has a potential for harm, and there are certain types of experiments that are "forbidden," such as the cloning of viral genes that might be involved in human cancer. In addition, a series of guidelines has been formulated to ensure that any potentially dangerous recombinants cannot escape from the laboratory and infect humans. It has been recognized, however, that the common laboratory strains of *E. coli* K12 used primarily in these experiments are unable to survive in nature. This inherent safety factor has led to a reduction in the safety requirements for many recombinant DNA experiments.

All in all, the advantages to humanity appear limitless. No one can predict the role of recombinant DNA technology during the next several decades. The commercial production of human insulin, human interferons, and various growth hormones, as well as vaccines for hepatitis B and foot-and-mouth disease are already in production from cloned antigens and, undoubtedly, more will be forthcoming. Various bacterial and viral antigens are also being cloned in vaccinia virus with the objective of using these hybrid viruses to immunize individuals against the respective products of these genes.

SUMMARY

Although the systematic study of genetic inheritance was begun by Mendel in 1865, an understanding of genetic mechanisms at the molecular level was initiated when it was shown that bacteria could be genetically transformed by the addition of DNA isolated from a closely related bacterium. It has been shown that some bacteria may be transformed only at a specific time of growth (i.e., late in the exponential stage of growth). Bacteria in this condition are said to be competent.

Transduction is a second mechanism for the transfer of genetic material from one bacterium to another closely related bacterium. In this case the bacterial DNA is transferred while enclosed in the protein coat of a bacterial virus. One can subdivide transducing phages into two broad types: those carrying out a restricted transduction and those involved in a generalized transduction. The former can transduce only the portion of the bacterial chromosome that lies immediately adjacent to the integrated genome of the prophage in a lysogenic bacterium. In this type of transduction, the transducing phage contains part of its own DNA and a part of the bacterial DNA. In generalized transduction, however, any portion of the bacterial DNA may be transduced. This apparently occurs as a mistake during the final assembly of the phage when a piece of bacterial DNA is enclosed in a phage coat. In this case the transducing phage may not contain any phage DNA, but it can inject the bacterial DNA into a second bacterium.

Bacterial conjugation occurs between donor cells containing a cytoplasmic sex factor called F and recipient cells called F^-. The F^- cells are rapidly converted to F^+ cells following conjugation, since the sex factor F is rapidly transferred during conjugation. However, the bacterial chromosome is not transferred during the conjugation of an F^+ and an F^- strain. This occurs only between Hfr strains and F^- strains, in which case the chromosome containing the inserted F factor is transferred to the F^- cell in a linear fashion, the F factor being the last to enter the F^- cell. Based on the time required for any specific gene to be transferred, it is possible to draw a map of the chromosome, using time of transfer for distance measurements between genes. Hfr strains originate by the integration of the F^+ sex factor into the circular bacterial chromosome. Because different Hfr strains may integrate the F factor at different positions in the circular chromosome and the chromosome always begins replication during conjugation at the position of the F factor, the initial gene introduced by the Hfr into the F^- strain will vary from strain to strain. Occasionally, the integrated F plasmid (in an Hfr strain) is excised from the chromosome. Abnormal excision may include bacterial genes on either side of the plasmid, resulting in an F plasmid carrying chromosomal genes. Such plasmids are designated F'.

Plasmids are known to encode for many functions, such as drug resistance, ultraviolet light resistance, toxins, metabolic enzymes, and transfer (tra) genes.

Transposons are areas of DNA that can move from one location to another. Many encode for toxins or drug-inactivating enzymes, and their mobility is responsible for the widespread occurrence of these traits among bacteria.

Restriction enzymes are enzymes that recognize and cleave at very specific sequences on a DNA molecule. Their normal function is to protect the cell against a takeover by foreign DNA, but they are indispensable for the construction of hybrid plasmids used in genetic engineering experiments. Using such enzymes, foreign DNA from any source can be inserted into a bacterial plasmid, which can subsequently be used to transform cultures of E. coli or the yeast S. cerevisiae. Such transformed strains are being used to produce many products, such as human insulin, human interferons, and growth hormones. Recombinant DNA can also be inserted into mammalian cells by incorporating the complementary mRNA into a retrovirus that is then used to infect the cell.

Many genetically engineered microorganisms have been designed to protect against frost damage, increase N_2 fixation, and interfere with diseases caused by fungi and viruses.

QUESTIONS FOR REVIEW

1. What is the true genetic material by which a species can pass on its morphological and biochemical characteristics to its descendants?
2. Explain how the answer to the first question was determined.
3. Explain why the transforming DNA must be closely related to the DNA of the recipient cell for transformation to occur.
4. How do restricted transduction and generalized transduction differ?
5. Define prophage, lysogeny, temperate phage, virulent phage.
6. With respect to the amount of DNA transferred, how does conjugation differ from transformation and transduction?
7. What morphological difference is there between donor and recipient cells?
8. What is a recombinant?
9. How can one map a chromosome?
10. What are plasmids? List some functions for which they encode.
11. What is a transposon? Give two examples of genes encoded in transposons.
12. What is a restriction enzyme and precisely how does it work to produce sticky ends in cleaved DNA?
13. What is meant by *genetic engineering*?
14. List the steps you would follow to construct a hybrid plasmid.
15. Why must one use cDNA when cloning eucaryotic DNA into a bacterium?
16. How could ice-minus bacteria provide protection against frost damage?

SUPPLEMENTARY READING

Anderson WF, Diacumakos EG: Genetic engineering in mammalian cells. *Sci Am* 245(1):106, 1981.

Avery OT, MacLoed CM, McCarthy M: Studies on the chemical nature of the substance inducing transformation of pneumococcal types. *J. Exptl Med* 79:137, 1944.

Barany F, Kahn M: Comparison of transformation mechanisms of *Haemophilus parainfluenzae* and *Haemophilus influenzae*. *J Bacteriol* 161:72, 1985.

Elwell LP, Shipley PL: Plasmid-mediated factors associated with virulence of bacteria to animals. *Annu Rev Microbiol* 34:465, 1980.

Gilbert W, Villa-Komaroff L: Useful proteins from recombinant bacteria. *Sci Am* 242(4):74, 1980.

Griffith F: The significance of pneumococcal types. *J Hyg* (Comb) 27:113, 1928.

Hardy K: *Bacterial Plasmids*, 2nd ed. Aspects of Microbiology Series, Washington, DC, American Society for Microbiology, 1986.

Lederberg J, Tatum EL: Genetic recombination in *Escherichia coli*. *Nature* 158:558, 1946.

Lewin R: Gene therapy—so near and yet so far away. *Science* 232:824, 1986.

Marshall E: Engineering crops to resist weed killers. *Science* 231:1360, 1986.

Mendel GJ: Versuche über Pflanzen-Hybriden. *Verh naturf Vereines Brunn* 4:3, 1866.

Stewart GJ, Carlson CA: The biology of natural transformation. *Annu Rev Microbiol* 40:211, 1986.

Zinder ND, Lederberg J: Genetic exchange in *Salmonella*. *J Bact* C4:679, 1952.

Chapter 9

Methods for the Control of Microorganisms

OBJECTIVES After study of this chapter, you should

1. Be able to define sterilization, disinfection, lyophilization, antiseptic, germicide, and disinfectant.

2. Be familiar with the term *nosocomial infections*.

3. Know the many ways in which heat is used to sterilize an object.

4. Be able to describe pasteurization and to differentiate it from sterilization.

5. Know the types of radiation used to control microorganisms and be able to describe how ultraviolet light kills a cell.

6. Know the type of filter that is most commonly used to remove bacteria from a solution.

7. Know how disinfectants act to kill microorganisms and what are the major variables involved in disinfection.

8. Be able to list the commonly used chemical disinfectants.

9. Understand the kinetics of the disinfection process.

10. Know the meaning of the term *phenol coefficient*.

INTRODUCTION TO MICROBIAL CONTROL

Medical history is filled with descriptions of epidemics that, up to a century ago, killed off vast numbers of people. Even minor surgery carried a high death rate because of postoperative infections, and complications due to infection following childbirth were indeed reason enough for a prospective father to pace the floor. However, in comparatively recent years the relationship between humans and pathogenic microorganisms has changed drastically in favor of humans. For the most part, this change has been due not to the disappearance of any pathogenic agents but rather to our increasing knowledge of how to control microorganisms. In other words, we are just as susceptible as ever to the ravages of infectious diseases, and any relaxation in the application of the principles of control could quickly change this to a less favorable relationship.

Our knowledge of microbial control did not come easily. Considering the generations of humans who have been subjected to the ravages of infectious diseases, it is, comparatively speaking, only very recently that we have had any real concept of the cause of these diseases.

Scarcely more than a century ago, a few scientific giants were subjected to ridicule for suggesting that infections were caused by organisms too small to be seen and, more importantly, that the doctors themselves were frequently responsible for carrying the infection from one patient to another. A list of these leaders would include Pasteur, who proved that microorganisms could cause disease; Semmelweis, who was persecuted by the medical profession for insisting that doctors wash their hands before surgery; Koch, who isolated many of the causative agents of disease and set up a series of rules (Koch's postulates) to follow in order to prove whether a particular organism caused a disease; Lister, who is given credit for the first antiseptic surgery, in which he sprayed dilute phenol (carbolic acid) into the air of the operating room during surgery. More names could be mentioned, but for our purposes it is sufficient to state that our knowledge of microbial control started about 100 years ago, and the procedures used today are the result of information gained through study and experience.

Today, in large part, many of the problems of the past century have been solved. The modern hospital operating room is a spotlessly clean place where surgical procedures can be carried out with reasonable safety from infections. The prospective father may still pace the floor, but not out of fear of subsequent infection that might prove fatal to his wife and newborn child.

Why? What has happened during the past 50 or 100 years? Why are our modern hospitals no longer to be feared, as were the hospitals of the past century? The answer is simple: we are learning how to control microorganisms through our expanding knowledge of

MICROBIOLOGY MILESTONES

During the 1850s, when Joseph Lister was beginning his surgical practice in Edinburgh, Scotland, the occurrence of pus in surgical incisions was so common that it was not only accepted, but was given the name "laudable pus," indicating that its presence was a good sign for wound healing. Many surgical procedures, however, were considered impossible because of the extremely high mortality rates from infections. As an example, the death rate from infection following the resetting of a compound fracture was so high that such injuries were routinely treated by amputation, a procedure that had only a 50 percent rate of success. Abdominal, chest, or skull surgery was not even considered.

Lister, who was familiar with Pasteur's experiments in fermentation, began to wonder if the "germs in the air" might not also be responsible for the putrefaction of wounds. If so, he concluded, all that was necessary was either to kill the germs infecting a wound or to prevent their access to the wound.

A number of chemicals were considered, but Lister's first attempts at antiseptic surgery employed a solution of carbolic acid (phenol) to treat a compound fracture in the lower leg of James Greenlees, an 11-year-old boy. Both wound and dressings were continually soaked in a dilute solution of carbolic acid and, in six weeks, the bones were united and the wound healed.

Lister continued the use of carbolic acid, even spraying it continuously in the air during surgery. He used it not only for compound fractures, but for all sorts of surgery, such as the removal of carbuncles, boils, and abscesses. He published his remarkable results in the medical journal *Lancet*. As one might expect, his concepts of antiseptic surgery were originally met with apathy and antagonism by his fellow surgeons, but during his lifetime he was showered with honors for his achievements. His gifts to medicine were well summed up by Dr. Irving Edgar, who wrote, "The present hospital system exists because of Lister, for without antisepsis, hospitals were houses of torture and death. ... Lord Lister's discoveries gave hospitals all over the world a new lease on life, making them houses of healing and cure."

what they are, where they are, how they live, and where their strengths and weaknesses lie.

We also know that there is much more to learn before our control can be absolute. For example, new lifesaving procedures have indirectly given rise to a whole category of hospital-acquired infections called **nosocomial infections.** Such infections occur most frequently in patients who are undergoing procedures that were undeveloped several decades ago; surprisingly, many are caused by organisms that are part of our normal flora. It has now become apparent that when organisms that are part of the normal intestinal flora, such as *Escherichia coli* or *Klebsiella pneumoniae,* are mechanically introduced into the bladder via an indwelling catheter, they will frequently cause a urinary tract infection. Other examples of nosocomial infections include respiratory infections resulting from contaminated respirators, blood infections from contaminated fluids or equipment used for administering intravenous fluids, and postoperative infections acquired from the patient's normal flora or from contact with hospital personnel carrying virulent organisms. Many patients who acquire a nosocomial infection either have an altered immune response (which may be caused by anticancer or antiinflammatory drugs) or they have had organisms mechanically implanted into areas that are normally sterile. Thus, although these new procedures have been lifesaving, they have also contributed to an added risk of infection. They have taught us, also, that any relaxation or shortcuts in the application of the principles of infection control may lead to infections that could have been avoided. We are now aware that the new miracle drugs and antibiotics are not able to compensate for inadequate prophylaxis. The present problem of controlling antibiotic-resistant staphylococci in hospitals all over the world emphasizes this fact. The microbiologist, the nurse, and the physician all know that the control of pathogenic microorganisms requires the everyday application of all the knowledge and techniques that have been learned through the years.

What are these methods of control, and what is their aim? They are techniques of killing microorganisms, and they are directed toward the complete exclusion of all microorganisms from any area where they might do harm. Bandages and surgical instruments are sterilized to keep a wound uninfected. Disinfection procedures are required for contaminated clothing or gowns, bed linens, and blankets. The aim of the **disinfection** process is the destruction of disease agents, whereas **sterilization** is an absolute term that means the killing of all forms of life in a given area. The term *sterilization* has a definite meaning to the microbiologist. No object can be almost sterile.

Sterilization procedures vary considerably, depending on such factors as the material of which the object is made and the circumstances incident to its use. New developments such as the use of indwelling intravenous polyethylene tubing and open-heart surgery continue to create new problems in sterilization. For many articles there is a variety of procedures whereby sterilization may be effected, while for other objects the choice may be limited. The major methods are (1) physical and (2) chemical.

This chapter is devoted to a detailed description of physical and chemical sterilization techniques; the principles discussed provide the basic knowledge necessary for continued success in the control and elimination of undesirable microorganisms.

PHYSICAL METHODS FOR THE CONTROL OF MICROORGANISMS

Heat

For most objects heat is the most practical and efficient method of sterilization. Although no one knows exactly how heat kills a specific microorganism, we do know that it is through the inactivation of one or more essential proteins such as enzymes. The amount of heat required to kill varies from one organism to another—in fact, for any organism one must consider both the amount of heat (i.e., the temperature) to be used and the length of time the material to be sterilized is maintained at a given temperature. The relationship of time to temperature in sterilization is the same as that in cooking or baking: the higher the temperature used, the shorter the time required.

Sterilization would not be too difficult in most cases if we did not have to contend with bacterial endospores that are formed by certain organisms (see Chapter 3). The bacterial endospore is probably the most resistant form of life known; some will survive the temperature of boiling water (100°C at sea level) for several hours. There is no standard pattern of heat resistance for endospores, since their resistance varies not only from species to species but also within the same species under different conditions of age or growth. *Clostridium botulinum,* the toxin of which causes a severe or even fatal type of food poisoning, and *Clostridium tetani,* the causative agent of tetanus, are two anaerobic species that form heat-resistant endospores. A number of factors have been regarded as contributing to the heat resistance of spores. Their low water content and thick spore coat are probably the most important.

Another important factor to be considered with heat sterilization is the environment of the microorganisms being destroyed (i.e., whether they are in a heavy blanket, in pus or feces, in a contaminated bandage, or in blood that has been allowed to clot in a syringe). Environment is important for two reasons: (1) in order to kill, the heat must reach the organisms, and (2) more heat than normal is required to kill organisms embedded in protein material such as pus, tissue, or tissue exudate. Viruses also demonstrate variation in their susceptibility to heat inactivation, but none possesses a resistance comparable to that of bacterial endospores.

From these major considerations it becomes obvious that we can set no standard temperature or time for all conditions of sterilization. Certainly, small tubes of media will become sterilized more quickly than will a large fabric bundle.

Methods of Sterilization by Heat

Moist Heat Steam under pressure is the most efficient sterilization agent and the chief means used for sterilizing surgical bandages, instruments, media, and contaminated material. The temperature of sterilizing is dependent on the pressure of the steam. Normally, the temperature of steam is 100°C and, although one could sterilize at this temperature, it would require a very long time—possibly hours. However, if the steam is confined in a closed vessel, its pressure is raised, and the temperature of the steam is elevated correspondingly. At 15 lb of pressure per square inch (1.05 kg/cm^2), the temperature of steam reaches 121°C; this is the temperature most commonly employed for sterilization.

It should be noted that if steam is to reach the expected temperature for a corresponding pressure (e.g., 121°C for 15 lb/in^2), the atmosphere must be free of air and contain only steam. These conditions are fulfilled in an autoclave, which is actually very similar to a large pressure cooker (see Figure 9.1). Most autoclaves are equipped with controls that automatically remove the air before the steam pressure is allowed to rise to 15 lb/in^2. However, in older autoclaves or small autoclaves in private offices, this must be done manually by first opening the vent and allowing the steam to force out the air.

Although a properly used autoclave is one of the surest means of sterilization, faulty operation can result in misplaced confidence in the sterility of equipment. Improper use of the autoclave is usually attributed to one of two mistakes: (1) failure to exhaust all the air before closing the exhaust vent and (2) overloading or improperly packing loads in the autoclave. It cannot be stressed too strongly that all the air must be forced out of the autoclave if the proper temperature is to be reached. Modern autoclaves automatically record a temperature graph during sterilization,

Figure 9.1 A modern autoclave.

and a glance at the chart before unloading will show whether the desired temperature was attained. Nonrecording autoclaves should be checked during the sterilizing period to be certain that the materials are being exposed to the correct sterilizing temperatures. The problem of loading involves a certain amount of knowledge and a lot of common sense. One must know that *in order to kill all microorganisms, the steam must actually penetrate through the entire load.* Thus, no object should be wrapped in a material such as a rubber sheeting, which is impermeable to steam, or sealed in a tight container. Common sense comes into play when one must decide on the length of time for the autoclaving process. Quite obviously, sterilizing a large load of fabric would require a considerably longer time than would sterilizing a load of instruments. Thus, if 20 min at 121°C is the recommended time for a small load, 60 min may be required for a larger load; if a flask containing 500 mL of solution must be exposed for 20 min, a similar one containing 1000 mL might require 25 min. These time intervals can be determined from tables made available by authoritative sources.

When using an autoclave, one can determine the effectiveness of the sterilization procedure by placing a container of heat-resistant endospores in the center of the load; this is then sent to the laboratory for culture to see if complete killing has taken place. Spores of *Bacillus stearothermophilus*, which are commercially available as spore-strip-set tests, are more resistant than any pathogen that may be involved. However, it is not feasible to include a test organism for laboratory culture in each load of material placed in the autoclave. The Centers for Disease Control recommend that an endospore test set be added to an autoclave load at least weekly. It is also common to use pressure- and temperature-sensitive tapes that change color when the proper temperature is reached (see Figure 9.2). These can be misleading, however, because a color change indicates only that the correct temperature was reached, not how long that temperature was maintained.

Boiling Water. Vegetative forms of pathogenic organisms are readily destroyed at the temperature of boiling water. Actually, they are usually killed within a few minutes at 80°C. However, some bacterial endospores show unusual heat resistance and may survive boiling temperatures for up to 20 h. These very highly resistant endospores have usually been isolated from foods, where they were protected. Mold spores are more readily killed than bacterial endospores, but one cannot trust boiling water to sterilize completely.

The killing effect of boiling water is greatly increased by the addition of 2 percent sodium carbonate or detergents. Spores found to be resistant to boiling water for 10 h can be killed at 98°C in 10 to 30 min when suspended in a solution of 2 percent sodium carbonate.

Dry-Heat Sterilization This is accomplished in an oven not unlike the one used in the home. In fact, the home oven—with the use of a household thermometer to maintain the proper temperature—can well be an improvised method of sterilization. Dry-heat sterilizers are usually heated by either gas or electricity and may or may not have a fan inside to circulate the air so as to effect a better distribution of the heat.

The effectiveness of dry heat depends on the penetration of the heat through the object to be sterilized; thus, it is possible to sterilize objects such as syringes that have already been assembled and sealed in a container. Dry heat is used to sterilize powders, special or petrolatum gauze dressings, and other items that would be damaged by steam or water. It is also effective for oily substances such as ointments that are insoluble in water and not permeable to moist heat.

Dry-heat sterilization requires considerably higher temperatures for complete effectiveness than does steam sterilization. As with steam sterilization, there is no definite standard time, but a temperature of 160 to 170°C for 2 h is probably used most commonly. It is necessary to add the time needed for the oven to reach that temperature and also to take into account the weight or bulk of the load to be sterilized.

Pasteurization

Pasteurization is not a form of sterilization, but it is a method to destroy disease-producing organisms. It was originally designed by Louis Pasteur to preserve wine by killing the bacteria that cause it to become sour. We are today more likely to associate the process with milk. It is usually accomplished by heating to 71.6°C for 15 s, although a holding method in which the milk is heated to 62.9°C for 30 min may also be used. Both procedures are followed by rapid cooling.

The major disease-producing organisms killed by pasteurization include those causing tuberculosis, brucellosis, typhoid fever, diphtheria, scarlet fever, and Q fever. Because many of the organisms that cause

Figure 9.2 Temperature-sensitive tape changes color at 121°C.

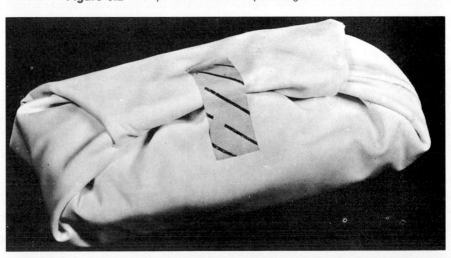

milk to sour are also killed, the storage life of pasteurized milk is also lengthened.

Desiccation

Because bacteria must grow in an environment of moisture, lack of water (desiccation) prevents multiplication. However, bacterial endospores are very resistant to drying, so the organisms are not immediately killed by desiccation and may remain viable for many years in a dried state. Some pathogens, such as the organism causing gonorrhea, survive drying for only an hour or so. Others, such as the tuberculosis bacterium, are more hardy and may remain viable for weeks, particularly if embedded in dried sputum. Much of the ability of a vegetative bacterium to withstand desiccation depends on the cell's immediate environment. For example, even a very sensitive bacterium is much more resistant if embedded in pus or feces. Viruses are usually quite resistant to desiccation but are less so than endospores.

Radiation

Radiation may be defined as the transmission of energy through space. All types of radiation can be injurious to microorganisms, causing either death or mutations. The two main groups of radiations that have been used for controlling microorganisms are ionizing radiations (x-rays, gamma rays, and cathode rays) and ultraviolet rays.

Ionizing Rays

The greatest advances in radiation application have been made with ionizing radiations.

X-rays (produced by generating machines and varying in wavelength from 0.1 to 40 nm) and gamma rays (originating from radioactive elements such as cobalt 60 and similar to short x-rays) are much more powerful and penetrating than ultraviolet light. Ionizing radiation may cause various effects in microorganisms; most investigators believe that the cellular DNA is the primary target. Moreover, it appears that the wide variations in sensitivity to radiation among different microorganisms is correlated with an organism's ability to repair damaged or broken DNA and not to any actual resistance to radiation damage.

One important effect of ionizing radiations is the formation of free radicals (^-OH and $^-HO_2$), which are produced as the high-energy radiation travels through water. These free radicals contribute to the lethal action of the ionizing radiation by forming peroxides that act as powerful oxidizing agents.

Ionizing radiations are used to sterilize certain pharmaceuticals and stable plastic items such as petri

dishes, catheters, and syringes. Low doses of gamma or x-rays are used also to preserve fresh fruits, vegetables, and pork. In a typical process, the food (which may be already packaged) is irradiated in a room with thick walls and, after irradiation, is removed from the chamber. For fresh foods, irradiation is used primarily to kill insects or bacteria that could cause spoilage. In pork, the main objective is to kill the organism that causes trichinosis.

Ultraviolet Light Ultraviolet light is, for the most part, invisible to the human eye. Its short wavelengths cover a range from partially visible to totally dark, that is, wavelengths of 390 to 40 nm, with the maximum killing effect at 260 nm.

Ultraviolet light is lethal because it is absorbed by the cells' nucleic acids (see Figure 9.3). When absorbed, cross-links in loops of a DNA strand are produced between neighboring thymine molecules, as

Figure 9.3 The black curve shows the bactericidal effectiveness of various ultraviolet wavelengths in killing bacteria. The red curve represents absorption by DNA. Note that both curves peak at 260 nm, indicating that bacterial death caused by ultraviolet light is due to absorption by DNA.

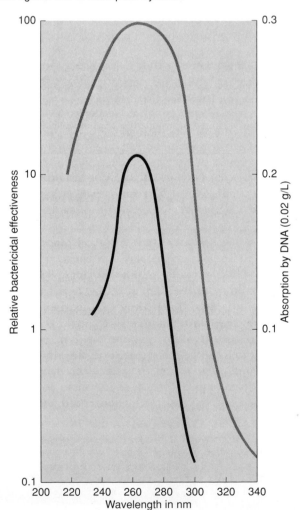

shown in Figure 9.4. These thymine dimers inhibit the normal replication of the DNA by blocking the passage of the replicating enzyme. In such cases a new, error-prone repair system is induced that has the ability to replicate through damaged DNA, although with considerable sacrifice in fidelity. Thus, the mutations induced by ultraviolet irradiation are the results of mistakes caused by this enzyme. There are also repair systems to repair such damaged DNA. One, photoreactivation, is a process whereby the visible light activates an enzyme that excises the thymine dimers while another enzyme, DNA polymerase, replaces the thymine dimer with individual thymine molecules, thus restoring the original DNA. Since these enzymes are activated by visible light, cells that have been killed by ultraviolet light can be photoreactivated by exposure to visible light. A second process, called excision repair, occurs in the dark; this system can also repair the DNA by eliminating the thymine dimers and replacing them with individual, noncovalently bound thymine. Thus, to be lethal, the amount of ultraviolet light absorbed by a bacterium must cause the formation of more thymine dimers than the cell can repair during a short period.

When used in sufficient intensity, ultraviolet light is effective in killing bacteria, but it has one major limitation: it does not penetrate ordinary glass, dirt, paper, pus, or for that matter, much of anything. Because it must be absorbed by the bacterium to be effective, the destructive effect is satisfactory only when the ultraviolet rays have direct contact with the organisms. Ultraviolet light has been used to reduce the microbial population of the air in operating rooms, nurseries, communicable disease wards, schoolrooms, bacteriology laboratories, bakeries, restaurants, and other food establishments. It should be emphasized that such germicidal lamps do not result in the sterilization of an area but only in the reduction of the number of bacteria in the air. Because ultraviolet light can cause severe inflamation if it is directed into the eyes, lamps must be placed high on the walls with the ultraviolet light directed toward the ceiling of the room or placed in ducts bringing air into a room.

Filtration

Certain substances that cannot be subjected to heat or chemical treatment without decomposition or other injury may be sterilized by the process of filtration. Bacteria are not killed during filtration but are physically separated from the fluid by this means. The filter proper, usually made of cellulose acetate, has many tiny pores through which the bacteria cannot pass. Many viruses are able to pass through filters that hold bacteria back.

Commonly Used Filters

Filtration has been used for almost 100 years to remove bacteria from liquids that would be destroyed by the normal heat sterilization procedures. Early filters, designed by Pasteur and Chamberlain in 1884, consisted of hollow candles composed of unglazed porcelain. Such candles could be made with various pore sizes, and those with the smallest pore size were quite effective in removing bacteria from a solution.

Later filters were made of materials such as diatomaceous earth (Berkefeld filters), asbestos pads (Sietz filters), and sintered glass filters. Essentially all of these have been replaced by membrane filters.

Membrane Filters These filters are the most widely used in both the laboratory and industry. The cellulose ester membranes are made with various pore sizes ranging from 8 to 0.025 μm (see Table 9.1). They can be purchased sterile from the manufacturer, although they can also be autoclaved at 121°C for 15 min. Because of their thinness (approximately 150 μm), their efficiency is based entirely on pore size; a

Figure 9.4 Formation of a thymine dimer by ultraviolet light. Repair takes place by one of several routes. In the light a photoenzyme may recognize and bind to the dimer in the strand of DNA. This complex then absorbs a photoreactivating wavelength of light, which results in the splitting of the dimer to restore the monomers. In the dark, excision repair consists of the removal from the strand of DNA of the dimer and several bases on either side by an excision enzyme, followed by replacement of the excised bases by DNA polymerase using the information on the complementary strand.

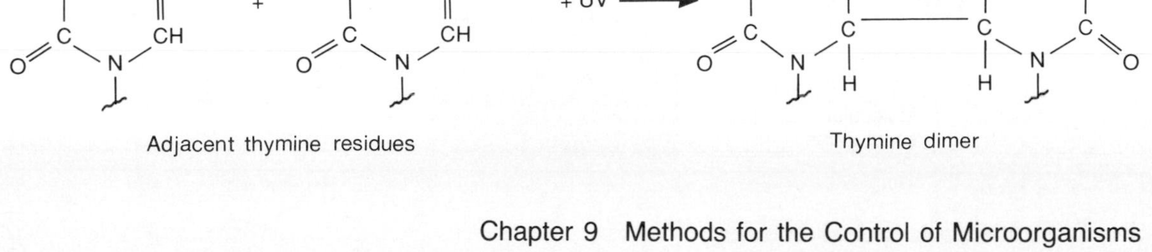

Adjacent thymine residues   Thymine dimer

TABLE 9.1 Pore Size and Filtration Rates for Membrane Filters

Membrane Filter Type	Pore Size	Rates of Flow	
		Water[a]	Air[b]
SC	8.0 μm $\pm$ 1.4 μm	950	55
SM	5.0 μm $\pm$ 1.2 μm	560	35
SS	3.0 μm $\pm$ 0.9 μm	400	20
RA	1.2 μm $\pm$ 0.3 μm	300	14
AA	0.8 μm $\pm$ 0.05 μm	220	9.8
DA	0.65 μm $\pm$ 0.03 μm	175	8.0
HA	0.45 μm $\pm$ 0.02 μm	65	4.9
PH	0.30 μm $\pm$ 0.02 μm	40	3.7
GS	0.22 μm $\pm$ 0.02 μm	22	2.5
VC	100 nm $\pm$ 8 nm	3.0	1.0
VM	50 nm $\pm$ 3 nm	1.5	0.7
VF	25 nm $\pm$ 2 nm	0.5	0.3

[a] (mL/min)/cm^2 ⎫
[b] (L/min)/cm^2 ⎬ Under 70 cm Hg differential pressure.

pore size of 0.22 μm is usually effective in removing all bacteria (see Figure 9.5). Also, because of this thinness and small pore size, they will easily become plugged if the liquid being filtered contains an appreciable amount of bacteria or particulate matter. It is common practice to force liquid through a deep prefilter composed of fibrous, granular, or sintered material to remove a large percentage of contaminating material before the liquid reaches the membrane filter. It should be noted that many viruses and some mycoplasmas may pass through these filters. In fact, membrane filters are used for the production of the live polio vaccine because they permit the virus to pass through the filter while retaining much of the cell debris from the monkey kidney cells in which the virus is grown.

Figure 9.5 Comparison of particle and pore sizes in Millipore membrane filters.

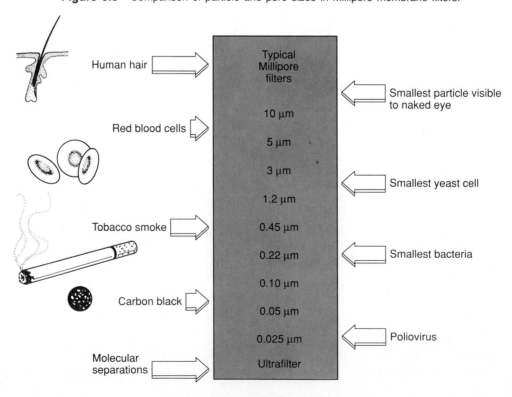

In general, basic membrane filters are small-volume systems that are used in laboratories or large-volume industrial systems that can filter up to thousands of gallons of liquid each day.

Membrane filters can also be used to quantitate the number of bacteria present in a sample of water or air. This is done by placing the membrane (after filtering the sample through it) directly onto suitable agar medium. The plate is then incubated, and each viable bacterium able to grow under the conditions used will produce a visible colony. Figure 9.6 shows a membrane filter of a size commonly used in laboratories and colonies that have grown on the filter after incubating the used membrane on an agar medium.

Membrane filters have been used to separate toxins from bacterial cells and to sterilize drugs and organic materials such as proteins (serum), antibiotics, and certain sugars. It has been suggested that membrane filters could be incorporated into intravenous tubing during manufacture. Fluids could then be pumped through the filter immediately before entering the patient's circulation. This would remove bacteria as well as any contaminating particles in the fluid.

Osmotic Pressure

If two solutions are separated from each other by a semipermeable membrane that permits the passage of water but not of dissolved material, water will pass from the less concentrated into the more concentrated solution. The rate at which water will pass from one solution to another is a function of the difference in concentration between the two solutions; it is known as the **osmotic pressure**. The bacterial cytoplasmic membrane functions in a similar manner. Therefore, if bacteria are placed into a highly concentrated solution of salt or sugar, water will flow from the bacterial cell into the salt or sugar solution. This, of course, prevents bacterial growth. The immersion of meats in brine (a concentrated salt solution) and fruits in thick syrups exemplifies the principle of osmotic pressure as used in food preservation.

Low Temperatures

Cold temperatures or freezing cannot be used as a means of sterilization because many bacteria survive at low temperatures for very long periods. In fact, low temperatures are used to preserve microorganisms. One such technique is called **lyophilization** or **freeze-drying**. In this procedure the organisms are frozen rapidly (usually in a dry-ice bath) and then dehydrated in high vacuum directly from the frozen state. Then they are stored under vacuum in sealed ampules in cold storage. Many bacteria may be kept alive for years in this manner.

Table 9.2 provides a summary of the physical methods used for the control of microorganisms.

CHEMICAL AGENTS FOR THE CONTROL OF MICROORGANISMS

In many of our everyday practices, chemical agents are used to control microorganisms. They are used in

Figure 9.6 (a) Membrane filter apparatus and membrane being laid on agar medium. (b) Typical coliform colonies with metallic "sheen" on a Millipore filter.

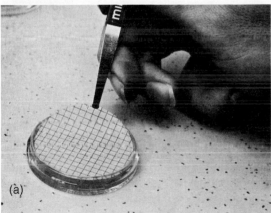

(a)

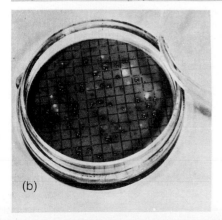

(b)

TABLE 9.2 Physical Methods for Control of Microorganisms

Agent	Action	Use
Moist heat		
Autoclave or steam under pressure	Coagulation	Sterilization of applicators, bacteriological media, drains, dressings and packs, instruments, intravenous equipment, rubber gloves, solutions, syringes, tongue depressors, transfusion equipment; preservation of foods in canning
Boiling	Coagulation	Destruction of vegetative cells on dishes, pitchers, basins, and various equipment
Live steam (tyndallization)	Coagulation	Fractional sterilization of bacteriological media
Dry heat		
Hot air	Oxidation	Sterilization of empty glassware (such as test tubes, petri dishes), instruments, needles, syringes, glycerin, paraffin, petrolatum, petrolatum gauze, sulfonamide powder, talcum powder, zinc oxide, and zinc peroxide
Open flame	Burning to ashes	Sterilization of inoculating loops, needles, etc.
Incineration	Burning to ashes	Complete destruction of paper cups, bags, wipes, soiled dressings, and animal carcasses
Cold temperatures	Not determined; there may be changes in cellular proteins; ice crystals may cause mechanical injury; chemical reactions are decreased	Bacteriostatic effect chiefly permits preservation of foods, drugs, and cultures of bacteria
Lyophilization	Dehydration	Preservation of bacterial cultures
Desiccation	Removes water	Effect chiefly bacteriostatic Preservation of various foods
Pasteurization	Coagulation—changes in cellular protein	Removing all pathogenic and some non-pathogenic organisms from milk
Radiation		
Ultraviolet	Formation of thymine dimers	Microbicidal effect, with limitations owing to lack of penetration; reduces airborne infections in hospitals, restaurants, and schoolrooms Destruction of organisms on surfaces, in water, etc.
X-rays	Ionization; peroxide formation	Research; used to induce mutations
Cathode rays	Ionization	Research; may be used for sterilizing effects in pharmaceutical houses and the food industry in the future
Filtration	Separation of bacteria from the suspending fluid	Sterilization of certain liquids injured by heat or chemical treatment; separation of bacteria from toxins, enzymes, etc.; measurement of the approximate size of some viruses
Osmotic pressure	Plasmolysis	Microbicidal effect in the preservation of various foods

medicine, agriculture, food preservation, and the bacteriology laboratory, as well as in many other areas. Some of the chemical agents are applied to human tissues only and some to inanimate objects, whereas others may be useful in both areas. Antibiotics are also chemical agents, but because of their specialized mechanism of action, the antibiotics will be discussed separately in Chapter 11.

Many of the terms used to describe the action of chemical agents on microorganisms are widely misapplied, having for many persons rather vague meanings. For instance, *antiseptic* and *disinfectant* are often confused. In practice, an antiseptic has come to mean an agent used for topical application to living tissue, primarily skin. A disinfectant, on the other hand, is thought of as a chemical agent used on nonliving or inanimate objects, such as a bedpan. Actually, certain chemical agents may be classed as antiseptics under one set of circumstances and as disinfectants under other conditions. The following definitions may help to clarify the matter.

An **antiseptic** is a chemical substance that is applied to the skin or mucous membranes to prevent growth by either inhibiting or destroying microorganisms, but normally does not sterilize an area.

A **disinfectant** will destroy vegetative cells, but this term is usually reserved for use on inanimate objects. It does not usually destroy endospores and may not inactivate viruses.

Combining forms indicate the type of action an agent exerts on microorganisms. The suffix *-stasis* refers to inhibition, *-cid(e)* to a killing effect.

A **bacteriostatic** agent acts by inhibiting growth; it does not necessarily kill the organism.

A **bactericide** is defined as an agent that kills bacteria. In actual practice only the rare bactericide is effective in killing endospores. **Germicide** is generally considered to be synonymous with bactericide.

In like manner one can define a **sporicide**, a **viricide**, and a **fungicide** as agents that will kill spores, viruses, and fungi, respectively.

A **sanitizer** is any agent that reduces bacterial numbers to safe levels according to public health requirements. Their chief use is for food-handling equipment and utensils for eating and drinking.

These terms are summarized in Table 9.3.

Mechanisms of Disinfectant Action

Although a disinfectant might be described as either bacteriostatic or bactericidal, such a designation is an oversimplification of its action. The same disinfectant may be bacteriostatic under one set of conditions and bactericidal under another set of conditions. The major factors that determine how a disinfectant acts are concentration of the disinfectant, time during which the disinfectant is allowed to act, temperature of disinfection, number and types of microorganisms present, and the nature of the material being disinfected. So you see, a number of factors must be kept in mind in order to do the best job possible under a given set of circumstances.

Just how does a disinfectant act in order to kill an organism or to prevent its growth? Quite obviously, for a disinfectant to have any effect it must influence some vital part of the cell. Parts of the cell most susceptible to the action of a disinfectant are the cytoplasmic membrane, certain enzymes, and structural proteins such as those found in the cell wall.

Action on the Cytoplasmic Membrane

All living cells have a semipermeable membrane that regulates the passage of substances into and out of

TABLE 9.3 Definitions of Terms Related to Microbial Control

Term	Definition
Sterilization	The destruction of *all* forms of life within a given area
Disinfectant	Use of a chemical to destroy the growth of potential disease-producing organisms. May not kill endospores or viruses. Term is usually restricted to chemicals used on inanimate objects
Antiseptic	Identical to disinfectant but may be only bacteriostatic; also, this term is usually restricted to chemicals used on skin and mucous membranes
Sanitizer	A disinfectant used on food-handling equipment
Bactericide	An agent that kills bacteria
Bacteriostatic agent	One that inhibits the growth of bacteria but does not kill the organisms

the cell. Injury to this membrane allows essential inorganic ions, nucleotides, coenzymes, and amino acids to leak out of the cell. In addition, such injury can prevent the entrance of essential materials into the cell because the cytoplasmic membrane also controls active transport into the cell. Thus the action of any substance that can inhibit these essential functions of the membrane will result in either death of the cell or its inability to grow. The end result depends on the severity of the damage to the membrane.

The cytoplasmic membrane is composed primarily of protein and lipid; as a result, the membrane is particularly susceptible to agents that lower the surface tension—surface-active agents. Essentially all soaps and detergents fall into this category, but some of those that are most effective and more commonly used as disinfectants include benzalkonium, phenol, cresols, and ethyl or propyl alcohol.

Action on Cellular Proteins

The so-called backbone of cellular structure is made up primarily of protein. In addition, all metabolic reactions of a cell are catalyzed by enzymes, which are made of protein. These metabolic reactions include both essential biosynthetic reactions and the essential energy-yielding reactions. Thus, any chemical agent that combines with proteins so as to prevent the protein from carrying out its normal function exerts either a bacteriostatic or a bactericidal effect.

Agents that injure the cell through their effect on proteins include acids, alkalies, phenol, cresol, alcohols, salts of heavy metals, formaldehyde, halogens, and other oxidizing agents such as hydrogen peroxide and potassium permanganate. Of these agents, the heavy metals are the strongest general inhibitors of the catalytic action of enzymes, and mercury is one of the most efficient. The sulfhydryl (—SH) groups of many enzymes are necessary for catalysis, and both mercury and the oxidizing agents produce their effect by combining with the SH groups.

Variables in Disinfection

The basic differences in bacteria possibly may account for the fact that certain agents are effective against one type of organism and not another. When the disinfecting process is aimed at a particular pathogen, the agent selected as a disinfectant must be known to be an effective bactericide against that organism. For example, iodine solutions or phenol—but not aqueous benzalkonium (Zephiran chloride)—can be used effectively against *Mycobacterium tuberculosis*.

Concentration

The concentration of disinfectant used will depend on the material being disinfected and on the organism to be destroyed. In general, a higher concentration will be bactericidal, whereas a weaker one may be only bacteriostatic. Alcohols form a general exception to this rule, since ethyl alcohol is most effective at 70 percent concentration and propyl alcohol at a concentration of 50 to 80 percent.

Time

The destruction of microorganisms by a disinfectant appears to be an orderly process. The time required may be influenced by many variables. In general, a wide margin of safety should always be ensured by allowing ample time for the disinfectant to act.

Temperature

It is a general rule that an increase in temperature speeds up the rate of a chemical reaction. Likewise, in disinfection, a rise in temperature usually hastens the process. It is not unusual for an increase of 10°C to double the rate of disinfection.

Nature of the Surrounding Medium

The pH of the medium and the presence of extraneous materials may greatly influence the disinfection process. In fact, pH alone may determine whether an agent is only inhibitory in action or is lethal. In other cases, some chemical agents may combine with the microbes, or certain agents may be unable to penetrate precipitated proteins and therefore will not produce the desired effect. Regardless of the organism or the nature of the chemical agent used, there is a slowing of the destructive effect of the disinfectant in the presence of organic materials. In practice, this is a point to be strongly emphasized, since a great deal of disinfection takes place amid various body fluids or secretions.

Phenol

Lister used phenol (carbolic acid) in his first attempts at antiseptic surgery, and its importance in reducing infections from surgical operations is common knowledge. It is bacteriostatic or bactericidal, depending on the concentration.

When used in high concentrations, phenol acts by totally disrupting the cytoplasmic membrane and precipitating the cell proteins. In concentrations of 0.1 to 2 percent, however, phenol distorts the cytoplasmic membrane, which permits the leakage of essential metabolites and, in addition, inactivates a number of bacterial enzyme systems.

Phenol is effective against vegetative forms of bacteria (including *M. tuberculosis*) and most fungi. Spores are not destroyed, and the effect of phenol on viruses varies according to whether or not the viral capsid is surrounded by a membrane envelope. Thus, viruses that bud from the host cell membrane (such as herpesviruses, poxviruses, and influenzaviruses) are quite susceptible to the action of phenol, whereas those existing as naked capsids are not particularly susceptible.

Derivatives of phenol are used as preservative ingredients in wood, paper, paints, textiles, and certain foods. Phenol has also been used as a standard for comparison with other disinfectants. While this has its limitations, it usually provides some useful information.

Substituted Phenols

Attempts to reduce the toxicity of the phenol compounds led to the synthesis of a number of substituted phenolic disinfectants. Two of the more common ones are hexylresorcinol and hexachlorophene. These substituted phenol compounds have been incorporated into soaps and detergents to enhance the germicidal activity of the surface-active agents. For example, the combination of hexachlorophene with soap is sold under trade names such as pHisoHex and Hexosan. These compounds are not toxic to skin and are highly bacteriostatic even in very low dilutions. As an example, the bacteriostatic end point of hexachlorophene against *Staphylococcus aureus* has been reported to be 1:2,500,000.

In 1972, the U.S. Food and Drug Administration (FDA) released laboratory findings that indicated hexachlorophene could cause brain damage in newborn animals. Furthermore, research with laboratory animals indicated that the practice of bathing newborn babies in hexachlorophene could result in the absorption of some of the disinfectant through the skin and into the circulation of the newborn. As a result, the FDA warned physicians that hexachlorophene should not be used for total body bathing, since absorption through the skin might well result in permanent brain damage. In the summer of 1972, 40 babies in France are believed to have died from a baby powder containing hexachlorophene. As a result of this tragedy and because of other scientific data, the FDA completely prohibited the use of hexachlorophene except by a physician's prescription. It also required the immediate recall of all cosmetics, powders, and so on, that contained more than 0.75 percent of hexachlorophene and prohibited the future manufacture of items containing any hexachlorophene.

Chlorhexidine has now largely replaced the chlorinated phenols as a disinfectant used in skin asepsis. Chlorhexidine is not a phenolic compound but, as the name suggests, it does contain chlorine. It is normally used in solutions containing 200 μg/ml, at which concentration it is rapidly effective against both gram-positive and gram-negative organisms. It is not absorbed through the skin, and its lack of toxicity results in its widespread use for hand scrubbing and preoperative skin preparation.

Table 9.4 includes the structures of phenol and some related phenolics.

Cresols

The destructive distillation of coal results in the production of not only phenol but also some compounds known as cresols. Cresols are effective as bactericidal agents, and their action is not seriously impaired by the presence of organic matter. However, they cause irritation to living tissues and hence are used primarily as disinfectants for inanimate materials. One percent Lysol (cresol mixed with soap) has been used on the skin, but concentrations greater than this cannot be tolerated.

The action of cresol is similar to that of phenol. However, its germicidal activity is greater, and the concentration usually used is 2 to 5 percent. Since the cresols are not soluble in water, they are usually mixed with soap solutions.

Alcohols

While ethyl alcohol is probably the most familiar and the most commonly used, isopropyl and benzyl alcohol are also antiseptics. Benzyl alcohol is used chiefly for its preservative effect.

The germicidal action of alcohol is apparently protein denaturation. The molecular weight of the particular alcohol has been shown to be related to the germicidal action. As the molecular weight increases, so does the germicidal activity. Propyl alcohol, then, is more effective than ethyl alcohol.

Reports concerning the destructive action of alcohol on vegetative cells are controversial; however, it has been verified that *M. tuberculosis* is quite readily destroyed by the action of alcohol. Alcohol is practically ineffective against bacterial endospores. Anthrax spores have been reported to have survived in alcohol for 20 years.

The most effective concentration of alcohol depends on the amount of moisture present; in the case of ethyl alcohol, a 70 percent solution seems to be the

TABLE 9.4 **Summary of Chemical Agents**

Agent	Chemical Nature	Mechanisms of Action	Practical Use
Acids	H^+ H_2SO_4 HCl HNO_3	Hydrolysis Coagulates proteins	Rarely used
Alkalies	OH^- KOH $NaOH$ NH_4OH	Hydrolysis Coagulates proteins	Rarely used
Phenol	OH (benzene ring structure)	Surface-active; disrupts membrane Inactivates enzymes Denatures proteins Toxic	Active against vegetative cells Relatively effective in presence of organic matter May be used for tubercular sputum Preservative Not generally effective against spores
Hexachlorophene	(chemical structure)		Phenolic compound used in skin antisepsis
Chlorhexidine	(chemical structure)		Antiseptic for handwashing
Cresols	Ortho- Meta- Para-cresol (structures)	Surface-active; disrupts membrane Inactivates enzymes Denatures proteins Toxic	Active against vegetative cells Relatively effective in presence of organic matter More active than phenol Disinfection of instruments Not generally effective against spores
Alcohols	Ethyl (C_2H_5OH) Isopropyl ($CH_3CHOH\text{-}CH_3$) Benzyl- (structure)	Denatures proteins (germicidal action increases with molecular weight)	Ethyl (skin antisepsis) Effective against *M. tuberculosis* Benzyl (preservative) Ethylene and propylene glycols as aerosols for air disinfection
Halogens 1. Chlorine and compounds 2. Iodine	$HClO$ (hypochlorous acid) $NaClO$ (sodium hypochlorite) $Ca(OCl)_2$ (calcium hypochlorite)	Oxidation Combines with protein to form protein halides	Chlorine (disinfects water) Hypochlorites (sanitizing utensils and dairy equipment) Iodine (active against spores, viruses, and fungi) May be used to disinfect various equipment

TABLE 9.4   Summary of Chemical Agents   *(Continued)*

Agent	Chemical Nature	Mechanisms of Action	Practical Use
Salts of heavy metals 1. Mercuric chloride 2. Silver nitrate	$HgCl_2$ $AgNO_3$	Oxidation Combines with proteins Toxic	Mercuric chloride (preservative) Silver nitrate (eye drops and lotion)
Dyes 1. Crystal violet, etc. 2. Acridine dyes	Amino derivatives of triphenylmethane	Believed to combine with proteins or interfere with reproductive mechanism May interfere with an oxidative process Acridine appears to interfere with enzymes	Inhibition of gram-positive bacteria in culture media Isolation of gram-negative pathogens Acridine (for treatment of wounds)
Quaternary ammonium compounds Detergent	Cationic 	Surface-active; disrupts cell membrane Inactivates enzymes Denatures proteins	Disinfection of utensils in dairies and restaurants No effect on spores
Formaldehyde	HCHO	Alkylating agent	May be used to kill *M. tuberculosis* in sputum and the fungus of athlete's foot in shoes Used in preparing vaccines Gas may be used to disinfect rooms Preservation of specimens Alcoholic solution for instruments
Hydrogen peroxide	H_2O_2	Oxidation	Cleansing of wounds
Potassium permanganate	$KMnO_4$	Oxidation	Antibacterial action on tissue surfaces
Ethylene oxide		Alkylating agent	Sterilization of heat-labile materials Effective against vegetative bacteria, spores and viruses
Betapropiolactone		Alkylating agent	Sterilization of bone, cartilage, artery grafts, and culture media Destroys hepatitis virus
Glutaraldehyde		Alkylating agent	Used primarily on inanimate objects Very effective against all forms of microbial life

most effective concentration for most purposes. Pure alcohols are less effective presumably because they dehydrate bacterial cells and thus enhance their survival.

Because alcohol does not destroy spores, it cannot be used as a sterilizing agent. When used on the skin, its contact is usually too short for much germicidal effect. However, it does remove oil and dust particles, and with them, no doubt, bacteria as well. The most common way to disinfect the skin prior to injection or withdrawal of blood is to (1) swab the area with alcohol to remove oil and surface debris; (2) swab the area with an iodine-alcohol solution to kill contaminating organisms; and (3) swab again with alcohol to remove most of the residual iodine. Propyl alcohol (80 percent) is considered to be particularly useful for skin antisepsis. The fact that alcohol will coagulate proteins limits its use where organic matter is present. One or two percent iodine enhances the effectiveness of alcohol.

Halogens

Of the halogens, chlorine and iodine are the only two that have been used to any extent for their antimicrobial properties. Chlorine has been widely used to sanitize drinking water, swimming pools, food handling and processing equipment, and sewage and waste water. There are a number of different disinfectants containing chlorine, ranging from chlorine gas to organic chloramines. All appear to exert their bactericidal effect by reacting with water to form hypochlorite. In the case of chlorine gas, the reaction is as follows:

$$Cl_2 + H_2O \longrightarrow HClO + HCl$$

The mechanism by which hypochlorite (HClO) destroys microorganisms is not completely clear. Hypochlorite is a very strong oxidizing agent, and evidence suggests that the bactericidal effect of chlorine results from the irreversible oxidation of sulfhydryl groups (—SH) on essential enzymes. This inactivates the enzyme by changing its tertiary structure.

Iodine has been used extensively for skin antisepsis and is germicidal against bacteria, fungi, spores, and viruses. Iodine may also be used for disinfecting various articles of equipment and for sterilizing certain instruments. It has certain advantages over chlorine, particularly in relation to its activity in different pH ranges, since acidity or alkalinity have less influence on its effectiveness.

The germicidal effect of iodine is believed to result from its reaction with the amino acid tyrosine, which prevents the normal function of tyrosine-containing enzymes. Iodine, however, like chlorine, also reacts with water to form hypoiodite (HIO); in all likelihood, the oxidizing ability of this compound exerts some killing effect.

Iodine may be combined chemically with large carrier molecules, such as polyvinylpyrrolidine, to produce iodophors (*iodo,* iodine; *phor,* carrier). Iodophors are readily soluble in water and when dissolved will liberate free iodine slowly from the complex. Commercially available iodophors such as Betadine, Isodine, Ioprep, and Surgidine are used as skin disinfectants, particularly for preparation for surgery and spinal taps.

Salts of Heavy Metals

Salts of heavy metals, such as mercury, silver, and arsenic, react with and inactivate many enzymes. Depending on the concentration, their action may be bacteriostatic or bactericidal. Silver nitrate has been used in eye drops to prevent gonococcal eye infections in newborns.

Like silver, the action of mercury results from a reaction between the mercury and the sulfhydryl groups (—SH) present on certain enzymes in the cell. Once mercury has reacted with these —SH groups, the enzymes can no longer function and the cell dies. Mercury inhibition can be reversed by the addition of other compounds that have —SH groups so that the enzyme-bound mercury reacts with this second compound (the antidote) and allows the enzymes to function again. British antilewisite (BAL) is one such antidote used for mercury poisoning.

Several inorganic mercury compounds are available. Of these, mercuric chloride is used as a preservative for wood, paper, and leather as well as for the control of fungal infections in seeds.

Several mercuric organic compounds are used rather commonly for skin disinfection, and some are used for their preservative effect. Merthiolate is used as a preservative in some vaccines and as a skin disinfectant. Merbromin (Mercurochrome) and nitromersol (Methaphen) are other compounds in use, chiefly for skin antisepsis.

Dyes

Certain dyes used for staining bacteria have a bacteriostatic action. They are most effective against the gram-positive bacteria, although some yeasts and molds have been inhibited or killed, depending on the concentration of the dye. It is believed that the dyes may combine with proteins or interfere with the reproductive mechanisms of the cell. Besides crystal violet (crude form, gentian violet), other dyes used as bacteriostats are malachite green and brilliant green.

Because of their selective inhibition of gram-positive bacteria, the dyes have been used in bacteriological culture media for the selective cultivation of the gram-negative bacteria. This has been especially useful in isolating gram-negative pathogens from the intestinal tract. In therapy, however, the dyes have generally been replaced by other agents.

Quaternary Ammonium Compounds

This group consists of a large number of compounds in which four carbon-containing substituents are covalently bound to a nitrogen atom (see Table 9.4). They may be either bacteriostatic or bactericidal, depending on the concentration used; in general, they are much more effective against gram-positive than gram-negative organisms.

They have been used to reduce bacterial counts in food-processing industries, on food and beverage utensils in restaurants, and on milk-handling and processing equipment. Quaternary ammonium compounds have been combined with nonionic detergents, providing both cleaning action and disinfection of large surface areas.

Their mode of action is not fully understood, but they appear to affect cell permeability by damaging the cytoplasmic membrane. They are also reported to inactivate enzymes and to denature proteins.

Soaps and Detergents

Soap acts primarily as a surface-active agent; that is, it reduces surface tension. This mechanical effect is important because the bacteria, along with oil and other particles, become enmeshed in the soap and are removed through the washing process. However, soap as ordinarily used cannot really be classed as a germicide. Even though soap is mildly bactericidal, its contact is usually too brief to produce much destructive effect. Nevertheless, fragile bacteria such as the gonococcus, the meningococcus, and the pneumococcus may be readily killed by the chemical action of soap.

Detergents include a large group of surface-active agents. They are usually more effective against gram-positive than against gram-negative organisms and appear to exert their disinfecting effect by disruption of membranes and denaturation of proteins. Those possessing a positive charge are called cationic detergents, whereas those having a negative charge are known as anionic detergents.

Formaldehyde

Formaldehyde is an excellent disinfectant when used as a gas. It is quite effective in closed areas as a bactericide and fungicide. In an aqueous solution of about 37 percent, formaldehyde is known as formalin.

Formaldehyde (or formalin) destroys spores of both bacteria and fungi; however, its very irritating vapor interferes with its use. It is used in the preservation of laboratory specimens and can be used in disinfecting shoes that carry the fungi of athlete's foot. In alcoholic solutions it may be used to sterilize instruments. Formaldehyde functions as an alkylating agent by attaching to proteins and nucleic acids, rendering them nonfunctional.

Glutaraldehyde also functions as an alkylating agent and is frequently used for sterilizing surgical instruments.

Hydrogen Peroxide

This agent owes its mildly antiseptic properties to its oxidizing ability. It is quite unstable but is frequently used in the cleaning of wounds, particularly deep wounds in which anaerobes are likely to be introduced. The action of the peroxide is limited since the enzyme catalase, present in tissue fluids, rapidly decomposes the peroxide to water and free oxygen. The liberation of the oxygen accounts for the foaming noted when peroxide is introduced into the mouth or open wounds. Usually it is used in a 3 percent solution.

Ethylene Oxide

Ethylene oxide boils at 10.8°C, so it will exist as a gas unless it is kept cold or tightly sealed. Since the mid-1920s it has been used as an insecticide, and in 1937 a patent was granted for its use as a germicide.

When used as either a gas or a liquid, ethylene oxide is a very effective killing agent for bacteria, spores, molds, and viruses. An important property that makes this compound such a valuable germicide is its ability to penetrate into and through essentially any substance that is not hermetically sealed. For example, it has been used commercially to sterilize barrels of spices without even opening the barrel. They are merely placed in a large drumlike apparatus and, after much of the air has been removed with a vacuum pump, the ethylene oxide is admitted.

It is not difficult to recognize the value of such an agent for sterilizing certain pieces of equipment used in the hospital operating room—particularly the pieces that are large, expensive, and delicate and cannot be subjected to heat or liquid sterilization. Heart pumps, respirometers, and intravenous catheters are easily sterilized with gas. It should also be noted that ethylene oxide is very irritating to skin and mucous membranes and must, therefore, be completely degassed from any material, such as catheterization tubing, before use.

The major disadvantage of ethylene oxide is that it is highly flammable in high concentrations. For safety, it is frequently mixed with an inert gas. The most commonly used mixture, called **carboxide**, consists of 10 percent ethylene oxide and 90 percent carbon dioxide. Another mixture, called **oxyfume**, is made up of 20 percent ethylene oxide and 80 percent carbon dioxide. Equipment is now available, however, that uses pure ethylene oxide for sterilization. These mixtures of ethylene oxide are used in specially constructed autoclaves in which the temperature and the humidity of the atmosphere are carefully controlled.

Ethylene oxide is a very active chemical and will react readily with free carboxyl, amino, sulfhydryl, and hydroxyl groups, replacing the labile hydrogen present in each of these groups with a hydroxyethyl radical ($—CH_2CH_2OH$). It is believed that it is this type of reaction with the structural and enzymatic components of the cell that causes the death of the cell.

Ethylene oxide is a very effective bactericidal compound and is used in special sterilizers. Heat-labile materials such as rubber and plastic can be sterilized effectively without the danger of injury that would be caused by heat. Bacteriological culture media also have been sterilized by the addition of liquid ethylene oxide. After the medium is warmed, the ethylene oxide is given off, and the medium may be used. It can be used on laboratory equipment and biological materials as well. It has also been used to sterilize valuable library and museum holdings. Aside from the explosive nature of pure ethylene oxide, its chief disadvantage is that it requires 4 to 12 h to sterilize an object effectively and another 4 to 12 h to permit the evaporation of any residue before use. Thus, it is an effective but time-consuming process.

Betapropiolactone (β-Propiolactone)

Betapropiolactone has many properties similar to those of ethylene oxide. It kills spores in concentrations not much greater than those required for killing vegetative bacteria. The effect is rapid, which is necessary, because betapropiolactone in aqueous solution undergoes fairly rapid hydrolysis to yield acrylic acid, so that after a few hours no betapropiolactone is left.

This instability is an advantage, since it permits addition of the substance to many materials, with the knowledge that it will disappear spontaneously in a few hours. Thus, betapropiolactone may be used to sterilize bone, cartilage, and artery grafts. Also, it may be added directly to serum to destroy hepatitis virus and will sterilize a culture medium when the two are mixed in liquid form. It has, however, been reported to produce tumors when applied repeatedly to mouse skin, and it is currently banned from interstate shipment if it is to be used as a bactericide. Because of its many advantages, one might predict that it will be used more in the future, as safety procedures and new applications are developed.

The Disinfection Process

For both physical and chemical methods of sterilization, the rate at which a microbe population dies can be plotted as a logarithmic function of the number of survivors at any given time. In other words, if we heat a suspension of vegetative bacterial cells to 65°C, they begin to die, but they do not die instantaneously.

Let us take a hypothetical example in which we begin with 100 million bacteria and heat the suspension at 70°C. Let us assume that after 1 min, 90 percent of the cells are killed, leaving only 10 million viable bacteria in the suspension. At first glance one would think that just a few seconds more of heating would result in a complete sterilization of the suspension. Not so! If 90 percent of the viable cells were killed during the first minute, 90 percent of the survivors will be killed during the second minute. Thus, although 90 million bacteria were killed during the first minute, only 9 million will be killed during the second minute, and only 900,000 will be killed during the third minute.

This type of reaction is called a first-order reaction because the rate of killing is dependent on the number of viable cells at any time. When this is plotted on a graph, we obtain a **death curve** that shows death to be a logarithmic function. Thus, a plot of the logarithm of survivors versus time will yield a straight line (see Figure 9.7). In general, the slope of the line (which will show the speed of the reaction) will vary according to the conditions of killing. At 60°C, it will take longer to kill than at 70°C, but both cases will yield a straight-line death curve.

In practical cases, the line may deviate from normal as the number of survivors becomes very small.

Figure 9.7 Graph showing exponential death of bacteria. Death occurs as a first-order reaction; thus, the actual rate of killing is a function of the number of survivors at any time.

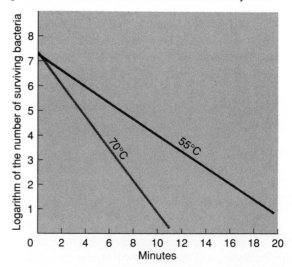

One must always allow a safety margin if sterilization conditions are mild and sterilization is necessary.

Tests for Evaluating Chemical Agents

No completely satisfactory method is available for determining the effectiveness of a disinfectant. However, certain tests are carried out that supply useful information. The limitations of all methods should be kept in mind when judging the efficiency of a disinfectant. At all times the practical application of a disinfectant is likely to differ considerably from the conditions of testing in the laboratory.

The Phenol Coefficient

The phenol coefficient has long been a standard method for testing chemical agents. The disinfectant to be tested is compared with phenol under identical conditions. Organisms to be used, according to the FDA, are specific strains of *Salmonella typhi*, *Pseudomonas aeruginosa*, and *Staphylococcus aureus*. These may be obtained from the American Type Culture Collection, Rockville, Maryland. Usually, however, a new disinfectant is tested against a wider variety of organisms. It must be emphasized that the phenol coefficient alone is of limited value. Its major significance is to relate other water-soluble phenolic compounds with phenol. When compared with emulsifiable phenolic compounds containing coal-tar hydrocarbons and other constituents or to other nonphenolic disinfectants, the use of the phenol coefficient may be misleading.

The numerical value of the phenol coefficient is determined by taking the highest dilution of a disinfectant that kills the organisms in 10 min (but not in 5 min) and dividing this by the highest dilution of phenol giving the same result. For example, disinfectant X at a dilution of 1:1000 kills cells in the same time as does a 1:90 dilution of phenol. The phenol coefficient is calculated by dividing 1000 by 90, giving a value of 11.1 for disinfectant X.

Various other methods have also been devised, such as the addition of 3 percent dried human feces to determine the effect of organic matter on the disinfectant being tested.

Use Dilution Test

For this method, the test organisms are dried on rings or glass rods, and these are then suspended in the disinfectants to be tested. At the end of specified periods, the rings or rods are transferred to a suitable culture medium. These results are interpreted in the same manner as the culture results in the phenol coefficient. This test is particularly useful for disinfectants such as the detergents.

SUMMARY

Physical methods of sterilization are effective either by killing all forms of life within a given area or by removing all forms by filtration. Heat is by far the most commonly used method of sterilization. Complete sterilization, which includes the killing of endospores, requires the use of steam under pressure. This is accomplished by autoclave sterilization with steam under a pressure of 15 lb/in^2, yielding a temperature of 121°C.

Dry heat is an effective sterilization method, although higher temperatures are required for complete effectiveness than are necessary in steam sterilization. Dry heat is used to sterilize powders, gauze dressings, glassware, and other things that should not come in contact with steam. Routinely, temperatures of 160 to 170°C for 2 h are used in dry-heat ovens.

Pasteurization is a mild heat treatment used only to destroy pathogenic vegetative cells. Today it is used primarily to kill pathogenic organisms in milk.

Radiation can be used for sterilization. Ultraviolet irradiation, which is commonly used, is effective by causing the formation of thymine dimers in the bacterial DNA. If the ultraviolet radiation is not too extensive, the cell can repair the damage by excision of the dimers; excision is accomplished either through a light-induced mechanism (photoreactivation) or by a dark reaction (excision repair).

Disinfection is an important process in the control of disease, because its aim is the destruction of pathogenic agents. Various terms are used in reference to the chemical agent according to its action or the specific organism affected. These include *disinfectant*, *antiseptic*, *bacteriostatic agent*, *bactericide*, *germicide*, *sporicide*, *viricide*, *fungicide*, and *preservative*.

The mechanism of disinfectant action may vary from one disinfectant to another. The effect may be caused by injury to the cell membrane or by actions on cellular proteins or on specific genes causing either death or mutation.

Factors that alter the rate of disinfection include the kind of agent, the concentration, the time and temperature, the number of organisms and their characteristics (e.g., species differences, spores, and capsules), and the nature of the surrounding medium.

Various tests are in use to evaluate the action of chemical agents. All supply a certain amount of useful information, but it is well to remember the limitations of the tests employed.

QUESTIONS FOR REVIEW

1. Give examples of materials that must be sterilized in the hospital.
2. Define *sterilization*.
3. What are the available methods of sterilization?
4. What pressure is ordinarily used in an autoclave? What temperature does this produce?
5. What is lyophilization?
6. For what is ultraviolet light used most? What are the disadvantages of ultraviolet light? Do you think it is effective for sterilizing a toilet seat in a public restroom?
7. What are thymine dimers? Explain photoreactivation.
8. List the range of pore sizes available in membrane filters. What pore size would you use to ensure the removal of all bacteria?
9. How may a disinfectant work to kill microorganisms?
10. In what way are alcohols an exception to the rule about concentration of disinfectants?
11. List the major variables in disinfection.
12. Name two substituted phenols used as antiseptics.
13. List the different kinds of alcohols used in disinfection and the specific use of each.
14. Make a list of the major uses for both chlorine and chlorine compounds, and for iodine.
15. What is the primary action of soap?
16. What is carboxide and for what kind of disinfection is it used?
17. Describe what is meant by a phenol coefficient.

SUPPLEMENTARY READING

Allman WF: Irradiated food. *Science '81* 2(8):14, 1981.

Block SS: *Disinfection, Sterilization, and Preservation*, 3rd ed. Philadelphia, Lea & Febiger, 1983.

Borick PM (ed): *Chemical Sterilization*. New York, Academic, 1972.

Edgar II: Modern surgery and Lord Lister. *J Hist Med Allied Sci* 16:145, 1961.

Gregor HP, Gregor CD: Synthetic membrane technology. *Sci Am* 239(1):112, 1978.

Hugo WB (ed): *Inhibition and Destruction of the Microbial Cell*. New York, Academic, 1971.

Phillips GB, Miller WS, (eds): *Industrial Sterilization*. Durham, NC, Duke University Press, 1973.

Upton AC: The biological effects of low-level ionizing radiation. *Sci Am* 246(2):41, 1982.

Chapter 10

Surgical and Medical Asepsis

OBJECTIVES A study of this chapter should provide a clear understanding of

1. Nosocomial infections and how they can be minimized.

2. The objectives and procedures of surgical asepsis.

3. The problems of disposal of waste material.

4. General procedures involved in medical asepsis.

5. The responsibilities and functions of an infection control committee.

A *sepsis* means the absence of sepsis or, to put it another way, the absence of any infectious agent. Aseptic techniques are procedures used by hospital personnel to protect the patient from infectious organisms. However, in spite of the precautions taken, 6 to 10 percent of all persons acquire an infection while they are patients in the hospital. These hospital-acquired infections have been given the special name of **nosocomial infections**; it has been estimated by the Centers for Disease Control (CDC) that nearly 2 million patients each year become infected during their hospital stay. Such infections contribute to about 80,000 deaths annually and carry an estimated yearly cost of nearly $2 billion. It is, therefore, the objective of this chapter to delineate some of the major procedures that must be followed to eliminate or reduce the number of nosocomial infections.

FACTORS INFLUENCING NOSOCOMIAL INFECTIONS

Many conditions contribute to the acquisition of a nosocomial infection, but the two most important considerations are the following: (1) because sick people are in the hospital, it is the most likely place to find virulent, disease-producing microorganisms, and (2) many hospital patients are particularly susceptible to infection, both as a result of hospital procedures that eliminate normal anatomical barriers to infection and because their normal immune responses may be impaired by drug therapy, malignancies, or the extremes of age (i.e., infancy or old age). Table 10.1 lists some of the predisposing factors leading to nosocomial infections. In the first case, control requires that patients

TABLE 10.1 Normal Host Defenses and Factors Predisposing to Nosocomial Infections

Altered Host Defense	Predisposing Factors
A. Protection afforded by normal flora-	1. Burns, trauma, other infection 2. Surgery 3. Hospitalization 4. Antibiotics
B. Anatomic barriers and secretions	1. Burns, trauma, surgery, other infection, inflammatory diseases 2. Extremes of age 3. Foreign body, prostheses 4. Diagnostic procedures 5. Urinary tract and intravenous catheters 6. Antimetabolites, irradiation
C. Inflammatory response	1. Diabetes mellitus, renal failure 2. Diseases of hematopoietic system 3. Antimetabolites, irradiation, corticosteroids, other drugs
D. Mononuclear phagocyte system (MNS)	1. Diseases involving the lymphoid or MNS, e.g., lymphoma, reticulum cell sarcoma 2. Antimetabolites, x-irradiation, corticosteroids
E. Immune response	1. Disease of the lymphoid tissues and MNS (e.g., multiple myeloma, chronic lymphatic leukemia) 2. Hodgkin's disease 3. Extremes of age 4. Debilitating diseases (e.g., liver disease, renal failure) 5. Antimetabolites, irradiation, corticosteroids

with communicable infectious diseases (like influenza, measles, typhoid fever, and whooping cough) be isolated to prevent the spread of these organisms to non-immune hospital patients. Surprisingly, however, the majority of nosocomial infections are caused by microorganisms that make up part of the patient's normal flora. Such microorganisms are termed *opportunists* because they normally produce infections under the following conditions: (1) when they are in a host whose immune system has been impaired, (2) when they can bypass anatomical barriers following burns or surgery, or (3) when they are implanted by contaminated catheters, syringes, or respirators. Table 10.2 lists some of the more frequent opportunists and the predisposing factors leading to nosocomial infections. Since nosocomial infections have reached almost epidemic proportions, let us explore some of the factors that contribute to their frequency and see what is being done to minimize their occurrence. Because the procedures involved differ somewhat for surgical and medical patients, they will be discussed separately.

SURGICAL ASEPSIS

One of our major safeguards against infection is the unbroken skin. If an individual loses this protection either by injury or by an operative procedure, this person is much more susceptible to infection. This is particularly true for patients with severe burns. Such persons are extremely susceptible to infection, and the nosocomial infection rate in many hospital burn units may routinely exceed 75 percent. To minimize nosocomial infections requires stringent surgical aseptic technique.

Surgical aseptic technique is aimed at preventing the infective agent from reaching the wound. This is a working example of the old adage that "an ounce of prevention is worth a pound of cure," since it is certainly more desirable to prevent an infection than to try to cure one after it has become established. This ounce of prevention starts before the patient enters the operating room and entails the cleansing and the disinfection, insofar as is possible, of the skin area to be opened. The procedure usually starts with the removal of any hair in the area, followed by a thorough cleansing of the area with alcohol or a commercial antiseptic. This removes most of the skin oils and permits better penetration of the antiseptic. Also, this procedure will result in the mechanical removal of a large percentage of the microorganisms present on the skin. Finally, an antiseptic is used to further decrease the microbial population. The antiseptic used may be an antibiotic or a chemical compound. A variety of

different antibiotics have been used, but a large study in England concluded that the topical application of ampicillin (a derivative of penicillin) or cephaloridine (one of the cephalosporins) is the most effective procedure in preventing infections following surgery. The use of antibiotics for topical application has some serious disadvantages—namely, the potential development of an allergic reaction to the antibiotic and the emergence of strains of bacteria that are resistant to the antibiotic. As a result, chemical disinfectants are much more widely used for skin disinfection than are antibiotics.

Providone-iodine (a water-soluble, chemically stable complex of iodine and the polymer polyvinyl pyrrolidone) is one of the more common antiseptics used to disinfect an area of skin prior to surgery. It is sold under a variety of trade names, such as Betadine, Isodine, Ioprep, and Surgidine.

It is also essential that the surgeon and others working in the operating room do not themselves become the source of an infection for the surgical patient. This entails the wearing of a sterile gown, the use of a mask to cover the mouth and the nose, the use of a cap or turban to cover the hair, the careful and thorough washing of hands, and the use of sterile rubber gloves when touching anything that will come in contact with the patient's wound.

The need for asepsis does not end when the operation is finished. The patient must be guarded against infection until the wound is sufficiently healed and infection is no longer a major problem. Dressings are usually changed at intervals during the convalescent period, at which time it is important to remember that bare hands or fingers should never touch the wound or any part of the dressing that comes in contact with the wound. Wearing sterile gloves is often inconvenient for this procedure, so dressings are removed with sterile forceps. A second pair is used to handle the fresh dressings. These instruments are usually referred to as transfer forceps and are used only for transferring sterile dressings from one sterile field to another.

Surgical asepsis is employed in obstetric procedures as well as in general surgery, because breaks in the mucous membranes (as well as the skin, in many cases) leave the patient very susceptible to infection.

MEDICAL ASEPSIS

A person who washes his or her hands, does the laundry, or scrubs the floor is involved in medical asepsis. Medical asepsis should be a part of the daily routine of every individual involved in hospital work or in

TABLE 10.2 Association of Specific Nosocomial Infections with Certain Predisposing Factors

Predisposing Factor	Frequent Opportunistic Invaders
Burns, trauma	*Pseudomonas* and other gram-negative bacilli *Serratia* *Staphylococcus* *Mucor*
Abdominopelvic surgery	Anaerobic streptococci plus *Bacteroides* Gram-negative bacilli, *Serratia-Enterobacter-Klebsiella* *Staphylococcus*
Cardiac surgery	Diphtheroids *Staphylococcus (aureus)* Aspergilli *Candida*
Intravenous catheter	*Acinetobacter* *Staphylococcus* *Candida* *Cryptococcus*
Urinary tract manipulation	*Pseudomonas, Proteus,* and other gram-negative bacilli *Serratia-Enterobacter-Klebsiella* *Staphylococcus (epidermidis)*
Diabetes	Gram-negative bacilli *Staphylococcus* *Candida* *Mucor* (with acidosis)
Renal failure	Bacilli *Bacteroides* *Serratia-Enterobacter-Klebsiella* *Staphylococcus* *Mucor* (with acidosis)
Liver failure	*Clostridia* Gram-negative bacilli, *Serratia-Enterobacter-Klebsiella* *Staphylococcus*
Diseases involving hematopoietic systems, MNS[a] lymphoid system	Diphtheroids *Listeria* *Pseudomonas* and other gram-negative bacilli *Serratia* *Staphylococcus* *Nocardia* *Aspergillus* *Candida* *Cryptococcus* *Mucor* Cytomegalovirus Herpes zoster *Pneumocystis*

[a] MNS—Mononuclear Phagocyte System.

Source: Klainer AS, Beisel WR: Opportunistic infection: a review. *Am J Med Sci* 258:431–456, 1969.

During the mid-1850s, long before the term *nosocomial infections* was coined, one's chance of dying following a normal delivery in the First Obstetric Clinic of Vienna's General Hospital was about 1 in 5. In fact, childbed fever (also known as puerperal fever) sometimes struck down every mother in row after row of beds.

It was at this time that Ignaz Philipp Semmelweis was appointed assistant of the First Obstetric Clinic. During his first few years in that position, Semmelweis noted that many of the physicians and medical students spent their mornings doing autopsies and then went directly to the hospital clinic to examine their obstetric patients and assist in child delivery. Semmelweis correctly deduced that these professionals were carrying the "poisons" directly from the cadavers to the childbearing women. As a result, he instituted the requirement that every physician and medical student who examined women in his clinic must first thoroughly scrub his hands in soap and water, then with a chlorinated lime solution, and then with clean sand until the odors of the dissecting room were removed.

Within two months, the mortality rate fell from about 20 percent to 1.2 percent and one would think that Semmelweis would have received praise and acclaim for his accomplishments. Not so! Many of his peers failed to appreciate his discovery, and he became an object of scorn and contempt by his colleagues. His own director, Professor Klein, refused to reappoint him and, after Semmelweis' departure, death from childbed fever again became prevalent, claiming hundreds of mothers admitted to the First Clinic.

Semmelweis countered by publishing a book on the cause of childbed fever but, as before, he was ridiculed by his fellow obstetricians. Both his physical and mental health suffered and, at the age of 47, he was committed to a mental sanatorium in Vienna. That same year he died—not from insanity, but from a blood infection he acquired in a finger wound while performing a surgical operation.

It is interesting to note that medical history books are filled with the accomplishments of Ignaz Philipp Semmelweis while the names of those who scorned him are lost forever.

---

any other occupation, including good housekeeping. Dust control measures are a part of medical asepsis, since reducing or eliminating dust reduces the chance for pathogenic organisms to be spread in the environment. The careful handling of dishes by procedures that ensure destruction of pathogens is carried out in restaurants, hospitals, and wherever food is served. Another common example of medical asepsis is covering the mouth and the nose with a handkerchief or tissue during coughing and sneezing to limit the spread of organisms in the air.

Awareness of the constant presence of microorganisms in our everyday environment should be sufficient indication of the desirability and, in most cases, the necessity of aseptic practices. Everyone carries a vast number of microorganisms on skin and body orifices. Most of these organisms are harmless and some are even beneficial—but only if they stay where they belong. Every organism must be treated as a potential pathogen, because even the most innocuous microbe may cause infection, particularly if it gets into a wound. Also, an individual may have an unrecognized infection with no symptoms of illness and may thus be distributing disease agents to others. Such individuals are called **carriers**. Some individuals become carriers after recovering from disease, while others may harbor a pathogenic organism without ever having symptoms of illness.

General medical asepsis should aim at destroying pathogens as well as at reducing the number of microorganisms in the environment. These aseptic techniques use both physical and chemical methods to control microorganisms. Physical methods include thorough general cleanliness measures, which help to prevent the transfer of microorganisms among hospital personnel, visitors, and patients. One of the important measures carried out to reduce this transfer is careful and thorough handwashing. A second very important practice is the use of individual equipment wherever possible.

All equipment used by the patient should be treated so as to destroy any pathogens present. In many cases, it is possible to identify certain body secretions, products, or areas as more highly contaminated than others. For example, poliomyelitis, hepatitis A, typhoid fever, and dysentery can be spread by contact with feces. In these cases articles that have been contaminated with urine or feces must be disinfected or sterilized after use.

A far more serious situation is present when handling blood or syringes that may be contaminated with hepatitis B virus or the causative agent of the acquired immunodeficiency syndrome—AIDS—(human immunodeficiency virus—HIV). There is no treatment for either of these infections and, as discussed in Chapter 27, a high percentage of HIV infections prove fatal. There have been several reported cases of HIV infection in health care workers who have acquired the virus by an accidental needlestick with a contaminated needle. Such tragic accidents emphasize the caution that health care workers must exercise in handling potentially infectious fluids.

Another major problem facing the modern hospital is disposal of the 10 to 40 lb of solid waste generated daily by each patient. Much of this waste is an accumulation of disposable items such as plastic syringes, sputum and stool containers, diapers, bedpans, catheters, gloves, caps, gowns, masks, dishes, silverware, baby bottles, and many more contaminated articles. Much of this waste is not particularly dangerous, but it has been estimated that 25 to 30 percent of the total amount may be contaminated with virulent organisms and, therefore, must be disinfected prior to disposal.

There are three general methods used to dispose of this large amount of material: (1) it may be incinerated, (2) it may be finely ground in large grinders and emptied into the municipal sewer system, or (3) it may be buried in a landfill. Each method presents special problems; the procedures used in each hospital are established by an infection control committee. In some cases, such decisions may be dictated by the availability of an adequate sewer system, incinerator, or landfill area.

GENERAL ASEPTIC PROCEDURES

Handwashing Methods

To be effective, handwashing requires careful cleaning and scrubbing with soap or detergent and running water, with repeated applications of the soap and frequent rinses. A quick dampening or rinsing of the hands has no place in hospital work (or anywhere else for that matter). Handwashing should be a very careful procedure, employed frequently—for example, before eating, before pouring medications, before serving trays to patients, before and after general care given to any patient, after toilet, and after handling soiled dressings and bedpans. Sterile gloves provide additional protection in caring for patients with certain communicable diseases.

Surgical scrubs require considerably more effort than does routine handwashing. In this instance, germicidal substances are required to reduce the number of resident skin bacteria. Effective agents include providone-iodine and chlorhexidine, all in combination with a suitable detergent. All of these can be purchased as sponges impregnated with the germicidal soap and sealed in individual packs.

Skin Antisepsis

The skin cannot be truly sterilized. Some resident bacteria have their habitat around the hair follicles and sweat glands; it is impossible for a chemical agent (at least one that could be used on the skin) to have direct contact with all the bacteria present. Preoperative skin washing with germicidal soaps followed by an alcohol wash does nevertheless drastically reduce the number of resident bacteria and therefore the chance of postoperative infections.

Care of Instruments

Instruments used in surgical procedures should preferably be sterilized in an autoclave. This ensures safety by destroying spores as well as vegetative cells. However, certain instruments, such as those with sharp cutting edges, may be damaged in the autoclave; therefore, in these cases, chemical treatment is recommended.

Table 10.3 lists a few common chemicals used to disinfect or sterilize instruments. It is important to note that many of these solutions will kill vegetative bacteria and fungi in 10 to 30 min, but the complete destruction of endospores and some viruses may require 3 to 12 h. It should be emphasized that all instruments should be cleaned of any blood or body secretions before any disinfection procedures are undertaken.

Care of Dressings

Dressings to be used in surgical procedures should be sterilized in an autoclave. There is no substitute for this method. In the home, a pressure cooker (or, less

TABLE 10.3 Chemical Compounds Used to Disinfect Instruments

Smooth, hard objects	Ethyl alcohol (70–90%) Iodophors (100–500 ppm iodine) Sodium hypochlorite Ethylene oxide
Rubber tubing and catheters	Iodophors (100–500 ppm iodine) Ethylene oxide Aqueous phenolic solution (1–5%)
Reusable thermometers	Ethyl alcohol (70–90%) containing iodine
Inhalation and anesthesia equipment	Ethyl alcohol (70–90%) Aqueous glutaraldehyde (2%)

effectively, an oven) can be used. Sterile dressings should be handled in a manner that precludes any contamination of the area that comes in contact with the patient's wound.

Soiled dressings should be autoclaved or burned in an incinerator. They should be handled with forceps and wrapped in paper prior to burning, or placed in an autoclavable bag.

Care of Infectious Body Discharges

Discharge that may contain pathogens should be incinerated when this is practical. If it is not possible, sputum or other discharges should be mixed with an effective chemical agent such as phenol or cresol. Cresol or chlorine compounds, such as chloride of lime, can be used to disinfect feces. Remember, *disinfection is always slowed down by organic material;* hence, the chemical agent should be mixed carefully with the infectious discharge and allowed sufficient time to be effective.

Care of Syringes and Needles

Because disposable commercial packets of syringes, needles, tubing, and so on, are used in most hospitals, the dangers of transferring infectious agents present in the blood from one patient to another is minimal. Individuals handling used syringes, however, must exercise extreme care so as not to stick themselves inadvertently with used needles; such accidents could cause hepatitis or AIDS, in addition to many other infectious diseases.

ISOLATION TECHNIQUES AND PROCEDURES

A sterile, protective barrier around each patient would probably be the most effective method of protecting patients and hospital personnel involved in patient care. Such a procedure, however, is neither possible nor practical; it is, therefore, necessary to use various isolation policies and practices that will specifically suit the patient and control the disease in question. Based primarily on how the disease is spread, various isolation policies have been established. Cards developed by the Public Health Service giving information about these specific isolation procedures are displayed prominently on the door of any hospital room housing a patient under an isolation procedure (see Figure 10.1). Those diseases requiring one of the isolation techniques described on the reverse side of each card are also shown in Figure 10.1.

INFECTION CONTROL COMMITTEE

The Joint Commission on Accreditation of Hospitals requires that each hospital establish a committee on

---

Figure 10.1 Hospital precaution signs used for the protection of patients, staff, and visitors.

CARD DESCRIBING ENTERIC PRECAUTIONS

ENTERIC PRECAUTIONS

Visitors—Report to Nurses' Station Before Entering Room

1. Private Room—*necessary for children only.*
2. Gowns—must be worn by all persons having direct contact with patient.
3. Masks—not necessary.
4. Hands—must be washed on entering and leaving room.
5. Gloves—must be worn by all persons having direct contact with patient or with articles contaminated with fecal material.
6. Articles—special precautions necessary for articles contaminated with urine and feces. Articles must be disinfected or discarded.

Front

Diseases Requiring Enteric Precautions*

1. Cholera
2. Diarrhea—acute illness with suspected infectious etiology
3. Enterocolitis, staphylococcal
4. Gastroenteritis caused by
   a. Enterotoxic (enteropathogenic) *Escherichia coli*
   b. Salmonella species
   c. Shigella species
   d. *Yersinia enterocolitica*
5. Hepatitis, viral type A or type B
6. Typhoid fever

* See *Isolation Techniques for Use in Hospitals* for details and recommended duration of isolation.

Back

(Continued on p. 156)

Figure 10.1 (*Continued from p. 155*)

CARD DESCRIBING RESPIRATORY ISOLATION

RESPIRATORY ISOLATION

Visitors—Report to Nurses' Station Before Entering Room

1. Private Room—*necessary;* door must be kept closed.
2. Gowns—not necessary.
3. Masks—must be worn unless person entering room is not susceptible to the disease.
4. Hands—must be washed on entering and leaving room.
5. Gloves—not necessary.
6. Articles—those contaminated with secretions must be disinfected.

Front

Diseases Requiring Respiratory Isolation*

1. Measles (rubeola)
2. Meningitis, meningococcal
3. Meningococcemia
4. Mumps
5. Pertussis (whooping cough)
6. Rubella (German measles), except congenital rubella syndrome
7. Tuberculosis, pulmonary—sputum positive or suspect

* See *Isolation Techniques for Use in Hospitals* for details and recommended duration of isolation.

Back

CARD DESCRIBING PROTECTIVE ISOLATION

PROTECTIVE ISOLATION

Visitors—Report to Nurses' Station Before Entering Room

1. Private Room—*necessary;* door must be kept closed.
2. Gowns—must be worn by all persons entering room.
3. Masks—must be worn by all persons entering room.
4. Hands—must be washed on entering and leaving room.
5. Gloves—must be worn by all persons having direct contact with patient.
6. Articles—see *Isolation Techniques for Use in Hospitals.*

Front

Conditions That May Require Protective Isolation*

1. Agranulocytosis
2. Certain patients with extensive noninfected burns
3. Dermatitis—noninfected vesicular, bullous, or eczematous disease when severe and extensive
4. Certain patients receiving immunosuppressive therapy
5. Certain patients with lymphomas and leukemia

* See *Isolation Techniques for Use in Hospitals* for details and recommended duration of isolation.

Back

CARD DESCRIBING STRICT ISOLATION

STRICT ISOLATION

Visitors—Report to Nurses' Station Before Entering Room

1. Private Room—*necessary;* door must be kept closed.
2. Gowns—must be worn by all persons entering room.
3. Masks—must be worn by all persons entering room.
4. Hands—must be washed on entering and leaving room.
5. Gloves—must be worn by all persons entering room.
6. Articles—must be discarded, or wrapped before being sent to Central Supply for disinfection or sterilization.

Front

Diseases Requiring Strict Isolation*

1. Anthrax, inhalation
2. Burn wound, extensive, infected with
   a. *Staphylococcus aureus,* or
   b. Group A streptococcus
3. Congenital rubella syndrome
4. Diphtheria
5. Disseminated neonatal *Herpesvirus hominis* (herpes simplex)
6. Plague, pulmonic
7. Pneumonia, infected with
   a. *Staphylococcus aureus,* or
   b. Group A streptococcus
8. Rabies
9. Skin infection, extensive, with
   a. *Staphylococcus aureus,* or
   b. Group A streptococcus
10. Vaccinia
    a. Generalized and progressive
    b. Eczema vaccinatum
11. Varicella and disseminated herpes zoster

* See *Isolation Techniques for Use in Hospitals* for details and recommended duration of isolation.

Back

WOUND AND SKIN PRECAUTIONS

Visitors—Report to Nurses' Station Before Entering Room

1. Private Room—desirable.
2. Gowns—must be worn by all persons having direct contact with patient.
3. Masks—not necessary except during dressing changes.
4. Hands—must be washed on entering and leaving room.
5. Gloves—must be worn by all persons having direct contact with infected area.
6. Articles—special precautions necessary for instruments, dressings, and linen.

Note: See *Isolation Techniques for Use in Hospitals* for Special Dressing Techniques to be used when changing dressings.

Front

Diseases Requiring Wound and Skin Precautions*

1. Burns with excessive purulent drainage except those with
   a. *Staphylococcus aureus,* or
   b. Group A streptococcus
2. Gas gangrene
3. Herpes zoster
4. Melioidosis
5. Plague, bubonic

6. Skin infection, extensive, that cannot be covered by a dressing except those infected with
   a. *Staphylococcus aureus,* or
   b. Group A streptococcus
7. Wound infection with excessive purulent drainage that cannot be covered by a dressing

* See *Isolation Techniques for Use in Hospitals* for details and recommended duration of isolation.

Back

Figure 10.1 (*Continued from p. 156*)

infection control. This committee should be headed by a hospital epidemiologist or infection control nurse and should include representatives from the hospital administration, nursing staff, and clinical departments. The general overall responsibilities of such a committee are to establish and operate a practical system for reporting and evaluating infections in patients, personnel, and discharged patients. This includes distinguishing nosocomial infections, identifying the source and method of transmission of each nosocomial infection, and preparing reports that are pertinent to infection control. Figure 10.2 and Table 10.4 are summaries of a nosocomial infection surveillance report from a large teaching hospital.

Many hospitals designate an infection control nurse who shares responsibility with the infection control officer. From the information presented, it can be seen that there is an increasing need for nurse or medical technician epidemiologists in hospitals today. In fact, it is important now more than ever that all allied health personnel understand the principles of surgical and medical asepsis, because only the proper practice of these established techniques can control the incidence of nosocomial infections.

Figure 10.2 Monthly nosocomial infection surveillance report from a large university medical school teaching hospital.

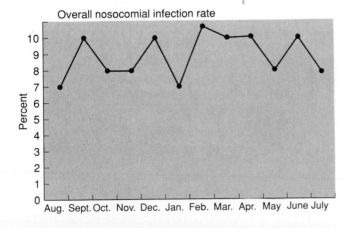

Overall nosocomial infection rate

TABLE 10.4 Mean Rates of Nosocomial Infcotions over a Three-Year Period in Each Service of a Large Teaching Hospital

Service	Mean Rate
General surgery	11%
Gynecology	4%
Medicine	7%
Neurology	5%
Neurosurgery	8%
Obstetrics	2%
Ophthalmology	0.5%
Otolaryngology	2%
Orthopedics	7%
Pediatrics	4%
Plastic surgery	11%
Burn unit	74%
Urology	8%

SUMMARY

Infections acquired during a hospital stay are called nosocomial infections. Six to ten percent of all patients admitted to a hospital in the United States acquire a nosocomial infection. This high rate stems from the fact that many disease-producing microorganisms are brought into a hospital by sick patients, and many patients undergo surgical procedures or are involved in drug therapies that leave them particularly susceptible to infection.

Surgical asepsis involves skin care to reduce bacterial numbers as well as the use of sterile equipment and sterile dressings. It is also concerned with the correct disposal of contaminated equipment and dressings to minimize the possibility of infecting other patients.

Medical asepsis is directed at the control of the environment to protect the patient. It involves aseptic procedures in handwashing as well as care of instruments, thermometers, syringes, needles, and other equipment.

Each hospital must establish isolation procedures to prevent the spread of infectious organisms to uninfected patients and hospital personnel. This is monitored by an infection control committee headed by an infection control officer and, in many cases, a nurse epidemiologist. It is the function of this committee to recognize nosocomial infections, identify the source of such infections, and prepare reports that are pertinent to infection control.

QUESTIONS FOR REVIEW

1. Define *medical asepsis* and *surgical asepsis*. How do they differ? Give an example of each.
2. What is an opportunist?
3. What is a carrier? Why is a carrier dangerous?
4. Find out all you can about how contaminated material is disposed of in a hospital. Can you suggest better ways to dispose of such material?
5. What is a nosocomial infection? How prevalent are nosocomial infections in U.S. hospitals?
6. What kinds of isolation procedures are used for hospital patients? Give several examples for each type of isolation procedure.
7. What are the duties of an infection control officer? A nurse epidemiologist?

SUPPLEMENTARY READING

American Hospital Association: *Infection Control in the Hospital,* 3rd ed. Chicago, American Hospital, 1974.

Bender GA: *Great Moments in Medicine.* Detroit, Northwood Institute Press, 1966.

Castle M: *Hospital Infection Control.* New York, Wiley, 1980.

McInnes B: *Controlling the Spread of Infection,* 2nd ed. St Louis, Mosby, 1977.

Perkins JJ: *Principles and Methods of Sterilization in Health Sciences,* 2nd ed. Springfield, IL, Charles C Thomas, 1982.

Spaulding EH, Groschel DHM: Hospital disinfectants and antiseptics. In: *Manual of Clinical Microbiology,* 4th ed. Lennette et al (eds), Washington, DC, American Society for Microbiology, 1985.

Chapter 11

Antimicrobial Agents in Therapy

OBJECTIVES Study of this chapter should provide a clear understanding of

1 The mechanism of action of the sulfonamides, trimethoprim, and isoniazid.

2 How various antibiotics inhibit cell wall synthesis.

3 The mechanism whereby penicillinase destroys penicillin.

4 How various antibiotics inhibit protein synthesis.

5 How polymyxin, nystatin, and amphotericin B affect cell membranes.

6 Antibiotics that bind to nucleic acids or inhibit their transcription.

7 The mechanism of transfer of drug resistances from one organism to another.

8 The problems associated with the use of subtherapeutic levels of antibiotics in animal feeds.

9 Laboratory tests for antibiotic susceptibility.

10 How some of the antiviral compounds exert their effects.

Sick people have been treated with herbs, bark, weeds, and a thousand other concoctions. However, there was no systematic approach to this problem of **chemotherapy** (treatment of disease with chemical compounds) until a series of experiments carried out in 1908 by Paul Ehrlich led to the development of an agent that would selectively destroy certain bacteria without serious injury to the host cells. This particular agent was called by several different names (e.g., arsphenamine, salvarsan, and 606), and it was used for many years for the successful treatment of syphilis. (The 606 referred to the fact that arsphenamine was the compound finally synthesized after 605 failures.)

Since 1935, a large number of chemotherapeutic agents have been developed. Many of these compounds are prepared synthetically in the laboratory, while others occur as byproducts of the metabolic activity of bacteria or fungi. This latter group of chemotherapeutic agents has been given the general name of *antibiotics*. The list of antibiotics that have been isolated and characterized is lengthy. Unfortunately, many of those that occur lack practical value because they are too toxic to the host to be used for the treatment of infectious diseases.

For any chemotherapeutic agent to be useful against infectious disease, several criteria should be met:

1. The drug should be low in toxicity to the host cells while destroying or inhibiting the disease agent. In other words, it must demonstrate a **selective toxicity** for the disease agent.
2. The host should not become allergic (hypersensitive) to it.
3. The organism should not readily become resistant to the drug.
4. The host should not destroy, neutralize, or excrete the drug too rapidly.
5. The drug should reach the site of infection.

As you will learn in this chapter, these criteria constitute an idealized list of properties for a chemotherapeutic drug. Even the best and most widely used drugs do not have all of these properties. For example, many drugs demonstrate some toxicity to the host (such as upset stomach, diarrhea, fever), and many individuals do become allergic to antibiotics. Let us see how chemotherapeutic agents exert their selective killing effect.

CHEMICAL SYNTHESIS OF CHEMOTHERAPEUTIC AGENTS

No one could have predicted in 1935 that the discovery of the sulfonamides would usher in an era of medicine in which the treatment of infectious diseases would be revolutionized.

Sulfonamides

The preparation that led to the discovery of the sulfonamides was a red dye known as Prontosil. In 1933 Prontosil was reported to cure a normally fatal bloodstream infection caused by staphylococci and to be highly effective against streptococcal infections as well. Subsequent research showed that the active part of Prontosil was actually sulfanilamide. Although it was effective against many microorganisms, it produced some toxic side effects. Dissatisfied, researchers sought to develop other sulfonamide compounds, some of which are considerably less toxic than the parent compound (see Figure 11.1).

The sulfonamides are outstanding in absorption properties, being absorbed almost completely when taken by mouth. However, they are excreted in the urine; and since they have a tendency to crystallize in an acid environment, they can damage the kidneys.

Figure 11.1 Structural formulas for some of the representative sulfonamides that are used as chemotherapeutic agents.

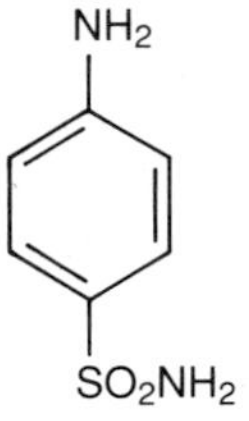

Sulfanilamide

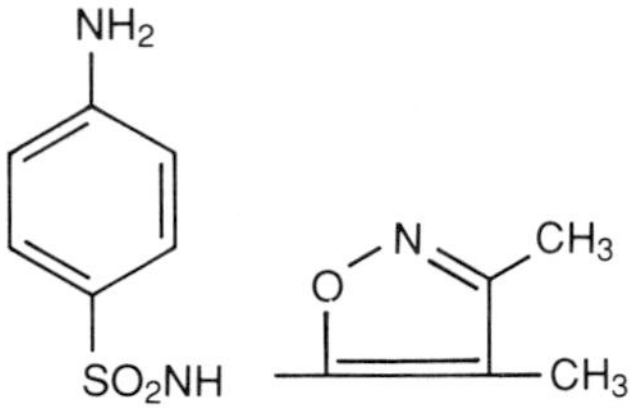

Sulfisoxazole (Gantrisin)

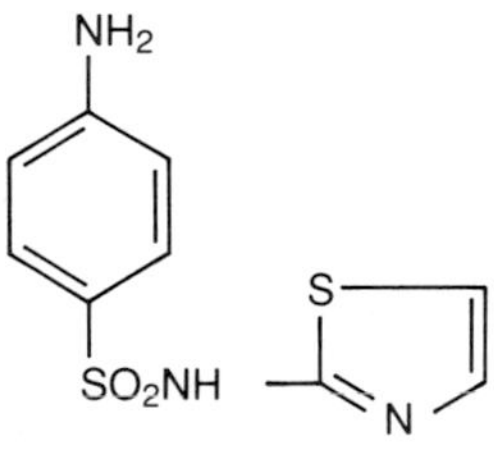

Sulfasuxidine

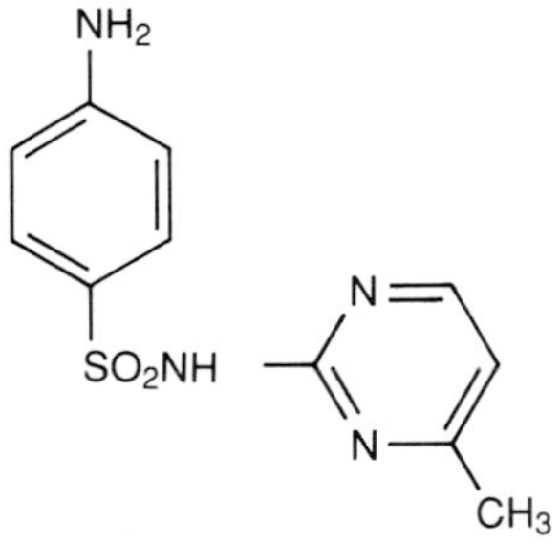

Sulfadiazine

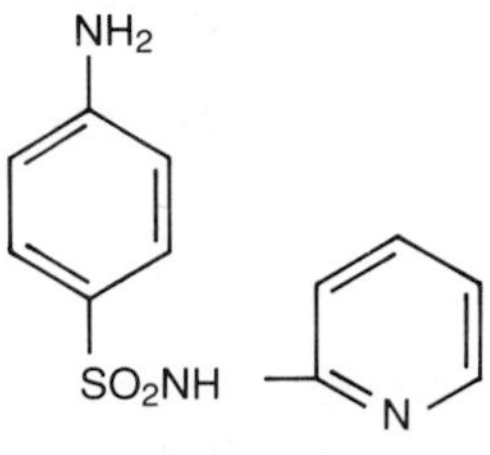

Sulfapyridine

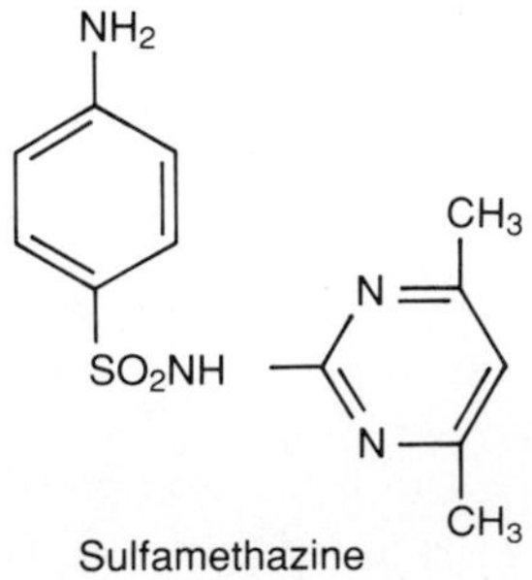

Sulfamethazine

Sulfisoxazole (Gantrisin) is a more soluble derivative and for this reason has been used for urinary tract infections.

The sulfonamides are effective in the treatment of diseases such as lymphogranuloma venereum and psittacosis. Other susceptible bacteria include gram-positive organisms such as the streptococci, the staphylococci, and the pneumococci and gram-negative organisms such as *Escherichia coli, Haemophilus influenzae, Enterobacter aerogenes,* and *Proteus vulgaris.*

The action of the sulfonamides is described as **competitive inhibition.** What actually happens can be explained most satisfactorily as follows: In order for the bacterium to synthesize folic acid (one of the B vitamins necessary for growth), it has to synthesize and put together several large molecules. One of these molecules, which makes up part of the folic acid molecule, is named paraaminobenzoic acid (frequently called PABA). The chemical structure of PABA, as shown below, is very similar to that of sulfanilamide:

$$NH_2 \qquad NH_2$$

COOH	SO_2NH_2
PABA	Sulfanilamide

Because of this similarity, the condensing enzyme attaches itself to the sulfanilamide instead of PABA (when the sulfanilamide is present in sufficient amounts) but then cannot make it fit into the folic acid molecule. Therefore, the cell is not able to make folic acid and hence is unable to grow. It must be kept in mind that this is *not* a killing effect, and treatment must be continued long enough to allow the body defenses to destroy and get rid of the infecting organism. Only organisms that synthesize their own folic acid are affected by the action of the sulfonamides. Resistance to sulfonamides occurs frequently, particularly in gram-negative organisms such as *E. coli, Neisseria gonorrhoeae,* and *N. meningitidis.* There are several mechanisms for resistance. Plasmid-coded resistance results from production of a new condensing enzyme that has a much lower affinity for sulfonamides relative to PABA. Chromosomally encoded mechanisms of resistance include decreased uptake, overproduction of PABA, or alteration of the normal enzyme so that sulfonamide binding is reduced.

Although most of the sulfonamides are readily absorbed through the intestines, there are several that are not absorbed. These are used to treat certain enteric infections and also for intestinal antisepsis in preparation for intestinal tract surgery. Sulfonamides that are not readily absorbed include succinylsulfathia-zole (Sulfasuxidine) and phthalylsulfathiazole (Sulfathalidine).

Trimethoprim

Trimethoprim inhibits another step in the metabolism of folic acid by binding to the enzyme dehydrofolate reductase. This enzyme, which reduces dihydrofolate to tetrahydrofolate, is essential for the biosynthesis of folic acid as well as being involved in the formation of thymine.

There are a number of dihydrofolate reductase inhibitors, including such drugs as aminopterin and amethopterin, each of which is inhibitory to both bacterial and eucaryotic cells. Trimethoprim, however, inhibits the bacterial form of this enzyme but does not significantly bind to the mammalian enzyme.

Trimethoprim acts synergistically with the sulfonamides since these drugs inhibit sequential steps in the same metabolic pathway (see Figure 11.2); therefore, it is frequently used in combination with sulfa drugs. Plasmid-coded resistance to trimethoprim results from the production of a dihydrofolate reductase that binds the drug less efficiently.

Ethambutol

Ethambutol, a synthetic drug, is used only for the treatment of tuberculosis, because other organisms are resistant to it. If administered alone, ethambutol-resistant strains of *M. tuberculosis* would soon become prevalent. It is therefore given together with isoniazid for the treatment of tuberculosis. Its precise mechanism of action is not known, but there is some evidence that it inhibits mycobacterial RNA synthesis.

Isonicotinic Acid Hydrazide

Isonicotinic acid hydrazide is frequently referred to as isoniazid, or INH. Like ethambutol, it demonstrates a highly selective activity against *M. tuberculosis,* although its effect against most other bacteria is negligible. In the treatment of tuberculosis, it is usually given concurrently with either streptomycin or ethambutol to reduce the development of resistant strains

Figure 11.2 Sequential steps in folic acid synthesis blocked by sulfonamides and trimethoprim.

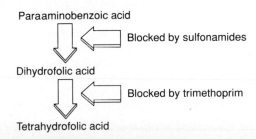

Paraaminobenzoic acid

Blocked by sulfonamides

Dihydrofolic acid

Blocked by trimethoprim

Tetrahydrofolic acid

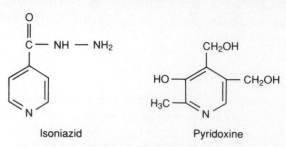

Figure 11.3 Competitive inhibition of pyridoxine biosynthesis is the result of the presence of a structurally related analogue, isoniazid.

of *M. tuberculosis.* INH is structurally similar to the B vitamin pyridoxine (Figure 11.3). As a result, its mode of action is considered to be competitive inhibition of pyridoxine-catalyzed reactions.

ANTIBIOTICS IN CHEMOTHERAPY

In 1929, Alexander Fleming reported that he had isolated a mold *(Penicillium notatum)* that secreted a substance inhibiting the growth of *Staphylococcus aureus.* The basic observation remained an idle curiosity for the following 10 years until the British scientist Howard Florey and his colleagues worked out the procedures to isolate and purify this unusual inhibitory mold product, subsequently named penicillin. By

1941, they produced enough penicillin to treat the first patient, a British policeman who had both a staphylococcal and a streptococcal infection. However, it took another 5 to 6 years of work by this group in the United States before the commercial production of penicillin achieved sufficient size to make it generally available.

In 1941, Selman Waksman and his associates reported the isolation of another antibacterial product, later named streptomycin, that was synthesized by the procaryotic actinomycete *Streptomyces griseus.* Probably no one, however, could have predicted at that time that the discovery of penicillin and streptomycin would usher in the greatest change in the treatment of infectious diseases since the advent of specific serum therapy. The next 30 years saw the isolation of literally hundreds of mold and bacterial products that exhibited a selective toxicity to many infectious agents but showed little or no toxicity toward mammalian cells. These metabolic products, termed **antibiotics,** have revolutionized the treatment of infectious diseases. The remainder of this chapter will be devoted to a discussion of some of the major antibiotics in use today, along with a discussion of how they can selectively kill bacteria or inhibit bacterial growth. It should be pointed out, though, that it is beyond the scope of this book to discuss every antibiotic used for the treatment of infectious diseases. Rather, we shall discuss selective groups of antibiotics based on their mechanism of antibacterial action, which can be categorized as follows: (1) inhibition of cell wall synthesis, (2) inhibition of protein synthesis, (3) injury to cell mem-

MICROBIOLOGY MILESTONES

During the hundreds of thousands of years that humans have inhabited the earth, it is only in the past century that there has been any successful use of chemotherapy for the treatment of infectious disease. Like Ponce de Leon's search for the fountain of youth, chemists dreamed of a "magic bullet" that would strike the invading disease-causing organisms without harm to the human host.

Paul Ehrlich was one of the early pioneers in this endeavor and his many achievements culminated in 1910 with the synthesis of a drug named salvarsan, the first reasonably nontoxic and effective treatment for syphilis. The next 25 years, however, yielded not a single drug having significant activity for the treatment of infectious diseases, and most medical personnel regarded "chemotherapy" as an impractical dream. Then came the synthesis of a dye named Prontosil by Gerhard Domagk in 1935, followed by the discovery that its antibacterial action was due to the sulfonamide moiety making up its structure. These results brought back into fashion a new wave of enthusiasm for chemotherapy.

Someone then remembered a paper published in 1929 by Dr. Alexander Fleming in which he described a mold that had contaminated a culture of staphylococci growing in a petri dish. Fleming reported that the presence of the mold prevented growth of the bacteria in the area surrounding the mold colony. He cultivated the mold and demonstrated that extracts of the culture medium prevented the growth of a number of other bacterial species. He named the growth-inhibiting substance, penicillin, and went on to demonstrate that it was "nontoxic to animals in enormous doses," suggesting that "it may be an efficient antiseptic for application to, or injection into, areas infected with penicillin-sensitive microbes."

The rest is history. Through the cooperative efforts of Howard Florey and Ernst Chain, plus many other collaborators, penicillin was eventually purified, opening the door to a revolution in the treatment of infectious diseases.

brane, and (4) inhibition of either DNA or RNA synthesis. Our discussions are designed to cover the general aspects of antibiotic action but, as you will see, any real comprehension of this field requires some detailed knowledge of the target reactions blocked by antibiotics.

Antibiotics That Inhibit Cell Wall Synthesis

A view of Chapter 3 will remind the reader that most procaryotic cells are surrounded by a rigid cell wall composed of a polymer of two carbohydrates. *N*-acetylglucosamine and *N*-acetylmuramic acid, as well as a small number of amino acids. The synthesis of this cell wall, termed **peptidoglycan**, involves a number of enzymatic steps, many of which are blocked by antibiotics. Since mammalian cells do not synthesize peptidoglycans, they do not possess an analogous biosynthetic pathway. So let us look at some of the biosynthetic reactions involved in peptidoglycan synthesis and see which steps are blocked by the antibiotics that inhibit cell wall synthesis.

Fosfomycin

One of the first steps in the synthesis of peptidoglycan is the production of the unique carbohydrate *N*-acetylmuramic acid. This is accomplished by enzymatically joining phosphoenolpyruvate to *N*-acetylglucosamine, as follows:

CH₂OH

N-Acetylglucosamine

$+$ $CH_2 = C - COOH$
$\quad\quad\quad | $
$\quad\quad\quad OPO_3H_2$

Phosphoenolpyruvate

$\longrightarrow$

CH₂OH

NHCOCH₃

O

HC — CH₃

COOH

N-Acetylmuramic acid

Fosfomycin is a structural analogue of phosphoenolpyruvate and consequently binds irreversibly to the muramic-acid-synthesizing enzyme, thus blocking the synthesis of muramic acid.

Fosfomycin is termed a **broad-spectrum antibiotic,** since it is bactericidal against both gram-positive and gram-negative bacteria. Resistance occurs primarily through a mutational loss of the transport system by which the antibiotic enters the cell.

Penicillin

We owe the discovery of penicillin to a fortunate accident in which a plate culture of *S. aureus* became contaminated with a mold, later identified as *P. notatum.* The clearing zone around the mold colony showed that the mold was producing some compound that was causing the bacteria to disintegrate. We now know that this was the antibiotic penicillin; today, however, the term *penicillin* refers not to a single product but to a group of very similar substances.

All the penicillins have the very distinct advantage of exerting a killing (bactericidal) action on susceptible organisms. However, the large variety of penicillins available for treatment is frequently confusing. It should be kept in mind that each preparation of penicillin possesses certain properties not shared by the other types. For example, benzyl penicillin is quick-acting but of short duration, requiring frequent injections to maintain high blood levels. The short duration of bactericidal activity stems from the drug's rapid excretion in the urine. A drug called probenecid (which does not itself possess any antibacterial activity) is frequently given concurrently with penicillin because probenecid blocks the excretion of penicillin, permitting higher serum levels and a longer duration of penicillin activity. If benzyl penicillin is mixed with equimolar amounts of procaine, a new penicillin called penicillin G procaine is formed. This preparation is not nearly as soluble as penicillin G alone and thus does not yield as high a concentration in the blood. However, it does maintain a lower concentration over a 24- to 96-h period, which is effective for many infections. Another long-acting penicillin is benzathine penicillin; one injection is effective for the prevention of rheumatic fever recurrences several weeks.

Penicillin (in its natural form) has several limitations that can be enumerated briefly as follows: (1) it is easily destroyed by stomach acids and, therefore, is of only limited value when taken orally; (2) it is destroyed by penicillinase, the enzyme formed by many bacteria (most notably the *Staphylococcus*) that makes them resistant to the effect of penicillin; and (3) it is, for the most part, effective only against gram-positive bacteria (although notable exceptions are the organisms causing gonorrhea, epidemic meningitis, and syphilis).

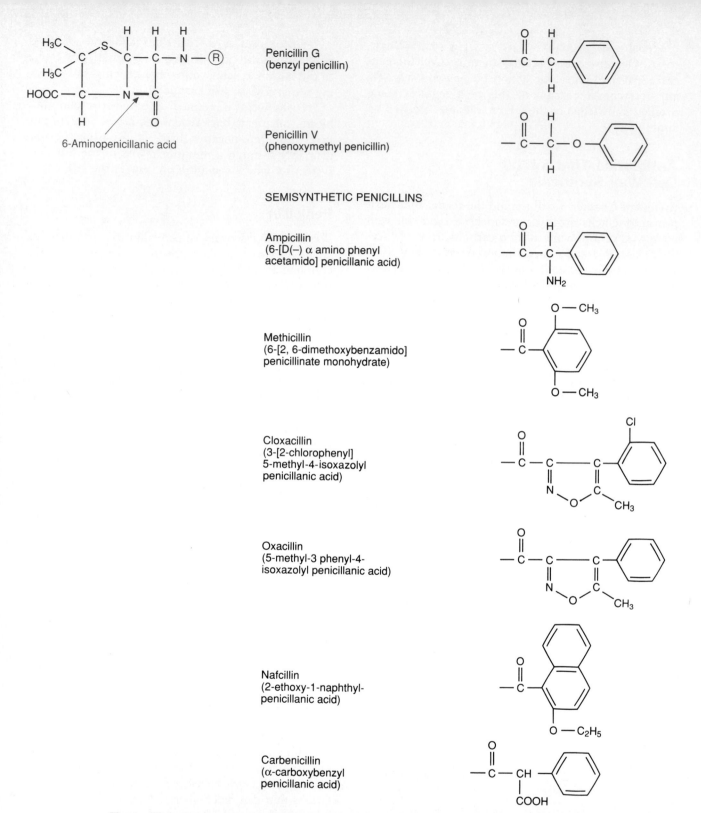

Penicillin G
(benzyl penicillin)

Penicillin V
(phenoxymethyl penicillin)

SEMISYNTHETIC PENICILLINS

Ampicillin
(6-[D(−) α amino phenyl
acetamido] penicillanic acid)

Methicillin
(6-[2, 6-dimethoxybenzamido]
penicillinate monohydrate)

Cloxacillin
(3-[2-chlorophenyl]
5-methyl-4-isoxazolyl
penicillanic acid)

Oxacillin
(5-methyl-3 phenyl-4-
isoxazolyl penicillanic acid)

Nafcillin
(2-ethoxy-1-naphthyl-
penicillanic acid)

Carbenicillin
(α-carboxybenzyl
penicillanic acid)

Figure 11.4 Structures of some natural and semisynthetic penicillins. Arrow indicates β-lactam bond that is hydrolyzed by penicillinase and (R) represents the various side groups present on the different natural and semisynthetic penicillins.

In recent decades, much research has been directed toward the problem of chemically overcoming the disadvantages of penicillin without losing its desirable features. Seemingly this has now been accomplished, at least in part, in two different ways. In the first case, one can alter the type of penicillin produced by the fungus by controlling the culture medium. For example, the addition of phenoxyacetic acid to the medium results in the production of phenoxymethyl penicillin, or, as it is usually called, penicillin V. Penicillin V is much more resistant than benzyl penicillin to the effects of acids and is therefore more effective when taken orally.

The real breakthrough came when it was discovered that all penicillins possess the same basic molecular structure and differ only in the types of side chains attached to this structure. It then became possible to produce semisynthetic penicillins either by interrupting the synthesis of penicillin so that only 6-amino penicillanic acid was produced or by enzymatically removing the side groups that the mold had attached to the 6-amino penicillanic acid and then chemically adding other side chains. Figure 11.4 shows the structures of a few natural and semisynthetic penicillins. The major advantage of the semisynthetic methicillin is its high resistance to the enzyme penicillinase. Penicillinase destroys penicillin by hydrolyzing the N—

CO bond (see arrow in Figure 11.4) of the β-lactam ring. The fact that penicillinase is not effective against methicillin makes this antibiotic extremely valuable for the treatment of organisms that produce penicillinase.

Ampicillin has overcome the problem of the limited spectrum of activity of the normal mold product. Ampicillin has the bactericidal properties of normal penicillin, but it is called a broad-spectrum antibiotic because it is effective against many gram-negative as well as gram-positive bacteria.

Chemistry of Penicillin To understand how penicillin exerts its bactericidal effect, it is necessary to know about the synthesis of the cell wall peptidoglycan. As you might surmise, anything that interferes with peptidoglycan synthesis will weaken the wall and cause the cell membrane to burst and liberate the cell contents. It may help to visualize the structure of peptidoglycan as similar to that of a barrel. In this analogy the long staves would be a repeating polymer of N-acetylglucosamine and N-acetylmuramic acid (see Figure 11.5). However, a barrel would not be able to hold together unless there were hoops around the staves. In peptidoglycan these hoops are composed of short peptides of amino acids, as represented by both the black and color circles in Figure 11.5. During growth, the bacterial cell synthesizes units of the cell

Figure 11.5 Schematic structure of the overall peptidoglycan of the cell wall of *S. aureus*. In this representation X represents *N*-acetylglucosamine and Y represents *N*-acetylmuramic acid. Color circles represent the four amino acids of the tetrapeptide L-alanyl-D-isoglutaminyl-L-lysyl-D-alanine. Black circles are pentapeptide bridges that interconnect polymers of *N*-acetylglucosamine and *N*-acetylmuramic acid. The shaded portion represents that part of the molecule that is transported to the cell wall. The arrow points to the peptide bond, the formation of which is inhibited by penicillin.

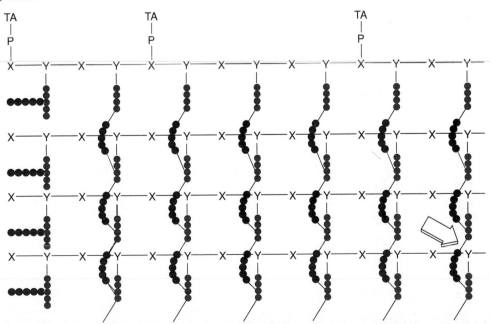

wall made up of (in the case of *S. aureus*) one *N*-acetylglucosamine and one *N*-acetylmuramic acid plus 10 amino acids that will provide the hoops to our bacterial barrel. This unit (shaded in Figure 11.5) is then transported to the growing section of the cell wall.

After incorporation of the *N*-acetylglucosamine and the *N*-acetylmuramic acid into the existing wall, the last step is to cross-link these linear polymers (staves) with a bridge of amino acids (hoops) to provide the required rigidity and strength. The mechanism of this cross-linking is shown in Figure 11.6, where it can be seen that two linear polymers will be cross-connected by a process called **transpeptidization.** In this process the terminal glycine from one polymer replaces the terminal D-alanine from an adjacent polymer, which results in a completed cross bridge and the liberation of one molecule of D-alanine. This point of joining is shown by the arrow in Figure 11.6. Herein lies the primary reaction inhibited by penicillin. Penicillin has a structure similar to that of the terminal D-alanyl-D-alanine, which must undergo transpeptidization (see Figure 11.7). Thus the enzyme transpeptidase reacts with the penicillin and is not available to complete the pentaglycine bridges between the linear polymers of *N*-acetylglucosamine and *N*-acetylmuramic acid. To return to our analogy, we end up with a barrel with very few hoops, and the bacterium, like the barrel, will burst. The amino acids that make up this cross-connecting bridge will vary from one genus to another, but the mechanism of penicillin ac-

Figure 11.7 Stereomodels of penicillin and of *(a)* the D-alanyl-D-alanine end of the peptidoglycan strand. *(b)* Arrows indicate the position of the CO—N bond in the β-lactam ring of penicillin (left) and of the CO—N bond in the D-alanyl-D-alanine. Hydrolysis of the β-lactam bond of penicillin by penicillinase changes the structure of the penicillinase so that it is no longer an analog of D-alanyl-D-alanine.

tion remains the same, because no matter which amino acids make up the connecting bridge, the transpeptidization always involves the terminal D-alanyl-D-alanine.

It should be noted that penicillin will also bind to other proteins in a bacterium. There are three to seven such penicillin-binding proteins (PBPs) in most species, each of which is involved in a different step in peptidoglycan synthesis. In *E. coli*, for example, one PBP is involved in septum formation, another in cell shape, and still another for elongation or overall wall growth. All, however, possess PBP1B or its equivalent, which has transpeptidase activity, and it is the binding of penicillin to this PBP that results in the lysis of the bacterium.

Cephalosporins

Cephalosporin C was isolated in 1952 as a product of the fungus *Cephalosporium*. This antibiotic is a β-lactam antibiotic structurally similar in many ways to

Figure 11.6 Transpeptidization to join two linear peptidoglycan strands by forming an interpeptide bridge of pentaglycine.

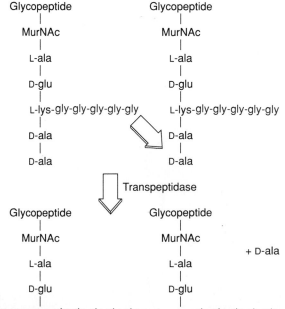

Glycopeptide | Glycopeptide
| |
MurNAc | MurNAc
| |
L-ala | L-ala
| |
D-glu | D-glu
| |
L-lys-gly-gly-gly-gly-gly | L-lys-gly-gly-gly-gly-gly
| |
D-ala | D-ala
| |
D-ala | D-ala

Transpeptidase

Glycopeptide | Glycopeptide
| |
MurNAc | MurNAc
| + D-ala
L-ala | L-ala
| |
D-glu | D-glu
| |

---gly-D-ala-L-lys-gly-gly-gly-gly-gly-D-ala-L-lys-gly-gly-gly-gly-gly---

penicillin. It was, in fact, superior to penicillin in that it was much more resistant to inactivation by penicillinase. Its major disadvantage, though, was that it was considerably less potent than penicillin.

Then, in 1961, a procedure was reported whereby the side groups of cephalosporin C could be removed chemically and, as with the semisynthetic penicillins, new groups could be added. This created a second generation of cephalosporins with greatly improved potency, particularly against gram-negative organisms. Figure 11.8 shows the structure of four of these semisynthetic cephalosporins.

In the late 1970s a third generation of cephalosporins came into being when chemists discovered how to exchange the atom of sulfur in the cephalosporin nucleus for an atom of oxygen. These cephalosporins, termed *oxacephalosporins* or *oxa-β-lactams*, have a much broader specificity and higher potency than did previous types of cephalosporins, particularly for gram-negative organisms. Major third-generation cephalosporins currently used included moxalactam, cefotaxime, cefoperazone, and ceftriaxone.

The mode of action of the cephalosporins appears to be identical to that of penicillin and, as one would guess, their action is bactericidal. They are inactivated by some β-lactamases (penicillinases) but are insensitive to others.

Synthetic Monobactams

During efforts to isolate new antibiotics produced by bacteria, a new class of monocyclic (possessing only one ring in their chemical structure) β-lactams were isolated. These antibiotics, given the general name of **monobactams,** have been isolated from such diverse genera as *Pseudomonas, Gluconobacter, Chromobacterium, Agrobacterium, Bacillus,* and others. They are highly active against most of the members of the Enterobacteriaceae, as well as against species in the genera *Neisseria* and *Haemophilus,* but have little or no activity against the facultative aerobic gram-positive bacteria or the anaerobic bacteria.

Alteration of the side chains of the monobactams has resulted in changes in their bioactivity as well as their resistance to hydrolysis by β-lactamases. Aztreonam, the first semisynthetic monobactam approved for use in patient trials, has proved to be highly successful for the treatment of a wide variety of infections caused by the facultative aerobic gram-negative bacteria. It has also proved effective for the elimination of the enterobacteria from the intestine when administered orally.

Bacitracin

Bacitracin is a polypeptide antibiotic produced by a spore-forming bacillus named *Bacillus licheniformis.* This antibiotic is bactericidal chiefly against gram-positive bacteria such as the staphylococci and streptococci. It is also effective against some gram-negative organisms such as the meningococci, the gonococci, and *Haemophilus influenzae.* Because of the toxic effect of bacitracin on the kidney, its use is limited. Bacitracin inhibits cell wall peptidoglycan synthesis by blocking the dephosphorylation (removal of phosphate) of a lipid carrier necessary to transport the

Figure 11.8 Structures of four "second-generation" cephalosporins. Arrow points to β-lactam bond. Shaded areas represent chemical groups that were added to the unshaded cephalosporin nucleus. "Third-generation" cephalosporins have substituted an oxygen atom for the sulfur atom.

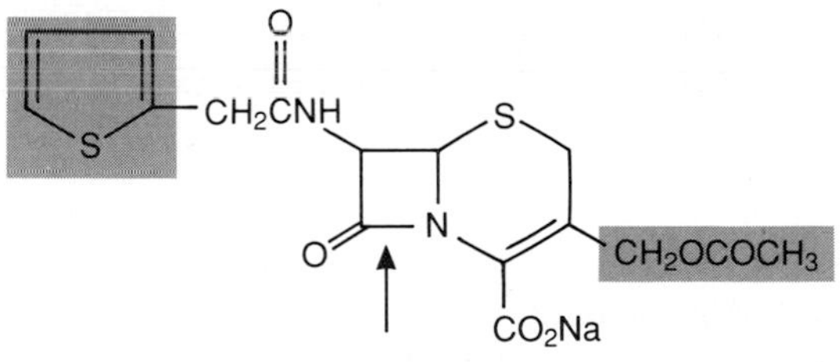

Cephalothin

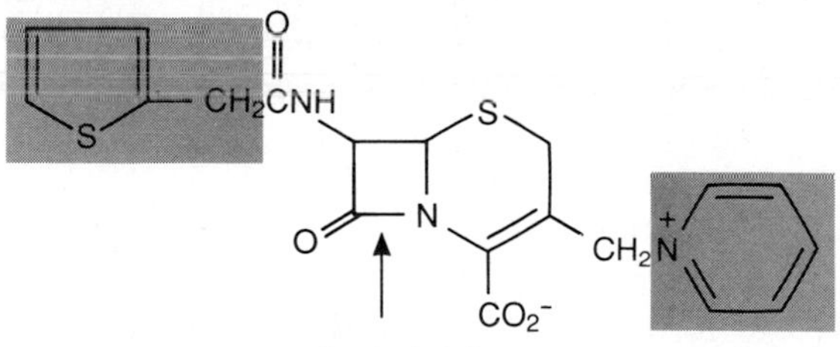

Cephaloridine

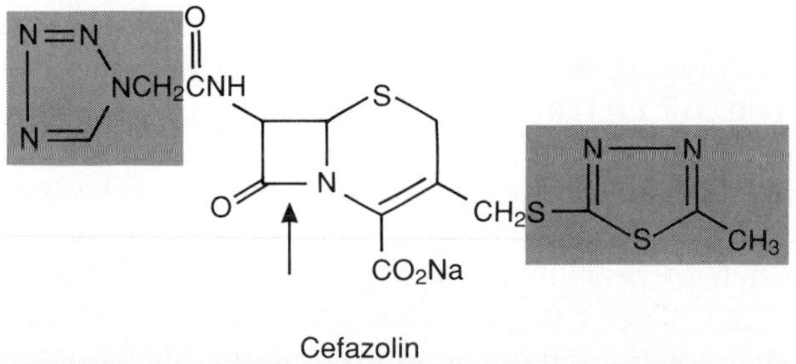

Cefazolin

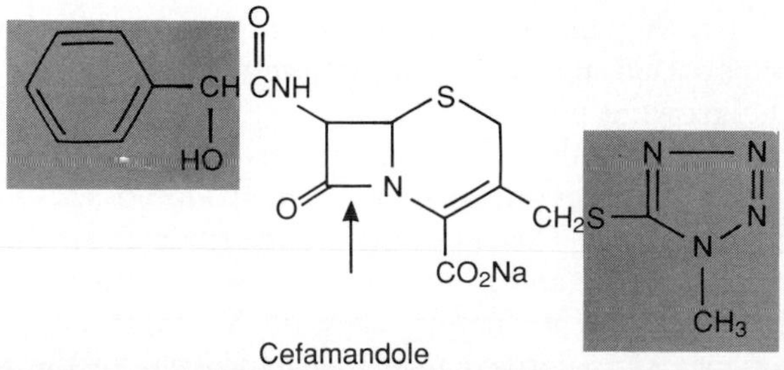

Cefamandole

newly formed cell wall components from the cytoplasmic membrane to the external wall. When this dephosphorylation is prevented (as in the case of cells treated with bacitracin), the lipid can no longer function as a carrier of the cell wall material and the cell dies in the same way as a cell killed by penicillin.

Vancomycin and Ristocetin

Both vancomycin and ristocetin are antibiotics bactericidal against gram-positive bacteria and spirochetes, and both act by inhibiting peptidoglycan synthesis. Their modes of action seem to be identical in that they bind tightly to the terminal D-alanyl-D-alanine present in the cell wall peptidoglycan and there prevent transpeptidization.

Antibiotics That Inhibit Protein Synthesis

Before we get into a discussion of how protein synthesis is blocked by antibiotics, we will review the major steps involved in protein synthesis. These were discussed in Chapter 7 and will only be briefly listed at this point.

1. The messenger RNA (mRNA) is transcribed from the DNA of the gene and carries the message from the DNA into the cytoplasm. In the cytoplasm are the structural units, the ribosomes, to which the mRNA must bind in order to be translated into protein. The ribosomes exist in the cytoplasm in two different subunits that combine to form a functional ribosome after reacting with an mRNA. The dissociated parts of the bacterial ribosome are called 30S and 50S subunits. The S refers to Svedberg unit, a measure of the rate at which the particles move when subjected to a centrifugal force. Thus, a 50S particle is heavier than a 30S particle, and when the two parts of the ribosome combine, they form a still heavier particle called a 70S ribosome. The mRNA binds to the 30S portion of the ribosome.

2. Each amino acid is first activated by reacting with adenosine triphosphate (ATP) and transfer RNA (tRNA) to form an amino acid-tRNA complex. This complex recognizes its specific codon on the mRNA and binds to the 50S portion of the ribosome.

3. A peptide bond is formed between the new amino acid and the carboxyl terminal amino acid of the growing peptide.

4. The ribosome then moves along the mRNA to the next codon, so that the next amino acid may be brought in. This movement of the ribosome is called translocation, and it occurs as a result of the enzyme translocase. The energy necessary for this ribosomal movement is derived from the hydrolysis of the terminal high-energy phosphate bond in guanosine triphosphate (GTP).

5. Finally, the ribosome reaches a termination codon on the mRNA, and the completed protein molecule is released from the ribosome.

As you can see, the blocking of any one of these steps would result in a cessation of protein synthesis and the eventual death of the cell. Let us see what steps are blocked by some of the common antibiotics in use today.

Chloramphenicol

Chloramphenicol is produced by the actinomycete *Streptomyces venezuelae,* which was originally isolated from a soil sample from Venezuela. Because of its comparatively simple structure, chloramphenicol is the only antibiotic completely synthesized chemically rather than isolated as a product of microbial metabolism.

Chloramphenicol is called a broad-spectrum antibiotic, since it is effective against both gram-positive and gram-negative bacteria. Unlike penicillin, its effect is bacteriostatic rather than bactericidal. It exerts this effect by reacting with the 50S portion of the ribosome, where it inhibits the enzyme peptidyl transferase. It is this enzyme that carries out step 3 by forming a peptide bond between the new amino acid, which is still attached to its tRNA, and the last amino acid of the growing peptide. As a result of this inhibition, all protein synthesis stops immediately.

The use of chloramphenicol is not without danger. In some people it suppresses the development of cells of the bone marrow and produces irreversible (and usually fatal) aplastic anemia (loss of stem cells that are precursors to red blood cells). Even though this is a rare complication, chloramphenicol is used only for serious infections caused by some anaerobes and for *H. influenzae* meningitis and typhoid fever. Plasmid-coded resistance results from the production of enzymes that alter the antibiotic so that it is no longer effective.

Macrolide Antibiotics

Members of the large family of macrolide antibiotics are characterized by possession of a large lactone ring in their structures. They include such antibiotics as erythromycin, oleandomycin, carbomycin, and spiromycin. All prevent protein synthesis by reacting with the 50S subunit of the ribosome, and all appear to possess a similar mechanism of inhibition. The bacterial spectrum of the macrolide antibiotics is similar to that of penicillin, and they are useful in treating patients allergic to penicillin. Macrolides are also effective for the treatment of mycoplasma infections.

Erythromycin is thought to interfere with the translocation step by blocking the movement of the ribosome. It is currently the drug of choice for the treatment of legionnaires' disease.

Lincomycin and Clindamycin

Like some other 50S subunit-binding antibiotics, lincomycin competes with chloramphenicol for a binding site on the ribosome. Thus, it also appears to inhibit peptide bond formation. Unlike chloramphenicol, however, lincomycin causes the breakdown of existing polysomes (multiple ribosomes on an mRNA molecule), and this results in the dissociation of the ribosomes into their 30S and 50S subunits.

Lincomycin is most effective against gram-positive bacteria and can replace penicillin in cases of allergy to penicillin. Because it is more toxic than erythromycin, it should be reserved for pathogens that are resistant to the macrolide antibiotics.

Clindamycin is a semisynthetic antibiotic that is synthesized from lincomycin. It apparently has a similar mode of action because it binds to the same receptor in the 50S subunits of the ribosome as erythromycin and lincomycin. Clindamycin is particularly effective for the treatment of infections caused by the obligately anaerobic bacteria, particularly those caused by *Bacteroides fragilis*, the most commonly encountered organism found in intraabdominal and pelvic infections.

The use of these two antibiotics is not without danger, though, because they may indirectly induce a very serious ulcerative colitis (intestinal ulceration) by permitting the overgrowth of a toxin-producing strain of *Clostridium difficile*.

Tetracyclines

The tetracyclines are a family of related antibiotics differing in the identity of several side chains, as shown in Figure 11.9. All have a very broad antibacterial spectrum, exerting a bacteriostatic effect on all bacteria except the mycobacteria. They inhibit protein synthesis by binding to the 30S subunit of the ribosome, thereby preventing the attachment of the amino-acid-carrying tRNA. Interestingly, the tetracyclines inhibit protein synthesis on both bacterial and eucaryotic ribosomes, but bacterial growth is selectively inhibited because bacteria accumulate high concentrations of these drugs by a transport process absent from eucaryotic cells.

Undesirable side effects include inhibition of much of the normal intestinal flora, resulting in the outgrowth of drug-resistant organisms. In addition, some of the tetracyclines are deposited in growing bones and teeth, causing staining and possible structural impairment, thus precluding their use during pregnancy or childhood.

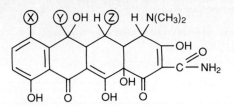

Antibiotic	X	Y	Z
Tetracycline	–H	–CH₃	–H
Oxytetracycline	–H	–CH₃	–OH
Chlortetracycline	–Cl	–CH₃	–H
Minocycline	–N(CH₃)₂	–H	–H

Figure 11.9 Structure of several important tetracyclines.

As with other antibiotics, the appearance of resistant organisms is a serious problem; nonetheless, the tetracyclines remain a very important collection of drugs for the treatment of infections caused by β-lactamase-producing bacteria as well as rickettsial and chlamydial infections.

Aminoglycoside Antibiotics

Streptomycin, one of the aminoglycoside antibiotics, exerts a bactericidal effect on a large number of gram-positive and gram-negative organisms. As a result of its very basic guanidino groups (see Figure 11.10), streptomycin causes a variety of nonspecific effects such as an efflux of potassium from cells or, in larger amounts, actual agglutination of the bacteria. However, its specific bactericidal effect depends on its ability to bind specifically to one of the proteins in the 30S subunit of the ribosome. This binding results in two major effects on protein synthesis: (1) it causes a misreading of the mRNA and (2) it prevents the movement of the ribosome after it has bound to the first amino acid to make up the protein. The result is that the mRNA binds to only a single ribosome at its initiation site, and that ribosome cannot move.

One might wonder which of these effects exerts the bactericidal action of streptomycin. The misreading has been shown to occur using synthetic polynucleotides as mRNAs. For example, it has been shown that, in the presence of streptomycin, an isoleucine is inserted instead of phenylalanine. Such misreading undoubtedly occurs in an in vivo situation and might contribute to the lethality of streptomycin, but it is generally assumed that the irreversible binding of streptomycin causing a freezing of the initiation complex accounts for most of its bactericidal activity.

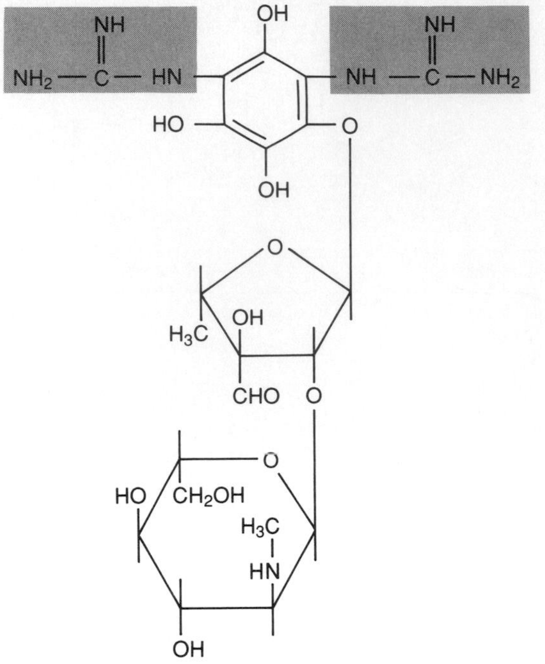

Figure 11.10 Structure of streptomycin. The two boxed-in areas with numerous amino groups are the basic guanidino groups and the bottom two-ring structures are carbohydrates; hence the general name of *aminoglycoside*.

Curiously, one can isolate mutant bacteria that are completely resistant to streptomycin as well as bacteria that are dependent on its presence for growth. The resistant organisms can be shown to have an altered protein in the 30S subunit of their ribosomes that will no longer react with streptomycin. Although the mechanism of streptomycin dependence is not readily apparent, one hypothesis suggests that the ribosomes in these organisms have undergone a conformational change that by itself causes misreading of the mRNA. The addition of streptomycin is thought to correct this malfunction, permitting the correct translation of the mRNA.

There are a number of other antibiotics synthesized by members of the genus *Streptomyces* that, like streptomycin, contain carbohydrates and basic amino groups in their structures. The more common of these aminoglycosides include neomycin, kanamycin, amikacin, tobramycin, gentamicin, and spectinomycin. All react with the 30S subunit of the ribosome and all except spectinomycin are bactericidal. Spectinomycin is bacteriostatic, probably because its reaction with the ribosome is reversible and it does not cause a misreading of the mRNA. The other aminoglycosides cause a greater degree of misreading than streptomycin and, since they bind irreversibly to the 30S subunit, their bactericidal effect appears to be similar to that of streptomycin. Unfortunately, the prolonged use of the aminoglycosides causes toxicity to the eighth cranial nerve (resulting in a hearing impairment); in some cases, they may also cause kidney damage.

The choice of which of these antibiotics to use is frequently based on the antibiotic susceptibility of the isolated pathogen. Thus, one may find organisms that are resistant to kanamycin and gentamicin but susceptible to amikacin (which is a semisynthetic antibiotic made from kanamycin) and tobramycin. It is also worth noting that a functioning electron transport system appears to be necessary for the uptake of aminoglycosides into the cell. As a result, they are essentially ineffective against anaerobic bacteria or even against sensitive, facultative bacteria that are growing under anaerobic conditions.

Antibiotics Affecting Cell Membranes

The cytoplasmic membrane is a semipermeable structure that controls the transport of many metabolites into and out of the cell. Thus, damage to this structure hampers or destroys its ability to act as an osmotic barrier and also prevents a number of necessary biosynthetic functions from taking place in the membrane. Many antibiotics that affect bacterial cell membranes are also injurious to mammalian cell membranes and, as a result, only a few such agents are clinically useful. Also, a number of the antibiotics affecting membranes are too toxic for parenteral use, and they are therefore used only for topical applications.

Polymyxins

Various polymyxins—designated A, B, C, D, and E—are produced by different strains of *Bacillus polymyxa*. Both B and E may be used effectively against certain gram-negative bacteria, but the other three are too toxic for general use. Polymyxin is a peptide in which one end of the molecule is soluble in lipids and the other end is soluble in water. When it gets into the cell membrane, the water-soluble end remains in the outer part of the membrane and the lipid-soluble end is dissolved in the interior area of the membrane. This results in a distortion between the layers of the membrane (see Figure 11.11), which allows substances free passage into and out of the cell.

Polymyxin is particularly effective against infections caused by members of the genus *Pseudomonas*. However, because toxic effects on the kidney and, to a lesser extent, the central nervous system may occur, it is used chiefly when the infective agent is resistant to other, less toxic antibiotics.

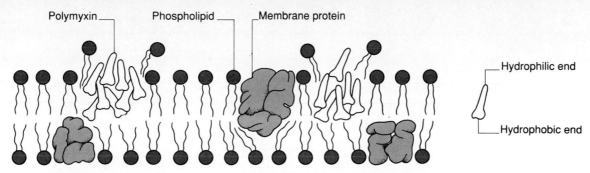

Figure 11.11 Schematic diagram illustrating how polymyxin distorts the cell membrane, destroying its effectiveness as a semipermeable barrier.

Nystatin and Amphotericin B

Both these antibiotics possess large ring structures and, because of the presence of a number of double bonds, they are frequently referred to as the polyene antibiotics.

The polyene antibiotics combine with sterols present in the cell membrane, causing disruption and leakage of the cytoplasmic contents. Therefore, they are specifically effective only against the sterol-containing mycoplasmas and the systemic eucaryotic fungal infections. They are ineffective for the treatment of other bacterial infections because bacteria other than mycoplasmas lack sterols in their membranes. They are used, however, for the treatment of a number of fungal infections.

Imidazole Derivatives

Several compounds have been synthesized by adding various chemical groups to imidazole, as shown in Figure 11.12. These compounds appear to inhibit the synthesis of ergosterol, which is an important constituent of fungal membranes. As such, these compounds are used primarily for the treatment of both dermatophyte and systemic fungal infections. Ketoconazole can be taken orally, whereas miconazole is effective only if used topically or intravenously. Clotrimazole is quite toxic, but it has been used successfully to treat oral candidiasis.

Antibiotics Affecting Nucleic Acid Biosynthesis

Any antibiotic that specifically reacts with DNA to prevent its replication or transcription will obviously inhibit cell growth and division. However, many antibiotics that possess this potential are not completely selective for the target microorganism and show considerable toxicity to the host's cells.

In general, antibiotics inhibit nucleic acid synthesis in one of two ways: (1) interaction with the strands of the DNA double helix in a manner that prevents subsequent replication or transcription or (2) combination with the polymerases involved in DNA or RNA biosynthesis.

Mitomycin and Actinomycin

Both mitomycin and actinomycin bind to DNA, preventing replication and transcription. These antibiotics have been useful research tools for the study of nucleic acid metabolism; however, they are too toxic to the host to be used for the treatment of infectious diseases. They are used as antitumor agents because they will affect rapidly dividing cells (tumor cells) more than slower-growing "normal" cells, and they are also effective against some of the DNA viruses.

Nalidixic Acid and Novobiocin

Both nalidixic acid and novobiocin exert multiple effects on susceptible cells, and the inhibition of DNA replication occurs immediately in their presence. Both drugs inhibit an enzyme (DNA gyrase) that puts supertwists into the closed, circular, double-helical DNA by opening the helix and unwinding it a few times. Superhelicity is necessary for DNA replication and for the expression of some, but certainly not all, genes.

Nalidixic acid is rarely used; it is primarily effective against gram-negative organisms and has been used for the treatment of enteric urinary tract infections. Novobiocin is bactericidal for gram-positive organisms but cannot enter most gram-negative bacteria.

The fluoroquinolones, such as ciprofloxacin, are substituted analogues of nalidixic acid and are hundreds of times more effective than is nalidixic acid. Such fluoroquinolones have a broad antibacterial spectrum (including some intracellular bacteria) and resistance to them has not yet become a significant problem.

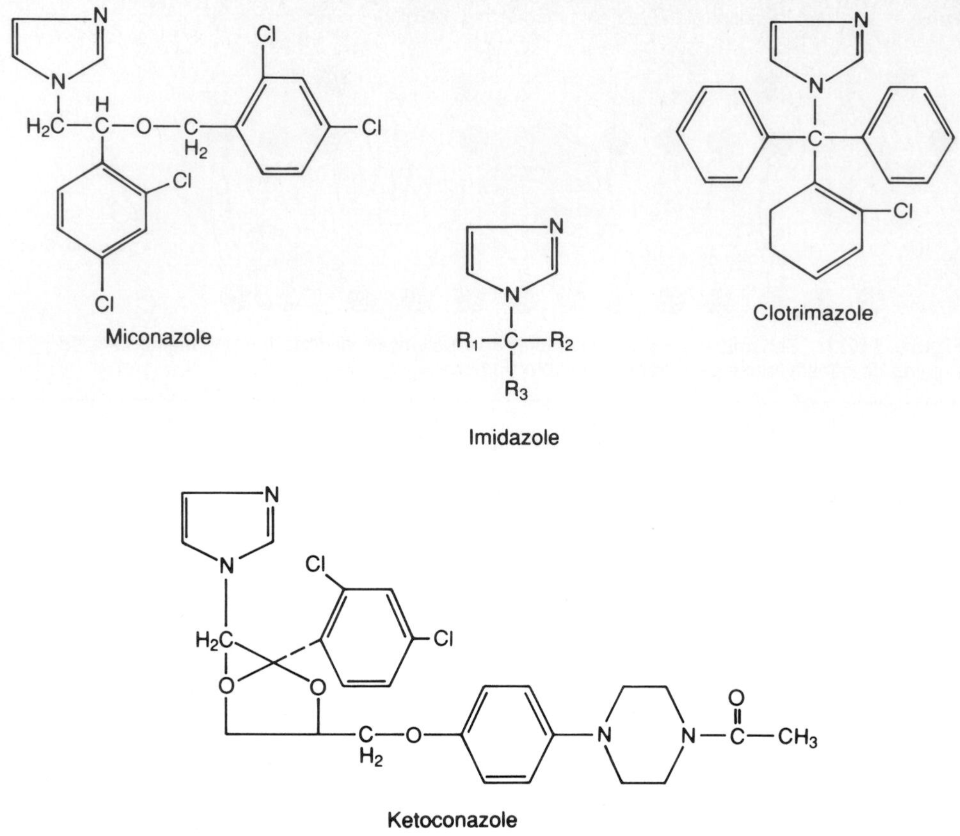

Figure 11.12 Chemical structure of the imidazole nucleus and three clinically useful derivatives: miconazole, clotrimazole, and ketoconazole.

Griseofulvin

Griseofulvin is used to treat superficial dermatophyte infections caused by fungi that destroy the keratin structures in the host. It inhibits growth only of fungi possessing cell walls containing chitin and has no effect either on fungi with cellulose cell walls or on bacteria or yeast. Its mode of action is unclear, but it has been proposed that griseofulvin inhibits DNA replication, although it is more likely that it blocks chitin synthesis.

Rifamycins

The rifamycins form a large family of drugs, many of which have been chemically modified to produce semisynthetic antibiotics. Rifampin, an effective semisynthetic rifamycin, inhibits DNA transcription by binding to the RNA polymerase that catalyzes the transcription of DNA. Rifampin is effective against many gram-positive and gram-negative bacteria and has proved particularly valuable for the treatment of tuberculosis and leprosy.

Table 11.1 provides a summary of some useful chemotherapeutic agents and their spectra of activity.

DRUG RESISTANCE

In the treatment of infectious diseases, one of the serious problems faced today is the development of bacterial resistance to the antibiotic used. An organism that has become resistant to a particular chemotherapeutic agent is said to be drug resistant.

The development of resistance may be due to the ability of an organism to destroy the antibiotic, to a mutation that allows the organism to bypass the sensitive step inhibited by the antibiotic, or to a mutation that causes the cell to become impermeable to the antibiotic.

An example of the first type of resistance (the destruction of the antibiotic) is the secretion of the enzyme penicillinase (also called a β-lactamase). This enzyme destroys penicillin by hydrolyzing one bond (the beta-lactam bond, see Fig 11.4) in the molecule and, even though the organisms may be sensitive to penicillin, the penicillin is rendered ineffective before it can exert its bactericidal effect. Interestingly, a comparative crystallographic study of the spatial arrangements of penicillinase and the D-alanyl-D-alanine pep-

TABLE 11.1   Major Spectrum and Target of Some Chemotherapeutic Agents

Spectrum	Target
Compete with PABA	
Sulfonamides	Enteric urinary tract infections
Paraaminosalicylic acid (PAS)	*Mycobacterium tuberculosis*
Trimethoprim	Broad-spectrum activity
Compete with pyridoxine	
Isonicotinic acid hydrazide (INH)	*Mycobacterium tuberculosis*
Inhibit cell wall peptidoglycan synthesis	
Penicillins	Mostly gram-positive bacteria and the *Neisseria;* some have broad-spectrum activity
Cephalosporins	Gram-positive bacteria and the *Neisseria*
Bacitracin	Gram-positive bacteria
Vancomycin	Gram-positive bacteria
Ristocetin	Gram-positive bacteria
Inhibit protein synthesis by binding to 50S subunit of ribosome	
Chloramphenicol	Broad-spectrum activity
Macrolide antibiotics	Mostly gram-positive bacteria, *Neisseria,* and spirochetes
Erythromycin	
Oleandomycin	
Carbomycin	
Spiramycin	
Lincomycin	Gram-positive bacteria
Clindamycin	Obligate anaerobes, especially *Bacteroides fragilis*
Inhibit protein synthesis by binding to 30S subunit of ribosome	
Tetracyclines	Broad-spectrum activity
Chlortetracycline	
Minocycline	
Oxytetracycline	
Tetracycline	
Aminoglycoside antibiotics	Gram-positive and gram-negative organisms
Streptomycin	
Amikacin	
Gentamicin	
Kanamycin	
Neomycin	
Tobramycin	
Disrupt cell membranes	
Polymyxins	Gram-negative bacteria, especially *Pseudomonas*
Polyene antibiotics	Fungi
Nystatin	
Amphotericin B	
Inhibit DNA synthesis	
Mitomycin and actinomycin	Not used to treat infections
Nalidixic acid	Gram-negative bacteria, especially enteric urinary tract infections
Ciprofloxacin	Fluoroquinolone with broad-spectrum activity
Novobiocin	Gram-positive bacteria, but rarely used
Griseofulvin	Fungi
Inhibit RNA synthesis	
Rifamycins	Gram-positive bacteria, but primarily *M. tuberculosis* and *M. leprae*
Inhibit purine synthesis	
Trimethoprim	Broad-spectrum activity

tidase involved in bacterial wall peptidoglycan synthesis reveals considerable similarity in the structure of their active enzymatic sites. As a result, it has been proposed that penicillinases and these peptidases share a common origin and that all of the various penicillinases arose during a divergent evolution of this ancestral molecule.

There are many genera able to produce penicillinase, of which one of medical importance is *Staphylococcus*. Fortunately, some of the semisynthetic penicillins are not destroyed by penicillinase, and these offer hope in overcoming this difficulty.

Resistance Involving Plasmids

An extremely important and interesting problem in drug resistance was first reported in 1956 in the genus *Shigella*, the cause of bacillary dysentery. The unusual nature of drug resistance in *Shigella* was that each resistant organism was resistant to several drugs, such as sulfonamide, streptomycin, chloramphenicol, and the tetracyclines. It was also observed that when patients who had been excreting drug-susceptible shigellae in their feces were treated with a single drug, such as chloramphenicol, they subsequently excreted organisms that were now multiple-drug resistant. It was not possible to induce multiple-drug-resistant strains from drug-susceptible shigellae grown in the laboratory.

To make a long story short, it was suggested that perhaps the multiple-drug resistance was transferred from resistant organisms to sensitive ones. Much research has shown this to be the case. Thus, if one mixes drug-resistant organisms with drug-sensitive organisms, all become resistant to the same drugs. It has been shown that for this multiple transfer to occur, direct cell-to-cell contact is necessary between susceptible and resistant organisms. Furthermore, drug resistance is usually transferred independently of host bacterial chromosomes. This has been interpreted to mean that although conjugation takes place, the multiple-drug-resistance factor that is transferred is not located on the chromosome but rather on a plasmid.

Many gram-negative bacteria contain these resistance factors, and they transfer them to other gram-negative bacteria by conjugation. The mechanism by which these plasmids impart resistance to the organism stems from their ability to encode different enzymes that specifically inactivate antibiotics such as streptomycin, kanamycin, and chloramphenicol. It has been possible to transfer streptomycin resistance from a resistant culture to a sensitive one by extracting the DNA from the resistant culture and mixing it with competent susceptible cells. Other antibiotics that are inactivated by enzymes encoded in plasmids include gentamicin, spectinomycin, and neomycin.

Gram-positive bacteria also possess plasmids that provide multiple antibiotic resistance. A staphylococcus may possess several plasmids capable of conferring resistance to penicillin, erythromycin, chloramphenicol, tetracycline, streptomycin, and others. However, unlike the resistance factors in gram-negative organisms, these plasmids are usually transferred to sensitive strains by transduction rather than conjugation.

Resistance Involving Chromosomal Mutations

Many mutations causing resistance to an antibiotic are alterations that change components so that they will no longer bind to the antibiotic. For instance, a mutation causing the substitution of a single amino acid in protein S-12 of the 30S ribosomal subunit of *E. coli* prevents the binding of streptomycin, thereby conferring resistance to that antibiotic. Similarly, an alteration in protein S-5 in the 30S ribosomal subunit provides resistance to aminoglycosides, such as neomycin and kanamycin. Alteration of a 50S ribosomal protein provides erythromycin resistance in *E. coli* and *B. subtilis*

Rifampin resistance results from an altered RNA polymerase that will no longer bind the antibiotic. In some cases, resistance may result from a mutation resulting in the inability of an organism to transport the antibiotic into the cell or by the acquisition of a new transport system that pumps the antibiotic out of the cell. Some types of tetracycline resistance occur by the latter mechanism.

Measures Used to Minimize Drug Resistance

How may the problem of drug resistance be solved? There is certainly no single answer, but several measures have been proposed for reducing the development of resistant strains. A most significant point is the avoidance of the indiscriminate use of antibiotics. Because antibiotics act as selective agents for resistant organisms, wide and indiscriminate use soon results in the death of all susceptible strains and leaves only resistant strains to infect new hosts. This particular fact has generated considerable controversy over the use of the subtherapeutic amounts of broad-spectrum antibiotics that are added to animal feeds to promote growth. Many millions of pounds of antibiotics are used annually in such feeds, and the resulting increased growth of chickens and livestock has produced an economic boon to both the farmers and the pharmaceutical companies. The emergence, however, of antibiotic-resistant bacteria in animals given the antibiotic food and their subsequent spread to humans has been well documented. For example, a review of

Salmonella-induced dysentery in the United States from 1971 to 1983 showed that food animals were the source of 69 percent of the outbreaks caused by resistant salmonellae. Serious human disease caused by a multiple-resistant *Salmonella typhimurium* originally found in cattle in the 1960s led to the ultimate ban of antibiotic-containing animal foods in England, followed by other countries in the European Economic Community. A subsequent publication in 1982 by Thomas O'Brien and his collaborators reported that one serotype-plasmid combination in the United States, which appeared to be endemic in cattle in 20 states, was also found in at least 26 persons in two states, indicating that resistance plasmids may be extensively shared between animal and human bacteria. Another report, in 1984 by Scott Holmberg and his associates, showed that 18 persons living in four midwestern states were infected with a multiple-antibiotic-resistant strain of *Salmonella newport*, which they acquired by eating hamburger originating from South Dakota beef cattle fed subtherapeutic chloramphenicol for growth promotion. It appears well established, therefore, that every animal taking an antibiotic (therapeutic or subtherapeutic) rapidly acquires resistant strains through selection of existing organisms, and that these resistant strains are readily passed to humans. In the United States, the unsettled controversy concerning the use of antibiotics in animal feeds revolves around the choice of accepting the emergence of antibiotic-resistant, disease-producing organisms, or accepting higher consumer meat costs associated with the cessation of the use of antibiotics in feed.

During treatment of a disease, the dose of antibiotic administered should be adequate to control the microbe population as quickly as possible. The faster the organisms are inhibited, the smaller the chance for resistant mutants to develop. Another measure that is sometimes effective is the use of a combination of two or more durgs simultaneously. Then, if a mutant to one drug does appear, the other drug still is able to inhibit its growth. However, because certain antibiotics tend to have mutually antagonistic effects, there should be clinical proof of their effectiveness in combination before use. A desirable combination is exemplified in the use of streptomycin and isoniazid or ethambutol in the treatment of tuberculosis.

When two chemotherapeutic agents are given simultaneously, the result may be a greater effect than would result from either alone. This is called a **synergistic effect**.

It is a matter of interest that the majority of antibiotic-resistant bacteria appear to be at some disadvantage in relation to their antibiotic-susceptible counterparts. Thus, if an antibiotic is no longer used in an environment that is heavily populated with antibiotic-resistant strains, susceptible strains will gradually replace the resistant strains, eventually becoming the predominant population.

LABORATORY TESTS FOR ANTIBIOTIC SUSCEPTIBILITY

A physician must know to which antibiotics a particular organism is most susceptible in order to make a decision concerning the treatment of a patient. Although many modifications are available, two general techniques are used in clinical laboratories to provide this information: (1) drug diffusion tests and (2) tube dilution assays.

Drug Diffusion Tests

There are a number of variations in the method of performing drug diffusion tests, but all modifications consist of inoculating a tube of melted agar with the test organism either by pouring the inoculated agar medium into a petri dish and allowing it to solidify or by swabbing the test organism onto the solidified agar in a petri dish. Dilutions of various antibiotics can then be applied to the surface of the agar medium to ascertain whether they will prevent the growth of the organism. The most common method of doing this is to use commercially available paper disks that have been impregnated with various concentrations of different antibiotics.

Kirby-Bauer Susceptibility Test

The Kirby-Bauer susceptibility test uses a single high-strength filter paper antibiotic disk that is placed on a Mueller-Hinton agar medium whose surface has been swabbed with the test organism. After incubation, the diameter of the zone of growth inhibition is measured with calipers (Figure 11.13). Different antibiotics diffuse at different rates, and it is, therefore, necessary to refer to a standard table to ascertain the degree of susceptibility of the test organism to the antibiotic in question (see Table 11.2). The results may be reported as susceptible (meaning that obtainable blood levels of the antibiotic would be effective) or resistant (indicating that obtainable blood levels would probably not be effective). Intermediate size zones may also be reported, but they are usually considered in the resistant category.

Tube Dilutions Assays

A more quantitative (and more time-consuming) approach is to use the tube dilution technique. In this procedure the antibiotic in question is diluted out in

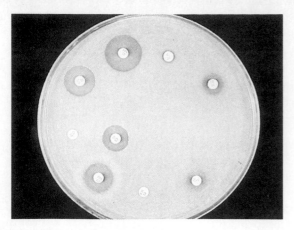

Figure 11.13 Kirby-Bauer disk diffusion plate of antimicrobial susceptibles. Numbers on disks refer to micrograms of antibiotic. Tobramycin (NN)—susceptible; kanamycin (K)—resistant; tetracycline (TE)—resistant; cephalothin (C)—resistant; ampicillin (AM)—resistant; carbenicillin (CB)—resistant; chloramphenicol (CL)—resistant; gentamicin (GM)—susceptible; and colistin (CS)—susceptible.

the growth medium in a series of twofold dilutions. Each well is then inoculated with the organism in question, and the minimal inhibitory concentration (MIC) is determined by looking at the wells after 24 h of incubation and observing the minimal concen-tration of the antibiotic that prevented growth (Figure 11.14). Since data are available concerning attainable blood levels for any specific antibiotic, a knowledge of the MIC provides direct evidence concerning the potential value of a particular antibiotic for treatment of a specific infection. One can also ascertain whether an antibiotic is bactericidal or merely bacteriostatic. This can be accomplished by subculturing a loopful of broth from wells that did not show growth into an antibiotic-free medium. Subsequent growth would in-dicate that the antibiotic tested exerted only a bac-teriostatic effect.

COMPLICATIONS OF CHEMOTHERAPY

A serious problem encountered with many of the chemotherapeutic agents is that patients may develop an allergy to them. This type of sensitivity, usually called hypersensitivity, elicits various reactions char-acteristic of allergic conditions. Skin rashes and fever are the most common manifestations, but there have been a number of deaths attributed directly to anti-biotic hypersensitivity.

TABLE 11.2 Zone Sizes and Their Interpretation for Frequently Used Chemotherapeutics

Antibiotic or Chemotherapeutic Agent	Disk Potency	Inhibition Zone Diameter to Nearest Millimeter		
		Resistant	*Intermediate*	*Sensitive*
Ampicillin				
S. aureus	10 μg	20 or less	21–28	29 or more
All other organisms	10 μg	11 or less	12–13	14 or more
Bacitracin	10 units	8 or less	9–12	13 or more
Cephalothin	30 μg	14 or less	15–17	18 or more
Chloramphenicol	30 μg	12 or less	13–17	18 or more
Colistin	10 μg	8 or less	9–10	11 or more
Erythromycin	15 μg	13 or less	14–17	18 or more
Kanamycin	30 μg	13 or less	14–17	18 or more
Lincomycin[a]	2 μg			17 or more
Methicillin	5 μg	9 or less	10–13	14 or more
Nalidixic acid[b]	30 μg	13 or less	14–18	19 or more
Neomycin	30 μg	12 or less	13–16	17 or more
Novobiocin[c]	30 μg	17 or less	18–21	22 or more
Oleandomycin	15 μg	11 or less	12–16	17 or more
Penicillin-G	10 units	20 or less	21–28	29 or more
Polymyxin-B	300 units	8 or less	9–11	12 or more
Streptomycin	10 μg	11 or less	12–14	15 or more
Sulfonamides[d]	300 μg	12 or less	13–16	17 or more
Tetracycline	30 μg	14 or less	15–18	19 or more
Vancomycin	30 μg	9 or less	10–11	12 or more

[a] Tentative standard.
[b] Standards apply to urinary tract infections only.
[c] Zone sizes not applicable when blood is added to medium.
[d] Any of the commercially available 300- or 250-μg sulfonamide disks may be used with the same standards of zone interpretation.

Source: Bauer AW, Kirby WMM, Sherris JC, Turck M: Antibiotic susceptibility testing by a standardized single-disk method. *Am J Clin Pathol* 45:493–496, 1966.

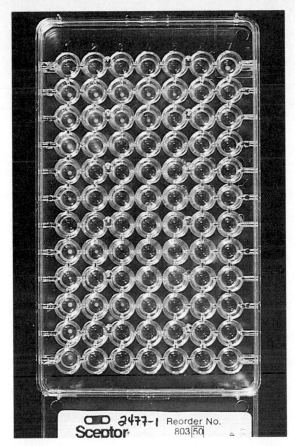

Figure 11.14 Minimal inhibitory concentration (MIC) microtiter system. Each horizontal row of growth media contains increasing concentrations of a particular antibiotic. Growth (indicating resistance at that antibiotic concentration) is shown by the presence of a button of cells in the bottom of the well. Results are reported as susceptible (S), moderately susceptible (MS), or resistant (R). Starting with the top row, this strain of *S. aureus* is reported as follows: (1) cephalothin—S; (2) penicillin—R; (3) ampicillin—R; (4) oxacillin—S; (5) amikacin—S; (6) gentamicin—S; (7) clindamycin—S; (8) erythromycin—MS; (9) tetracycline—S; (10) chloramphenical—S; and (11) vancomycin—S. Row 12 indicates that the organism produces a β-lactamase and is sensitive to nitrofurantoin.

Undoubtedly penicillin is the antibiotic most frequently administered; therefore, it is not surprising that it has been responsible for more side reactions than any other drug. Paradoxically, penicillin is one of the least toxic of the antibiotics, and it can be taken by most people in enormous quantities with no undesirable results.

The allergic reactions resulting from penicillin may be grouped into a number of specific types. The majority of reactions to penicillin are not life-threatening and generally consist of skin rashes or hives (urticaria). However, symptoms similar to serum sickness (fever and joint pain) are not uncommon. More severe manifestations result in anaphylactic (unusual or exaggerated) reactions and not infrequently in death.

In addition to the hypersensitivity reactions, severe toxic reactions may be caused by several chemotherapeutic agents. For example, the toxic effects of streptomycin taken over a prolonged period may result in damage to the hearing mechanism or in dizziness. A toxic effect on the kidney has been noted with several agents, particularly the polymyxins.

The long-term use of the broad-spectrum antibiotics provokes another side effect. In these cases one may find that the normal flora of the body is destroyed by the antibiotic and, as a result, organisms such as *Candida albicans* may take over and flourish. Pseudomembranous colitis (a severe ulceration of the intestine) occurs occasionally after the destruction of the normal intestinal flora by antibiotics, especially clindamycin and ampicillin. Such destruction permits an overgrowth of toxin-producing *Clostridium difficile*. When certain urinary tract infections are treated with the broad-spectrum antibiotics, it is not unusual to find that antibiotic-resistant strains of *Proteus* may take over and be extremely difficult to eradicate. Such infections are sometimes called **superinfections.**

Table 11.3 summarizes some of the adverse effects that, although far from universal, do present problems in many individuals undergoing antibiotic therapy.

ANTIVIRAL COMPOUNDS

Antimetabolites that inhibit the growth of viruses are not nearly so common as the antibiotics discussed in this chapter. There are, however, a few such substances that are effective for the prevention or treatment of certain viral diseases. A few are inhibitory for a wide variety of viruses while most are restricted to a single genus or group of viruses. Such antiviral compounds can be categorized as (1) naturally occurring compounds synthesized by a virus infected cell, (2) structural analogues of purines and pyrimidines that block viral DNA replication, (3) antiviral drugs that are acted on by viral enzymes or will block the activity of viral enzymes, and (4) compounds that prevent the release of viral nucleic acids from the endosome into the cytoplasm of the infected cell.

Interferon is a family of natural products that arise in most, if not all, cells when they are infected by a virus. Once produced, interferon prevents the subsequent infection of noninfected cells. We shall discus the mechanism whereby interferon causes this inhibition in Chapter 17. Interferon has been demonstrated to be effective in eliminating viruses from infected cell cultures, and it is now being used for clinical trials in animal models as well as certain life-threatening infections in humans.

There are also a number of compounds that are structural analogues of the normal purines and pyrim-

TABLE 11.3 Adverse Reactions and Major Contraindications for the Use of Various Antibiotics

Antibiotic	Most Common Adverse Reaction or Contraindication for Use
Penicillins	Hypersensitivity shown by about 5% of Americans
Probenecid	Hypersensitivity, headaches, gastrointestinal symptoms, and hemolytic anemia in some cases
Cephalosporins	Similar to penicillin
Chloramphenicol	Irreversible aplastic anemia
Erythromycin	Relatively nontoxic; jaundice in about 0.4% of cases where used over 10 days
Lincomycin and clinda-mycin	Diarrhea; severe colitis
Tetracyclines	Permanent staining of teeth and bones if given during last half of pregnancy to 8 years of age; increased photosensitivity in some adults; gastrointestinal irritation
Aminoglycosides	Eighth-nerve damage (may be irreversible), skin eruptions, dizziness, toxic to kidney
Polymyxins	Toxic to kidney
Nalidixic acid	Gastrointestinal upset; rash; photosensitivity; headache
Trimethoprim	Rash; fever; kidney and liver damage; aplastic anemia (rare)
Sulfonamides	Similar to trimethoprim

idines, which form the building blocks of nucleic acids. Figure 11.15 shows the structure of a few such compounds. Note that most differ from normal nucleosides either by having halogens such as iodide or fluoride attached to the ring structure or by being attached to the sugar arabinose instead of the normal sugar deoxyribose. Because they are structurally similar to the normal nucleosides, they prevent viral reproduction by inhibiting DNA synthesis. All, however, are quite toxic to mammalian cells. Thus, these nucleic acid analogues are used topically to treat herpesvirus eye infections that might result in blindness if untreated. They have also been injected and given orally for the treatment of herpesvirus encephalitis, cytomegalovirus infections in newborns, and herpes zoster in immunosuppressed patients.

Acyclovir, an analogue of guanosine, is at present the most effective and widely used antiviral agent for the treatment of herpesvirus infections. It is administered intravenously for the treatment of herpes encephalitis or herpesvirus infections in newborns and is given orally for the treatment of genital herpesvirus infections. It inhibits viral replication because it is specifically phosphorylated by the herpesvirus thymidine kinase and the resulting acyclovir phosphate effectively inhibits herpesvirus DNA replication.

As is described in Chapter 14, retrovirus replication requires that the viral RNA be first transcribed into DNA. The viral enzyme that catalyzes this step is called a **reverse transcriptase** and, since this enzyme has no role in eucaryotic cell metabolism, it provides a unique target for antiviral compounds directed against retroviruses. Retroviruses are the causative agents of a number of animal tumors and leukemias but in humans are known only to cause two different T-cell leukemias (HTLV [human T-cell lymphotrophic virus]-1 and HTLV-2) and acquired immunodeficiency syndrome (AIDS). It is this latter disease, caused by human immunodeficiency virus (HIV), that has received billions of dollars worth of research funds during the past decade. Much of this effort has been directed toward methods of inducing an immune response that will prevent an initial infection by HIV and also toward a treatment that could eliminate the virus from individuals who are already infected. This latter endeavor has uncovered several drugs that inhibit reverse transcriptase, but only one of these, azidothymidine (AZT), is sufficiently nontoxic for human use. This drug, under the tradename of zidovudine, may cause severe anemia or neurological symptoms, but it is currently being used to treat HIV-infected patients and, although it cannot cure the infection, it does appear to prolong the latent period between the time of infection and the appearance of the overt symptoms of AIDS.

Finally, there are several antiviral compounds that prevent very early steps in viral replication and, hence, are most effective if given prophylactically before viral infection occurs. One, termed **amantadine,** is particularly effective against influenzavirus type A. It exerts its inhibitory effect by preventing the release of viral RNA from the endosome into the cytoplasm. Others, such as the purine nucleoside analogue **ribavirin** inhibit viral replication by blocking the synthesis of

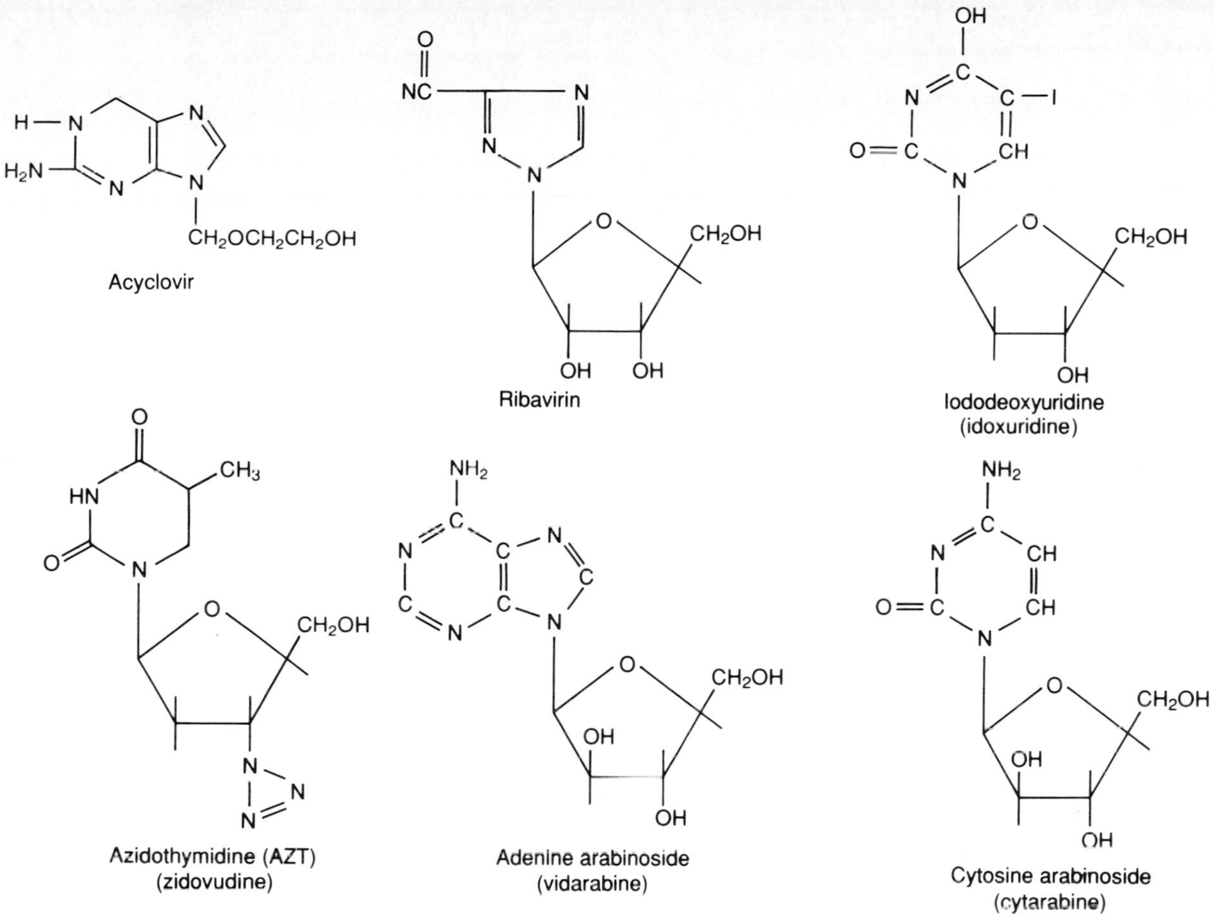

Figure 11.15 Structural formulas of antiviral agents.

GMP, an essential step in nucleic acid synthesis, which interferes with the synthesis of viral single-stranded RNA.

Undoubtedly, the future will see the advent of many new antiviral compounds, and it seems possible that one day viral infections will be treated as effectively as are bacterial infections today.

Table 11.4 lists the major antiviral compounds and the mechanism by which each inhibits viral replication.

TABLE 11.4 Summary of Antiviral Compounds and their Mechanisms of Action

Compound	Action
Interferon[a]	Antiviral compound synthesized by virus-infected host cells
Purine and pyrimidine analogues Idoxuridine Vidarabine Cytarabine Acyclovir	Interfere with viral DNA synthesis; particularly useful for the treatment of herpesvirus infections
Amantadine	Inhibits release of viral RNA from the endosome; used prophylactically for influenzavirus type A infections
Ribavirin	Inhibits nucleic acid synthesis by interfering with the formation of GMP
Azidothymidine (AZT)	Inhibits action of reverse transcriptase

[a] Described in greater detail in Chapter 14.

SUMMARY

Chemotherapy is the treatment of disease by chemical agents. The chemotherapeutic agents may be synthesized chemically, or they may result from microbial synthesis. Metabolic products that are the product of microbial action and that inhibit or destroy other microorganisms are called antibiotics.

The sulfonamides are the chief antibacterial agents produced by chemical synthesis. They function as competitive inhibitors by blocking the incorporation of PABA into folic acid. Their antibacterial spectrum is rather wide, but they are effective chiefly against gram-positive bacteria and some gram-negative organisms. Trimethoprim also blocks the synthesis of folic acid, but at a different step from that of the sulfonamides. Paraaminosalicylic acid and isonicotinic acid hydrazide are also produced by chemical synthesis and act as competitive inhibitors. They are used primarily in combination with streptomycin against *M. tuberculosis*.

Antibiotics in general exert their bactericidal or bacteriostatic effect on susceptible organisms by (1) inhibiting cell wall synthesis, (2) injuring the cytoplasmic membrane, (3) inhibiting protein biosynthesis, or (4) inhibiting nucleic acid synthesis.

There are a number of different penicillins; all are bactericidal by virtue of their ability to inhibit cell wall peptidoglycan synthesis. Specifically, they inhibit the formation of cross bridges between the polymers of *N*-acetylglucosamine and *N*-acetylmuramic acid by inhibiting a transpeptidization reaction that would link these linear polymers. The cephalosporins and the monobactams appear to act in a manner similar to that of the penicillins. Bacitracin, fosfomycin, ristocetin, and vancomycin also inhibit cell wall biosynthesis, but at different steps from those affected by penicillin.

Polymyxin acts against susceptible bacteria by altering the semipermeability of the cell membrane. Other antibiotics causing membrane injury are nystatin and amphotericin B. These last two antibiotics react with the sterols in the membrane and are hence used as antifungal agents.

Chloramphenicol is a broad-spectrum antibiotic that inhibits protein synthesis by inhibiting peptide bond formation between the incoming amino acid and the growing peptide. Erythromycin and the other macrolide antibiotics act by preventing translocation of the ribosome. Lincomycin and clindamycin also react with the 50S subunit of the ribosome and appear to act similarly to chloramphenicol and erythromycin by inhibiting peptide bond formation. The tetracyclines appear to prevent protein synthesis by reacting with the 30S portion of the ribosome, thus preventing the binding of the amino acid-tRNA complex to mRNA.

Streptomycin has a dual effect on protein synthesis. It causes misreading of the mRNA and inhibits the initiation of protein synthesis. It is apparently this latter effect that causes the bactericidal action.

Other antibiotics that inhibit protein synthesis include neomycin, kanamycin, amikacin, tobramycin, spectinomycin, and gentamicin. Antibiotics such as rifampin, nalidixic acid, novobiocin, griseofulvin, actinomycin, and mitomycin inhibit DNA replication or RNA synthesis.

Antibiotic resistance in bacteria may result from the ability of an organism to destroy or alter the antibiotic, or from a mutation that bypasses the inhibited step or causes the cell to become impermeable to the drug. Enzymes involved in the destruction or alteration of an antibiotic are usually encoded in transmissible plasmids.

The selection of antibiotic-resistant strains of bacteria is facilitated by the presence of subtherapeutic levels of an antibiotic. Such a condition is routinely maintained in many of our chicken, turkey, and livestock farms through the use of antibiotic-containing feeds. The use of high levels of antibiotic for treatment of disease as well as using combinations of two or more drugs reduces the possibility of selecting out resistant strains.

There are a number of compounds that are structural analogues of the normal purines and pyrimidines that are used to treat viral infections, particularly those caused by herpesviruses. Other antivirals interfere with viral enzymes or the release of viral nucleic acids into the host-cell cytoplasm.

QUESTIONS FOR REVIEW

1. What is meant by the term *chemotherapeutic agent*? What is an antibiotic?
2. What is competitive inhibition? How does it work with the sulfonamides?
3. Against which normal metabolites do PAS and INH compete?
4. What bacteria, in general, are susceptible to penicillin?
5. How does penicillin work to kill bacteria?
6. What is the mechanism of action of the polymyxins? Nystatin? Amphotericin B?
7. Which step in protein synthesis is blocked by chloramphenicol? Erythromycin? Tetracyclines? Spectinomycin?
8. How does streptomycin affect protein synthesis?
9. Describe two general ways in which a bacterium becomes resistant to an antibiotic.
10. By what mechanism is antibiotic resistance transferred among gram-negative organisms? gram-positive organisms?

11. Describe two mechanisms used to test for antibiotic sensitivity.
12. How do antiviral compounds inhibit viral replication?

SUPPLEMENTARY READING

Abraham EP: The beta-lactam antibiotics. *Sci Am* 244(6):76, 1981.

Davis, BD: The mechanism of bactericidal action of aminoglycosides. *Microbiol Rev* 51:341, 1987.

Dolin R: Antiviral chemotherapy and chemoprophylaxis. *Science* 227:1296, 1985.

Goldstein FW, et al: Plasmid-mediated resistance to multiple antibiotics in *Salmonella typhi*. *J Infect Dis* 153:261, 1986.

Holmberg SD, et al: Drug-resistant salmonella from animals fed antimicrobials. *N Engl J Med* 311,617: 1984.

Kelley JA, et al: On the origin of bacterial resistance to penicillin: Comparison of a beta lactamase and a penicillin target, *Science* 231:1429, 1986.

Kerridge D: Mode of action of clinically important antifungal drugs. *Adv Microbiol Physiol* 27:1, 1986.

Levy SB, Burke JP, Wallace CK (eds): Antibiotic use and antibiotic resistance worldwide. *Rev Infect Dis* 9:S231, 1987.

Levy SB: Tetracycline resistance determinants are widespread. *ASM News* 54:418, 1988.

Lorian V (ed): *Antibiotics in Laboratory Medicine*, 2nd ed. Baltimore, Williams & Wilkins, 1986.

O'Brien TF, et al: Molecular epidemiology of antibiotic resistance in *Salmonella* from animals and human beings in the United States. *N Engl J Med* 307:1, 1982.

Streissle G, Paessens A, Oediger H: New antiviral compounds. *Adv Virus Res* 30:83, 1985.

Chapter 12

Classification Schemes for Common Procaryotes

OBJECTIVES Study of this chapter should provide a comprehension of

1 Past and present attempts to classify living organisms.

2 The five-kingdom system proposed by Whittaker.

3 The three-kingdom scheme proposed by Woese.

4 The use of rRNA sequences to establish phylogenetic relationships among living cells.

5 The properties of the archaebacteria that differentiate them from the eubacteria.

6 Procedures used for the identification of bacteria.

7 What types of biochemical information can be used for identification purposes.

8 What is meant by numerical taxonomy.

9 How specific antibodies or bacteriophages can be used to type microorganisms.

10 The use of genetic classifications such as plasmid exchange or DNA homology studies

11 How the current *Bergey's Manual* has divided the procaryotes into 33 different sections and the major characteristics for each section.

Taxonomy is the branch of biology that places organisms into a systematic classification, making it possible not only to identify a particular specimen, but also to establish the relationship of one organism with other closely related groups. For many centuries, morphology and reproductive compatibility were the primary criteria used to establish taxonomic relatedness. As more was learned about the metabolic pathways of microorganisms, biochemical properties became important in defining taxonomic groups. This is essentially where the science of taxonomy stood until the late 1900s.

Then, a virtual revolution took place when techniques became available to compare nucleic acids using radioactive probes or to determine the actual base sequences of ribosomal RNA. Such information has made it possible to propose evolutionary relationships that may have begun billions of years ago.

But, we will come back to that later in this chapter. Let us first look at how taxonomic groups are defined and what types of information may be necessary to place an unknown organism into a specific taxonomic niche.

TAXONOMIC SCHEMES

Quite obviously, if we are going to establish an orderly and systematic classification of anything, we must give each entity a name. The naming of organisms is termed **nomenclature** and the system of nomenclature in use today was first proposed in 1753 by the Swedish naturalist, Carolus Linnaeus. This system uses a **binomial nomenclature,** in which each taxonomic entity is assigned a two-part name consisting of the **genus** and the **species.** Both names are usually Latin or latinized words with the generic name always a noun and beginning with a capital letter and the species name usually an adjective and never capitalized. We thus have names such as *Homo sapiens* (modern humans), *Ixodes dammini* (a tick that transmits Lyme disease

to humans), and *Borrelia burgdorferi* (the bacterium that causes Lyme disease).

To be a workable system, we also must establish a taxonomic hierarchy in which related organisms are grouped together into taxons that separate them from other groups or organisms. This is accomplished by assigning related genera into a single family. Similar families are then grouped into a single order, and related orders are assigned to a class. Related classes are grouped into a division (called a phylum in zoology) and divisions are combined to make up the final unit in our hierarchy, the kingdom. Table 12.1 provides a summary of how this system would be used to pinpoint a specific species.

How Many Kingdoms

When Linnaeus proposed his system of taxonomy in 1753, he established only two kingdoms, **Animalia** and **Plantae.** The assignment of organisms to one or the other of these kingdoms was generally based on photosynthesis, nonmotility, and rigid cell walls for plants, and nonphotosynthesis, motility, and the absence of cell walls for animals. This two-kingdom scheme proved adequate until the mid-1800s when Haeckel suggested that unicellular organisms be placed in their own kingdom since some appeared to possess properties associated with both animal and plant cells. Thus, in addition to the Animalia and Plantae, Haeckel proposed the addition of a third kingdom, the **Protista,** which would include the unicellular algae and the bacteria, fungi, and protozoa. This also seemed adequate, but as more was learned about cell structure, it seemed unrealistic to lump both eucaryotic and procaryotic organisms into the same kingdom. To overcome the difficulty, H. R. Whittaker proposed a five-kingdom system in 1969. His classification scheme retained the kingdoms Animalia and Plantae, but subdivided Haeckel's Protista into three additional kingdoms— the **Protista** (unicellular algae and protozoa), the **Myceteae** (eucaryotic yeasts and molds), and the **Monera** (containing all of the procaryotic organisms).

TABLE 12.1 Taxonomic Breakdown from Kingdom to Species

Kingdom	Animalia	Plantae	Procaryotae
Division or Phylum	Chordata	Tracheophyta	Gracilicutes (thin cell walls)
Class	Mammalia	Angiospermae	Scotobacteria (nonphotosynthetic)
Order	Primates	Fagales	Rickettsiales
Family	Hominidae	Fagaceae	Rickettsiaceae
Genus	*Homo*	*Quercus*	*Coxiella*
Species	*Homo sapiens* (humans)	*Quercus alba* (white oak)	*Coxiella burnetii* (cause of Q fever)

One would think that this ought to be enough but, as is described in the section on phylogenetic relationships, modern technology provided information that led to yet another proposal for the classification of living organisms. Using the base sequences occurring in ribosomal RNA, Carl Woese proposed in 1978 that the procaryotes should really be divided into two kingdoms, namely, the **Archaebacteria** and the **Eubacteria**. All other organisms would be placed in a kingdom termed the **Eucaryotes**. Note that whereas earlier systems of classification were based primarily on morphologic considerations, Woese used morphology to differentiate procaryotes from eucaryotes, but employed biochemical data to subdivide the procaryotes into two kingdoms. Table 12.2 summarizes the historic development of these classification schemes.

Phylogenetic Relationships Among Procaryotes

Charles Darwin's theories of evolution were based on the concepts that all forms of life had their origin in a common ancestor and that by observing fossils of prehistoric animals and plants along with contemporary organisms, one could construct an evolutionary pattern. This type of classification is termed a **phylogenetic classification,** and it has proved to be very valuable in establishing relationships among higher animals and plants.

But how does one construct a phylogenetic relationship among procaryotes? Few fossils are available, and morphologic differences are not sufficient to establish a realistic evolutionary sequence. In fact, most microbiologists had given up the idea that such relationships could be established until the exciting work of Carl Woese and his colleagues demonstrated that the nucleotide sequences occurring in ribosomal RNA (rRNA) could be used to suggest phylogenetic relationships among procaryotes. In other words, Woese reasoned that changes in rRNA sequences that still permitted protein synthesis would occur so slowly that such changes could be used as a clock to compare one organism with another. Thus, if six different genera of bacteria all contained the same specific changes in a segment of their rRNA, one could suppose that all had a common ancestral origin. To put it another way, the more widespread the existence of a specific rRNA sequence, the longer ago it occurred.

Why are rRNA sequences so highly conserved? Keep in mind that such RNA must maintain a fairly restricted secondary structure in order to function in protein synthesis and that this structure is dependent on base pairing within the RNA molecule. Any mutational change incompatible with such secondary structure will be lethal to the organism. In addition,

TABLE 12.2   Kingdom Proposals for Living Organisms

	Kingdoms
Linnaeus 1753	Animalae 　　Nonphotosynthetic 　　Absence of cell walls 　　Motile Plantae 　　Photosynthetic 　　Rigid cell walls 　　Nonmotile
Haeckel 1866	Animalia 　　Multicellular animals Plantae 　　Multicellular plants Protists 　　All unicellular organisms
Whittaker 1969	Animalia 　　Multicellular animals Plantae 　　Multicellular plants Fungi 　　Yeasts and molds Protista 　　Protozoa and unicellular 　　　algae Monera 　　Procaryotic organisms
Woese 1978	Eucaryotes 　　All eucaryotic organisms Eubacteria 　　Procaryotes with cell walls of 　　　peptidoglycan 　　Membrane lipids attached 　　　through ester linkages. Archaebacteria 　　Procaryotes without peptido- 　　　glycan 　　Membrane lipids attached 　　　through ether linkages

rRNA must be able to bind to proteins within the ribosome, and any changes that interfere with such binding would also be lethal.

Woese's sequence analyses led to the conclusion that all living organisms could be fitted into one of three distinct kingdoms. One, comprising all of the eucaryotic cell types was termed the **Eucaryotes** and the other two kingdoms, containing procaryotic cells, were designated the **Eubacteria** (true bacteria) and the **Archaebacteria** (ancient bacteria). Moreover, it was subsequently shown that the rRNA existing in mitochondria and chloroplasts (obtained from eucaryotic cells) is closely related to that found in the Eubacteria. In addition, at least one protein sequence occurring in eucaryotic cells is closely related to a similar protein found in the Archaebacteria. Thus, eucaryotic cells

One would think that all of the new biotechnology would tend to make classification easier but the status of *Pneumocystis carinii* refutes the old adage that states: If it looks like a duck, walks like a duck, and quacks like a duck, it must be a duck.

P. carinii is a very common respiratory parasite that causes a mild or asymptomatic respiratory infection in essentially 100 percent of humans. In fact, by age 4, between 75 and 90 percent of healthy children have developed circulating antibodies against this organism and, it appears, the organisms remain as saprophytes in the lungs of humans and a variety of other animal species. During the last decade, however, *P. carinii* has acquired a new notoriety by causing the most common opportunistic infection in individuals with AIDS as well as being responsible for significant morbidity and mortality in immunosuppressed persons.

So, just what kind of an organism is *P. carinii?* Based on morphology and growth cycle, it has been classified as a protozoan in the class Sporozoea. It has a complex reproductive cycle involving trophozoites, cysts, and sporozoites which has a number of similarities to the reproduction of the sporozoan that causes malaria. Everyone was happy with this classification until the ribosomal RNA sequence data of *P. carinii* was compared with different classes of organisms. These data showed that the rRNA of *P. carinii* had over a 90 percent homology with certain fungi and only about a 40 percent homology with members of the class Sporozoea. So, what is it, a protozoan or a fungus? There is as yet no agreement. The biotechnologists state that *P. carinii* is more closely related to fungi than to protozoa and thus should be classified as a fungus. Those who disagree say that rRNA homology is not sufficient for taxonomic placement. What do textbook authors do? They place it in a category termed "unclassified"!

---

appear to contain genes derived from both eubacteria and archaebacteria as well as distinct gene sets that have evolved in eucaryotic cells.

Archaebacteria

So, we are back again to the old question of what came first, the chicken or the egg? It seems at least reasonable to conclude that procaryotic cells preceded eucaryotic cells and that eucaryotic mitochondria and chloroplasts, as they exist today, may well have originated as endosymbionts of eubacteria within the emerging eucaryotic cell. It also appears plausible that the archaebacteria may have transferred some of their genes to an evolving eucaryotic cell.

The conclusion that the evolution of the archaebacteria and the eubacteria diverged very early is based on major differences in their cell composition. For example, archaebacteria do not contain peptidoglycan in their cell walls, and their membrane lipids are composed of branched carbon chains attached to glycerol via an ether linkage. Eubacteria, on the other hand, contain peptidoglycan, and their membrane lipids consist of straight-chain fatty acids attached to glycerol via an ester linkage.

Information suggesting that archaebacteria may have been the original life forms is based mainly on the fact that their form of metabolism could be supported by conditions thought to exist during the evolution of life on earth. Metabolically, they fall into the following three groups: (1) the obligately anaerobic methanogens that reduce carbon dioxide (CO_2) to methane (CH_4); (2) the extreme halophiles, which grow only in the presence of high concentrations of salt; and (3) the thermoacidophiles that can grow at temperatures as high as 90°C at a pH as low as 2. Figure 12.1 illustrates Woese's proposed evolutionary tree beginning with an ancestral procaryote. The ancestral eucaryotic cell that existed before the acquisition of mitochondria and chloroplasts is termed a **urkaryote**.

Bacteriological Keys

Each procaryotic cell type has further evolved to adapt to the millions of niches available on this earth. Thus, there are bacteria that grow in hot springs at temperatures approaching 100°C, whereas others grow best at temperatures near 0°C. Some bacteria will grow only in the presence of oxygen, others only in its absence, and still others can grow in the presence or absence of oxygen. At a finer level, some bacteria are found only in very specific niches such as cow's milk or seawater, while others occur only as parasites on and within various animals, including humans. Moreover, bacteria have evolved so that some are found exclusively in the mouth and throat, others on the skin, and yet others in the intestine or genitourinary tract. Some rarely, if ever, cause disease while others routinely induce a harmful infection.

To the student who is beginning the study of microbiology, all of this diversity among procaryotes would seem to make the identification of a specific bacterium a formidable task. We shall see, however, that a number of standard procedures have been developed which are used to characterize an unknown

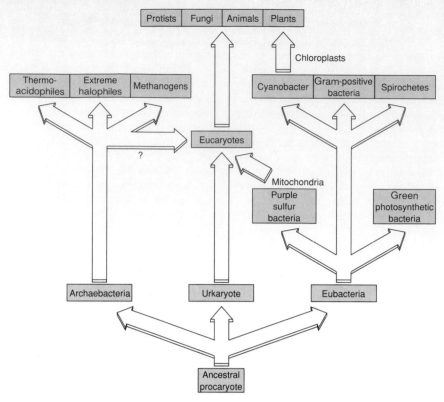

Figure 12.1 Proposed evolutionary development of eucaryotes, eubacteria, and archaebacteria. (Adapted from Woese CR: Archaebacteria. *Sci Am* 244(6):98, 1981.)

organism. The objective of the remaining sections of this chapter is to outline the techniques used to show relatedness of one bacterium to another so as to provide a systematic scheme for the identification of an unknown organism.

PROCEDURES FOR THE IDENTIFICATION OF BACTERIA

The first step in establishing the identity of an unknown bacterium is to determine its shape and Gram reaction. This, along with a minimum of additional information (such as oxygen requirements for growth and colonial appearance), is often sufficient to place the organism into a specific genus. Further identification, however, may require the use of biochemical, serological, or genetic techniques to obtain the information necessary to assign a species or subspecies designation.

Biochemical Classifications

The ability to metabolize specific sugars has long been a primary basis for species identification. This is fre-

quently accomplished by testing for the production of acid by incorporating an acid-base indicator into the sugar-containing medium.

Simple methods to test for the production of other metabolic end products such as carbon dioxide and hydrogen sulfide are also available. Gas-liquid chromatography can give a generic identification of many anaerobic bacteria by providing a rapid qualitative and quantitative evaluation of fatty acid endproducts formed during fermentation. Also existent are computer-controlled gas chromatographic systems that will automate an analysis of the fatty acid composition of an organism. The computer will compare the results of such a determination with data obtained from known organisms. Using such a technique, it is possible, in just a few hours, to identify an organism from an overnight culture.

Differences among various species or genera may also be distinguished by the use of colorimetric assays to detect the presence of various enzymes. Examples include the production of high levels of urease by members of the genus *Proteus* in contrast to other enteric bacteria, the production of β-galactosidase by lactose-fermenting enteric organisms, and the ability to transport citrate into the cell. Figure 12.2 is part of a key taken from *Bergey's Manual* that shows how some of these tests are applied to complete an identification.

	g.	Do not produce ß-glucosidase
	11.	*Bacteroides oris*
gg.		Produce ß-glucosidase
	12.	*Bacteroides buccae*

ff. No growth at 45°C, isolated from ruminants

     21a *Bacteroides ruminicola* subspecies

             ruminicola

ee. Xylose no acid

    f. Arabinose acid

      21b *Bacteroides ruminicola* subspecies *brevis*

    ff. Arabinose no acid

      g. Salicin no acid, pigment produces on

            blood agar

        13. *Bacteroides loescheii*

      gg. Salicin acid, no black pigment on blood agar

        14. *Bacteroides oralis*

  dd. Cellibiose no acid

    e. Esculin acid

      15. *Bacteroides denticola*

    ee. Esculin no acid

      16. *Bacteroides melaninogenicus*

bb. Giant cells greater than 2 μm in diameter

    22. *Bacteroides hypermegas*

Figure 12.2 Part of a key from *Bergey's Manual* showing how species of *Bacteroides* are differentiated from one another.

Numerical Taxonomy

Relatedness among a large number of organisms can be estimated by a numerical taxonomy (also termed *Adansonian analysis*) in which 100 to 300 biochemical and morphologic characteristics are compared for each strain. A similarity index is then calculated for each pair of organisms, using the formula

$$S = \frac{NS}{NS + ND},$$

where S is the similarity index, NS is the number of characteristics common to both organisms, and ND is the number of characteristics not shared. The higher the value of S, the more related are the strains. Examination of a large number of parameters reduces the bias inherent in less extensive classification schemes that are based on the presence or absence of only a few key enzymes. Needless to say, the use of numerous parameters to compare many strains of bacteria would not be feasible without the aid of a computer. Figure 12.3 illustrates how data from this type of analysis can be plotted to show similarities among strains.

SEROLOGIC CLASSIFICATIONS

Serologic classifications are commonly used to distinguish differences among species within a given genus. These tests employ antibodies as very sensitive and specific probes for the presence of various antigenic conformations present on the bacterial surface. The commercial availability of monoclonal antibodies directed against single, antigenic determinants has greatly increased the specificity for many such anti-

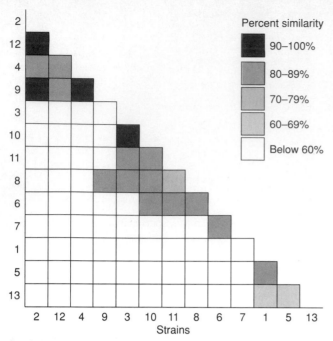

Figure 12.3 Thirteen different strains of bacteria are arranged according to their similarity indices. Each intersecting square is then shaded to show the degree of similarity existing between each pair. With this type of matrix, one can readily see clusters of similarity such as shown here among strains 2, 12, 4, and 9 and among strains 3, 10, 11, 8, 6, and 7.

Bacteriophage Typing

Bacteriophages (also called phages) are bacterial viruses that frequently are very specific for the strain of bacteria they will infect. As a result, the similarity among various strains of a bacterial species is a reflection of their susceptibility to infection and lysis by the same phage. Phage typing is used primarily to identify strains of *Staphylococcus aureus, Pseudomonas aeruginosa, Vibrio chlolerae, Salmonella typhi,* and the mycobacteria.

Phage typing has proved to be a particularly valuable tool for the epidemiologist to locate the origin of a strain causing an epidemic of nosocomial infections. For example, following the spread of staphylococcal infections that appeared to be associated with an operating room, all personnel affiliated with that operating room would be cultured for *S. aureus.* Positive cultures would be phage typed and compared with the phage type of the cultures isolated from the infected patients.

Phage typing is carried out by placing a drop of each diluted phage onto a lawn of bacteria inoculated onto a plate of nutrient medium. The phage numbers causing lysis are noted and, as shown in Figure 12.4, that culture would be designated "*Staphylococcus aureus,* phage type 6/42E/54."

GENETIC CLASSIFICATIONS

Among eucaryotic organisms, the ability to interbreed has long been used as a measurement of relatedness. The exchange of genetic material between two procaryotes also indicates relatedness, but it is much more difficult to quantitate such observations than are the

genic probes. In addition, antibodies directed toward specific surface components can be linked to fluorescent dyes so that a positive identification can be achieved by flooding a smear of organisms with a solution of the fluorescently labeled antibody. After washing off unreacted antibodies, the slide can be viewed using a fluorescence microscope to ascertain if the antibodies bound to the cells.

Figure 12.4 Bacteriophage typing shows that this strain of *Staphylococcus aureus* is lysed by phages 6, 42E, and 54, thus providing another parameter for the identification of this organism.

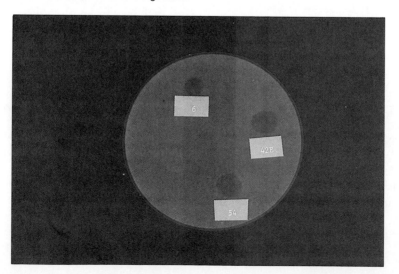

biochemical procedures for nucleic acid homology. For example, the fact that *Escherichia coli* could exchange plasmids with both *Klebsiella* and *Erwinia* does not indicate just how closely related these three organisms might be to each other. The following methods, however, would provide a numerical value to express their genetic relatedness.

DNA Homology

DNA homology is a measure of relatedness at a molecular level and, as such, provides a much better criterion for genetic relatedness than can be obtained with living cells. In the most direct test for DNA ho-

mology, the DNA from one organism is made radioactive by growing the organisms in a medium containing a radioactive component such as phosphorus. It is then mixed with an excess of small DNA fragments obtained from the test organism. The mixed samples are heated to separate the strands of DNA

Figure 12.5 The relatedness of two organisms can be determined by measuring the degree of DNA homology existing between the organisms. The reference organism is grown in a medium containing tritiated thymidine so that all of its DNA will be radioactively labeled with tritium. The test organism is grown in a medium containing no radioactively labeled components. The DNA is isolated from each organism and treated as described. Only the fragments of DNA from the test organism possessing sequences complementary to the reference DNA will reanneal to form double-stranded DNA.

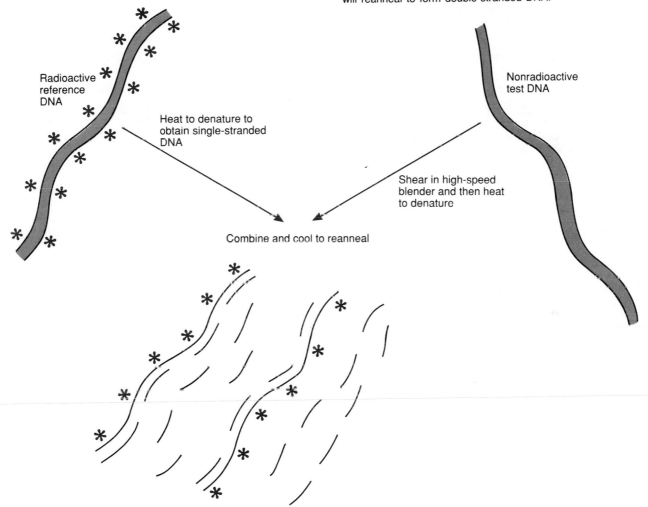

Radioactive reference DNA

Heat to denature to obtain single-stranded DNA

Nonradioactive test DNA

Shear in high-speed blender and then heat to denature

Combine and cool to reanneal

Hydrolyze residual single-stranded DNA with pancreatic DNAse; collect and count radioactivity remaining in double-stranded DNA

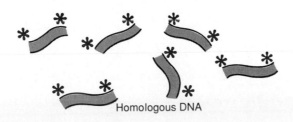

Homologous DNA

and then are slowly cooled to allow the re-formation of the double-stranded regions by any homologous portions of the two DNAs (see Figure 12.5).

The degree of DNA homology between the two organisms is revealed by the rate and extent to which the test DNA will form stable double-stranded duplexes with that from the reference organism. There are several ways of determining the amount of radioactivity present in the double-stranded DNA species such as the chromatographic separation of double-stranded DNA from single-stranded DNA or by the preferential degradation of the single-stranded molecules by the action of the S1 nuclease enzyme. Figure 12.6 shows relatedness among several enteric bacteria using this technique.

The percentage of guanosine plus cytosine (G + C) also may be a measure of relationship. This value can be estimated from the temperature at which the paired strands of DNA separate ("melting point" of DNA), or from the buoyant density determined by high-speed centrifugation of the bacterial DNA in a gradient of cesium chloride. It must be emphasized, however, that even though a large difference in G + C content between two strains certainly indicates unrelatedness, the reverse is not true, that is, a similar G + C content does not by itself prove similarity between two bacterial strains.

PRESENT APPROACHES TO CLASSIFICATION

The major system of classification of bacteria that is used routinely in the United States is outlined in a series of four volumes entitled *Bergey's Manual of Systematic Bacteriology*. This is actually a continuation of the *Bergey's Manual* that was originally published in 1923 and underwent eight revisions before the current edition was published. This latest edition, however, has a new format and it is published as the first edition of this classification. In it, the procaryotic organisms are assembled into a number of different sections based on their Gram reaction, morphology, and, in some cases, the mechanism by which they obtain their energy. For example, all aerobic, gram-negative rods and cocci are placed in one section while the facultatively anaerobic gram-negative rods are placed in a different section. Phylogenetic relationships are not a consideration in compiling the various sections of this manual. In fact, rRNA sequence data indicate that many organisms in the same section are only distantly related while others that are placed in separate sections appear to be closely related phylogenetically. The exclusive purpose of *Bergey's Manual* is therefore to provide keys (such as shown in Figure 12.2) that can be used for the identification of bacteria.

Because many of the bacteria described in *Bergey's Manual* are discussed elsewhere in this book, plus the fact that many microbiologists are unfamiliar with a large number of these organisms, no attempt will be made to list or describe the hundreds of genera included in this publication. Rather, a brief list of the contents of each volume is provided in Tables 12.3 through 12.6. One representative genus is included for each section.

An indication of the rapidly changing taxonomy of bacteria is illustrated by the observation that Volume IV of *Bergey's Manual* has an addendum listing 48 new species of actinomycetes that were validly described after Volume IV went to press.

Figure 12.6 DNA relatedness among Enterobacteriaceae. The numbers represent approximate percentage of relatedness. (From *Bergey's Manual of Systematic Bacteriology,* Vol. 1, page 411 (1984).)

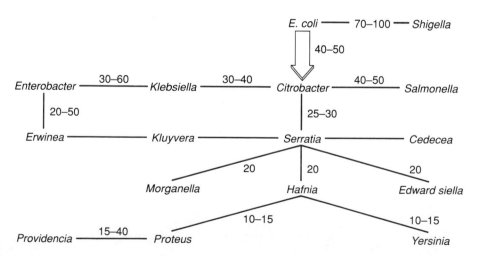

TABLE 12.3 *Bergey's Manual,* Volume I: Gram-Negative Bacteria of General, Medical, or Industrial Importance

Section	Representative Genus
1. Spirochaetales	*Treponema*
2. Motile helical/vibrioid	*Spirillum*
3. Nonmotile helical	*Spirosoma*
4. Aerobic rods and cocci	*Pseudomonas*
5. Facultatively anaerobic rods	*Escherichia*
6. Anaerobic rods	*Bacteroides*
7. Dissimilatory SO_4 or S-reducing	*Desulfovibrio*
8. Anaerobic cocci	*Veillonella*
9. Rickettsias and chlamydias	*Rickettsia*
10. Mycoplasmas	*Mycoplasma*
11. Endosymbionts	*Blattabacterium*

TABLE 12.4 *Bergey's Manual,* Volume II: Gram-Positive Bacteria

Section	Representative Genus
12. Cocci	*Staphylococcus*
13. Endospore-forming	*Bacillus*
14. Regular nonsporing rods	*Lactobacillus*
15. Irregular nonsporing rods	*Actinomyces*
16. Mycobacteria	*Mycobacterium*
17. Nocardioforms	*Nocardia*

TABLE 12.5 *Bergey's Manual,* Volume III: Archaebacteria, Cyanobacteria, and Remaining Gram-Negative Bacteria

Section	Representative Genus
18. Anoxygenic phototrophs	*Rhodospirillum*
19. Oxygenic phototrophs	*Oscillatoria*
20. Aerobic chemolithotrophs	*Nitrobacter*
21. Budding and appendaged bacteria	*Gallionella*
22. Sheathed bacteria	*Sphaerotilus*
23. Nonphotosynthetic nonfruiting gliders	*Cytophaga*
24. Fruiting gliders	*Myxococcus*
25. Archaebacteria	*Methanobacterium*

TABLE 12.6 *Bergey's Manual,* Volume IV: Actinomycetes

Section	Representative Genus
26. Nocardioform actinomycetes (Update of section 17)	*Nocardia*
27. Multilocular sporangia	*Frankia*
28. Actinoplanetes	*Actinoplanes*
29. Streptomycetes	*Streptomyces*
30. Maduromycetes	*Actinomadura*
31. Thermomonospora	*Thermomonospora*
32. Thermoactinomycetes	*Thermoactinomyces*
33. Other genera	*Glycomyces*

SUMMARY

Bacteria are named according to a binomial nomenclature consisting of a two-part name—the genus and the species. Various attempts to categorize all living matter has resulted in proposals using two, three, or five different kingdoms. The most recent system, using sequence data from rRNA, divides the procaryotes into two kingdoms, the Archaebacteria and the Eubacteria. All eucaryotic organisms are placed in a single kingdom, Eucaryote.

Procaryotic cells can be divided into three basic cell types: (1) gram-positive cells, (2) gram-negative cells, and (3) the Archaebacteria. Each cell type has evolved into multiple biological types.

Identification of an organism requires that one know its shape and gram reaction. Additional information may necessitate the biochemical testing for end products of fermentation; for the ability to ferment specific sugars; for the presence or absence of specific enzymes; or for the fatty acid composition of the organisms. A system of classification, termed *numerical taxonomy*, or *Adansonian analysis*, uses a similarity index that is calculated after comparing 100 to 300 varied characteristics for each pair of organisms.

Serological classification is commonly used to distinguish differences among species within a given genus. Similarly, phage typing frequently is used to identify strains occurring within a single species. Genetic classification may merely determine the ability of an organism to exchange genetic material with another organism, or it may measure relatedness by determining the DNA homology that exists between two organisms. Ribosomal RNA sequences have been shown to be the most conserved of any nucleic acid sequences and, as such, can be used to establish phylogenetic relationships.

The primary system of classification of bacteria, used routinely in the United States, is outlined in a series of volumes entitled *Bergey's Manual of Systematic Bacteriology*. Currently, all procaryotes are placed into one of 33 different sections based on morphology, Gram reaction, and oxygen requirements for growth.

One of these groups, termed the *archaebacteria*, differs from the other bacteria in that they lack peptidoglycan in their cell walls, possess unusual lipids in their membranes, and have different biochemical characteristics in their transfer RNAs and RNA polymerase subunit structure. Based on rRNA sequence data, it is postulated that the archaebacteria and the eubacteria arose from a phylogenetic split that occurred eons ago.

QUESTIONS FOR REVIEW

1. How does the five-kingdom system proposed by Whittaker differ from Woese's three-kingdom proposal?
2. What are the first kinds of information one needs to begin to identify a bacterium?
3. What types of biochemical data could be useful for the identification process?
4. Describe how antibodies or phages can be used to identify an organism.
5. What types of genetic information would be of value in comparing two organisms?
6. How can rRNA sequence data be used to establish a phylogenetic classification?
7. What criteria are used in *Bergey's Manual* to establish the different sections of procaryotes?
8. What properties differentiate the archaebacteria from the eubacteria?

SUPPLEMENTARY READING

Fox GE, et al: The phylogeny of the procaryotes. *Science* 209:457, 1980.

Holt JG (ed): *Bergey's Manual of Systematic Bacteriology*. Baltimore, Williams & Wilkins, 1984.

Lewin R: RNA catalysis gives fresh perspective on the origin of life. *Science* 231:545, 1986.

Olsen GJ, et al: Microbial ecology and evolution: A ribosomal approach. *Annu Rev Microbiol* 40:337, 1986.

Pace NR, Olsen GJ, Woese CR: Ribosomal RNA phylogeny and the primary lines of evolutionary descent. *Cell* 45:325, 1986.

Pace NR, et al: Analyzing natural microbial populations by rRNA sequences. *ASM News* 51:4, 1985.

Schwartz RM, Dayhoff MO: Origins of procaryotes, eucaryotes, mitochondria, and chloroplasts. *Science* 199:395, 1978.

Stoffler-Meilicke M, et al: Structure of ribosomal subunits of *M. vannielii*: Ribosomal morphology as a phylogenetic marker. *Science* 231:1306, 1986.

Van Valen LM, Maiorana V: The archaebacteria and eucaryotic origins. *Nature* 287:248, 1980.

Woese CR: Archaebacteria. *Sci Am* 244(6):98, 1981.

Chapter 13

Unusual Procaryotic Cells

The organisms discussed in this chapter have been arbitrarily grouped together because of one or more properties that make them "atypical" when compared with what have been described as typical procaryotic cells. Nevertheless, they are procaryotic cells, and they are described in *Bergey's Manual of Systematic Bacteriology*. Let us see just what makes these particular bacteria atypical.

RICKETTSIAE

The rickettsiae (*sing.* rickettsia) are smaller than most bacteria. In fact, most of them are just barely within the range of visibility of the ordinary light microscope. However, it is not their size that sets them apart from typical bacteria but rather that, with one exception (*Rochalimaea quintana*), they grow only inside the cells of animals; that is, they are obligate intracellular parasites. They cannot be grown in the laboratory on artificial media as can most bacteria; they require living cells for their growth.

Rickettsiae appear to be structurally related to the gram-negative bacteria. Typically, they are rod-shaped, with average dimensions of 0.3 to 0.7 μm by 1.5 to 2.0 μm (see Figure 13.1). Electron microscopy reveals a cell wall consisting of an inner membrane and an outer membrane. Furthermore, when stained with the Gram stain, they appear as gram-negative cells. Rickettsiae have been shown to contain peptidoglycan, a substance found only in procaryotic cells, and diaminopimelic acid, a substance found only in the cell walls of gram-negative bacteria. It seems that rickettsiae are really a special type of gram-negative bacteria.

We may wonder, then, why they are obligate intracellular parasites. If they originally evolved from gram-negative bacteria, what have they lost that will not allow them to grow extracellularly? The answers to these questions are not completely known, but recent research has provided information that makes it possible at least to speculate. Once rickettsiae are removed from their intracellular environment, they lose viability and infectivity. This seems to be correlated with the loss of intracellular metabolites. The organisms can be stabilized or even restored to infectivity by the addition of metabolites such as nicotinamide adenine dinucleotide (NAD), coenzyme A, and adenosine triphosphate (ATP) and of inorganic ions such as K^+ and Mg^{2+}. The surprising thing about these observations is that substances such as NAD, coenzyme A, and ATP cannot usually get into a normal bacterial cell. The fact, however, that these metabolites can either stabilize or restore infectivity to the rickettsiae certainly indicates that they must get into the rickettsial cell. Thus, one can postulate that the unusual transport systems of the rickettsial cell membrane account for its obligate intracellular habitat and that only in this environment can it maintain its cellular integrity.

A great deal remains to be learned about the catabolic and anabolic abilities of the rickettsiae. In general, however, they appear able to synthesize their own proteins, nucleic acids, and other macromolecules necessary for the cell structure. They are also able to oxidize glutamic acid, as well as a number of intermediates of the citric acid cycle, and to trap the released energy as ATP. It would appear that they have a functioning citric acid cycle. However, even though rickettsiae have been shown to possess some of the enyzmes necessary to metabolize glucose, none is able to utilize glucose as a substrate.

Culture Techniques

In view of all this, how can we propagate rickettsiae in the laboratory? One way is to infect susceptible animals, such as guinea pigs. Rickettsiae can also be grown in tissue cultures of living animal cells in test tubes. The most common method for growing rickettsiae used during the past several decades is to in-

Figure 13.1 Electron micrographs of four species of rickettsiae.

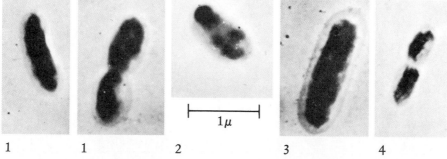

1   1   2   3   4

1μ

oculate them into chick embryos. For this purpose, fertile hens' eggs are incubated at 35°C for 7 to 10 days, at which time the embryo is well along in its development. The organisms are then injected into the yolk sac, where they enter and grow in the cells of the membrane surrounding the yolk sac. After an additional 4 to 5 days of incubation, the egg is opened and the yolk sac membrane is removed. The rickettsiae can then be partially purified by a number of techniques that usually involve disruption of the yolk sac membrane cells followed by differential centrifugation.

Rickettsiae can also be grown in monolayer cultures of chick or duck embryo cells. This procedure is currently used for the production of certain rickettsial vaccines. It results in a vaccine that is free of yolk sac contaminants and that, therefore, causes fewer allergic reactions.

Pathogenesis

There is one other rather distinctive difference between the rickettsiae and most other bacteria—the manner in which humans (or other animals) are infected. With one exception (Q fever), humans are infected only by the bite of an infected tick, louse, flea, or mite. Thus, the rickettsiae are injected directly into the blood by the bite of an arthropod vector. (Ticks, lice, fleas, and mites are members of the phylum Arthropoda.) The word **vector** means a carrier of disease-producing agents from one host to another.

A knowledge of the mechanism by which the rickettsiae are maintained in nature is helpful for an understanding of how humans become infected with these organisms. The major reservoir of the rickettsiae is in the wild animal population. The organisms are spread from animal to animal by the bite of an infected arthropod vector. If a tick or flea takes a blood meal from an infected rat or rabbit, the rickettsiae will infect the tick or flea. Then when the tick or flea subsequently bites an uninfected animal, this second animal becomes infected. Hence, in the normal sequence of events, we have a cycle of infection going from animal to arthropod and back to animal. With one exception (the human body louse, which carries epidemic typhus), the infected arthropod vector appears not to be harmed. Moreover, the rickettsiae are passed along a transovarian route by the vector through the eggs, from mother to offspring. Thus, even the wild animal reservoir is not necessary for the maintenance of these organisms in nature. Furthermore, in diseases in which the vector is not the human body louse, humans become infected only when they accidentally enter the animal-to-arthropod cycle of infection. Table 13.1 lists the rickettsiae and their arthropod vectors that cause disease in humans. The specific diseases will be discussed in detail in Unit III.

CHLAMYDIAE

The chlamydiae are another group of obligate intracellular parasites. Prior to 1966, when the name *chlamydiae* was adopted, these organisms were frequently referred to as the large viruses, or the psittacosis-lymphogranuloma-trachoma group. We know now that the chlamydiae are not viruses but are procaryotic cells.

Morphology and Reproduction

The characteristic that does most to delineate the chlamydiae as a distinct group of organisms is their rather complex method of reproduction. The sequence of events occurring during the reproduction of these organisms has been referred to as the developmental cycle and goes like this: (1) a small, dense cell (called

TABLE 13.1 Infectivity Cycles of Rickettsiae

Species	Normal Host	Mode of Transmission to Humans	Disease
Rickettsia prowazekii	Human	Louse and flying squirrel	Epidemic typhus fever
Rochalimaea quintana	Human	Louse	Trench fever
Rickettsia typhi	Rats and mice	Flea	Endemic typhus fever
Rickettsia tsutsugamushi	Several rodents	Mite	Tsutsugamushi or scrub typhus
Rickettsia akari	Mice	Mite	Rickettsialpox
Rickettsia rickettsii	Many small mammals and ticks	Tick	Rocky Mountain spotted fever
Coxiella burnetii	Many mammals	Primarily through respiratory route	Q fever

an elementary body), about 0.3 μm in diameter, is taken into a host cell through the cell membrane (see Figure 13.2); (2) the small cell undergoes a "reorganization" into a large, less dense cell called an initial or reticulate body; (3) the large cell grows in size and multiplies by binary fission; (4) after 24 to 48 h, the initial bodies (which by themselves are noninfectious) are reorganized into the small, dense, infective elementary bodies, thus completing the development cycle; (5) the host cell bursts and liberates the small, dense, infective cells. It is not known what triggers this reorganization of large cells back into small cells. It is known, however, that this reduction in size is accompanied by the loss of a great deal of RNA from the large cell, since the RNA-to-DNA ratio in the small cells is approximately 1:1, whereas that of the large initial bodies is about 4:1. The high content of RNA in the dividing large cells is undoubtedly responsible for the high rate of reproduction—and hence protein synthesis—taking place in the large cells. Because the small cells do not divide, a high concentration of RNA is not necessary for their maintenance.

In spite of their rather complex method of reproduction, the chlamydiae are probably descendants of the gram-negative bacteria. Their cell walls contain peptidoglycan, they reproduce by binary fission, and their DNA is not surrounded by a nuclear membrane. We are still faced, however, with the question of why these organisms can grow only inside a host cell. The answer appears to lie in their inability to synthesize their own ATP. Their enzyme systems seem to be moderately complete in that they synthesize their own proteins, lipids, and macromolecules necessary for their cell structure. But the synthesis of all these molecules requires energy, and the chlamydiae are metabolically defective in that they have no mechanism for the trapping of energy. Furthermore, the chlamydiae must have unusual transport systems, since they can take in not only molecules like ATP but also large protein molecules. Quite obviously, such an organism would find it impossible to survive in the world of extracellular life.

Figure 13.2 Microcolony of *C. psittaci* in a McCoy cell (about ×34,000). In this electron micrograph both small, dense, infectious elementary bodies (two are at the lower right) and larger, thin-walled, noninfectious initial bodies are shown. The initial body on the left is dividing.

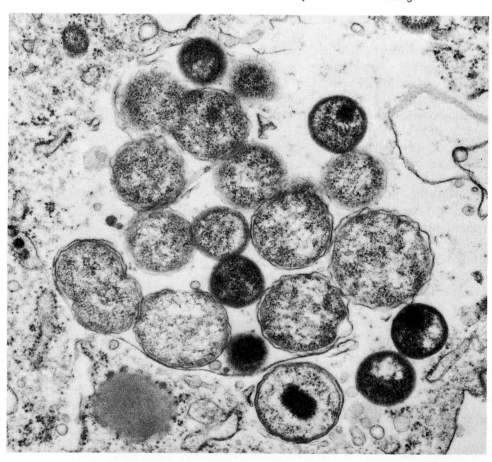

Endosymbionts comprise an unusual group of microorganisms that routinely grow in the cytoplasm or nucleus of eucaryotic cells. Most have never been cultured, and interestingly, it is not usually possible to infect cells through the medium. In other words, the endosymbionts are unable to leave the parasitized cell and enter a new host cell. Thus, transmission occurs only as a result of cellular heredity.

Endosymbionts have been described as occurring in insects, fungi, algae, and protozoa. In many cases they are essential for cell survival in that antibiotic treatment to kill the endosymbiont also results in the death of the host cell.

Little is known concerning the contribution of the endosymbiont to the welfare of the host cell. It is known, however, that many will confer on the host the ability to produce toxins. Some such toxins, termed *killers*, are liberated into the medium, where they will kill cells that are similar to the host cell but are not infected with an endosymbiont. Thus, the endosymbiont in some way confers resistance to the host cell to any toxin produced. Others induce the formation of a toxin that can kill only after cell-to-cell contact. Such organisms have been designated *mate killers*.

A number of the endosymbionts resemble organelles of eucaryotic cells, leading to speculation of the role of procaryotic organisms in the evolutionary origin of mitochondria and chloroplasts. We may never know for certain but rRNA sequence data, as well as the fact that ribosomes from eucaryotic cells are 70S, certainly support this postulation.

Classification

During the past several decades, there have been numerous attempts to classify this very complex group of organisms. Early schemes assumed that the chlamydiae were large viruses, and classification schemes, for the most part, named them after the diseases they caused. Later, taxonomists tended to group the chlamydiae with the rickettsiae, because both groups consist of procaryotic organisms that are obligately intracellular parasites. However, it is apparent that the complex developmental cycle of the chlamydiae clearly separates this group of organisms from all other procaryotic cells; no other bacterial organism produces daughter cells that must undergo a morphologic reorganization (such as an initial body to an elementary body) before they are able to infect a host cell.

The chlamydiae are divided into two subgroups based on the morphologic appearance of their intracellular inclusion bodies, the presence or absence of glycogen in the inclusion, their sensitivity to sulfonamides, and the extent of DNA homology between related organisms. Organisms in group A (which contain the causative agents of trachoma, inclusion conjunctivitis, nongonococcal urethritis, epididymitis, female pelvic inflammatory disease, and lymphogranuloma venereum) have been placed by many investigators into a single species named *Chlamydia trachomatis*. These organisms form compact inclusion bodies containing glycogen (see Figure 13.3) and are inhibited by sulfonamides. The organisms in group B, which include *Chlamydia psittaci* and *Chlamydia pneumoniae*, are resistant to the presence of sulfonamides and produce a diffuse inclusion body that does not contain glycogen. *C. psittaci* causes a disease in birds called psittacosis (ornithosis) and humans may develop a severe pneumonia if infected bird feces are inhaled. *C. pneumoniae* also causes a pneumonia in humans but this organism appears to infect only humans and, as a result, is transmitted directly from one person to another.

MYCOPLASMAS

Mycoplasmas are the smallest organisms known capable of growth and reproduction outside of living host cells. Because of the pleomorphism of mycoplasmas, the actual size of the individual cells is variable.

Figure 13.3 *(a)* Compact inclusion body of *C. trachomatis* in a McCoy cell. The glycogen-containing inclusion appears dark after staining with 5 percent iodine-potassium iodide solution (I-KI). *(b)* Diffuse inclusion of *C. psittaci* in a mononuclear mouse cell. The chlamydiae are the small dark bodies distributed about the cytoplasm in this phase-contrast micrograph of a fresh wet mount.

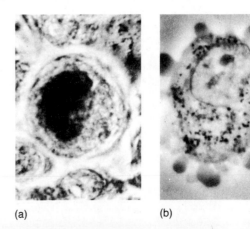

(a)   (b)

TABLE 13.2   Partial List of Mycoplasmas that Infect Animals

Organism	Host	Disease
M. bovigenitalium	Cattle	Mastitis and vulvovaginitis
M. bovirhinis	Cattle	(?)Respiratory disease
M. mycoides	Cattle	Pleuropneumonia
M. ovipneumoniae	Lambs	Pneumonia
M. conjunctivae	Goats and sheep	Conjunctivitis
M. gallisepticum	Poultry	Respiratory disease and encephalitis
M. pulmonis	Mice and rats	Respiratory disease
M. arthritidis	Rats	Polyarthritis
M. hyorhinis	Swine	Arthritis
M. hyosynoviae	Swine	Arthritis and synovitis
M. orale	Humans	Parasite of oropharynx
M. pneumoniae	Humans	Atypical primary pneumonia
M. salivarium	Humans	Parasite of oropharynx
Ureaplasma (T strains)	Humans and other animals	(?)Urethritis

However, it is generally agreed that they range from 0.12 to 0.25 μm in diameter.

The first mycoplasma isolated was an organism that caused a pleuropneumonia in cattle. It was originally thought to be a virus, because it would pass through a filter that would hold back bacteria. Subsequently, it was shown not to be a virus, because it would grow on artificial media. We now know that there are many similar organisms that cause disease in animals other than cattle (see Table 13.2). Because the original organism isolated from cattle was called a pleuropneumonia organism, all similar organisms isolated afterward were referred to as pleuropneumonialike organisms, or more commonly by the abbreviation PPLO. This abbreviation was widely used until 1967 when these organisms were placed in a new class, Mollicutes ("soft skins"). This class is divided into three families and a number of genera, as shown in Table 13.3.

Morphology and Reproduction

The major morphologic difference between the mycoplasmas and other procaryotic cells is that mycoplasmas completely lack a cell wall. Because there is no cell wall, the major components of a mycoplasma cell are simply the plasma membrane, the cytoplasm (with its inclusions), and chromosomal DNA. No cell wall components, such as peptidoglycan or diaminopimelic acid, are synthesized. As might be expected for cells so poorly protected from the osmotic vagaries of their environment, the shapes of mycoplasmas vary over a wide range, from ultramicroscopic coccoid cells to long filaments that may or may not be branched.

There is still some confusion concerning the details of mycoplasma reproduction. It appears that a coccoid cell routinely elongates into a long filament. This is followed by the formation of coccoid structures within the filament, which are subsequently released by fragmentation (see Figure 13.4). Reproduction by budding also may occur.

Mycoplasmas also vary considerably in their growth requirements, but all can be grown on artificial media. Members of the family Mycoplasmataceae require cholesterol for growth. The exact function of the cholesterol is not known, but it is adsorbed to a constituent in the cell membrane and is thought to be necessary for this membrane's pliability and tensile

TABLE 13.3   Generic Classification of Mollicutes

Families and Genera	Major Characteristics
Mycoplasmataceae	All require cholesterol
Genus 1: *Mycoplasma*	Pathogenic for animals and humans; colonies may be 600 μm diameter
Genus 2: *Ureaplasma*	Proposed name for urea-utilizing T strains; pathogens or parasites; tiny colonies 10 to 30 μm diameter
Acholeplasmataceae	None requires cholesterol
Genus 1: *Acholeplasma*	Saprophyte or parasite for mammals and birds.
Genus 2: *Thermoplasma*	Saprophyte having an optimum growth temperature of 59°C at pH 1 to 3
Genus 3: *Anaeroplasma*	Saprophyte obligately anaerobic isolated from rumen of cattle and sheep
Spiroplasmataceae	
Genus 1: *Spiroplasma*	Plant/animal pathogen possessing helical and branched filaments

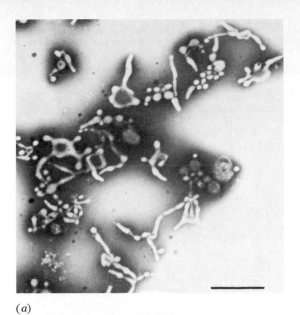

(a)

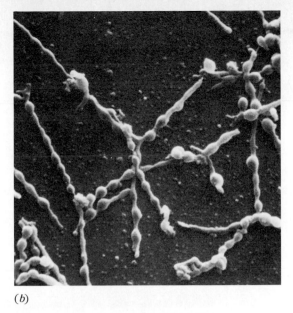

(b)

Figure 13.4 *(a)* Electron micrograph of a negatively stained preparation of *M. pneumoniae.* Scale = 1 μm. The variable morphology, ranging from rings with lobes to beaded filaments, is evident. *(b)* Scanning electron micrograph of *M. pneumoniae* (×10,000); note the coccoid structures within the filaments.

strength. Members of the family Acholeplasmataceae do not require added sterols for growth, but they do synthesize carotenoids, which are deposited in the cell membrane and may serve a function similar to that of cholesterol.

Growth of *Mycoplasma* on a solid medium routinely results in the formation of colonies too small to be seen with the naked eye. When cultures are viewed under the low power (×10) of the light microscope, diphasic colonies with a "fried-egg look" are frequently seen. Such colonies are formed when the organisms in the center of the colony grow down into the medium (see Figure 13.5).

T strains of mycoplasma are so named because they grow as exceptionally tiny colonies, usually 10 to 20 μm in diameter. Such colonies cannot be seen with the naked eye and (as shown in Figure 13.6) are much smaller than normal mycoplasma colonies.

T strains require both cholesterol and urea for growth; the incorporation of 10 percent horse serum into a medium will satisfy both requirements. Because of their absolute requirements for urea, the generic name of *Ureaplasma* is now used for these organisms.

Pathogenic Significance

The mycoplasmas are widespread in both animals and humans. They have been shown to be the causative agents of pleuropneumonia in cattle, arthritis in rats, and a neurologic disorder of mice called rolling disease. They are found commonly in the human mouth

and frequently in the human genital tract, especially T-strain mycoplasmas. In the latter case, it has been postulated but not proved that they may cause an inflammation of the genitourinary tract. Mycoplasmas have been isolated from the joints of patients with arthritis; again, it has been postulated but not proved that they are involved in the arthritic syndrome. How-

Figure 13.5 Colonies of *Mycoplasma fermentans* show "fried-egg" appearance; growth was on agar for 14 days (about ×1885).

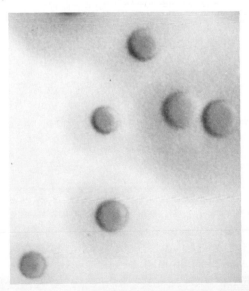

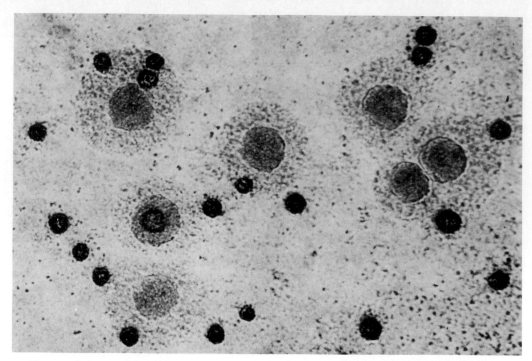

Figure 13.6 This micrograph demonstrates the size difference between colonies of the strain mycoplasma *Ureaplasma urealyticum* (small, dark colonies) and the "fried-egg" colonies of the classic large-colony *Mycoplasma hominis* (×195). This is a standard agar culture on which the direct spot test for urease was applied: the dark urease-positive strain colonies show the specific development of manganese reaction product, which positively identifies them as those of *Ureaplasma*. The larger, lighter, urease-negative *M. hominis* colonies have been counterstained with a blue dye.

ever, in spite of the frequent appearance of mycoplasmas in humans, there is only one human disease that is unequivocally caused by these organisms. This infection, primary atypical pneumonia, will be discussed in detail in Unit III.

A great many diseases of plants that had only recently been thought to be viral diseases are now believed to be caused by mycoplasmalike organisms. It seems quite probable that future research on these many plant diseases will provide a much more complex classification of the causative organisms than we now have.

We may conclude that mycoplasmas are another example of procaryotic cells. Furthermore, if they were originally derived from bacteria, they must have appeared early in evolution.

L FORMS

In 1935, Emmy Klieneberger-Nobel was doing research with a bacterium called *Streptobacillus moniliformis*, the causative agent of rat-bite fever. She found that no matter how carefully she purified her culture, there were always two colony types: one made up of the normal streptobacilli and the other a very tiny colony composed of very tiny organisms that did not appear to possess rigid cell walls. Klieneberger-Nobel eventually came to the conclusion that these very tiny organisms were being formed spontaneously from the normal streptobacilli and, furthermore, possessed the potential to revert to the normal streptobacillus form. Since she was working in the Lister Institute in London, she named these bacterial derivatives **L forms**.

We now know that many bacteria can be induced to form L forms by the addition of such chemicals as lithium salts. Also, any substance, such as penicillin or bacitracin, that interferes with the synthesis of the cell wall peptidoglycan can induce L forms of bacteria. Although many of the L forms are capable of reverting to the parental bacterial form, others are extremely stable and apparently never revert. There has always been controversy concerning whether the stable L forms are essentially the same as mycoplasmas. Based on such things as a comparison of the guanosine-cytosine content of their DNA, the incorporation of cholesterol into the membranes of most mycoplasma, and the considerably smaller size of the mycoplasma cell, it is generally agreed that the L forms and the mycoplasmas are not closely related.

Although bacterial L forms have not been directly implicated in the production of disease, there is some indirect evidence that they may be an important survival mechanism for pathogenic organisms. Thus, treatment of an infection with an antibiotic such as penicillin may induce the formation of L forms in the host. Since L forms have no cell wall, they are not susceptible to the effect of penicillin. After the cessation of penicillin therapy, the L forms would then be capable of reverting to the parent form. Although evidence for this complete cycle of events is not absolute, it is true that L forms have been isolated from patients undergoing penicillin therapy.

BDELLOVIBRIOS

The bdellovibrios were first described in the 1960s and represent a very unusual procaryotic organism in that they parasitize other bacteria. The bdellovibrios are 1 to 2 μm in length but only about 0.35 μm in diameter. Thus, they are only slightly larger than the limit of resolution of the light microscope. Their parasitic reproduction appears to go through the following stages:

1. While traveling at a fantastic speed (i.e., as fast as 100 cell lengths per second), the bdellovibrio collides with a bacterium. It then rotates rapidly around its long axis at speeds up to 100 revolutions per second. The effect of this "boring" action is not understood, but the result is the formation of a pore in the bacterium's cell wall. However, it appears likely that the formation of this pore is also facilitated by enzymatic action of the bdellovibrio.

2. The bdellovibrio enters through the pore and lodges between the cell wall and the cytoplasmic membrane. It does not appear to enter the cytoplasm of the bacterium.

3. The bdellovibrio, receiving its nutrients from the parasitized bacterium, continues to elongate until it is 5 to 10 times longer than the original infecting bdellovibrio. The elonged cell finally segements into a number of daughter cells that then leave the "ghost" bacterium.

A large number of bdellovibrio strains that can parasitize many of the enteric bacteria and pseudomonads have been isolated from sewage and soil. It is also possible, occasionally, to select host-independent strains that are able to grow in the absence of a susceptible bacterium. A large number of bacterial viruses (bacteriophages) that can infect and destroy the bdellovibrios have also been isolated. Thus, again we see an illustration of the saying "big fleas have little fleas upon their backs to bite them, and little fleas have lesser fleas, and so on, *ad infinitum.*"

SHEATHED BACTERIA

One moderately large group of bacteria grows as individual gram-negative cells that are enclosed in an extracellular, tubular sheath. Most such organisms are found in aquatic habitats and many possess a holdfast at one end of the sheath that anchors it to a solid surface.

Members of the genus *Sphaerotilus* grow profusely in fresh water contaminated with sewage or wastewater from agricultural industries. In such cases, their tangled filaments may plug wastewater pipes or prevent the settling of sludge during wastewater treatment. Members of the genus *Leptothrix* deposit ferric and manganic oxides in their sheaths, giving them a yellow to dark brown appearance. As a result, if one examines microscopically the reddish-yellow deposits occurring in iron-containing streams, one will see numerous long tubular filaments (almost always empty) that were formed by leptothrix.

In all cases, motile swarmer cells are released from one end of the sheath, which will eventually attach to a solid matrix before forming a new sheath.

ACTINOMYCETES

Actinomycetes is the collective name for eight different families of bacteria that grow as many-branched long or short filaments of cells. They divide by binary fission and may or may not produce external spores. By far the majority of these organisms are soil and water saprophytes (organisms living on decaying organic matter) and are exceedingly important for their roles in the cycles of nature, such as the decomposition of organic material and the fixation of nitrogen.

The actinomycetes are moldlike in outward appearance and in many texts are discussed along with the eucaryotic fungi. They are, however, true bacteria, as judged by all the criteria for a procaryotic cell. They contain peptidoglycan in their cell walls, they lack mitochondria, they contain 70S ribosomes (eucaryotic cells possess 80S ribosomes in their cytoplasm), they lack a nuclear envelope, the diameter of their cells ranges from 0.5 to 2.0 μm, and they are killed or inhibited by many bacterial antibiotics.

Of the many different genera currently classified in the order Actinomycetales, only a few produce disease in humans; we shall describe only these medically important actinomycetes.

Actinomyces

The genus *Actinomyces* contains obligately anaerobic or microaerophilic gram-positive organisms that grow in a mass of branched filaments (called a mycelium). The tangled mycelium readily breaks up into bacillary and coccoid forms. Members of the genus *Actinomyces* do not form spores, and only two species are of major medical importance. Both—*Actinomyces bovis*, the causative agent of lumpy jaw in cattle, and *Actinomyces israelii*, the etiologic agent of actinomycosis in humans—cause similar types of infections. Interestingly, *A. israelii* occurs as part of the normal flora in the crypts of the tonsils, in dental caries, and occasionally in the intestinal tract or the lungs. Thus, infections caused by this organism (cervicofacial and abdominal actinomycosis) originate from an endogenous source, and the initial lesions, which usually contain a mixture of *Actinomyces* with other endogenous bacteria, occur in the cervicofacial, abdominal, or lung tissue.

Nocardia

Members of the genus *Nocardia* are distributed worldwide in the soil but are not considered to be part of the normal flora of the human body. Only two species, *Nocardia asteroides* and *Nocardia brasiliensis,* are considered valid pathogens of humans.

These organisms grow aerobically on simple media, forming long filaments that easily fragment into rather pleomorphic bacillary and coccoid cells. They are gram-positive, and the pathogenic species are partially acid-fast. Because these organisms are frequently involved in lung infections, the fragmented mycelium can be mistaken for tubercle bacilli. Human actinomycotic infections will be discussed in Unit III.

Streptomyces

The genus *Streptomyces* includes an extremely large group of organisms distributed worldwide. Like the other actinomycetes, they grow in long branched filaments; unlike members of the genus *Actinomyces* and *Nocardia*, they form long chains of aerial spores called conidia.

Members of the genus *Streptomyces* are only rarely pathogenic, but they have achieved prominence as a result of their ability to produce antibiotics. Streptomycin and actinomycin were the first to be isolated and characterized, but since 1940 over 500 different antibacterial compounds, including many of the antibiotics in current use, have been isolated and characterized from organisms classified as *Streptomyces*.

APPENDAGE BACTERIA

In addition to flagella and fimbriae, there are over a dozen genera of bacteria that form appendages (termed prostheca). These appendages are small outward extensions of the cell wall.

Within this group, members of the genus *Caulobacter* have been the most extensively studied. As shown in Figure 13.7, *Caulobacter* possesses a holdfast, located on the tip of the appendage, that anchors the bacterium to a solid surface. During division, the prosthecate cell undergoes transverse binary fission to produce a daughter cell that possesses a single polar flagellum. After separation, the daughter cell moves off and secretes a holdfast prior to a new prostheca, and begins a new cycle of replication.

Because these events are steps in the differentiation of the cell, considerable effort has gone into the study of their genetic control in *Caulobacter*. The appendage bacteria are found primarily in freshwater ponds and streams, and none is known to cause disease.

GLIDING BACTERIA

The gliding bacteria comprise a very large heterogeneous group of gram-negative bacteria that lack flagella but move with a gliding motion when in contact with a solid surface. The mechanism of their motility is not well understood, but may result from the excretion of slime material by the cell.

Many of the gliders are large, multicellular bacteria, whereas others are unicellular organisms classified in the order Myxobacteriales. These latter or-

Figure 13.7 Dividing cell of *Caulobacter crescentus*. Note that the cell with appendate is adhering to a solid surface whereas the newly formed cell has a flagellum that permits it to swim to a new location before forming its own appendage and hold fast.

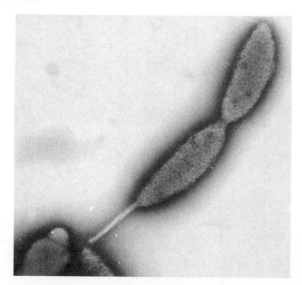

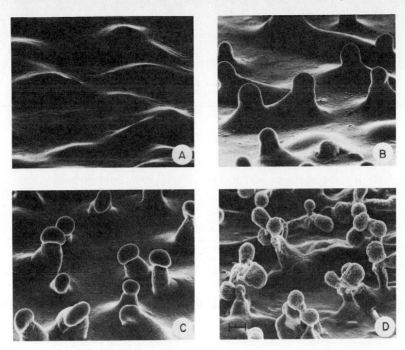

Figure 13.8 Stages of fruiting body formation of the myxobacterium *Stigmatella aurantiaca. (a)* Early aggregates of vegetative cells; *(b)* early stalks; *(c)* late stalks; *(d)* mature fruiting bodies, containing mature myxospores.

ganisms are all rod-shaped vegetative cells, many of which respond to certain unknown environmental factors in a coordinated action to produce an elaborate fruiting body containing rounded cells referred to as myxospores (see Figure 13.8). The resulting myxospores are quite resistant to desiccation but are not appreciably more resistant to heat than are the vegetative cells.

The myxobacteria are, for the most part, found in decaying vegetation or in manure piles. A few do cause infections in fish, but none has been implicated in human disease.

The best-known examples of the multicellular gliding bacteria are found in the genera *Beggiatoa, Thiothrix,* and *Leucothrix.* These organisms grow in long filaments and are frequently found in waters that are rich in H_2S such as sulfur springs or waters heavily contaminated with sewage. These organisms, sometimes referred to as the filamentous sulfur bacteria, obtain their energy by the oxidation of H_2S to elemental sulfur, which is deposited inside of the cell as sulfur globules. In the absence of H_2S, this intracellular sulfur can be further oxidized to sulfate, providing energy for the anabolic activities of the cell.

SUMMARY

Rickettsiae are small procaryotic organisms that appear to have evolved from the gram-negative bacteria. With the exception of one species, *R. quintana,* all rickettsiae are obligate intracellular parasites. An explanation for the obligate parasitism of the rickettsiae is their unusual transport systems; in an extracellular environment, these organisms lose a large part of their essential nutrients. The rickettsiae are unusual in that they are (again, with one exception—Q fever) transmitted from host to host only via the bite of an arthropod vector such as the tick, louse, flea, or mite.

The chlamydiae are another group of obligate intracellular, parasitic, procaryotic organisms. These organisms possess a rather complex cycle of reproduction that involves a small, dense, infective particle and a larger reproductive vegetative cell. Chlamydiae are also differentiated from the rickettsiae by virtue of the fact that they are transmitted from host to host without the intervention of an arthropod vector. Their obligate parasitism appears to result from their leaky membrane and inability to synthesize ATP. Major human diseases caused by the chlamydiae include psittacosis, lymphogranuloma venereum, trachoma, and inclusion conjunctivitis.

The mycoplasmas are the smallest known organisms capable of growth and reproduction outside living host cells. These organisms do not possess a rigid cell wall. Their cells are, therefore, composed only of a plasma membrane, cytoplasm, and a strand of DNA. They are widespread in the animal kingdom but have thus far been shown to cause only a few diseases.

L forms are organisms that are derived from bacteria and are devoid of cell walls. Their formation can be induced by several toxic salts or by antibiotics such as penicillin that inhibit cell wall biosynthesis. Although L forms possess many similarities to the mycoplasmas, the consensus is that the mycoplasmas are not directly derived from L forms of bacteria.

Bdellovibrios are procaryotic organisms that parasitize other procaryotes by forming a pore in the parasitized cell and lodging between the cell wall and the cytoplasmic membrane.

Sheathed bacteria are characterized by growing inside a tubular sheath. They are found primarily in aquatic habitats.

Actinomycetes include bacteria that characteristically grow in straight and branched filaments of cells. Two genera of this large group of organisms, *Actinomyces* and *Nocardia,* are known to cause disease in humans. Members of the genus *Streptomyces* are important sources of antibiotics.

Appendage bacteria are characterized by an extension of the cell wall into a structure that is morphologically narrower than the main body of the cell. The appendage may or may not be associated with a holdfast that anchors the bacterial cell to a solid surface.

Gliding bacteria are so named because they exhibit a gliding motility when in contact with a solid surface. Members of the Myxobacteriales may form a resting cell termed a *myxospore.* Such spores result from a rounding up of the vegetative cell usually within a fruiting body. Filamentous gliders oxidize sulfur compounds as their source of energy.

QUESTIONS FOR REVIEW

1. What is an obligate parasite?
2. What observations can you list to explain why rickettsiae grow only intracellularly in susceptible host cells?
3. Give three methods by which rickettsiae can be grown in the laboratory. Which is the most common method?
4. Where are rickettsiae usually found, and how do humans become infected with them?
5. What is the transovarian passage of rickettsiae?
6. How do the chlamydiae differ from the rickettsiae?
7. What explanation has been offered for the obligate intracellular growth of chlamydiae?
8. List the steps involved in the reproduction of the chlamydiae.
9. Name one human disease known to be caused by mycoplasmas.
10. How would you explain the fact that mycoplasmas are less fragile than protoplasts?
11. What is an L form? How does it differ from a mycoplasma?
12. What is the importance of L forms in human disease?
13. How do the parasitic bdellovibrios obtain their nutrients?
14. What are sheathed bacteria and where might one find them?
15. How would you differentiate between members of the genus *Actinomyces* and the genus *Nocardia?*
16. What is the normal habitat for the pathogenic *Actinomyces? Nocardia?*
17. Why are members of the genus *Streptomyces* important?
18. What are appendaged bacteria and what is their normal habitat?
19. What is a myxospore?

SUPPLEMENTARY READING

Burchard RP: Gliding motility of prokaryotes: Ultrastructure, physiology, and genetics. *Annu Rev Microbiol* 35:497, 1981.

Kalakoutskii LV, Ogre NS: Comparative aspects of development and differentiation in actinomycetes. *Bacteriol Rev* 40:469, 1976.

Maniloff J: Molecular biology of mycoplasma. In: Schlessinger D (ed): *Microbiology—1978.* Washington, DC, American Society for Microbiology, 1978, p. 390.

Pachas WN, Madoff S: Biological significance of bacterial L-forms. In: Schlessinger D (ed): *Microbiology—1978.* Washington, DC, American Society for Microbiology, 1978, p. 412.

Razin S: Physiology of the mycoplasmas. *Adv Microbiol Physiol* 10:2, 1973.

Schachter J, Caldwell HD: Chlamydiae. *Annu Rev Microbiol* 34:285, 1980.

Staley JT (ed): *Bergey's Manual of Systematic Bacteriology,* 9th ed. Vols. 3 & 4. Baltimore, Williams & Wilkins, 1989.

Williams JC, et al: Molecular biology of rickettsiae. In: Leive L, Schlessinger D (eds): *Microbiology—1984.* Washington, DC, American Society for Microbiology, 1984, p. 239.

Wyrick PB, Newhall WJ: Biology and pathogenesis of *Chlamydiae.* In: Leive L (ed): *Microbiology—1986.* Washington, DC, American Society for Microbiology, 1986, p. 80.

Chapter 14

Viruses

In this chapter we shall be concerned with biological entities that do not by themselves possess life, because a virus manifests life as measured by reproduction only after it has entered a susceptible host cell. Thus, viruses exist in the foggy semantic area between "living" and "nonliving," their status depending on whether they are reproducing within a susceptible cell or whether they are in an extracellular state.

There are probably few cells, eucaryotic or procaryotic, that cannot be infected by some virus. Once a virus begins to replicate, a host cell does not usually continue to function as a normal, uninfected cell. In fact, death is probably the most common fate of an infected cell. This is dramatically illustrated when nonregenerating nervous tissue is infected with viruses such as polio or rabies; the destruction of the infected nerve cells results in either permanent disability or death of the entire host organism.

Other viruses may cause the proliferation of infected cells, resulting in such manifestations as warts. Still other viruses may cause changes that transform the normal host cell into a cancer cell, no longer affected by the controls that regulate normal growth. In other instances, the effect of a virus may be so subtle that one can only postulate the mechanism by which it damages its host. Such a case is provided by the infection of women with rubella virus during their first trimester of pregnancy. When the virus infects the fetus, its major effect is to slow down cell growth and, as a result, to interfere with cell differentiation, causing either the death of the fetus or a large variety of fetal abnormalities. After cell differentiation is essentially completed (by the beginning of the second trimester of pregnancy), infection of the fetus is usually uneventful and is followed by complete recovery. Finally, it should be pointed out that some viruses do not cause any noticeable damage to the infected cell. They appear to have reached the ultimate state of parasitism wherein both the virus and the host cell continue to replicate in peaceful coexistence.

Before we can hope to understand what viruses do, however, we must learn what they are, how they replicate, and how they can be distinguished from each other. It is the purpose of this chapter to answer these questions.

.

THE NATURE OF VIRUSES

Viruses are incredibly simple structurally compared with procaryotic or eucaryotic cells. They can exist with a simple structure because, like several other groups of organisms we have discussed, they are obligate intracellular parasites. But here the comparison of the viruses with rickettsiae or chlamydiae ends. For

example, someday someone might devise a nutrient medium that will sustain rickettsiae in spite of their leaky membranes. Such a thing is impossible in the case of viruses, because a virion (virus particle) lacks certain components absolutely essential for its own replication and must depend on the host cell in which it is replicating to provide these missing factors. One component missing from all viruses is an adenosine triphosphate (ATP)-generating system. Biological synthesis requires energy, and this energy is provided in ATP as chemical energy in the form of high-energy phosphate bonds (see Chapter 4). However, for independent life, a cell must carry out oxidations to provide energy to regenerate the high-energy phosphate bonds used for biosynthetic reactions. No virion possesses this regeneration system; hence, it must rely on the ATP-generating system present in the infected host cell. A second component that viruses lack, and that the host cell must provide, is the structural component for protein synthesis, that is, ribosomes. The synthesis of any protein requires that a ribonucleic acid (messenger RNA) be attached to a ribosome so that the individual amino acids can be joined to form the protein. The virion does carry its own ribonucleic acid (RNA) or deoxyribonucleic acid (DNA), but, as far as is currently known, all viruses must use the host cell ribosomes for protein synthesis. Another characteristic that is peculiar *only* to viruses is that, whereas all other forms of life contain both RNA and DNA, *viruses contain only one type of nucleic acid;* in some cases it is RNA, and in others, DNA, but never both.

SIZES AND SHAPES

Viruses vary considerably in size but, in general, they are well below the limit of visibility of the light microscope. In early literature, they were often referred to as filterable viruses, because they would go through filters that would not allow most bacteria to pass. In fact, errors concerning the etiology of a disease have occurred because a virus was held to be the causative agent merely because it was "filterable."

Virus sizes (they fall into the range of 20 to 250 nm) may be measured by several techniques. Three basic techniques used to determine virus size are (1) filtration through graded membranes in which the pore size of the membrane is known; (2) high-speed centrifugation (greater than 100,000 times gravity), in which the size of a virus can be calculated by determining the rate at which the virus particles settle to the bottom of the centrifuge tube; and (3) direct observation with an electron microscope.

One can think of a virus in its simplest form as nothing more than nucleic acid (either DNA or RNA) surrounded by a protein overcoat called a *capsid*. Many viruses, however, are more complex in that they may contain enzymes necessary for the replication of their nucleic acid. Also some may have carbohydrates bound to their protein coat (glycoproteins); others may be enclosed in a membrane acquired as they budded from the host cell.

Viral Capsids

In essence, the viral capsid protects the enclosed nucleic acid from both physical destruction and enzymatic hydrolysis by host cell nucleases. It possesses binding sites that enable the virus to attach to specific receptor sites on the host cell. Finally, the capsid is responsible for the ultimate shape of the virion. Electron microscopy has revealed that, with the exception of the relatively complex poxviruses, all animal viruses are either icosahedral or helical in shape.

Icosahedral viruses have 20 facets (or faces), each an equilateral triangle. Schematic and electron microscopic views of an icosahedral virus are shown in Figure 14.1. As depicted in this figure, the completed capsid is made up of repeating morphologic units called capsomers. These capsomers are visible by electron microscopy as small protein structures. One means of classifying icosahedral viruses is based on the total number of capsomers present in the viral capsid. For example, all herpesviruses possess 162 capsomers, whereas the capsids of all adenoviruses contain 252 capsomers per intact virion.

In the case of the helical viruses, the viral nucleic acid is closely associated with the protein capsid, forming a coil-shaped nucleocapsid that becomes enclosed in a membrane as it buds from the host cell. All helical animal viruses are RNA viruses and, although the exact nature of the nucleic acid interaction with the capsid is not well understood, the protein does protect the RNA from enzymatic hydrolysis (by RNase) while still allowing the RNA to be transcribed from the intact nucleoprotein. Figure 14.2(a) depicts a schematic representation of the helical nucleocapsid of a rhabdovirus, and Figure 14.2(b) shows an electron micrograph of a similar rhabdovirus virion in which the striations of nucleic acid can be seen wound in a helical pattern in the interior of the particle. Figure 14.2(c) shows a virus budding from a host cell membrane.

With the exception of the bullet-shaped rhabdoviruses, other helical animal viruses exist as spherical virions containing a helical-shaped nucleocapsid surrounded by a membrane.

Viral Nucleic Acid

The nucleic acid for each family of viruses is characteristic for that taxonomic group. Some possess double-stranded DNA (dsDNA), while others contain single-stranded DNA (ssDNA), dsRNA, or ssRNA. DNA virus nucleic acid is always in one piece; some RNA viruses, however, contain a segmented genome in which their nucleic acid exists in a number of separate pieces. For example, each reovirus has been shown to contain 10 different molecules of dsRNA, rotoviruses contain 11 segments of dsRNA, and influenza viruses possess 8 separate segments of ssRNA.

Figure 14.1 *(a)* Electron micrograph of a negatively stained adenovirus virion. *(b)* Model of the same virion showing 252 spheres in icosahedral symmetry.

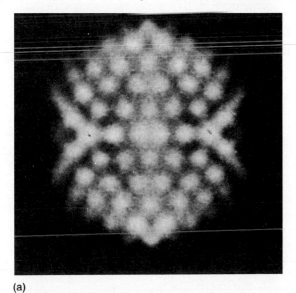

(a)

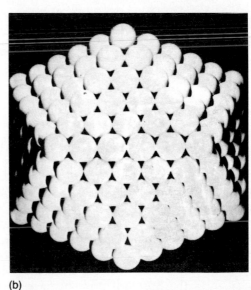

(b)

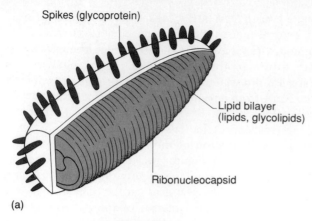

Spikes (glycoprotein)

Lipid bilayer
(lipids, glycolipids)

Ribonucleocapsid

(a)

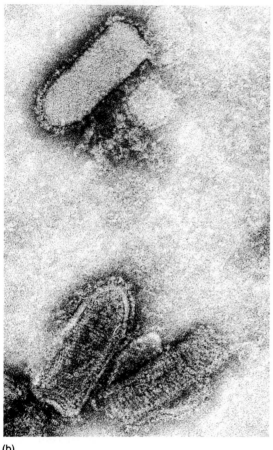

(b)

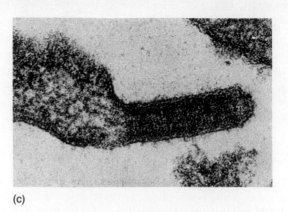

(c)

Figure 14.2 *(a)* Schematic representation of the rhabdovirus, vesicular stomatitis. *(b)* Virions of vesicular stomatitis virus (VSV); the helical nucleocapsid is visible as cross striations (×207,000). *(c)* VSV budding from a cell (×147,000).

teins (proteins linked to a carbohydrate moiety). The glycoprotein usually occurs as projections or spikes on the outer surface of the envelope. Some of these glycoproteins have special functions, particularly the attachment to host cell receptors to initiate the entrance of the virion into the cell. Some glycoprotein spikes will also attach to receptors on red blood cells, linking the cells together and causing them to clump. Others possess enzymatic activity and will cleave neuraminic acid from host cell glycoproteins. It is not surprising, therefore, that antibodies capable of neutralizing viral infectivity are frequently specific for the viral glycoprotein spikes.

In addition to protein and glycoprotein, the viral envelope contains 20 to 30 percent lipid, totally derived from the host cell membrane. Thus, the viral envelope is structurally similar to the host cell membrane, differing only in that it contains virus-coded proteins and glycoproteins. Figures 14.2*(c)* and 14.3

Figure 14.3 A C-type RNA retrovirus budding from an infected cell (×168,000).

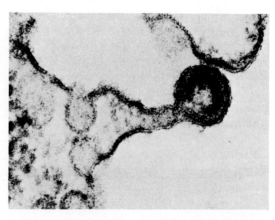

Viral Envelopes

The combination of capsid and nucleic acid is referred to as the **nucleocapsid,** and, in the case of some animal virus groups, this constitutes the completed virion. A large number of animal viruses, however, are surrounded by an additional envelope that they acquire during their final stage of replication as they bud through special areas of the host cell membrane. In these special areas the host cell proteins have been replaced by viral-coded polypeptides and glycopro-

show electron micrographs of a virus budding from a host cell membrane.

CELLS USED TO GROW ANIMAL VIRUSES

Growth of viruses requires susceptible host cells capable of replicating the virus. The studies of early virologists were restricted to the use of virus obtained from the blood or tissues of whole animals. As scientists learned to grow animal cells in culture, the use of whole animals has become much less important.

Types of Cell Cultures

Animal cell cultures can be classified either as primary cell cultures, diploid cell cultures, or permanent cell lines. **Primary cell cultures** used for the propagation of viruses may be obtained directly from the animal organ as follows: (1) the organ (such as monkey kidney) is minced and treated with a proteolytic enzyme to separate the cells; (2) the cells are dispersed into sterile tubes or dishes and covered with a buffered growth medium containing serum; (3) after the cells have adhered to the surface of the container, they can be infected with the virus to be propagated. Unlike a bacterial culture that can be transferred indefinitely, primary cell cultures can be subcultured at most five to six times before they die.

The second type of cell culture, **diploid cell strains**, can be derived from human embryonic tissue. They are morphologically normal and possess a normal chromosome number. Diploid cell cultures can be grown for a number of generations; they will die after 40 to 50 subcultures.

During the growth of diploid cell strains, rare mutations may occur that give rise to a **permanent cell line**, capable of being cultivated for an unlimited number of generations. Stable cell lines of this type, however, possess an altered morphology from the parent cell culture and are more like cells derived from malignant tumors.

Cells derived from cancer tissue or those that have been transformed into cancer cells by chemicals or oncogenic (cancer-producing) viruses can also be cultivated for an unlimited number of generations. Permanent cell lines of this type differ from diploid cells in that they will grow in suspension, and they will frequently produce tumors if injected into susceptible animals. Table 14.1 lists some stable, representative cell lines that are commercially available and the original source of the tissue from which they were isolated. Such cell lines (as well as diploid strains) may be kept frozen for very long periods; after thawing, they can again be grown.

TABLE 14.1 Selected Commercially Available Cell Lines

Cell Line	Source
HeLa	Human carcinoma of the cervix
HEp-2	Human carcinoma of the larynx
L-132	Human embryonic lung
Raji	Human Burkitt's lymphoma
RPMI 8226	Human myeloma
WI-38	Human normal diploid female
MDCK	Dog kidney
BHK-21	Hamster kidney
BS-C-1	African green monkey kidney
LLC-MK$_2$	Rhesus monkey kidney
MOPC-31-C	Mouse plasmacytoma
3T3	Mouse embryonic fibroblast
RTG-2	Rainbow trout gonadal tissue
LLC-RK$_1$	Rabbit kidney
P1-1-Ut	Racoon uterus
XC	Rat sarcoma
IgH-2	Iguana heart

Growth Medium for Cell Cultures

For animal cells to remain viable and grow in a cell culture, they must, like bacteria, be immersed in a growth medium. Numerous such media have been developed and many are available from commercial sources. In general, they contain most, if not all, of the amino acids and a number of vitamins as well as salts, glucose, and a buffer. The buffer is usually a bicarbonate buffer designed to have a pH of 7.2 when the cells are cultured in special incubators under an atmosphere of 5 percent carbon dioxide and 95 percent air. Phenol red (a pH indicator) is also added to permit visual monitoring of the pH and, prior to use, 5 to 10 percent sterile serum is added to the medium. Cell cultures are usually incubated at 37°C; under such conditions the number of cells will double every 24 to 48 h.

Cells (which are usually attached to the surface of the culture flask) will continue to divide until they completely cover the bottom of the flask, forming a monolayer of contiguous cells. Except in the case of malignant tumor cells, cell culture growth then stops because of a phenomenon known as **contact inhibition.** It is at this point that the medium can be removed and the cells inoculated with a suspension of virus. After adsorption of the virus to the cells, the cell culture is again covered with medium to sustain the cells during virus replication.

Chick embryos and, to a lesser extent, duck embryos are also used for the propagation of certain viruses. Figure 14.4 shows a 2-wk-old chick embryo immersed in a fluid enclosed by the amniotic membrane; this, in turn, is enclosed in a cavity bounded by the chorioallantoic membrane. Viruses can be in-

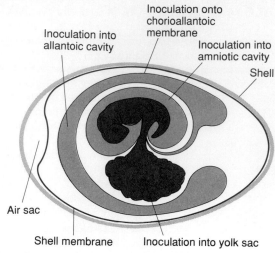

Inoculation into
allantoic cavity

Inoculation onto
chorioallantoic
membrane

Inoculation into
amniotic cavity

Shell

Air sac

Shell membrane

Inoculation into yolk sac

Figure 14.4 A chick embryo may be inoculated with virus suspension into the yolk sac, the amniotic cavity, the allantoic cavity, or directly onto the chorioallantoic membrane. In all cases the viruses grow in the membrane cells surrounding the cavity inoculated.

oculated directly into the allantoic or amniotic cavity, where they will grow in the corresponding membrane cells. After 4 to 6 days, the released virus can be harvested from the respective fluid. Several viral vaccines, including influenza and mumps vaccines, are produced in this manner. Some viruses, such as herpesviruses and poxviruses, can be placed directly on the chorioallantoic membrane, where they will produce pocklike lesions.

REPLICATION OF ANIMAL VIRUSES

Virus multiplication in each infected cell occurs in a series of independent events that culminate in the assembly of the viral nucleic acid, viral protein, and any other viral components. The details of these reproductive steps vary for each type of virus, but the following steps describe the general sequence of events:

1. Adsorption of the virion to specific receptor sites on the host cell surface is probably the most specific reaction between virus and host cell. A cell lacking such receptor sites is resistant to infection by the virus.

2. Penetration occurs either by engulfment of the intact virion or by fusion of the viral envelope with the host cell membrane so that only the nucleocapsid is allowed to enter the cell.

3. Uncoating releases the viral nucleic acid from the capsid, making it accessible to the host enzymes that will transcribe, translate, and replicate it.

4. The nucleic acid is eventually translated, to produce naked capsids, and replicated, to produce more viral nucleic acid.

5. Assembly of the various viral components into nucleocapsids occurs shortly after the replication of the viral nucleic acid. This appears to be a self-assembly process in which the virion is completed when the nucleic acid becomes enclosed within the capsid.

6. Release of the completed virions is the final step in virus multiplication. Viruses that exist as naked nucleocapsids may be released by the lysis of the host cell, or they may be extruded by a sort of reverse phagocytosis. Enveloped viruses are released by budding through special areas of the host cell membrane where proteins and glycoproteins coded by the virus have replaced those normally present in the host cell membrane.

Bearing in mind that all protein synthesis requires mRNA, there will be obvious differences between the replication of RNA viruses and DNA viruses. For example, viral DNA must first be transcribed by host cell polymerases to form mRNA. RNA viruses, on the other hand, may contain ssRNA that can function directly as mRNA without being transcribed. Such viruses are referred to as **plus strand viruses.** In others, however, their ssRNA must first be transcribed into a complementary RNA that then functions as the mRNA. Such viruses are called **negative strand viruses,** and they must carry their own RNA polymerase within the virion in order to synthesize mRNA from their negative strand of RNA. Figure 14.5 summarizes the steps involved for a DNA virus and plus strand RNA virus.

Assay of Viruses

There are a number of ways to enumerate infectious virions. One of the more common techniques is very similar to the dilution procedures for counting bacteria described in Chapter 2. In general, a suitable dilution of a viral suspension is added to a layer of susceptible tissue cells that are growing in a single layer in a bottle or petri dish. After adsorption and penetration, the entire surface is covered with a medium containing agar to prevent the free movement of released virions. However, this does not prevent infection of cells immediately adjacent to the originally infected cell. After a few rounds of replication, the viruses that kill their host cells will show small plaques of dead cells, which can be counted and multiplied by the virus dilution factor to provide the number of infectious virions in the original suspension. Figure 14.6 illustrates such plaques occurring on a monolayer of duck embryo cells.

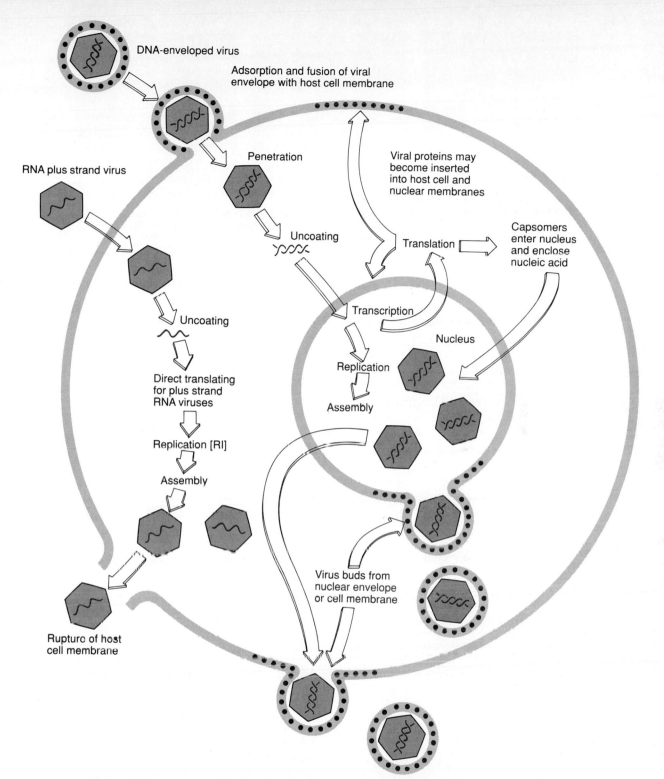

Figure 14.5 Schematic drawing illustrating the replication of an enveloped DNA virus and an RNA plus strand virus. In this and similar drawings of viral replication in the chapters to follow, the outer membrane of the nuclear envelope has been omitted for clarity.

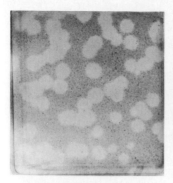

Figure 14.6 In the left and center flasks, areas of duck embryo cells have been killed by viruses, as shown by the clear areas surrounded by living cells stained with neutral red. On the right is an uninfected culture of stained duck embryo cells.

Reaction of the Host Cell to Viral Infection

In the introduction to this chapter, some of the reactions of the host cell to viral infection were enumerated. These are summarized in Table 14.2, and it can be seen that some viruses may cause more than one effect on a host cell.

In addition, some viruses produce intracellular inclusions in the infected host cell. These **inclusion bodies** are frequently but not always areas of viral assembly, and their intracellular location and appearance are constant for a particular virus. As a result, many inclusion bodies have been given specific names, and their appearance within a cell is a diagnostic criterion for a specific infection. Figure 14.7 shows one such body.

TABLE 14.2   Reaction of Host Cell to Viral Infection

Reaction	Representative Viruses Causing These Effects
Death of host cell Proliferation of host cells	Most viruses Poxvirus Papovaviruses Papillomavirus
Fusion of membranes of adjacent cells to form multinucleate hybrid cells	Respiratory syncytial virus Measles virus Sendai virus Herpesvirus
Transformation of normal cells into malignant cancer cells	Polyomaviruses Herpesvirus Adenovirus RNA retroviruses Papillomaviruses
No histologic change in host cell appearance for several weeks	Rubella virus Some adenoviruses

Host Cell Metabolism in Infected Cells

In general, viruses that kill their host cells do so by inhibiting cellular synthesis at the expense of virus production. Details of how this is accomplished are unexplained, but it has been established that infection by cytolytic viruses (those causing death and lysis of the host cell) results in a cessation of cellular macromolecular synthesis. For example, 2 to 3 h after infection of HeLa cells with poliovirus, the rate of cellular protein synthesis is about 20 percent that of uninfected cells. Cellular RNA and DNA synthesis are also rapidly inhibited. In some cases, such effects may be due to an excess of viral mRNA competing for host cell ribosomes and enzymes. In other cases, as reported for poliovirus infections, the virus appears to synthesize a product that inhibits chain initiation for the synthesis of host cell proteins. Regardless of the mechanisms involved, the effect is the channeling of the cellular biosynthetic activities toward the production of virus, ultimately resulting in the death of the cell.

Latent Virus Infections

Recovery from a primary infection by some viruses may be followed by an asymptomatic latent infection. The physical state of the virus in such cases is unknown, but it probably exists as naked nucleic acid within the nucleus of the infected cell. Herpesviruses are a common cause of latent infections, particularly those that may recur many times during the life of an infected individual. Recurring fever blisters around the mouth are probably the most common example of the reactivation of a latent herpesvirus infection but, as we shall learn later, genital herpesvirus infections have also become epidemic throughout the world. Other herpesviruses may also cause latent infections, but overt recurrences are not routine.

Measles virus may also cause a latent infection

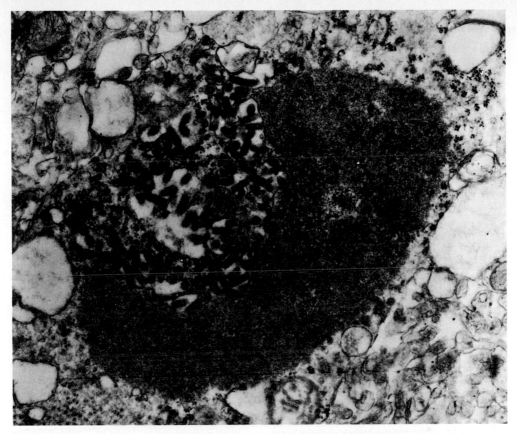

Figure 14.7 Rabies virus inclusion body in fox brain tissue (×50,000). Note the individual virions in close proximity to the inclusion.

that, on rare occasions, manifests itself by recurring as a central nervous system infection.

CLASSIFICATIONS OF ANIMAL VIRUSES

Based on structural and chemical criteria, most of the viruses infecting mammalian hosts have been assigned to one of a number of virus families, and these families have been subdivided, in turn, into one or more genera. Table 14.3 lists the current classification for the major animal DNA viruses, and Table 14.4 provides this information for the RNA-containing viruses. Figure 14.8 on page 216 is a drawing showing the relative sizes and shapes of animal viruses of the major taxonomic groups.

It should be emphasized that, although a generic type of classification of viruses is essential to provide a systematic handle for the animal viruses, some of the generic names are new and some may be used only rarely, especially by the medical virologist, who is more likely to refer to the name of a virus by the name

of the disease for which it is responsible. In Unit III our discussions will, for the most part, utilize the virus names that are more commonly used by the medical virologist.

BACTERIOPHAGES

Viruses that infect bacteria are called bacteriophages or merely phages. Although the spectrum of bacteria that any one phage can infect is limited, the enormous number of phages in existence makes it likely that there is at least one phage for every type of bacterium.

Chemical Composition and Structure

Bacteriophages occur in assorted shapes and, like animal viruses, all have a protein capsid that encloses the phage nucleic acid. Some also possess complex structures that are used to attach the phage to a susceptible cell. Figure 14.9(a) is a drawing of a T4 phage (which infects certain strains of *E. coli*) and Figure

TABLE 14.3 Chemical and Physical Properties of Animal DNA Viruses

Families and Genera of DNA Viruses	Capsid Shape (Symmetry)	Number of Capsomers	Presence of an Envelope	Mol. Wt. (in Millions of Daltons) and Physical Properties of Virion Nucleic Acid[a]	Approx. Diameter of Virion (nm)
Adenoviridae					
Mastadenovirus (infects mammals)	Icosahedral	252	No	20–25 ds	70
Aviadenovirus (infects birds)	Icosahedral	252	No		
Herpetoviridae					
Herpesvirus (human herpes)	Icosahedral	162	Yes	100 ds	150[b]
Hepadnaviruses					
Hepatitis B virus	Icosahedral	42	Yes	1.6[c]	42
Poxviridae					
Orthopoxvirus (vaccinia and smallpox)	Very complex, brick-shaped	Not applicable	Yes	100–200 ds	300 × 240
Papovaviridae					
Polyomavirus (rodents and primates)	Icosahedral	72	No	3 ds	45
Papillomavirus (produces papillomas in humans as a sexually transmitted disease)	Icosahedral	72	No	5 ds	55
Parvoviridae					
Parvovirus (infects vertebrate hosts)	Icosahedral	32	No	1.2–1.8 ss	20

[a] ds, double-stranded; ss, single-stranded.
[b] Includes envelope; naked capsid is approximately 100 nm in diameter.
[c] Second strand is incomplete and varies in length.

14.9(b) shows an electron micrograph of a similar phage attached to a bacterial cell wall.

The phage head, which contains the nucleic acid, is frequently icosahedral, but there are several other shapes, including round and cylindrical. Not all phages possess a tail; among those that do, there is considerable variation in its structure.

Replication

The sequence of events occurring during phage replication is similar for most phages; it is, however, sufficiently different from that described for animal viruses to justify a separate description.

Adsorption

In the case of the DNA-tailed phages, adsorption occurs tail first at specific receptor sites in the bacterial cell wall (see Figure 14.9). Infection cannot occur in the absence of adsorption; thus, if a bacterium loses the ability to synthesize specific phage receptors, it will become resistant to infection.

Filamentous DNA phages, containing ssDNA, adsorb specifically to the tip of the F pilus (see Chapter 8). As a result, such phages are male specific. Following adsorption, the entire phage particle travels to the cytoplasmic membrane, after which the DNA enters the cytoplasm and the capsid proteins are incorporated into the cytoplasmic membrane.

Penetration

Following adsorption, the phage injects its nucleic acid into the bacterial cytoplasm while the protein capsid remains outside the bacterium. Note that this procedure is unlike that followed by the animal viruses, which must enter their host cell before uncoating occurs.

TABLE 14.4 Chemical and Physical Properties of Animal RNA Viruses

Families and Genera of RNA Viruses	Capsid (Symmetry)	Number of Capsomers	Envelope	Mol. Wt. (in Millions of Daltons) and Physical Properties of Virion Nucleic Acid	Approx. Diameter of Virion (nm)
Picornaviridae					
Enterovirus (enteric)	Icosahedral	32	No	2.6 ss	27
Rhinovirus (respiratory)	Icosahedral	32	No	2.6 ss	28
Calicivirus (possible genus swine infection)	Icosahedral	32	No		35–40
Togaviridae					
Alphavirus (group A arboviruses)	Icosahedral	32	Yes	3 ss	50–70
Flavivirus (group B arboviruses)	Icosahedral	32	Yes	3 ss	50
Rubivirus (rubella)	?		Yes	2.5–3 ss	50–80
Orthomyxoviridae					
Influenzavirus (influenza)	Helical		Yes	5 ss	100
Paramyxoviridae					
Paramyxovirus (Newcastle disease)	Helical		Yes	6.5–7.5	150–300
Morbillivirus (measles)	Helical		Yes		
Pneumovirus (respiratory syncytial)			Yes		
Rhabdoviridae					
Lyssavirus (rabies group)	Helical		Yes	4 ss	70 × 170
Vesiculovirus (vesicular stomatitis group)	Helical		Yes	4 ss	
Arenaviridae					
Arenavirus (lymphocytic choriomeningitis)	?		Yes	1–3	
Bunyaviridae					
Bunyavirus (Bunyamwere supergroup)	Helical		Yes		90–100
Coronaviridae					
Coronavirus	Helical(?)		Yes	?	120
Hetroviridae (tumors & AIDS)	?			9–12 ss	100
Reoviridae (dsRNA viruses)					
Reovirus	Icosahedral	120	No	15 ds	75–80
Orbivirus	Icosahedral	32	No	15 ds	65–80
Rotavirus	Icosahedral		No	?	?

Phage DNA Transcription

The steps involved in the replication of the phage DNA vary considerably from one phage to another, but in the usual sequence of events the injected nucleic acid codes for the synthesis of a number of enzymes. These enzymes cause the destruction of the bacterial nucleic acid and direct the synthesis of more phage nucleic acid and more phage capsids.

Assembly and Release

When the synthesis of both nucleic acid and structural proteins is well under way, the phage begins to assemble into mature phage particles, as illustrated in Figure 14.10. After the assembly of 100 to 200 phage particles, the bacterial cell lyses due to the formation of a phage-coded lysozyme that hydrolyzes the bacterial peptidoglycan, releasing the completed phages.

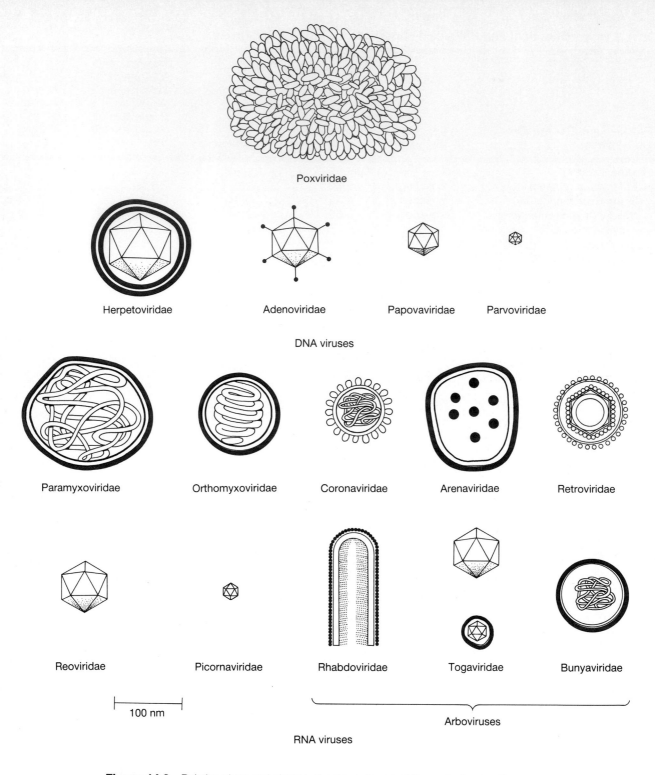

Poxviridae

Herpetoviridae Adenoviridae Papovaviridae Parvoviridae

DNA viruses

Paramyxoviridae Orthomyxoviridae Coronaviridae Arenaviridae Retroviridae

Reoviridae Picornaviridae Rhabdoviridae Togaviridae Bunyaviridae

100 nm

Arboviruses

RNA viruses

Figure 14.8 Relative sizes and shapes of animal viruses of the major taxonomic groups.

Release of the filamentous DNA phages occurs at points of adhesion between the cytoplasmic membrane and the outer membrane. In this case, cell lysis does not occur and the infected bacterium continues to multiply while producing phage.

RNA-Containing Phages

The small ssRNA phages are icosahedral and, like the ssDNA phages, adsorb specifically to the F pilus. Following adsorption, a maturation protein in the capsid

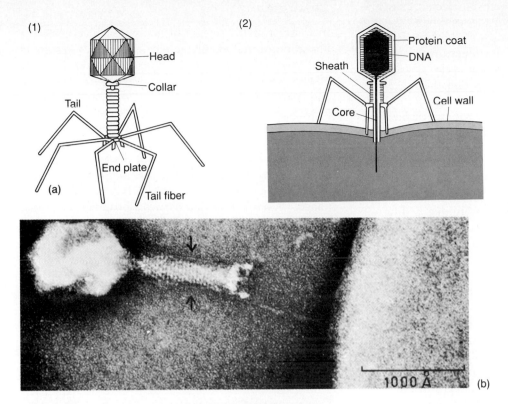

(1)
Head
Collar
Tail
End plate
Tail fiber
(a)

(2)
Protein coat
DNA
Sheath
Core
Cell wall
(b)

1000 Å

Figure 14.9 (a1) T-4 bacterial virus is an assembly of protein components. The head is a protein membrane, shaped like a kind of prolate icosahedron with 30 facets and filled with deoxyribonucleic acid (DNA). It is attached by a neck to a tail consisting of a hollow core surrounded by a contractile sheath and based on a spiked end plate to which six fibers are attached. (a2) The sheath contracts, driving the core through the wall, and viral DNA enters the cell. (b) Beginning phase of adsorption of bacteriophage T2 on *Escherichia coli*. Note the phage tail fibers adhering to the bacterial cell wall. Two retracted tail fibers are against the tail sheath (indicated by arrows) (×280,000).

Figure 14.10 The morphogenic pathway of phage maturation has three principal branches leading independently to formation of heads, tails, and tail fibers, which then combine to form complete phage particles. The numbers refer to the T-even phage genes, whose products are involved at each step. The solid arrows indicate the steps that have been shown to occur in extracts.

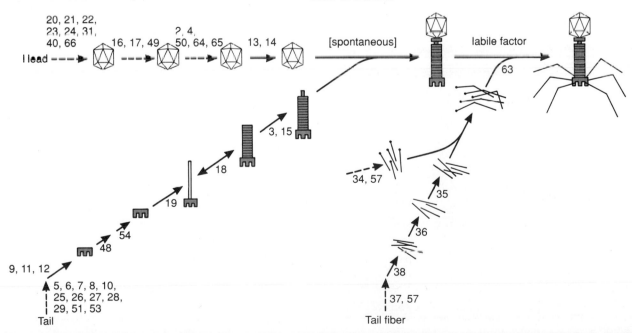

20, 21, 22,
23, 24, 31,
40, 66   16, 17, 49   2, 4,
                                50, 64, 65   13, 14   [spontaneous]   labile factor
Head ----→   ---→   ---→   ---→   63

                                                          3, 15
                                             18
                                        19
                                   54
                              48   34, 57
9, 11, 12   35
                                                         36
5, 6, 7, 8, 10,   38
25, 26, 27, 28,   37, 57
29, 51, 53
Tail   Tail fiber

is cleaved, allowing the RNA to be ejected from the capsid into the pilus. The RNA moves down the pilus and, after entering the cytoplasm of the bacterium, it is replicated and translated to form mature phage particles. Near the end of the infection cycle, a lysis protein is produced, causing cell lysis and the release of mature phage particles.

Many such phages are specific for enteric bacteria and, as a result, can be found in high concentrations in sewage and feces.

Assay

Bacteriophages can be counted by mixing a dilution of the phage with a large excess of susceptible bacteria. The large excess of bacteria is necessary so that each cell is infected by only one phage. After adsorption has taken place, the entire mixture is plated on a suitable solid medium to form a bacterial lawn. As each phage reproduces and lyses its host bacterium, the progeny infect adjacent bacteria on the plate. The result is a clear area, or plaque, occurring for each phage present in the original dilution (see Figure 14.11). By counting the plaques and multiplying by the dilution factor, the number of phages in the original material can be calculated; the method is the same as the one that was explained in detail in Chapter 5 for enumerating bacteria.

Figure 14.11 Bacteriophages can be counted by mixing a dilution of the phage with a large excess of susceptible bacteria and then plating the entire mixture on a suitable medium. As each phage lyses the original bacterium it has infected, the progeny infect adjacent bacteria on the plate. The result is a clear area, or plaque, occurring for each phage present in the original dilution. By counting these plaques and multiplying by the dilution factor, the number of bacteriophages in the original material can be determined.

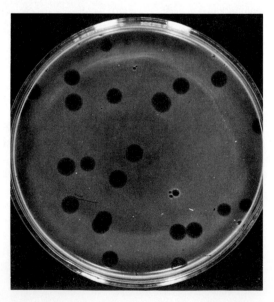

Lysogeny

In contrast to the release of many animal viruses, the liberation of mature phage particles usually results in the lysis and death of the infected bacterium. The phages whose infection is followed by reproduction and bacterial lysis are called "virulent" bacteriophages. Many phages, however, are able to infect a bacterium without inducing the production of more phage or the lysis of the infected cell. In such cases, the phage DNA becomes a part of the genetic material of its host bacterium. Phage DNA that exists and replicates along with the bacterial DNA is called "prophage." A bacterium that carries a prophage is termed "lysogenic," indicating that it possesses the potential to lyse, and a phage that can produce lysogeny is called a "temperate" phage. Of course, unless a prophage occasionally became virulent and produced mature phage particles, it would be difficult to know whether or not any particular bacterial strain was lysogenic. However, an occasional lysogenic bacterium does go through the lytic cycle, and the liberated phage can be assayed by plating the supernate on a second bacterial culture lysed by the liberated temperate phage.

Lysogenic Conversion

Even though a prophage does not transcribe all its DNA (unless it enters the lytic cycle), all the prophage genes are not necessarily blocked. As a result, part of the prophage DNA may be transcribed to form proteins new to the bacterium. These products may be enzymes that induce a change in the structure of a bacterium's outer membrane lipopolysaccharide or toxins that are excreted by the bacterium into the surrounding environment. These excreted toxic proteins, called exotoxins, are the basis for such serious diseases as diphtheria, scarlet fever, and botulism. Interestingly, if such organisms lose their lysogenic state, they are no longer able to cause disease. The acquisition of new characteristics coded for by prophage DNA is called lysogenic conversion.

ONCOGENIC VIRUSES

A number of DNA and RNA viruses are able to transform normal cells into cancer cells. Such viruses are termed **oncogenic viruses** because of their ability to produce malignant tumors in susceptible animals. In many cases the molecular events induced by a virus leading to the establishment of a cancer cell result from an activation or mutation of certain normal cellular genes termed **protooncogenes**. A study of the RNA tumor viruses, known as retroviruses, has provided considerable insight into the mechanism of malignant transformation.

The **retroviruses** constitute a large group of tumor-producing RNA viruses. The first of these was described in the early 1900s, when it was shown that leukemias and sarcomas of chickens could be transmitted to newborn, healthy chickens using cell-free extracts of the tumors. At that time, the phenomenon was an intellectual curiosity; few scientists realized its implications in relation to the role of viruses in cancer. It is now known that the RNA tumor viruses are widespread in nature, having been isolated from tumors of many animals such as birds (especially chickens), mice, cats, hamsters, and primates.

Replication of Retroviruses

The replication of the RNA tumor viruses begins with the interaction of the viral envelope glycoprotein spikes with cellular receptors on the host cell membrane. Following the fusion of the viral envelope with the host cell membrane, the viral core is released into the cell. At this point, the replication of the RNA tumor viruses differs from that of all other viruses *because the first step is the transcription of the viral RNA into complementary single-stranded DNA*. This reaction is catalyzed by a viral enzyme called RNA-dependent DNA polymerase; however, because this transcription uses RNA as a template to synthesize complementary DNA, the enzyme is frequently called a **reverse transcriptase** (the basis of the name *retroviruses*). A model for the replication of the retroviruses would be as follows:

1. DNA is transcribed from viral RNA and is then replicated to form a double-stranded DNA, which is subsequently transported to the nucleus of the host cell.

2. One or two such DNA molecules integrate into the host genome.

3. The integrated viral DNA is transcribed by the host cell RNA polymerases.

4. Viral proteins (made in the cytoplasm), including the reverse transcriptase and structural proteins, migrate to specific areas of the host cell membrane, where viral glycoproteins have been incorporated into the cell membrane.

5. The assembly of RNA, viral enzymes, and structural proteins occurs at the cell membrane, and the virus is released by budding from the membrane.

Figure 14.12 summarizes the replication of a retrovirus.

Retrovirus Gene Structure

A good deal is now known of the mechanisms whereby the retroviruses induce malignancies. To understand these mechanisms, it is necessary to learn a few facts concerning the gene structure of these viruses.

Retroviruses that can replicate independently within a host cell are termed **nondefective,** whereas retroviruses that require a helper virus for replication are designated **defective** viruses. On the basis of their ability to transform a host cell into a malignant cell and to replicate independently within a susceptible cell, retroviruses can be divided into three categories: (1) nondefective viruses that are also capable of causing a rapid transformation of host cells; (2) nondefective viruses that are incapable of transforming cells in culture but will cause malignancies in animals months or years after infection; and (3) defective viruses that can readily transform susceptible cells. As we shall see, these properties are correlated with the retrovirus gene structure.

Rous sarcoma virus (RSV) (a nondefective retrovirus) has been the most intensively studied of the RNA tumor viruses; we will use its genomic structure as a model for the gene structure of the retroviruses. From Figure 14.13(a), which includes a schematic diagram of one linear strand of the RSV genome, it can be seen that it contains only four genes.

The *gag* gene encodes for group-specific proteins found in the nucleocapsid. The *pol* gene encodes for the reverse transcriptase, and the *env* gene carries the genetic information for two glycoproteins that are inserted as spikes into the envelope of the mature virion. The *src* (pronounced "sarc") gene is completely unnecessary for viral replication; it does, however, encode for a protein kinase that is responsible for the malignant transformation of the host cell.

The nondefective leukemia viruses of virtually all retroviruses possess a genome like that in Figure 14.13(b). The genome of these viruses appears identical to that of RSV except that it lacks the *src* gene. Later, we shall postulate how these viruses induce leukemia.

The genome shown in Figure 14.13(c) is representative of defective transforming viruses from many different animal species. These viruses contain a transforming gene termed an oncogene (*onc*), which is analogous to the *src* gene of RSV. Such viruses, however, lack functional genes for either all or part of *gag*, *pol*, or *env* and, as a result, are unable to replicate unless the host cell is simultaneously infected with a competent helper leukemia virus that can provide these gene functions for the defective virus. Note, however, their *onc* gene is complete and they can, therefore, induce cell transformation in the absence of replication.

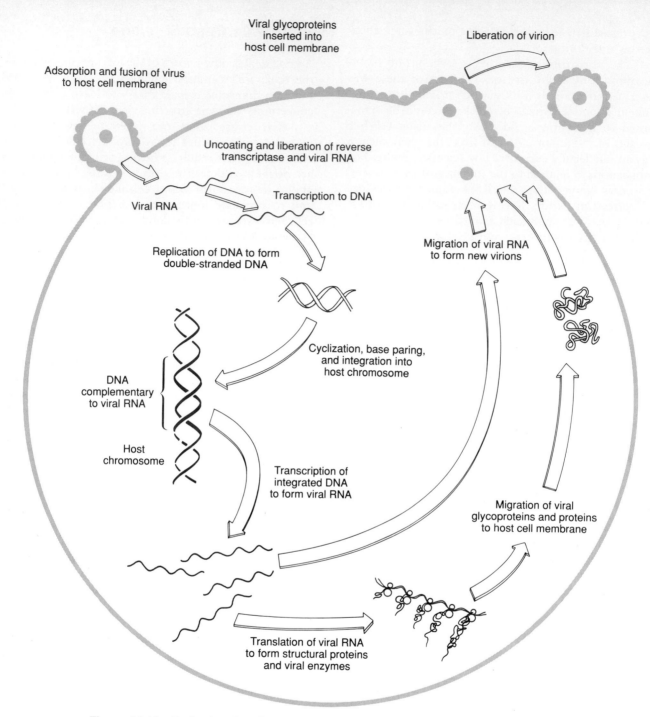

Viral glycoproteins
inserted into
host cell membrane

Liberation of virion

Adsorption and fusion of virus
to host cell membrane

Uncoating and liberation of reverse
transcriptase and viral RNA

Transcription to DNA

Viral RNA

Replication of DNA to form
double-stranded DNA

Migration of viral RNA
to form new virions

DNA
complementary
to viral RNA

Cyclization, base paring,
and integration into
host chromosome

Host
chromosome

Transcription of
integrated DNA
to form viral RNA

Migration of viral
glycoproteins and proteins
to host cell membrane

Translation of viral RNA
to form structural proteins
and viral enzymes

Figure 14.12   Replication of an RNA tumor virus. Note that the viral RNA must be transcribed
to a complementary DNA, which becomes integrated into the host genome before viral replication
can occur.

Occurrence of Oncogenes in the Normal Host Cell Genome

We have now learned that there is such a thing as an oncogene that encodes for a protein that can transform normal host cells into malignant cells. In a search for oncogenes in normal chicken cells, J. Michael Bishop and his collaborators prepared highly radioactive

DNA that was complementary to the *src* gene of RSV. It was then possible to use this complementary DNA as a probe to see if it would hybridize (i.e., form double-stranded DNA) with normal host cell DNA. Surprisingly, it was found that not only chicken DNA but also the DNA of all vertebrate cells investigated (including human cells) contained sequences homologous to *src*-complementary DNA. Using the same

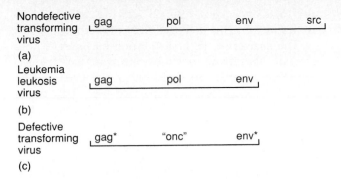

Nondefective transforming virus

(a)

| gag | pol | env | src |

Leukemia leukosis virus

(b)

| gag | pol | env |

Defective transforming virus

(c)

| gag* | "onc" | env* |

Figure 14.13 Coding regions of various RNA tumor viruses. *(a)* RSV genome group antigens (gag), reverse transcriptase (pol), and envelope (env) make up the required part of the RSV genome. Src encodes for a protein kinase that causes a malignant transformation of the host cell. *(b)* Nondefective genome of avian leukosis virus, which is essentially identical to RSV but lacks src gene. *(c)* This represents all sarcoma genomes except RSV. Such viruses are defective and cannot replicate in the absence of a helper virus. Notice that onc gene (analogous to src) occurs in the middle of the genome. Genes with an asterisk may be incomplete and, hence, nonfunctional.

technique, normal vertebrate DNA was also shown to hybridize with the oncogenes of other sarcoma viruses and acute leukemia viruses. The presence of related oncogene sequences within all vertebrates indicates that such oncogenes arose at the time of emergence of vertebrates and have been kept essentially unchanged during the course of evolution and the development of vertebrate species.

Mechanism of Transformation by Acute Leukemia and Sarcoma Viruses

An early understanding of the mechanisms of retrovirus transformation stemmed largely from studies of RSV. The protein encoded by the *src* gene of RSV (designated *v-src*) appears to be identical to one encoded in the normal host cell genome designated as *c-src*. Both have been shown to be enzymes that will transfer the phosphate group from ATP to tyrosine residues in a number of different proteins.

The paradox of this situation is that the normal host cell, *c-src*, is essentially indistinguishable from *v-src*. Why, then, does infection with RSV result in the malignant transformation of the infected cell? The answer to this question is not definite, but it has been postulated (and it seems probable) that transformation may result from a dose effect of the *v-src* enzyme. Thus, the normal cell might well require a low activity of *c-src* to maintain its metabolic and structural integrity, but the addition of large amounts of *v-src* results in excessive protein phosphorylation of some membrane protein or proteins, causing structural and metabolic changes characteristic of malignant cells.

Transformation by Nonacute Leukemia Viruses

Referring again to Figure 14.13*(b)*, it can be seen that these leukemia viruses completely lack any gene product that could cause a host cell transformation. Such viruses are the nonacute leukemia viruses, characterized by the fact that they will induce leukemia only after a latent period of many months. The mechanism by which these viruses induce leukemia appears to involve the insertion of their DNA adjacent to an endogenous host cell protooncogene. Under normal circumstances, these genes are transcribed at low levels, perhaps because they lack an efficient promoter for the attachment of the RNA polymerase to transcribe the DNA. Integration of the viral DNA, however, provides a highly efficient promoter; analysis of a number of transformed cells induced by avian leukosis virus has shown that the cellular oncogene is expressed as a result of *downstream promotion*. That is, the viral promoter permits the transcription of the gene immediately "downstream" from where it integrated into the cellular chromosome. The latent period is so long because it is only when such viruses integrate adjacent to a host cell oncogene that cellular transformation occurs, whereas in the majority of cases the position of viral integration does not result in the downstream promotion of a host cell oncogene.

Chromosomal Translocations

There have been more than 20 different cellular protooncogenes discovered in mammalian chromosomes, and several mechanisms of cellular transformation by these genes have been described. One of these mechanisms involves a chromosomal translocation in which a part of the chromosome containing the protooncogene is broken off and translocated to another chromosome. In illustration, about 90 percent of all cases of Burkitt's lymphoma contain a reciprocal translocation of a portion of chromosome 8 to chromosome 14; thus, the protooncogene *c-myc*, which is normally located on chromosome 8, is translocated to chromosome 14, the chromosome encoding for immunoglobulin heavy chains. In other cases, *c-myc* is translocated to chromosome 2 (the chromosome encoding for kappa light chains) or to chromosome 22 (where the genes for lambda light chains reside). An identical translocation involving *c-myc* has been observed in mouse plasmacytomas, a B-cell malignancy. The possible role of virus infection in such cases is not yet explained, but these observations provide strong evidence that malignancy can result from activation of a protooncogene by chromosomal rearrangements, presumably by increasing the rate of gene transcription.

Protooncogene Activation

Evidence has been presented of the involvement of other protooncogenes in human malignancies. Studies have demonstrated that the isolated DNA from some human tumor cells could be used to transfect mouse cells in a manner analogous to bacterial transformation. Using molecular cloning techniques, it has been shown that the human DNA sequences responsible for the mouse cell transformation correspond to a human protooncogene termed *c-ras,* supporting the concept that these human tumors contained activated protooncogenes that were responsible for the original malignancy. Nucleotide sequence analysis of such activated genes revealed a small but significant alteration in the protooncogene-encoded protein, suggesting that a mutagenic event had given rise to the production of a functionally altered gene protein. To date, several human tumors (bladder carcinoma, lung carcinoma, and neuroblastoma) have exhibited activated protooncogenes.

Current published evidence suggests that there are four possible pathways for tumor initiation resulting from activated oncogenes.

1. Acute retrovirus infection;
2. Promoter insertional activation of a protooncogene;
3. Activation of a protooncogene by chromosomal translocation;
4. Mutations leading to the activation of a protooncogene.

Normal Function of Cellular Protooncogenes

It would appear very strange that this large series of genes have been conserved throughout hundreds of millions of years of evolution if their sole function were to cause cancer. Instead, it has now been shown that many, if not all, protooncogenes encode for factors that regulate normal cellular growth. Thus, some act as growth factors, others as receptors for growth factors, and others (such as the protein kinases) regulate cellular activities by a more indirect route. The important concept to remember is that any event, whether chemical or viral, that alters the amount or changes the function of a protooncogene product, may transform the cell into one that is no longer subject to normal regulation.

Retroviruses Oncogenes

Since we have now established that protooncogenes are components of normal eucaryotic cells, we should now ask two questions: (1) How did retroviruses acquire their oncogenes, and (2) Are the retroviral oncogenes identical to their cellular counterparts?

The answer to the first question becomes clear if we recall that retroviruses must integrate into the cellular DNA in order to replicate. It has been shown that on very rare occasions part of the retrovirus is incorrectly transcribed together with a protooncogene, resulting in the incorporation of the oncogene into the viral capsid. The retrovirus has thus "captured" all, or part, of a cellular protooncogene.

We also know that although the retroviral oncogenes are very similar to the protooncogenes, they are not always identical. For example, the amino acid sequence of *v-ras* differs by one or two amino acids from that of *c-ras,* a change that is sufficient to permit *v-ras* to transform normal cells. Other examples include *v-erbB,* which is a truncated version of a growth factor receptor, and *v-sis,* which forms part of a normal growth factor. Thus, we see that many "captured" oncogenes either are incomplete or have undergone mutational changes so that they somehow interfere with the products of the normal cellular protooncogenes.

Retroviruses in Human Cancer

There is no doubt that retroviruses are involved in human malignancies, but experimental proof such as that provided by Koch's postulates is not possible. Two different retroviruses have, however, been isolated from human lymphomas. They are now assigned to be the etiologic agents of these malignancies, but the precise mechanism by which they cause the malignancy is unknown. These viruses have been named **human T-cell lymphoma virus I and II** (usually abbreviated as HTLV-I and HTLV-II). A third retrovirus, designated as human immunodeficiency virus (HIV), is known to cause **acquired immune deficiency syndrome** (better known as AIDS), but because this is not really a malignancy, it will be discussed in more detail with the sexually transmitted diseases.

Tumor-Suppressor Genes

There appears to be little doubt that the overexpression or mutation of protooncogenes can result in the occurrence of malignant tumors. But now a second category of genes, tumor suppressors (also called antioncogenes), has been implicated in tumor formation. The existence of such genes was first suspected when it was observed that the fusion of normal cells with most tumor cells yielded a normal hybrid. This was interpreted to show that normal cells contained a product that could suppress the malignant growth of a tumor cell. In addition, epidemiologic and genetic studies of a heritable childhood tumor of the eye,

The existence of an antioncogene was first established through studies of a malignant eye cancer known as retinoblastosis (RB). Genetic studies revealed that the majority of cases of RB were familial with about one half the children from afflicted families acquiring the malignancy during early childhood and the other half remaining disease-free. Thus, it was apparent that RB was being transmitted genetically from parent to offspring, but the figures were not compatible with either a dominantly or recessively transmitted trait.

This puzzle was eventually solved when it was shown that RB tumor cells were missing a portion of their chromosome 13. This, and other data, led to the following conclusion: (1) When one parent is heterozygous for the Rb1 gene, half of that parent's offspring will receive one defective Rb1 gene. The remaining children will receive a normal Rb1 gene from both parents. (2) If a mutation occurs in the normal Rb1 gene in a child who already has one defective Rb1 gene, that person will develop RB. Persons with two normal Rb1 genes would require that mutations occur in both genes before developing the malignancy.

It is not known precisely why the loss of the Rb1 gene product should result in a malignancy, but it is known that it is a 105-kd protein that is found in the nucleus, and one could propose that its normal function is to regulate DNA transcription.

Are antioncogenes (tumor-suppressing genes) involved in other types of cancer? Probably. But our knowledge in this field is still fragmentary. Activated oncogenes can be detected in only about 20 percent of human cancer cells, suggesting that many cancers might have arisen because of a loss rather than through the acquisition or activation of an oncogene. Moreover, it seems well established that the majority of malignancies arise only after two or more mutational events, which is also conducive with the loss of a gene product. In any event, it seems highly probable that more such genes will be described, providing a better understanding of the molecular changes that transform a normal cell into a malignant one.

called retinoblastoma, revealed that the tumor resulted from a *loss* of a gene. More specifically, it was shown that retinoblastoma tumor cells always contained a deletion in chromosome 13 (not seen in normal cells), which has subsequently been designated the Rb1 locus. Human osteosarcoma (bone cancer) has also been shown to result from the inactivation of the Rb1 locus. Introduction of the cloned, normal retinoblastoma gene into either retinoblastoma tumor cells or osteosarcoma cells reversed the malignancy. In addition, most, if not all, human small-cell lung cancers show a deletion of the Rb1 gene, reinforcing the concept that the Rb1 gene product is essential for normal cell growth.

Another tumor-suppressor gene product, which has been designated as p53, has also been shown to be deleted in a large number of lung cancers, colon cancers, and osteogenic sarcomas. The normal function of such tumor-suppressing gene products is as yet unknown, but it seems quite clear that the absence or removal of these "antioncogene" products results in the transformation of a normal cell into one that is malignant.

Oncogenic DNA Viruses

There are a number of DNA viruses that are known to produce tumors in animals and several, such as the herpesviruses, the papillomaviruses, and hepatitis B virus, are strongly implicated as causes of human cancer (see Unit III). In no case, however, has it been shown that a DNA virus contained a cellular protooncogene as was described for the retroviruses. Whatever gene product is responsible for cell transformation is also essential for DNA virus replication and, therefore, unlike the retroviruses, it is not possible to separate virus replication from cell transformation. It is possible, however, to pinpoint certain genes in the oncogenic DNA viruses that are involved in malignancy. Let us look at several examples that provide some insight into the mechanism whereby DNA viruses might induce oncogenic changes.

SV40 and polyomavirus (both are papovaviruses) will produce tumors in some animals and will transform certain animal cells in culture. Both produce several antigens known as "tumor antigens" (T-Ags), which are essential for replication but are also involved in cell transformation. Certain adenoviruses also produce malignant tumors in animals, and the gene products associated with their oncogenicity are termed E1A and E1B. Our final example is the papillomaviruses, some of which are strongly associated with cervical and genital cancers in humans. The gene products of papillomaviruses involved with malignancy are designated E6 and E7.

What have all these DNA virus "oncogenes" in common? First, most are early viral gene products that are involved in the regulation of viral nucleic acid transcription, and second, most of them appear to form stable complexes with tumor-suppressor gene products. Thus, SV40 T antigens, adenovirus E1A, and papillomavirus E7 all bind with the retinoblastoma gene product. SV40 T antigens and adenovirus E1B bind to p53. One could postulate, therefore, that these viruses induce malignancies through their ability to remove a tumor-suppressing gene product.

When the story is finally known, it seems highly likely that it will be considerably more complex than presented here. A complete understanding will also have to include the mechanism whereby hepatitis B virus induces liver cancer and why some viruses produce cancer in one type of cell and a productive infection in another. Hopefully, in the next edition of this book we will be able to present answers to these questions.

VIRUS MUTATION

Viruses, like living cells, occasionally form mutants; that is, a new virus will be formed that differs from its parent and in turn will produce offspring like itself (i.e., the mutant virus) and not like the original parent strain. These mutants may differ in many ways, but perhaps the most interesting and useful mutants are those that have lost the ability to produce disease but retain the other characteristics of the parent virus. Such mutants can still infect but can no longer produce symptoms of disease. Mutant viruses are used to immunize against pathogenic strains; for example, live measles vaccine is made from a nonpathogenic measles virus mutant.

Control of Viral Infections

Immunization with attenuated (nonvirulent) or killed viruses is still the foremost method in use today for the control of human and animal virus infections. Perhaps the newest innovation is the production of immunogens from cloned recombinant DNA derived from the virus. Using recombinant DNA techniques (see Chapter 8), it is possible to produce a purified protein or glycoprotein that, when injected into an individual or animal, will stimulate the production of antibodies and thus protect against infection. Cloned viral products are being used as immunogens to prevent hepatitis B, foot-and-mouth disease, and possibly soon rabies and influenza.

In Unit III, we shall also discuss specific chemicals that are effective for the treatment of certain viral infections, as well as interferon that is produced by virally infected cells.

PLANT VIRUSES

There are a large number of viruses that cause plant diseases, which range from essentially asymptomatic infections to ones that rapidly kill the plant. The majority of such viruses contain ssRNA as their genomic nucleic acid, which may exist as either plus or minus strands. There are also plant viruses that contain dsRNA, ssDNA, and dsDNA.

Nomenclature and Taxonomy of Plant Viruses

Plant viruses have, in general, been named according to the plant they normally infect and the disease resulting from such an infection. This has yielded names such as tobacco mosaic virus, tomato bushy stunt virus, and turnip yellow mosaic virus. Plant virologists have not, however, assigned the plant viruses to families as was done for the animal viruses but, rather, have usually placed closely related viruses into groups that have been designated by syllables taken from the name of the disease entity caused by the virus. Thus, as shown in Table 14.5, tobacco mosaic virus belongs to the tobamovirus group, tomato bushy stunt virus to the tombusvirus group, and turnip yellow virus to the tymovirus group.

Transmission and Replication of Plant Viruses

Plant viruses do not appear to possess specific cell receptors such as occur on animal viruses. Initial infection, therefore, follows injury to the leaf, permitting entry of the virus. Under normal circumstances, such injury is done by insects such as aphids or leaf hoppers, which feed on plant leaves and carry the virus from one plant to another. Laboratory infections are frequently accomplished by rubbing the leaf lightly with sand at the time of adding the virus.

Much is yet to be learned about the replication of plant viruses, but there are unusual characteristics of some of these agents that are not seen in animal viruses. For example, many of the RNA-containing plant viruses possess a segmented genome consisting of two to four strands of RNA. The unusual aspect of some of these divided-genome viruses is that each RNA molecule is packaged in a separate capsid. Viruses such as alfalfa mosaic and cucumoviruses package their RNA in four separate capsids, and the subsequent infection of a new plant therefore, requires the entry of multiple capsids. It is also noteworthy that most plant cells contain an RNA-dependent polymerase and, thus, it is probably unnecessary for the

TABLE 14.5 Examples of Selected Plant Viruses and the Groups into Which They Are Placed

Virus	Number of Viruses in Group	International Group Name
Turnip yellow mosaic virus	16	Tymovirus
Barley yellow dwarf virus	28	Luteovirus
Tomato bushy stunt virus	8	Tombusvirus
Carnation latent virus	21	Carlavirus
Potato virus Y	73	Potyvirus
Potato virus X	24	Potexvirus
Tobacco mosaic virus	15	Tobamovirus
Tobacco ringspot virus	18	Nepovirus
Cowpea mosaic virus	10	Comovirus
Tobacco rattle virus	2	Tobravirus
Cucumber mosaic virus	4	Cucumovirus
Brome mosaic virus	3	Bromovirus
Tobacco streak		Ilarvirus
Barley stripe mosaic		Hordeivirus
Beet yellow		Closterovirus
Cauliflower mosaic		Caulimovirus

RNA plant viruses to encode for their own polymerase.

Structure of Plant Viruses

Plant viruses exist in both helical and isometric shapes, but there is great variation in the capsid size of the various groups. Figure 14.14 illustrates the comparative sizes and shapes of a few representative plant viruses. Note that an infectious unit for viruses such as tobacco rattle, alfalfa mosaic, and tobacco streak consists of multiple capsids of varying sizes. Cucumber mosaic and brome mosaic viruses also are segmented into multiple capsids but in these cases they are found in similar-sized icosahedral capsids.

Of the DNA-containing plant viruses, the geminiviruses have the most unusual structure. These agents, such as bean golden mosaic and maize streak viruses, consist of two icosahedral capsids occurring in pairs, with each pair containing two molecules of ssDNA. This appears to be the only known example of a segmented DNA genome.

Response of Plants to Virus Infection

It is common for plant viruses to cause a mosaic disease in which various areas of the leaf are light green or yellow in color because of a destruction of chloroplasts and, hence, a deficiency of chlorophyll. Other viruses, such as the plant reoviruses, may induce tumor formation while some cause necrosis of plant cells, presumably by stimulating the plant to form excess enzymes such as phenol oxidase. One group, exemplified by tomato bushy stunt virus, causes plants to remain abnormally small. Such viruses are believed to act by interfering with growth hormone production by the plant.

VIROIDS AND PRIONS

Small molecules of naked RNA that cause several plant diseases have been isolated and characterized. These very unusual infectious molecules do not possess a protein coat and exist solely as short, infectious molecules of RNA. Such molecules of free, infectious RNA are called **viroids.** They have never been observed in the animal kingdom, probably because such unprotected nucleic acid would be quickly destroyed by the many nucleases present in the cell.

Potato spindle tuber viroid has been sequenced and shown to contain 359 nucleotides, which are joined to form a covalently closed circle. The mechanism whereby viroids cause plant diseases is unknown, but there is no dispute that naked RNA is the etiologic agent for viroid-induced plant diseases.

Prions is the term given to the infectious agents that cause a variety of neurological diseases in humans and animals. Such agents are included among the slow viruses because of the very long incubation periods that occur between the time of infection and the appearance of symptoms. These infectious agents are incredibly resistant to heat, withstanding an hour in

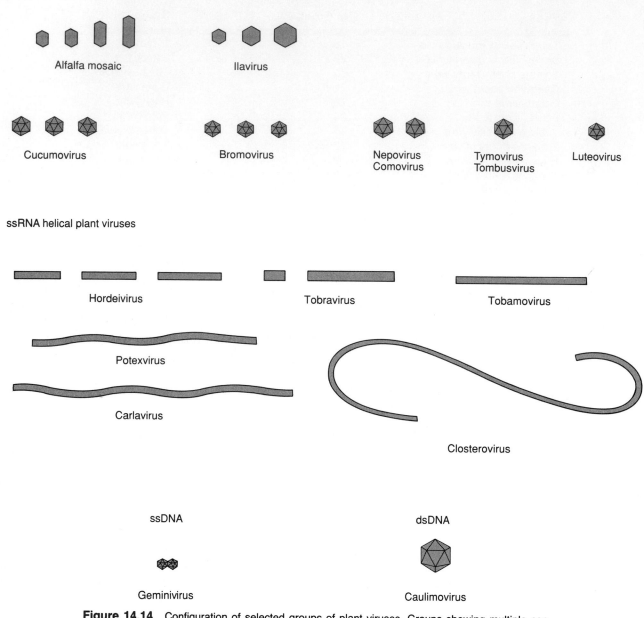

ssRNA isometric plant viruses

Alfalfa mosaic

Ilavirus

Cucumovirus

Bromovirus

Nepovirus
Comovirus

Tymovirus
Tombusvirus

Luteovirus

ssRNA helical plant viruses

Hordeivirus

Tobravirus

Tobamovirus

Potexvirus

Carlavirus

Closterovirus

ssDNA

Geminivirus

dsDNA

Caulimovirus

Figure 14.14 Configuration of selected groups of plant viruses. Groups showing multiple capsids indicate that different segments of the virus genome are packaged in separate capsids. Infection by such viruses requires that the plant cell receive at least one of each of the different capsids.

an autoclave. They are also resistant to destruction by nucleases but are destroyed by proteinases. Thus, the term *prion* was coined by Professor Stanley Prusiner to represent proteinaceous infectious particles. It is very important to keep in mind, however, that no one has ever isolated or grown a prion, and many scientists find the concept of a self-replicating, infectious protein impossible to conceive.

Many diseases have been attributed to prions, but the only ones of these that are actually known to be infective are **kuru** and **Creutzfeldt-Jakob disease** in humans and **scrapie** in sheep. All of these diseases lead to dementia and death, and no cures are known. It will obviously take much more research before we know the nature of these infectious agents and whether there actually is such a thing as a prion.

SUMMARY

Viruses can grow only within susceptible living cells because they must use the host cell's ATP-generating system for energy and its ribosomes to synthesize viral proteins. Viruses contain either DNA or RNA, but not both. Their nucleic acid is enclosed in a protein overcoat, called a capsid, and in many instances the capsid is surrounded by an envelope acquired as the virus budded from the host cell envelope. Viruses are usually grown in cultures of animal cells, although chick embryos are used for some viruses.

Animal viruses are taken intact into susceptible cells, and there the nucleic acid is liberated by uncoating the virus. The nucleic acid is transcribed and translated to produce naked capsids and replicated to form more viral nucleic acid. The various components of the virus are then assembled and the virus is released from the host, either by budding from the membrane or by lysis of the cell.

When some animal viruses are growing on a monolayer of animal cells, each infectious unit of virus may destroy a small area of the cell culture. These areas may be counted and used to calculate the number of infectious virions in a specimen. After initiating replication in a susceptible cell, some viruses form characteristic forms, called inclusion bodies, inside the cell.

Bacteriophages are bacterial viruses most of which infect the host bacterium by adsorbing to the cell and injecting their nucleic acid into the bacterium. In most cases, the bacteriophage nucleic acid is transcribed, translated, and replicated, resulting in the formation of several hundred mature phage particles. At this point, the bacterium bursts and liberates the bacteriophages. In some instances phage DNA can remain in the bacterium, replicating along with the bacterial DNA but not forming mature phage particles. This phage DNA is called prophage, and a culture containing prophage is referred to as lysogenic. If the prophage imparts new properties to the bacterium, the process is called lysogenic conversion.

Many viruses, called oncogenic viruses, will transform their host cell into a malignant cell. Oncogenic RNA viruses are called retroviruses because, in order to replicate, their RNA must first be transcribed into DNA by a viral enzyme called *reverse transcriptase*. The transcribed DNA is then integrated into the host cell chromosome where it, in turn, is transcribed to produce more virus and/or to transform the host cell into a malignant cell.

Some retroviruses carry in their nucleic acid an oncogene that, when transcribed and translated, forms a protein responsible for the cell transformation. Others lack such a gene product and appear to transform by integrating into the host cell's chromosome adjacent to an endogenous oncogene. This provides a promoter region for the RNA polymerase, permitting transcription of the endogenous oncogene.

Some tumors appear to result from the activation of the host's normal protooncogenes. Such activation may follow the translocation of the cellular oncogene to another chromosome where it is more efficiently transcribed; other protooncogenes appear to become activated through mutational events to induce tumors.

Normal cells are also known to possess tumor-suppressing genes such as Rb1 and p53. The loss of the product of such genes, either through mutation or by binding to a DNA tumor gene product, can result in the transformation from normal to malignant growth.

Viruses may also infect a variety of plants. Such viruses are usually transmitted from one plant to another by insects, such as aphids or leaf hoppers. Some plant viruses have a segmented RNA genome and will enclose each segment of RNA in a separate capsid. Plant viruses may cause an assortment of effects on infected plants, such as mosaics of discoloration, tumor formation, necrosis, or inhibition of normal growth.

Viroids are molecules of circular, infectious RNA that cause certain plant diseases. Prion is a term that has been given to the infectious unit of certain neurological diseases such as kuru, Creutzfeldt-Jakob disease, and scrapie. No one, however, has ever seen or isolated a prion so that the true nature of these agents remains unknown.

QUESTIONS FOR REVIEW

1. Cite two reasons why it is not possible to grow viruses outside living cells.
2. What is a viral capsid? List two functions of the capsid.
3. How does a virus acquire an envelope?
4. List three types of cell cultures. How do they differ from each other?
5. List the steps involved in the replication of a bacteriophage.
6. What is a lysogenic culture?
7. What is an oncogenic virus?
8. How does the replication of a retrovirus differ from that of other RNA viruses?
9. What is an oncogene?
10. How do the nonacute leukemia viruses that lack an oncogene transform their host cells?
11. By what methods can a protooncogene be activated to produce a malignant cell?
12. What are the normal functions of protooncogenes?
13. What are tumor-suppressor genes? Name two.
14. Postulate how a DNA virus such as papillomavirus can induce tumors.
15. How can recombinant DNA techniques be of value in controlling viral infections?
16. How are plant viruses named and classified?
17. What is unusual about the RNA packaging in alfalfa mosaic virus?
18. What is a viroid? A prion?

SUPPLEMENTARY READING

Bishop JM: Viral oncogenes: A review. *Cell* 42:23, 1985.

Croce CM: Chromosomal translocations oncogenes and B-cell tumors. *Hosp Pract* 20:41, 1985.

Croce CM, Klein G: Chromosome translocations and human cancer. *Sci Am* 252(3):54, 1985.

Diener TD: Viroids. *Sci Am* 244(1):66, 1980.

Gallo RC: The first human retrovirus. *Sci Am* 255(6):88, 1986.

Gallo RC: The AIDS virus. *Sci Am* 256(1):46, 1987.

Huang HS, et al.: Suppression of the neoplastic phenotype by replacement of the RB gene in human cancer cells. *Science* 242:1563, 1988.

Lennette EH (ed): *Laboratory Diagnosis of Viral Infections*. New York, Dekker, 1985.

Marx JL: How DNA viruses may cause cancer. *Science* 243:1012, 1989.

McKinley MP, et al: Molecular characteristics of prion rods purified from scrapie-infected hamster brains. *J Infect Dis* 154:110, 1986.

Spriggs DR: Viral oncogenesis: Variations on a theme. *J Infect Dis* 153:179, 1986.

Takahashi T, et al.: p 53: A frequent target for genetic abnormalities in lung cancer. *Science* 246:491, 1989.

Weinberg WA: Finding the anti-oncogene. *Sci Am* 259(3):44, 1988.

Chapter 15

Eucaryotic Microorganisms

OBJECTIVES After study of this chapter, you should

1. Be able to list one or more properties for each of the six divisions of algae.

2. Be familiar with the term phytoplankton and know their role in our ecology.

3. Be able to differentiate between a yeast and a mold

4. Know the types of sexual and asexual spores produced by each major class of fungi.

5. Be able to list important products that are the result of yeast or mold metabolism.

6. Be cognizant of what are slime molds and be familiar with their fruiting body and spore formation.

7. Be able to define a lichen.

8. Know the general morphologic properties of each class of protozoa.

9. Be able to list one or more diseases caused by each class of protozoa.

10. Be familiar with morphology and replication of each of the parasitic classes of helminths.

This chapter is designed to acquaint the reader with a few of the properties of eucaryotic microorganisms, some of which form complex cellular structures that are readily seen without the aid of a microscope. Our goal is to provide a general concept of each of the four major divisions of eucaryotic microorganisms, the algae, the fungi, the protozoa, and the helminths. Diseases caused by these eucaryotic organisms are discussed in Unit III.

Before proceeding, the reader is encouraged to review, in Chapter 1, the properties that differentiate eucaryotic and procaryotic cells.

ALGAE

Eucaryotic algae comprise organisms that contain one or more types of chlorophyll plus additional pigments known as carotenoids and biloproteins (also called phycobilins). Carotenoids are yellow, orange, or red water-insoluble linear hydrocarbons; biloproteins are blue or red water-soluble pigment-protein complexes.

The color of the different groups of algae is dependent on the ratio of these pigments that occur within the cell. All, of course, contain chlorophyll, but in some groups the green is masked by the carotenoids, giving a brown or red color to the algae.

All algae obtain their energy from the type of photosynthesis described in Chapter 6 for green plants. Thus, water serves as a reductant for nicotinamide adenine dinucleotide phosphate (NADP), and oxygen is evolved from the photolysis of water as shown below:

$$CO_2 + 2H_2O \longrightarrow CH_2O + H_2O + O_2.$$

The algae comprise a group of comparatively simple plants that are differentiated from the seed plants by the fact that they lack both vascular tissue (xylem and phloem) and airborne spores. Moreover, algae do not possess multicellular reproductive structures as found in higher plants.

Algae are widely distributed in nature, existing almost anywhere there is sufficient light, moisture, and nutrients to sustain their growth. Many grow as microscopic forms that float on or near the surface in both fresh and salt water. Such organisms, called **phytoplankton,** provide the major source of food for many aquatic animals. Because they can fix carbon dioxide into organic sugars using light as their sole source of energy, they constitute the beginning of an essential food chain. The depth at which algae may be found in such cases will vary according to their pigments and to the intensity of the sunlight, but many grow well at several hundred feet below the surface. Algae

are also found in damp soil and other terrestrial niches where conditions permit growth.

Cellular Organization

Many algae grow solely as single cells whose size may vary from several micrometers to more than a centimeter in diameter. Others, however, form multicellular colonies that may contain morphologically identical cells or may consist of a complex of differentiated cells. Interestingly, some colonies containing thousands of individual cells behave as a single organism in that reproduction occurs only in an intact colony and the movement of the colony results from the coordinated action of all cells. Still others may grow to lengths of greater than 100 m. In general, however, their individual cells are surrounded by a rigid cell wall. Within the cell are found the usual components of a eucaryotic organism, such as mitochondria, storage granules, a nucleus enclosed within a nuclear membrane, and one or more chlorophyll-containing chloroplasts. Some of the large multicellular algae produce special air-filled structures, termed **bladders,** which serve to maintain the organism at a water depth most conducive to photosynthesis. Because there is such diversity among the algae, let us briefly examine the major characteristics of each of the divisions listed in Table 15.1.

Chlorophyta

The Chlorophyta, commonly called the green algae, are considered by some to be the ancestors of the first land plants. Members of this division exist as single cells, colonies of cells, and filamentous forms that produce holdfasts to attach the organisms to a solid surface. The cell walls of the green algae are composed of cellulose, but many produce a two-layered wall in which the outer layer consists of pectin (a polymer of galacturonic acid, galactose, and arabinose). Most are found in fresh water, and they are characterized by storing starch as a reserve food material.

Chlamydomonas is one of the more common unicellular green algae. Each cell normally possesses two flagella and a single chloroplast, which, in addition to chlorophyll, contains a red spot that is thought to be involved in light perception. *Chlamydomonas* reproduces asexually by longitudinal fission to form two to eight daughter cells, which are then released from the parent organism (see Figure 15.1). Sexual reproduction also occurs through the fusion of opposite mating types.

Volvox is a green alga that exists only in a rather complex colony. Each colony consists of several hundred to many thousands of individual cells that are cemented together by a gel-like substance, forming a hollow sphere with each cell so oriented that its

TABLE 15.1   Characteristics of Eucaryotic Algae

Major Divisions	Common Name	Method of Reproduction	Nature of Cell Wall	Nature of Reserve Food Storage	Major Habitat
Chlorophyta	Green algae	Asexual: fission, motile zoospores Sexual: fusion	Cellulose and pectin	Starch	Fresh water; soil
Euglenophyta	Euglenids	Asexual: fission	No true cell wall	Fat and carbohydrate	Fresh water
Chrysophyta	Yellow-green algae Golden-brown algae Diatoms	Asexual: fission	Pectin, frequently impregnated with silica	Oils	Fresh and salt water; soil
Phaeophyta	Brown algae	Asexual: motile zoospores Sexual: motile gametes	Cellulose and pectin; alginic acid	Carbohydrates, fats	Salt water
Pyrrophyta	Dinoflagellates	Asexual: fission Sexual (rare): fusion	Cellulose and pectin	Starch and oil	Salt water
Rhodophyta	Red algae	Asexual: nonmotile spores Sexual: nonmotile gametes	Cellulose and pectin	Starch	Salt water

flagella face outward (Figure 15.2). Although the individual cells appear similar to each other, only a few are destined to give rise to daughter colonies. In some species of *Volvox* each individual colony is either male or female, and sexual reproduction occurs when the released male sperm fertilize the egg within a female colony.

Euglenophyta

The euglenids comprise motile, single-celled microorganisms that reproduce by longitudinal binary fission. *Euglena*, one of the most studied genera, possesses many animal cell characteristics in addition to being photosynthetic. *Euglena* does not possess a rigid

Figure 15.1 *Chlamydomonas reinhardtii*. Note the two flagella occurring on this common unicellular green alga.

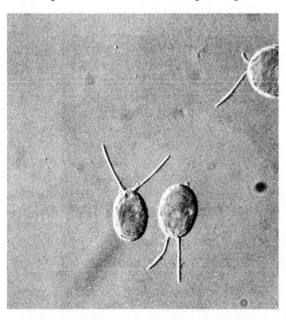

Figure 15.2 A mature *Volvox carteri* spheroid containing 16 asexually produced daughter spheroids. When released, the daughter spheroids will expand in size and the 16 large cells inside each spheroid will cleave to form new identical daughter spheroids. Each spheroid contains approximately 1000 somatic cells (similar in structure to *Chlamydomonas*) and 16 gonida or asexual reproductive cells.

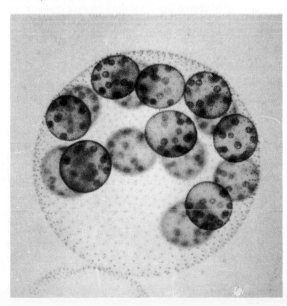

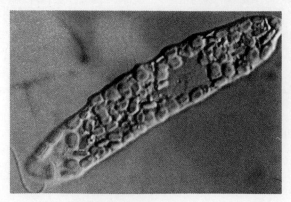

Figure 15.3 *Euglena gracilis* (×600). Note the single polar flagellum on this photosynthetic organism.

cell wall and does have an anterior gullet, but this is not normally used for the ingestion of food (Figure 15.3). Some, however, are facultative chemoheterotrophs, permitting them to grow in the dark by ingesting organic matter. These attributes indicate that the euglenids probably descended from animal-like ancestors. They can be readily found in both soil and water.

Chrysophyta

The Chrysophyta are sometimes referred to as the golden-brown algae. Two of the most characteristic features of these organisms are the storage of oils as a reserve food and the possession of large amounts of the carotenoid pigment carotene (which imparts the golden color). Most forms are unicellular, and many (the **diatoms**) have cell walls that exist in two portions, one overlapping the other, resembling the two parts

of a petri dish. These unicellular algae are abundant in both fresh and marine waters. There are at least 40,000 species of diatoms occurring in a variety of shapes, many with intricate designs, and all consisting of opposing cell walls containing silica (Figure 15.4). In some areas of the world there are huge deposits of these shells; the resulting diatomaceous earth is used commercially as a cosmetics base, as a polishing material, and for clarifying fruit juices.

Phaeophyta

The Phaeophyta contain large amounts of the brown pigment fucoxanthin and, consequently, are often called the brown algae. They are the most complex of the algae, with some (the kelps) growing to lengths of over 100 m. They characteristically occur in marine waters and in the Sargasso Sea, an area of the North Atlantic where many thousands of square miles are covered with a brown alga called *Sargassum*.

Pyrrophyta

This division contains a large group of motile, unicellular algae known as the dinoflagellates. They are characterized by cell walls composed of interlocking plates of cellulose and pectin. Two red-pigmented dinoflagellates, *Gonyaulax catanella* and *Ptychodiscus brevis* (older name, *Gymnodinium breve*), occasionally multiply to very high concentrations in coastal waters. This phenomenon, known as the "red tide," is a potential danger to humans because of toxic reactions following ingestion of shellfish that have fed on these dinoflagellates. Muscle weakness is the most distinctive symptom, and in the event of paralysis of the

Figure 15.4 Representative diatoms. *(a)* Marine diatom after acid treatment, leaving only siliceous components. *(b) Fragilaria sp.*, a common marine diatom. Individual cells of this diatom adhere to each other at their corners by gelatinous pads.

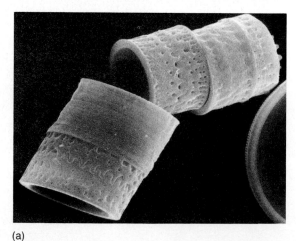

(a)

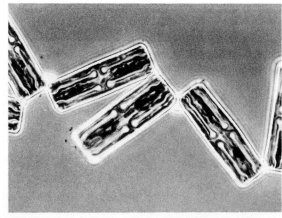

(b)

respiratory muscles, the toxemia may terminate in death.

In the United States red tides appear on the Pacific Coast and in the Gulf of Mexico; occasionally, they invade the coastal waters of New England and eastern Canada. During such outbreaks, shell fishing is banned.

Rhodophyta

These are the red algae found primarily in marine environments. They may be unicellular or filamentous, with the larger forms growing to lengths slightly greater than 1 m. Calcium carbonate from seawater is deposited on the surfaces of some species and over many centuries, this has resulted in the formation of algal reefs.

One species, commonly called Irish moss, produces a gelatinous material called carrageen, which is used to thicken puddings, evaporated milk, and ice cream. In Japan, organisms in the genus *Porphyra* are harvested for food, and without species of *Gelidium*, the bacteriologist would have to find a substitute for agar, since these organisms are the source for this material.

Ecological Role of Algae

Algae of all kinds, particularly the unicellular forms known as phytoplankton, provide the chief source of food for fish; such phytoplankton grow in vast numbers in both fresh and marine waters. In addition, algae are extremely important for releasing oxygen formed during photosynthesis. In areas where brown algae are plentiful, the plants are ground and used to fertilize the soil. Some algae are used in special foods, and it is conceivable that some could be produced commercially to provide a source of protein for both humans and domestic animals.

Water that is polluted with sewage or other organic matter provides an ideal environment for the growth of many algae and, as a result, **blooms** of algae are frequently indicative of water pollution. The unfortunate consequences of such blooms occur when the algae die and decompose, depleting the dissolved oxygen in the water.

It is also of interest that much of our current petroleum and natural gas deposits are the result of the partial decomposition of plankton that was covered by sediments many millions of years ago.

FUNGI

Fungi are eucaryotic organisms that lack chlorophyll, are nonmotile (with the exception of certain spore forms), and may grow as single cells (yeast) or as long, branched, filamentous structures composed of many cells (molds). Included among the fungi also are such macroscopic organisms as mushrooms and puffballs. Some fungi are termed **dimorphic** because they will grow as single-celled yeast under one set of conditions or as a filamentous mold under different circumstances. It is estimated that over 100,000 diverse species of fungi participate in the cycles of nature and, fortunately, only a few cause disease in humans.

Molds

Although patches of mold are visible, the individual cells are microscopic. Molds are composed of long filaments of cells joined together, end to end. The filaments are called **hyphae** (*sing.* hypha). Many molds have cross walls (**septa**) in their hyphae that divide each hypha into many different cells with individual nuclei. This type of arrangement is referred to as a **septate hypha** [Figure 15.5(*a*)]. In some classes of fungi the filaments do not contain septa and thus appear to be one long cell containing many nuclei. This type of hypha is called a **coenocytic hypha** [Figure 15.5(*b*)].

Figure 15.5 Septate and coenocytic hyphae. *(a) Phialophora verrucosa* showing hyphal strands clearly divided by septa. *(b) Absidia corymbifera* produces coenocytic hyphae, resulting in the free movement of nuclei within the hyphal strand.

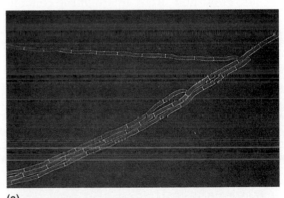

(a)

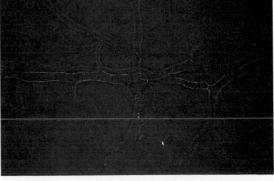

(b)

The size of the cells that make up the hypha varies from one mold to another. The larger may have diameters of 10 to 20 μm (contrasted with bacterial cells, which have an average diameter of 1 μm); smaller ones may be nearly as small as bacterial cells. The length of the filaments may vary a great deal, depending on a number of factors such as the manner in which the mold is grown. Moreover, hyphae are usually very fragile, and any handling of the mold will result in broken hyphae and filaments shorter than normal.

As a mold grows, the hyphae intertwine to form a mass of filaments (the mass is referred to as a **mycelium**) that is large enough to be seen with the naked eye. It is this fuzzy mycelium that allows molds to be recognized easily. The various pigments observed on molds are present only after the spores develop.

Conditions Affecting Growth

Molds require abundant moisture, a supply of organic matter, and a supply of oxygen for growth. Most are **saphrophytes**, that is, they live on dead or decaying organic matter. Although molds will grow best at about normal room temperature, many will grow at refrigerator temperatures. In general, however, a warm, humid environment accelerates mold growth. This is particularly noticeable during the summer, when it is not unusual to find mold on shoes and clothing that have been left in a closet.

Molds grow well in an environment containing large amounts of sugars (producing a high osmotic pressure) and under acidic conditions that would be unfavorable to bacteria. It is these properties that enable molds to develop on the surface of jellies, jams, or pickles kept in a refrigerator.

Growth and Reproduction

A hyphal filament elongates by cell division, but the principal reproductive mechanism of molds is through spore formation. Mold spores are usually produced in very large numbers and are easily spread by wind and insects. Under favorable conditions the spore will germinate, giving rise to a new mold colony.

Although all fungi reproduce by spores, the means by which the spores are produced can be either sexual or asexual. Sexual spores in fungi are formed following fertilization (fusion of two cells) and meiotic cell division (meiosis). As a result, the spores resulting from a fusion are genetically different from the parental types. Asexual spores result from mitotic cell division and are, as a rule, genetically identical to the parental cells.

Classification of Fungi

Classification of the fungi continues to undergo changes as more is learned about their molecular biology. For purposes of this overview, however, we shall restrict our comments to a brief discussion of the major classes of fungi, with emphasis on those of medical importance.

Based on differences in their sexual and asexual spores (see Table 15.2), the fungi have been separated into three large divisions as shown in Table 15.3. The

TABLE 15.2 Fungal Spore Types

Name	Type	Fungal Groups	Morphologic Characteristics
Arthroconidia	Asexual	Geotrichum, Coccidioides	Produced by fragmentation of septate hyphae into single, slightly thickened cells
Chlamydoconidia	Asexual	Found in all groups	Thick-walled spores formed either within segments of hyphae or as terminal spores of a hyphal filament
Blastoconidia	Asexual	Found in all yeasts	Appear as buds on mother cell
Microconidia	Asexual	Found in most fungi except Zygomycetes	Produced by constriction of hypha—borne on a stalk (conidiophore) and may occur singly or in chains
Macroconidia	Asexual	Found in dermatophytes	Large multicelled spores borne on a stalk (conidiophore)
Sporangiospores	Asexual	Zygomycetes	Spores formed within a sac (sporangium) that is borne on a sporangiophore
Ascospores	Sexual	Ascomycetes	Spores that are formed within a saclike structure called an ascus
Zygospores	Sexual	Zygomycetes	Resting spore resulting from the fusion of two morphologically similar cells
Basidiospores	Sexual	Basidiomycetes	Spores that are borne externally on a club-shaped cell called a basidium

TABLE 15.3   Summary of the Major Divisions of Fungi

Classification	Description
Kingdom: Fungi	
Division: Gymnomycota	Vegetative cells do not possess cell walls
Class: Acrasiomycetes	Vegetative cells are uninucleated amoeba
Class: Myxomycetes	Vegetative form consists of multinucleated plasmodium
Division: Mastigomycota	Motile asexual spores
Class: Chytridiomycetes	Aquatic fungi; haploid hyphae
Class: Hyphochytridiomycetes	Flagellated zoospores; coenocytic hyphae
Class: Plasmodiophoromycetes	Aquatic fungi; diploid hyphae
Class: Oomycetes	Flagellated zoospores; coenocytic hyphae
Division: Amastigomycota	Nonmotile asexual spores
Class: Zygomycetes	Primarily terrestrial; sexual spore is a zygospore; asexual spore is a sporangiospore. Some are parasitic on insects, nematodes, protozoa, and plants. Human infections are routinely caused by opportunistic parasites that are normally saprophytic organisms.
Class: Ascomycetes	Produce sexual spores within a sac called an ascus. Yeasts are important for fermentation and baking. Some produce serious plant infections (Dutch elm disease, rusts); some cause human diseases.
Class: Basidiomycetes	Sexual spores are borne externally on a basidium. Some cause plant diseases (rusts and smuts). Many mushrooms in this class are extremely toxic if ingested.
Class: Deuteromycetes	A convenience group. Includes fungi for which no sexual stage has been demonstrated; also called the Fungi Imperfecti. Many human and animal pathogens are found in this class.

Gymnomycota contains the slime molds, while all fungi producing motile, asexual spores are placed in the Mastigomycota. None of the fungi in these two divisions has ever been involved in human disease. The third division, Amastigomycota, is characterized by production of nonmotile, asexual spores and conidia, and it is in this division that we find the human and animal pathogens. This division comprises four classes, namely, the Zygomycetes, the Ascomycetes, the Basidiomycetes, and the Deuteromycetes.

Zygomycetes   These organisms lack regular septa in their hyphal filaments, resulting in the presence of many nuclei in each cell filament. Their sexual spores, called zygospores, are formed by the fusion of two similar cells. Asexual spores, called sporangiospores, are formed by mitosis inside a saclike structure known as a sporangium (see Figure 15.6). Rupture of the sporangium liberates hundreds of sporangiospores. None of the Zygomycetes are frank pathogens for humans, but several, such as *Rhizopus* and *Mucor*, are opportunists that can cause a serious, often fatal, pneumonia.

Ascomycetes   Sometimes known as sac fungi, ascomycetes form one or more (usually eight) sexual

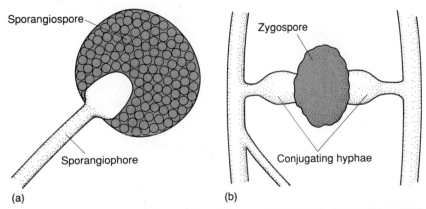

Figure 15.6   *(a)* Mature sporangium containing the asexual sporangiospore produced by the Zygomycetes. *(b)* Mature sexual zygospore formed by the fusion of two hyphal tips.

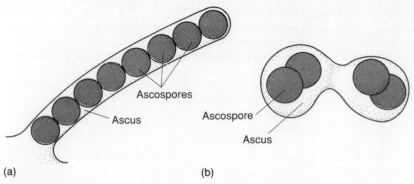

(a)   (b)

Figure 15.7 *(a)* Mature ascus, formed by filamentous fungus, containing eight ascospores. *(b)* Four ascospores within an ascus that is made up of two conjugating yeast cells.

spores within a saclike cell called an ascus (see Figure 15.7). The asexual spores produced by the Ascomycetes are frequently single-celled microconidia. Microconidia may be produced as macroaleuroconidia, as shown in Figure 15.8(*a*), or in long chains extending from a specialized aerial hypha called a conidiophore (meaning "conidia bearing"), as shown in Figure 15.8(*b*).

The Ascomycetes produce regular septa that divide the mycelium into a large number of individual cells. Each septum, however, has a "hole" in it, which permits the free flow of cytoplasmic and nuclear material between the cells. Many Ascomycetes grow only as single-celled yeasts, and by far the best-known yeasts in this class are those in the genus *Saccharomyces,* on which both the baking and the alcoholic beverage industries are totally dependent.

Basidiomycetes Basidiomycetes form their sexual basidiospores externally on club-shaped cells called basidia (see Figure 15.9). Asexual reproduction may occur by budding, through microconidia, or by fragmentation of the hyphal filament. The hyphae are usually septate. Very few human diseases are caused by members of the Basidiomycetes, although organisms in this class do cause several plant diseases. Note that mushrooms, some of which may be highly toxic, belong to this class.

Deuteromycetes Also called Fungi Imperfecti, these comprise a large group of fungi for which no sexual state has been demonstrated. During the past decade, however, some well-established members have been shown to produce sexual spores when mixed with the correct mating type and, as a result, have acquired two names, one for the sexual classification and the other for the older asexual name. Because many of the human fungal pathogens belong to the Deuteromycetes, the tendency has been to retain the older name even though a sexual stage has been shown to exist. Most mycologists, however, believe the sexual-state name is the preferred term and should eventually be the only name. Many species of this class produce

Figure 15.8 *(a)* Tuberculated macroconidia from *Histoplasma capsulatum (Ajellomyces capsulata). (b)* Chains of conidia radiating from special cells on the swollen end of a conidiophore of *Aspergillus.*

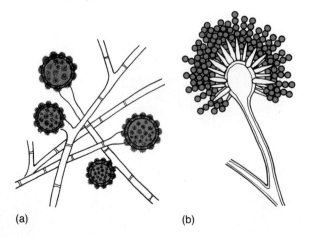

(a)   (b)

Figure 15.9 *(a)* Immature basidium showing spore nuclei migrating to form basidiospores. *(b)* Mature basidiospores borne externally on the basidium.

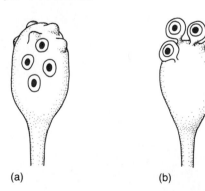

(a)   (b)

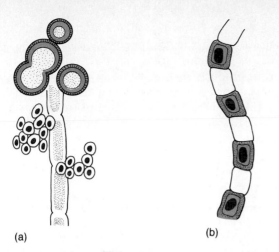

(a)   (b)

Figure 15.10 *(a)* Large chlamydoconidia borne on pseudo-hyphae of *C. albicans* with small blastoconidia. *(b)* Thick-walled arthroconidia alternating with empty cells in the hyphae are characteristic of *Coccidiodes immitis.*

both microconidia and macroconidia, as well as chlamydoconidia, arthroconidia, and blastoconidia (see Figure 15.10).

Diseases caused by the Deuteromycetes include (1) superficial infections that are of concern only because of their cosmetic appearance; (2) cutaneous infections caused by the dermatophytes that are restricted to the keratinized tissues such as the nails, hair, and stratum corneum of the skin; and (3) the subcutaneous and deep-seated systemic infections, resulting in debilitating and fatal diseases.

Yeasts

Yeasts, like molds, are microscopic fungi, but unlike every other type of fungi, they exist as simple, independent cells. These are usually round or ovoid but may have other shapes. Yeast cells differ from bacteria in that they are eucaryotic cells; they are usually larger than the average bacterium, and they reproduce by different mechanisms. Yeasts, then, are simpler than molds, but their cellular structure is more complex than that of the bacteria. All classes of fungi contain yeast.

Reproduction

Like molds, most yeasts reproduce both sexually and asexually. A common method of sexual spore formation is that in which two yeast cells fuse to a single saclike cell called an ascus. Within the ascus one to eight discrete spores are formed. The ascus then breaks open to liberate the individual spores (**ascospores**). Under suitable conditions for growth each of the ascospores subsequently will germinate to form a new yeast cell.

The usual method of yeast reproduction is by an asexual process called **budding.** Microscopic examination of a growing culture of yeast will reveal bulges of various sizes protruding from the mother cell. These bulges (buds) continue to enlarge, usually until they are about the same size as the mother cell. As the bud enlarges, it is gradually pinched off by the continued stricture of the cell wall between it and the mother cell until finally it is severed completely. The new cell is called the daughter cell (see Figure 15.11). Budding should not be confused with binary fission; it is an entirely different process. All yeasts reproduce asexually, but not all yeasts are able to reproduce sexually. Those that reproduce only asexually are classed as Deuteromycetes or Fungi Imperfecti.

Figure 15.11 Yeast—a one-celled fungus—shown budding. CW—cell wall; CM—cell membrane; M—mitochondria; N—nucleus; IM—internal membrane.

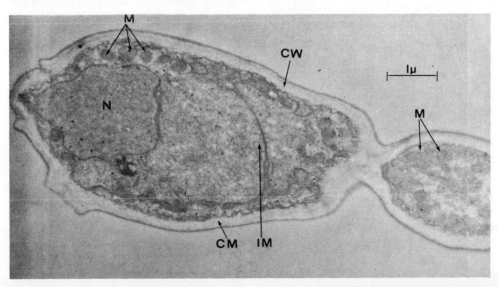

Like molds, yeasts are rather widely distributed in nature. They are found in water, soil, and dust and are commonly present on many fruits and vegetables. While all molds are described as aerobic, many yeasts are able to grow facultatively. In fact, in his early studies on fermentation, Pasteur recognized that yeasts could grow in the absence of air. Under anaerobic conditions, glucose can be converted to alcohol and carbon dioxide by yeast enzymes via the Embden-Meyerhof pathway. Under aerobic conditions, glucose can also be oxidized completely to carbon dioxide and water by some yeasts via the citric acid cycle. Yeasts are able to synthesize vitamins and proteins; thus, yeast cells provide a good source of nutrients.

Physiologic Activities of Yeasts and Molds

Scientists have developed many uses for yeasts and molds. The production of drugs, organic acids, and enzymes, all of which are products of yeast or mold metabolism, makes fungi particularly important to biochemists and physicians. Perhaps the best-known mold product used in medicine today is the antibiotic penicillin, which is produced by *Penicillium chrysogenum* or *P. notatum*. Certain cheeses are the result of mold action on milk products. The caves of Roquefort, France, are said to be particularly amenable to the growth of *Penicillium roqueforti* in sheep's milk; the result is Roquefort cheese. Camembert cheese is another mold-ripened cheese (*P. camemberti*). Alkaloids from the ergot fungus have been used to produce contractions of the uterus following childbirth. Toxic effects have been attributed to some 30 or 40 species of mushrooms. Because fungi were used in various foodstuffs in the Orient more than 2000 years ago, their exploitation is not a recent endeavor.

In addition, yeasts in the class Ascomycetes, have been used since the beginning of history both for the fermentation of fruit juices (in the manufacture of wines) and in the brewing of beer. Fermentation is the process by which sugar is decomposed to form alcohol and carbon dioxide. The yeast cakes familiar to every housewife consist of yeast in a cornstarch medium. When this is introduced into dough, the yeasts ferment the sugar in the dough. The resultant carbon dioxide bubbles up through the dough to form the small holes that give the bread its lightness. It is interesting to note that a species of yeast best for wine production usually would not be good for baking bread, and vice versa.

As they exist in nature, the fungi are important because they help decompose dead plants and animals and their wastes, thus making these complex substances again available for plant growth. However, from our point of view, some species of molds are highly destructive. They are among the chief causes of disease in cultivated plants (wheat rust, potato blight) and also bring about considerable loss in stored seeds. In addition, molds cause deterioration of numerous products such as wood, leather goods, rubber articles, paper, fabrics, and even glass lenses. Finally—and this is what mainly concerns us—yeasts and molds cause many diseases of humans and domestic animals.

SLIME MOLDS

Slime molds have a particularly interesting growth pattern and are fascinating to study. They seem to resemble protozoa more than the true fungi, although in one stage of development they do form spores. They can be categorized either as plasmodial slime molds (Myxomycetes) or cellular slime molds (Acrasiomycetes). Both classes of slime molds have a complex life cycle in which the vegetative cells form complex fruiting bodies containing spores.

The vegetative form of a plasmodial slime mold consists of a large (several centimeters in diameter) multinucleated mass of protoplasm called a plasmodium. The plasmodium moves along a surface, ingesting bacteria and other small particles. If conditions become unfavorable, the plasmodium forms many small, brightly colored fruiting bodies, each consisting of a stalk bearing a number of sporangia. Each sporangium contains many uninuclear spores that, under favorable conditions, can germinate to form a new plasmodium.

The cellular slime molds differ from the plasmodial type in that the vegetative phase consists of uninuclear amoeboid cells that move along a surface by the extension of pseudopodia. Under favorable conditions, these cells secrete a hormone (cyclic AMP) that causes many of the amoeba to aggregate, forming a slug. Through a series of very elaborate migrations of the individual amoeba within the slug, an elevated fruiting body is formed that contains the spores. After being released from the fruiting body, these spores will germinate to form the ameboid cells. Figure 15.12 illustrates a slime mold fruiting body.

Slime molds, like the true fungi, have a role in the decay process; they are found in damp areas, particularly on logs and stumps. Certain species are parasitic on some of the higher plants. They apparently have no role in human or animal infections.

FUNGAL SYMBIOSIS

A number of fungi (mostly ascomycetes) are able to grow only in a symbiotic relationship with an alga.

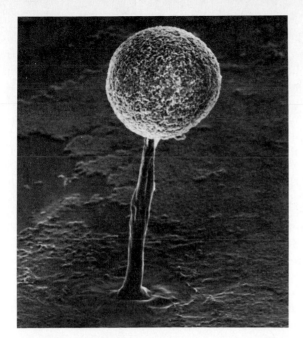

Figure 15.12 Fruiting body of the slime mold, *Lamproderma* (×240).

The resulting plant, comprising a fungus and an alga, is termed a **lichen**. The fungus moiety of the lichen gains carbohydrates formed during the photosynthesis of the alga component and the alga gains other nutrients and minerals made available by the enzymatic activity of the fungus. The alga also is protected from desiccation by the fungus and is anchored to a solid stratum by the fungal holdfast.

Approximately 30 different genera of symbiotic algae have been described in lichens. The majority belong to the green algae, but the procaryotic cyanobacteria are also found as components of lichens. Usually only one species of alga is found associated with each lichen, but two have been described.

Lichens in the subarctic are a prime source of food for reindeer and caribou; the Laplanders graze their herds of reindeer on the lichen fields of northern Scandinavia and harvest lichens to be used as winter fodder. Berber tribes in Libya graze their sheep on lichens, and it has been suggested that the fabled "manna" of the Israelites was a lichen. These remarkable plants occupy an important niche in world ecology.

PROTOZOA

It is generally agreed that protozoa are neither plants nor animals but belong instead in a separate kingdom. Hence, the designation of a phylum for all protozoa has been abandoned in favor of a kingdom, the Pro-

tista, which includes all of the unicellular eucaryotes. This belated recognition of the great diversity of the various groups has resulted from studies of their ultrastructure and biochemistry. The brief presentation of the protozoa in this chapter does not permit a detailed discussion of the current status of their taxonomy; however, the reference list at the end of this chapter will provide additional information.

Morphology

Although protozoa (*sing*. protozoan) are usually described as unicellular, many exist in groups called colonies. Each cell of a colony usually represents an independent unit, although certain specialized functions are observed in some.

No effort of imagination could create more diverse patterns than this group has provided. Some are spherical to oval, some are elongated, and some even change their shape as they move along a surface. Overall size is relatively constant within a species but varies considerably in different species. Some may be as small as 10 μm in length and some may measure 200 μm. Actually, certain species may measure 1 to 2 mm and are large enough to be seen with the unaided eye.

In general, the structure present in protozoa does not differ essentially from the structure of other animal cells. A typical protozoan cell consists of a tough covering, or pellicle; a plasma membrane; an outer layer of cytoplasm (called the ectoplasm); and an inner, more granular cytoplasm (called the endoplasm). Within the cytoplasm are found one or two nuclei, mitochondria, food vacuoles (spaces in the cytoplasm containing food for digestion), and one or more contractile vacuoles, which pump out excess water. The protozoa also possess structures for motility and for securing food.

Life Processes

For the most part, protozoa are aquatic organisms with habitats ranging from droplets of moisture around soil particles to large bodies of water such as rivers and oceans. Some live only as obligate parasites in animals. The distribution of the protozoa is worldwide, and many produce chronic to acute diseases in humans.

Nutrition

Unlike bacteria, which obtain their food by the diffusion of dissolved substances into the bacteria cell, most protozoa are **holozoic**. They ingest their food as solid particles through a mouth opening or cytosome. The food ingested by protozoa is usually bacteria, algae, or other protozoa. After ingestion, the food is contained in a food vacuole. Enzymes are secreted into the vacuole to speed the breakdown of complex materials into simpler substances that can be dissolved by the water inside the food vacuole. Once dissolved, they enter the cytoplasm, where they are used for the synthesis of cell material or are broken down further to provide energy for the cell. Any food material not digested in the vacuole is either expelled from the cell body through an anal pore or brought to the surface, where the vacuole breaks open to free the undigested material from the cell. Unlike the free-living forms, some of the parasitic protozoa obtain their nutritive substances from their host in the same manner as bacteria, by the diffusion of soluble material into the cell.

Reproduction

Almost all protozoa are able to reproduce both asexually and sexually. Asexually, most reproduce by splitting into two cells of equal size. However, unlike bacteria, which divide by simple transverse fission (crosswise), most protozoa divide longitudinally (lengthwise). Unequal fission, or budding, also is found in some species.

Sexual reproduction occurs when two different cells join (conjugate) and exchange nuclear material.

The Cyst or Encystment

Under certain adverse environmental conditions, many protozoa may become encysted; that is, the cell assumes a fairly round or oval shape and secretes a heavy protective coating around itself. At this stage, the protozoan may survive when there is a lack of food or moisture, during adverse temperature changes, and while in contact with toxic chemical agents. The cyst stage is particularly valuable in permitting parasitic forms to survive outside the host and to transfer to a new host. Under favorable conditions, water is absorbed into the cyst; the protozoan then emerges and resumes its growth.

In addition to the free-living forms, parasitic species are found in all classes of the protozoa. As would be expected, the parasitic forms are found most frequently in association with the particular host that they parasitize. In some rare cases the protozoan exists in its host in what is referred to as a **symbiotic** relationship, that is, one in which the host benefits from the presence of this parasite and the parasite benefits from the host. One notable example of this is the common termite, which could not live if a protozoan in its intestine were not present to digest the wood ingested by the termite. On the other hand, this particular protozoan cannot live if it is removed from the intestine of the termite. Thus, a mutually beneficial relationship exists between the termite and its protozoan parasite.

Most animals infected by protozoa are not this fortunate, and the result of the infection is either an acute or a chronic disease. Parasitic infections are much more common in tropical areas, where conditions seem to be more favorable for the growth and the spread of the parasite.

Classification

The protozoa are divided into four classes based on their mechanism of locomotion (see Table 15.4). Of the approximately 40,000 species of protozoa that have been described, only a few cause disease in humans. These pathogenic species are distributed in all four classes; thus, a study of the medically important species requires a general knowledge of each of the classes.

Protozoa are not routinely grown by the diagnostic laboratory. To diagnose a protozoan infection, it is necessary to locate the organisms in material taken directly from the host. Therefore, it is imperative that careful attention be paid to the specimens sent to the laboratory—particularly stool specimens, which should be taken to the laboratory while fresh.

TABLE 15.4 Properties of Protozoa Classes

Class	Description
Mastigophora	Flagellated, elongated cell which is surrounded by a pellicle; may have mouth opening and a gullet; major pathogens are *Trypanosoma, Leishmania, Trichomonas,* and *Giardia.* Some, like *Giardia,* form cysts.
Sarcodina	Flexible amoebas; may form cysts; major pathogen is *Entamoeba histolytica,* the cause of amoebic dysentery.
Sporozoa	All are nonmotile, animal parasites with a complex life cycle that may require different hosts for sexual and asexual reproduction; do not engulf particulate matter; major pathogens are *Plasmodium* (malaria) and *Toxoplasma* (toxoplasmosis)
Ciliata	Highly developed cells covered with short, hairlike projections called cilia; shape of cells highly variable, but constant within a species; *Paramecium* is best-studied free-living ciliate; *Balantidium coli* causes dysentery in humans. Some form cysts.

Class Mastigophora

Flagella (*sing.* flagellum = whip) are characteristic of the class Mastigophora. One or more of these whiplike projections provides the means of locomotion. These organisms have a tough, flexible covering, or pellicle, surrounding the protoplasm, which prevents the changes of form observed in amoebae.

A typical cell is usually elongated in shape, with flagella extending from the forward (anterior) end. Some have distinct food-getting devices, such as a mouth opening and a gullet (see Fig. 15.13*a*). Reproduction is accomplished by longitudinal binary fission.

Many of the organisms in this group possess properties that are characteristic of both the plant and the animal kingdoms. *Euglena viridis* is an example of such an organism. It contains chlorophyll and thus can obtain its energy from light, but, on the other hand, it is morphologically similar to many non-chlorophyll-containing members of this class.

Among the other members of this class are found a number of parasites of humans. A few of the more common diseases and the disease-producing organisms are listed below.

Giardiasis *Giardia lamblia* causes an intestinal infection that is characterized by cramps and diarrhea. It is usually acquired by the ingestion of water contaminated with feces, but person-to-person transmission can occur. The organism is easily recognized because of its bilateral symmetry (Figure 15.13*a*).

Trichomoniasis *Trichomonas vaginalis* is the causative agent of a common nonfatal infection of the genitourinary tract. It is primarily a sexually transmitted disease, since it is almost always acquired through sexual intercourse; these organisms occur throughout the world.

Trypanosomiasis *Trypanosoma gambiense* and *T. rhodesiense* are the causative agents of a very important disease known as African sleeping sickness. *T. cruzi* causes Chagas' disease, which is found in the western hemisphere.

Leishmaniasis This is a disease caused by a flagellated protozoan of the genus *Leishmania*. Some forms are fatal if untreated, and they are found throughout much of the world (including South America and Mexico, but not the United States).

Class Sarcodina

The class Sarcodina contains the amoebas (genus *Amoeba*), which move by sending out fingerlike projections called pseudopodia (meaning false feet) from the cell. These fingerlike projections provide not only for the movement of the animal but also the means of securing food. The cell body is a flexible mass of protoplasm except in the encysted stage. Because of this characteristic, amoebas lack a definite shape, as this changes with the motion of the cellular contents (see Figure 15.13*b*). The cells multiply by binary fission.

Some free-living amoebas secrete a protective layer external to the cytoplasmic membrane. An example of this is seen in the *Foraminifera*; these ocean-dwelling organisms secrete a chalky covering containing many small pores that still allow the protoplasm to flow through and provide movement. The famous White Cliffs of Dover are composed largely of the remains of these particular organisms.

Although the majority of the protozoa in this class are free-living, several parasitic forms are known. Perhaps the most common pathogen in this class is *Entamoeba histolytica*. This organism causes a human intestinal disease called amoebiasis or amoebic dysentery. The infection may be asymptomatic, may produce slight abdominal discomfort and diarrhea, or it may cause abscesses of the liver and other organs. The disease can, in some instances, be fatal.

Class Sporozoa

Although many of the Sporozoa are harmless, all are parasites of some animal host. They differ from the

---

Figure 15.13 Examples of three groups of protozoa. (a) *Giardia*, a flagellate; (b) *Entamoeba*, an ameba; (c) *Balantidium*, a ciliate.

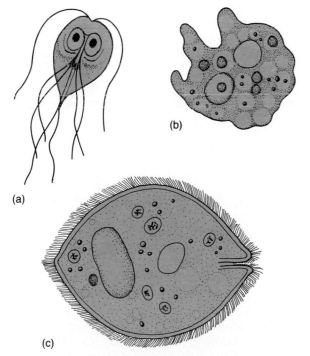

(a)

(b)

(c)

other classes of protozoa in several respects: (1) the mature organism is not motile; (2) they do not engulf solid particles of food but obtain their food by the diffusion of soluble nutrients into the cell; and (3) they have a rather complex life cycle that includes both asexual and sexual reproduction. In some instances, both the asexual and the sexual stages can take place in a single host; in others, two hosts are required (i.e., one host for the sexual stage and a second, different, host for the asexual stage).

Malaria The sporozoans that cause malaria (several species of *Plasmodium*) infect an estimated 150 million persons each year, resulting in approximately two million deaths. The organisms reproduce asexually within the human and sexually within the mosquito. The disease is spread by the bite of infected mosquitoes.

Toxoplasmosis *Toxoplasma gondii* has infected about 100 million adults in the United States. The disease is usually mild or asymptomatic except when acquired congenitally or by an immunocompromised host. The parasites reproduce asexually in many animals, but sexual fusion occurs only when they infect a member of the cat family.

Pneumocystis carnii This parasite, *Pneumocystis carnii*, rarely, if ever, causes disease in a healthy individual, but it has become an important cause of pneumonia in immunosuppressed individuals, particularly those with AIDS. This organism appears to occupy a taxonomic niche in which it shares certain properties with both protozoa and fungi. Current information on spore formation and sequence data of its ribosomal RNA, however, suggest that *P. carinii* is a fungus, belonging with the Ascomycetes.

Class Ciliata

Cilia—short, hairlike projections—provide the organisms of the class Ciliata with structures for movement (see Figure 15.13c). The ciliates are the most highly developed protozoa. The shape of the cell varies considerably among different species, ranging from a simple ovoid to trumpet, kidney, vase, and other shapes. The outer covering is sufficiently sturdy to maintain a definite shape within a given species.

Each cell contains two nuclei, one small nucleus called a micronucleus and a second, larger one called a macronucleus. The micronucleus functions in the reproduction of the cell, and the macronucleus apparently controls the metabolic activities within the cell. The members of this class may reproduce either asexually by transverse fission or sexually by conjugation, in which two individual cells come together and exchange nuclear material.

Paramecium caudatum is a free-living protozoan that is frequently used for study because its characteristics are typical of organisms of this class. Specialized structures in this organism include an oral groove, a gullet, an anal pore, and two contractile vacuoles.

Some members of this class are parasites of various animals, but *Balantidium coli* is the only species recognized as being infective in humans. It is an intestinal parasite that causes a severe dysentery. Table 15.4 summarizes the properties of each class of protozoa.

HELMINTHS

Helminths (from Greek *helmins*—worm) are classified in the phyla Platyhelminthes (flatworms) and Aschelminthes (roundworms). Many live only in the intestinal tract of a parasitized host, whereas others invade internal organs such as the liver, lungs, blood, subcutaneous tissue, and brain, and yet others are free-living and nonparasitic. Most are macroscopic, but identification frequently requires a microscopic examination of their eggs.

The helminths are placed in one of three classes: the Cestoda and Trematoda, both flatworms, and the Nematoda, a group of roundworms. (*Note:* not all nematode worms are parasitic. Some are free-living in soil or water.)

Platyhelminths

The platyhelminths, or flatworms, are the most primitive of the helminths in that they characteristically have either no digestive tract or only a rudimentary one. They are flat, having a shape roughly like a leaf or a measuring tape, and are, in most cases, hermaphroditic (both male and female organs occur in the same animal). Many of the parasitic species have complicated life cycles that require an alternation of hosts such as humans and cows or pigs or fish.

Cestodes

The habitat of the adult cestode (tapeworm) is the intestinal tract, and the animal in which the larval stage develops into an adult worm is termed the **definitive host.** Animals in which the eggs develop into the larval stage are called **intermediate hosts.**

Adult tapeworms are usually long and ribbonlike, and are divided into segments called proglottids (Figure 15.14). The scolex (head) at the anterior end is provided with suckers and, in some cases, hooks that attach the worm to the intestinal wall. Posterior to the scolex is the neck region, followed by immature

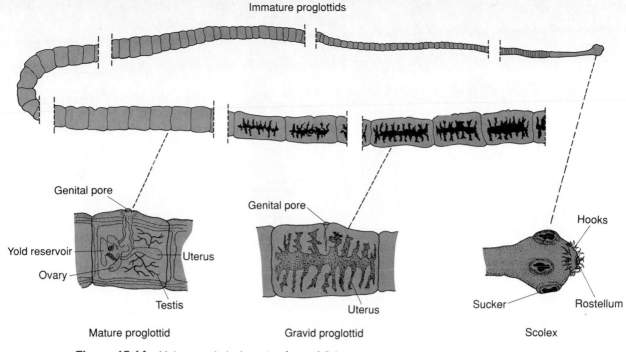

Immature proglottids

Genital pore
Yold reservoir
Uterus
Ovary
Testis

Mature proglottid

Genital pore
Uterus

Gravid proglottid

Hooks
Sucker
Rostellum

Scolex

Figure 15.14 Major morphologic parts of an adult tapeworm.

proglottids, mature proglottids containing male and female sex organs, and fertilized segments, called gravid proglottids, containing fertilized eggs.

Humans are a definitive host for *Taenia saginata* (beef tapeworm), *T. solium* (pork tapeworm), *Diphyllobothrium latum* (fish tapeworm), and *Hymenolepis nana* (dwarf tapeworm). Humans also can serve as an intermediate host, resulting in a disseminated infection, for *T. solium* (pork tapeworm) and for *Echinococcus granulosus,* (the causative agent of hydatid disease), if the eggs instead of the larvae of these tapeworms are ingested.

Additional details of these infections and the life cycles of their respective tapeworms are described in Unit III.

Trematodes

Trematodes are also referred to as flukes. Adult worms may vary in size from 1 mm to several centimeters, and all possess suckers used for attachment to host tissue.

Trematodes have a worldwide distribution, and each species is found in association with the specific species of snail required as its intermediate host.

Comprehension of the epidemiology of trematode infections requires a general knowledge of the complex events that occur between the excretion of the egg by the definitive host and the formation of the final larval stage. During this period, four different larval forms occur (except in the blood flukes, which have only three):

1. The hatching of the egg usually occurs only in fresh water. Within the egg is formed a **miracidium,** a small ciliated larva. The miracidium escapes from the egg and swims about until it finds a snail of the right species for its first intermediate host.

2. The miracidium bores into the snail's tissues, after which it loses its cilia and undergoes a metamorphosis to form a long tubular larva called a **sporocyst.**

3. The sporocyst migrates to the snail's liver, where it continues to form masses of germ cells within a saclike structure. With the exception of the blood flukes, the sporocyst then undergoes another morphological change to become a more differentiated larva possessing a mouth and rudimentary digestive tract. This larval stage is called a **redia.**

4. Within each redia, germ cells develop into more rediae, but eventually a final larval change occurs when **cercariae** begin to develop within the rediae. A cercaria resembles the adult worm possessing suckers and a rudimentary digestive and excretory system. It also possesses a tail for locomotion after leaving the snail. The cercaria can infect a new intermediate host such as a freshwater fish, a crab, another snail, or aquatic vegetation. In many cases, after infection the cercaria loses its tail and secretes a cyst wall around the larva. This cyst form is called a **metacercaria** and, except for the blood flukes, *humans are infected only*

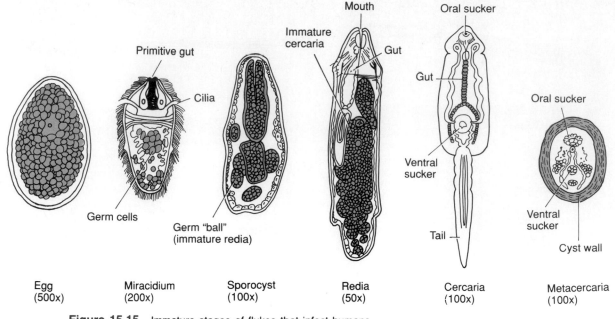

Figure 15.15 Immature stages of flukes that infect humans.

Egg (500x)	Miracidium (200x)	Sporocyst (100x)
Redia (50x)	Cercaria (100x)	Metacercaria (100x)

by the ingestion of metacercaria. The morphological appearance of each of these larval stages is illustrated in Figure 15.15.

As described in Unit III, the flukes are divided into four types based primarily on the type of infection caused, namely, the intestinal flukes, the liver flukes, the lung flukes, and the blood flukes. Table 15.5 summarizes the important platyhelminth infection of humans.

Aschelminths

Nematodes are round, elongated worms; the sexes are separate and, unlike the flatworms, adult nematodes possess complete digestive systems that include both a mouth and an anus (Figure 15.16). Human infections by these roundworms are usually divided into two large categories: the intestinal roundworms and the blood and tissue roundworms.

TABLE 15.5 Platyhelminth Infections in Humans

	Disease	Intermediate Host	Definitive Host	Stage Passed to Humans
Cestodes				
Taenia saginata	Beef tapeworm	Cattle	Humans	Larvae ingested with infected meat
Taenia solium	Pork tapeworm	Pork	Humans	Larvae ingested with infected meat
Diphyllobothrium latum	Fish tapeworm	Crustacean Fish	Humans, others	Larvae ingested with infected fish
Hymenolepis nana	Dwarf tapeworm	None required	Humans, mice, rats	Fecal-oral transfer of eggs
Echinococcus granulosus	Hydatid disease	Sheep, cattle	Humans, dogs, foxes	Hand-to-mouth transfer of eggs from contaminated soil or dog fur
Trematodes				
Fasciolopsis buski	Intestinal fluke	Snails	Humans, pigs	Ingestion of metacercariae
Fasciola hepatica	Liver fluke	Snails	Humans, sheep, cattle	Ingestion of metacercariae
Paragonismus westermani	Lung fluke	Snails	Humans	Ingestion of metacercariae
Shistosoma	Blood flukes	Snails	Humans	Penetration of skin by cercariae

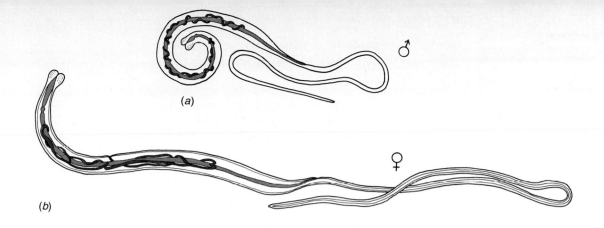

Figure 15.16 The nemotode Trichuris Trichina A, male B, Female.

Intestinal Nematode Infections

In these infections, the roundworms exist in the intestine during their adult stage, but the larval stages of some may be widely distributed throughout the body of an infected person.

The details of the major nematode infections are described in Unit III. Suffice it to state here that most of the intestinal infections are acquired by the ingestion of the larvae or eggs of the adult worms. Table 15.6 summarizes some of these infections, including the mechanism by which humans usually become infected.

Blood and Tissue Nematode Infections

Unlike intestinal roundworms, nematodes infecting blood and tissue are carried from human to human by the bite of an arthropod vector.

With the exception of *Dracunculus medinensis* (guinea worm), the remaining blood and tissue nematodes belong to the superfamily **Filarioidea** and the resulting human infection is called **filariasis.** Adult worms generally range from 2 to 30 cm long, with the female about twice the size of the male. One property distinguishing the filariae from other nematodes is that the female does not lay eggs, but instead gives birth to prelarval forms called **microfilariae.** The ingestion of the microfilariae by blood-sucking vectors provides for the transmission of the filariae from one person to another.

As with the other helminthic infections, details are discussed in Unit III. Table 15.7, however, summarizes the important aspects of the epidemiology of the blood and tissue nematode infections.

TABLE 15.6 Intestinal Nematode Infections in Humans

Organism	Disease	Stage Passed to Humans	Location in Humans
Trichinella spiralis	Trichinosis	Larvae in meat, especially pork	Intestine and muscles
Trichuris trichiura	Whipworm disease	Ingestion of larvae in soil or contaminated food	Cecum
Ascaris lumbricoides	Ascariasis	Ingestion of eggs in soil or contaminated food	Small intestine
Enterobius vermicularis	Pinworms	Ingestion of eggs	Large intestine
Nector americanis	Hookworm disease	Penetration of larvae thru skin	Small intestine

It is probable that few persons who read this book have ever heard of the disease "onchocerciasis" or, as it is frequently called, "river blindness." In spite of the unfamiliarity of the Western World to this affliction, over 18 million persons are infected with this parasite (mostly in Africa), and approximately 500,000 individuals become permanently blinded each year from this infection.

The worms that cause onchocerciasis are transmitted to humans by the bite of infected black flies that breed along the banks of rivers and streams. The prelarval forms (called *microfilariae*) move throughout the dermis, causing a severe inflammation, which after long-established, chronic disease, can be seen as thick, wrinkled, hyperpigmented skin. The most serious lesions, however, occur when the microfilariae migrate into the eyes, leading to eventual blindness.

After untold centuries of suffering, it now appears possible to rid the world of this disease. Merck and Company have developed a drug termed "mectizan," a derivative of another drug, ivermectin. Two doses per year provide complete protection with minimal side effects. Mectizan kills the microfilariae in the body and prevents the adult worms from producing new microfilariae. Thus, if an entire population can be treated for a period of about 10 years, all adult worms should have died and the disease eliminated from the area.

To make this possible, Merck and Company has offered to donate the drug free to developing countries. The World Health Organization, along with a review committee set up by Merck, will establish distribution and reporting systems. Hopefully, the sight of children leading their blind elders to the fields where they serve as human scarecrows, will someday be a thing of the past.

TABLE 15.7   Blood and Tissue Nematode Infections in Humans

Organism	Disease	Vector
Wuchereria bancrofti	Bancroftian filariasis	Mosquito
Brugia malaya	Malayan filariasis	Mosquito
Loa loa (African eye worm)	Loiasis	Deer flies
Onchocerca volvulus	Onchocerciasis (River blindness)	Black flies
Dracunculus medinensis (Guinea worm)	Dracontiasis	Ingestion of small crustaceans

SUMMARY

Eucaryotic algae comprise organisms that contain one or more kinds of chlorophyll in addition to other pigments and carry out a green plant-type photosynthesis resulting in the photolysis of water and the evolution of oxygen. They are divided into six classes on the basis of their cellular structure and pigment composition. Many exist as single cells or colonies of cells, but some, such as the brown algae, may grow to lengths of over 100 m. Unicellular forms of algae, known as phytoplankton, are widely distributed in both fresh and marine waters, where they serve as a major source of food for aquatic animals. The cyanobacteria are procaryotic cells, but they carry out the same type of photosynthesis as do the eucaryotic algae.

The fungi (molds and yeasts) are eucaryotic cells. Characteristically, molds grow as long filaments of cells called hyphae. A mass of filaments is called mycelium. Mycelia may be coenocytic or septate. The major method of reproduction of fungi is by asexual spore formation, although most can produce spores by sexual means. Classification of the fungi is based primarily on the appearance and the mechanism of spore formation.

Yeasts differ from molds in that they exist as single cells. They reproduce asexually by a process called budding. Most yeasts produce sexual spores called ascospores.

Both molds and yeasts are used widely in industry. Molds are used in the production of certain antibiotics and in the ripening of some cheeses.

Slime molds are not true fungi. They possess an interesting life cycle resembling that of both protozoa and fungi.

A lichen is a plant structure comprising an alga or a cyanobacterium and a fungus living in symbiosis. Lichens are extremely widespread and provide a major source of food for many animals, particularly in arctic regions or under desert conditions, where many other plants are unable to thrive.

The protozoa consist of unicellular and colonial cells that exhibit some characteristics typical of animal life. Some protozoa may possess special structures such as an oral groove, a gullet, an anal pore, and contractile vacuoles in addition to the structures of a typical cell.

In general, protozoa are classified on the basis of locomotive structures. Members of the class Sarcodina move by means of pseudopodia, those of Mastigophora by flagella, and those of Ciliata by cilia. Mature members of the class Sporozoa have no means of locomotion.

All classes of protozoa contain some species that cause disease in humans. Such infections are most common in the tropics, but some occur in temperate zones.

Many protozoan infections are transmitted through the bite of an insect vector, although the intestinal parasites are transmitted via fecal contamination, and some others, such as trichomoniasis, result from direct person-to-person contact.

Helminths are parasites that can be divided into platyhelminths (flatworms) and aschelminths (roundworms). Some live only in the intestinal tract of infected humans (mostly tapeworms) and others may be disseminated to various parts of the body such as muscle, lungs, liver, and blood. The cestodes and trematodes are acquired by ingestion of an egg, larva, or cyst (except for the blood flukes, which penetrate the skin), and the nematodes are spread to humans by the bite of an arthropod vector (except for the guinea worm, *Dracunculus medinensis*).

QUESTIONS FOR REVIEW

1. List the common name of each of the classes of algae and give at least one major property that defines each class.
2. What is phytoplankton and what two functions does it provide?
3. What is diatomaceous earth?
4. What causes the red tide and why is it dangerous to humans?
5. Contrast yeasts and molds.
6. What is a sexual spore? An asexual spore?
7. List all the types of sexual and asexual spores and describe how they are formed.
8. What are coenocytic hyphae?
9. Name some mold products that are beneficial to humans.
10. How are slime molds similar to fungi? To protozoa?
11. What is a lichen?

12. Discuss the size of protozoa as compared with bacteria.
13. How does the nutrition of protozoa differ from that of bacteria?
14. How are protozoa classified? What classes are recognized? What distinguishes each class?
15. What is meant by the terms *definitive host* and *intermediate host*? Give an example.
16. How do humans become infected with the various types of flukes?
17. How do nematodes differ from cestodes and trematodes?
18. How do humans acquire an infection with the blood and tissue nematodes?

SUPPLEMENTARY READING

Alexopoulos CJ, Mims CW: *Introductory Mycology*, 3rd ed. New York, Wiley, 1979.

Beaver PC, Jung RC, Cupp EW: *Clinical Parasitology*. Philadelphia, Lea & Febiger, 1984.

Bold HD, Wynne MJ: *Introduction to the Algae: Structure and Reproduction*. Englewood Cliffs, NJ, Prentice-Hall, 1978.

Blumenthal DS: Intestinal nematodes in the United States. *N Engl J Med* 297:1437, 1977.

Brown DH, et al: *Lichenology: Progress and Problems*. New York, Academic, 1976.

Dworkin M, Kaiser D: Cell interactions in myxobacterial growth and development. *Science* 230:18, 1985.

Holland HD, Lazar B, McCaffrey M: Evolution of the atmosphere and oceans. *Nature* 320:27, 1986.

Marples MJ: Life on the human skin. *Sci Am* 220(1):108, 1969.

Moore-Landecker E: *The Fundamentals of Fungi*, 2nd ed. Englewood Cliffs, NJ, Prentice-Hall, 1982.

Most H: Trichinosis: Preventable yet still with us. *N Engl J Med* 298:1178, 1978.

Nanduri J, Kazura JW: Clinical and laboratory aspects of filariasis. *Clin Microbiol Rev* 2:39, 1989.

Ross IK: *Biology of the Fungi*. New York, McGraw-Hill, 1979.

Trainor FR: *Introductory Phycology*. New York, Wiley, 1978.

Vidal G: The oldest eucaryotic cells. *Sci Am* 250(2):48, 1984.

Warren KS, Mahmoud AAF: Liver, intestinal and lung flukes. *J Infect Dis* 135:692, 1977.

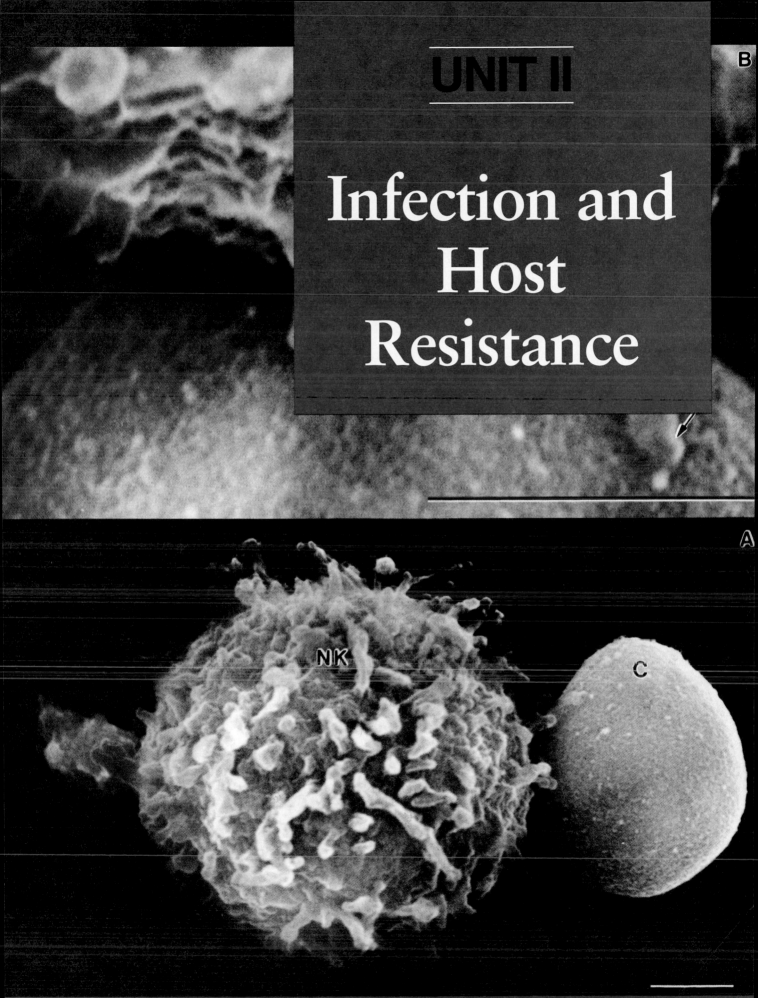

UNIT II

Infection and Host Resistance

Chapter 16

Infection and Bacterial Invasiveness

OBJECTIVES

After a study of this chapter, you should

1. Be able to differentiate among acute, chronic, local, systemic, and inapparent infection.

2. Be familiar with how communicable diseases are transmitted to a susceptible individual.

3. Know how to use Koch's postulates to establish the causes of a disease.

4. Be able to recognize any properties possessed by a bacterium that contribute to its ability to produce disease.

5. Know what is meant by such terms as *portal of entry, virulence,* and *vector.*

All of us are infected from our birth until our death. Our normal flora is actually an infection by microorganisms that have adapted to a specific habitat in our bodies where they can replicate without harming their host. In fact, our normal flora provides us with protection by interfering with the growth of organisms that may be harmful. We see examples of this in individuals who have had part of their normal flora destroyed by long-term antibiotic treatment, resulting in the overgrowth of the yeast *Candida albicans* in the intestinal tract, in the mouth, or on the skin. However, should any one species of the microbes comprised by our normal flora invade and grow in an area of our bodies where it is not adapted to a harmless coexistence with its host, it will cause **disease.** For instance, the majority of urinary tract infections are caused by organisms that are normal flora of the intestinal tract. In addition, we frequently harbor organisms that, although they result in no disease to us, may provide a source of infection and disease to those with whom we come in contact, particularly those who are already ill or debilitated, as in a hospital environment. Many of us may peacefully coexist with staphylococci in our nose and throat, but if these staphylococci infect a burn or a wound, a fatal disease may result.

This difference between infection and disease is important to the physician who must identify the causative agent of an infectious disease. To make this decision, he or she must know what organisms should be expected to occur in the various areas of the body. For example, one would not be surprised to isolate *Escherichia coli* from a normal stool specimen, but if a large number of *E. coli* were isolated from a catheterized urine specimen, it would unequivocally indicate a urinary tract infection and hence disease. Similarly, the isolation of α-hemolytic streptococci from the throat would be a normal expectation, whereas isolation of the same organism from the blood could indicate serious disease.

In this chapter we discuss the general characteristics of organisms that are able to cause disease, how they are transmitted to humans—and, insofar as possible—what characteristics these disease-producing microorganisms possess that allow them to invade and cause disease.

The division between disease-producing bacteria and those unable to cause disease is not, however, a sharp demarcation. Some organisms routinely cause illness when they infect an individual, whereas others may cause symptomatic infections in some persons but not in others. There are terms to describe such situations, but it is difficult to assign quantitative values to them. For example, **pathogenicity** is defined as the ability of a microorganism to cause disease, while **virulence** refers to the extent of pathogenicity. We can then describe some pathogenic organisms as highly virulent while others that only occasionally cause disease are spoken of as being weakly virulent.

TRANSMISSION OF INFECTIOUS DISEASES

Not all infectious diseases are spread from person to person. For example, in tetanus and also in gas gangrene, the infective agent is introduced through the skin and into the muscle following a wound such as a puncture by a nail or a gunshot. Other infectious diseases may be acquired from animals, either by direct contact (such as handling infected meat) or by the ingestion of contaminated meat or milk products. On the other hand, many diseases are transmitted, either directly or indirectly, from one individual to another. This type of disease is called a **communicable** or **contagious disease.** We may also categorize infectious diseases on the basis of how often they occur. If a disease is constantly present within a stated geographic area, it is said to be **endemic.** The common cold is usually endemic within most localities. An **epidemic,** on the other hand, is a disease occurring in excess of normal expectancy. It is difficult to assign numbers to an epidemic because we are concerned with the occurrence over normal expectancy. For example, four or five cases of yellow fever in a village in South America might not be considered an epidemic, but these same cases in a city in the United States would certainly be labeled as one. Diseases such as influenza and measles often spread as epidemics. When an epidemic of a disease becomes worldwide, it is given the special title of **pandemic.** Pandemics of influenza have caused the deaths of millions of people. If a particular disease occurs only occasionally, it is said to be **sporadic.**

When a microorganism capable of producing disease gains entrance into a host, a period of time elapses before there is any manifestation of illness. This time interval between infection and the appearance of the first symptoms is called the **incubation period.** For some diseases it is constant and predictable (e.g., 11 days for measles), while in others it may vary (e.g., from 15 to 40 days for hepatitis A). The incubation period is followed by the period of **illness,** and this, in turn, by the period of **convalescence.**

Even though a person has no symptoms of disease during the incubation period and the period of convalescence, infectious disease organisms may be spread by that person. In some instances, an individual who has fully recovered from a disease may continue to carry and spread the causative organisms for months or even years. Such people are designated as **carriers.** They have apparently acquired sufficient immunity to

Completely asymptomatic carriers are responsible for the spread of a variety of diseases, particularly gastrointestinal infections such as dysentery and typhoid fever. Probably the most famous carrier in modern history became known as Typhoid Mary.

Her story began in 1868 when a typhoid epidemic raged on a ship carrying immigrants from Europe to America. Mary was about 13 years old and, although she recovered from typhoid fever, both of her parents died. It is believed that the ship's cook, Sean Mallon, took charge of Mary after her parents died and that he taught her how to cook. Because she became a very good cook, she had little trouble getting jobs as a domestic in the New York area and, on at least one occasion, she worked in the kitchen of a New York hotel.

An unusual series of tragedies, however, seemed to follow her. Often, within a few weeks or months, one or more of her employers died of typhoid fever. And the hotel that employed her was forced to close after it was established that several deaths from typhoid fever apparently originated there.

It is probable that by this time Mary Mallon (who had taken the name of the ship's cook) was aware that she was the cause of the typhoid fever and the ensuing deaths, but the alternatives to being a cook did not appeal to her so she kept moving from job to job as the disease continued to strike the households where she was employed. She changed her name frequently and is believed to have cooked in both Boston and Philadelphia only to return to New York when typhoid fever struck again. By the turn of the century, the newspapers had dubbed her "Typhoid Mary" and health authorities were desperately searching for the cook deemed to be a typhoid carrier. Of course, one must consider that during the early 1900s, typhoid fever was quite common, especially in large cities. For example, New York reported 3467 cases of typhoid fever with 700 deaths during the summer of 1906.

Mary was finally apprehended in 1907 and was placed in isolation in a bungalow at Riverside Hospital. She remained there for 3 years until the courts finally decided that her incarceration was unconstitutional. When she was released, she signed a promise not to engage in any kind of cooking or handling of food products.

She immediately broke her promise and, after many different positions (followed by cases of typhoid fever), was finally apprehended again after more than 20 cases of typhoid fever occurred in a women's hospital where Mary was employed as a cook. This episode ended Mary's tragic role as a source of typhoid fever, and she spent the remainder of her life in an isolated bungalow where she finally died on November 11, 1938.

There are a lot of people today who would fit into a category analogous to Typhoid Mary to whom we might give names such as gonorrhea Jane, hepatitis B Tom, or AIDS Joe, indicating that many infectious diseases are spread by asymptomatic carriers who may or may not be aware of their infection.

---

the disease-producing agent to prevent the occurrence of symptoms, but they are unable to eliminate the agent from their bodies. The most common type of carriers are those who carry disease-producing agents in their intestines, resulting in the spread of these organisms via contaminated food and water. Some infectious agents that are commonly spread by such carriers include those causing typhoid fever, bacillary dysentery, amebic dysentery, bacterial diarrhea, and viral hepatitis. Other infectious agents may linger in the bloodstream of people who, though not ill themselves, continue to carry the disease for months, years, or even a lifetime. This type of carrier is particularly dangerous if unwittingly used as a blood donor. Viral hepatitis and AIDS are, without doubt, the most common examples of disease agents transmitted via blood products and body secretions.

Infectious diseases may also be spread by people who have mild symptoms of an undiagnosed infection. One typical example of this is seen in the spread of tuberculosis. This disease usually begins with a mild but persistent cough, causing the organisms in the lungs to be expelled in an aerosol. By the time such an individual is diagnosed as having tuberculosis, the organisms may have been spread to numerous contacts.

We shall learn in Unit III that many infectious disease agents exist in a reservoir (e.g., wild animals or ticks) from which they are transmitted to humans. For instance, many rickettsial diseases exist in the wild animal population; the organisms causing plague, tularemia, and brucellosis also exist primarily in both wild and domestic animals. Birds constitute the chief reservoir for a number of agents causing viral encephalitis, and, of course, humans are the reservoir for many bacterial and viral diseases.

Acute and Chronic Infections

Acute infections are those that develop rapidly, are of short duration, and usually result in a high fever. By far, the majority of hospital patients with infectious diseases are acutely ill. In contrast, a chronic infection develops more slowly, with milder but longer-lasting symptoms. Of course, a chronic infection may become acute, and an acute infection may become chronic. Tuberculosis and leprosy are examples of infections that are more frequently chronic, whereas scarlet fever or toxic shock syndrome are generally acute infections.

Local and Systemic Infections

A local infection is one in which the causative agent is limited to one locality of the body. A boil (furuncle) is a localized infection. A systemic infection is one in which the infecting agent spreads throughout the body, as in Lyme disease or mumps. Of course, the clinical picture is never quite as simple as the examples given here, because a local infection may cause general symptoms similar to those resulting from a systemic infection. Local infections with general symptoms are usually the result of the elaboration of poisonous substances (toxins) by the microorganism. Such is the case with diphtheria, in which the organism remains localized in the throat and nasopharynx but produces a powerful exotoxin that is transported throughout the body. Tetanus and pertussis are also diseases of this type.

Occasionally an infection occurring in one body area (particularly an abscess) may act as a nucleus for its spread to other sites. An example of this type of focal infection is an abscess of a tooth or tonsils that continues to seed organisms into the blood. It is frequently necessary to find the area of focal infection and remove it surgically.

Primary and Secondary Infections

A primary infection is the initial infection causing the illness. To accomplish this, the microorganism requires a certain amount of what we may vaguely term **invasiveness**—that is, the property that enables an organism to overcome the normal body defenses and cause an overt infection. Many organisms lack this ability and are able to cause infections only after a primary infection or suppression of the immune system has weakened the body's defenses. These opportunistic bacteria, or **opportunists,** are said to cause secondary infections. A good example is staphylococcal pneumonia, which frequently follows primary respiratory infections such as measles and influenza. Another example is the frequently fatal pneumonia caused by the sporozoan *Pneumocystis carnii*, which occurs almost exclusively in persons with acquired immunodeficiency syndrome (AIDS).

Inapparent Infection

It is difficult to differentiate between an inapparent infection and a mild, undiagnosed one. By **inapparent** infection we mean an illness in which the symptoms are either absent or so mild that it goes undetected and thus undiagnosed. This category, in all likelihood, covers most of our illnesses. Poliomyelitis and hepatitis A are excellent examples of diseases in which most cases go undiagnosed. We know this is true because most people who have no history of having either infection have protective antibodies against both of the viruses causing these illnesses.

Bacteremia refers to the presence of bacteria in the blood, and **septicemia** implies that the bacteria are multiplying in the bloodstream. **Viremia** has a similar meaning applied to viruses. Most infectious organisms, with the exception of those causing strictly local infections, are transported via the blood during some part of the disease process. The presence of microbes in the blood may be only transient, or these microbes may be always present during the acute stage of the disease. Because of the presence of toxic substances that make up part of the cell wall of gram-negative bacteria (endotoxins), the presence of large numbers of these organisms in the blood can result in a general toxic condition known as toxemia (toxins in the blood). Toxemia could also occur in the absence of organisms.

PROOF OF THE CAUSE OF A DISEASE

The isolation of a particular organism from an infected person does not establish proof that it is the causative agent of the disease. It may exist in or near a lesion merely as normal flora or as a transient contaminant. Robert Koch faced this problem in the 1870s during his work to establish the cause of anthrax. Consequently, Koch laid down a series of experimental steps that should be followed to establish unequivocally a causal relationship between an organism and a specific disease. These rules, known as **Koch's postulates,** can be summarized as follows:

1. The same organism must be found in all cases of a given disease.
2. The organism must be isolated from the infected person and grown in pure culture.
3. The organisms from the pure culture must reproduce the disease when inoculated into a susceptible animal.
4. The organism must then again be isolated from the experimentally infected animal.

Although these postulates are effective in determining the causative agents of most diseases, there are a few exceptions. For example, neither *Mycobacterium leprae* nor *Treponema pallidum* has been grown in the laboratory and, therefore, fit only the first postulate. Some of the causes of viral hepatitis also fall into the category in which the last three postulates are not fulfilled.

Occasionally it is the third postulate that is difficult to fulfill, particularly in a human disease for which there is no readily available experimental animal. Such a case was seen in 1978 in investigations to establish the causal agent of a seemingly new malady called toxic shock syndrome (discussed in Unit III). *Staphylococcus aureus* was isolated in almost pure culture from the vaginas of the majority of women suffering from this syndrome, leaving little doubt as to its role in this disease. The final proof, however, came when it was shown that these strains of *Staph. aureus* produced an exotoxin that, when injected into rabbits, produced many of the symptoms of this disease.

Legionellosis (legionnaires' disease) is another recent example of the use of Koch's postulates to establish a cause-and-effect relationship between a microorganism and a disease. This disease, first described in 1978, yielded an organism that had not been previously described. Its causal role was established when it was shown always to be associated with the disease. In addition, it was grown in pure culture in the laboratory; it caused a pneumonia in guinea pigs similar to the human disease; and it could be reisolated from such infected guinea pigs.

Probably the most famous disease to fall into this gray area of causative agents is AIDS. As is discussed in Unit III, this disease has an incubation period that may be as long as 10 years, but once symptoms of immunodeficiency develop, it is uniformly fatal. The etiologic agent appears to be a retrovirus that has been named **human immunodeficiency virus** (HIV), but there are still a few scientists who believe that a rigorous proof has not yet established that HIV is the proven etiologic agent of AIDS.

HOW PATHOGENS ENTER AND LEAVE THE BODY

Secretions and excretions from infected areas often contain the infective microbe. In some illnesses (such as malaria) there may be no obvious exit, since the organisms are present in the blood and require a mosquito for transmission. However, whether it is obvious or not, each organism capable of producing disease has its own portal or portals of entry as well as a means of escape (portal of exit) from the host. The infectious agent in discharges from infected areas must be destroyed to prevent the transmission of the agent to a new host.

Microorganisms have been found to enter the body through the following areas:

1. Respiratory tract via nose and mouth. This is the portal of entrance of all microbes causing respiratory diseases such as the common cold, influenza, measles, pneumonia, and tuberculosis.

2. Gastrointestinal tract via mouth. Examples include agents responsible for diseases such as typhoid fever, paratyphoid fever, dysentery, cholera, poliomyelitis, and hepatitis as well as food illnesses discussed in Chapter 26.

3. Skin and mucous membranes. Although the skin provides an effective protective barrier, minor breaks are undoubtedly always present that may allow the entrance of certain organisms. The staphylococcus that causes boils (furuncles) is one of the more frequent organisms entering in this way; however, the streptococci also may cause spreading skin infections. Tularemia is an example of a severe systemic disease that may be contracted by handling infected animals.

4. Genitourinary system. The mucous membranes of the genital tract are the common site for invasion by sexually transmitted disease agents such as those causing syphilis, AIDS, and gonorrhea. In addition, the urinary tract may be infected by microorganisms in the blood or by their introduction into the bladder during catheterization.

5. Blood. Organisms that must be introduced directly into the blood in order to cause disease usually are transmitted from one individual to another by insect bites. The best-known examples of diseases in this category are malaria and yellow fever—both transmitted by the mosquito. Others include the rickettsial diseases (ticks, fleas, lice, and mites), bubonic plague (rat fleas), tularemia (deer flies and ticks), Lyme disease (deer ticks), relapsing fever (soft ticks), and the viral encephalitides (mosquitoes). Unfortunately, our advancing civilization has added another way for direct blood inoculation: blood transfusions (AIDS and hepatitis) or inadequately sterilized syringes and needles.

The portals of exit for a disease agent may be the same as the portals of entry. Thus, diseases of the respiratory tract are spread by way of secretions and excretions of the respiratory tract and the mouth. Enteric infections (typhoid fever, poliomyelitis, dysentery, etc.) enter the body at one end of the gastrointestinal tract and leave the body via the other end. Skin or wound infections are spread by the drainage from these areas either directly to another person or through contamination of some inanimate object. Blood infections, which are spread by insects or contaminated needles and syringes, usually leave the individual in a similar manner, that is, through direct contamination of a needle or syringe during the withdrawal of blood or the injection of drugs, or the ingestion of the microorganism by a biting insect.

VIRULENCE FACTORS

For an organism to be virulent, to invade and cause disease, it must possess certain characteristics or properties that are not possessed by the saprophytic organisms. In some cases the properties conferring virulence on an organism are either unknown or poorly understood. However, many bacteria are known to possess special structures that protect them from the host's defenses, or they may be capable of secreting certain substances that contribute to their virulence. In the following discussion are described some of the factors believed to contribute to the pathogenicity of a microorganism.

Capsules

Certain pathogenic organisms secrete large capsules around their cell walls. The ability of these organisms to produce disease depends on the presence of this capsule. Loss of the capsule as a result of a mutation invariably results in the inability to produce disease. The chemistry of some of these capsules is discussed in Chapter 3, but this knowledge does not explain why virulence is associated with the presence of a capsule. Experiments have demonstrated, however, that the capsule prevents phagocytosis, that is, the engulfment and digestion of the organisms by the host's leukocytes. The exact reason for this antiphagocytic activity is not known, but it seems to be a surface phenomenon in which the phagocyte is unable to form a sufficiently intimate contact with the microorganism to allow phagocytosis to occur. The binding of antibody to the capsular material provides receptors on the capsule and allows phagocytosis to take place. Thus, immunity to pneumonia caused by *Streptococcus pneumoniae* depends only on an antibody to the microorganism's capsule. In the presence of this antibody, the invading organism is rapidly engulfed and destroyed by the host's leukocytes.

Fimbriae

Just as nonpathogenic bacteria may possess capsules, so may non-disease-producing bacteria possess fimbriae (also called pili; see Chapter 3). This obviously demonstrates that neither capsules nor fimbriae are sole determinants for the virulence of microorganisms. It is known, however, that the possession of fimbriae endows an organism with an enhanced ability to adhere to other bacteria and to the membranes of the host's cells and phagocytes, thereby impairing ingestion by the phagocyte.

There are a number of examples in which the possession of fimbriae appears to be required for an organism to cause disease. For instance, *E. coli* causes about 75 percent of all urinary tract infections, and essentially all possess fimbriae that bind the bacterium to the epithelial cells of the urinary tract. *Neisseria gonorrhoeae* (the cause of gonorrhea) as well as *N. meningitidis* (the cause of epidemic meningitis) lose their virulence if they lose their fimbriae. Enterotoxigenic *E. coli* also are unable to cause diarrheal disease if they lose their fimbriae, even though they continue to secrete the enterotoxins that are directly responsible for the symptoms of the disease. It is also noteworthy that specific antibodies directed to the homologous fimbriae are protective in all of the above cases.

Exotoxins and Enzymes

Because many organisms do not possess antiphagocytic capsules, their virulence must be due to other characteristics that permit them to overcome normal host defenses. Many of these characteristics remain unknown to us, but some organisms, for example, produce disease by excreting a toxic substance that kills or injures host cells, including leukocytes. Toxic substances excreted from the bacterial cell are called **exotoxins.** Exotoxins are responsible for the symptoms of diseases such as diphtheria, pertussis, cholera, dysentery, anthrax, tetanus, gas gangrene, and scarlet fever. The food poisonings caused by *Clostridium botulinum, S. aureus,* and *Bacillus cereus* are due to the liberation of an exotoxin into the food. The ability to produce an exotoxin is the only property that allows many organisms to produce disease. Thus, the loss of this ability results in their complete loss of pathogenicity.

However, many pathogenic organisms do not possess antiphagocytic capsules and do not produce exotoxins. It is somewhat more difficult to pinpoint the properties they possess that are responsible for their invasiveness. There are, however, a host of substances—some of which are enzymes—that are secreted by various bacteria and are believed to enhance the invasiveness of the organism and, as a result, the spread of disease within the host. Most of these substances have not been isolated, purified, or characterized and have been named based on observations with crude materials.

Hemolysins

A number of different hemolysins are produced by bacteria that induce the lysis of the host's red blood cells. Hemolytic anemia, however, is not a normal result of an infection even by organisms producing large amounts of hemolysin. Such hemolysis does, however, provide iron essential for growth from the liberated hemoglobin and, as a result, contributes to the virulence of the hemolysin-producing organism.

Many of these hemolysins are also capable of causing membrane damage to many different kinds of eucaryotic cells. One could assume that damage to phagocytic cells would enhance the survival of the invading organism. In addition it is frequently important to know whether an organism produces a hemolysin and the type of action of the hemolysin in order to identify the bacterium in the laboratory.

Leukocidins

Leukocidins—little-studied, nondescript substances—kill the host's leukocytes (white blood cells). It is difficult to evaluate their role in the invasiveness of the organism, but because leukocytes are so important in the defense mechanisms of the body, any substance toxic to them will give an advantage to the invading organism.

Hyaluronidase

Various pathogenic bacteria excrete an enzyme capable of breaking down hyaluronic acid, the intracellular material of connective tissue. The hydrolysis of this material would seem to help in the spread of the organisms and, based on this postulation, hyaluronidase sometimes has been referred to as the "spreading factor." Experimentally, it is difficult to quantitate the value of hyaluronidase in microbial invasiveness, but it certainly is probable that it does aid in penetration and in the spread of certain organisms in host tissues.

Streptokinase

By activating a proteolytic enzyme called plasmin, which is normally present in the host's plasma, streptokinase causes the dissolution of blood clots and thus allows the spread of the organism. An essentially identical substance produced by the staphylococci is referred to as **staphylokinase.**

Coagulase

The major organism producing coagulase is the familiar staphylococcus. The end result of the action of coagulase is the coagulation of plasma, which produces a fibrin clot. It is postulated that this ability allows the organisms to lay down a thin fibrin layer around each individual cell, thus preventing or inhibiting phagocytosis by the host's leukocytes. Furthermore, it is thought by many investigators that co-agulase is responsible for the fibrin barrier, typical of an abscess, that walls off the local infection from the host's defense mechanisms. In any case, it is certainly true that of all the properties associated with the virulence of staphylococci, the production of coagulase is the most constant.

Collagenase

Collagenase is formed by some of the clostridia that cause gas gangrene. Collagenase causes the breakdown of collagen, which is the ground substance of bone, skin, and cartilage. Thus, collagenase helps in the spread of the organism from the initial site of the infection.

A few other damaging substances are a necrotizing factor (causing necrosis of tissue cells), a hypothermic factor (affecting body temperature), and an edema-producing substance. Like those already mentioned, these substances may well be involved in the invasive ability of pathogens.

Siderophores

Vertebrates have evolved some ingenious ways of retaining iron in a soluble state so that this highly insoluble metal is available for the synthesis of new cells. This is accomplished by binding iron to very-high-affinity glycoproteins, which not only keep it soluble, but serve to transport iron throughout the body. One of these, termed **lactoferrin,** is found primarily in secretions such as milk, tears, saliva, mucus, and intestinal fluid. The other major iron transporter is called **transferrin,** and it is found mostly in plasma. The important thing for us to remember, however, is that microorganisms also require iron for growth, and unless they are able in some way to "steal" the bound iron from transferrin or lactoferrin, they would be unable to reproduce within the body. And, that is exactly what many pathogenic microorganisms are able to do through the production of **siderophores** and membrane receptors for the siderophore-iron complex.

Siderophores are low-molecular-weight compounds that have a very high affinity for iron. In fact, many have higher affinities than transferrin and thus are able to capture iron, permitting growth within the host. After accepting a molecule of iron, the siderophore then binds to a specific membrane receptor on the bacterium, releasing iron for microbial growth. There are many different siderophores, but they all fall into one of two types: (1) catechols and (2) hydroxymates. In either case, the ferric ion is chelated between two hydroxyl groups or, in some cases, between a hydroxyl group and an amino group.

Thus, the ability to produce a siderophore that can successfully compete for host iron is an important virulence factor. Many organisms synthesize more than one siderophore, and many others are able to bind to siderophores that they are unable to synthesize. Their importance has been clearly demonstrated in many experiments with animals and a large variety of microorganisms, which showed that a genetic loss in the ability to synthesize a siderophore is correlated with loss in virulence and vice versa.

Exotoxins Associated with Specific Diseases

The symptoms of many diseases are directly attributable to one or more specific exotoxins produced by the infecting organisms. The mechanism of action of such exotoxins is described in Unit III; here we will only categorize the major types that are associated with human disease.

Enterotoxins

Enterotoxins are exotoxins that interact with the gastrointestinal system to cause diarrhea or dysentery (bloody diarrhea). There are several different enterotoxins but, in general, they can be grouped into two types: (1) those that cause an electrolyte imbalance resulting in a flow of water into the intestine, and (2) those that inhibit protein synthesis and, as a result, kill the epithelial cells lining the intestine. Enterotoxins causing an electrolyte imbalance are characteristic of *Vibrio cholerae* (cholera), *E. coli* (traveler's diarrhea), and possibly staphylococcal food poisoning. Enterotoxins blocking protein synthesis are produced by *Shigella* species (dysentery) and certain strains of *E. coli* (enterohemorrhagic *E. coli*).

Neurotoxins

Tetanus toxin causes a spastic paralysis through its ability to block inhibitory nerve impulses in the central nervous system, causing the somatic motor neuron to fire continuously. Botulism toxin, on the other hand, causes a flaccid paralysis by preventing the release of acetylcholine at the neuromuscular junction.

cAMP-Producing Toxins

Cyclic adenosine monophosphate (cAMP) is termed a second messenger because its concentration is governed by certain hormones. Once formed, cAMP is involved in the activation of a number of protein kinases that, in turn, control the activity of a large number of different enzymatic reactions. It is easy to see, therefore, that any increase in cAMP above physiologic levels will result in a detrimental reaction within the cell. There are several bacterial exotoxins that induce the formation of cAMP. These include the en-terotoxins that cause diarrhea by creating an electrolyte imbalance in the intestinal epithelium as well as pertussis toxin (whooping cough) and anthrax toxin.

Protein-Synthesis–Inhibiting Toxins

We have already listed the *Shigella* enterotoxin (Shiga toxin) that causes dysentery through its ability to block protein synthesis in intestinal epithelial cells. Shiga toxin accomplishes this by inactivation of the 60S ribosomal subunit. Diphtheria toxin also prevents protein synthesis by blocking the movement of the ribosome along its mRNA.

Table 16.1 provides a summary of human diseases in which the major symptoms are the result of exotoxin production by the infecting bacteria.

Endotoxins

These substances are large molecules of lipopolysaccharide (composed of lipid and carbohydrate) that are normal components of the outer membrane of all gram-negative organisms. They are sloughed off in small amounts during the growth of the cell, but also may be able to manifest their toxicity while still attached to the bacterial cell wall. The biological effects of endotoxins are manifold, but the two most pronounced effects that result from an infection with a gram-negative organism are fever and shock (a severe drop in blood pressure). Humans are particularly sensitive to endotoxin and respond with fever to minute amounts of endotoxin or a mild gram-negative bacterial infection. The production of irreversible shock requires somewhat larger amounts of endotoxin but does happen during massive gram-negative bacteremia.

TABLE 16.1 Major Diseases Associated with Exotoxins

Exotoxins that increase cAMP levels
  Vibrio cholerae enterotoxin
  Escherichia coli enterotoxin
  Anthrax toxin
  Pertussis toxin
Exotoxins that block protein synthesis
  Shigella enterotoxin (Shiga toxin)
  Corynebacterium diphtheriae toxin
Neurotoxins
  Clostridium botulinum toxin
  Clostridium tetani toxin
Miscellaneous toxins whose detailed mechanism of action is unknown
  Streptococcus pyogenes scarlet fever toxin
  Staphylococcus aureus
    Scalded skin syndrome toxin
    Toxic shock syndrome toxin
    Food poisoning enterotoxins
  Clostridium perfringens gas gangrene toxins

The real conundrum was just now a molecule like lipopolysaccharide could evoke the symptoms attributed to endotoxin. The answer to this puzzle appears to have been solved in part by the observation that endotoxins cause macrophages to secrete a cytokine whose hormonelike effects are responsible for many of the symptoms associated with endotoxin. Because this cytokine had been previously associated with the lysis of malignant tumors, it was given the name of **tumor necrosis factor alpha** (TNF-α). This factor is also responsible for a syndrome associated with chronic infection and severe weight loss, known as cachexia. For this reason, TNF-α has also been termed **cachectin.**

The gene for TNF-α has been cloned and it has been shown experimentally that the injection of TNF-α into a susceptible animal will reproduce the severe hypotension (low blood pressure) and lethal shock characteristic of gram-negative bacterial infections. Moreover, antibodies to TNF-α are able to prevent endotoxin-induced lethality.

The mechanism whereby TNF-α induces its effects is a little less clear, but it is known that TNF-α induces the release of another cytokine from macrophages and epithelial cells (interleukin-1) and that this cytokine together with TNF-α has multiple effects on polymorphonuclear neutrophils in addition to their interaction with the temperature centers of the brain to produce prostaglandins (see Figure 16.1), and it appears that prostaglandins are the "second messengers" of TNF-α-induced toxemia.

Table 16.2 lists some microbial factors that appear to be involved in the production of disease.

FACTORS IN THE DEVELOPMENT OF DISEASE

We can now see that a number of different factors come into play in order for a disease agent to produce an infection.

Portal of Entrance

The portal of entrance must be suitable for the particular pathogen to cause disease. In many cases the organism is restricted to only one portal of entry—for example, the typhoid fever organism, which must be swallowed. On the other hand, the staphylococcus can cause pneumonia via the respiratory route, boils and furuncles via the skin, internal abscesses via the blood, or food poisoning via the gastrointestinal tract.

Ability to Flourish Outside the Body

The exit from the body and the ability of the organisms to survive in the outside world are important factors in disease development. Some organisms, such as the meningococcus and the gonococcus, are extremely sensitive to drying and will die in several hours outside the host. However, organisms such as the tubercle bacillus may survive in dried sputum for weeks and still maintain its pathogenicity. Certain protozoan par-

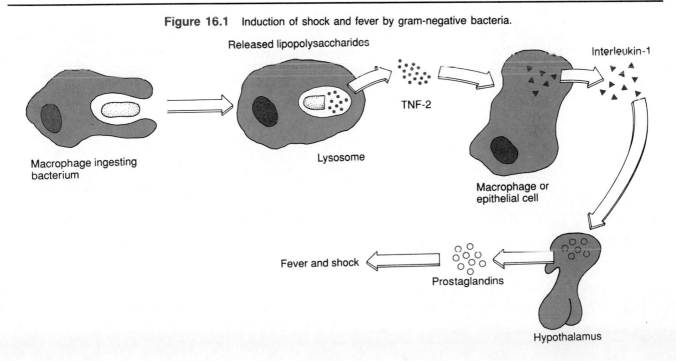

Figure 16.1   Induction of shock and fever by gram-negative bacteria.

Released lipopolysaccharides

Interleukin-1

TNF-2

Macrophage ingesting bacterium

Lysosome

Macrophage or epithelial cell

Fever and shock

Prostaglandins

Hypothalamus

TABLE 16.2 Toxic Factors of Bacteria

Factor	Effect	Examples of Pathogens that Excrete Damaging Substances
Exotoxins	Toxic effect varies with the species	*Bacillus anthracis* *Bordetella pertussis* *Clostridium botulinum* *C. perfringens* *C. tetani* *Corynebacterium diphtheriae* Enterotoxigenic *E. coli* *Shigella dysenteriae* *Staphylococcus* sp. *Vibrio cholerae*
Endotoxins	Liberated in small amounts during autolysis of gram negative bacteria	*Salmonella typhi* *Shigella flexneri* *Neisseria gonorrhoeae* *N. meningitidis*

Hemolysins
  Streptolysin (streptococci)
  Staphylolysin (staphylococci)
	Cause destruction of red blood cells as well as other eucaryotic cells	Staphylococci
		Streptococci
		Str. pneumoniae
		C. perfringens
		A number of saprophytic bacteria

Leukocidins	Cause destruction of leukocytes	Staphylococci Streptococci *Str. pneumoniae*
Hyaluronidase	Increases permeability of tissues to bacteria and their products; contributes to their spreading through the tissues	Staphylococci Streptococci *Str. pneumoniae* Certain obligate anaerobes *C. perfringens*

Kinases
  Streptokinase (streptococci)
  Staphylokinase (staphylococci)
	Dissolve fibrin	Streptococci
		Staphylococci

Coagulase	Causes plasma to clot	Staphylococci (occasional strains of *Pseudomonas aeruginosa, Serratia marcescens, Bacillus subtilis, E. coli*)
Collagenase	Breaks down collagen Muscular disintegration follows destruction of ground substances Contributes to the spread of organisms in the tissues	*C. perfringens*
Necrotizing factor (necrotoxin)	Kills tissue cells	Staphylococci
Hypothermic factor	Lowers body temperature	*Shigella dysenteriae*
Edema-producing substance	Edema	*Bacillus anthracis*
Pyrogenic toxins	Fever and shock	Streptococci and Staphylococci

asites secrete a dense covering around their vegetative cells, forming cysts that enable them to survive after leaving the body. Similarly, most of the flatworms in the class Trematoda are able to form cysts, called metacercaria, which protect the organism until it is ingested.

Vector

Vectors are necessary for the transmission of certain diseases. Without mosquitoes, there would be no malaria or yellow fever and without ticks, no Rocky Mountain spotted fever or Lyme disease.

Number of Pathogens

The number of organisms required to cause disease is certainly a very important variable. It may require many thousands of staphylococci to cause infection in a clean cut, but a few hundred may infect a suture. On the other hand, *Franciscella tularensis* (the organism causing tularemia) can cause severe disease even when only three or four organisms invade the skin. Thus, it is easy to see that the division between pathogens and saprophytes is a vague line and that it is only at the extreme ends that one can see clear-cut examples of obligate pathogens (tularemia, syphilis) or obligate saprophytes such as *Bacillus megaterium* and *B. subtilis*.

As our study of diseases becomes involved with specific infections (Unit III), we shall see that some diseases are found only in humans (typhoid fever, cholera, syphilis, epidemic meningitis); hence, humans are necessary in the continued life of these organisms. We shall see also that there are many infectious diseases that humans acquire from wild or domestic animals only as accidental hosts. Examples of this category include yellow fever, typhus, Rocky Mountain spotted fever, tularemia, and brucellosis. A third category of infectious organisms, the soil saprophytes, live and grow in the soil without requiring any host for survival, but, if introduced into a susceptible host, they will cause serious illness and frequently death. These include such diseases as tetanus, gas gangrene, anthrax, and the fatal food poisoning caused by *C. botulinum*.

SUMMARY

Disease is an undesirable host-parasite relationship involving the entry and multiplication of microorganisms.

Koch's postulates establish proof of the association of a bacterial species with a specific disease. However, proof of the cause of certain diseases has been established without fulfilling all the steps of the postulates.

Pathogens may enter the body through the respiratory tract, the digestive tract, the skin and mucous membranes, the genitourinary system, and the blood. Pathogens may leave the body through secretions and excretions from the respiratory tract and the mouth, through feces and urine, and through infected areas such as wounds and boils, the mucous membranes, and blood.

Diseases may be transmitted by direct and indirect contact. Various vectors in disease transmission include hands, food, contaminated inanimate objects, droplets, dust, infected persons or carriers, human blood and its derivatives, and animals, including insects.

Pathogenicity is the ability of an organism to cause disease; virulence refers to the extent of pathogenicity. For an organism to be pathogenic, it must be invasive and/or toxic. To be invasive, an organism must be able to resist the host's defense mechanisms. Many organisms accomplish this through the production of capsules, exotoxins, siderophores, and enzymes. In many others, however, we do not know what property makes them invasive.

Major bacterial exotoxins can be categorized as enterotoxins, neurotoxins, cAMP-producing toxins, and protein-synthesis-inhibiting toxins. Endotoxins are lipopolysaccharides present in the outer membrane of gram-negative bacteria. They are able to induce fever and shock through their ability to cause macrophages to secrete tumor necrosis factor alpha and interleukin-1.

Several factors are of considerable significance in the development of disease. These include the portal of entry, the virulence of the organisms, exit from the host and the ability to survive in the outside world, agents of transmission (such as insect vectors), the number of invaders, and the ability of the host to combat the invaders.

QUESTIONS FOR REVIEW

1. What is meant by *infection*?
2. What is an infectious disease?
3. Differentiate between *endemic, epidemic, pandemic,* and *sporadic*.
4. What is the meaning of *incubation period*?
5. Contrast the following:
   a. Acute and chronic infection
   b. Local and systemic infection
   c. Primary and secondary infection
6. What is a focal infection?
7. What is the significance of Koch's postulates?
8. List the possible portals of entry for infectious agents.
9. List the possible avenues of exit from the body for infectious agents.
10. In what ways may infectious diseases be transmitted?
11. How do capsules serve as a virulence determinant? Fimbriae? Siderophores?
12. List four categories of exotoxins that are associated with specific diseases.
13. What is endotoxin? By what mechanism do they induce fever and shock?

SUPPLEMENTARY READING

Costerton J, Geesey G, Gheng K: How bacteria stick. *Sci Am* 238(1):86, 1978.

Holt SC: Bacterial adhesion in pathogenesis. In Schlessinger D (ed): *Microbiology—1982*, Washington, DC, American Society for Microbiology, 1982, p. 261.

Old LJ: Tumor necrosis factor. *Sci Am* 258(5):59, 1988.

Roitt I: *Essential Immunology*, 6th ed. Oxford, Blackwell Scientific Publications, 1988.

Roth JA (ed): *Virulence Mechanisms of Bacterial Pathogens*. Washington, DC, American Society for Microbiology, 1988.

Stephen J, Pietrowski RA: *Bacterial Toxins*, 2nd ed. *Aspects of Microbiology Series*. Washington, DC, American Society for Microbiology, 1986.

Tracey KJ, Lowry SF, Cerami A: Cachectin: A hormone that triggers acute shock and chronic cachexia. *J Infect Dis* 157:413, 1988.

Weinberg ED: Cellular regulation of iron assimilation. *Q Rev Biol* 64:1, 1989.

Chapter 17

Nonspecific Host Resistance

OBJECTIVES   This chapter is designed to introduce you to

1 The role of mechanical and chemical mechanisms in our first line of defense.

2 The nature of interferons, including their induction and mode of action.

3 The kinds of cells involved in the phagocytosis of invading organisms.

4 The mechanism by which phagocytic cells kill microorganisms.

5 Genetic defects in the ability to carry out phagocytosis.

The complex reactions that a host animal undergoes after contact with microorganisms may be grouped under the rather broad heading of **resistance.** Many facets of host resistance to infection are poorly understood; on the basis of our present knowledge, however, we may categorize resistance as being of two major types: (1) nonspecific or natural resistance and (2) specific resistance that is acquired and directed against specific microbes.

Resistance or susceptibility (lack of resistance) to infections may vary tremendously from one species of animal to another. For example, white mice are extremely susceptible to infection by *Streptococcus pneumoniae*. The introduction of just two or three organisms into the healthy mouse will almost invariably kill it within 24 to 36 h. Humans, on the other hand, are relatively resistant to pneumococcal infections, as is evidenced by the large percentage of persons who carry the organism in their respiratory tracts without symptoms of infection. Another example is the extreme susceptibility of the guinea pig to human tuberculosis, as opposed to the relatively high resistance to the tubercle bacillus by humans.

We have no complete answer as to why resistance varies so widely from one species to another. It is simple to state the obvious—that cold-blooded animals are resistant to infection from microbes that cannot grow at their body temperature; but it is more difficult to understand why the mouse is so susceptible to pneumococcal infections, why humans are so susceptible to tularemia, or, for that matter, why both are relatively resistant to infection by streptococci, staphylococci, and a host of other microorganisms.

The fact that some people suffer a disease mildly and others suffer severely emphasizes the variation in resistance between individuals. For the most part, we are unable to explain this variation in nonspecific or natural resistance. There are genetic or racial factors that make certain ethnic groups of people more susceptible or resistant to a particular infection than others. (Examples are the inordinate susceptibility of the American Indian to tuberculosis and the increased resistance of American and African Blacks to malaria.) While the observation of increased susceptibility appears to be correct, the interpretation may well be wrong. Most cases seem to be found in a group or race of people whose low socioeconomic status brings additional variables into the picture. Thus, such factors as diet and environment must be evaluated. A number of laboratories are currently directing their research activities toward a better understanding of this problem. However, that some genetic factors are involved in our resistance to infection is supported by the observation that if one identical twin contracts tuberculosis, there is a 75 percent chance that the other will also get tuberculosis. On the contrary, among fraternal twins, there is only a 33 percent chance that the second twin will become infected by the tubercle bacilli. Quite obviously, there is some genetic determinant that provides most of us with resistance to tuberculosis as well as to many other organisms.

THE FIRST LINE OF DEFENSE: MECHANICAL AND CHEMICAL MECHANISMS

Skin

The intact skin and mucous membranes provide a mechanical barrier that prevents the entrance of most microbial species. However, even though the structure of the skin itself undoubtedly provides a great deal of protection, considerably more important are the fatty acids secreted by the sebaceous glands in the skin and the propionic acid produced by the normal flora of the skin. Secretions by the sebaceous glands contain both saturated and unsaturated fatty acids, which are lethal for many bacteria and fungi. A striking example of this type of resistance to infection is seen in the case of the fungus causing ringworm of the scalp. This infection is difficult to cure in young children, but after puberty it disappears without treatment, presumably as a result of a change in the amount and kinds of fatty acids secreted by the sebaceous glands.

Mucous Membranes

The beating action of the cilia that extend from the mucous membranes of the respiratory tract provides for the continuous movement of a fluid layer of mucus. Particles of dust or microorganisms adhere to this mucus and are moved to the exterior. In this way, the lungs are kept remarkably free of microorganisms. The mucous membranes of the genitourinary tract also provide a defense barrier through mucus secretions.

Chemical Factors

The washing action of tears helps to eliminate injurious microbes in the eye. Also, tears are very rich in lysozyme, an enzyme that hydrolyzes the bond between *N*-acetylmuramic acid and *N*-acetylglucosamine in the peptidoglycan part of the bacterial cell wall. As a result, many organisms are completely lysed (dissolved) in the presence of lysozyme. Lysozyme is also present in the blood and tissues, where it is undoubtedly effective in eliminating invading microorganisms. Acids of the stomach, vagina, and skin also provide important resistance to infection. A notable example of this is the greatly reduced number of organisms required to cause typhoid fever if the acid in the stomach is neutralized before ingestion of the organisms.

Interferon-type Interference

Interferon-mediated resistance results from the production of a family of proteins by a virus-infected host cell or an antigen-stimulated T cell. This type of interference was originally described when it was discovered that allantoic fluid obtained from a chick embryo infected with influenza virus would prevent the subsequent infection of uninfected chick embryos. In other words, something was produced during the initial infection that, when injected into other chick embryos, made them resistant to a subsequent viral infection. Interferons were later shown to be soluble proteins produced by cells infected with almost any animal virus.

An interesting property of most interferons is that they are not virus-specific; rather, they are host-specific. This means that the interferon produced by chick embryo cells infected with an influenza virus is effective in preventing the infection of other chick embryo cells with almost any virus. On the other hand, interferon produced in a mouse as a result of virus infection is effective in protecting other mice from a virus infection but of little value when used with chicken or human cells. Thus, one can generalize by stating that interferon inhibits viral replication most effectively in the species in which it was produced and that it is nonspecific with respect to the types of viruses it can inhibit.

Types of Interferon

Interferons originally were named according to the cell type from which they were produced, for example, leukocyte interferon, fibroblast interferon, and immune interferon (also called T-cell interferon). It was subsequently determined that a single cell line could produce more than one species of interferon; the current classification is shown in Table 17.1.

Molecular cloning and DNA sequencing of human interferon genes revealed that the interferons are encoded by a structurally related multigene complex.

To date, 14 distinct human IFNα genes have been identified in addition to single genes for IFNβ and IFNγ. DNA sequence analysis has shown that the IFNα are structurally related, sharing 80 to 95 percent sequence homology. In contrast, IFNβ and IFNγ are more distantly related, sharing only about 30 to 40 percent sequence homology with IFNα. IFNγ can be differentiated from IFNα and IFNβ serologically and by the observation that it is rapidly destroyed at pH 2, whereas IFNα and IFNβ are stable at that pH.

Induction of Interferon

The synthesis of IFNα and IFNβ occurs when a cell becomes infected with a virus. Essentially any virus is effective, although the amount of interferon produced will vary depending on both the cell line and the species of virus used as an inducer. The properties shared by this wide array of interferon-inducing viruses are not yet completely understood, but some light was shed when it was discovered that double-stranded RNA alone would act as an inducer of interferon synthesis. It is now generally believed that double-stranded RNA is the actual inducer of interferon synthesis during a viral infection.

Synthesis of IFNγ follows the addition of an antigen to T cells that have been sensitized to that antigen, such as the addition of tuberculin to cells from a tuberculin-positive individual. It can also be induced by the addition of a mitogen (which causes cells to divide), such as endotoxin, to normal, unprimed T cells.

The overall importance of IFNγ has been difficult to assess, but it has been reported that it regulates natural-killer-cell activity and enhances macrophage activation. Natural killer cells appear to be lymphocytes whose main function is to kill tumor cells that may arise in the body. As a result, IFNγ can be considered to be a lymphokine (secreted by lymphocytes), which exerts a protective effect by regulation of nonspecific host immunity.

Mechanism of Antiviral Effect

Interferons are small protein molecules secreted in exceedingly limited amounts by the virus-infected cell.

TABLE 17.1 Types of Human Interferon

Inducer	Producing Cell	General Name	Current Classification
Virus	Leukocytes	Le interferon	IFNα
Virus	Fibroblast	F interferon	IFNβ
Antigen	Primed T cells	Immune interferon	IFNγ
Mitogens	Unprimed T cells	Immune interferon	

They are, by themselves, not active against viruses and only after a molecule of interferon leaves the cell in which it was produced and binds to an uninfected cell can it exert its antiviral effect. It does this by derepressing at least two host cell genes, resulting in the synthesis of two new enzymes by the interferon-treated cell. One, a protein kinase, transfers a phosphate group from adenosine triphosphate (ATP) to an initiation factor necessary for protein synthesis. This results in a greatly reduced ability to translate messenger RNA. The other interferon-induced enzyme catalyzes the formation of a short polymer of adenylic acid. This oligoadenylate acts as a positive effector to activate an inactive endonuclease, RNase L, which then degrades both viral and host cell messenger RNA (Figure 17.1).

Therapeutic Use of Interferon

Human interferons have been purified to homogeneity, and all of the interferon genes have been individually cloned in *Escherichia coli* and yeast. Thus, through genetic engineering technology, large quantities of each of the interferons can now be produced in *E. coli* or yeast, making significant amounts available for clinical trials. Toxic side effects following the intramuscular or intravenous administration of interferon include fever, severe fatigue, headache, nausea, and a dry mouth.

Recent studies using human IFNα produced in *E. coli* have shown that colds can be prevented by the internasal instillation of IFNα in high dosage at frequent intervals prior to exposure to the virus. Inter-

Figure 17.1 Mechanism of interferon action. Viral infection of Cell 1 induces that cell to synthesize interferon. The released interferon enters an uninfected cell, causing that cell to synthesize two enzymes, protein kinase and polyadenylate synthetase. The protein kinase inhibits the translation of viral mRNA and the polyadenylate induces the formation of an endonuclease that degrades viral mRNA.

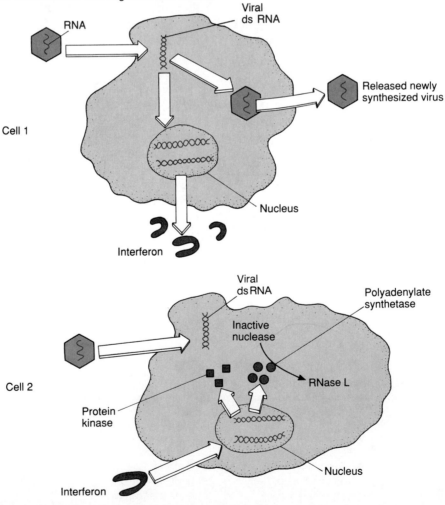

feron was not, however, particularly effective in curing a cold, and its value for treatment of the common cold is probably impractical.

Interferon treatment may have an important role in the treatment of certain severe viral infections. For instance, chronic hepatitis B and some herpesvirus infections appear to be amenable to prolonged treatment with high doses of IFNα or IFNβ. Interferon has proved extremely useful in treating juvenile laryngeal papilloma, a severe recurrent viral infection requiring surgical treatment for the removal of growths from the larynx. However, these growths tend to recur after cessation of interferon treatment. Interferon has also been used successfully to treat other types of human papilloma infections, including common warts and genital warts. In addition, studies have shown that interferon is effective in protecting animals after challenge with rabies virus, and it will undoubtedly be used following human exposure to this virus.

Antitumor Effects of Interferon

The value of interferon in the treatment of cancer remains to be firmly established. Several groups have reported promising results using interferon to treat osteosarcoma (bone cancer) multiple myeloma, Hodgkin's disease, and acute lymphoma. On the other hand, several other carefully conducted trials have yielded discouraging results, indicating that it is not yet possible to evaluate the antitumor effects of interferon. Obviously, more evidence on the value of interferon as an anticancer agent will be forthcoming during the next decade.

THE SECOND LINE OF DEFENSE: PHAGOCYTOSIS AND INFLAMMATION

In spite of all the barriers that prevent the entrance of microorganisms into our bodies, probably no day passes when bacteria do not enter the bloodstream. This undoubtedly occurs any time we cut ourselves, brush our teeth, shave, or for that matter, even chew gum vigorously. Since the blood provides such an excellent growth medium, we may then ask why we all are not dead.

We survive these daily attacks of bacteria because several mechanisms destroy and eliminate the offending agents in the circulatory system.

Circulating Phagocytic Cells

Leukocytes (white blood cells), in general, are the scavengers in the blood; that is, they remove debris, in-

cluding bacteria, by engulfing or devouring foreign particles. This process of engulfing particulate matter is called phagocytosis; the white cell itself is called a **phagocyte.** Frequently, the leukocytes themselves are killed in the process and accumulate at the site of an infection. This accumulation of dead cells, along with bacteria and serum, is known as pus.

There are several types of leukocytes circulating in the blood, but by far the most prevalent is the **polymorphonuclear neutrophil** or **granulocyte** (usually called a **PMN** or **poly**). PMNs normally make up 60 to 70 percent of the total leukocyte count. This type of white cell contains granules that stain with neutral dyes, and the nuclei are divided into large lobes (see Figure 17.2). In many infections there is an explosive increase in the number of these leukocytes in the blood, a condition known as **leukocytosis.** However, in other infections, particularly those caused by gram-negative bacteria, there may be a decrease in the number of circulating leukocytes as a result of their sticking to capillary walls. This is referred to as **leukopenia.** Thus, a leukocytosis may be good evidence for a bacterial infection, but the absence of any change, or a leukopenia, does not help in determining whether or not an infection is present. It was the early work of

Figure 17.2 Electron micrograph of a mature neutrophilic granulocyte (PMN) from rabbit bone marrow, showing two lobes of its dense nucleus and cytoplasmic granules. There are two types of granules: the larger, denser azurophil granules contain peroxidase, lysozyme, and lysosomal enzymes and the smaller, less dense specific granules that lack lysosomal enzymes but contain alkaline phosphatase, lysozyme, and lactoferrin (×25,000).

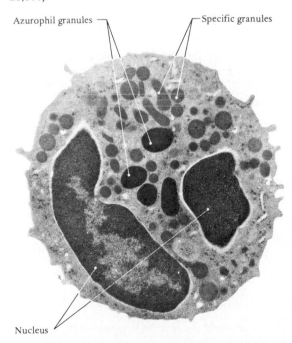

Azurophil granules — Specific granules

Nucleus

It became well established during the last half of the 19th century that fresh normal serum could kill a variety of bacteria, which gave rise to the concept of "humoral immunity," meaning that the immunity existed in a fluid nature in blood serum.

A new concept of immunity was proposed in 1884 by Elie Metchnikoff, a Russian-born zoologist who worked extensively in Russia and France. Metchnikoff was studying mobile cells in transparent starfish larva when he was struck by the idea that such cells might serve in the defense against foreign intruders. In his initial experiments, Metchnikoff placed small rose thorns under the skin of the starfish larva, and by observing the mobilization of white cells around the thorn, he formulated his theory of phagocytosis as a defense against infection.

The next 25 years of his life were devoted to proving that inflammation and phagocytosis did indeed comprise a defense mechanism that could be separated from the humoral immunity functioning in the absence of cells. It was not easy! Many experiments and many animal types were required. Moreover, it was during this period that Behring discovered antitoxins, favoring the humoral theory of immunity and dealing a blow to the cellular concept of phagocytosis. But, Metchnikoff persevered in his studies and eventually laid the groundwork establishing the existence of both a humoral and a cellular type of immunity. In 1908 he was awarded the Nobel Prize for his achievements.

Elie Metchnikoff did not, however, retire after receiving the Nobel Prize. He became interested in prolonging life and became convinced that, in most cases, death was the result of an autointoxication from aromatic compounds, such as indole and phenol, which were liberated by organisms in the large intestine, particularly *Escherichia coli* and *Clostridium perfringens*. He found that if he grew these organisms in the presence of lactic acid bacteria, no such aromatic compounds were formed. This led to his proposal that the ingestion of milk soured with Lactobacillus would eventually replace much of the normal flora of the intestine, resulting in a prolongation of life. Unfortunately, it is not possible to supplant the normal intestinal flora by drinking sour milk, but many stores still sell "acidophilus" milk (soured by *Lactobacillus acidophilus*). While it may be healthful, there is no data supporting the contention that this milk will prolong life.

Elie Metchnikoff, a Russian zoologist, that laid the foundation for the study of the important part played by the white cells in defending body tissues from disease organisms.

A second phagocytic cell, called a **monocyte,** is also produced in the bone marrow and released into the bloodstream; after one or two days, it migrates through the vessel walls into the surrounding tissues. The monocytes then begin to differentiate into large, phagocytic cells called **macrophages** or **mononuclear phagocytes,** since they are also capable of surrounding and engulfing microorganisms (see Figure 17.3).

Other circulating leukocytes include the **basophils,** whose granules stain with basic dyes, and **eosinophils,** which contain granules that stain with the acidic dye eosin. These two cell types are present in small numbers and are primarily involved in allergic reactions.

Lymphocytes make up 25 to 35 percent of the white cell population. We shall learn much more about these cells when we study their role in antibody synthesis and foreign cell destruction.

Table 17.2 summarizes the types of leukocytes and their primary function.

Mononuclear Phagocyte System

The mononuclear phagocyte system (sometimes called the reticuloendothelial system) comprises macrophages that are not free to move around because they are fixed in certain organs. They are found lining the blood vessels in the liver, the spleen, the bone marrow, and the lungs and also in other tissues such as lymph nodes. Like PMNs, they function in removing bacteria and other particulate matter from the circulating fluids.

Antimicrobial Systems of the Phagocyte

Because the ingestion and digestion of foreign material by the host's phagocytic cells is such an important process in host resistance to bacterial infections, let us examine briefly just how the leukocyte is able to accomplish this task.

The Inflammatory Response

The inflammatory response is the overall reaction of the body to injury. It is in this response that the individual actions of the phagocytes, the bactericidal substances in the blood, and the body's attempt to localize the infection all come into play. It has been suggested that the various stages of the inflammatory response occur as follows: (1) After a momentary constriction, the blood vessels gradually dilate, resulting in an increased blood flow to the injured area. At the same time there is an increase in vascular permeability,

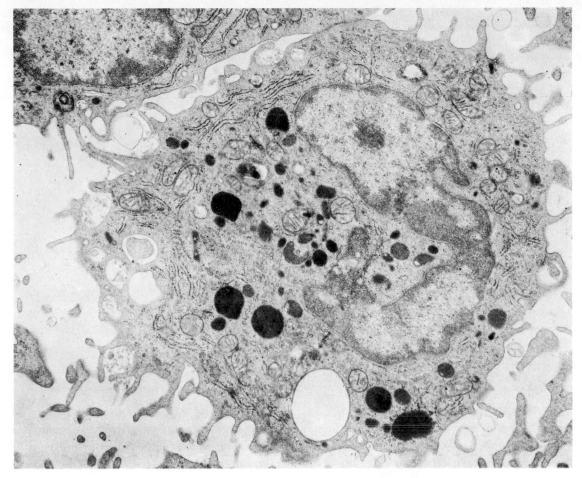

Figure 17.3 Electron micrograph of a peritoneal macrophage from a mouse. These cells were obtained from a peritoneal cavity stimulated with a lipid emulsion. Cytoplasmic processes are prominent, some clearly fingerlike. Lysosomal dense bodies are numerous—some small and homogeneous, others large and heterogeneous (×12,000).

permitting plasma to enter the inflamed area. The resulting swelling is referred to as **edema.** This vasodilation and increased vascular permeability result from the action of various amines released by the injured phagocytes. These vasoactive amines, such as **hista-**mine, **serotonin,** and **bradykinin,** are discussed in greater detail in Chapter 19. (2) As the blood flow slows, leukocytes begin to stick to the sides of the blood vessel closest to the injury. (3) From there they migrate through the walls of the blood vessel to enter the injured area and phagocytose particulate matter such as bacteria. During this time many of the PMNs will die and disintegrate. They are soon replaced by the larger mononuclear macrophages, which engulf not only infecting particles but also the dead PMNs. Humans are also exquisitely sensitive to many bacterial products, particularly endotoxin, and will respond with an elevated temperature (see Chapter 16). It is difficult to evaluate precisely the role of fever, but it is generally agreed that a few degrees of fever may be beneficial by enhancing the reactions involved in tissue repair.

One interesting question is: What stimulates the leukocytes to migrate to an injured area of the body? This process, called **chemotaxis,** is not fully under-

TABLE 17.2 Types of Leukocytes

Name	Function
Polymorphonuclear phagocyte (also called PMN, poly, or neutrophil	Phagocytosis
Monocyte (circulating)	Phagocytosis
Macrophage (in tissues)	Phagocytosis
Eosinophil	Allergic reactions
	Some phagocytosis
Basophil	Allergic reactions
Lymphocytes	Antibody synthesis
	Cellular immunity

stood, but it is well established that a large number of different chemotactic factors are involved. These factors include (1) a number of bacterial products (which are not well characterized); (2) the products from damaged tissue cells (including the supernates from virus-infected cells); (3) certain components produced during the activation of the complement system (see Chapter 18); and (4) an array of compounds that include such things as antibody-antigen reactions, aggregated proteins, and many bacterial and host cell enzymes. From the foregoing, it is apparent that the rapid accumulation of PMNs at an infected site follows the release of a variety of chemotactic factors from that site.

Attachment and Ingestion by the Phagocyte

For the leukocyte to surround and engulf an infecting microorganism, there must be very close contact between the two (see Figure 17.4). It appears that the effectiveness of this contact can be related to the cell surface properties of the invading microorganism. For example, there are a number of pathogenic microbes that produce capsules, which seem to make these organisms too slippery to phagocytose. *Streptococcus pneumoniae, Haemophilus influenzae, Klebsiella pneumoniae,* and *Bacillus anthracis* are all examples of encapsulated organisms that lose their ability to produce disease when they lose the ability to form a capsule. Others, such as *Neisseria gonorrhoeae,* possess fimbriae that bind to the external surface of the phagocyte, preventing the intimate contact necessary for phagocytosis to occur.

We shall see later in this unit how the action of a specific antibody against the capsular material or fimbriae works by providing receptor sites that enable the phagocyte to come into intimate contact with the microorganism and then engulf and destroy it.

Lysosomes

In 1955, Christian de Duve discovered that many of the small granules within the cytoplasm of animal cells contained hydrolytic enzymes. These granular structures have been named lysosomes because they contain enzymes capable of hydrolyzing the major constituents of living cells—that is, proteins, nucleic acids, fats, and complex carbohydrates. Lysosomes are present in the cytoplasm of essentially all animal cells, and they probably have many roles, including maintenance of our tissue cells by facilitating normal turnover of cell constituents. However, here we shall be concerned with the function of lysosomes in the host's phagocytic cells, where they function to kill bacterial invaders.

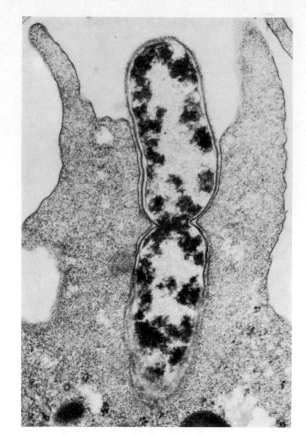

Figure 17.4 PMN engulfing a microorganism. When a bacterium is internalized by a PMN, the plasma membrane invaginates to form the wall of a phagosome. Note that the plasma membrane follows precisely the contour of the bacterium (×35,000).

Immediately after phagocytosis occurs, lysosomes begin to fuse with the membrane of the phagosome and to release their contents into it. Because the lysosomes exist as small granules within the phagocyte, this process is called degranulation and the resulting phagosome is now referred to as a phagolysosome (see Figure 17.5). Over 60 different enzyme activities have been found within lysosomes, and it appears likely that a number of these enzymes are involved in the killing and digestion of phagocytosed microorganisms. However, we shall limit our discussion to several that have been shown to be of primary importance.

Mechanism of Phagocyte Killing

At the time of phagocytosis, there is a burst of metabolic activity within the phagocyte in which the stores of glycogen in the cytoplasm of the PMN are metabolized. This metabolic activity results in the formation of excess reduced nicotinamide adenine dinucleotide

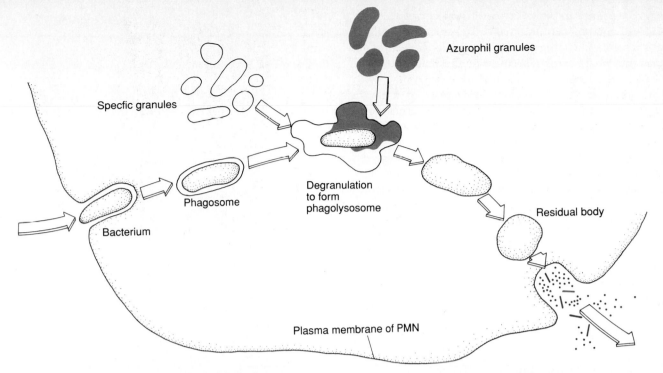

Figure 17.5 After engulfment, lysosomal enzymes from the azurophil granules, along with enzymes in specific granules, enter the phagosome by fusion of the granule membranes with the phagosome membrane. The ingested bacterium within the phagolysosome is killed and either digested or retained within a residual body, from which it can be excreted by fusion with the plasma membrane of the PMN.

phosphate (NADPH), which is oxidized as shown in the following equation:

$$NADPH + H^+ + O_2 \xrightarrow{\text{NADPH oxidase}} NADP + H_2O_2$$

The hydrogen peroxide (H_2O_2) diffuses from the cytoplasm into the phagolysosome, where, in conjunction with the lysosomal enzyme myeloperoxidase (MPO) and chloride ion present in the phagolysosome, it takes part in the following reaction:

$$H_2O_2 + Cl^- \xrightarrow{\text{MPO}} ClO^- + H_2O$$

Hypochlorite (ClO^-) is a strong oxidizing agent and is probably directly involved in the microbicidal (killing) effect.

A number of additional enzymes are emptied into the phagolysosome during degranulation of the PMN and undoubtedly are involved in the antimicrobial action of the leukocyte. A few of the better-known factors are lysozyme (which hydrolyzes the peptidoglycan cell wall of certain sensitive bacteria), lactoferrin (which binds iron present in the phagolysosome, pre-venting growth of the phagocytosed microorganism), and basic proteins such as phagocytin and leukin (which have been shown to exert antimicrobial activity, although their mechanism of action or contribution to the killing effect is not known). In addition, acid produced by the metabolic activity of the PMN lowers the pH of the phagolysosome to about 3 or 4, which alone may cause the death of many microorganisms.

Microbial Resistance to Phagocyte Killing

Obviously, if all invading organisms were to be phagocytosed and killed, all infections would be mild or inapparent. We know, however, that this is untrue, indicating that many microorganisms have developed ways to circumvent our second line of defense. These defensive mechanisms fall into two general categories: (1) the ability to resist phagocytosis and (2) a mechanism that prevents lysosomes from fusing with the phagosome, thus inhibiting degranulation and the subsequent phagocyte killing.

Resistance to phagocytosis can be correlated with the presence of certain surface structures present on a microbe. The presence of a capsule, for example, prevents phagocytosis because of the inability of the phagocyte to make a sufficiently intimate contact with the organism. Similarly, the occurrence of fimbriae results in binding of the microbe to the external membrane of the phagocyte, preventing intimate contact between the phagocyte and the microbe.

A number of pathogenic organisms possess the ability to inhibit degranulation, especially within macrophages and monocytes. Such organisms, therefore, multiply intracellularly, where they are protected from the host's immune system. Some examples of diseases caused by such organisms include tuberculosis, leprosy, brucellosis, tularemia, and histoplasmosis. The precise property that such organisms possess that inhibits lysosomal fusion with the phagosome is unknown.

Defects in Intracellular Killing

The importance of the phagocyte as a major defense against disease became more obvious with the discovery that some individuals produce leukocytes that are defective in their ability to phagocytose and destroy invading microorganisms. Such people experience repeated episodes of infection in spite of the presence of high levels of circulating specific antibodies. We shall describe only two such conditions.

Chronic granulomatous disease (CGD) is a frequently fatal genetic disorder that is characterized by repeated bacterial infections—most commonly with chronically infecting organisms such as *S. aureus* or gram-negative rods. These individuals possess PMNs that phagocytose invading bacteria normally but are unable to kill many of the ingested microorganisms, as shown in Figure 17.6. The major defect in the PMNs of these patients has been shown to be related to the inability of the phagocyte to produce hydrogen peroxide. This seems to result from a specific enzymatic defect involving NADPH oxidase, the enzyme that catalyzes the major hydrogen-peroxide-generating reaction.

Individuals with Chediak-Higashi syndrome (CHS) suffer recurrent infections similar to those described for persons with CGD. However, this defect results from the presence of giant lysosomes that do not fuse readily with a cytoplasmic phagosome (see Figure 17.7). Such cells have other defects also, including a defective chemotactic response to infection.

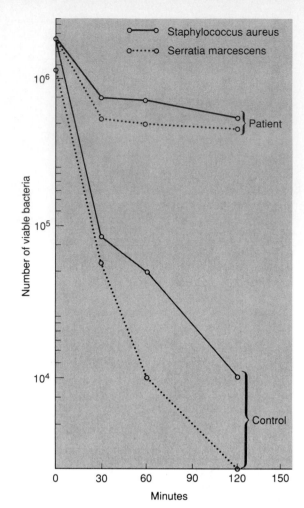

Figure 17.6 *S. aureus* and *S. marcescens* were incubated with PMNs from a patient with chronic granulomatous disease and with PMNs from a normal control. The inability of the patient's PMNs to kill either species of bacterium is readily apparent from the number of viable bacteria.

Of course, the fact that bacteria cause disease in normal people in spite of their leukocytes demonstrates that some bacteria have developed adaptations that enable them to survive lysosomal attack. Some have capsules that resist phagocytosis; others secrete substances (leukocidins) that destroy leukocytes; still others, such as the tubercle bacillus, have thick, waxy coats that resist attack by lysosomal enzymes. The remainder of this unit will discuss the role of specific factors produced by the body that aid in the prevention of disease.

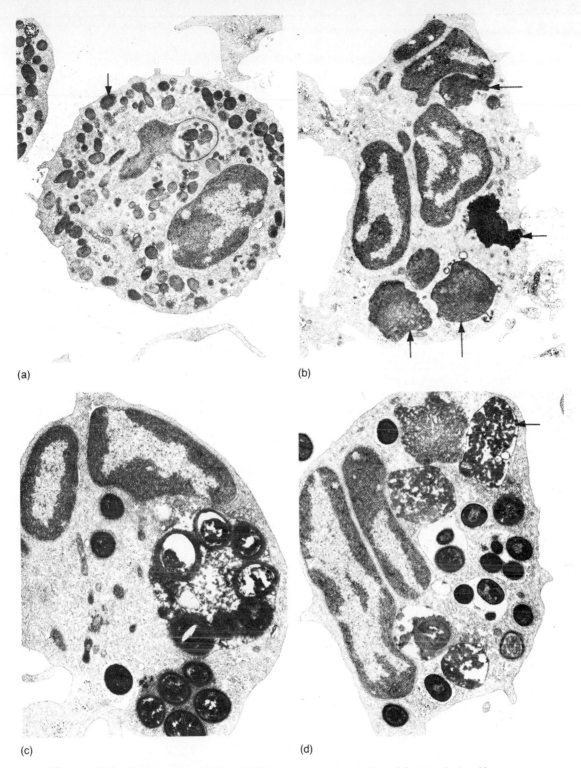

Figure 17.7 Electron micrographs of PMNs from normal controls and from patients with Chediak-Higashi syndrome (CHS). PMNs have been stained histochemically for peroxidase, causing the azurophil lysosomes to appear dark. *(a)* Normal PMN with arrow pointing to one of the many lysosomes. *(b)* PMN from patient with CHS. Arrows point to the very few, large lysosomes. *(c)* Normal PMN 60 min after mixing with staphylococci, showing phagosomal fusion with peroxidase activity in the phagolysosomes and a lack of peroxidase-positive granules in the cytoplasm (compare with a). *(d)* CHS-PMN 60 min after mixing with staphylococci. Note the persistence of structurally intact peroxidase-positive giant lysosomes (arrow).

SUMMARY

Our first line of defense against infection is a nonspecific one; it is directed against all invading organisms. This type of defense seems to vary considerably from one animal species to another and even between individual humans. Much of this may well reflect genetic differences. The skin and mucous membranes provide a mechanical barrier to infecting microorganisms. However, much of the effectiveness of the skin in preventing infection lies in the secretion of unsaturated fatty acids by the sebaceous glands. These acids are bactericidal for many microorganisms. Tears contain lysozyme, an enzyme capable of lysing, or killing, many bacteria. Interferon is a substance produced by a virus-infected cell that can prevent the subsequent infection of other cells by many different viruses. However, even though it is nonspecific for the viruses it inhibits, it is usually species-specific. Thus, only interferon produced in a human is effective in humans. Interferon production can be induced by double-stranded RNA.

Leukocytes provide a second line of defense by engulfing invading microorganisms. There are several different kinds of leukocytes, but all act to destroy cells by releasing enzymes from small granules called lysosomes into the phagosome. Phagocytosed bacteria are killed by several mechanisms, of which the action of myeloperoxidase on hydrogen peroxide and chloride ion to form hypochlorite has been the most studied. Other enzymes released from the lysosome are capable of breaking down proteins, fats, nucleic acids, and carbohydrates. The inflammatory response is an overall reaction of the body aimed at localizing and destroying invading microorganisms. As one part of this response, chemotactic substances are released from the infected site, inducing the accumulation of leukocytes in the injured area.

QUESTIONS FOR REVIEW

1. How do mechanical and chemical mechanisms function in our defense?
2. What evidence is there that genetic factors, which differ from one individual to another, are important in the prevention of disease?
3. In addition to providing a mechanical barrier, how does the skin function to prevent infection?
4. What is lysozyme?
5. What is meant by the statement that interferon is cell-specific and not virus-specific?
6. How does interferon act to inhibit viral replication?
7. How is interferon produced commercially?
8. What is phagocytosis? What part does it play in disease control?
9. What is the mononuclear phagocyte system?
10. What is leukocytosis? Leukopenia?
11. What are lysosomes, and of what importance are they in nonspecific immunity?
12. What is the inflammatory response? Explain its importance in localizing infectious agents.
13. What is myeloperoxidase? What is its role in killing by the phagocyte?

SUPPLEMENTARY READING

Boxer GL, et al: Polymorphonuclear leukocyte function. *Hosp Pract* 20(3):69, 1985.

Di Sabato (ed): *Chemotaxis and Inflammation. Methods in Enzymology,* vol. 163. San Diego, Academic Press, 1988.

Edelson RL, Fink JM: The immunologic function of skin. *Sci Am* 252(6):46, 1985.

Elsbach P: Degradation of microorganisms by phagocytic cells. *Rev Infect Dis* 2:106, 1981.

Kleinman R, et al: Natural killer cells: Killing of microorganisms. In Leive L and Schlessinger D (eds): *Microbiology—1984.* Washington, DC, American Society for Microbiology, 1984, p. 319.

Kornfeld S, Sly WS: Lysosomal storage defects. *Hosp Pract* 20(8):71, 1985.

Larsen G, Henson P: Mediators of inflammation. *Annu Rev Immunol* 1:335, 1983.

Michna H: *The Human Macrophage System: Activity and Functional Morphology.* New York, Karger, 1988.

Oldham RK: Biologicals for cancer treatment: Interferons. *Hosp Pract* 20(12):71, 1985.

Snyderman R, Goetzl EJ: Molecular and cellular mechanisms of leukocyte chemotaxis. *Science* 213:830, 1981.

Weiss L: *Cells and Tissues of the Immune System.* Englewood Cliffs, NJ, Prentice-Hall, 1972.

Chapter 18

Antigens and Antibodies

OBJECTIVES

After studying this chapter, you should be familiar with the

1. Nature of specific or acquired immunity.

2. Properties a substance must possess to be an antigen.

3. Definition of determinant group and hapten.

4. General structure of each class of antibody.

5. Principal functions of each class of antibody.

6. Origin and role of B cells and T cells in antibody synthesis.

7. Role of the macrophage in antibody synthesis.

8. Kinetics of antibody synthesis.

9. Meaning of adjuvants and monoclonal antibodies.

10. Several models of autoimmune diseases.

In the previous chapter, we considered the nonspecific mechanisms by which the body protects itself from infections. However, because occasionally almost everyone becomes ill, common sense indicates that these mechanisms alone are insufficient protection from all microbial invaders. It is general knowledge that, following recovery from one of many common diseases, there probably will not be a recurrence and we consider ourselves immune to that particular disease. In other words, we do not anticipate a second attack of measles, mumps, or chickenpox. The state of being highly resistant to a specific pathogenic organism is called **immunity,** and the type we are here concerned with is referred to as **specific** or **acquired,** since it is specific against particular organisms and must be acquired during life.

What is the nature of this specific acquired immunity? What have our bodies done that will prevent a subsequent attack by the same organisms? Acquired immunity is understood best by visualizing in higher animals a system whereby they are able to distinguish between materials that make up their own bodies and those that are foreign. They do not react against their own substances but will set in motion a series of complex reactions that aid in the elimination or neutralization of foreign substances, such as cells from other animals, bacteria, viruses, and toxins. It is this ability to recognize **self** versus **nonself** that forms the basis for specific or acquired immunity.

To understand the formation of these protective reactions, it is necessary to know something of the nature of the foreign substances that induce their synthesis.

ANTIGENS

Foreign materials that gain entrance to the body and induce a specific immune response are called immunogens or **antigens** (Ag). This resulting specific immune response may take the form of proteins called **antibodies** (Ab) that circulate in the bloodstream (**antibody-mediated immunity**), certain cells that have acquired an increased ability to destroy other cells (**cell-mediated immunity**), or both. In any case, this acquired immunity enables the body to destroy or neutralize invading organisms or toxins.

Chemically, antigens are usually composed of protein, polysaccharide, or combinations such as glycoproteins (carbohydrate linked to protein), nucleoproteins, or glycolipids (carbohydrate linked to lipids). Many of the antigens of concern to us occur as components of various membranes or as structures surrounding bacteria, viruses, or even cancer cells. Examples of such antigens include bacterial capsules, the cell wall lipopolysaccharides of gram-negative bacteria, specific glycoproteins occurring in cell membranes, or the attachment sites on viruses that interact with specific receptors on mammalian cells. Other antigens, however, may be soluble substances such as bacterial toxins or various snake and insect venoms. And individuals who suffer from any one of a number of allergies know that pollens, foods, and house dust also act as antigens. Some antigens are more effective than others in inducing a specific immune response, or, stated another way, there are strong antigens and there are weak antigens. Some proteins, such as diphtheria or tetanus toxoids, are strong antigens when injected and induce a good antibody response, whereas foreign antigens occurring on the surface of a cancer cell may be closely related to self-antigens and, as a result, will elicit a poor immune response or none at all.

Properties of Antigens

An antigen may be a soluble substance such as horse serum proteins or a bacterial toxin, or it may exist on particulate matter such as a red blood cell, a bacterial cell, or a virus. In any case, it must have a molecular configuration with certain consistent characteristics to be able to induce antibody synthesis. Let us look at some of the properties a molecule must possess to be an effective antigen or, for that matter, to be any kind of an antigen.

First and foremost, an antigen must be foreign to the host. We can easily see that unless this is true, an animal would produce antibodies against antigens on its own cells and subsequently destroy these cells. Sir Macfarlane Burnet, an Australian immunologist, proposed the theory that during the prenatal period the developing lymphoid system learns to distinguish "self" from "nonself" by a mechanism in which cells capable of making antibody to "self" are destroyed. Thus, one develops a permanent self-tolerance to antigens that are normal components of one's own cells. We shall learn later that occasional changes in the structure of normal antigens can induce the immune system to respond toward antigens that are self, resulting in autoimmune diseases.

The second property a substance must have to be capable of inducing an antibody response is that it must be a reasonably large molecule. It is rare for any compound with a molecular weight of less than 6000 daltons to act alone as an antigen. This seems to be unusual because the antibodies formed in response to the presence of an antigen react with only a small portion of the antigen. The portion of the antigen that specifically combines with antibody is spoken of as its **determinant group** or **epitope.** If we could degrade the antigen—that is, chemically break it down and isolate a determinant group (which would have a molecular weight varying from 200 to 1000 daltons)—we would find that this isolated portion of the antigen still reacts with antibodies formed in response to the original antigen but by itself is unable to induce the production of antibodies when injected into an animal.

Considerable insight into the nature of an antigen and the specificity of an antibody to its antigenic determinant evolved from experiments carried out by Karl Landsteiner and his collaborators over 50 years ago. They prepared synthetic antigens by linking a number of small organic compounds to carrier proteins, as shown in Figure 18.1, for the coupling of metaaminobenzene sulfonic acid to chicken globulin. When these synthetic antigens were injected into rabbits, antibodies were made that were specific for the attached organic compounds. The injection of these same organic compounds by themselves, however, failed to induce any antibody response. It was apparent, therefore, that antibodies could bind to these determinants on the antigen, but that alone they were unable to induce antibody formation. Such molecules of known chemical composition are called **haptens,** which, by definition, are too small to induce antibody synthesis but, when conjugated to a large carrier molecule, become the determinant groups toward which the antibody is directed. Moreover, if we examine most antigens, we would see that they possess multiple epitopes, each of which may induce an antibody response. For example, tetanus toxoid (used to immunize individuals to tetanus) will induce a family of over 20 different antibodies, each capable of binding to a different epitope on the toxoid molecule. We shall see how this occurs later in this chapter.

We have now covered the two important properties a compound must possess to be an antigen: (1) it must be foreign to the host and (2) it must be a reasonably large molecule.

ANTIBODIES

We have referred to antibodies as molecules synthesized in response to the presence of an antigen that, once formed, combine with that antigen. Let us now look at some of the properties of antibodies.

First, we can make the generalization that all antibodies are proteins containing a small amount of carbohydrate attached to the molecule. Some types have a greater tendency to become attached to tissues than others, but all the ones discussed in this section are found in serum.

Figure 18.1 Schematic diagram illustrating the coupling of an aromatic hapten to a carrier protein. The final protein will induce antibody synthesis to both the carrier protein and the aromatic hapten.

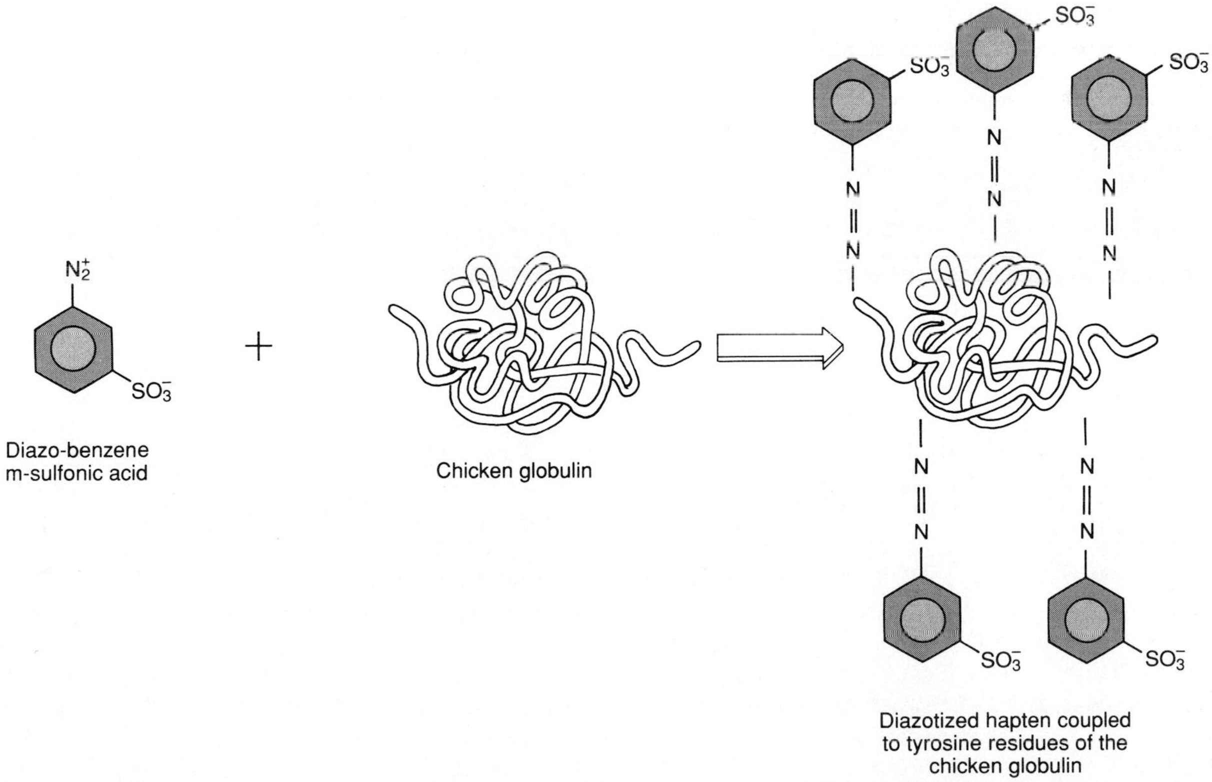

Diazo-benzene
m-sulfonic acid

Chicken globulin

Diazotized hapten coupled
to tyrosine residues of the
chicken globulin

In the early days of immunology, before the complexity of serum was known, serum proteins were classified according to their solubility as either albumins or globulins. It was observed that the circulating antibodies were found only in the globulins and that, if the globulins underwent electrophoresis (were put in an electrical field), they separated into three major components (see Figure 18.2). These components were designated alpha (α) globulin, beta (β) globulin, and gamma (γ) globulin. Almost all circulating antibodies were found in the γ globulin fraction, and, as a result, the term *gamma globulin* has been used as a synonym for circulating antibodies. However, since there are exceptions, the more proper name **immunoglobulins** (abbreviated Ig) is now used as a general name for antibodies.

A vast amount of research during the past couple of decades has been directed toward understanding the structure and workings of antibodies. It is beyond the scope of this text to do more than list and summarize some of the major current concepts concerning antibodies.

Structure

Figure 18.3(*a*) is a schematic drawing of the basic structure of an antibody molecule. One can see that it is made up of four polypeptide chains joined by disulfide (—S—S—) bonds. This basic structure has a molecular weight of approximately 150,000. If the disulfide bonds joining these polypeptides are broken, two identical polypeptides with molecular weights of approximately 25,000 and two larger polypeptides having molecular weights of about 50,000 are obtained from each antibody molecule. The smaller polypeptide chain is called the *light* chain (L chain), and the larger one is called the *heavy* chain (H chain).

Note in Figure 18.3(*b*) that the antibody molecule can be split by the enzyme papain into two major regions called the **Fab** (because it is the fragment containing the antigen combining site) and the **Fc** (because this fragment has been crystallized). We shall see that the Fc portion contributes a number of properties to the antibody molecule.

Classification

Use of various procedures such as zone electrophoresis and analytical centrifugation, has proved that immunoglobulins are heterogeneous; that is, they are not identical. A source of only one kind of immunoglobulin became available when it was recognized that individuals with a malignancy of the lymphoid system called **multiple myeloma** produced large amounts of a single kind of immunoglobulin. Using purified immunoglobulins from many different patients with multiple myeloma, it was possible to study the chemical and antigenic characteristics of a series of different, pure immunoglobulins.

Figure 18.2 (*a*) Electrophoresis scale diagram of rabbit serum containing antibodies to egg albumin. Peaks for alpha, beta, and gamma globulin are shown. (*b*) Same diagram after absorption of serum with egg albumin. Note the dramatic decrease in size of the gamma globulin peak, demonstrating that the anti-egg albumin antibodies were gamma globulins.

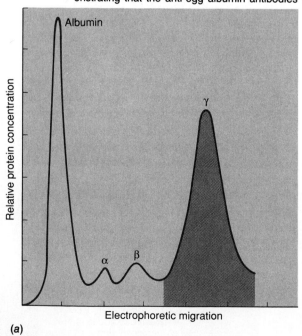

(a)

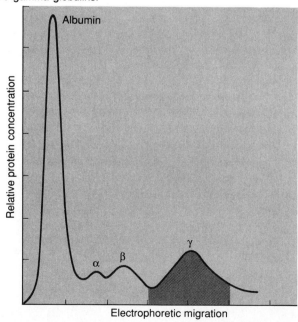

(b)

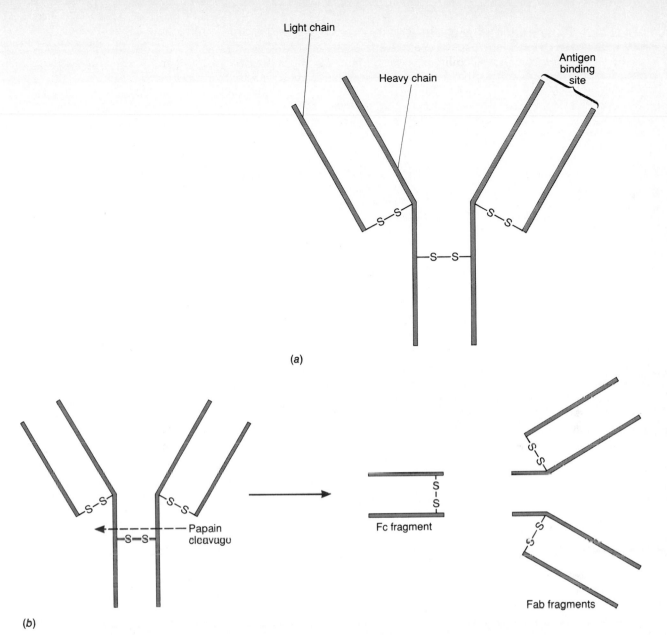

Figure 18.3 (a) Schematic stick model of IgG showing two heavy and two light chains held together by interchain disulfide bonds. (b) Schematic illustration of the digestion of IgG with papain (a proteolytic enzyme) to yield one Fc fragment and two Fab fragments.

Initial studies revealed that immunoglobulins could be divided into different classes based on their antigenic properties. In other words, when immunoglobulin (i.e., antibody) is injected into a species of animal different from the one from which it was obtained, the receiving animal will produce antibodies to the foreign immunoglobulin. By using these antibodies to compare the immunoglobulins from a large series of myeloma patients, it was found that immunoglobulins could be divided into five major classes. Also, it was discovered that the antigenic differences among these five classes exist in their heavy chains. The five classes of immunoglobulins have been named IgG (for immunoglobulin G), IgA, IgM, IgE, and IgD. It is now known that all normal individuals possess varying amounts of immunoglobulin of each of the five classes. Table 18.1 lists the major characteristics for each class of immunoglobulin.

Similar studies of light chains revealed that there are two major, antigenically different types, which have been named kappa (κ) and lambda (λ). Both κ and λ chains, however, are found in all five classes of immunoglobulin. Thus, you can see that each major class of immunoglobulin can exist in two forms, containing either two λ light chains or two κ light chains. Also, there are subclasses of some heavy chains, but

TABLE 18.1 Antibody Characteristics

	IgG	IgA	IgM	IgD	IgE
Molecular weight	150,000	170,000 and 390,000	900,000	170,000	200,000
Sedimentation coefficient	7S	7S and 11S	19S	7S	8S
Concentration in serum	12 mg/mL	3.5 mg/mL	1.5 mg/mL	0.03 mg/mL	0.0003 mg/mL
Antibody combining sites	2	≥2	5 or more	2	2
Ability to fix complement	Yes	No	Yes	No	No
Ability to pass placenta in human	Yes	No	No	No	No

these subclasses represent only minor variations of the major antibody classes.

Considerable information concerning the function of various parts of the antibody molecule has been obtained by partially breaking the molecule apart with proteolytic enzymes and then observing what changes occur in the antibody properties. Pepsin chews up the Fc piece of the molecule but leaves the remainder of the molecule intact with its two antigen binding sites (see Figure 18.4). This pepsin-treated molecule is still capable of causing agglutination or precipitin reactions but is now unable to fix complement (see Chapter 19), or, in the case of IgG, to pass through the placental wall from mother to fetus. Thus, we can see that the intact Fc piece of the molecule is extremely important for the total normal physiological function of antibody.

Functions

A single exposure to an antigen can stimulate the immune system to synthesize different classes of immuno-globulins, each of which possess specific characteristics.

IgG is quantitatively the most important class of immunoglobulins in human serum because it accounts for more than 80 percent of the total immunoglobulins. It is the only immunoglobulin that can freely pass the placental wall and is therefore extremely important in providing antibody immunity to the newborn.

IgA occurs both in serum and in mucous secretions. There are two forms of this antibody. One appears to be very similar to IgG; it is found only in the serum and is essentially the same size as IgG. The other is more than twice as large as serum IgG (11S, molecular weight, 390,000) and is composed of two molecules of serum type IgA together with a β-globulin component called the secretory component. This larger IgA is found primarily in body secretions, such as saliva, tears, and seminal fluid and in the gastrointestinal and genitourinary tracts. It is present also in human milk and undoubtedly is of value to the newborn infant in the regulation of its gastrointestinal tract flora. The secretory IgA is thought to be partic-

Figure 18.4 Schematic illustration of the digestion of IgG with pepsin to yield one F(ab')$_2$ fragment plus small peptides.

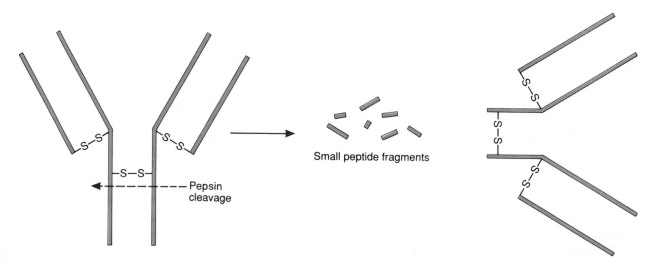

Small peptide fragments

Pepsin cleavage

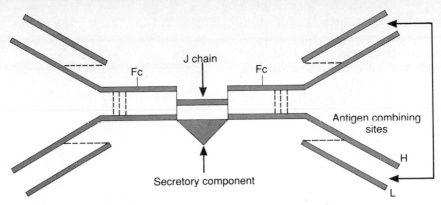

Figure 18.5 Model of secretory IgA showing J chain and secretory piece.

ularly valuable as our first line of antibody defense against pathogenic bacteria and viruses. Thus, pathogenic microorganisms entering via the respiratory tract come in contact with the secretory IgA before it is possible for antibodies in the serum to be effective. Figure 18.5 shows a schematic model of secretory IgA in which the secretory component is attached to the Fc portion of each of two molecules of serum IgA. Secretory IgA also contains a polypeptide that has been named the J chain. The J chain is responsible for linking the monomeric units of IgA to form dimers, whereas the secretory piece appears to function mainly in the transport of the IgA molecules to the external secretions. The secretory piece is believed also to protect the secreted IgA from proteolytic digestion by extracellular proteases present in gastrointestinal and other secretions.

IgM is the largest of the immunoglobulins and is composed of five basic units attached to one another by way of disulfide bridges between their Fc pieces, as shown in Figure 18.6. If one gently breaks these disulfide bonds, one obtains five subunits and a J chain similar to that found in IgA. Thus, the intact IgM molecule is composed of 10 light chains (either κ or λ) and 10 heavy chains, which are antigenically distinct for IgM. One would predict, therefore, that IgM has 10 antigenic binding sites, but this is not always true. Because IgM is such a compact molecule, it appears that many antigens are just too large to fit physically into each of the 10 potential antigen binding sites, and one frequently finds that IgM is capable of binding only five molecules of antigen. Apparently, depending on the size of the antigen, one half of the potential sites may be sterically restricted, leaving only five functional antigen binding sites. If one follows the appearance of an antibody after injection of an antigen, IgM is the first antibody to be detected. However, IgM synthesis is not usually prolonged, and IgG antibodies become the most prevalent. Since IgM antibodies have a potent bactericidal effect on gram-negative bacteria in the presence of complement, they

are undoubtedly extremely important for protection very early after infection (see Chapter 19).

IgD is present in serum at a level of about only 0.2 percent of the concentration of IgG, and because of this low concentration and the unusual lability of the molecule during purification, it has been difficult to assign specific functions of this class of immunoglobulin. It is known, however, that IgD is present on the surface of B lymphocytes (cells that differentiate into antibody-producing cells), and it has been proposed that IgD acts as a cell surface receptor for antigens, regulating the differentiation of the B lymphocyte into the plasma cells that synthesize the immunoglobulins.

IgE antibodies are normally present in even smaller amounts than are IgD antibodies, but we do know a little more about this class of immunoglobulins. First of all, it is IgE immunoglobulins that are

Figure 18.6 Schematic stick model of IgM. The dashed lines represent interchain disulfide bonds. It is not certain whether the J chain is linked to one or two heavy chains.

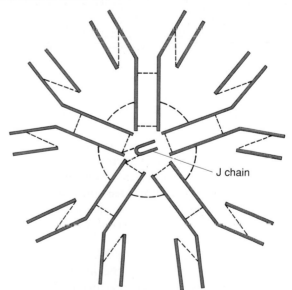

responsible for the many different allergies from which certain people suffer: food, ragweed and other pollens, dust, and almost any material imaginable. Although structurally they are similar but a bit larger than IgG, they exhibit some rather unusual properties. They exist in the allergic individual in considerably higher concentration than in the nonallergic person. In addition, they exist bound tightly to the surface of circulating mast cells and leukocytes. Mast cells exist in connective tissue adjacent to capillaries throughout the body. There are especially high concentrations of mast cells in the lungs and skin as well as in the gastrointestinal and genitourinary tracts. When a specific antigen (such as ragweed pollen) induces the synthesis of IgE antibodies, the IgE binds tightly to the surface of mast cells. Then, if an individual is subsequently exposed to the same antigen, the reaction takes place on the surface of the cell to which the IgE is fixed. This results in the release of pharmacologically active substances (such as histamine) from the mast cell, which in turn dilate capillaries, alter permeability, and cause serious reactions such as bronchial constriction, resulting in an acute asthmatic attack. One of the major differences between IgE and the other classes of immunoglobulins resides in the Fc portion of the IgE molecule, and it is this part of the molecule that fixes so tightly to cell membranes. You can see why people suffering from allergies obtain relief from their symptoms by taking antihistamines, substances that react with histamine before it can exert any pharmacological effect.

Table 18.2 summarizes the functions of each antibody class.

Antibody Diversity

Using homogeneous IgG preparations from myeloma patients, it has been possible to determine the primary structure of a number of light and heavy chains. Determination of the amino acid sequence of many different immunoglobulins disclosed that light chains were comprised of two approximately equal halves, of which the amino terminal half (about 110 amino acid residues) was highly variable and the carboxy terminal half was remarkably constant. These parts are termed *variable-light* (V_L) and *constant-light* (C_L).

Similar studies with heavy chains revealed four parts (termed *domains*), of which the amino terminal domain (about 110 amino acid residues) was highly variable while the remaining three domains were quite constant for any one class of immunoglobulin. These heavy chain domains are referred to as V_H, C_H1, C_H2, and C_H3.

It is now known that the antibody combining site on the molecule is formed by the interaction of the V_L and V_H domains and that the incredible diversity

TABLE 18.2   Functions of Each Antibody Class

Class	Primary Function
IgG	Principal circulating Ab; passes placental membrane to newborn; fixes complement and lyses gram-negative bacteria; binds to gram-positive bacteria and provides receptors for phagocytes
IgM	First Ab synthesized after Ag stimulation; more efficient than IgG in its ability to fix complement and to lyse gram-negative bacteria
IgA	Occurs in serum as a monomer and is found externally as a dimer bound to a secretory piece and a J chain; first line of Ab defense against organisms entering the body through the mucous membranes; major Ab in breast milk and provides newborn with preformed antibodies.
IgD	Occurs almost exclusively on the surface of B cells; is believed to be involved in the regulation of B cell differentiation
IgE	Binds very tightly to mast cells and leukocytes; when cross-linked by their specific Ag, mast cells release pharmacologically active substances (such as histamine), which result in various allergic reactions to the antigen in question

of antibody specificities is a result of the amino acid sequences making up the variable domains of both the light and heavy chains. These studies also revealed that the region between C_H1 and C_H2 possessed a high content of the amino acid proline, which provided the flexibility necessary for the spreading of the Fab arms. This part of the antibody molecule has been designated as the **hinge** region. Figure 18.7 shows a schematic illustration of an IgG molecule.

The Antigen Binding Site

The antigen binding site of an antibody is a convoluted surface that is complementary to the antigenic determinant to which it binds. Thus, wherever there is a depression or protrusion in the antibody binding site, there is a protrusion or depression, respectively, on the antigenic determinant. As a result, a very close fit between antigen and antibody is achieved.

ANTIBODY SYNTHESIS

In most cases, antibody synthesis requires the interaction of at least three types of cells: (1) B cells, (2) T cells, and (3) an antigen presenting cell. The interaction of each of these cell types induces the secretion

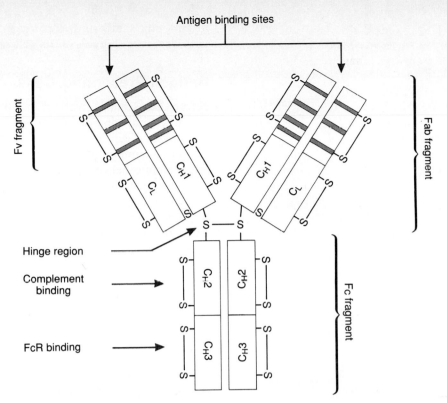

Figure 18.7 Diagrammatic representation of the IgG antibody molecule showing its basic structural features. This model shows the two-chain structure of the IgG molecule. The position of the Fab and Fc regions of the molecule are shown. The Fv fragment is another fragment that can be generated by digestion with the proteolytic enzyme trypsin. This fragment contains the variable region domains of the heavy and light chains. Each Fv fragment has a simple antigen combining site. Also shown are (1) the four domains of heavy chain (V_H, C_H1, C_H2, and C_H3) and the two domains of light chain (V_L and C_L) each with its intrachain disulfide bond; (2) the hinge region; and (3) the heavy-chain domains involved in fixation of complement and binding to Fc receptors on phagocytic cells (monocyte binding).

of one or more soluble factors, which stimulates the division and/or the differentiation of the lymphocytes involved in antibody synthesis. In addition, we shall see that T cells can recognize an antigenic determinant only when it is presented on the surface of an antigen presenting cell in association with a self antigen.

Let us take a look at each of these cell types and then put them all together to show the interactions occurring during antibody synthesis. But first, we will see what types of antigens define "self."

Transplantation

Transplantation of tissue from one individual to another on a random basis invariably leads to rejection of the transplant by the immune system of the recipient. This rejection is specific for cell surface antigens on the transplanted tissue. Such antigens are called **alloantigens** and are encoded by any one of a number of genetic loci within the genome. In mammals, however, one set of closely linked genes, termed the **major histocompatibility complex (MHC)**, is of primary im-

portance in determining the outcome of transplantation as well as having a central role in controlling immune responses to essentially all antigens.

Two different classes of self antigens (Classes I and II) are encoded in the MHC. In humans three loci exist for Class I antigens (designated as HLA-A, HLA-B, and HLA-C) and four loci for Class II antigens (termed HLA-D, HLA-DR, HLA-DQ, and HLA-DP). Keep in mind that there are many alleles for most of these loci, resulting in an incredible diversity of transplantation antigens. Only identical twins share exactly the same MHC antigens and, as a result, only in the case of twins can tissue be freely transplanted from one person to another.

Role of the B Cell

A chicken has an organ in the lower gut called the **bursa of Fabricius** (see Figure 18.8). If it is removed from a newly hatched chick, the animal will be unable to form circulating antibodies. Thus, this organ appears to be important in the generation and dissem-

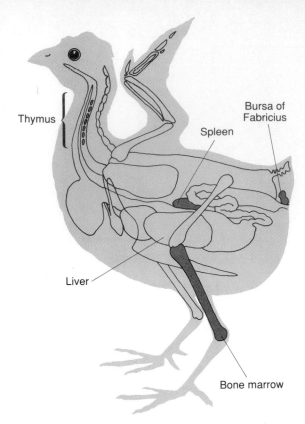

Figure 18.8 The location of the bursa of Fabricius and thymus in the chicken. The multilobed thymus is the origin of T lymphocytes, and the bursa (found only in birds) is the organ from which B lymphocytes originate. After leaving the thymus or the bursa, the lymphocytes may colonize the spleen or bone marrow.

ination of antibody-producing cells. The cells apparently leave the bursa of Fabricius very early in life and become implanted primarily in the spleen, lungs, and bone marrow, where they exist as small lymphocytes. These cells, commonly called **B cells** (for bursa or bursa-equivalent), synthesize immunoglobulins that remain bound to the cell membrane through the Fc part of the molecule, leaving the antibody binding sites exposed. All of the antibodies on the surface of any one B cell have identical antigen binding sites; however, there are literally millions of B cells, each possessing membrane-bound antibodies with different antigenic specificities. Then, when an antigen enters the body, it binds only to B cells that carry on their surface membrane antibodies specific for a determinant on that antigen. After binding, the antigen is taken into the B cell, where it is degraded into small peptides, which are then redeposited on the surface of the B cell in association with their MHC class II antigen.

Mammals do not possess a bursa of Fabricius, and their B cells are thought to arise in the fetal liver and mature in the bone marrow.

We shall now discuss the other cells taking part in antibody synthesis before returning to our B cell with degraded antigen on its cell surface.

Antigen-Presenting Cells

Before discussing the role of T cells in antibody synthesis, we must first describe how T cells interact with an antigen. Keep in mind that the B cell binds intact antigen to immunoglobulin receptors on its surface and then internalizes and degrades that antigen before redepositing individual antigenic determinants back on its surface.

The T cell cannot interact with an intact antigen and thus requires that an antigen-presenting cell first process the antigen and then present it in a manner that the T cell can recognize. Both macrophages and dendritic cells serve this function by internalizing an antigen by endocytosis, degrading it into smaller determinants, and then repositioning the fragments on their cell surface in association with their MHC class II antigens.

We shall return in the following section to discuss why this antigenic fragment must be deposited in association with self antigens of the antigen-presenting cell.

Role of the T Cell

T cells are so named because their maturation from a precursor cell occurs within the thymus (an organ existing as two lateral lobes extending into the neck on the front and side of the trachea). Based on their ultimate function and the presence of certain surface proteins, T cells can be divided into four distinct subsets: (1) helper T cells (Th), which "help" in the antigen-specific activation of B cells; (2) delayed-type hypersensitivity T cells (Tdth), which are involved in cell-mediated immune responses; (3) cytotoxic T cells (Tc), which are also involved in cell-mediated immune responses and lyse target cells by a direct cell-cell contact; and (4) suppressor T cells (Ts), which downregulate immune responses. In this section we shall be concerned only with Th cells and will return to the other subsets of T cells in later sections.

T cells do not produce antibodies, nor do they have antigen-specific immunoglobulins on their cell surface. In fact, a small percentage of antigens can induce antibody synthesis in the absence of any T-cell help. Such antigens are termed *T-independent antigens*, and they are characterized by being poorly metabolized and by usually possessing repeating antigenic determinants. Moreover, such antigens can induce only an IgM response. A number of capsular polysaccharides fall into this category.

The nature of the antigen-specific T-cell receptor on Th cells was far more difficult to characterize than that of the B-cell receptor because Th cells do not bind to most free antigens as do B cells but instead recognize a foreign antigen in association with cell surface molecules encoded by genes in the MHC. They do, however, have antigen-specific receptors, which consist of a 90,000-dalton glycoprotein that will specifically bind to an antigenic determinant when it is presented in association with a Class II MHC surface antigen.

So, now let us see what happens when the body is confronted with a T-dependent antigen (such as diphtheria toxin).

T-Dependent Antibody Synthesis

Antibody synthesis to a T-dependent antigen can be summarized as follows:

1. The antigen first binds to an antigen-presenting cell (a macrophage or dendritic cell), where it is internalized, degraded, and fragments are redeposited on its surface in association with a Class II MHC molecule.

2. The antigen also binds to immunoglobulins on the surface of the B cell, where it is internalized, degraded, and fragments are redeposited back on the surface of the B cell in association with a Class II MHC antigen in the B cell surface.

3. A Th cell binds through its specific receptor to the processed antigen, which is associated with the Class II antigen on the antigen-presenting cell (APC) (Figure 18.9). This stimulates the APC to secrete a soluble factor termed **interleukin-1.**

4. Interleukin-1 induces the Th cell to divide and to synthesize and secrete other soluble growth factors (known as lymphokines), which result in the proliferation of the Th cells specific for the epitope presented on the APC.

5. The activated Th cell then binds to the *same antigenic determinant* associated with a Class II molecule on the surface of the B cell. This results in the secretion of several factors by the activated Th cell at the surface of the antigen-presenting B cell. These lymphokines drive the B cell to divide and differentiate into antibody-producing plasma cells and memory cells. The plasma cell then secretes about 2000 molecules per second of antibody, which possesses a specificity identical to that of the surface immunoglobulin of the progenitor B cell (see Figure 18.10).

Kinetics of Antibody Synthesis

The appearance of antibodies in the blood serum following initial exposure to an antigen is known as a primary antibody response. Both the rate and the extent of the primary response depend on the nature of the antigen, size of the antigenic dose, and route of administration. Figure 18.11 depicts a typical primary

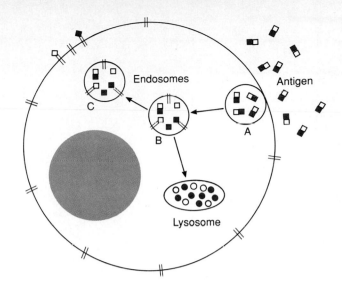

| | Fully-degraded | | Antigen-derived |
| Class II molecule ● & ○ | antigen | □ & ■ | peptides |

Figure 18.9 Antigen processing by the antigen-presenting cell (APC). The activation of helper T cells (T$_H$) and delayed type hypersensitivity T cells (Tdth) depends on the uptake and processing of soluble foreign antigen molecules by a nonspecific APC. The antigen is endocytosed by the cell and degraded in acidic intracellular compartments (thought to be the light endosomes). Fragments of the antigen are taken up by Class II molecules and presented as peptide-Class II complexes on the surface of the APC. It is thought that the association between the peptide antigen and the Class II molecule first occurs in the intracellular compartment prior to appearance of the complex on the cell surface. Class II–restricted T cells specifically recognize and interact with this complex. A number of cell types can perform this antigen-presentation function.

response curve. In a usual case, measurable antibodies appear after about 5 days and reach a peak in 2 to 3 weeks. The duration of antibody in the serum depends on a continued stimulation of antibody production by antigen and on the rate of antibody turnover, which varies for each class of immunoglobulin.

As shown in Figure 18.11, a second exposure to an antigen, months or even years after the primary response, results in an almost immediate appearance of antibodies (within 1 to 3 days), reaching levels that may be 10 to 15 times higher than that occurring during the primary response. Since antibody-secreting cells are short-lived, this secondary type of response cannot be attributed to a reactivation of dormant plasma cells but, rather, to an activation of memory B and T cells that were formed during the primary response. Such memory cells are very long lived and are already programmed to differentiate rapidly into functional Th cells and antibody-producing plasma cells. As a consequence, when they come into contact with their specific antigenic determinant, antibody production begins almost immediately.

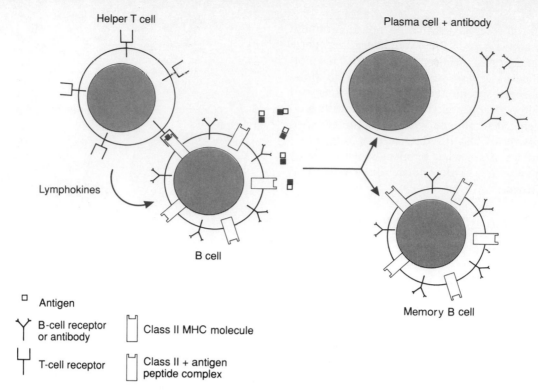

Helper T cell

Lymphokines

B cell

Plasma cell + antibody

Memory B cell

☐ Antigen

Y B-cell receptor or antibody

Y T-cell receptor

☐ Class II MHC molecule

☐ Class II + antigen peptide complex

Figure 18.10 Induction of an antibody response to a T-dependent antigen. Antigen is recognized and bound by the B cell immunoglobulin antigen-specific receptor. Receptor-mediated endocytosis of the antigen (and the immunoglobulin receptor itself) leads to processing of the antigen as described for APC in Figure 18.9. A fragment of the antigen is bound by a Class II molecule and is presented on the B-cell surface as a peptide-Class II complex. A helper T cell, which has been preactivated by interaction with an identical peptide-Class II complex on an APC, interacts with this complex on the cell. This Th cell is then activated and delivers both proliferative and differentiative signals (many in the form of secreted lymphokines) to the B cell, causing the formation of memory B cells and plasma cells, which synthesize and secrete antibody. On subsequent contact with antigen and Th, the memory B cell will give rise to more memory B cells and to more plasma cells secreting antibody.

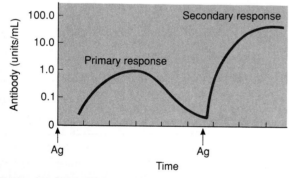

Figure 18.11 Antibody response curves. Because of the large variability for different antigens (Ags), time is shown as arbitrary units. However, a maximum primary response might require 10 to 20 days, whereas a secondary response occurs in 1 to 3 days.

Adjuvants

The immunizing effect of many soluble antigens can be increased by mixing the antigens with insoluble materials that act to keep the antigen in the tissues for much longer periods than if the soluble antigen were injected alone. The insoluble materials are called **adjuvants.** For instance, diphtheria toxoid is frequently mixed with aluminum hydroxide or aluminum phosphate to form an alum-precipitated toxoid. This type of preparation causes a very slow release of antigen from the injection site and, thus, a much longer stimulation of antibody synthesis. Water and oil emulsions are also very effective—particularly if mixed with killed tubercle bacilli. This mixture is called **Freund's adjuvant;** but, because of the inflammation that occurs at the site of the injection, it is not used in humans.

Monoclonal Antibodies

As shown in Figure 18.12 for sperm whale myoglobin, most proteins possess a number of different antigenic determinants. As a result, serum from an animal or human producing antibodies to a protein or cellular constituent will contain a complex mixture of antibodies. This mixture contains antibodies to all determinants as well as antibodies of different classes. If one could isolate and culture a single antibody-producing cell, it would provide a source of homogeneous

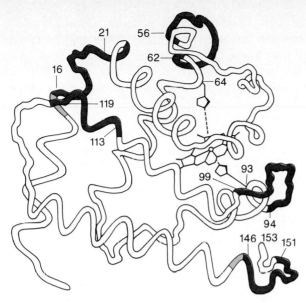

Figure 18.12 A schematic diagram showing the mode of folding of sperm whale myoglobin. The five solid color portions represent entire antigenic reactive regions. The light gray segments, each corresponding to only one amino acid residue, can be part of the antigenic reactive regions with some antisera. The white regions represent parts of the molecule that reside outside the reactive regions. The numbers identify landmark amino acid residues.

monoclonal antibodies that would bind to only a single, antigenic determinant.

Unfortunately, plasma cells cannot be propagated in vitro, but in 1975 Geörges Köhler and Cesare Milstein used a clever trick to overcome this obstacle. They took spleen cells (containing B cells and plasma cells) and mixed them with myeloma cells (malignant lymphocytes) in the presence of polyethylene glycol. This caused the cell membranes of a number of antibody-producing spleen cells to fuse with the myeloma cells. Such fused cells, called **hybridomas,** can be grown in vitro, and single clones of these fused cells can be isolated and propagated to produce **monoclonal antibodies** (see Figure 18.13).

Monoclonal antibodies are now available for hundreds of different determinants and are being used as research tools to study protein structure as well as virus and toxin neutralization and to isolate specific proteins from complex mixtures. Many commercially available monoclonal antibodies are being used in extremely sensitive and specific techniques for the diagnosis of various infections, for example, hepatitis B.

In addition, monoclonal antibodies directed against specific tumor antigens on leukemia cells have been isolated. Such monoclonal antibodies have been used to make immunoconjugates, that is, antibodies linked to drugs. Successful immunoconjugates have been synthesized by covalently linking monoclonal antibodies to the toxic moiety of ricin (plant toxin de-

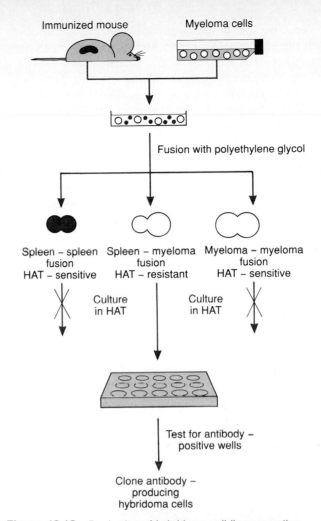

Figure 18.13 Production of hybridoma cell lines secreting monoclonal antibodies. The procedure for producing monoclonal antibodies is shown. Activated B cells from an immunized individual (e.g., spleen cells from immunized mice) are fused with malignant plasma cells isolated from plasmacytomas and adapted to tissue culture. The myeloma cell has a mutant gene that renders it sensitive to the drug aminopterin. The activated B cells, although resistant to aminopterin, have a very limited lifetime in culture, and die naturally. The B-cell–myeloma cell hybrid is resistant to aminopterin because the B cell provides the missing genes. Therefore, the B-cell–myeloma cell hybrid (the "hybridoma") is the only fusion product that can survive in the hypoxanthine-aminopterin-thymidine (HAT) selective culture medium used. The hybrids are distributed into a large number of culture wells in multiwell culture plates and are allowed to grow from a short period. The culture supernatant of these wells is then tested for the desired antibody. The cultures that are positive are cloned, and the hybridoma cell producing the desired antibody is propagated and used as a source of the monoclonal antibody.

rived from castor beans) or to specific drugs that will kill a cell after the monoclonal antibody binds to the cell. Thus, you can see that such monoclonal antibodies act as a very specific delivery system by binding only to cells possessing their specific antigen.

Monoclonal antibodies, however, do not need to be conjugated to drugs or toxins to be useful for immunotherapy. When these antibodies plus complement or the toxin-coupled monoclonals are mixed with bone marrow cells from leukemia patients, they exhibit a very efficient killing of the malignant cells. Hopefully, this will permit a surgeon to withdraw bone marrow cells, rid the preparation of leukemia cells, and reinject the leukemia-free marrow after the patient has undergone whole-body irradiation to kill all white cells.

In 1986, monoclonal antibodies that bind specifically to human T cells were approved for human use by the U.S. Food and Drug Administration to prevent kidney transplant rejection. Such a procedure is still experimental, but in clinical trials the use of these antibodies reversed approximately 94 percent of the rejections.

It seems very probable, therefore, that the use of monoclonal antibodies for research, diagnosis, and treatment of cellular disorders will become more common each year.

AUTOIMMUNE DISEASE

Our discussion of the immune system has described a mechanism for the generation of antibody-producing cells that provides them with an almost limitless potential to bind to various molecular configurations.

Yet with all of this vast potential for reactivity, a functional immune system must be able to differentiate self from nonself. Unfortunately, this occasionally goes awry, and the destruction caused by this inability to distinguish self from nonself is termed **autoimmune disease.**

There are several fairly well established models to explain some types of autoimmune diseases, whereas the mechanism of other autoimmune diseases remains obscure. We shall discuss three documented models.

Sequestered Antigens

During the development of the embryo, while the immune system is learning to distinguish self, there are certain antigens that are sequestered from the circulatory system and, insofar as the adult immune system is concerned, are always regarded as nonself.

One such antigen is found on the surface of mature sperm, which arise later in the development during both embryogenesis and normal adult life. As a result, antibodies do not normally arise to this sperm antigen and, if they should form, do not appear to cause autoimmune disease in humans.

Myelin, the covering surrounding nerve fibers, also appears to be a sequestered antigen. Under conditions still unexplained, an individual may mount an

---

MICROBIOLOGY MILESTONES

Our experience with the acquired immunodeficiency syndrome (AIDS) has taught us that a functional immune system is essential and that any defect in the expression of immunity can lead to death from a myriad of infectious agents. We also know that this same immune system can become programmed to react with innocuous environmental substances such as pollens, foods, plants, and insect bites, leading to symptoms of discomfort or even death. Defects in our immune system also can occur, one of which is manifested in the inability to discriminate between self and nonself. This is referred to as an **autoimmune disease**.

There are many autoimmune diseases; one of the most common is called **systemic lupus erythematosus** (SLE). This autoimmune disease is found mostly in young women (80 to 90 percent of cases) and is characterized by fever, facial rash, and usually some arthritis. Within the serum of such patients are found antibodies to many self antigens, the most common of which are directed toward the patient's DNA. As a result, the serum contains large amounts of complexes, composed of antigen, antibody, and complement, that are deposited in the kidney, leading to kidney failure and death. The cause of SLE is still unknown, but the

observation that when one identical twin has SLE, there is a 60 percent chance that the other twin will also acquire the disease suggests that it occurs as a heritable defect.

Other autoimmune diseases occur when individuals form antibodies against their own (1) red blood cells (hemolytic anemia), (2) immunoglobulins (rheumatoid arthritis), (3) platelets (thrombocytopenic purpura), (4) brain cells (allergic encephalitis), or thyroglobulin (Hashimoto's disease). Juvenile diabetes also appears to be an autoimmune disease in which one's T cells destroy the insulin-producing cells in the pancreas (islets of Langerhans).

There are a number of other autoimmune diseases but the few mentioned here should emphasize the importance of being able to discern self from nonself. There are probably numerous causes for the failure to do so, but for the most part they can be divided into two categories: (1) heritable abnormalities in the immune system, and (2) mutations in cells of the immune system that accumulate during one's lifetime and eventually reach a level (usually in the late stages of life) at which they are expressed as a defect.

---

immune response to myelin, causing an autoimmune encephalitis (inflammation of the brain).

Eye lens constitutes another sequestered antigen that, following injury to the eye, may induce an immune response resulting in a severe inflammation in the eye.

Altered Self

An excellent example of how a minor alteration in self can lead to an autoimmune disease is seen in human thyroiditis, in which an individual makes antibodies to thyroglobulin (a hormone produced by the thyroid gland). In brief, it has been shown that all normal people possess B cells capable of binding to thyroglobulin, but helper T cells that can bind to the hormone are nonexistent; thus, antibodies to thyroglobulin cannot be produced. Occasionally, however, a metabolic defect results in the production of an altered thyroglobulin expressing a determinant that is recognized by a helper T cell as foreign, resulting in the formation of antibodies that destroy the hormone.

Lack of Suppressor T Cells

As was described earlier in this chapter, there are several subsets of T cells that carry out functions other than those of helper T cells. One such subset is referred to as **suppressor T cells.** Such cells regulate the immune system by specifically blocking the action of helper T cells. Because suppressor T cells possess different surface antigens than do helper T cells, it is possible to remove them by using antibodies directed against these specific determinants. When this is done in mice, they develop an autoimmune hemolytic anemia in which they destroy their own red cells. Other systems have shown that suppressor T cells prevent the formation of antibodies to self; it thus appears that some autoimmune diseases result from a dysfunction in these necessary suppressor T cells.

DEFECTS IN ANTIBODY SYNTHESIS

There are disorders in which an individual is unable to respond to an antigenic stimulation. Defects in antibody synthesis go by various names, such as *antibody deficiency syndrome, hypogammaglobulinemia,* and *agammaglobulinemia.* Many of these disorders are heritable defects, and only during the past few decades has it been possible to keep many such individuals alive long enough to determine the clinical nature of

these syndromes. We can now broadly categorize immunodeficiency diseases as defects in B cells, T cells, or both.

B-Cell Deficiencies

If we consider all of the differentiative steps involved in the formation of a mature B cell from a precursor bone marrow stem cell, it is no wonder that occasional disorders resulting in an immunodeficiency disease occur.

Selective IgA deficiency is the most common disorder of this type, occurring in about 2 out of every 1000 individuals. Such people do not appear to suffer from major overt diseases but do experience frequent respiratory infections. The mode of the IgA deficiency is unknown, but 40 percent of such persons possess anti-IgA circulating antibody. Moreover, they have normal levels of B cells with surface IgA present, and one might therefore assume either that such cells are incapable of being stimulated by antigen or that the IgA antibody produced is rapidly being removed from circulation.

Another deficiency is characterized by the complete absence of mature B cells. This disorder, termed **sex-linked agammaglobulinemia,** is restricted to males and is characterized by recurrent bacterial infections.

T-Cell Deficiencies

The failure of the body to develop a functional thymus gland leads to a class of T-cell immunodeficiency diseases characterized by a failure of T-cell maturation. Such people usually lack a cell-mediated immune system (see Chapter 20) but may have a fairly functional humoral immune system because of a small amount of functional thymus that can supply helper T cells.

Severe Combined Immunodeficiency Disease

Severe combined immunodeficiency disease (SCID) is characterized by the complete absence of both B and T cells, and the disease is usually first noted because of frequent, recurrent infections starting early in infancy and involving all exposed surfaces of the body. SCID appears to result from a defect in a stem cell that normally would differentiate to form B and T cells. As you might surmise, most persons with SCID die very young. Recently, a case was publicized involving a boy with all the features of SCID who survived by living his entire life in a sterile bubble environment. He died, however, at age 13 after an unsuccessful bone marrow transplant designed to supply him with functional B and T cells.

SUMMARY

The state of being resistant to a specific pathogenic organism is called specific or acquired immunity because it results from the presence of specific antibodies that were induced by antigens present on or secreted by the organism. Antigens are substances capable of inducing antibody synthesis. To be an antigen, a substance must be foreign to the host and it must be a reasonably large molecule.

The immunoglobulins formed in response to an antigen are proteins composed of two different polypeptide chains; the large chain is called the heavy chain and the smaller, the light chain. Based on antigenic differences in their heavy chains, immunoglobulins can be divided into five classes designated IgG, IgA, IgM, IgD, and IgE. There are two antigenic types of light chains called kappa and lambda, and both of these types are found in all classes of immunoglobulins. An antibody molecule can be degraded into parts that have been designated as Fab and Fc. There are two Fabs for each molecule, and it is this portion of the molecule that contains the antigen binding site.

Each class of immunoglobulin possesses unique properties. IgG is quantitatively most important because it constitutes about 80 percent of the circulating immunoglobulins. IgA is found primarily in body secretions such as saliva and urine and in the gastrointestinal and genitourinary tracts. IgM is composed of 10 heavy chains and 10 light chains; it is usually the first antibody class detected after induction by an antigen. IgM is extremely effective in killing gram-negative bacteria. Very little is known about IgD, but it is found mostly on the surface of B cells, and it is postulated that it may play a role in the regulation of antibody synthesis. IgE is responsible for many types of allergies to food and pollens. It exists tightly bound to mast cells in an individual and, when it reacts with its antigen, several pharmacologically active substances (such as histamine) are released.

Three types of cells are involved in the synthesis of most immunoglobulins: (1) B cells, which are lymphocytes that differentiate into plasma cells, the actual antibody-producing cells; (2) T cells, which must recognize and react with the antigen in order to induce the B cell to differentiate into an antibody-producing cell; and (3) macrophages or dendritic cells, which are necessary to present the antigenic determinants to the T cell.

Th cells can bind to antigen only after it has been processed and redeposited on the cell surface of an antigen-presenting cell in association with a Class II MHC antigen. Once bound to such processed antigen, both the antigen-presenting cell and the Th cell secrete various lymphokines that activate the Th cell to divide and differentiate into T memory cells. After binding to processed antigen on B cells, such activated Th cells secrete additional lymphokines, which drive the B cell to divide and to differentiate into antibody-producing plasma cells and memory cells.

Following the initial contact with an antigen, there is a primary response during which B cells differentiate into plasma cells and both B and T cells differentiate into many memory cells. Measurable amounts of antibody cannot usually be detected in the serum before 5 to 10 days. A second injection of the same antigen weeks or months later results in a secondary or anamnestic response during which memory cells react with the antigen, resulting in a very rapid production of antibody.

Monoclonal antibodies can be obtained by isolating and propagating single cells of a hybridoma. Hybridomas are constructed by the fusion of antibody-producing cells with myeloma cells, resulting in a cell line that secretes antibodies and can be grown in vitro.

Autoimmune diseases result when the immune system fails to recognize self and mounts an immune response against self antigens. This can occur when antigens that were sequestered from the immune system during embryogenesis enter the circulatory system. It may also result from changes in an antigen that provide a new determinant then recognized as nonself. Finally, autoimmune disease may result from the loss or dysfunction of suppressor T cells that normally function to prevent an immune response to self antigens.

There are a number of defects in antibody synthesis that can be categorized as B-cell deficiencies, T-cell deficiencies, or both.

QUESTIONS FOR REVIEW

1. Differentiate between specific and nonspecific immunity.
2. What is a hapten?
3. What properties must a substance possess to be an antigen?
4. List the five major classes of immunoglobulin. What basis is used to divide the immunoglobulins into these classes?
5. Give two properties that appear to belong to the Fc portion of antibody. Give one property of the Fab portion.
6. How do the two types of IgA differ chemically? How does this affect their distribution in the body?
7. What property does the Fc portion of IgE have that the other classes of immunoglobulins do not?
8. Which class of immunoglobulin is both the largest and the first to be detected?
9. Outline the steps involved in the synthesis of antibody. What cell types seem to be involved? What appears to be the role for each cell type?
10. What is a secondary or anamnestic response?
11. What are hybridomas and how are they made?
12. List three models to explain autoimmune disease.
13. List three major types of immune deficiency diseases.

SUPPLEMENTARY READING

Abbas AK: A reassessment of the mechanisms of antigen-specific T-cell-dependent B-cell activation. *Immunol Today* 9:89, 1988.

Collier JR: Immunotoxins. *Sci Am* 251(1):56, 1984.

Davies D, Metzgar H: Structural basis of antibody function. *Annu Rev Immunol* 1:87, 1983.

de Macario EC, Macario AJL: Monoclonal antibodies for bacterial identification and taxonomy. *ASM News* 49:1, 1983.

Fink PJ, et al: Correlations between T-cell specificity and the structure of the antigen receptor. *Nature* 321:219, 1986.

Gordon J, Guy GR: The molecules controlling B-lymphocytes. *Immunol Today* 8:339, 1987.

Jeske, DJ, Capra JD: Structure and function. In Paul WE (ed.): *Fundamental Immunology*. New York, Raven Press, 1984.

Kohler G, Milstein C: Continuous cultures of fused cells secreting antibody of predefined specificity. *Nature* 256:495, 1975.

Leder P: Genetic control of immunoglobulin production. *Hosp Pract* 18(2):73, 1983.

Levy R: Biologicals for cancer treatment: Monoclonal antibodies. *Hosp Pract* 20(11):67, 1985.

Marrack P, Kappler J: The T-cell and its receptor. *Sci Am* 253(2):36, 1986.

Nowinski RC, et al: Monoclonal antibodies for diagnosis of infectious diseases in humans. *Science* 219:637, 1983.

Pollock RR, et al: Monoclonal antibodies: A powerful tool for selecting and analyzing mutations in antigens and antibodies. *Annu Rev Microbiol* 38:389, 1984.

Rennie J: The body against itself. *Sci Amer* 106:263(6), 1990.

Roit I: *Essential Immunology*, 7th ed. London, Blackwell Scientific Publications, 1991.

Tonegawa S: The molecules of the immune system. *Sci Am* 253(4):122, 1985.

Unanue E: Antigen-presenting function of the macrophage. *Annu Rev Immunol* 2:395, 1984.

Chapter 19

Measurement of Antibodies and Their Role in Immunity and Hypersensitivity

OBJECTIVES   The completion of this chapter should make the reader familiar with

1 A number of techniques used in the laboratory to detect and measure specific antibodies.

2 Terms such as opsonins, neutralizing antibodies, and quellung reactions.

3 The mechanism whereby antibodies protect the body from toxins, viruses, gram-negative bacteria and gram-positive bacteria.

4 The nature of the complement system and its role in the immune system.

5 The ABO system of classification of red blood cells and the role of alloantibodies in transfusion reactions.

6 The nature of the Rh factor and the conditions under which a newborn might be subjected to anti-Rh antibodies received from its mother.

7 A method for assaying anti-Rh antibodies.

8 The differences between active and passive immunity.

9 The types of allergic reactions resulting from immediate-type hypersensitivity.

10 The series of events leading to anaphylaxis.

11 The role of antibodies and complement in the induction of serum sickness or the Arthus reaction.

It has been said that there are more experimental methods for studying antigen-antibody reactions than there are ways of cooking potatoes. Space precludes a detailed description of all the methods currently used, but we shall discuss some of the techniques commonly employed to measure these reactions.

MEASUREMENT OF ANTIBODY

The study of antigen-antibody reactions in the laboratory is called **serology.** Using standard serologic tests, a known antigen (Ag) can be tested with a person's serum to find a specific antibody (Ab) or, conversely, serum containing known antibody may be reacted with organisms isolated from a patient to confirm the identification of the organisms. Serological techniques are required in blood banks for typing blood, for typing tissue before transplant operations, for typing immunoglobulins, and for many other clinical tests.

A number of methods can be used to determine the amount of antibody in an individual's serum. The method used depends mostly on whether the antigen is a whole cell, a toxin, a virus, or an encapsulated organism. Because all antibodies have two or more identical combining sites and most antigens have considerably more than two reactive sites, a reaction between antibody and antigen forms a latticework of many antigen-antibody molecules that in time becomes quite large. If the antigen is on a particulate material such as a bacterial cell, the end result is a clumping of the cells, or **agglutination.** A soluble antigen builds up into an antigen-antibody complex too large to stay in solution, and the result is a precipitation, or **precipitin reaction.** Other reactions merely measure the gain or loss of a property of the antigen. An example of this is a reaction with a bacterial toxin to neutralize it; such antibodies are given the general name **antitoxin.** Still other reactions neutralize viruses so that they cannot infect cells (**neutralizing antibody**) or react with bacterial cells to make them more easily phagocytosed by leukocytes (**opsonins**).

The multiplicity of Ab names does not, however, indicate different types of antibodies for each reaction; each name merely denotes the type of reaction being measured. Thus, if sufficient antitoxin is mixed with a soluble toxin, it will produce a precipitin reaction, or, if this soluble toxin is adsorbed to particles such as styrene beads, the same specific Ab will agglutinate the beads or enhance their phagocytosis. Therefore, specific Ab names such as *precipitin, agglutinin, antitoxin,* and others are used only to indicate the physical state of the Ag or the type of Ag-Ab reaction being measured. With this in mind, we may now consider some specific techniques for measuring Ag-Ab reactions.

The Precipitin Reaction

The precipitin reaction is a grossly visible one. A clear sample of serum containing the antibody is mixed with a solution of antigen. Within a few hours, sometimes in minutes, the solution becomes opalescent, and flocculent particles precipitate from what had been a clear solution. It is important to understand the following facts: (1) the reaction is specific in that only the antigen or compounds closely related in structure can combine and precipitate with the antibodies; (2) the reaction is reversible, since excess antigen dissolves the precipitate; (3) the *proper combination* of antigen and antibody causes precipitation from solution of all the antibody and all the antigen. This is called the equivalence point. This phenomenon results from reactants in the precipitin reaction having multiple reactive sites, which permits the development of innumerable combinations of antigen (Ag) and antibody (Ab). The earliest reaction is

$$Ag + Ab \rightleftharpoons Ag \cdot Ab$$

While the reaction is proceeding, the product Ag · Ab can react with other Ag · Ab, with Ag, or with Ab to form a variety of larger antigen-antibody complexes. In turn, these products can interact to form still larger complexes. Figure 19.1 shows a precipitin curve and a schematic illustration of the complexes formed as a function of Ag:Ab ratios.

Modifications of the Precipitin Reaction

A number of modifications of the classic precipitin reaction have been developed. Most such changes are designed either to increase the sensitivity of the reaction or to identify specific Ag-Ab reactions occurring in a system containing multiple Ags and Abs.

Double Diffusion (Ouchterlony Technique)
When soluble Ag and soluble Ab are placed in separate small wells punched into agar that has solidified on a slide or glass plate, the Ag and the Ab will diffuse through the agar. The wells are located only a few millimeters apart; Ab and Ag will interact in the area in which they are in optimal proportions to form a line of precipitate. Because different Ags will diffuse at different rates and also may require different concentrations of Ab for optimal precipitation, the position of the precipitin band will usually vary for each Ag. Figure 19.2(*a*) is a schematic illustration of double

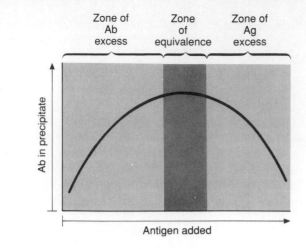

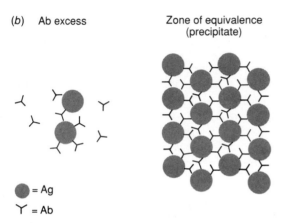

(a)

(b)   Ab excess

Zone of equivalence
(precipitate)

= Ag

= Ab

Figure 19.1 (a) Precipitin curve illustrating the effect of Ag concentration on the total amount of antibody precipitated as insoluble complexes of Ab and Ag. (b) Schematic illustration of the complexes formed as a function of Ag:Ab ratios.

diffusion in which there are several identical and nonidentical Ags. Thus you can see that this technique can, in some cases, determine the number of Ags present in a solution. Figure 19.2(b) is an example of actual double diffusions showing identities, partial identities, and nonidentities.

Immunoelectrophoresis The procedure of immunoelectrophoresis combines electrophoresis and double diffusion for the separation of Ag-Ab reactions in agar. In essence, one places a small drop of solution containing the Ags (usually proteins) into a small hole punched out of solidified agar on a slide. The slide is then placed in an electric field to allow the electrophoretic migration of the Ags. Different Ags will migrate at different rates or even in different directions depending on their size and charge. After the completion of the electrophoresis, a trough is removed from the agar along one or both sides of the slide and the appropriate antiserum is placed into the troughs. As in double diffusion, the Ags and Abs diffuse toward each other, which results in the formation of precipitin bands. As is shown in Figure 19.3, this procedure can be used to separate many antigens as well as to indicate the potential purity of an Ag.

Other modifications of the precipitin reaction are radial immunodiffusion and radioimmunoassays. For

Figure 19.2 Double diffusion in agar. In each case, well 1 contained antiserum and wells 2 and 3 contained one or more antigens. (a) The lines of precipitate formed by the Ag and Ab are confluent, indicating the Ag in the two wells was identical; (b) here a spur formation in addition to confluent lines signifies partial identity, i.e., well 2 contained one Ag that was identical to that in well 3 but it also contained an Ag not present in well 3; (c) both lines of precipitate cross, showing no confluence and, hence, complete non identity. *Credit:* David Normansell, Department of Pathology, University of Virginia.

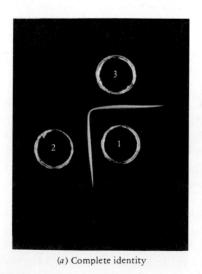

(a) Complete identity

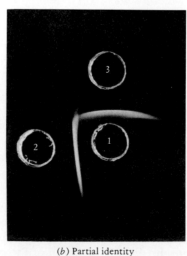

(b) Partial identity

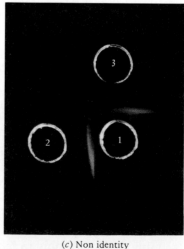

(c) Non identity

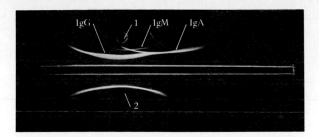

Figure 19.3 Immunoelectrophoresis: Normal human serum was placed in well 1 and purified human IgG placed in well 2. After 2 h of electrophoresis, the electric current was turned off and purified antiserum containing Abs to heavy chains γ, α, and μ and to light chains κ and λ was placed in the trough. During the ensuing 24 h, a line of precipitate formed as Ag and Ab diffused toward each other. Note that the whole serum shows well separated lines of IgG, IgM, and IgA. *Credit:* David Normansell, Department of Pathology, University of Virginia.

information concerning these assays, the reader is referred to books on immunology listed in the supplementary reading section at the end of this chapter.

The Agglutination Reaction

The agglutination reaction is a very sensitive serological technique for the detection and approximate quantitation of serum antibodies. It is widely used in clinical laboratories for the tentative diagnosis of infectious diseases, for blood grouping, and for the diagnosis of noninfectious diseases such as rheumatoid arthritis and lupus erythematosus.

The major difference between the agglutination reaction and the precipitin reaction is that in the former the antigens are on the surface of a particulate substance (bacteria, red blood cells, antigen coated on latex particles, etc.). A reaction between bivalent antibody and two particles containing antigen links the two visible particles, and a complex of visibly agglu-

tinated particles is built up much as in the precipitin reaction. Much less antibody is required to produce a visible agglutination pattern than is required to produce a precipitin reaction with soluble antigen. This makes the agglutination reaction a very sensitive assay for Ag-Ab reactions.

The usual agglutination reaction is actually only a semiquantitative measure of Ab, because one cannot normally determine the number of milligrams or moles of Ab that are attached to a clump of bacterial cells, as one can in the precipitin test. Instead, serial dilutions of the serum are made, as is shown schematically in Figure 19.4. A constant amount of cells is then added to each tube and, following several hours of incubation at 37°C, clumping is recorded by visual inspection. The titer of the antiserum is given as the reciprocal of the highest dilution that causes clumping. Thus, an antiserum that caused agglutination at a dilution of 1 to 128 (but not at 1 to 256) would be reported to have a titer of 128. Note that this is not an absolute value but rather an expression of a relative concentration of a specific antibody present in the serum.

Opsonins

Many bacteria are able to resist phagocytosis by the host's white cells because of the presence of capsules or fimbriae that prevent intimate contact between the bacterium and the phagocyte. In the presence of antibodies that are specific for these structures, however, ingestion by the phagocyte readily occurs. The antibodies inducing this increased rate of phagocytosis are referred to as **opsonins,** and they function by providing a substrate to which the phagocyte can bind. In other words, when an antibody binds to a bacterium, its Fc fragment projects out from the cell. Most phagocytic cells possess Fc receptors on their cell surfaces; these

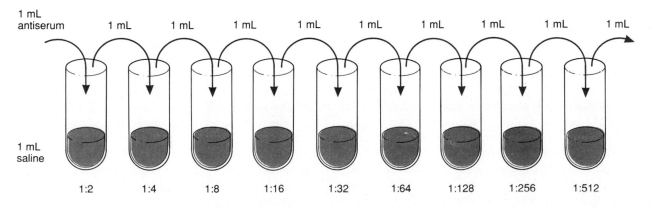

Figure 19.4 Double dilutions of Ab carried out in an agglutination reaction. Each tube contained 1 mL of saline before the double dilutions with antiserum were begun. The numbers under each tube represent the dilution of serum in that tube.

1 mL antiserum   1 mL   1 mL   1 mL   1 mL   1 mL   1 mL   1 mL   1 mL

1 mL saline

1:2   1:4   1:8   1:16   1:32   1:64   1:128   1:256   1:512

provide a mechanism for a close contact between the phagocyte and the antibody-coated bacterium.

Neutralizing Antibodies

Antibodies are involved in neutralizing the toxic effects of bacterial toxins as well as in preventing viruses from infecting host cells. A detailed model of exactly how they accomplish this neutralization is unavailable, but it is known that both toxins and viruses must interact with specific receptors on the host cell. Specific antibody binds to the toxin or virus and prevents this interaction either by sterically blocking the binding or by inducing a change in the molecular configuration of the toxin or virus so that it can no longer bind to the host cell receptors.

Special Serological Tests

There are a number of other serological tests that are specialized and sometimes difficult to process. The **hemagglutination inhibition test** (red cell agglutination inhibition) is used to identify certain viruses that themselves routinely cause the agglutination of red blood cells (RBCs) when the virus and RBCs are mixed together in a test tube. If, however, specific antibodies to the virus are mixed with the virus before the RBCs are added, agglutination is inhibited. Thus, by using antiserum of known specificity, one can sometimes identify an unknown virus.

Another specialized procedure is to attach chemically a fluorescent dye to an antibody. If cells are then treated with this fluorescently labeled antibody, the antibody combines with its specific antigen, and the entire complex becomes visible when viewed with a special fluorescence microscope.

The **fluorescent antibody** technique can also be used in an indirect method as follows: Human immunoglobulin is injected into an animal, which will make antibodies to the human immunoglobulin. These antibodies, called *antihuman immunoglobulin,* are made fluorescent by the attachment of a fluorescent dye. They can be used to detect very rapidly a specific human antibody. For example, let us suppose we want to check a patient's serum for the presence of antibodies to the organism that causes typhoid fever, *Salmonella typhi.* We would mix a dilution of the patient's serum with the *S. typhi* organisms and, after a short interval, wash off the excess serum and add the fluorescently labeled antihuman immunoglobulin. If the patient's serum contained antibodies to *S. typhi,* these antibodies would have reacted with the antigens on the surface of the cell. The subsequent addition of fluorescently labeled antihuman immunoglobulin would then react with the human immunoglobulin on the *S. typhi,* causing the cells to fluoresce. On the other

hand, if the patient's serum did not contain specific antibodies to the *S. typhi* organisms, no human immunoglobulin would react with the organisms; the subsequent addition of the fluorescent antihuman immunoglobulin would not react with the cells and would not result in the presence of fluorescent cells. As you can see, this same technique can be, and is, used for many different types of antibody reactions, since the fluorescent antihuman immunoglobulin is actually specific for human antibodies; as a result, it will react with all human antibodies. This indirect method of determination is sometimes referred to as the indirect fluorescent antibody technique or the fluorescent sandwich technique. Using an antiserum of known specificity, this same technique can be used for rapid identification of an unknown bacterium. The indirect fluorescent antibody technique is used routinely in diagnostic laboratories to test for the presence of antibodies to *Treponema pallidum,* the causative agent of syphilis. Figure 19.5 summarizes these fluorescent antibody techniques.

Neufeld Typing, or the Quellung Reaction

This test is based on the observation that if encapsulated bacteria are mixed with their specific antibodies, the capsule appears to swell to a size considerably larger than normal. It is on the basis of this test that one can divide the pneumococci into about 100 types. Other organisms that can be typed on the basis of quellung reactions include *Neisseria meningitidis,* the causative agent of epidemic meningitis; *Haemophilus influenzae,* a frequent cause of childhood meningitis; and *Klebsiella pneumoniae,* a gram-negative rod that can cause severe pneumonia.

Passive Agglutination Reactions Because the difference between a precipitin reaction and an agglutination reaction is based on whether the Ag is soluble or particulate, a precipitin reaction can be measured as an agglutination reaction by adsorbing a soluble antigen to the surface of a cell or particle. One may use synthetic particles such as polystyrene beads, but RBCs can be made to adsorb many soluble Ags readily; therefore they are frequently used for this purpose.

Most polysaccharides or lipopolysaccharides will spontaneously adsorb to the surface of red blood cells. Proteins do not adsorb well unless the RBCs are first treated with tannic acid. These "tanned" cells are then mixed with a soluble protein Ag prior to their use in passive hemagglutination.

Such tests are frequently carried out in plastic trays in which twofold dilutions of antiserum can be accomplished using a standardized loop. The addition of tanned RBCs possessing the adsorbed Ag to the Ab-containing wells results in agglutination patterns as shown in Figure 19.6.

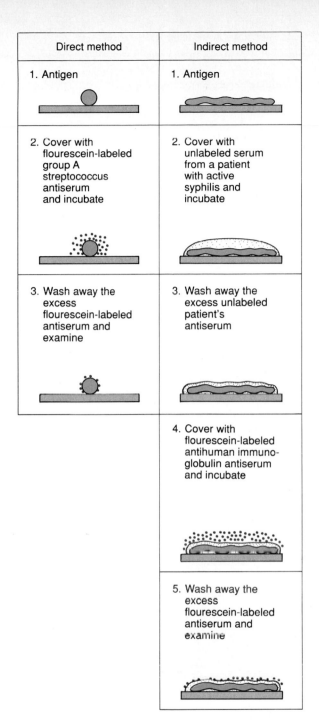

Direct method	Indirect method
1. Antigen	1. Antigen
2. Cover with flourescein-labeled group A streptococcus antiserum and incubate	2. Cover with unlabeled serum from a patient with active syphilis and incubate
3. Wash away the excess flourescein-labeled antiserum and examine	3. Wash away the excess unlabeled patient's antiserum
	4. Cover with flourescein-labeled antihuman immuno-globulin antiserum and incubate
	5. Wash away the excess flourescein-labeled antiserum and examine

Figure 19.5 Fluorescein-labeled antibody test methods. Note that one must have fluorescein-labeled antibody that is specific for the organism in question in order to use the direct method; however, the indirect method can be adapted to any organism using fluorescein-labeled antihuman gamma globulin as the only labeled antiserum.

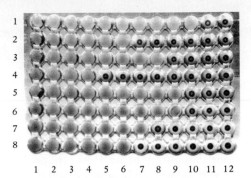

Figure 19.6 Hemagglutination patterns. The wells in each horizontal row contain double dilutions of antiserum (0.05 mL) to which 0.05 mL of appropriate red blood cells are added. Horizontal rows 1 through 4 contain sheep's red blood cells (SRBC) and serum from rabbits immunized with SRBC. Rows 5 through 8 contain SRBC that have been coated with human gamma globulin (HGG) and serum from rabbits immunized with HGG. Positive hemagglutination is seen as diffuse clumping of the red blood cells, while a negative reaction appears as a small button of red blood cells that have settled to the bottom of the well, as in the right-hand wells here.

MECHANISM OF IN VIVO ANTIBODY PROTECTION

The reactions between antibodies and antigens that are carried out in the laboratory do not, generally, occur in the body. In other words, the protective effect of antibodies is not due to precipitin or agglutination reactions or to any of the other procedures described for the measurement of antibodies. So we might ask just what antibodies do in an in vivo situation to prevent disease. This is fairly straightforward in the case of viral infections and diseases due directly to bacterial toxins (such as diphtheria and tetanus). In these cases specific antibodies act as neutralizing antibodies. The antibodies prevent infection by reacting with receptors on the virus to prevent the virus from entering a host cell. Antibodies neutralize toxins by binding to the molecule, thereby masking any toxic effect or preventing the toxin from binding to its host cell receptor.

The mechanism by which antibodies protect against bacterial infections is a bit more indirect; antibody alone is essentially ineffective. In other words, with the exception of the neutralization of toxins or of virus infectivity, the major function of an antibody is only to recognize a foreign antigen and to bind to it. By doing so, it can (1) present the Fc portion of the bound Ab to the Fc receptors on phagocytes, resulting in the rapid engulfment and destruction of the cell, or (2) provide a site for the initiation of the reactions of the **complement system.** It is the activation of this system that (1) leads to the lysis of foreign cells, (2) enhances the phagocytosis of invading microorganisms, and (3) causes a local inflammation that stimulates the chemotactic activity of the host's leukocytes.

Classical Pathway of Complement Activation

Complement is an incredibly complex multicomponent system composed of at least 16 proteins (see Table 19.1). It is present in normal serum and is effective only when it has been activated by a series of ordered reactions. In the classical pathway of activation, this cascade is initiated when the first component of complement binds to the Fc region and cross-links two or more molecules of antibody that have already bound to their respective antigens. The steps involved in the complement-mediated destruction of a foreign cell are immensely complex, but we can briefly categorize them into three steps.

Only the first component, C1, is involved in the **recognition unit.** This component, as indicated in Figure 19.7, is actually composed of three different proteins that interact with each other on the Ag-Ab complex. The reaction begins when C1q (the recognition subunit of C1) binds to the antigen-antibody complexes. C1r then binds to antibody-C1q, which initiates a partial hydrolysis of C1r, converting it into an active enzyme (**C1r**) whose only known substrate is C1s. Once activated, **C1r** converts **C1s** to an active enzyme, which initiates the assembly of the activation unit.

The assembly of the **activation unit** involves a series of enzymatic reactions that sequentially cleave

ACTIVATION OF RECOGNITION UNIT
C1 + Ab → Ab C1 (Ab·C1q·C1r·C1s) → ***C1s***

ASSEMBLY OF ACTIVATION UNIT
C1s
↓
C4 → C4a + C4b
C1s
↓
C2 → C2a + C2b
C4b + C2a → ***C4b2a*** C3 convertase
C4b2a
↓
C3 → C3a + C3b
C4b2a + C3b → ***C4b2a3b*** C5 convertase

ASSEMBLY OF MEMBRANE ATTACK COMPLEX
C4b2a3b
↓
C5 → C5a + C5b
C5b + C6 + C7 → C5b67
C5b67 + C8 + C9 → $C5b6789_n$

Figure 19.7 The classical pathway of complement activation. All activated enzymes are in bold italics.

the C4, C2, and C3 components of complement. The resulting activated subunits bind together to form yet another new enzyme (**C4b2b3b**), whose sole function is to cleave C5 into C5a and C5b.

Once C5b is formed, the remaining components spontaneously bind together to form the **membrane attack complex** (C5b6789). Figure 19.8 summarizes

TABLE 19.1   Components of the Human Complement System

Protein	Serum Concentration (μg/ml)	Molecular Weight	Number of Chains
Classical pathway			
C1q	180	410,000	18
C1r	50	83,000	1
C1s	40	83,000	1
C2	25	117,000	1
C3	1600	185,000	2
C4	640	206,000	3
C5	70	191,000	2
C6	64	120,000	1
C7	56	110,000	1
C8	55	151,000	3
C9	59	71,000	1
Alternate pathway			
Factor B (B)	200	93,000	—
Factor D (D)	1	24,000	—
Factor H (H)	500	150,000	—
Factor I (I)	34	88,000	—
Properdin	20	224,000	—

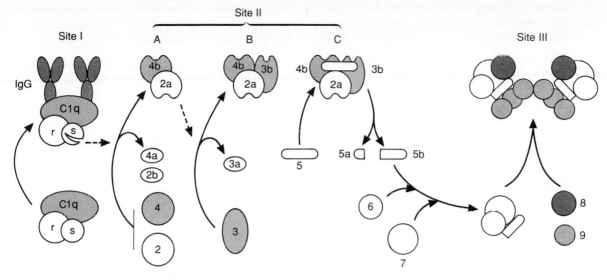

FLUID PHASE

Figure 19.8 The three-site model of antibody-dependent complement activation at a cell surface. The C1 complex reversibly interacts with antibody molecules that have bound to cell surface antigen molecules. Through an internal reaction, the C1s subunit of C1 is activated. This activated C1s protease cleaves first the C4 and then the C2 complement components. The C4b cleavage product binds to appropriate sites on the membrane (Site II) as shown in A above. The C2a cleavage product binds to the membrane-bound C4b, resulting in a new proteolytic enzyme called *C3 convertase*. The C3 convertase hydrolyzes the third component of complement (C3), producing C3a and C3b. The C3b portion then associates with the C4bC2a enzyme (see B in Site II) resulting in an enzyme, C5 convertase, with a new substrate specificity. The fifth component of complement, C5, binds to the C4b and C3b portions of the C5 convertase, is cleaved and released into the fluid phase. The C5b product then associates with C6 and C7, forming a complex that is stabilized by interaction with the membrane at Site III. Components C8 and C9 then associate with the C5b67 complex on the membrane, leading to membrane destabilization and lysis.

the reactions involved in the formation of the attack complex. This final complex binds to a site on foreign cells and gram-negative bacteria, causing membrane destruction and lysis of the cell. The mechanism by which this attack complex causes the lysis is not clearly understood, but an electron micrograph of such a cell shows that the attack complex causes the formation of a number of small, doughnut-shaped holes in the membrane (see Figure 19.9).

Because of their thick peptidoglycan cell walls, gram-positive bacteria are not lysed by the attack complex of complement. However, one of the components formed during the activation of complement (called C3b) binds tightly to the gram-positive cell. Normal leukocytes possess binding sites for C3b; hence they bind to the C3b component that is attached to the gram-positive cell wall. This results in an intimate contact between bacterium and leukocyte, leading to phagocytosis of the invading organism.

Finally, two other components that are formed during complement activation (C5a and C5b67) are chemotactic, resulting in an increased infiltration of leukocytes into an area of inflammation. As you might guess, individuals with defects in their complement

Figure 19.9 Sheep erythrocyte membrane treated with antibody and human complement. Subsequent treatment with trypsin was used to remove other proteins but did not affect the round lesions caused by complement ($\times$320,000).

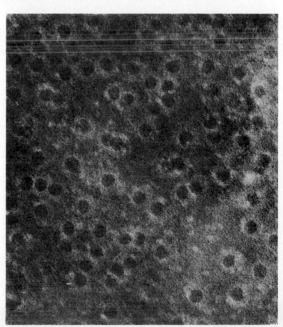

system are subject to recurrent infections, depending, of course, on the nature of the defect.

Note that each of these three steps takes place at a different site on the membrane as shown in Figure 19.8.

Alternate Pathway of Complement Activation

The alternate pathway of complement activation (the properdin pathway) does not require the presence of specific antibodies for initiation and, as a result, provides a mechanism of nonspecific resistance to infection. Moreover, this pathway does not use C1, C4, or C2, which are the early reactants in the classical pathway of activation. It should be kept in mind, however, that the overall result of this pathway is the same as that of the classical pathway: C3 is split into C3a and C3b, and C5 is cleaved into C5a and C5b, thus permitting the spontaneous formation of the C5b-9 membrane-attack complex. The enzymes catalyzing these conversions are, however, entirely different from the C3 and C5 convertases described for the classical pathway of complement activation.

The alternate pathway of complement activation uses three normal serum proteins that, together with

C3, form a functional C3 convertase and a C5 convertase. These are **factor B, factor D,** and **properdin.** To comprehend this pathway fully, you should keep the following facts in mind:

1. These are normal serum proteins, and the alternate pathway is continually being activated in the absence of any stimulus.

2. In the absence of initiators (also termed *activator surfaces*) such as bacterial membranes (lipopolysaccharide), yeast cell walls (zymosan), inulin (polyfructose), some RBCs, and neuraminic acid–poor or neuraminidase-treated membranes in general, the initial complexes are rapidly destroyed by two normal serum proteins known as factor H and factor I.

3. In the presence of these activator surfaces, such complexes are stabilized, and complement is activated to form the identical membrane-attack complexes described for the classical pathway (see Figure 19.10).

Complement Fixation

A very sensitive technique for the detection of either a specific antibody or a specific antigen is to determine whether complement was used up (fixed) by the reaction. This type of assay has been used in the past primarily to detect whether an individual possessed

Figure 19.10 Activator and nonactivator cell surfaces. Not all cell surfaces are capable of stabilizing C3b following complement activation. This diagram shows a model explaining the mechanisms responsible for this difference. C3b is inactivated by a proteolytic enzyme called *C3b inactivator* or *factor I.* The activity of this enzyme is promoted by interaction of another protein, factor H, with the C3b molecule. Binding of C3b to an "activator" surface (*a*) is thought to result in a conformational change in the C3b molecule that prevents binding of factor H that, in turn, fails to promote cleavage by factor I. However, as shown in (*b*), binding of C3b to a nonactivator surface fails to induce this conformational change. Thus, factor H can still bind and promote inactivation of C3b by factor I.

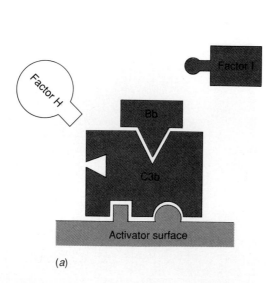

(a)

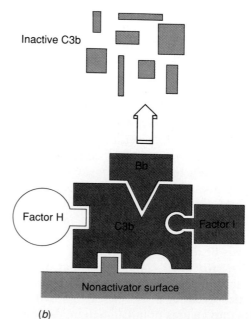

(b)

antibodies to a specific microorganism. The Wasserman test for the detection of antibodies related to syphilis was probably the best known of the complement fixation tests.

In general, a dilution of a person's serum (heated at 56°C for 30 min to inactivate any complement present) is mixed with a known antigen plus a known amount of complement. After allowing time for the reaction to go to completion, one then determines whether the added complement was "fixed" during the reaction or if it still remains free in the reaction mixture. To do this, a second antibody-antigen system that has an absolute requirement for complement is added. The indicator system routinely used consists of sheep's RBCs plus antibody to sheep's RBCs, which, in the presence of complement, will lyse the RBCs. If the complement had been fixed in the first reaction, no lysis would occur in the indicator system. If, however, there were no antibodies to the antigen used in the person's serum, complement would not have been fixed and the indicator system would lyse. Figure 19.11 summarizes the complement fixation reactions.

ALLOANTIBODIES

Early in the use of blood transfusions, it was observed that the recipient had to have blood compatibility with the donor if a serious reaction was to be avoided. It was soon learned that the basis of incompatibility was immunological in nature and that some individuals possess naturally occurring antibodies in their serum that can lyse other types of red blood cells. This led to the discovery of the ABO system of classification for red blood cells.

Karl Landsteiner observed that if serum from a large number of individuals was mixed with RBCs obtained from these people in all possible combination, the RBCs were agglutinated by some sera and not by others. On the basis of these findings, he determined that RBCs could be classified into four major types designated A, B, O, and AB. This classification is based on the presence or absence of two antigens, A and B, which may be present on RBCs together, singly, or not at all. These RBC antigens have been termed **alloantigens** because of the variations among members of the same species.

Unexpectedly, Landsteiner found that each individual's serum contained **alloantibodies** against the A and B alloantigens that were not present on the person's own RBCs. Thus, as shown in Table 19.2, type A persons possess antibodies to type B antigen on RBCs; type B persons possess antibodies to type A antigen on RBCs; type O persons have both anti-A and anti-B antibodies; and type AB individuals have neither of these antibodies.

A sometimes fatal transfusion reaction with fever, prostration, and renal insufficiency may occur if a patient receives blood from a different blood group; this results from the hemolysis of the transfused cells by the patient's own alloantibodies and serum complement. Careful tests to identify the blood group in both the donor and recipient, as well as compatibility tests to ascertain possible agglutination of donor cells (cross-matching), must be conducted prior to transfusion to avoid this hazard. It should be noted that these alloantibodies are mostly of the IgM class, and that this class of antibody cannot pass from the mother to the fetus. One exception occurs in type O individuals who may have anti-A and anti-B antibodies that are partly of the IgG class. Thus, type O mothers who are carrying a type A or B fetus may be stimulated by fetal red blood cells to produce a high titer of IgG class anti-A or anti-B. These antibodies can then cross the placental wall, causing the destruction of the fetal RBCs. This form of ABO-hemolytic disease of the newborn is rare, but severe cases requiring exchange transfusion do occur.

Rh Factor

The Rh factor, present in about 85 percent of the population, is another antigen found on red blood

Figure 19.11 Complement-fixation test. In the positive reaction, C was fixed when Ab bound to Ag. As a result, the Ab to sheep's red blood cells (SRBC) in the indicator system was unable to lyse the SRBC. In the negative reaction, C was not fixed because there was no Ab. Thus, C was still available to cause lysis of the indicator system.

Positive test

$$Ag + Ab + C \xrightarrow{\text{wait 12 hours}} \begin{array}{l} SRBC \\ + Ab\ to\ SRBC \end{array} \longrightarrow No\ lysis$$

Negative test (no Ab Present)

$$Ag + C \xrightarrow{\text{wait 12 hours}} \begin{array}{l} SRBC + C \\ + Ab\ to\ SRBC \end{array} \longrightarrow Lysis$$

TABLE 19.2 Alloantigens and Alloantibodies Associated with the ABO Classification of Red Blood Cells

Blood Type	Alloantigen on RBC	Alloantibody in Serum
A	A	B
B	B	A
AB	AB	None
O	None	AB

cells. The expression *Rh* comes from *rhesus monkey,* the species used in the early laboratory studies in this field. Individuals whose blood cells contain this factor are said to be Rh-positive; those without it are Rh-negative. Unlike the A, B, and O system, no alloantibodies to Rh antigen are formed without RBC stimulation. It is vitally important, however, that the Rh status be determined prior to transfusion, since an Rh-negative recipient of Rh-positive blood will develop antibodies within a few weeks' time, and these will lyse Rh-positive red blood cells. Thus, in a subsequent transfusion from an Rh-positive donor, all transfused red blood cells will be hemolyzed, resulting in a severe reaction and perhaps death from mass hemolysis. Note also that anti-Rh antibodies are of the IgG class.

The greatest importance of the Rh factor is the danger encountered during pregnancy when the mother is Rh-negative and the father is Rh-positive. If the fetus is Rh-positive (inheriting the factor from the father) and any fetal blood gets into the mother's circulation at the time of delivery or through a placental defect, the mother will form antibodies that are directed against the foreign Rh factor in the red blood cells of the fetus. Since her anti-Rh IgG antibodies can cross the placental wall, they will react with the RBCs of the fetus and produce hemolytic disease of the newborn (a condition known as erythroblastosis fetalis) and sometimes death. Once these antibodies are formed, there is increased danger of the same tragic result in any future pregnancy. Similarly, it can occur if the Rh-negative mother has been previously transfused from an Rh-positive donor as mentioned above.

Exposure of an Rh-negative mother to the Rh-positive cells of her baby, however, does not normally occur in sufficient concentrations to induce a primary response until the time of delivery. A first child, therefore, is not usually in danger. However, if the mother makes a primary response to the Rh Ag at the time of delivery, when some blood may be exchanged through trauma, a secondary (anamnestic) response can be induced at a later date with much smaller amounts of Rh Ag. Thus, a subsequent Rh-positive fetus might trigger the mother to form anti-Rh antibodies of the IgG class, which would then cross the placenta and destroy the fetal red blood cells. To avoid this, we can take advantage of the fact that an excess of injected antibody to the Rh-positive red blood cells will react with and destroy such cells, preventing the mother from making a primary response to the Rh Ag. It is now routine at the time of delivery to inject anti-Rh antibodies (known as RhoGAM) into Rh-negative mothers with Rh-positive babies.

RhoGAM is also given routinely to all Rh-negative women who have an abortion, whether spontaneous or induced. To be effective, it must be given within 72 h after delivery or abortion.

Coombs Antiglobulin Test

Anti-Rh antibodies are of the IgG type, but they will not normally agglutinate Rh-positive RBCs. The best explanation for this observation is that there are insufficient Rh factor sites to permit the IgG antibodies to overcome the normal electrostatic repulsion between RBCs. A clever way to measure these nonagglutinating Abs was described by British immunologist R.R.A. Coombs. He reasoned that if RBCs were coated with anti-Rh Abs, the addition of Abs that were directed against the Rh Abs would cause the erythrocytes to agglutinate. Accordingly, he injected human immunoglobulin into rabbits to induce them to make antihuman immunoglobulin (also known now as Coombs reagent). The antiglobulin test for Rh Abs could then be accomplished by (1) reacting Rh-positive RBCs with the serum to be tested and (2) after a short incubation period, adding antihuman immunoglobulin. If the serum in question contained Rh Abs, the RBCs would then agglutinate, as shown schematically in Figure 19.12.

Figure 19.12 Coombs antiglobulin test used to detect nonagglutinating Rh Abs.

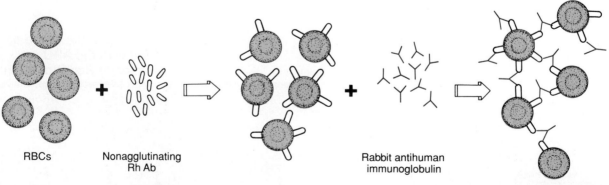

RBCs   Nonagglutinating Rh Ab   Rabbit antihuman immunoglobulin   Agglutinated RBCs

THE ROLE OF ANTIBODIES IN IMMUNITY

As an example of specific acquired immunity, let us consider the childhood disease chickenpox. During this illness the body forms antibodies that react with and neutralize the chickenpox virus.

After recovery, antibodies continue to circulate in the blood, and any subsequent exposure to chickenpox virus results in the inactivation of the virus. Although we are interested primarily in the antigens present in pathogenic microorganisms, it is important to realize that many chemical substances (not part of any microorganism) may be antigens. If an animal is injected with egg white, for example, it will form specific antibodies that would react with the egg white.

The types of immunity resulting from antibody production may be categorized on the basis of the origin of the antibodies in the following manner.

Natural Active Immunity

A person acquires natural active immunity through the experience of having a given disease. During the course of the disease, antibodies against the disease organism are formed in the body. In some cases these antibodies may provide lifelong immunity to reinfection, in particular to such viral diseases as measles, chickenpox, yellow fever, and smallpox. In other cases the antibodies may provide immunity against reinfection for a shorter time—possibly a few years. Examples in this group are influenza, diphtheria, and possibly tetanus. Sometimes the antibodies formed seem not to be protective at all, since reinfection can occur almost immediately; such an apparent lack of protective immunity may be explained by the fact that many different antigenic variants of the same organism cause identical symptoms. An example of this type of disease is the common cold; although the body does produce protective immunity to one type of rhinovirus, it can still remain susceptible to other types.

Not infrequently an infection is so mild that it goes unrecognized, but the body still responds by developing protective antibodies to the causative agent. This process is very common in diseases such as poliomyelitis and mumps. Although many of us have never had clinical symptoms of either of these illnesses, by early adulthood most of us have protective circulating antibodies in our blood directed against them.

Artificial Active Immunity

Artificial active immunity is acquired by the individual who has received a vaccine or inactivated toxin (toxoid). Thus an individual is able to build up protective antibodies against an organism without ever having had the disease. It must be remembered that the formation of antibody by the individual is a slow process, and at least 7 to 10 days may be required before effective antibody levels are reached. The antigens used to stimulate artificial active immunity may be killed

---

MICROBIOLOGY MILESTONES

Acquired immunity to diseases has been known to exist since the beginning of recorded history, but a comprehension of the cause of these diseases and the resulting acquired immunity took a long time to formulate. Let us look at a few ideas that were prevalent at several stages of mankind's scientific development.

Primitive societies were dominated, in large part, by spirits and demons, and everyday lives were governed by sets of taboos. Disease occurred as a punishment for some sin or the disregard of a taboo. Prevention of disease involved the wearing of amulets and the offering of sacrifices to please the gods. The Old Testament provides numerous examples of God causing disease as a punishment for transgressions. Acquired immunity after recovery was viewed as a reward following a punishment for minor sins. Other theories ascribed major epidemics of plague and syphilis in Europe to an influence of the stars and planets.

Girolama Fracastora presented one of the more unusual concepts of smallpox and its acquired immunity. Fracastora correctly believed that disease was spread from person to person by the passage of small seeds.

He believed that different seeds had affinities for various organs and tissues and that the seed causing smallpox had an affinity for that small amount of "menstrual" blood that each individual acquired in utero. After infection with smallpox, that trace of menstrual blood would rise to the skin and force itself out via smallpox pustules. Acquired immunity followed because, after recovery, the menstrual blood was purged and could no longer support the seed of smallpox.

Another ingenious explanation for acquired immunity was proposed in the early 1700s. This theory suggested that each person is born with a variety of innate seeds, termed "ovula" and that disease occurred when the ovule for a specific disease was fertilized. On recovery, the ovule was destroyed and, hence, it was not possible to be reinfected with the same disease.

There were many other proposed concepts of disease and acquired immunity, but these few examples illustrate that we have come a long way in the past few centuries in understanding disease and acquired immunity.

---

organisms (typhoid fever), live but attenuated (having lost the ability to produce disease) organisms (small-pox, yellow fever, live poliomyelitis vaccine, bacille Calmette–Guérin), or toxoids (tetanus, diphtheria). In any of these cases the body forms protective antibodies so that in the event of exposure to the disease organism itself, it is able to neutralize or inactivate it.

Passive Immunity

A person who is given preformed antibodies (antibodies formed by another person or an animal) acquires passive immunity. Passive immunity can be subdivided into what we may call **artificial passive immunity** and **natural passive immunity.** In artificial passive immunity, serum containing antibodies formed by one person or animal is injected into another individual. One of the more frequent uses of artificial passive immunity is the administration of tetanus antitoxin (antibodies against tetanus toxin, usually formed in other humans) to an individual with a dirty wound. Individuals not previously immunized with the toxoid do not have antibodies in their blood. The importance of using preformed antibodies is that this provides immediate protection, which is not provided by the administration of an antigen that stimulates artificial active immunity.

Under certain conditions pooled gamma globulin is used to give passive immunity. It is obtained from large pools of human blood and contains antibodies representative of the general population. It is used primarily to give immunity to an individual exposed to an infection from which most of the population has already recovered. Until recently, gamma globulin was commonly used to prevent active disease in a person exposed to measles or poliomyelitis. However, now that effective vaccines are available for both diseases, gamma globulin is not as commonly used for prevention or attenuation of these diseases. It is still used to confer passive immunity on people exposed to hepatitis A and for individuals who have been exposed but not previously immunized to measles or polio. It is also given to individuals who lack immunoglobulins from birth. The important thing to remember is that *passive immunity is always temporary.* Adults maintain this immunity at an effective level for only about 4 to 6 weeks, after which they are again susceptible to infection.

Natural passive immunity refers to the passage of preformed antibodies from the mother to the fetus or the newborn. In the human they pass through the placenta into the unborn fetus as well as being present in breast milk. In other animals, such as the cow, the antibodies are passed only in the first milk. In either case the newborn is provided with the mother's circulating antibodies. Again, this is only temporary.

However, the newborn does not metabolize or get rid of passively transferred antibodies as rapidly as the adult, and hence they are usually effective for 4 to 6 months.

ALLERGY AND HYPERSENSITIVITY

So far in this chapter we have pointed out the protective functions of antibodies in warding off infectious diseases. This is only part of the story because some antibodies can also take part in Ab-Ag reactions that result in minor or major illnesses. This harmful reaction of an antibody with its antigen is referred to as **allergy** or **hypersensitivity** (too much sensitivity). The overall field of allergy is too vast for any comprehensive discussion in this book, but a few types of allergy are particularly important. Though the terminology of allergic reactions is inadequate, we are stuck with it for the time being; hence it is necessary to define the two types of allergy.

In this chapter we shall confine ourselves to allergic reactions resulting from circulating antibodies. This type of allergy is commonly called **immediate hypersensitivity.** In the next chapter we shall discuss a different type of response, which is associated with a **cell-mediated hypersensitivity.**

Types of Allergic Reactions

Before an allergic reaction can occur, an individual must have been sensitized to the antigen involved. This sensitization normally follows contact with the antigen in question (frequently called the **allergen**), but it can also be induced passively by the injection of preformed antibodies to the allergen. In either case, subsequent contact with the allergen causes the release of pharmacologically active mediators from certain host cells. The primary effect of these mediators is to cause smooth muscle contraction and to dilate capillaries. Table 19.3 divides the immediate-type hypersensitivities into three groups on the basis of the class of antibody involved, the requirement for complement, and the type of cell destruction that releases the pharmacological mediators.

Anaphylaxis

The most serious hypersensitivity is called **anaphylaxis.** This extremely severe reaction frequently causes death within 2 to 3 min after the antigen is injected into a highly susceptible person or animal. Anaphylaxis is mediated by IgE-type antibodies (bound to circulating mast cells) when these antibodies are cross-

TABLE 19.3   Immediate-type Hypersensitivities

Group	Ab Involved	Complement Involved	Examples	Mediator Cells
Anaphylaxis	IgE	No	Anaphylactic shock Cutaneous anaphylaxis Hives Asthma Hay fever Drug allergies	Primarily mast cells buy may also include basophils
Immune complex syndromes	Any class, but primarily IgG or IgM	Yes	Serum sickness Arthus reaction Glomerulonephritis Rheumatoid arthritis Systemic lupus erythematosus	Primarily neutrophils and platelets
Autoimmune reactions	Any class, but primarily IgG	Yes	Transfusion reactions Rh incompatibility Hemolytic anemia	None

linked with antigen (see Figure 19.13). This may occur, for example, following the injection of penicillin into a highly susceptible individual. Another quite common cause of death is the anaphylaxis occurring in an individual who is highly sensitive to insect stings. The reaction occurs almost immediately; as a result, large quantities of vasoactive amines (histamine, serotonin, bradykinin, and others), which damage the host cells, are released. The actual major pharmacological mediator in humans appears to be histamine, but in animals such as mice and rats, it may be serotonin. These agents are extremely toxic and cause contraction of smooth muscles in the body. In humans this occurs primarily in the lungs and results in suffocation. The symptoms can be relieved by the immediate injection of epinephrine or antihistamines when an individual is going into shock. In summer many people carry insect sting kits with a syringe containing epinephrine, which they can inject into themselves in the event of a sting. The ready availability of such a kit has undoubtedly saved many lives.

This same type of immediate hypersensitivity may also be manifested locally when the antigen enters the body through the mucous membranes of the digestive tract. The antibody type is still IgE, but the antigen is one of many present in air or food that provoke allergies such as hay fever, asthma, ragweed, and food allergies. It is frequently possible to identify the offending antigen with skin tests of the allergic individual. In this case, extracts from a series of possible allergens (ragweed, dust, feathers, penicillin) are scratched into the patient's skin. Any antigen to which the patient is highly sensitive will provoke a wheal-and-flare reaction at the site of the scratch in 10 to 15 min. A wheal-and-flare reaction is a sharply circumscribed, elevated blanched area surrounded by a flare spreading from the center. Because some people are exquisitely sensitive to small amounts of certain allergens, such tests should be conducted only by a competent allergist.

In addition to the release of the preformed mediators already described, several secondary chemical mediators are released during mast cell immunological activation that appear to be byproducts of the reaction. One of the foremost among these has been termed **slow-reacting substance of anaphylaxis** (SRS-A). Chemically, SRS-A is known to consist of a family

Figure 19.13 Antigen-induced mediator release from mast cells. IgE antibody bound to mast cells by IgE Fc receptors is cross-linked by specific antigen, resulting in an influx of Ca⁺⁺ followed by a rise in intracellular cyclic AMP. This induces degranulation and mediator release. It also results in activation of phospholipase A₂, which in turn catalyzes production of arachidonic acid from membrane phospholipids.

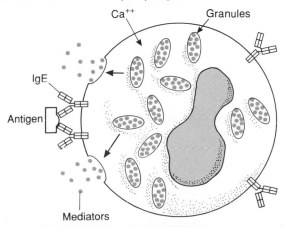

of four or more oxidation products of arachidonic acid that have been termed **leukotrienes**. The presence of SRS-A has long been thought to be a principal mediator of human asthma; indeed, it has now been demonstrated that leukotriene C_4 (LTC_4) and LTD_4 are spasmogens that are more than 1000 times as potent as histamine on a molar basis. Moreover, these two leukotrienes will elicit a wheal and flare at approximately one tenth of the concentration required for histamine. In vitro studies have shown that another leukotriene, LTB_4, is a strong chemotactic factor for certain phagocytic cells.

Once the allergen is determined, a patient can frequently be desensitized by injections of very small amounts of the allergen. The dose may be increased over a period of months and years. The actual mechanism of desensitization is believed to be the result of the formation of large amounts of IgG- or IgA-type antibody to the allergen. Then, when the individual comes in contact with the offending allergen (such as ragweed pollen), the allergen is neutralized by the IgG or IgA before it can react with IgE-type antibody. Since IgG and IgA are not fixed to mast cells, reaction of these types of immunoglobulin with their antigen does not result in the release of large amounts of histamine. Because they act to intercept the allergen from the IgE, they are frequently called **blocking antibodies**.

Immune Complex Syndromes

This type of circulating-antibody-type allergy also results from IgG or IgM, plus complement, but in this case the cellular damage appears to be due to the presence of large amounts of circulating primary Ag-Ab complexes. This occurs whenever an individual who has a high antibody level is placed in contact with a large amount of antigen. It can be manifested as either a generalized or a localized reaction. The generalized reaction is referred to as serum sickness and the localized injury as an Arthus reaction.

The **Arthus reaction** occurs when an antigen is injected into the skin of an individual who already possesses a high antibody level against that antigen. This reaction is characterized by a marked local inflammation and appears to occur as follows: (1) Intradermally injected antigen reacts with antibody (IgG or IgM) to form a precipitate that penetrates the local blood vessel walls. (2) These complexes fix complement resulting in the formation of the active chemotactic factors C5a and C5b67. The attracted polymorphonuclear leukocytes ingest the immune precipitates, causing the release of lysosomal enzymes (see Chapter 17). (3) The lysosomal enzymes digest adjacent cells, resulting in more inflammation.

The time course of the Arthus reaction is considerably slower than that described for anaphylactic-type reactions. Slight swelling and fluid accumulation (edema) may begin within 1 to 2 h, and the usual reaction will reach a maximum in about 4 h. Because the intensity of the inflammation is a function of the number of Ag-Ab complexes, the rate of healing depends on the severity of the reaction.

Clinically, Arthus reactions characterized by lung lesions have been described in individuals who inhale spores from the mold *Aspergillus* (whose normal habitat is decaying vegetation) or in those who inhale mold spores used in the manufacture of certain cheeses.

Serum sickness gets its name from the type of reaction that sometimes follows the injection of a foreign antiserum (such as tetanus antitoxin from a horse). Since the horse serum contains foreign proteins, the patient begins to form antibodies to the antigens in the horse serum. After about a week, antibodies are rapidly being synthesized and immediately react with any residual circulating horse proteins in the patient's blood. The resulting Ab-Ag complexes fix complement (particularly with a large antigen excess) and may cause fever and joint pains that persist until the last bit of horse protein has been eliminated. Usually serum sickness does not result from the first injection, because by the time any appreciable amount of antibody has been formed the foreign proteins have been largely destroyed by other metabolic processes in the patient. On a second injection—even years later—the patient may start forming antibodies much more rapidly, usually within a day or two.

The reactions described for serum sickness can also occur as a result of other Ab-Ag reactions, and there are a few cases in which infectious agents that initiate this type of allergic disease have been identified. Group A streptococcal infections, particularly those caused by type 12, have long been known to initiate an acute glomerulonephritis (an inflammation of the capillaries in the glomeruli of the kidney). The specific nature of the streptococcal antigen involved in this immune complex disease is not known, but several laboratories have demonstrated the presence of IgG, complement, and streptococcal antigens in the granular deposits along the glomerular capillary walls.

Rheumatoid arthritis is characterized by a chronic inflammation of the joints in which there are high levels of immunoglobulins, complement, and complement-fixing aggregates of immunoglobulins. The resulting inflammation may well be similar to that described for the Arthus reaction. The greatest mystery concerning rheumatoid arthritis is the nature of the antigen involved. There is, however, circumstantial evidence that aggregated IgG might function as the inducing antigen.

The obligatory role of complement in both the Arthus reaction and immune complex diseases (such as serum sickness) indicates that these allergic reactions have much in common. In fact, the overall pathology of serum sickness is not unlike a disseminated Arthus reaction in which the immune complexes have been deposited in the blood vessel walls, kidneys, and tissues throughout the body, where they are ingested by leukocytes.

SUMMARY

Antibodies can be measured in a number of ways—for example, agglutination, precipitation, the use of opsonins or neutralizing antibodies, and complement fixation. In addition, one can use several specialized techniques, including double diffusion in agar, immunoelectrophoresis, and using fluorescently labeled antibodies.

An individual's blood type—A, B, AB, or O—is based on antigens on the red blood cells. All individuals have alloantibodies to the A or B antigens not present on their red blood cells. Thus, type A has anti-B, type B has anti-A, and type O has anti-A and anti-B. In addition, one's red blood cells may or may not have an antigen called the Rh antigen.

Based on whether an individual forms antibodies as a result of an antigenic stimulation or receipt of preformed antibodies, the resulting immunity can be classified as active or passive. Active immunity may be lifelong, whereas passive immunity is effective in an adult for only 4 to 6 weeks. The in vivo mechanism of antibody protection from viruses or toxins results from neutralization of infectivity or toxicity, respectively. Bacteria are destroyed by the activation of the complement system, followed by membrane destruction in gram-negative organisms or phagocytosis of gram-positive organisms. Some components of the activated complement system also enhance the chemotaxis of leukocytes.

Many allergies appear to result from the reaction of an allergen with IgE, a reaction that releases histamine and other mediators from the mast cell on which the IgE was bound. This may result in anaphylaxis or, in the case of pollen allergies, in bronchial spasms. Identification of the offending allergen can be accomplished by scratching minute amounts of possible allergens into the skin and observing for a skin reaction, called a wheal and flare, that will rise in 10 to 15 min if the patient is hypersensitive to a particular allergen. Serum sickness occurs as a result of the injection of foreign serum. Antibodies to the serum are formed and react to form immune complexes that are toxic to the patient. The Arthus reaction is a localized version of an immune complex disease, and acute glomerulonephritis and rheumatoid arthritis may be immune complex diseases of bacterial origin.

QUESTIONS FOR REVIEW

1. List six ways to detect antibodies by serological techniques.
2. What is the basis for the different blood groups? Explain.
3. What is the Rh factor? What complication may result to a fetus if the mother is Rh-negative and the father is Rh-positive? If the father is Rh-negative and the mother is Rh-positive?
4. Differentiate between naturally acquired active immunity and artificially acquired active immunity.
5. Differentiate between naturally acquired passive immunity and artificially acquired passive immunity.
6. Discuss the duration of immunity resulting from active versus passive immunity.
7. What is allergy? What is a synonym for the word *allergy*?
8. Give two examples of anaphylactic reactions. How may the symptoms be relieved?
9. How may skin tests be used to identify suspected allergen?
10. Discuss the in vivo mechanism of antibody protection against viruses; against bacteria.
11. What is serum sickness and how does it differ from the Arthus reaction?

SUPPLEMENTARY READING

Austen KF: Tissue mast cells in immediate hypersensitivity. *Hosp Pract* 17(11):98, 1982.

Berzofsky JA, Berkower IJ: Antigen-antibody interaction. In Paul W (ed.): *Fundamental Immunology.* New York, Raven, 1984.

Doring NHG, Schiotz PO: The role of immune complexes in the pathogenesis of bacterial infections. *Annu Rev Microbiol* 40:29, 1986.

Hill HR, Matsen JM: Enzyme-linked immunosorbent assay and radioimmunoassay in the serologic diagnosis of infectious disease. *J Infect Dis* 147:258, 1983.

Chapter 20

Cellular Immunity

OBJECTIVES   The material in this chapter will familiarize you with

1. The types of cells taking part in cell-mediated immunity.

2. The nature and types of lymphokines involved in delayed-type hypersensitivity.

3. The mechanisms whereby delayed-type hypersensitivity cells provide their immune function.

4. The way in which cytotoxic T cells kill foreign target cells.

5. The difference between natural killer cells and killer cells.

6. The cellular aspects of transplantation rejection.

7. The role of tumor antigens in immune surveillance.

8. The mechanisms of immunological tolerance.

E ach normal vertebrate possesses two systems of immunity. The humoral system, composed of soluble immunoglobulins, was discussed in the previous chapter. The second system is one that depends on the presence of living cells that have become sensitized to a specific antigen. Let us compare some of the things we know about the cell-mediated immunity system with what we have already learned concerning the humoral system of immunity.

Cell-mediated immunity is known to be mediated by two distinguishable groups of lymphocytes: T cells and large granular lymphocytes (LGL). The T-cell group is composed of (1) Tdth cells, which on interaction with foreign antigens presented on the surface of macrophages secrete a number of biologically active factors that stimulate other cell types involved in the delayed-type hypersensitivity (DTH) reaction, and (2) Tc cells, which specifically kill target cells bearing non-self antigens, such as tumor cells, virus-infected cells, or foreign cells, by release of lysins following specific recognition of foreign antigens on the target cell surface.

The LGL group kill target cells by mechanisms similar to those used by the Tc cell. This group is composed of (1) natural killer cells, which kill certain tumor cells in spite of lacking antigen-specific receptors for such target cells, and (2) killer cells, which possess Fc receptors on their cell surface and, as a result, recognize antibody-coated target cells.

DELAYED-TYPE HYPERSENSITIVITY

Delayed-type hypersensitivity (DTH) derives its name from an immune reaction that develops many hours after contact with a specific antigen. DTH reactions provide the prime mode of defense against many diseases, particularly viral, fungal, and chronic bacterial infections. But, like immediate hypersensitivity, DTH reactions also may be induced by a number of apparently innocuous substances that can result in allergic reactions causing considerable damage to the responding individual.

The Role of DTH in Immunity

Although unaware of the nature of the inflammatory response, German bacteriologist Robert Koch described cellular hypersensitivity in 1891. He observed that if tubercle bacilli (the causative agent of tuberculosis) were injected into the skin of a guinea pig that had previously been infected with tubercle bacilli, an intense area of inflammation would develop in a day or two at the site of the injection. This response did not occur when tubercle bacilli were injected into guinea pigs that had not previously been infected with these organisms. Because it required 24 to 48 h for this inflammation to reach a peak, Koch referred to it as delayed hypersensitivity; later, when other, similar systems were described, these were called delayed-type hypersensitivities.

The puzzling aspect of this reaction was that, unlike Arthus-type sensitivity, it could not be passively transferred using serum from a sensitized animal. The first breakthrough in our understanding of this type of hypersensitivity came in 1942, when Karl Landsteiner and Merril Chase showed that DTH could be passively transferred from a sensitized to a nonsensitized guinea pig using washed, intact, living white blood cells. This observation confirmed that DTH is mediated by cells that function in the absence of an antibody. Some of the chief properties that differentiate DTH from immediate-type hypersensitivity are listed in Table 20.1.

The Tdth cell has the same cell-surface phenotype as does the Th cell, and its activation occurs in a manner identical to that described for the Th cell in Chapter 18. The antigen must be processed by an antigen-presenting cell (APC), such as a macrophage, and be presented to the Tdth cell in association with a Class II molecule on the macrophage cell surface. This interaction between the macrophage and the Tdth cell results in the release of interleukin-1 by the macrophage, which, along with interleukin-2 produced by an antigen-specific Th cell, induces the Tdth cells to proliferate and differentiate into activated Tdth cells and memory Tdth cells. Note that, just as in antibody synthesis, the activation of a Tdth cell requires an interaction with a Th cell.

TABLE 20.1 Differences in Delayed- and Immediate-type Hypersensitivity

Property	Delayed-type	Immediate-type
Mediators	Cells	Antibodies
Time for skin reactions to reach maximum	24–48 h	Minutes to several hours
Type of cells involved	T lymphocytes and macrophages	B lymphocytes
Passively transferred with:	T cells	Soluble antibodies

Soluble Factors Mediating DTH

DTH is an inflammation that is mediated by soluble, biologically active factors released by activated Tdth cells and macrophages. Such lymphocyte-released factors have been given the general name of **lymphokines,** whereas those released by macrophages are termed **monokines.** We shall see that although the specificity of this form of cell-mediated inflammation is due to the reaction of the Tdth cell with its antigen, *it is through a subsequent stimulation by released lymphokines that the macrophage becomes activated to produce the actual inflammatory response.*

Tdth-Produced Factors (Lymphokines)

Lymphokines are soluble factors that stimulate cell proliferation, influence cell motility, and induce killing activities in macrophages. Many different lymphokine activities have been described, but here we shall be concerned only with those that are mediators of DTH.

Migration Inhibition Factor One of the early techniques for the in vitro determination of DTH measured the inhibition of migration of white cells in the presence of a specific sensitizing antigen. This procedure is generally carried out by placing white cells from a sensitized animal into a capillary tube. Under normal conditions, macrophages will migrate out of the open end of the capillary tube when it is placed into a cell culture medium. If, however, the antigen to which the animal had been sensitized is added to the culture, the migration is inhibited, as shown in Figure 20.1. This inhibition is highly specific for the antigen to which the cells have been sensitized. If one obtains macrophages from a nonsensitized animal and mixes them with lymphocytes from a sensitized animal, migration of these macrophages is also inhibited in the presence of the specific antigen to which the lymphocytes are sensitized. Moreover, it requires only about 1 sensitized lymphocyte for each 99 macrophages to demonstrate this inhibition of migration. If the T cells are destroyed using antilymphocytic serum plus complement, the inhibition of macrophage migration is completely abolished.

It is now known that in the presence of antigen, sensitized T cells secrete a lymphokine called **migration inhibition factor** (MIF). This factor acts to keep the macrophage localized in the area of inflammation and, since it is the macrophage that inhibits the multiplication of an infectious agent, it is easy to see that MIF has an important role in our cellular immune system.

Figure 20.1 Effect of ovalbumin and diphtheria toxoid on the migration of peritoneal cells obtained from guinea pigs exhibiting delayed-type hypersensitivity to ovalbumin or diphtheria toxoid. Photographs were taken after 24-h incubation of each cell type with the respective antigen. Migration was inhibited only by the antigen to which the guinea pig exhibited a delayed-type hypersensitivity.

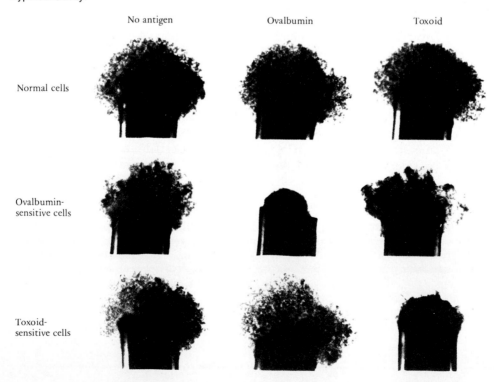

Chemotactic Factor for Macrophages

Observations of cell types found at the sites of delayed-type skin reactions have shown the major type of cell present to be a rapidly dividing monocyte that eventually differentiates into a mature macrophage. The assumption that antigen-induced lymphocytes release a factor that attracts monocytes has been proved by in vitro experiments using chambers separated by a micropore filter. Monocytes were placed in one chamber and lymphocytes plus antigen in the other chamber. Chemotactic activity was evaluated by counting the number of monocytes that migrated through the filter into the lymphocyte chamber. The results clearly demonstrated that a chemotactic factor is released when lymphocytes react with antigen and, furthermore, that the reaction is specific for the antigen to which the lymphocytes are sensitized. A partial characterization of this chemotactic factor for macrophages indicates that it is a protein capable of being separated from MIF on gel electrophoresis.

Macrophage Activating Factor

Acquired immunity to diseases such as tuberculosis and brucellosis has been clearly shown to require activation of macrophages. Observations to support this conclusion have repeatedly demonstrated that macrophages from immunized animals have acquired an enhanced bacteriostatic or bactericidal activity not seen in control macrophages obtained from nonimmune animals. The changes that take place are spoken of as **macrophage activation** and, in general, include (1) increased phagocytic activity, (2) increased metabolic activity, and (3) increased numbers of lysosomes and lysosomal enzymes. (In fact, because of the increased activity of such stimulated macrophages, some investigators have referred to them as "enraged or activated macrophages.")

The molecular events culminating in the enhancement of macrophage function are not known. In vitro experiments have shown that the supernatant solution from a mixture of stimulated lymphocytes and macrophages will mediate this activity. Attempts to purify the macrophage activating factor indicate that it is a glycoprotein similar to the same mediator that inhibits the migration of macrophages (MIF).

Lymphotoxin

We have discussed the antigen-mediated release of lymphokines that influence the behavior of macrophages. Also included in this catalog of substances are soluble mediators that kill target cells. The term **lymphotoxin** (LT) has been used to describe one such mediator.

LT is released by lymphocytes on stimulation by the specific antigen to which they have been sensitized. It appears to kill cells by interacting with the plasma membrane of a variety of normal and malignant cells.

Interferon

The role of interferon in the immune system is discussed in detail in Chapter 17. Here it will only be pointed out that immune interferon (IFNγ) is a lymphokine released by T cells following interaction with their sensitizing antigen. Immune interferon appears to enhance the activity of yet another type of lymphocyte termed **natural killer** cells (NK). The prime target of NK cells is thought to be tumor cells, particularly those that have been virally induced, but the mechanism by which NK cells kill tumor cells is unknown.

A number of additional lymphokines have been described but are not yet well characterized. Most of these have been given names describing their observed effects, such as leukocyte inhibitory factor, mitogenic factor for lymphocytes, T-cell replacing factor, osteoclast activation factor, and colony-stimulating factor. One lymphokine, termed *transfer factor*, has been reported to permit the transfer of DTH from one human to another using soluble extracts. All of these lymphokines, however, will require considerable additional study before we will be able to characterize definitively their biological activities on a molecular basis.

Table 20.2 lists the principal lymphokines involved in DTH, together with their role in the cellular immune response.

Role of DTH in Human Disease

It is difficult to predict accurately just what type of invading organism will preferentially stimulate a DTH response over an antibody-mediated immunity. In general, however, fungal infections routinely induce a

TABLE 20.2 Lymphokines and Their Function

Lymphokine	Function
Migration inhibition factor	To keep macrophages localized at the site of inflammation
Chemotactic factors for macrophages	To attract monocytes to the site of inflammation
Macrophage activating factor	To increase phagocytic activity, metabolic activity, numbers of lysosomes
Transfer factor	Possible factor that can transfer DTH to nonsensitized T cells
Lymphotoxin	To kill target cells
Immune interferon	To stimulate natural killer cells

DTH response against one or more of their structural proteins, and the intradermal injection of such products will elicit a local DTH response in a sensitized individual. Most viruses that reproduce by budding from the host cell membrane also induce a DTH response against the membrane-associated viral antigen. It seems probable that this response is important for the destruction of virus-infected cells before the mature virions can be released from the host cell.

DTH responses are also induced by many different bacterial infections, but the immune function of a DTH reaction is probably of value primarily in infections in which the bacterium grows intracellularly in host cells. Such infections include tuberculosis, leprosy, typhoid fever, brucellosis, tularemia, and listeriosis as well as a few less common diseases.

Measurement of DTH

In vivo measurements of DTH have been used for years as an aid in the diagnosis of certain diseases. Their value is based on the observation that a delayed-type inflammation to many infectious agents can be invoked by either the intradermal injection of a small amount of microbial products (as shown in Table 20.3) or by a test in which these products are applied to a patch that is allowed to remain in contact with the skin for approximately 24 h. One must remember, however, that DTH may last for years after clinical recovery from a disease. A single positive skin test can therefore be interpreted only as evidence of a past or present infection.

In spite of the use of such skin reactions for more than half a century, it was not until techniques for in vitro measurements of DTH became available that a molecular comprehension of these reactions unfolded.

The Role of DTH in Allergic Contact Dermatitis

It is well established that a single antigen may induce both humoral antibody synthesis and a specific DTH.

TABLE 20.3   Some Microbial Products Used to Elicit Delayed-Type Hypersensitivity Reactions in Sensitive Individuals

Product	Disease
Tuberculin	Tuberculosis
Brucellergin	Brucellosis
Lepromin	Leprosy
Histoplasmin	Histoplasmosis
Blastomycin	Blastomycosis
Coccidioidin	Coccidioidomycosis
Killed mumps virus	Mumps

However, in the latter case, it appears that the lower limits of antigen size are considerably larger than that required for the induction of immunoglobulin synthesis.

One apparent contradiction to this is seen in a delayed-type skin response known as allergic contact dermatitis. The allergens in this case generally are low-molecular-weight chemicals such as the catechols of poison ivy, cosmetics, drugs, or antibiotics. Sensitization occurs when these substances combine with cell surface skin proteins to form the antigens necessary to induce DTH. Once an individual is sensitized, subsequent skin contact with the allergen—generating the same hapten-protein conjugate—will induce a delayed-type reaction that reaches maximum intensity in about 24 to 48 h. This sequence of events is depicted schematically in Figure 20.2.

CYTOTOXIC T CELLS

The existence of cytotoxic T lymphocytes (Tc) was first suspected when it was reported that lymphocytes obtained from dogs that had rejected kidney transplants were able to destroy donor kidney cells in vitro. This cytotoxicity was demonstrated in the absence of antibody-forming cells, and it was completely eliminated when such cells were treated with antibody and complement to destroy T cells, confirming that the cytotoxicity resided in the T-cell population.

Most Tc cells are able to recognize their specific antigen on the surface of a target cell only when that antigen is associated with either Class I or Class II MHC-encoded "self" antigens. In fact, different clones of Tc cells are specific for an antigenic determinant on a viral antigen in conjunction with each of the HLA-A, HLA-B, and HLA-C gene products. For example, for an individual who is homozygous at each of these three loci, three types of virus-specific Tc cells may exist: one for virus plus HLA-A, one for virus plus HLA-B, and one for virus plus HLA-C. In an individual who is heterozygous at each of these MHC loci, six types of Tc cells would be theoretically present, one for virus at each of the two alleles at each of the three loci.

Assay for Tc Cells

The end result of Tc-cell action is the lysis of the target cell. It can be quantitated by measuring the release of radioactive chromium-51 (^{51}Cr) from the lysed target cells. In general, target cells are prelabeled with ^{51}Cr in the form of sodium chromate by allowing the labeled chromium to diffuse across the cell membrane

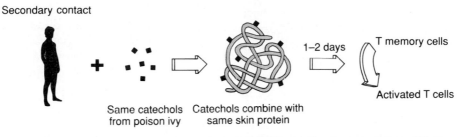

Primary sensitization

Low-molecular-weight catechols from poison ivy

Catechols combine with high-molecular-weight skin protein

7–10 days

T cells

T memory cell

Secondary contact

Same catechols from poison ivy

Catechols combine with same skin protein

1–2 days

T memory cells

Activated T cells

Figure 20.2 Schematic illustration showing the acquisition of allergic contact dermatitis to poison ivy. Dermatitis would not normally occur as a result of a primary sensitization because the inducing Ag would be gone before sufficient T cells became available. On secondary contact, however, T memory cells are rapidly converted to activated T cells.

of the target cells. Once inside, the $^{51}CrO_4^{2-}$ is reduced to $^{51}Cr^{2+}$, which binds tightly to cellular proteins. The radioactively labeled target cells are then incubated with the Tc cells, and the rate of target-cell lysis is measured by determining the amount of radioactivity released into the medium.

Physiological Role of Tc-Cell Response

It is a little difficult to imagine that millions of years ago an immune response evolved with the sole purpose of rejecting organ grafts. It seems more likely that the

MICROBIOLOGY MILESTONES

The ability to isolate monoclonal antibodies possessing an almost limitless repertoire of binding specificities has spawned the new technology of immunotoxin synthesis. Immunotoxins are molecules designed to bind to specific cells and, once bound, deliver their conjugated toxin and kill the cell.

The first step in the synthesis of an immunotoxin is to isolate cells making a monoclonal antibody that has the desired specificity. These may be directed against cancer cells, virally infected cells, or even against subsets of normal cells (such as cytotoxic T cells) to prevent the rejection of an organ transplant. The primary objective is to select a hybridoma that makes monoclonal antibodies that are specific only for the cell type you wish to destroy.

Once this is accomplished, a toxin that will kill the target cell must be conjugated to the monoclonal antibody. Ricin, a toxin produced by beans of the plant *Ricinus communis*, has been widely used in such studies, but other plant toxins such as abrin, as well as diphtheria toxin, have been tried. All such toxins are able to kill cells through their ability to prevent protein synthesis.

The potential value of immunotoxins is their use in cancer therapy but both animal experimentation and human trials have yielded variable results. In general, solid tumors have proven to be more refractory to killing than are circulating cells, probably due to accessibility of the target cells. In one clinical trial involving 46 patients with advanced melanoma, there was one complete remission and three partial responses. Immunotoxins directed against circulating donor T cells were much more effective, in that 12 of 15 patients showed significant reduction in their graft-versus-host disease.

Future research with immunotoxins will be directed along several lines, such as (1) the use of Fab fragments instead of whole antibody molecules to increase accessibility, (2) the use of new linkers to conjugate toxin and antibody to make more effective toxins, and (3) the use of genetically cloned but altered toxin molecules that are not rapidly cleared by the liver. There seems little doubt that we shall hear a great deal more about immunotoxins in the future.

true physiological role of the Tc cell is to destroy self cells that have acquired nonself surface antigens. This is exemplified by the observation that many virus-infected cells or virus-induced tumor cells are lysed by Tc cells. Thus, Tc cells probably evolved to prevent successful viral replication by lysing the virally infected cells early in the replication cycle (after viral antigens were inserted into the host cell membrane), before the intracellular assembly of the completed virions occurred.

It is only recently that some insight has been gained concerning the mechanism whereby Tc causes the lysis of a target cell. The data indicate that when the Tc cell attaches to an appropriate target cell, granules within the Tc cell move to the point of contact between the two cells and fuse with the membrane of the target cell as shown schematically in Figure 20.3. This occurs within 5 to 10 min after contact between the two cells, after which the Tc cell moves away unharmed and the target cell undergoes lysis. Figure 20.4 shows phase contrast micrographs of this series of events.

Figure 20.3 The cytotoxic T lymphocyte (CTL) lytic process. The process leading to lysis of target cells by cytotoxic T lymphocytes begins by specific recognition of the antigen-MHC complex on the target cell by the T-cell receptor. Other molecules on the T-cell surface then promote adhesion of the T cell to the target cell. Soon thereafter, cytoplasmic granules containing a number of active molecules including perforin begin to move from the distal portion of the CTL and accumulate at the interface between the CTL and the target cell. Degranulation at the interface results in release of perforin and other active components. Although not completely understood, it is quite clear that perforin aggregates on the surface of the target cell, forming complexes similar to those seen with antibody and complement. Assembly of this polyperforin then presumably leads to destabilization of the membrane, rapid ion exchange, and lysis. The CTL is not affected by this process and lives to seek another target cell.

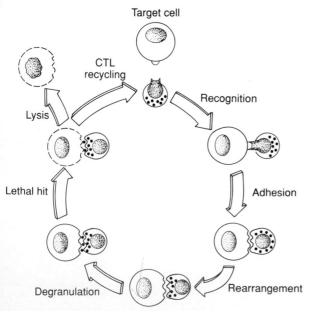

Suppression of Cell-Mediated Response

With the advent of organ transplants has come the necessity to suppress cell-mediated reactions in order to prevent graft rejection. The success of such suppression is greatly enhanced if the donor and recipient cells are immunologically similar, which is most likely to occur among siblings.

Both hormones and chemicals are used to suppress the cell-mediated response. Each immunosuppressive agent has some disadvantages and each significantly reduces the patient's ability to ward off infection by any number of microorganisms. Some of the more commonly used immunosuppressive agents are briefly discussed below.

Steroids

Prednisolone or prednisone is frequently given prior to transplantation, beginning with relatively high doses and gradually tapering off to maintenance doses. Infections, high blood pressure, and peptic ulcers are some of the possible side effects.

Azathioprine

Azathioprine is a purine analog and it prevents lymphocyte proliferation during alloantigen stimulation. Unfortunately, it also prevents proliferation of bone marrow cells, resulting in a very narrow range between effective and toxic doses.

Cyclosporin

Cyclosporin is a cyclic peptide that was originally isolated from several species of fungi and is now chemically synthesized. Cyclosporin appears to be a potent immunosuppressive agent, apparently preventing lymphocyte activation. Thus, the drug seems to be most effective if given before or at the time of organ transplantation.

Antilymphocyte Serum

Antisera to human lymphocytes can be prepared in animals. The major reactivity of such antilymphocyte serum (ALS) is against T cells, but its usefulness is controversial because it does not appear to prolong graft survival beyond that obtained with steroids or other drugs alone.

Monoclonal Anti-T-Cell Antibodies

The general efficacy of this class of reagents is yet to be determined, but a monoclonal antibody directed against an antigen common to all T cells has been used successfully to reverse an acute renal episode.

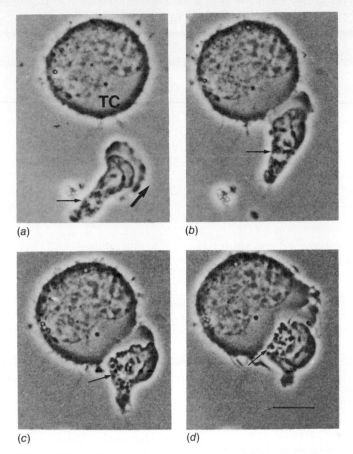

Figure 20.4 Phase contrast micrographs of a Tc approaching and interacting with a target cell (TC), which it can kill. The small arrows indicate Tc granule position at the various times. Prior to contact, the large arrow indicates the direction of Tc movement. (a) The Tc has a granule-filled uropod in the rear. (b) The Tc has made contact with the target cell. (c) Two minutes after contact, the Tc has begun to round up and initiate granule rearrangement. (d) Ten minutes after contact, the Tc granules occupy a position in the zone of Tc-target cell contact.

One might speculate that in the near future a panel of monoclonal reagents might be available that could selectively destroy the various T-cell subsets.

LARGE GRANULAR LYMPHOCYTES

Large granular lymphocytes are comprised of two cell types which have been termed (1) natural killer (NK) cells and (2) killer (K) cells.

Natural Killer Cells

NK cells are lymphocytes that lack most of the cell surface antigens that are characteristic for B cells and T cells. Their lineage is uncertain, but they are derived from bone marrow and their distribution is restricted to lymphoid tissue.

The major property that differentiates NK cells from Tcs is that NK cells are not induced by antigen. Thus, there is no specific recognition of target cells (at least in the same sense as Tc recognition); however, the mechanism of target cell lysis by NK cells appears to be very similar or identical to that described for Tc cells.

The putative function of NK cells is to kill malignant cells that have acquired nonself surface antigens. The mechanism by which such cells are recognized is not understood.

Killer Cells

K cells have an Fc receptor on their surface and therefore will bind and kill any cell that is coated with

TABLE 20.4   Cells Directly Involved in Cell-Mediated Immunity

Cell Type	Target
T lymphocytes	
Tdth cells	Fungal infections
	Virus-infected cells
	Bacteria that normally grow intracellularly
Tc cells	Tumor cells
	Virus-infected cells
	Organ transplants
Large granular lymphocytes	
Natural killer cells	Tumor cells
Killer cells	Foreign cells that are coated with antibodies

antibody. This activity is referred to as **antibody-dependent cell-mediated cytotoxicity (ADCC).** Their mechanism of cytotoxicity is unknown. It is independent of complement but is entirely dependent on the presence of IgG antibody bound to the target cell.

Table 20.4 summarizes cell types directly involved in cell-mediated immunity.

CELLULAR ASPECTS OF TRANSPLANTATION REJECTION

It has been known for many years that only under very restricted circumstances can one successfully transplant skin or organs from one individual to another. When such transplants are made, the foreign tissue is usually killed and rejected within 10 to 14 days after transplantation. If a second transplant from the same donor is applied, it is rejected much sooner than was the first transplant. A second transplant from a different, unrelated donor, however, requires essentially the same time period for rejection as did the first transplant. These observations clearly establish the following facts: (1) rejection is the result of an immunological reaction; (2) a second transplant from the same donor calls forth an anamnestic response, which results in a more rapid rejection; and (3) rejection is directed by specific tissue antigens, which in most cases differ from one individual to another.

Tissue Transplant Terminology

Because graft rejection is the result of an immunological reaction, tissue that is antigenically identical to that of the recipient would not be expected to be rejected. Thus, **autografts,** which are transplants from one area to another in the same person, or **isografts,** which are transplants between genetically identical people (identical twins), would not be rejected. This

is also observed in strains of mice and guinea pigs that have been inbred so that all offspring are genetically similar.

Allografts (older terminology: *homografts*) are transplants between genetically nonidentical animals of the same species—that is, from human to human or mouse to mouse. **Xenografts** are grafts that cross species lines. Because the intensity of a rejection is a function of the degree of antigenic difference between donor and recipient, one might correctly expect that xenografts are usually vigorously rejected, whereas the reaction of an allograft depends on the relatedness between the donor and the recipient. This terminology is summarized in Table 20.5.

Cell-Mediated versus Antibody Rejection

Allograft destruction is characterized by the infiltration of lymphocytes into the foreign tissue. In fact, the transfer of lymphocytes from an animal that has rejected an allograft will confer on the recipient the ability to respond with an accelerated rejection of tissue from the same allograft donor, indicating that immunological memory resides in the lymphocytes. Moreover, if one removes the thymus from a newborn

TABLE 20.5   Types and Examples of Tissue Grafts

Type	Example
Autografts	From one area of an individual's body to another
Isografts	Between genetically identical animals, such as identical twins or inbred strains of animals
Allografts	Between members of the same species, such as human to human
Xenografts	From one species to another, such as monkey to human

mouse, it will usually be unable to reject allografts, but it will acquire this ability if given lymphocytes from a genetically identical animal. It therefore seems to be established that allograft rejection is a function of cell-mediated immunity. In fact, the passive transfer of serum containing antibodies to the foreign tissue does not normally result in an accelerated allograft rejection.

Antigens Responsible for Allograft Rejection

There are a number of gene loci that code for cellular antigens involved in allograft rejection. For example, about 40 such genes have been detected in the mouse, and it would appear likely that humans are no less complex. In all vertebrate species so far examined, however, the antigens evoking the strongest response have been shown to be encoded in a single chromosomal segment termed the **major histocompatibility complex (MHC)**. The MHC in the human has been given the specific name of the **HLA complex**, derived from the term *human leukocyte antigens.*

Scientists studying the HLA complex used human sera collected from thousands of individuals and tested for antibodies capable of agglutinating or killing donor lymphocytes in the presence of complement. Consequently, a large number of human cell antigens have been described, and sera directed against such antigens are available for tissue typing. Such typing has become important in assessing the potential for successful organ transplants.

TUMOR ANTIGENS

As might be expected, when a normal cell undergoes transformation to a malignant cancer cell, it also acquires new antigens on its cell surface. A large number of viruses are now known that, under appropriate conditions, will produce cancer in animals. Furthermore, the tumor antigens occurring on these virus-induced malignancies are constant for any one virus. In other words, polyomavirus can induce bone cancer, liver cancer, or any other organ cancer in the mouse, but the tumor antigens occurring on the cancer cells of any polyoma-induced cancer are always the same. Thus, we can conclude that in virus-induced malignancies, the tumor antigen is coded for by the virus. Table 20.6 lists some of the important viruses known to induce cancer in animals.

We might wonder just what the importance to us of these tumor antigens is. The answer lies in the fact that our bodies normally respond to the presence of a tumor cell exactly as they do to newly transplanted

TABLE 20.6   Viruses Producing Tumors in Animals

Virus	Animal
Polyomavirus	Mouse
SV40 virus	Baby hamster
Herpes virus	Chickens, monkeys, frogs, rabbits, and hamsters
Adenovirus 12 and 18	Baby hamster
Shope papilloma	Rabbit
Mammary-tumor virus	Mouse
Leukemia (many different viruses)	Many different animals (including humans)
Rous sarcoma	Chickens

tissue. A cellular immune response is produced that results in the selection of lymphocytes specifically sensitized to the foreign antigens on the tumor cell. This, in turn, results in the destruction of the foreign tumor cell. It appears probable that many of us may develop cancer cells during our lifetimes that are rapidly destroyed by our cellular immune systems. This system for the destruction of malignant cells has been termed **immune surveillance.** You can also understand now why people who have received organ transplants and, as a result, have had their cellular immune system suppressed have a higher than normal incidence of some types of cancer.

IMMUNOLOGICAL TOLERANCE

Immunological tolerance is most simply defined as the lack of a specific immune response resulting from a previous exposure to the inducing antigen. The most notable example is immunological tolerance to self. Recall that this recognition of self versus nonself was proposed by Burnet to occur during fetal life by a mechanism in which cells capable of responding to self were destroyed. This concept has been termed **clonal deletion,** and it involves the specific removal or inactivation of B or T cells following contact with antigen at the right time, in the correct dosage, and by the correct route. Such a mechanism could account for the continuous elimination of clones directed toward self that constantly arise from stem cells throughout life.

A second mechanism of immunological tolerance has been shown to be the effect of the presence of **suppressor T (Ts) cells,** which prevent both an antibody and a cellular response to specific antigens such as self. The fact that such Ts cells exist for self antigens can be demonstrated in the following manner. If the thymus (a source of Ts cells) is removed from lethally irradiated mice and such mice are then given a transplant of bone marrow cells and T cells in which the Ts cells have been specifically destroyed, these mice will produce autoantibodies to their own red blood cells. Moreover, if T lymphocytes from a mouse made tolerant to sheep's red blood cells are transferred to a second mouse, that mouse becomes incapable of forming antibodies to sheep's red blood cells. Additional evidence for the role of Ts cells in tolerance to self can be seen in certain strains of mice that routinely develop autoimmune diseases in adult life. Such diseases can be prevented or delayed if these mice are given repeated injections of thymocytes (containing precursors of Ts cells) from young genetically similar mice.

It therefore appears very likely that at least part of our inability to recognize self results from the presence of specific Ts cells that prevent both humoral and cellular responses to self antigens. As one might surmise, any defective production of Ts cells, or the intentional destruction of such cells to prevent organ transplant rejection, may result in the occurrence of an autoimmune disease.

SUMMARY

Each individual possesses two separate but interrelated immune systems, one humoral and one cellular. The cell-mediated immunity is more frequently directed against antigens occurring on the surfaces of cells such as fungi, bacteria, virus-infected cells, foreign cells, and tumor cells. Such cell-mediated immunity is mediated by two different subsets of cells: (1) Tdth cells, which, on interaction with foreign antigens presented on the surface of macrophages, secrete a number of biologically active factors (called lymphokines) that stimulate other types of immune cells, particularly macrophages; and (2) Tc cells, which specifically kill target cells bearing nonself antigens such as tumor cells, virus-infected cells, or foreign cells.

Delayed-type hypersensitivity can be measured by the injection of microbial products into the skin of a sensitized person or by the in vitro demonstration of a lymphokine termed *migration inhibition factor*. Other lymphokines secreted by Tdth-activated cells include chemotactic factors for macrophages, macrophage activating factor, lymphotoxin, and interferon.

Cytotoxic T cells function by causing the lysis of cells bearing nonself antigens. Thus, they are primarily responsible for the rejection of organ transplants and for the destruction of virally infected cells. It is also thought that they may carry out an immune surveillance, influencing the destruction of any emerging tumor cells, since such tumor cells bear antigens that may be recognized as foreign by the Tc cells.

Natural killer cells lyse target cells independent of antigen recognition, whereas killer cells are effective only after binding to the Fc portion of antibody-coated target cells.

Tissue transplants are termed autografts, isografts, allographs, or xenografts. The chief antigens that induce a cell-mediated response resulting in rejection are encoded in a single chromosomal segment termed the major histocompatibility complex; using specific antibodies directed against human leukocyte antigens, it is possible to carry out tissue typing to assess the potential for successful organ transplants.

Immunological tolerance is defined as the lack of a specific immune response due to a previous exposure to the inducing antigen. The most notable example is immunological tolerance to self. It appears to occur by one of two mechanisms: (1) clonal deletion, in which immature B or T cells are destroyed following contact with an antigen; and (2) the formation of Ts cells, which prevent both an antibody and a cellular response to antigens such as self. Defects in the ability to make such Ts cells are believed to provoke various types of autoimmune diseases.

QUESTIONS FOR REVIEW

1. How are the antigens that induce cellular immunity different from those that induce a humoral antibody response?
2. What type of cell is responsible for DTH? What is the origin of this cell?
3. How can DTH be measured in vivo? In vitro?
4. What are lymphokines?
5. Explain two mechanisms whereby a sensitized lymphocyte can result in the destruction of an invading bacterium.
6. How can cellular immunity be passively transferred from one individual to another?
7. Name and define four types of tissue transplants.
8. List several techniques by which the cellular immune response can be suppressed.
9. What problems are involved in the long-term use of immunosuppressive agents?
10. What are tumor antigens?
11. Why are the recipients of organ transplants more susceptible to cancer than they would be normally?
12. Give two mechanisms to explain immunological tolerance.

SUPPLEMENTARY READING

DiNome MA, Young JD: How lymphocytes kill tumor and other cellular targets. *Hosp Pract* 22:59, 1987.

Henkart PA: Mechanisms of lymphocyte-mediated cytotoxicity. *Annu Rev Immunol* 3:31, 1985.

MacDonald HG: Differentiation of cytolytic T lymphocytes. *Immunol Today* 3:183, 1983.

Marx JL: How killer cells kill their targets. *Science* 231:1367, 1986.

Moller G (ed): Mechanism of action of cytotoxic T-cells. *Immunological Reviews*, vol. 72. Copenhagen, Munksgaard, 1983.

Nabholz M: Cytotoxic T lymphocytes. *Annu Rev Immunol* 1:272, 1983.

Rose NR: Autoimmune diseases. *Sci Am* 244(2):80, 1981.

Weiss S, Dennert G: T-cell lines active in the delayed-type hypersensitivity reaction (DTH). *J Immunol* 126:2031, 1981.

Chapter 21

Antisera and Vaccines

OBJECTIVES

A study of this chapter should familiarize you with

1. The general types of vaccines presently in use.

2. How recombinant DNA techniques are being exploited for the preparation of safer vaccines.

3. The expected duration of immunity from the administration of killed versus living attenuated vaccines.

4. Some of the hazards associated with the use of the current pertussis and polio vaccines.

5. What situations would require the use of specific antiserum rather than a vaccine for disease prevention.

6. The source of specific antisera and the potential hazards associated with its employment.

7. The duration of passively acquired immunity as compared to actively acquired immunity.

Throughout history we have used many devices to protect ourselves from disease. Probably the first major attempt to induce immunity was a process called variolation. Practiced in China and India in ancient times, variolation involved inoculating susceptible individuals with dried scabs obtained from mild cases of smallpox. The hope was that the inoculated individuals would have a mild disease and, as a result, develop immunity to smallpox. Although it appears that this crude technique killed 2 to 3 percent of the patients, the death rate in epidemic smallpox was 20 to 30 percent.

We have come a long way in our knowledge and ability to apply the principles of immunology toward the elimination of disease. The principles underlying the immune response were presented in the earlier chapters of this unit. In this chapter, we shall discuss the methods by which we attempt to apply these principles to the prevention and treatment of infectious disease in humans and domestic animals.

VACCINES

The broad area in which humans require protection ranges from the toxic products of bacteria (tetanus, diphtheria) through the infections produced by protozoa, fungi, bacteria, rickettsiae, and viruses. An ideal vaccine would be one that would (1) stimulate lifelong immunity, (2) be completely safe to use, (3) require only one administration, (4) be relatively easy to produce, and (5) be stable under a variety of storage conditions. Many vaccines in use today do not possess all of these characteristics.

There are no effective vaccines for many infectious diseases. When vaccines are used, they are of different types. These types are discussed below.

Types of Vaccines

Vaccines may be grouped according to the type of antigen used and the method of preparation. In general, **vaccines** consist of killed or inactivated pathogenic organisms or viruses, living avirulent (attenuated) organisms or viruses, inactivated exotoxins (toxoids), or cloned fragments of antigens from pathogenic organisms.

On the basis of specific content, vaccines may be classified as described below.

Bacterial Vaccines

Bacterial vaccines contain bacterial cells that are either avirulent (attenuated) or killed. Inactivated bacterial vaccines in use today are used to stimulate immunity

---

MICROBIOLOGY MILESTONES

It was not many years ago when infectious diseases were the number one cause of death in the United States (as well as the rest of the world). Within the lifetime of many persons living today, childhood diseases such as measles, mumps, rubella, and whooping cough were almost inevitable. Other serious diseases that were commonplace during the first half of the 20th century included typhoid fever, scarlet fever, and polio. And, if we go back to the 10th century, we would have to add smallpox and diphtheria to our list.

All of these diseases, except smallpox, are still around and many still exact a high mortality and morbidity in Third World countries. In the Western World, however, essentially all have been drastically reduced, and mortality from infectious diseases now lags far behind heart disease and cancer.

How were these diseases curtailed? Some, such as scarlet fever, owe their lower incidence to antibiotics. Others, such as typhoid fever, have become moderately rare because of modern-day sewage disposal and water purification techniques. Control of all the other diseases listed is through immunization with vaccines consisting of killed agents, living attenuated agents, or inactivated toxins. Probably the crowning achievement of vaccines has been to eliminate smallpox, with the result that smallpox vaccine is no longer used.

Does this mean that vaccines are always good? Good for society—definitely yes! Good for the individual—sometimes no. The oral polio vaccine and the pertussis vaccine provide examples of vaccines that are definitely good for society but in extremely rare situations may not be good for the individual.

During the past 15 years there have been 225 cases of paralytic polio in the United States, most of which resulted from mutations in the live vaccine. This amounts to 15 cases a year as compared to 4000 to 5000 such cases before use of the vaccine.

Pertussis vaccine is very effective for the prevention of whooping cough, and before it was available many thousands of young children died from this disease. The vaccine, however, is not without side effects. These include a temperature of 40°C (1 out of 330); continuous crying for more than 3 hours (1 out of 100); convulsions or episodes of limpness and paleness (1 out of 1750); and permanent brain damage (1 out of 310,000).

Should these vaccines be discontinued? This has been tried in some countries with the pertussis vaccine, resulting in considerably more deaths than would be expected from the vaccine. The best solution, of course, will be to develop vaccines that are both safe and effective.

---

against many of the gram-negative organisms. This includes such diseases as pertussis, typhoid and paratyphoid fever, and plague. A vaccine designed to protect against the more common types of *Streptococcus pneumoniae* is now available; it contains capsular material from the virulent organisms. A similar type of vaccine is used for protection against Group A and C meningococci (see Unit III).

Living attenuated bacteria are used to stimulate immunity to such diseases as tuberculosis and, to a very limited extent, tularemia. Attenuated anthrax and brucella organisms have been used in farm animals.

Toxins and Toxoids

Toxins are poisonous metabolic products; therefore, they cannot be administered safely unless they are treated to reduce their harmful effects. This is accomplished by treating the toxin with formaldehyde, which destroys its toxicity without changing its antigenicity. This inactivated toxin is called a **toxoid**. Toxoids are frequently prepared by precipitating the toxoid on alum (aluminum sulfate). Because this results in an insoluble antigen, it takes longer for the body to destroy and eliminate it. As a result, the antigen remains effective for a longer period and is thought to produce a higher level of antibody response in the individual than the toxoid alone.

Toxoids are used for active immunization against tetanus and diphtheria.

Rickettsial Vaccines

Rickettsial vaccines contain the rickettsial organisms as the antigenic components. The rickettsiae are usually grown in the yolk sacs of chick embryos or in cell cultures, and killed by treatment with formaldehyde. Vaccines prepared in cell cultures are highly effective, and because they are free of any egg yolk contaminants they can be used safely in individuals who are allergic to eggs. Although rickettsial vaccines are not used widely, they do offer protection to people who are more likely to be infected by way of tick bites. Vaccines have been used to protect against Rocky Mountain spotted fever, typhus, and certain other rickettsial diseases.

Viral Vaccines

Viral vaccines may be composed of inactivated or active attenuated virions that will induce an immune response against the virulent organism. One major inactivated viral vaccine used for humans is for the prevention of influenza. In the past an inactivated vaccine was used for both poliomyelitis and measles, but these have now been largely supplanted by the use of active attenuated viral vaccines, since active vaccines may induce a much longer-lasting immunity, even lifelong—as does the actual disease. As a result, virologists interested in viral immunity are always searching for viral mutants that will induce immunity without causing disease. Such active attenuated virus vaccines are now used for immunization against all three types of poliomyelitis, red measles (rubeola), German measles (rubella), yellow fever, and mumps. It was once postulated that a similar vaccine could be developed for the common cold, but owing to the multiplicity of viral agents responsible for this malady, it seems unlikely that a viral vaccine will be beneficial. Active attenuated rabies virus has been used for years to immunize dogs, but it is not used in humans. The rabies vaccine used in humans contains inactivated rabies virus that has been grown in cell culture.

One new, promising method for the production of vaccines makes use of recombinant DNA techniques to produce viral components that can serve as antigens to induce immunity to a viral disease. This is accomplished by cloning part of the viral DNA in *Escherichia coli,* which then makes the DNA-encoded protein. Such preparations have been designated as split or subunit vaccines because the cloned segment encodes only for one or two viral antigens. The putative advantage of such vaccines is that they will induce a protective immunity, but they lack the toxic components that may be present in the whole virion preparation. Subunit vaccines that contain specific viral antigens have been cloned in both bacteria and yeast. Some of those being currently evaluated in animal studies include a surface antigen for several influenza strains (the hemagglutinin), and surface antigens from rabies virus, herpesvirus, and foot-and-mouth disease. The surface antigen from hepatitis B (HB_sAg) also has been cloned in both bacteria and yeast cells, and in 1986 the U.S. Food and Drug Administration approved the recombinant yeast product to be used to immunize humans against hepatitis B. Undoubtedly, approval for use of other recombinant vaccines will follow this decision.

Another innovative recombinant type of vaccine has been produced by the incorporation of DNA fragments into vaccinia virus. Vaccinia virus is an avirulent poxvirus that has been used in the past to immunize individuals against smallpox. Using recombinant DNA technology it is possible to insert 22 to 25 kilobases of foreign DNA into vaccinia virus without diminishing its ability to replicate. Various hybrid vaccinia virus strains have been made that encode for the surface antigens for hepatitis B virus, influenza viruses, and herpesviruses. Animals infected with such vaccinia virus strains form protective antibodies against the specific antigens whose DNA has been cloned into the vaccinia virus genome. It seems very probable that these vaccines will soon be widely used in the human population.

Other novel vaccines produced include a recombinant vaccinia virus that contains a gene for the immunogenic glycoprotein of rabies virus. This recombinant virus expresses the rabies glycoprotein in its viral envelope, and immunization of experimental animals led to a complete protection against intracerebral injection of native rabies virus.

It is also interesting to note that there are still a lot of cases of polio occurring in Third World countries because vaccines are not readily available there. One helpful solution was reported by Karen Burke and her colleagues in England. Using recombinant DNA technology, they succeeded in constructing a single hybrid poliovirus that contained the epitopes of the three serotypes of poliovirus. Experimental animals immunized with this hybrid have been shown to produce antibodies reactive with all three serotypes. Similar methods may be used to produce improved vaccines against a number of picornaviruses, including hepatitis A.

More recently, the gene encoding p120, the surface glycoprotein of human immunodeficiency virus (the cause of the acquired immunodeficiency syndrome) has been cloned and the expressed protein is currently undergoing clinical trials for use as a vaccine.

Table 21.1 summarizes the types of vaccines in use, along with selected examples of diseases controlled by each type of vaccine.

Duration of Immunity

The best we can hope for from the administration of a vaccine is that it will stimulate active immunity that will be as effective and as lasting as the immunity acquired by recovery from the natural disease. After recovering from a staphylococcal infection, one does not apparently get a lasting immunity; we would not, therefore, expect a vaccine to stimulate lifelong protection.

What kind of immunity can we expect following the administration of a vaccine? There is no single answer to that question because of the wide variation in both the effectiveness and the duration of immunity provided by different vaccines. In general, antibody will reach measurable levels within 7 to 10 days after administration of the vaccine. This level may continue to rise for a week or two and then level off, maintaining a certain antibody titer. This antibody level may be maintained for 6 months or a year and then usually will start to decline. In many cases it will reach very low levels within 1 to 2 years. However, if a second injection of the same vaccine (a booster dose) is then administered, antibody production starts very rapidly and may reach a high level within 2 to 3 days.

If one correlates the duration of immunity with the type of vaccine used, the live or active viral vaccine for the most part provides a longer immunity than does an inactivated vaccine. One may postulate several reasons for this: (1) the living organism may stimulate cellular immunity, which is not stimulated by the inactivated vaccine; (2) the living organisms may grow in the individual and thus provide a much greater antigenic mass than would an injection of killed organisms; (3) the living organisms may be able to live and grow in the vaccinated individual and thus continue to stimulate antibody formation over a long period.

With this in mind, it might seem logical always to use live attenuated organisms in place of killed vaccines. Unfortunately, live avirulent strains of virulent organisms are not always available. Also, if one considers the source of the active avirulent vaccine, one finds that they have been selected from a virulent strain of the disease organism. For example, the yellow

TABLE 21.1   Types of Vaccines Commonly Used to Stimulate an Active Immune Response

Vaccine	Selected Examples
Bacterial	
Killed whole cells	Typhoid fever, pertussis, rickettsial diseases
Living attenuated cells	Brucellosis,[a] anthrax,[a] tuberculosis
Capsular material	Pneumococcal pneumonia, Haemophilus influenzae, Neisseria meningitidis
Inactivated toxins	Diphtheria, tetanus
Viral	
Killed whole virions	Influenza, adenoviruses, polioviruses, rabies
Living attenuated virions	Measles, rubella, polioviruses, mumps, yellow fever
Split virion components[b]	Influenza, rabies, herpesviruses, hepatitis B, foot-and-mouth disease

[a] Not used in humans.
[b] May not yet be commercially available.

fever vaccine is an avirulent mutant of virulent yellow fever virus. In like manner, the active poliomyelitis vaccine came originally from virulent poliomyelitis strains. The same is true for the vaccines used for tuberculosis, tularemia, anthrax, and measles. Thus, there remains the possibility that these organisms may revert to the virulent form and cause serious disease.

Practical Considerations in the Use of Vaccines

In addition to using aseptic procedures in the administration of vaccines, certain other precautions must be taken to ensure their effectiveness. The proper storage temperature must be maintained and the expiration date checked, particularly for live attenuated vaccines, because storage at room temperature may result in inactivation of the organism. The lot number is important in the event of undesirable reactions. This makes possible the recall of the entire lot if necessary.

Some vaccines may contain products that produce severe allergic reactions in the patient. For example, organisms grown in embryonated eggs (rickettsiae and influenza virus) contain trace amounts of egg material, which may cause a severe reaction in an individual allergic to eggs. In like manner, one must remember that most viral vaccines, whether active or inactivated, were grown in the presence of small amounts of penicillin. Thus, an individual who is very sensitive to penicillin might show a serious allergic reaction following the administration of a viral vaccine.

One major concern of public health officials is the difficulty of getting parents to have their children immunized against a disease that, for all practical purposes, seems to have disappeared. The principal diseases in this category include measles, poliomyelitis, pertussis, and diphtheria. If, as appears possible, barely enough people take the vaccines to retard the spread of diseases such as measles and poliomyelitis, the result would be that a very large number of people, even though not immunized, will reach older ages without acquiring immunity to these diseases. Thus, the stage might be set for a serious epidemic among these people at an age when the disease is considerably more serious than it would have been in childhood. The major hope for reversing this trend of indifference seems to be extensive publicity coupled with increased medical care of poor people in the large cities.

Table 21.2 lists the normal recommended immunization schedule for persons living in the United States. It has been suggested that such immunizations could begin at one month of age if the infant resides in an area where these diseases are endemic.

HAZARDS OF IMMUNIZATION

Although the application of immunology through the use of vaccines has virtually eliminated many serious diseases from the Western world, their use has not

TABLE 21.2   Recommended Schedule for Active Immunization of Normal Infants and Children

Recommended Age	Vaccines	Comments
2 mo	DTP#1[a], OPV#1[b]	OPV and DTP can be given earlier in areas of high endemicity
4 mo	DTP#2, OPV#2	6-wk to 2-mo interval desired between OPV doses
6 mo	DTP#3	An additional dose of OPV at this time is optional in areas with a high risk of poliovirus exposure
15 mo	MMR[c], DTP#4 OPV#3	Completion of primary series of DTP and OPV
18 mo	HbOC[d]	Conjugate preferred over polysaccharide vaccine
4–6 yr	DTP#5, OPV#4	At or before school entry
14–16 yr	Td[e]	Repeat every 10 yr throughout life

[a] Diphtheria, tetanus, and pertussis.
[b] Oral polio vaccine.
[c] Measles, mumps, and rubella virus vaccine.
[d] *Haemophilus influenzae* b polysaccharide antigen conjugated to a protein carrier.
[e] Tetanus and diphtheria toxoids: contains the same amount of tetanus toxoid as DTP or DT but a reduced dose of diphtheria toxoid.

been without hazard to the recipients. It is certainly not within the scope of this book to list the occasions when immunization has resulted in death as a sequel to contamination of the vaccine with virulent or live organisms. Suffice it to say that most of these situations occurred during the early use of diphtheria immunizations. The prevention of tragedies of this type rests solely in the continued and careful supervision of all vaccine production.

The most serious aspects of artificial immunizations are those that are associated with the toxic properties of the vaccine itself, as shown by the pertussis vaccine or by the rare reversion of an attenuated vaccine, as occurs in the polio vaccine.

The pertussis (whooping cough) vaccine used in the United States consists of killed phase-1 organisms that are usually incorporated with diphtheria and tetanus toxoids (commonly called DTP). This vaccine contains pertussis toxin, adenylate cyclase, and filamentous hemagglutinin, as well as structural components of the cell such as endotoxin. Unfortunately, it causes a local reaction in many infants and, on rare occasions, can cause serious systemic toxic reactions, including brain damage and death. For example, a government report indicates that over the past 10 years, 5 to 20 children die in the United States annually following DTP immunizations, and approximately 50 suffer permanent brain damage. These figures are statistically very small when compared with the many thousands of infant deaths that occurred before the advent of pertussis immunization, but they have been the cause of considerable concern for parents, vaccine producers, and legislators.

In 1984, two of the three vaccine manufacturers in the United States announced that, because of legal liabilities, they were ceasing the production of the pertussis vaccine. There has also been substantial pressure for the U.S. government to underwrite all vaccines and thus accept all liabilities associated with their use.

On the other side of the argument, however, are the epidemics of pertussis that ensued in Japan and England when they stopped using DTP because of fears of severe reactions. England experienced outbreaks of pertussis between 1977 and 1979 that resulted in 36 deaths, giving rise to a mortality rate significantly higher than when the vaccine was in use.

The best solution, of course, is to produce a less toxic vaccine that would be tolerated by everyone. Some such experimental vaccines are now available in Japan and their utilization is discussed in Chapter 24.

The use of an active attenuated viral vaccine for the prevention of polio is also not without some risk. Although there is no dispute that the use of this vaccine has prevented many thousands of cases of fatal and paralytic polio, there has been a very small number of cases of vaccine-associated poliomyelitis. During the period 1972 to 1983, 278.8 million doses of the oral vaccine were distributed, and 87 vaccine-associated cases of polio were reported. It must be kept in mind, however, that the viruses causing polio are still widespread, and failure to administer the vaccine could result in the return of large numbers of cases of paralytic polio.

ANTISERUM

The first use of antiserum to treat disease occurred in the 1890s, when serum from immunized animals was used in treating diphtheria. The observation that animals could be immunized with either bacteria or their toxic products and their immune serum used successfully against disease opened up a new era of medicine. During the next 50 years, immune serum was the only therapy for diphtheria, tetanus, gas gangrene, botulism, pneumococcal pneumonia, meningitis, and some other infections. The major disadvantage of antiserum treatment was its specificity. This presented no problem in the case of diphtheria or tetanus, since there is only one antigenic type of diphtheria toxin and only one type of tetanus toxin. However, there are many different antigenic types of pneumococci that cause pneumonia, and several antigenic types of meningococci that cause meningitis. In these cases it was necessary for the physician to ascertain which antigenic type was the causative organism before the correct antiserum could be administered. Today, the status of antiserum treatment has changed as a result of the discovery of antibiotics. The ease of antibiotic therapy has eliminated the time-consuming procedures of isolation, identification, and serological typing.

The administration of antiserum still maintains an important though limited place in modern medicine. The speed with which effective levels of antibody can be given often saves a life. For the most part, present-day uses of antiserum are restricted to the temporary prevention of certain viral infections and for the neutralization of certain potent bacterial toxins.

Source

An antiserum is a serum that contains antibodies against an antigen such as a toxin, a virus, or a bacterial cell. Although many different animals have been used for this purpose, the horse, because of its size and availability, has long been the animal of choice. Today antiserum is obtained by one of several general techniques. In one case, the antigen (toxin, virus, foreign lymphocytes) is inoculated into an animal or human, which, in turn, forms antibodies in response to the foreign material. The animal is bled, and the separated serum is used as the antiserum.

In a second technique, human serum is pooled and chemically fractionated to obtain the gamma globulin fraction, which contains the antibodies. Remember that pooled human gamma globulin is effective only against the diseases from which the majority of the population have recovered and thus would be ineffective in the prevention of such diseases as rabies. In this case, the rabies virus must be grown in cell cultures and then inoculated into an animal from which the antiserum is later obtained.

Because of the possibility of serum sickness or other allergic reaction to foreign serum, it is becoming more common to immunize human volunteers to a variety of toxins (and a few viruses) and to use this human serum as a source of antiserum for human injection.

Current Therapeutic Uses

As has been pointed out, the major advantage of antiserum is that it provides the patient with circulating antibodies that are effective immediately after its injection. It is used for both the prevention and treatment of certain diseases when one cannot wait for the individual to form his or her own antibodies.

Probably the most frequent use of antiserum today is for the prevention of disease. People who have been exposed to certain diseases are given passively transferred antibodies. Before the availability of an effective measles vaccine, a small amount of pooled human gamma globulin was given to an individual who had been exposed to measles. The injection of a limited amount of antiserum greatly lessened the severity of the disease. It did not prevent the individual from having a mild case of measles, and, as a result, allowed a buildup of a permanent immunity by the patient's own immune response. Today, a nonimmune person who had been exposed to measles would probably be given sufficient gamma globulin to prevent the disease entirely, and then, as the passively transferred antibodies disappeared, would be actively immunized using the live measles vaccine. A similar situation would exist in nonimmunized people who had been exposed to poliomyelitis. Hepatitis A is another disease for which pooled human gamma globulin has been successfully used to prevent active disease in exposed people. However, in this case there is not yet available an effective vaccine with which one can induce active immunity.

Because most of the bacterial infections in which antiserum would be of value are now treated with antibiotics, its use is limited to bacterial infections in which the illness is a result of an exotoxin. Therapy is aimed at neutralizing the toxin as it is produced by the invading organism. Tetanus and diphtheria are the major bacterial diseases in which specific antitoxin is used. For example, an individual who has a deep wound and has not been previously immunized against tetanus must be protected by the administration of specific tetanus antitoxin.

Gas gangrene and botulism are also diseases in which the symptoms are the result of exotoxins and for which specific antiserum is used. A more recent result of our medical and technological development is the use of antilymphocyte serum in individuals who have received organ transplants.

Practical Considerations of Antiserum Therapy

Adverse Reactions

For many years the horse and the rabbit have been the animals of choice for the production of protective sera. However, the use of animal serum has created several problems. As was discussed in Chapter 19, many people will form antibodies against the proteins in the animal serum and, as a result, develop serum sickness. This reaction has been partially eliminated by administering antiserum made in cows or humans to those who are allergic to horse or rabbit serum.

Frequently, a skin test is performed to determine whether a patient is allergic to a particular antiserum. A small amount of the antiserum is injected into the skin. If the individual is allergic, inflammation at the site of injection will appear, usually within 10 to 15 min. However, even this can be dangerous, because an exquisitely sensitive individual may have a sudden, severe allergic reaction following the injection of even a minute amount of antiserum.

Hepatitis A and hepatitis B are two infections in which the virus is present in the blood during a long, asymptomatic incubation period; in the case of hepatitis B, it may remain in the blood for months or years after the patient has recovered from the overt disease. Because these diseases are fairly common, any large pool of human serum is likely to be contaminated with one or the other. This problem has been greatly reduced by routinely testing individual donor serum for antigens to hepatitis B virus or by purifying the gamma globulin from pooled human sera, since these viruses are eliminated in the fractionation procedure.

The virus causing acquired immunodeficiency syndrome (AIDS) also has become a major problem in the use of human serum and human blood products, and all donor sera are now routinely assayed for human immunodeficiency virus by looking for the presence of antibodies to the AIDS virus. However, in at least one reported case, the virus was passed by a donor who had only recently been infected and had not yet developed measurable antibodies to the AIDS virus.

Duration

Antiserum therapy provides temporary immunity in an adult at effective levels for only about 4 to 6 weeks.

Thus, active immunization for such diseases as tetanus, diphtheria, and poliomyelitis should be carried out simultaneously with antiserum therapy, so that by the time the passively transferred antibodies are gone, the individual's immune system will be forming antibodies.

SUMMARY

Vaccines may be made up of inactivated bacterial toxins (toxoids), killed bacteria or viruses, or living attenuated bacteria or viruses. They are used to induce artificial active immunity toward specific diseases. Passive immunity as obtained by the injection of antiserum provides only 4 to 6 weeks of effective protection in the adult. Active immunization, however, may result in lifelong immunity, particularly in cases in which live attenuated viral vaccines are used. One of the major concerns of public health officials is to get parents to have their children immunized to a disease that no longer seems threatening. The present apathy could produce an adult population that is largely nonimmune to such diseases as measles, poliomyelitis, and diphtheria. Although artificial immunizations have virtually eliminated many serious diseases, there are certain hazards involved in the use of vaccines. The most important of these result from an abnormal reaction by the recipient to the vaccine or the reversion of an attenuated vaccine to a virulent strain.

Antisera obtained from immunized animals or humans or from gamma globulin from pooled human serum contain preformed antibodies. These preformed antibodies are used to provide artificial passive immunity in order to prevent or modify disease. Some of the more common usages of antiserum are for the prevention of tetanus following a deep wound, the prevention of hepatitis A following exposure, and treatment of botulism or gas gangrene.

QUESTIONS FOR REVIEW

1. Name four general types of vaccines. Describe briefly how each is made.
2. In general, what can we say about the duration of effectiveness of vaccines? How is the duration affected by the use of live versus killed vaccines?
3. Where do the attenuated organisms used as vaccines come from?
4. Even though we have effective vaccines for measles and poliomyelitis, public health officials are concerned that in the not too distant future we may have a very large adult population that is not immune to these diseases, thus setting the stage for some major, severe epidemics. What is the reason for this, and how would you prevent this from occurring?
5. What are some of the hazards of the use of vaccines?
6. Why is antiserum therapy used much less frequently today than it was 30 years ago?
7. What kinds of antisera are used and from where do they come?
8. What are the limitations of human gamma globulin? Name two diseases in which you would expect it to be effective. Name two in which it would be expected to be ineffective.
9. What is serum sickness?
10. Would you expect to find cases of serum sickness following the use of human gamma globulin? Why?
11. What is the duration of effectiveness of antiserum therapy?

SUPPLEMENTARY READING

Brown F: Synthetic viral vaccines. *Annu Rev Microbiol* 38:221, 1984.

Kolata G: Litigation causes huge price increase in childhood vaccines. *Science* 232:1339, 1986.

Lerner RA: Synthetic vaccines. *Sci Am* 248(2):66, 1983.

MacDonald GA, et al: Cloned gene of *Rickettsia rickettsiae* surface antigen: Candidate vaccine for Rocky Mountain spotted fever. *Science* 235:83, 1987.

Sun M: The vexing problems of vaccine compensation. *Science* 227:1012, 1985.

Sun M: Vaccine compensation proposals abound on Capitol Hill. *Science* 233:415, 1986.

Yelverton E, et al: Rabies virus glycoprotein analogs: Biosynthesis in *Escherichia coli*. *Science* 219:614, 1983.

UNIT III

Pathogenic Micro-organisms

Chapter 22

Introduction to the Pathogens

OBJECTIVES   After a study of this chapter, you should be able to

1 Discuss the changing patterns of disease over the past century and list the factors responsible for these alterations.

2 Explain why infections by normal flora or organisms usually considered to be nonpathogens are more prevalent now than 30 to 50 years ago.

3 Define epidemiology and outline the types of problems that the epidemiologist is called on to solve.

4 Describe the major responsibilities of the Centers for Disease Control.

5 Discuss the different methods for diagnosing an infectious disease.

6 Describe the procedures for the collection of various specimens and be aware of any precautions that must be taken to preserve them until they are delivered to the diagnostic laboratory.

In this unit the emphasis shifts from a study of the features of microorganisms in general to a consideration of the complex changes that occur when a particular organism invades and produces an infectious disease in the human host. This chapter is designed to introduce you to some of the variables that influence our exposure and susceptibility to infectious diseases. Subsequent chapters discuss many of the details of these specific host-parasite relationships.

CHANGING PATTERNS OF DISEASE

It has been just a little over 100 years since the germ theory was accepted by scientists. Since then, the relationship between humans and pathogenic microorganisms has changed dramatically, mostly in our favor, but the conquest of one problem frequently gives rise to the appearance of new challenges in the control of infectious organisms. Our hospitals are no longer the death traps of the nineteenth century, and new lifesaving techniques have resulted in types of infections unheard of several decades ago.

During the past century, we learned that microorganisms cause diseases such as cholera, tuberculosis, typhoid fever, and anthrax. Moreover, it was found that a rather precise series of events must occur in order for such microorganisms to be transmitted from an infected person or animal to a noninfected host. This concept gave rise to the science of epidemiology, and the epidemiologist learned where to break the link in the chain of transmission. Thus, better sanitation of drinking water drastically curtailed typhoid fever and dysentery; control of mosquitoes eliminated urban yellow fever; and scrubbing, disinfectants, and sterile techniques reduced surgical infections.

In this century of rapidly developing knowledge, the science of immunology began with the discovery that individuals could be protected from many infectious diseases by artificially inducing a specific immunity. By the middle of the twentieth century, diseases such as diphtheria, pertussis, tetanus, and polio could be prevented and, during the past quarter century, effective vaccines have become available for many other infectious diseases.

The next era in the conquest of infectious diseases began in the 1930s and 1940s, when chemotherapeutic drugs and antibiotics became available and bacterial infections seemed destined to be eliminated. However, it soon became apparent that many microorganisms can mutate to become antibiotic-resistant forms and that the microbiologist must battle constantly to develop new antibiotics to replace those that are no longer effective.

Where do we stand now? Many dreaded organisms have been curtailed to the point at which they no longer cause widespread diseases, particularly in the Western world. However, recent medical knowledge and techniques have led indirectly to the occurrence of many previously rare or unknown infections caused either by organisms considered to be "normal flora" or by organisms previously thought of as "nonpathogens." These opportunistic infections occur when host resistance is reduced to such a degree that normal flora or nonpathogens are able to grow and flourish in areas where they otherwise could not penetrate or, if they did invade, would be killed by immune mechanisms. A few situations leading to opportunistic infections include (1) the use of immunosuppressants to prevent the rejection of organ transplants or infection by the acquired immunodeficiency syndrome virus, which destroys helper T cells; (2) the use of cancer chemotherapy or irradiation, which also kills many cells in the immune system; (3) the use of kidney dialysis machines and heart pumps; (4) the frequent use of urethral catheters, leading to urinary tract infections; (5) the use of antibiotics, which disturbs the ecological balance of our normal flora; and (6) poor nutritional status. In fact, any procedure that destroys the protection offered by an intact skin and mucosa, interferes with the immune system or cough reflexes, or changes the normal flora is conducive to serious infections by "harmless" organisms.

ROLE OF THE EPIDEMIOLOGIST

Just how does our society keep track of the current status of infectious diseases? How do we know that an epidemic of influenza is beginning? If a number of cases of typhoid fever occur in a certain area, how can we locate the source of these infections in order to eliminate them? The study comes from **epidemiology,** the study of disease occurrence in human populations. The epidemiologist attempts to answer the following questions:

1. Who gets a particular disease?
2. How does an individual acquire the disease?
3. How can the spread of this disease be prevented?

The epidemiologist is concerned with all diseases, including heart disease, cancer, diabetes, and congenital abnormalities, as well as animal and plant diseases.

Another important function of the epidemiologist is to locate the source of an agent that is causing an unexplained number of infections. An example of this type of detective work can be seen in tracing the source of a typhoid fever epidemic that occurred in 1981 in San Antonio, Texas. This epidemic resulted in over 50 cases of typhoid fever, and epidemiologists from the Centers for Disease Control (CDC) converged on San Antonio to stop the outbreak by finding how these organisms were being spread within the city. Because typhoid fever is usually spread via contaminated food or water, a detailed questionnaire concerning foods eaten during the past month was given to each patient. To make a long story short, at the end of 3 or 4 days, the CDC epidemiologists had located a restaurant in which at least two employees were asymptomatic carriers, harboring the typhoid bacillus in their intestinal tracts. Such carriers invariably carry the organisms on their hands, resulting in the contamination of any food products they handle.

Another interesting example involved a large number of cases of diarrhea caused by *Salmonella muenchen* in 1981 over a four-state area. Epidemiologists questioned many of the patients and found that a common link was the fact that most admitted to personal or home exposure to marijuana. Samples of marijuana obtained from patients' households were subsequently shown to contain as many as 10 million *S. muenchen* per gram.

Finally, the epidemiologist may be called on to find the source of an infectious agent that causes a previously undescribed disease. Such was the case in 1976, when a number of fatal cases of pneumonia occurred in individuals attending an American Legion convention in Philadelphia. The disease, subsequently named legionnaires' disease, brought panic to the city of Philadelphia and resulted in the closing of one of that city's largest hotels (where the convention had been held). All of this occurred because no one could determine the cause of legionnaires' disease, and everyone had a personal hypothesis to explain this new malady. The causative agent has now been grown and shown to possess characteristics setting it apart from previously described bacteria. The surprising aspect of legionnaires' disease is not the severity of the pneumonia but rather that the causative agent remained unknown until 1976 in spite of its apparent worldwide distribution. Because it appears to be acquired as a respiratory disease, we shall discuss legionnaires' disease more completely in Chapter 24. The point to be made here is that there are still a number of diseases of unknown cause (such as multiple sclerosis and human cancer), and it is certainly possible that new infectious agents will continue to be found.

Centers for Disease Control

In the United States, each state requires physicians and hospital authorities to report the occurrence of any of about 50 different bacterial, viral, fungal, or parasitic infections. This information is then transmitted to a federal agency located in Atlanta, Georgia, called the Centers for Disease Control (abbreviated CDC). The CDC thus becomes quickly aware of any increase in the occurrence of a specific disease, such as measles or influenza, and can set up immunization programs in an attempt to stop its spread.

DIAGNOSIS OF INFECTIOUS DISEASES

Many infectious diseases have characteristic features that implicate a particular illness. Symptoms common to many are malaise, headache, fever, chills, cough, and exhaustion. Diagnosis of infectious diseases is at several levels. The first is a clinical diagnosis, which frequently is sufficient in itself. It is hardly necessary to attempt to isolate the virus from a child with chickenpox. The clinical features of chickenpox, mumps, and several other infectious diseases are adequate for making a definitive clinical diagnosis. More-specific diagnoses are made by one of two techniques. In the first, a demonstration of a specific antibody response during the course of the illness will usually provide a retrospective diagnosis. However, the best specific diagnosis results from the isolation and identification of the organism from the infected individual. Subsequent chapters of this unit will describe how many of these disease-producing organisms are identified; here we will emphasize only one important aspect of these procedures.

Collection of Specimens

Frequently, the weakest link in making a bacteriological diagnosis lies in the collection of material to be used as a source for the isolation and its subsequent delivery to the diagnostic laboratory. Thus, the collection of an adequate amount of the correct specimen is useless if the time between collection and culturing allows the disease-producing organism to die or to be overgrown by normal flora that may contaminate the specimen. It is of utmost importance that specimens be transported quickly to the diagnostic laboratory, where they can be promptly processed to ensure the best possible chance of growth, isolation, and identification of the disease-producing agent. In some situations, it is recommended that the growth medium be inoculated immediately after the specimen is obtained from the patient and that the inoculated medium then be taken to the laboratory to be grown and the bacteria identified. In other cases, material should first be dispersed in a nutrient broth or buffered transport medium, and then transported to the laboratory to be grown. In all instances, quick processing of specimens aids in providing the fastest and most reliable identification of the disease-producing organism.

Collection of a Blood Specimen

In taking a blood specimen for culture, one should be aware that, although blood is normally sterile, the skin that must be penetrated is not. Routinely, the skin should be first cleansed with 70 to 95 percent alcohol to remove dirt, lipids, and fatty acids. The site should then be scrubbed with a circular, concentric motion (working out from the starting point) using a sterile gauze pad soaked in 2 percent iodophor. The iodine should be allowed to remain on the skin for at least one minute before it is removed by wiping with a sterile gauze pad soaked with 70 to 95 percent alcohol. It must be emphasized, however, that all of this will be useless if the person drawing the blood palpates the vein after the cleansing process, thereby contaminating the very site that had been cleaned.

After cleansing the penetration site, the blood may be withdrawn using either a sterile needle and syringe or a commercially available, evacuated blood-collection tube. Approximately 10 mL of blood should be inoculated into 100 mL of the appropriate medium at the bedside.

Specimen Collection from the Respiratory Tract

Because of the myriad normal resident flora in the upper respiratory tract (mouth, nose, and throat), the isolation of lower respiratory tract infectious agents can be difficult and sometimes confusing. Therefore, one must be certain that lower respiratory tract specimens represent sputum that has been brought up by a deep cough. In a very young child, a debilitated older person, or someone who is comatose, a specimen may be obtained by transtracheal aspiration. This procedure uses a needle and tube about 30 cm long that is inserted through the trachea into the lung.

Organisms causing upper respiratory infections that appear as vesicles or ulcers in the throat should be obtained with a cotton swab and streaked on a blood agar plate as soon as possible. Nasopharyngeal cultures usually are obtained with a cotton swab on a bent wire that can be passed either through the nose or by way of the mouth, carefully bypassing the tongue and oropharynx. If unable to streak media immedi-

ately, swabs should be kept moist until delivered to the laboratory.

Collection of Specimens from Wounds and Abscesses

Whenever possible, a sterile syringe and needle should be used to collect specimens from wounds and abscesses. The use of a swab is generally unsatisfactory because of the limited amount of material collected by this method and because many organisms die as the swab dries. It is also important to remember that wounds and abscesses are commonly infected with obligately anaerobic bacteria, which quickly die on a swab that is exposed to the atmosphere. Therefore, all aspirates should be transported to the laboratory in commercially available bottles containing oxygen-free gas.

Specimens from Intestinal Contents

When one wishes to culture intestinal contents, there is no wide choice as to the material to be taken from the patient. If a specimen contains blood or mucus, these should be included in the material to be sent to the laboratory. When a sterile swab is used to collect a fecal specimen, the swab must be inserted past the anal sphincter and rotated several times before being withdrawn.

It is a common misconception that the microorganisms found in feces are rather hearty and that special precautions to preserve the viability of suspected pathogens are not required. Nothing could be further from the truth. Unless fecal specimens can be taken directly to the laboratory for culturing, they must be placed in a stool preservative containing a buffer that will maintain the pH near neutrality. Failure to use a preservative will result in the death of many enteric pathogens. When rectal swabs are used, they should be immersed in a commercially available, gram-negative enrichment broth until they can be inoculated onto appropriate media.

Collection of Urine Specimens

The majority of all urinary tract infections are caused by normal-flora enterics, among which may be species of *Escherichia, Klebsiella, Enterobacter, Proteus, Pseudomonas,* and *Enterococcus.* Urethral catheterization may yield samples with minimal contamination, but the danger of introducing organisms from the urethra into the bladder contributes some risk to this procedure. Hence, voided samples frequently are used after careful cleansing of the external genitalia.

In 1989, the journal *Science* selected the polymerase chain reaction (PCR) as the major scientific development of the year. Just what is the PCR and why was it accorded this accolade?

The PCR is a method used to amplify a specific segment from impure DNA as much as a millionfold in a matter of a few hours. The reaction is based on the repetitive cycling of three reactions: (1) heat denaturation of the sample DNA to yield single-stranded molecules, (2) the addition of two short DNA primers that will anneal to the opposite ends of the target DNA, and (3) the addition of a DNA polymerase and an excess of deoxyribonucleoside triphosphates. The polymerase catalyzes the synthesis of a new strand of DNA between the two primers. The series of reactions is then repeated, and since each newly synthesized segment is able to bind to new primers for each cycle of the reaction, the number of newly synthesized DNA strands increases exponentially. As a result, one can look for a particular DNA sequence in a single cell. All one really needs to amplify minuscule traces of a specific DNA segment is the two small primers that anneal to the 3' and 5' ends of the target DNA.

The applications of the PCR are enormous. One primary use will be to amplify and detect genetic material of infectious disease agents present in blood, cells, or food. In fact, it is currently used to detect the presence of a number of different viruses, particularly the acquired immunodeficiency syndrome virus and viruses associated with cancer such as HTLV-1 and papillomaviruses. It can also be used to compare the DNA from blood, hair, or sperm found at the scene of a crime with that of a suspect. Determination of paternity, as well as matching donor and recipients of transplants, are made more easily and more quickly using the PCR.

There is little doubt, therefore, that the use of the PCR for infectious diseases will continue to expand as more and more specific sequences of DNA become known for a multitude of infectious disease agents and their products.

One must be aware, however, that all voided urine samples will contain some bacteria; therefore, a quantitative assay for the number of bacteria present must be carried out. Urine should be collected in a sterile cup only after the first 20 to 25 mL have been voided, because the flushing action of the initial flow removes many of the organisms present in the urethra.

Considerable experimental data have resulted in the formation of the following rules:

1. 10^5 bacteria or more per milliliter from a clean, voided specimen indicate a urinary tract infection;

2. A value of 10^3 to 10^4 bacteria per milliliter in a symptomatic patient requires a repeat culture; and

3. 10^3 bacteria or less per milliliter are not usually considered significant in a voided sample.

Diagnosis of Nonbacterial Infections

Many of the procedures described in this chapter are equally applicable to the isolation of viruses, fungi, and parasites. But final identification of these agents varies considerably for each group. Therefore, the definitive identification of all infectious agents is described in conjunction with the diseases they cause.

ORGANIZATIONAL PRESENTATION OF THE PATHOGENS

There are several approaches to the method of presenting the pathogenic organisms to the student. This book relies primarily on the portal of entry, even though it is recognized that in many instances an organism may have several different portals of entry rather than a single one. It is also recognized that in many cases the site of the disease is far from the site of entrance. Nevertheless, the increasing importance of understanding the natural resistance to the infecting organism makes it necessary to focus again on its initial point of implantation. This may provide significant clues as to what happens in the short period after the pathogen passes through a particular portal of entry. In fact, this period may be decisive in determining the outcome of an infection.

The remaining chapters in this unit will deal with specific host-parasite relationships that result in disease. Table 22.1, which is based on *Bergey's Manual of Systematic Bacteriology*, lists some of the more common disease-producing bacteria and the diseases they produce.

TABLE 22.1 Classification of Common Disease-Producing Bacteria

Classification	Some Resulting Infections
Gram-negative aerobic rods	
Brucella abortus	
Brucella melitensis	Brucellosis
Brucella suis	
Bordetella pertussis	Whooping cough
Francisella tularensis	Tularemia
Pseudomonas aeruginosa	Urinary tract, wounds, burns, lungs
Legionella pneumophila	Legionellosis
Gram-negative facultative rods	
Escherichia coli	Diarrhea, urinary tract, meningitis in children
Enterobacter aerogenes	Urinary tract
Klebsiella pneumoniae	Pneumonia, urinary tract
Proteus vulgaris	Urinary tract, wound
Salmonella typhi	Typhoid fever
Salmonella paratyphi	Paratyphoid fevers
Salmonella typhimurium	Food infection
Shigella dysenteriae	
Shigella flexneri	Bacillary dysentery
Shigella sonnei	
Campylobacter jejuni	Diarrhea
Yersinia enterocolitica	Diarrhea
Helicobacter pylori	Chronic gastritis
Yersinia pestis	Plague
Vibrio cholerae	Cholera
Vibrio parahaemolyticus	Food poisoning
Vibrio vulnificus	Food poisoning & sepsis
Haemophilus influenzae	Childhood meningitis
Haemophilus aegypticus	Conjunctivitis
Gram-negative anaerobic rods	
Bacteroides fragilis	Abscesses
Fusobacterium necrophorus	
Gram-positive cocci	
Staphylococcus aureus	Boils, carbuncles, food poisoning
Streptococcus pyogenes	Scarlet fever, tonsillitis, pharyngitis
Streptococcus pneumoniae	Pneumonia, meningitis
Gram-negative cocci	
Neisseria meningitidis	Meningitis
Neisseria gonorrhoeae	Gonorrhea
Non-spore-forming gram-positive rods	
Listeria monocytogenes	Meningitis, septicemia, conjunctivitis
Endospore-forming rods	
Bacillus anthracis	Anthrax
Clostridium tetani	Tetanus
Clostridium perfringens	Gas gangrene, food infection
Clostridium botulinum	Food poisoning (botulism)
Actinomycetes and similar organisms	
Corynebacterium diphtheriae	Diphtheria
Mycobacterium tuberculosis	Tuberculosis
Mycobacterium leprae	Leprosy
Actinomyces israelii	Actinomycosis
Actinomyces bovis	
Nocardia asteroides	Nocardiosis
Spirochetes	
Treponema pallidum	Syphilis
Treponema pertenue	Yaws
Treponema carateum	Pinta
Treponema vincentii	Vincent's angina (trench mouth)

TABLE 22.1 *(Continued)*

Classification	Some Resulting Infections
Leptospira interrogans	Infectious jaundice
Borrelia recurrentis	Relapsing fever
Borrelia burgdorferi	Lyme disease
Rickettsias	
Rickettsia rickettsiae	Rocky Mountain spotted fever
Rickettsia akari	Rickettsialpox
Rickettsia prowazekii	Epidemic typhus
Rickettsia typhi	Endemic typhus
Rickettsia tsutsugamushi	Scrub typhus
Rochalimaea quintana	Trench fever
Coxiella burnetii	Q fever
Ehrlichia canis	Ehrlichiosis
Chlamydlae	
Chlamydia trachomatis	Trachoma, inclusion conjunctivitis, urinary tract infections
Chlamydia psittaci	Psittacosis, ornithosis
Mycoplasmas	
Mycoplasma pneumoniae	Atypical pneumonia
Mycoplasma hominis	Vaginitis (?), urinary tract infections (?)
Ureaplasma sp.	Urethritis (?)

SUMMARY

Major changes in sanitation practices, the use of vaccines, and the availability of chemotherapeutic agents have greatly changed the infectious disease picture during the past century. The use of immunosuppressants as well as more complicated surgical techniques have also contributed to a changing pattern of infectious diseases. However, infectious organisms are also changing—mutants that are resistant to drugs are developing rapidly, and organisms that in the past caused disease only rarely are being seen more frequently.

Epidemiology is the study of disease occurrence in any population. Physicians send epidemiological information to the Centers for Disease Control for analysis from everywhere in the United States. The CDC also is involved in locating the source of the agent causing any unexplained outbreak of infections.

Most infectious diseases produce characteristic symptoms in the host—malaise, headache, fever, chills, cough, and fatigue. Diagnosis of infectious diseases may be based on clinical appearance, serological tests, or by the isolation and identification of the offending microorganism. The latter requires that the correct material be collected from the patient and sent to the diagnostic laboratory.

Special techniques and precautions are required for the collection of each type of specimen to be sent to the diagnostic laboratory. For example, blood samples must be collected so that they are free of skin contaminants; lower respiratory tract samples must originate from the lungs and not from the upper respiratory tract; material from wounds or abscesses should be kept under anaerobic conditions while transporting it to the laboratory; intestinal specimens must be kept in a buffered medium to maintain a neutral pH; and voided urine samples must be collected so as to avoid gross contamination from the external genitalia.

QUESTIONS FOR REVIEW

1. During the past century, there have been three major types of changes that altered the incidence of infectious diseases in humans. These changes can be categorized as (1) environmental, (2) immunological, and (3) therapeutic. For each category, list as many changes as you can and correlate these changes with an alteration of disease patterns.
2. Name as many situations as you can that would be conducive to disease caused by our normal flora, or by organisms that are considered to be nonpathogens.
3. Define epidemiology and list the specific functions of an epidemiologist. Give an example in which the epidemiologist has answered each function you have listed.
4. What is the CDC and what are its major functions?
5. How may an infectious disease be diagnosed? Which procedure is best?
6. What is the main source of contaminants in a blood sample?

Most organisms in the external environment apparently do not find the body to be a favorable habitat. Characteristic features of the various body areas such as temperature, nutrients available, and pH influence the population that is able to survive and establish itself. Because these conditions vary from site to site in the body, different sites acquire considerably different organisms as their normal flora. Once the normal flora is established, it will actually be beneficial to the body by preventing the overgrowth of undesirable organisms. Destruction of the normal flora frequently disrupts the status quo, resulting in the growth of harmful organisms. This can be seen following the prolonged administration of broad-spectrum antibiotics. For example, if the normal flora of the intestinal tract and vagina are in large part destroyed, the yeast *Candida albicans,* which is resistant to these antibiotics, may grow unchecked to become the major organism in these areas. It may then infect the mucous membranes and the skin. A more frequent complication of antibiotic therapy is a severe gastroenteritis known as pseudomembranous colitis. This syndrome has been associated with a number of antimicrobial agents, but the antibiotics clindamycin and lincomycin have been incriminated most often. The mechanism of this severe diarrhea—at least for the clindamycin- and lincomycin-induced gastroenteritis—was elucidated in 1978, when it was observed that the use of these antibiotics resulted in an overgrowth in the intestine of an organism identified as *Clostridium difficile.* This organism produces an enterotoxin that causes the gastroenteritis. The important concept to be learned is that *C. difficile* can exist as normal flora in the intestine but is able to cause severe disease only when antibiotic therapy destroys much of the other normal intestinal flora, permitting the uncontrolled growth of *C. difficile* in the intestine.

In addition, physiological mechanisms in certain body areas tend to limit or prevent the entrance of microorganisms. Among these are the ciliary action in the respiratory tract, which keeps mucus (with adherent bacteria) moving outward, and the acid pH of the gastric contents and the vagina, which is unfavorable for most microorganisms. The outward flow of urine also tends to prevent deep penetration of microbes in the urinary tract.

The Skin

The pH of the skin is usually about 5.6. This factor alone may be responsible for inhibiting the establishment of many bacterial species, although the skin (particularly on the hands) is subject to a considerable transient population of organisms. Moist areas of skin, such as the axilla, are far more densely populated than are relatively dry areas, such as the forearm. For example, it has been reported that under a moist dressing on the arm, the number of bacteria per square centimeter rose in 4 days from an initial value of 1000 to almost 10 million bacteria.

Although some bacteria occur very superficially on the skin surface, much of the bacterial flora is located in the openings of the hair follicles. Consequently, to be effective, an antiseptic must enter the openings of the hair follicles.

Even though the skin is exposed to the external environment, the number of genera routinely found are fewer than one might expect. *Propionibacterium acnes* constitutes part of the deep-seated skin flora, and the production of propionic acid by this organism appears to exert a bacteriostatic effect on many other organisms. Probably the greatest microbial population on the skin is *Staphylococcus epidermidis; S. aureus* is present to a much lesser extent.

Other organisms commonly found on the skin include *Corynebacterium* sp., *Micrococcus* sp., *Peptostreptococcus* sp., and various *Neisseria* sp. Certain potentially pathogenic yeasts such as *Candida albicans, Pityrosporum ovale,* and *Pityrosporum orbiculare* may also be present. Organisms from upper-respiratory-tract secretions may be deposited on the skin, particularly during an overt infection. The external ear, the axillae, and the genital region also usually harbor nonpathogenic acid-fast mycobacteria.

The Eye

It is surprising that the eye is so seldom infected, since, at first thought, one would think of it as being particularly susceptible to disease. One explanation for this may lie in the fact that the secretions of the eye are particularly rich in the enzyme lysozyme (found in other parts of the body as well as in tears), which causes the lysis (dissolving) of many bacteria. This action, plus the washing action of tears, helps to eliminate many organisms from the eye.

A number of skin organisms may, however, be found in the normal conjunctiva. These frequently include *S. epidermidis, Haemophilus influenzae, Neisseria* sp., and *Corynebacterium xerosis.* Other organisms occasionally isolated from the healthy conjunctiva include *Streptococcus pneumoniae,* enteric gram-negative rods, and the viridans streptococci (normal flora of the mouth).

The Respiratory Tract

The Upper Respiratory Tract

In the course of normal breathing, many kinds of microbes are inhaled into the nose. Among these are normal soil inhabitants as well as pathogenic and potentially pathogenic bacteria. Some will be filtered out by the hairs in the nose or will land on the moist surface of the nasal mucous membranes. Others will get back to the nasopharynx and will take up residence there.

Bacteria routinely found in these areas include *Lactobacillus* sp., *Corynebacterium pseudodiphtheriticum*, and viridans streptococci. Obligate anaerobes include *Bacteroides* sp., *Fusobacterium* sp., *Bifidobacterium* sp., nonpathogenic *Treponema*, *Actinomyces israelii*, and *Veillonella* sp. Organisms that may cause severe disease but are carried by many persons as normal flora include *S. pneumoniae, S. pyogenes, S. aureus, H. influenzae,* and *N. meningitidis.* Other organisms found less frequently in the upper respiratory tract are included in Table 23.1.

TABLE 23.1   Summary of Common Microbial Flora of the Human Body

Body Area	Microbial Flora
Skin	
General	*Acinetobacter* sp.
	Candida albicans and other yeasts
	Micrococcus sp.
	Neisseria sp.
	Peptostreptococcus sp.
	Propionibacterium acnes
	Staphylococcus epidermidis
External ear	*Corynebacterium* sp.
	Mycobacterium sp.
	Staphylococcus aureus
	Staphylococcus epidermidis
Axilla and groin	*Corynebacterium* sp.
	Mycobacterium smegmatis
	Staphylococcus epidermidis
Eye	
Conjunctiva	*Corynebacterium* sp.
	Haemophilus influenzae
	Neisseria sp.
	Staphylococcus epidermidis
	Viridans streptococci
Respiratory	
Mouth and tonsils	*Actinomyces* sp.
	Bacteroides sp.
	Bifidobacterium sp.
	Borrelia refringens
	Candida albicans and other yeasts
	Corynebacterium sp.
	Coliforms
	Fusobacterium sp.
	Haemophilus sp.
	Lactobacillus sp.
	Micrococcus sp.
	Mycoplasma sp.
	Neisseria sp.
	Peptostreptococcus sp.
	Staphylococcus aureus
	Staphylococcus epidermidis
	Streptococcus pneumoniae
	Treponema denticum
	Veillonella sp.
	Viridans streptococci
Nose and nasopharynx	*Corynebacterium* sp.
	Haemophilus sp.
	Neisseria sp.
	Staphylococcus aureus
	Staphylococcus epidermidis
	Streptococcus pneumoniae

TABLE 23.1 *(Continued)*

Body Area	Microbial Flora
Larynx, trachea, bronchi and lungs, and accessory nasal sinuses	No permanent normal flora; transient flora destroyed
Gastrointestinal tract	
Stomach	No permanent normal flora; transient flora destroyed by high acidity
Intestines	
Duodenum	Usually none
Jejunum and upper ileum	Usually none or very few
Lower ileum and large intestine	*Achromobacter* sp.
	Bacteroides sp.
	Bifidobacterium sp.
	Candida albicans and other yeasts
	Clostridium sp. including *C. tetani* and *C. perfringens*
	Enteric organisms included in the Enterobacteriaceae
	Eubacterium sp.
	Fusobacterium sp.
	Lactobacillus sp.
	Peptostreptococcus sp.
	Pseudomonas aeruginosa
	Staphylococcus aureus
	Viridans streptococci
Genitourinary tract	
Bladder	None
Urethra (anterior)	*Acinetobacter* sp.
	Candida albicans and other yeasts
	Corynebacterium sp.
	Enteric organisms included in the Enterobacteriaceae
	Mycobacterium sp.
	Mycoplasma sp.
	Neisseria sp.
	Trichomonas vaginalis
Vagina	*Candida albicans*
Prior to puberty and after menopause	*Corynebacterium* sp.
	Enteric organisms included in the Enterobacteriaceae
	Micrococcus sp.
	Staphylococcus epidermidis
	Viridans streptococci
Between puberty and menopause	*Acinetobacter* sp.
	Bifidobacterium sp.
	Candida albicans and other yeasts
	Clostridium sp.
	Corynebacterium sp.
	Fusobacterium sp.
	Group B streptococci
	Haemophilus vaginalis
	Lactobacillus acidophilus and other lactobacilli
	Mycobacterium sp.
	Mycoplasma sp.
	Peptostreptococcus sp.
	Trichomonas vaginalis
	Viridans streptococci

The Lower Respiratory Tract

With the exception of an occasional organism that passes into the larynx, no normal inhabitants are found in the larynx, trachea, bronchi, or lungs. One reason for this (as was pointed out above) is that organisms are filtered out by the upper respiratory tract. Thus, when the upper tract is bypassed, as in a tracheostomy (an incision of the trachea and insertion of a tube to facilitate breathing), it is very important to realize that a major defense against infection of the lower tract is no longer operating. Hence, care must be exercised so that pathogenic or potentially pathogenic organisms do not enter and contaminate the trachea.

The Digestive Tract

The Mouth and Oropharynx

Microorganisms are taken into the mouth with food and drink, the hands, and various objects carrying dust and bacteria. Others are transmitted by a chewed pencil or a finger.

Because of difficulties encountered in cultivation, some organisms present in the mouth probably have not been grown in the laboratory or even identified. However, microbes recognized as normal inhabitants of the mouth include lactobacilli, spirochetes, various cocci (particularly the viridans streptococci), spore-forming bacilli, coliforms, vibrios, and the fusiform bacillus. It is important to realize that the viridans streptococci are normal inhabitants of the mouth and the oropharynx, whereas the other streptococci are not normal flora (even though occasionally present) and should be eradicated if found.

As can be seen from the extensive list above, the mouth contains many microorganisms present as normal flora. Food left on and between the teeth undoubtedly causes a large increase in the numbers of oral organisms. We shall show in Chapter 24 how the fermentation of ingested sugar by our oral bacteria causes dental caries through the production of acids that dissolve the tooth enamel.

The Stomach

Although many microorganisms get past the gastric barrier to become established in the intestinal tract, the healthy stomach has no natural flora. Its acid content is either inhibitory or destructive to swallowed organisms. Those that do pass into the intestinal tract probably had only short contact with the gastric juice at a time when the acidity was temporarily lessened by food intake.

In certain disease conditions, such as carcinoma of the stomach, some bacteria are able to multiply in the stomach. It has also been recently reported that *Helicobacter pylori*, a curved gram-negative bacterium, can colonize the mucus-secreting epithelial cells of the human stomach, causing peptic ulcers or a chronic gastritis.

However, aside from the possible route of infection, the transitory flora of the stomach is sometimes of great importance in the diagnosis of pulmonary infections. This is particularly true in the diagnosis of tuberculosis, when swallowed lung secretions can provide the acid-fast bacilli that identify the condition. However, the mere presence of acid-fast rods in stomach washings is not sufficient evidence for a diagnosis of lung tuberculosis. Many edible foods contain saprophytic acid-fast organisms, so the bacteria must be cultured and identified by other means in order to provide a definite diagnosis.

The Intestines

Certain external factors may influence the composition of the intestinal flora. Although diet may be one factor, another important factor is antibiotic therapy, particularly if of long duration. In this case, the normal flora may be so changed that organisms (such as *Candida albicans* or *Clostridium difficile*) normally present in small numbers may become the predominant microbe present. However, in the absence of some unusual factor, the intestine has a fairly characteristic bacterial population.

The Small Intestine The stomach contents pass into the duodenum, or first part of the small intestine, which, like the stomach, has no natural flora. Organisms begin to appear in the jejunum and the ileum (the second and the third segments), but it is not until near the lower end of the ileum that large numbers of bacteria occur. The major bacteria found in this area include α-hemolytic streptococci (particularly *Streptococcus faecalis*), staphylococci, lactobacilli, *C. perfringens*, *Veillonella* species, and, occasionally, *Escherichia coli*. Yeasts also may be present.

The Large Intestine This body area contains a great deal of digested food and food wastes and provides excellent growth conditions for a diverse population. As a result, the large intestine contains an incredible diversity of microbes. The largest single group consists of the obligately anaerobic gram-negative organisms. Of these, *Bacteroides* species account for the vast majority, followed in frequency by *Fusobacterium* species. Many gram-positive anaerobes are also found, such as *Clostridium tetani*, *C. perfringens*, *C. sporogenes*, *C. putrificum*, *C. difficile*, *C.*

histolyticum, *Bifidobacterium* sp., and *Eubacterium* sp.

The lower intestine also contains a number of facultative gram-negative organisms, which are much easier to grow than are the obligate anaerobes but constitute only about 5 percent of the intestinal flora. The most common of these include *E. coli*, *Enterobacter aerogenes*, *K. pneumoniae*, and *P. aeruginosa*. Also found are *Enterococcus faecalis*, viridans streptococci, *Mycoplasma* species, and lactobacilli. The major fungi present include *Candida*, *Geotrichum*, *Cryptococcus*, *Penicillium*, and *Aspergillus*. *Vibrio*, *Spirillum*, *Borrelia*, *Treponema*, and even strains of pathogenic *Salmonella* and *Shigella* have been isolated from some healthy individuals. So you see that there are great numbers of microorganisms growing in the intestinal tract, most of which can be thought of as opportunists.

A few (even though still opportunists) can also be thought of as mutualistic symbionts as long as they remain in the large intestine. This category includes organisms that synthesize excess vitamin K and B-complex vitamins, which can be used by the host.

The Genitourinary Tract

The Urethra

The urinary tract, except for the external urethra of the male, is normally sterile, although a few non-pathogenic cocci may be present in the female urethra (see Table 23.1). In suspected cases of gonorrhea, one must be careful not to confuse the short gram-negative diplobacilli in the genus *Acinetobacter* with the causative agent of gonorrhea, *N. gonorrhoeae*. In actual cases, many of the *Neisseria* will be found inside or adhering to the white cells making up the exudate so common in this disease.

Mycoplasma organisms have been isolated from the urethras of both male and female patients. It is believed that mycoplasmas account for many of the inflammatory infections of the urethra; it is also true that they are found in the urethras of many healthy individuals. *Mycobacterium smegmatis*, an acid-fast rod, is present on the external genitalia in both sexes. This organism is morphologically similar to the tubercle bacillus but can be differentiated easily on the basis of its rapid growth on laboratory media.

The Vagina

Between puberty and menopause, the vaginal area has an acid pH, whereas at other times (prior to puberty and after menopause), the secretions are alkaline. This change in pH results in changes in the normal flora of the vagina, as shown in Table 23.1. In an acid environment, lactobacilli predominate, but corynebacteria, yeasts, staphylococci, streptococci (α- and β-hemolytic) and anaerobic cocci are frequently found in lesser numbers.

The Blood

As a rule, in the absence of disease, microorganisms are not found in blood or healthy tissue. Although it is true that bacteria may get into the blood through cuts or abrasions or even by way of food through the intestine, these microorganisms are quickly engulfed and destroyed by the white cells in the blood.

This is not always the case with viral organisms. In the first place, most viral diseases have an incubation period (the time from infection until the appearance of symptoms) of about a week, although some incubation periods are as short as a day and others as long as a month. During this incubation period the viruses may be circulating in the blood and thus are a potential hazard in blood transfusions. In certain viral infections—particularly hepatitis B—the virus may be found in the blood for months or even years after recovery from the disease. Blood banks now routinely test all blood for the presence of a surface antigen on the hepatitis B virus in order to minimize the risk of transmitting this disease through blood transfusions. However, as we shall discuss in Chapter 26, a different virus (called non-A, non-B hepatitis or hepatitis C) now accounts for about 90 percent of posttransfusional hepatitis infections. The virus causing AIDS is believed to have an incubation period of as long as 10 years, during which time it can be spread via blood products. As a result, all blood to be used for transfusions is checked for the presence of antibodies to the AIDS virus. Blood banks also refuse blood from anyone who has had malaria, syphilis, and certain other infections. There is always the risk that a blood donor may be in the incubation period of some disease and not be aware that the donated blood may cause serious disease in a recipient.

SUMMARY

Our normal flora is made up of organisms that inhabit some area of the human body. Many are always present; others may persist for a few weeks or months before disappearing. Organisms can be categorized as helpful (mutualistic symbionts), harmless (commensals), or potentially harmful (opportunists). Most must be considered as opportunists because almost all can cause disease in individuals whose normal barriers to infection or immune mechanisms are nonfunctional. Some organisms that are considered to be serious pathogens may be carried asymptomatically by a large percentage of the population. Examples in the group include *N. meningitidis,* the cause of epidemic meningitis, *S. pneumoniae,* a major cause of lobar pneumonia, and *S. pyogenes,* a cause of scarlet fever and septic sore throat.

Areas of the body that have an extensive microbial flora include the skin, eye, upper respiratory tract, mouth and oropharynx, intestines, and genitourinary tract. Major organisms routinely found in each of these body areas are listed in Table 23.1.

QUESTIONS FOR REVIEW

1. Define *commensalism, mutualism, opportunism,* and *parasitism.*
2. Why are most of the organisms that make up the normal bacterial flora of the body referred to as opportunists?
3. What factors influence the population in various body areas?
4. Why does the blood have no natural flora?
5. What are some organisms found to be normal inhabitants of the various body areas?
6. What is the significance of the finding of *M. smegmatis* in the urine?
7. What is lysozyme and what is its significance?
8. Under what conditions do organisms multiply in the stomach?
9. How does the bacterial flora of the large intestine function to benefit the host?
10. How may bacteria gain access to the bladder? What usually tends to prevent invasion?
11. What factors contribute to preventing the invasion of bacteria in the trachea and the lungs?
12. Describe the mechanism for clindamycin-induced gastroenteritis.

SUPPLEMENTARY READING

Aronsson B, Mollby R, Nord CE: Antimicrobial agents and *Clostridium difficile* in acute enteric disease. *J Infect Dis* 151:476, 1985.

Gorbach SL, Goldin BR: The intestinal microflora and the colon cancer connection. *Rev Infect Dis* 12(S2):S252, 1990.

Mackowiak PA: The normal microbial flora. *N Engl J Med* 307:83, 1982.

Savage DC: Microbial ecology of the gastrointestinal tract. *Annu Rev Microbiol* 31:107, 1977.

Skinner FA, Carr JG: *The Normal Microbial Flora of Man.* New York, Academic, 1974.

Chapter 24

Bacteria and Fungi That Enter the Body via the Respiratory Tract

OBJECTIVES

This chapter is designed to acquaint you with

1 The classification and characteristics of group A streptococci.

2 The types of infections caused by group A streptococci.

3 A characterization of the late, nonsuppurative complications that may follow group A streptococcal infections.

4 A description of group B streptococcal infections and their danger to the newborn.

5 The role of streptococci in dental caries and periodontal disease.

6 A description of *Streptococcus pneumoniae* and its role in human disease.

7 The serological classification of *Neisseria meningitidis* and the epidemiology and control of meningococcal meningitis.

8 The molecular basis and control of whooping cough.

9 The classification of *Haemophilus influenzae* and the types of infections caused by these organisms.

10 The mechanism of disease production and control of *Corynebacterium diphtheriae*.

11 The epidemiology, control, and diagnostic procedures for tuberculosis and leprosy.

- **12** A characterization of infections caused by *Mycoplasma pneumoniae* and *Legionella pneumophila*.

- **13** The systemic mycoses, which include blastomycosis, histoplasmosis, coccidioidomycosis, and paracoccidioidomycosis.

- **14** The occurrence of infections by opportunistic fungi such as *Geotrichum* and *Aspergillus*.

- **15** The type and source of infections caused by *Cryptococcus neoformans*.

- **16** The epidemiology of avian-borne and non-avian-borne chlamydial pneumonia.

- **17** The reason why *Coxiella burnetii* is more resistant to heat and desiccation than are the other rickettsia.

For a proper perspective in understanding the respiratory tract, two facts should be borne in mind. First, the upper end of the respiratory tract is in intimate contact with the air, which we know to be contaminated with microorganisms. Second, the other end of the respiratory tract is a large surface with very thin walls that is in intimate contact with blood vessels. This is an ideal situation for setting up many types of infection. However, the fact is that bacteria from the air only rarely reach the lower end of the respiratory tract to produce disease.

One explanation of how the lower respiratory tract is kept free of microorganisms centers on the lining of the tract, with its cilia and mucus secreting cells. The combined action of the mucus secretion and the ciliary beating tend to produce a mucociliary "escalator" that effectively removes any bacteria or other particles that may have gained access to the lower respiratory tract.

This chapter places emphasis on microorganisms that enter mainly via the respiratory tract. Some of these organisms may have other portals of entry as well. There are still other organisms that occasionally enter the body and produce disease via the respiratory tract that are not included in this chapter because of their rarity as disease producers.

STREPTOCOCCI

Streptococci are important pathogens both because of the many severe infections they produce and because of the complications that may occur after recovery from the acute infection. Complications following streptococcal infections include rheumatic fever and acute glomerulonephritis.

The bacteria classified in the genus *Streptococcus* share certain morphological and biochemical characteristics. The most characteristic property of these organisms is their appearance. They are more or less spherical cells that grow as chains. The organisms divide in only one direction, but instead of breaking apart into individual cocci, they have a tendency to remain together and form a chain of cocci (see Figure 24.1). The length of the chain you are likely to see when staining these organisms depends to some extent on whether the organisms were grown on a solid or a liquid medium and how roughly they were handled in the process of making a smear. The longest chains are seen in a wet mount of a liquid culture. The streptococci are all gram-positive and nonmotile.

Figure 24.1 Chains of *Streptococcus mutans* (×5000). As is discussed later in the chapter, this streptococcus utilizes sucrose to produce an acid capable of eroding tooth enamel.

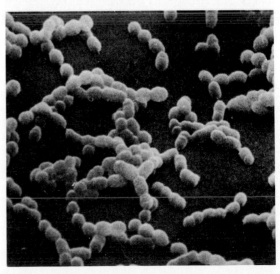

For the most part, streptococci are aerotolerant organisms, but a few obligate anaerobes do exist. They are demanding in their growth requirements because they have lost the ability to synthesize many of the nutrients they need. For example, some streptococci require up to 15 different amino acids, all of the known B vitamins, and some purines and pyrimidines plus asparagine or glutamine for growth. Quite obviously such a complex medium would be used only in a research laboratory. For routine culturing of streptococci, a complex medium containing peptones, meat infusion, salts, glucose, and agar to solidify it is used. To this is added 5 percent sterile defibrinated blood before the medium is poured into petri dishes. The resulting plates are usually referred to as blood agar plates. You will see later in this section that blood agar plates are of value both for growing and for identifying the streptococci.

Classification

Until the 1930s the classification of the streptococci was in a confused state. Many streptococci were considered specific for the disease entity from which they were isolated and were given species names based on the type of infection they caused. Examples of this were such names as *Streptococcus erysipelatis* for an organism isolated from the skin infection erysipelas and *S. scarlatinae* for one isolated from a case of scarlet fever. Biochemical classification, proposed by J.M. Sherman (see Table 24.1), proved valuable in the overall classification of this genus, but since essentially all acute infections were caused by organisms in Sherman's pyogenic group, this classification was of little value to the medical epidemiologist. Our present classification of this genus still uses the criteria proposed by Sherman but in addition uses antigenic criteria to subdivide the hemolytic streptococci into many groups and types. This latter classification was originally proposed by Rebecca Lancefield in 1933, and it was a result of her classification that it became obvious that a single streptococcal species could be responsible for a variety of disease entities. One important pathogen,

S. pneumoniae, does not fit into either the Sherman or Lancefield classification schemes. However, as will become clear later in this chapter, *S. pneumoniae* possesses many properties that warrant its inclusion in the genus *Streptococcus*. However, before we can discuss these classification schemes, we must learn about the hemolysins associated with the streptococci.

Hemolysins

As streptococci grow, they secrete a large number of toxins and enzymes into the surrounding medium. For the most part, we do not know the role of these products in the production of disease, but we do make use of them for identification of the streptococci and, in some cases, for diagnosis of a past streptococcal infection. Among these secreted products may be one of several hemolysins that cause the lysis of red blood cells present in the medium. On the basis of the type of red cell destruction caused by the hemolysin, the streptococci can be divided into three groups.

The α(alpha)-hemolytic streptococci produce a hemolysis of the red blood cells that results in a greenish-brown discoloration surrounding the colony. The green discoloration is caused by the formation of an unidentified reduced product of hemoglobin. The streptococci producing α-hemolysis are also called **viridans** streptococci.

The β(beta)-hemolytic streptococci cause a hemolysis of red blood cells surrounding the colony, resulting in a completely clear zone in which no color of any kind remains (see Figure 24.2). We now know that β-hemolysis by the streptococci results from the secretion of one or both of two different hemolysins by the cells. These hemolysins have been named **streptolysin S** and **streptolysin O**.

Streptolysin S was so named because early work with this hemolysin showed that it could be extracted from cells with serum. However, because it is stable in the presence of atmospheric oxygen (streptolysin O is not), the S might represent *stable* as well as *serum-extractable*. Because streptolysin O is inactivated in the presence of atmospheric oxygen, the hemolysis one

TABLE 24.1 **Major Criteria Used in Sherman's Biochemical Classification of the Streptococci**

Group	Characteristics
Pyogenic (all Lancefield groups except D and N)	Mostly β-hemolytic, will not grow at 45°C or in the presence of 6.5% NaCl
Viridans (not classifiable in Lancefield's classification)	α-hemolytic, will grow at 45°C but not in the presence of 6.5% NaCl
Lactic (Lancefield's group N)	Nonhemolytic, will not grow at 45°C or in the presence of 6.5% NaCl, will grow at 10°C in the presence of 0.1% methylene blue in milk
Enterococcus (Lancefield's group D)	Usually not hemolytic, will grow at 45°C in the presence of 6.5% NaCl, will grow at pH 9.6

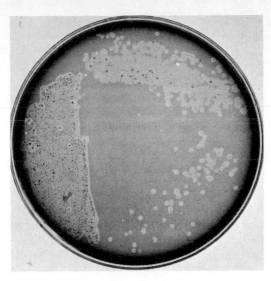

Figure 24.2 Streak plate of β-hemolytic streptococci, showing a zone of complete hemolysis around each colony.

sees on an aerobically incubated blood agar plate is the result of streptolysin S rather than streptolysin O activity. If, however, the cells are grown anaerobically, both hemolysins produce β-hemolysis.

Streptolysin O binds to cholesterol in the cell membrane and causes lysis of erythrocytes. However, because anemia is not a usual result of streptococcal infections, it is not this function that contributes to the virulence of the organisms; it is the ability of streptolysin O to react with sterols in the host's leukocyte membranes and cause a release of the enzymes in the cell's lysosomes, resulting in degranulation and death of the leukocyte (see Figure 24.3). Furthermore, strep-

Figure 24.3 Fragment of rabbit erythrocyte lysed with streptolysin-O (SLO). The toxin polymers form channels through the cell membrane. *Credit:* Sucharit Bhakdi. Institute of Medical Microbiology, University of Gussen.

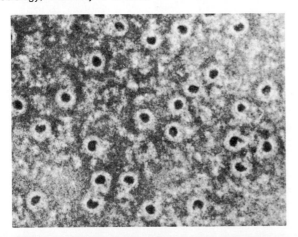

tolysin O is capable of suppressing chemotaxis and leukocyte mobility. The fact that streptolysin O is antigenic and that antistreptolysin O (ASO) will prevent the hemolysis of erythrocytes provides a very important tool for determining a recent streptococcal infection. Such information is a propitious aid in the diagnosis of the late complications of streptococcal infections, namely, rheumatic fever or acute glomerulonephritis.

Streptolysin S is a small polypeptide which, in addition to causing β-hemolysis, is able to inhibit chemotaxis and phagocytosis as well as exert a cytotoxic effect on certain types of eucaryotic cells. It acts by binding to phospholipid in the target cell membrane, and the mere external contact of group A streptococci with leukocytes is sufficient to kill the phagocytic cells. Unlike its O counterpart, streptolysin S is not antigenic—probably because of its small size.

The third group of streptococci produces no hemolysins and hence has no effect on the blood cells in the agar medium. Members of this group of streptococci sometimes are called the γ(gamma)-hemolytic streptococci, although the term is really a misnomer, because they are not at all hemolytic.

Antigenicity

Now back to our classification of the streptococci. The major classification of these bacteria—particularly those of medical importance—is based on the presence of antigens in or on the cell walls of the streptococci. Lancefield found that if streptococci were placed in dilute acid (pH 2) and heated at 100°C for 10 min, a soluble carbohydrate antigen was extracted from their cell walls. This carbohydrate, which she called C carbohydrate, could also be extracted with formamide by heating the cells at 150°C for 15 min. All streptococci except the viridans groups (see Table 24.1) possessed this C carbohydrate. When the C carbohydrate from many different streptococcal isolates was categorized using antibodies directed against the streptococci from which they were extracted, it was found that there were 13 different antigenic C carbohydrates. Based on these antigenic differences in their C carbohydrate, Lancefield divided the streptococci into groups that have been designated as groups A, B, C, D, E, F, G, H, K, L, M, N, and O. Thus, the organisms in group A all possess the same antigenic C carbohydrate. The organisms in group B also all possess the same C carbohydrate, but a different one from that found in group A organisms. It soon became apparent that each group had a usual habitat; group A were primarily human pathogens, group B cattle and human pathogens, and so on (see Table 24.2). Though common, these habitats are not rigid.

TABLE 24.2 Lancefield's Group Classification and Normal Habitat of Streptococci

Group	Normal Habitat
A	Humans
B	Cattle and humans
C	Wide variety of animals and humans
D	Intestinal tract of humans and animals (enterococci)
E	Swine
F	Humans
G	Humans and dogs
H	Humans
K	Humans
L	Dogs
M	Dogs
N	Dairy products (never hemolytic on blood agar)
O	Humans

In addition to the carbohydrate used to classify the streptococci into groups, other antigens are present in each group. The characterization of these antigens allows the further subdivision of each group of organism into types. The type-specific substance of some groups is a carbohydrate (different from the group carbohydrate), whereas in other groups it is a cell wall protein.

Group A (*S. pyogenes*) is by far the most important to us, since it contains most of the streptococci that produce acute disease in humans. These organisms are for the most part β-hemolytic. Group A can be further divided into types on the basis of a protein in the cell wall called the M protein (see Figure 24.4). Serological differences in the M protein allow for the subdivision of group A streptococci into over 50 different types. The M protein is toxic to host cell leukocytes and, as a result, prevents phagocytosis of the invading organisms. The rare streptococcal mutants that have lost their M protein are rapidly phagocytosed and killed. It is interesting to note that although antibodies are produced against both the C carbohydrate and the M protein in the group A streptococci, only antibodies against the M protein are important in preventing recurrent disease. This is because specific antibodies to the M protein neutralize its toxicity to host cell leukocytes and promote the phagocytosis and destruction of the organism. It can be readily seen that acquired immunity is type specific because it is directed against the type-specific M protein. Following recovery from one type of infection, a person would be immune only to reinfection by that type of streptococci but would not possess immunity to other types of group A streptococci.

Toxigenicity

In addition to the various hemolysins produced by the streptococci, several other toxins or enzymes may be secreted in varying amounts by different strains of group A streptococci. These extracellular products include enzymes that break down nucleic acids (DNase), hydrolyze nicotinamide adenine dinucleotide

Figure 24.4 (*a*) Layers of the cell wall and capsule of a group A streptococcus. (*b*) Section through a streptococcal cell showing fimbriae of M protein (×100,000). (*c*) Section through a type of streptococcal cell without M protein. (×80,000; compare with *b*).

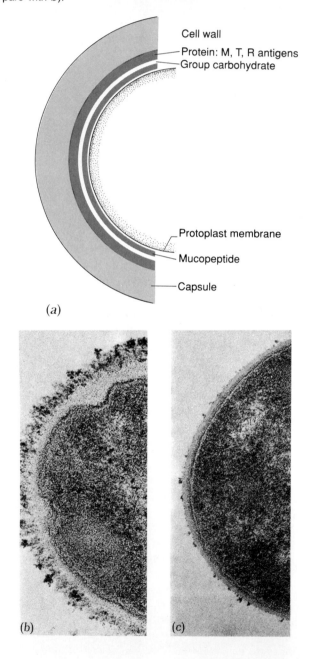

Cell wall
Protein: M, T, R antigens
Group carbohydrate
Protoplast membrane
Mucopeptide
Capsule

(*a*)

(*b*)   (*c*)

(NADase), hydrolyze hyaluronic acid (hyaluronidase), lyse blood clots (streptokinase), cause a skin rash (erythrogenic toxin), kill leukocytes (leukocidins), and kill mice (lethal toxin). Little is known concerning some of these products, whereas others have been more extensively studied and may well be important factors in the ability of the streptococci to produce disease. For example, streptokinase (once called fibrinolysin) causes the lysis of a blood clot by activating an inactive proteolytic enzyme (plasminogen), which is normally present in blood plasma, to an active form (plasmin), which is then able to dissolve the blood clot. Streptokinase is produced commercially and is used by many hospitals to dissolve blood clots occurring within the heart following a heart attack. This has now, however, been largely superceded by the use of a cloned human tissue plasminogen activator. The prompt removal of such clots reduces damage to the heart, resulting in the more rapid recovery of the patient. Also, there are indications that the excreted enzyme, NADase, can kill leukocytes. Hyaluronidase has been called the **spreading factor** by some investigators because the ability to hydrolyze the hyaluronic acid in the host's connective tissue is thought to enable the organism to spread more rapidly. It is noteworthy that many streptococci form large hyaluronic acid capsules, which appear to prevent the host's leukocytes from phagocytosing and destroying them.

Group A β-Hemolytic Streptococcal Infections

A survey of the pediatric practice in Rochester, New York, showed that approximately 10 percent of all office visits were related to streptococcal disease. By far the most common types of group A streptococcal infections are pharyngitis (sore throat) and impetigo (infection of the skin).

Streptococcal Sore Throat Clinical features of streptococcal infection vary. The most frequent type is an infection of the tonsils and the pharynx. In children particularly, a streptococcal sore throat may be acute. The mucous membranes are usually red and swollen, with a purulent exudate. Cervical lymph nodes may be enlarged, and the temperature is usually high. The white blood cell count also is usually increased. The incubation period varies from 1 to 3 days and the infection is usually self-limiting, with symptoms lasting less than 5 days. Epidemics of this disease are usually the result of personal contact with either infected people or healthy carriers. Epidemiologic studies have shown that it is usually the school child who brings these infections home and spreads them in the family.

Impetigo Impetigo (also called streptococcal pyoderma) is an infection of the skin that occurs most often in young children, particularly those living in crowded, low-socioeconomic-level conditions. Streptococcal impetigo is characterized by the occurrence on the skin of small vesicles that eventually form a thin, amber crust. The lesions are painless and healing occurs without scarring.

Scarlet Fever Scarlet fever may be caused by any of the types of group A streptococci that are able to secrete one of the **erythrogenic toxins.** There are three distinct types of this toxin, which are also termed **streptococcal pyrogenic exotoxins,** each of which will cause a skin rash. There are considerable data suggesting that the actual rash is a result of a hypersensitivity reaction to the toxin. Thus, scarlet fever is a streptococcal infection (i.e., sore throat) in which a pyrogenic toxin-producing strain is involved. It is now known that, like many exotoxin-producing bacteria, the pyrogenic toxin-producing streptococci are lysogenic, and toxin production is a result of their lysogenicity, or lysogenic conversion. The streptococcus itself is usually confined to the throat and the nasopharynx, but on some occasions it may invade the blood to cause a streptococcal blood infection. After the onset of the sore throat, the scarlet fever rash usually appears within two days.

It should be noted that the streptococcal pyrogenic toxins are part of a large family of toxins that includes staphylococcal pyrogenic toxins A and B, staphylococcal enterotoxins A through E, and staphylococcal toxic shock syndrome toxin. All of the toxins in this group share a number of biological properties such as pyrogenicity (induction of fever), immunosuppression, mitogenicity for lymphocytes, and the ability to greatly enhance the lethality of endotoxic shock. It is not surprising, therefore, that a number of cases of typical toxic shock syndrome (see staphylococcal infections) have been ascribed to infections caused by group A streptococci. In one study of severe group A streptococcal infections associated with a toxic shock-like syndrome (mortality 30 percent), cultures from such patients predominately produced pyrogenic toxin A.

It is also noteworthy that a streptococcal shock-like syndrome is not always a throat or wound infection. Muppeteer Jim Hensen's sudden death in 1990 resulted from a pneumonia caused by a pyrogenic toxin-producing group A streptococcus. Such infections may result in death in a matter of hours after the onset of symptoms.

Immunity to scarlet fever results from the presence of antitoxin to the pyrogenic toxin in the blood. It must be remembered that the body then is protected only against the scarlet fever toxin and that other types of streptococci can still invade and cause other infections.

Other Group A Streptococcal Infections

Puerperal sepsis is an infection of the uterus that has claimed the lives of many women following childbirth. Fortunately, modern aseptic techniques have eliminated much of this type of infection in developed countries. Streptococci may also occasionally spread to the sinuses and middle ear.

Wound infections and puerperal sepsis may also be caused by the obligately anaerobic streptococci that are part of the normal flora of the intestinal tract. Such infections usually occur in conjunction with some of the other obligately anaerobic organisms from these areas, for example, *Bacteroides*.

Late Nonsuppurative Complications

Two serious complications of group A streptococcal infections are rheumatic fever, involving the heart and joints, and glomerulonephritis, involving the kidney. These complications are referred to as late nonsuppurative (non-pus-forming) sequelae because symptoms for either one do not begin until 1 or 2 weeks after apparent recovery from the acute streptococcal infections. Also, during the course of these complications, it is not possible to isolate streptococci from the patient.

Rheumatic Fever

Rheumatic fever follows a small percentage of untreated β-hemolytic group A streptococcal infections. It may occur as a sequela of any type of group A infection. There is at present no evidence that streptococcal infections developing into rheumatic fever are any different from those that do not. Recovery from rheumatic fever occurs with no permanent damage to the joints, but the involvement of the heart is the most important part of this disease, because it is in this organ that permanent damage may occur. The mechanism by which the streptococci produce rheumatic fever is still obscure, but a great deal of circumstantial evidence indicates that rheumatic fever is the result of an immunological reaction. For example, infection with group A streptococci might result in the deposition of streptococcal antigens (such as the M protein) into the tissues of the joints and heart. When the patient's antibodies form and react with these antigens, damage results. Other reports claim that streptococci have an antigen that induces the synthesis of antibodies that will damage the heart. This concept is strongly supported by the isolation of a series of murine monoclonal antibodies that bind to both *S. pyogenes* and human heart tissue. The antigens involved have been shown to possess at least three different cross-reacting epitopes (antigenic determinants) common to the group A streptococcus and the heart. The concept that rheumatic fever is the result of an immunological reaction is also supported by postmortem studies that showed large deposits of immunoglobulins and the C3 component of complement within the damaged heart. Thus, although there are gaps in our knowledge of the specific antigen involved, it is clear that the reaction of an antibody induced by the streptococcus—either directly with heart tissue or with an antigen deposited in the heart—would activate the complement cascade as described in Chapter 19; this would result in the formation of anaphylatoxins (C3a and C5a) that cause localized tissue damage.

Glomerulonephritis

Glomerulonephritis, an inflammation of the glomeruli in the kidney, is less frequently a consequence of streptococcal infection than is rheumatic fever. Many cases of glomerulonephritis follow group A, type 12 streptococcal infections, but a few have been reported to follow types 4, 18, 25, 49, 52, and 55. Interestingly, a number of cases of glomerulonephritis follow streptococcal skin infections, a situation not seen with rheumatic fever. Bloody urine (hematuria) is the predominant symptom, often accompanied by high blood pressure. Glomerulonephritis is believed to be an autoimmune disease in which either the streptococci possess or synthesize an antigen that induces antibodies that react with glomerular basement membranes of the kidney or the streptococci deposit antigen-antibody complexes on the basement membranes. In either case, the activation of the C3 and C5 components of the complement system would lead to tissue destruction.

Diagnosis

To prevent late nonsuppurative complications, it is essential that the β-hemolytic streptococcal infections be diagnosed promptly and treated adequately. Although there are several techniques for the diagnosis of streptococcal infections, the only positive way to make a diagnosis is to isolate and identify the causative organism. To accomplish this, material from the patient (usually a throat swab) is streaked on a blood agar plate. After approximately 24 h of incubation at 35°C, the plate is examined for the presence of β-hemolytic streptococci, indicated by tiny, compact, dull colonies surrounded by areas of clear hemolysis. If a Gram stain of such a colony reveals gram-positive cocci in chains, a positive diagnosis is made.

Since group A streptococci are more sensitive than other groups of streptococci to the antibiotic bacitracin, presumptive evidence that a β-hemolytic streptococcus belongs to group A can be obtained by using disks containing a calibrated amount of bacitracin,

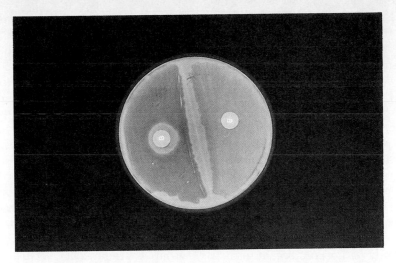

Figure 24.5 Effect of bacitracin on the growth of group A streptococci. Note the clear area around the bacitracin disk where the group A streptococci are unable to grow as compared to growth immediately adjacent to the disk by non–group A streptococci.

which will inhibit group A streptococci but not other groups of streptococci (see Figure 24.5). Fluorescent antibodies against the C substance of group A may also be used for a definitive identification of these organisms.

Probably the most common method uses commercially available kits that contain antibodies to the C carbohydrate of group A streptococci bound to latex beads. In this procedure, the throat is swabbed and the group A carbohydrate is extracted from the swab into about 150 μL of extraction solution. Depending on the intensity of the infection, this provides sufficient group A carbohydrate to cause agglutination of the antibody-coated latex beads. A major advantage is that this procedure requires less than 30 min to complete, allowing the physician to prescribe an appropriate antibiotic before comparing these "quick" results with culture results 24 h later.

It has been estimated that approximately 3 percent of untreated group A streptococcal infections culminate in rheumatic fever. A prompt diagnosis of a streptococcal sore throat is therefore imperative, because rheumatic fever can be prevented if adequate antibiotic therapy is initiated within 1 week after the initial symptoms of the infection.

Prevention of glomerulonephritis is not so clearcut, perhaps because many such cases follow streptococcal skin infections that are relatively mild and so do not routinely receive early treatment.

By the time a patient experiences the symptoms of either rheumatic fever or glomerulonephritis, the individual has usually recovered from the streptococcal infection and it is no longer possible to find the organisms. Consequently, in order to make a diagnosis, it is necessary to show that the individual has recovered from a recent streptococcal infection. This can be done by measuring the amount of antibodies in the patient's serum that are directed toward streptococcal products. For example, one can determine the amount of antibody a patient has that is directed against streptolysin O. If the antistreptolysin O titer is high, it indicates a recent infection by β-hemolytic streptococci. Another such test, termed the streptozyme test, uses sheep's red blood cells that have been coated with a crude mixture of many streptococcal products. When such cells are mixed with a dilution of patient's serum, the presence of any antistreptococcal antibodies will cause the agglutination of the sheep's red blood cells.

Clinical Management of Group A Infections

Fortunately, almost all streptococci except some enterococci are sensitive to a wide variety of antibiotics. Penicillin is still the antibiotic of choice, but most experts agree that therapeutic levels of penicillin must be maintained for at least 8 to 10 days to ensure complete eradication of the organisms. It is extremely important to remember that adequate treatment during the acute infection will prevent the complications of rheumatic fever. In spite of the ability of penicillin to destroy the streptococci, even intensive antibiotic therapy does little to shorten the course of a throat infection. In general, most symptoms last about 5 days whether or not penicillin is administered.

Individuals who have recovered from rheumatic fever are particularly vulnerable to a recurrence of rheumatic fever in the event they again become infected with a group A streptococcus. As a result, these persons are usually kept on low levels of oral penicillin for many years in order to prevent such an infection. People who have recovered from glomerulonephritis are usually not given prophylactic penicillin because there are only a few types of streptococci that might cause a recurrence of their symptoms.

It is essentially impossible to control the spread of streptococcal infections because there are so many asymptomatic, healthy carriers of β-hemolytic streptococci. Depending on the season of the year, anywhere from 5 to 10 percent of apparently normal people may carry these potentially pathogenic organisms.

Group B Streptococci

Group B streptococci (*S. agalactiae*) are the causative agents of bovine mastitis and until recently were not believed to cause serious human disease. However, it is now established that group B streptococci are present in the vaginal flora of approximately 25 percent of all women and, although these organisms rarely cause overt disease in adults, cases of bacteremia (bacteria in the blood) and meningitis (infection of the membrane surrounding the brain and spinal cord) have occurred in diabetics and in individuals on immunosuppressive drugs. In the United States, these organisms are also responsible for about 45,000 cases per year of postpartum endometritis (infection of the endometrium) following childbirth. Group B streptococci also have been frequently isolated from rectal swabs. The organisms are probably spread from person to person both by indirect contact and as a sexually transmitted disease.

It is in the newborn that serious infections most commonly occur. About 1 percent of children born to mothers infected with group B streptococci will develop bacteremia and pneumonia within the first 5 days of life, and such infections carry a 50 to 70 percent mortality rate in spite of intensive antibiotic therapy. Moreover, about 50 percent of infants surviving a group B streptococcal infection will succumb to permanent neurological handicaps, such as mental retardation, cerebral palsy, or deafness. Meningitis also occurs, with a mortality rate of approximately 15 percent, but this complication is usually seen between the 10th and 60th day of life. Altogether, about 7000 to 10,000 infants are infected each year in the United States. All five serotypes of group B streptococci are equally involved in the early infections. However, type III is much more frequently associated with the later meningeal infections of neonates. These infants come from mothers having no type III antibody, suggesting that the mother is not the source of the infection.

Antibiotic therapy of expectant mothers who are colonized with group B streptococci has been disappointing, perhaps because the organisms are difficult to eradicate from the lower intestinal tract, which can act as a reservoir to infect the mucous membranes of the vagina. The most promising procedure for the prevention of neonatal group B streptococcal infections was reported in 1986 by Kenneth Boyer and Samuel Gotoff. They found that if expectant mothers with group B streptococcal colonization were given intravenous ampicillin during labor, only 9 percent of infants born to such mothers were infected with the group B streptococci versus 51 percent of infected infants from untreated controls.

It has also been suggested that if a vaccine inducing type-specific antibodies were available for expectant mothers, its use would provide the newborn with protective maternal antibodies. Results from experimental trials using type III capsular polysaccharide to immunize expectant mothers is encouraging.

Other Streptococcal Infections

Subacute Bacterial Endocarditis

The α-hemolytic streptococci include a number of species that are normal flora in the throat and nasopharynx of humans and *Enterococcus* (Lancefield group D), which are normal inhabitants of the large intestine. Neither produces acute infections, but they can act as secondary invaders.

The one infection in which these streptococci are unquestionably involved is **subacute bacterial endocarditis.** This infection may be caused by a number of different bacteria, but the viridans streptococci and the enterococci are the most common etiological agents. It occurs in people who already have injured heart valves (congenitally formed or possibly damaged by rheumatic fever). On reaching the damaged valve, the organisms multiply, forming vegetative growths that further impede the functioning of the valve. The characteristic findings in such cases are fever, heart murmurs, enlarged spleen, and anemia. Once the disease is suspected, the diagnosis is made by obtaining blood cultures from which the bacteria are isolated. Left untreated, this condition is almost invariably fatal.

The α-hemolytic streptococci are sensitive to penicillin; usually an intensive course of penicillin therapy for 2 weeks is effective. When an enterococcus is the infecting organism, a combination of penicillin and streptomycin is commonly used.

Dental Caries

The observation that dental caries are the result of a bacterial infection is now well established. The demineralization of tooth enamel requires the presence of dense masses of bacteria that adhere tightly to the tooth. Such masses are termed **dental plaques.** The first step in the formation of a dental plaque is the adsorption of glycoproteins from the saliva. This film, called the **acquired pellicle,** is formed within minutes on a clean tooth. Saliva contains approximately 10^8 bacteria per milliliter and, of these, *S. sanguis* appears to possess the greatest affinity for the acquired pellicle. Other gram-positive cocci and rods (predominantly other species of streptococci and lactobacilli) soon adhere to the acquired pellicle and to each other. As these organisms grow in the early plaque, small niches of anaerobiosis are formed, permitting the growth of the obligate anaerobes such as *Actinomyces, Viellonella, Bifidobacterium, Bacteroides, Fusobacterium,* and spirochetes.

The mechanism by which these organisms bind to the tooth surface is not yet fully understood. However, it is apparent that many, if not all, of the bacteria making up the dental plaque possess specific receptors for either the glycoproteins of the acquired pellicle or for each other. These receptors have been shown in many cases to exist as fimbriae, which have the lectinlike properties of binding to various carbohydrates present on the bound salivary glycoproteins or to polysaccharides excreted by bacterial residents of the plaque. Thus, if not mechanically removed, a tightly adhering mature plaque develops on the tooth surface, consisting of millions of bacteria per square millimeter, all enmeshed in salivary glycoproteins and excreted bacterial polysaccharides.

The formation of dental caries results from the dissolution of the tooth enamel by bacterially produced organic acids. It seems probable that many of the organisms comprising the dental plaque contribute to caries formation but the principal etiological agent is *S. mutans.*

S. mutans secretes an extracellular glucosyltransferase, also called dextransucrase, that specifically forms a large insoluble polymer of glucose (i.e., a glucan), as is shown by the following equation:

$$n \text{ Sucrose} \xrightarrow{\text{dextransucrase}} (\text{glucan})n + \text{fructose}$$

The glucan (also termed a glycocalyx) adheres tightly to the surface of the tooth and to the bacterium, bringing viable streptococci into very close association with the tooth enamel. As the streptococci, which are embedded in the plaque, ferment the fructose that was cleaved from the sucrose, lactic acid is formed. As a result of its close contact with the tooth, this lactic acid causes decalcification, which allows subsequent decay in the dentin area of the tooth. Moreover, it should come as no surprise that countries with a high per capita sugar consumption also suffer from a high rate of dental caries. This is indicated in Figure 24.6, which shows that children in countries such as China have many times fewer cavities than do those in the United States.

Controlled studies using 436 adults in a mental institution in Sweden demonstrated that the incidence of dental caries was related not only to the amount of sugar consumed but also to the frequency of sugar ingestion and the nature of the sugar-containing food. Thus, individuals who consumed sugar only at mealtime experienced fewer caries than those consuming equal amounts of sugar largely between meals. Such results are no doubt related to the fact that sugar ingested during meals is more rapidly cleared from the mouth than is sugar consumed between meals. It is also noteworthy that surveys in Europe and Japan showed that the number of caries was dramatically

Figure 24.6 Cumulative dental decay prevalence, expressed as decayed, missing, and filled (DMF) permanent teeth in children ages 11 to 12; corresponding 1959 per capita sucrose utilization data from 18 countries and the state of Hawaii, from the Food and Agricultural Organization of the United Nations.

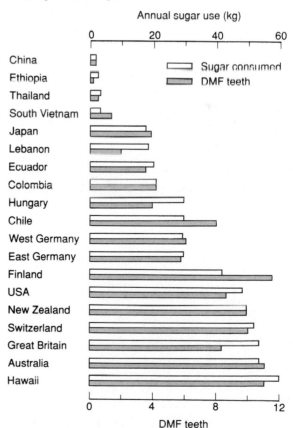

reduced during the period of wartime restrictions of sugar and sugar products.

You can see, therefore, that although sugar is the indirect cause of dental caries, bacterial fermentation is the direct cause of such decay. As a result, prevention of caries should be directed toward good dental hygiene (brushing and flossing) to remove any dental plaques before extensive fermentation occurs.

Because this enzyme functions only on sucrose, the avoidance of candies and foods rich in sucrose is effective in reducing tooth decay.

Experimental immunization of monkeys by the injection of *S. mutans* cells and cell products into the vicinity of the salivary and parotid glands has been reported to cause the formation of IgA antibody, resulting in the formation of fewer dental caries in immunized animals than in control animals. Subcutaneous immunization with whole cells or with isolated proteins derived from cultures of *S. mutans* has also been used successfully to reduce the occurrence of dental caries in rhesus monkeys.

Periodontal Disease

Periodontal disease derives its name from the fact that it is a disease that occurs around (peri-) the tooth (dontal). It is thus an infection that involves the gingiva (gums) and the alveolar bone, which supports the teeth. Periodontal disease is a slow, chronic, painless infection that affects about 80 percent of teenagers and adults. It is mostly a preventable disease and is the major cause of tooth loss.

Anatomy of Tooth Support A somewhat simplistic view of tooth attachment is shown in Figure 24.7. Here it can be seen that the gingiva forms a tight collar around the neck of the tooth and that a structure called the **periodontal ligament** attaches the tooth to the alveolar bone.

Plaque Formation in Periodontal Disease As described for dental caries, the same plaque formation is the initiating event in periodontal disease. Only, in this case, the overall process is considerably more complex than simple decalcification by organic acids.

Periodontal disease begins when the plaque extends into the gingival sulcus. If this plaque is not removed by brushing or flossing, calcium is deposited on the plaque surface, forming a very rough, stony crust, termed **calculus** or **tartar,** which binds tenaciously to the tooth surface. More plaque continues to be laid down on calculus surfaces, eventually forming a thick layer of calcified plaque.

Pathogenesis of Periodontal Disease It should be kept in mind that there are two primary pathological events occurring during periodontal disease. First, the extension of the calculus to the bottom of the gingival sulcus creates conditions that lead to inflammation of the gingiva. Such gingivitis may be chronic for years and is mainly characterized by a small amount of bleeding that occurs while brushing the teeth. This bleeding results from a loss of attachment of the gingival epithelium to the tooth surface, and its subsequent tearing by the cutting action of the calculus, which, as it progresses, allows pockets of inflammation to form in the gingival sulcus. The depth of these pockets, as measured by a periodontal probe, provides an indication of the extent of the periodontal disease (Figure 24.8).

The second event of periodontal disease is the resorption of the alveolar bone structure surrounding

Figure 24.7 Normal tooth showing support through linkage of the periodontal ligaments to the alveolar bone.

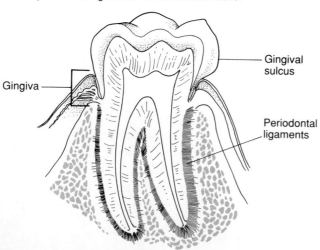

Figure 24.8 Note loss of bone and support of tooth in moderately advanced periodontal disease.

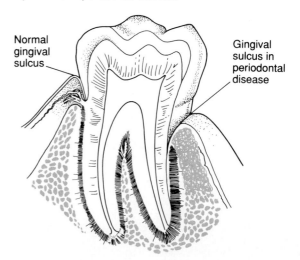

the roots of the tooth. As this progresses, the tooth becomes loose and falls out. The plaque organisms do not actually penetrate the gingival tissue. Instead, the pathogenicity of plaque bacteria appears to result from (1) destruction by secreted enzymes, (2) inflammation, and (3) allergic reactions to bacterial products diffusing from the plaque.

Prevention and Control of Periodontal Disease The prevention of periodontal disease requires the mechanical removal of plaque that has accumulated in the gingival sulcus. This can be accomplished by brushing and flossing. However, once this plaque becomes calcified, it can be removed only by scraping the tooth, a procedure that requires professional skill. In more advanced cases of periodontal disease, it may be necessary to cut back the gingiva to permit the removal of accumulated calculus.

Streptococcus pneumoniae: Pneumococcal Pneumonia

Organisms in the species *S. pneumoniae* are most commonly referred to as the **pneumococci.** They are lancet-shaped cocci that are usually arranged in pairs, although short chains also may be seen (see Figure 24.9). They are gram-positive, nonmotile, and non-spore-forming but are encapsulated. On the basis of the antigenic nature of the polysaccharide capsule, it is possible to subdivide this species into approximately 80 types.

Nutritionally the pneumococci are very similar to the other streptococci; thus, they are usually grown on a blood agar medium. When growing on blood medium, the pneumococci produce a hemolysin, which is always of the alpha type. Other types of extracellular toxins or enzymes known to be produced by the pneumococci are leukocidins and hyaluronidase.

Since the pneumococci very closely resemble the viridans streptococci in colonial appearance and morphology, it is necessary to do additional tests to differentiate between the two. Two laboratory procedures are used for this purpose. The first is based on the fact that the pneumococci will dissolve in bile salts. In fact, any surface-active agent (such as soaps or detergents) seems to activate their autolytic enzyme system, resulting in the complete lysis of the pneumococci. This is usually spoken of as **bile solubility.** Since the viridans streptococci are not bile soluble, this test serves to distinguish between the two.

Because a bile solubility test requires prior growth of the organisms, this determination has been supplanted by a much simpler method: commercially available optochin disks (impregnated with ethylhydrocupreine hydrochloride) are laid down on the surface of an agar plate that has been inoculated with the unknown organism. The pneumococci are exceedingly susceptible to this compound and will fail to grow in the proximity of the disk. The α-hemolytic viridans streptococci are insensitive to Optochin and will grow adjacent to the implanted disk. A third laboratory procedure is used less frequently and is based on the pathogenicity of the pneumococci for white mice. The intraperitoneal injection of only a few pneumococci (as contrasted with the streptococci) into a white mouse will result in death, usually within 18 to 24 h. Type XIV pneumococcus appears to be the one exception to this rule.

Pathogenicity

It appears that the major, or perhaps sole, reason why the pneumococcus can survive and produce disease in the host is related to its capsule. Once a strain loses its capsule, it can very readily be phagocytosed and destroyed. Thus a rough unencapsulated strain of pneumococcus is no longer virulent. The conversion of one type of *S. pneumoniae* to a new type by the use of purified DNA extracted from organisms of the new type opened a new area of genetic understanding.

Pneumococcal pneumonia is an acute inflammation of the lungs that is most commonly lobar, that is, involving the tissues in one or more lobes of the lungs. The disease is usually sudden in onset and is characterized by chills, fever, and pleural pain. The alveoli fill with exudate, and in about 25 percent of cases, bacteremia is found early in the course of the

Figure 24.9 *S. pneumoniae* (about ×46,000). Note the diploid arrangement of cells and the lancet shape of individual cells. Capsules are not visible in this electron micrograph.

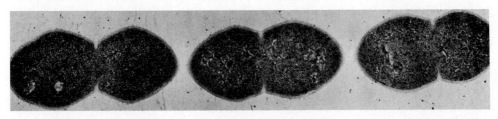

disease. The incubation period for pneumococcal pneumonia is usually 1 to 3 days.

The estimated incidence of pneumococcal pneumonia varies considerably with an individual's age, conditions of crowding, and occupation. In the United States, the incidence is thought to be 300,000 to 400,000 cases per year, with an overall mortality of 15,000 to 60,000. Military personnel living in a closed group have an incidence of 10 to 20 times higher per 1000 population; the rates of pneumococcal pneumonia in African gold miners is about 100-fold greater than that for the general population.

S. pneumoniae is frequently found as part of the normal flora of the upper respiratory tract in healthy individuals. We do not understand the mechanism that prevents these carriers from getting pneumonia, but it does seem that they are a source for the dissemination of the organisms. The infected patient may also act as a source for the spread of these bacteria but usually has less opportunity to do so than the healthy carrier.

The pneumococci also have the dubious honor of being the second most common cause of bacterial meningitis in adults. Meningitis may arise as a complication of pneumonia or a sinus infection in which the bacteria reach the meninges by way of the bloodstream, or it may result from a skull fracture or other injury permitting organisms from the nasopharynx to enter the meninges.

Finally, we should note that at least 75 percent of all children have middle ear infections (otitis media) by age 6; the pneumococcus is the etiological agent for about half these infections. Such infections are rare in adults. Children experiencing an initial middle ear infection during the first year of life are likely to have recurrent infections during early childhood, resulting in some hearing loss.

Clinical Management

Collection and examination of sputum are very important. Direct smears of sputum can be made to observe the gram-positive encapsulated diplococci. One rather specific test that aids in establishing the identity of the pneumococci is the **Neufeld (quellung) test.** This test is based on the fact that if encapsulated bacteria are mixed with type-specific antibodies, the capsule becomes more refractile and appears to swell to a size considerably larger than normal. This capsular swelling is diagnostic (see Figure 24.10). Actually, since the advent of penicillin, the Neufeld test is used only as an epidemiological tool; such typing is no longer routinely done. Thus one must rely mainly on the morphological appearance and optochin sensitivity for the identification of the pneumococci.

In cases of suspected pneumococcal meningitis, essentially the same diagnostic procedures are followed except that a spinal tap is performed and the

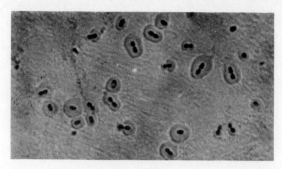

Figure 24.10 Encapsulated cells of *S. pneumoniae*. The capsules around the pairs of cells are swollen by use of type-specific antibody.

spinal fluid is observed rather than the sputum. It is also routine to withdraw and culture blood from a patient suspected of either pneumonia or meningitis in order to provide another means of isolating the organism and, hence, arriving at a definitive diagnosis.

The most effective treatment of pneumococcal infections is the administration of antibiotics such as penicillin, the tetracyclines, or erythromycin. Of these, penicillin is probably the best and most frequently used. However, occasional rare penicillin-resistant strains have been isolated, particularly from patients who are on extended long-term therapy with penicillin or ampicillin because of immune deficiencies or for rheumatic fever prophylaxis. In such cases erythromycin or chloramphenicol has been effective. Sulfonamides will also exert a bacteriostatic effect on the pneumococcus but are not as effective in treatment.

Antiserum could be used, but in this case it is necessary to type the organism first so that the proper specific antiserum can be administered. Actually, since antibiotics have been available, antiserum therapy is no longer used.

Because the organisms are so widely distributed in the human population, control measures are difficult. The case fatality rate for pneumococcal pneumonia averages between 5 and 10 percent, and for individuals older than 70 years of age, it approaches 60 percent. The observation that about 90 percent of all pneumococcal disease is caused by only 23 different types of pneumococci has prompted the production of a 23-type pneumococcal vaccine. This vaccine, licensed in the United States in 1978, consists of purified polysaccharide capsular material from types 1, 2, 3, 4, 5, 8, 9, 12, 14, 17, 19, 20, 22, 23, 26, 34, 43, 51, 54, 56, 57, 68, and 70. Field tests have indicated that a single injection of this vaccine results in a rise in antibody to the capsular polysaccharides and a decrease of the incidence of pneumococcal pneumonia.

Mass immunizations are not proposed at this time, but the vaccine is currently used for individuals having an enhanced risk of pneumococcal disease, such as elderly people in nursing homes, people with chronic diseases such as diabetes mellitus, persons who have lost their spleens, and any group with a high risk of infection with influenza virus.

OTHER RESPIRATORY INFECTIONS

Neisseria meningitidis: Epidemic Meningitis

Bacteria in the genus *Neisseria* are characteristically gram-negative diplococci (see Figure 24.11). Several nonpathogenic species can be routinely isolated from normal saliva. The most common of these is *N. sicca*. All *Neisseria* are nonmotile, and all species are able to oxidize dimethyl- and tetramethylparaphenylene diamine hydrochloride. This property can be used to distinguish colonies of *Neisseria* from colonies of most other bacteria growing on the same plate, but it does not distinguish one species of *Neisseria* from another. In actual practice the test, called the **oxidase test,** is carried out by flooding the colonies on a plate with either dimethyl- or tetramethylparaphenylene diamine hydrochloride and observing for color changes in the colonies. Colonies that are oxidase-positive (*Neisseria*) first turn pink, then dark red, and finally black. The reagent eventually kills the cells, but if the colony is restreaked on fresh media before it turns dark black, the culture can be grown again. The reagent can also be placed on a piece of filter paper and part of a

suspected bacterial colony smeared with a platinum loop or toothpick on the wet filter paper. A positive reaction will be indicated by the development of a dark purple color within 10 to 15 s.

Only two species in this genus are pathogenic for humans. They are *N. meningitidis,* the causative agent of epidemic meningitis, and *N. gonorrhoeae,* the cause of gonorrhea. *N. gonorrhoeae* is discussed in Chapter 27. Both of these pathogenic species possess certain characteristics not found in nonpathogenic *Neisseria* that are part of our normal flora. The first of these is the inability of pathogenic *Neisseria* to grow at a temperature below 30°C. All nonpathogenic species grow at room temperature, around 20 to 22°C. The second major difference is seen in the growth requirements of the pathogens. The nonpathogens grow well on ordinary laboratory media exposed to air. The pathogens, however, are extremely sensitive to fatty acids and trace metals that are present in peptones and agar of ordinary media. This inhibitory effect can be eliminated by the addition of serum or blood to the medium. Moreover, if the blood is heated to 80°C for 10 min, it is even more effective. This procedure turns the agar medium a dark brown, and the resulting medium is commonly called chocolate blood agar. Also, the pathogens require for optimal growth more carbon dioxide than is present in the atmosphere. It is customary, therefore, to culture pathogenic *Neisseria* plates in a jar in which a candle has been lit before it is closed. The candle uses up about half the oxygen and releases carbon dioxide, resulting in a final concentration of about 10 percent carbon dioxide. In large laboratories, a carbon dioxide tank is used to supply a mixture of air containing excess carbon dioxide.

N. meningitidis (commonly called the meningococcus) is responsible for 3000 to 4000 cases of acute bacterial meningitis per year in the United States.

Antigenic Classification Meningococci can be subdivided into a number of groups based on common antigens. Current serogroups are designated A, B, C, D, X, Y, Z, L, W135, and 29E. Members of groups X and Y have occasionally been isolated from cases of meningitis, but in most cases X, Y, and Z are found in carriers and have not been involved in large epidemics of meningitis. It is groups A, B, and C that are the major causes of epidemics of meningitis.

The group-specific antigen for the meningococci is a polysaccharide capsule. Some groups of meningococci are further subdivided into distinct serotypes based on the antigenicity of their outer membrane proteins.

Pathogenicity The meningococcus is a strict parasite of humans and can be routinely cultured from

Figure 24.11 *N. meningitidis* (×1540). Note how cells appear in pairs.

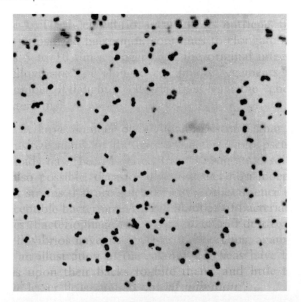

the nasopharynx of asymptomatic carriers. In periods between epidemics, civilian carrier rates range from 2 to 8 percent, depending on the season, whereas in a closed society such as a military camp, the average carrier rate has been reported to exceed 40 percent.

It is, therefore, obvious that the organisms are spread from person to person by the respiratory route. Even in a situation in which a large number of individuals are carrying the virulent organisms in their nasopharynx, usually very few get sick.

If we follow the fate of the infecting organism, meningococcal infections can be categorized into three stages. The first stage is characterized by a nasopharyngeal infection, which is usually asymptomatic but may show symptoms of a minor inflammation. This stage may last for days to months and will induce the formation of protective antibodies within a week, even though the infection remains asymptomatic.

In a small percentage of cases the meningococci may enter the bloodstream. This second stage, called meningococcemia, may be explosive, resulting in death within 6 to 8 h, or may begin more gradually with fever, malaise, and a skin rash containing the cocci.

In the third stage of meningococcal infections, the organisms may cross the blood-brain barrier and infect the meninges, causing the major symptoms of severe headache, stiff neck, and vomiting, accompanied by delirium and confusion.

An unusual property of the pathogenic meningococci is that all isolated strains secrete a proteolytic enzyme whose only known substrate is the IgA1 class of immunoglobulins. (Based on minor antigenic differences in their heavy chains, IgA can be divided into two subclasses that have been designated as IgA1 and IgA2.) This enzyme cleaves the IgA1 into two fragments, thereby largely destroying any effectiveness possessed by the antibody. One can easily imagine that this property could aid in the colonization of mucous membranes, where specific IgA is a major defense mechanism of the host.

Clinical Management Diagnosis is confirmed by direct microscopic examination of smears (see Figure 24.12) and by cultures of spinal fluid, blood, nasopharyngeal swabs, and skin lesions. In direct microscopic examination, the meningococci are frequently observed to be inside polymorphonuclear cells. Throat swabs are usually streaked on blood or chocolate blood agar, but the selective Thayer-Martin medium is frequently used because it contains polymyxin, vancomycin, and nystatin. These antibiotics inhibit the growth of many bacteria and yeast contaminants while allowing the growth of the pathogenic *Neisseria*.

Final identification of the meningococcus requires both sugar fermentation tests (Table 24.3) and serological tests using group-specific antisera for the agglutination of the unknown isolate.

Until the early 1960s, the sulfonamides were the drug of choice for treatment of infections caused by

Figure 24.12 *N. meningitidis,* the causative agent of epidemic meningitis. Note the characteristic paired cocci near the center of the preparation.

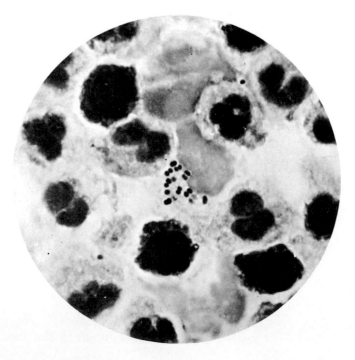

TABLE 24.3 Growth Characteristics and Sugar Fermentations by *Neisseria* and Morphologically Similar Human Parasites

Organism	Growth in the Absence of Blood	Glucose	Maltose	Sucrose	Lactose	Fructose
N. meningitidis	–	+	+	–	–	–
N. gonorrhoeae	–	+	–	–	–	–
N. flavescens	+	–	–	–	–	–
N. sicca	+	+	+	+	–	+
Branhamella catarrhalis	+	–	–	–	–	–

N. meningitidis because they readily penetrate the blood-brain barrier. However, the incidence of organisms resistant to sulfonamides has increased to the point at which this treatment is rarely used. Fortunately, penicillin is effective and is now considered the drug of choice. Erythromycin and chloramphenicol are also effective and can be used in individuals hypersensitive to penicillin.

Vaccines consisting of purified capsular polysaccharide are now commercially available for groups A, C, Y, and W135. In 1974 a group A and C vaccine was used to stop a major epidemic in São Paulo, Brazil, where over 20,000 cases and 3000 deaths were caused by both group A and group C meningococci. The efficacy of these vaccines has also been shown in American and Finnish military recruits. The group A vaccine has been shown to be effective in children as young as 3 months of age but, unfortunately, the group C vaccine does not induce protective antibodies in children under the age of 2 years.

Purified group B polysaccharide is not immunogenic in humans, and there is as yet no commercial vaccine for group B meningococci. Group B organisms can be subdivided into approximately 15 serotypes based on the antigenicity of their outer membrane proteins. Using animal models and human volunteers, it has been shown that antibodies directed against these type-specific outer membrane proteins are protective, demonstrating the potential for a type-specific group B vaccine. Furthermore, because serotype 2 is known to cause at least half of group B infections, the use of serotype 2 vaccine would considerably reduce group B infections. Experimental vaccines in which the purified group B polysaccharide was complexed with an outer membrane protein appear to enhance the immunogenicity of the group B polysaccharide, suggesting that such vaccines could be effective for the entire group B meningococci.

In the past few years meningitis caused by group Y and group W135 has become more prevalent, with group W135 accounting for 10 to 15 percent of meningococcal disease in Europe and group Y strains have steadily increased as a cause of sporadic infections in the United States.

Bordetella pertussis: Whooping Cough

Bordetella pertussis is the causative agent of pertussis, or whooping cough. This organism was first isolated and described by Bordet and Gengou. It is a small gram-negative coccobacillus that is nonmotile and non-spore-forming but may be encapsulated.

Enriched media, usually containing blood, are used for the primary isolation. These organisms are readily killed by drying and survive for only short periods outside the body.

Pathogenicity   An acute, noninvasive infection of the ciliated epithelial cells of the respiratory tract (see Figure 24.13) begins after an incubation period of 7 to 10 days with a catarrhal stage characterized by sneezing and a mild but irritating cough. After 1 to 2 weeks, the disease enters the spasmodic stage, in which the cough becomes very violent. These episodes of violent coughing are frequently followed by a whoop on inspiration, and it is this characteristic whoop that gives the disease its name. The paroxysms of coughing may be so severe that cyanosis, vomiting,

Figure 24.13   *Bordetella pertussis*. Note that organisms are attached directly to the cilia and not to the cell membrane of the epithelial cells.

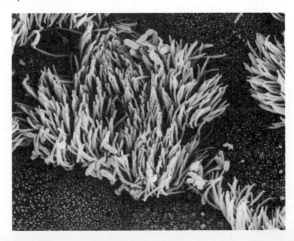

and convulsions follow and completely exhaust the patient. The spasmodic state persists for about 2 weeks and is followed by a convalescent stage, which lasts an additional 2 weeks.

The respiratory discharge from an infected person is the source of the infectious agent. It may be transmitted either by direct contact, by droplets, or through freshly contaminated articles. The major spread probably occurs in the catarrhal stage before the disease is diagnosed. In this stage the organisms are present in the nasopharynx, and when the individual coughs, even though the cough is not very severe, the organisms are expelled and sprayed through the air. Unfortunately, as a rule it is only after the person begins to whoop that he or she is isolated.

The organisms do not invade the bloodstream but remain localized in the respiratory tract. Healthy carriers have not been recognized but it is generally believed that adults with mild cases of whooping cough (due to a partial immunity remaining from a childhood immunization) are the primary source of infection for children in an immunized population.

Few organisms have been studied as intensively as *B. pertussis* in attempts to understand virulence at a molecular level. The results have been the isolation and characterization of four primary virulence factors: (1) pertussis toxin, (2) extracytoplasmic adenylate cyclase, (3) filamentous hemagglutinin, and (4) tracheal cytotoxin. Other factors, including the gram-negative endotoxin, may also be involved in the symptomatology of the infection.

Pertussis Toxin   This exotoxin possesses an enzymatic activity that hydrolyzes the coenzyme nicotinamide adenine dinucleotide (NAD) into adenosine diphosphate-ribose (ADP-ribose) and nicotinamide. It then transfers the resulting ADP-ribose to a host protein whose normal function is to regulate the activity of the cellular adenylate cyclase. This, in essence, destroys its regulating ability, permitting an uncontrolled activation of adenylate cyclase. This, in turn, hydrolyzes ATP to form large amounts of cyclic adenosine monophosphate (cAMP). This increased level of cAMP results in the activation of protein kinases, which phosphorylate certain host cell proteins. The major consequences of such phosphorylation include: (1) a histamine-sensitizing effect, resulting in a high susceptibility to anaphylaxis; (2) an islet-activating effect, resulting in an increased insulin synthesis and a consequent hypoglycemia; and (3) a lymphocytosis-promoting effect manifested by the presence of an excess of lymphocytes in the blood. The events leading to cAMP synthesis are summarized in Figure 24.14. The pertussis toxin is also involved in the specific adherence of the organisms to the cilia of the respiratory epithelium.

Extracytoplasmic Adenylate Cyclase   In addition to the activation of the host's adenylate cyclase by pertussis toxin, these organisms also secrete an extracellular bacterial adenylate cyclase. This enzyme is taken into the host cells where it functions to produce still more cAMP. One important effect of the increased cAMP levels is to inhibit the killing function of neutrophils, monocytes, and natural killer cells. This occurs, presumably, through the activation of protein kinases, resulting in the blocking of chemotaxis, prevention of the myeloperoxidase-dependent production of hydrogen peroxide, and the inhibition of the generation of superoxide.

Filamentous Hemagglutinin   This virulence factor exists as filamentous rods extending from the cell surface of the bacterium. It appears that the filamentous hemagglutinin acts together with pertussis toxin to bind the cells to the cilia of the respiratory epithelium.

Tracheal Cytotoxin   One of the characteristic symptoms of an infection by *B. pertussis* is the severe cough from which the disease gets its name. The severity of the cough is believed to be correlated with the destruction of the ciliated respiratory cells by a bacterial product termed *tracheal cytotoxin*, resulting

Figure 24.14   The ultimate effect of pertussis toxin can be divided into two general steps: (1) an inactivation of an adenylate cyclase regulating protein, and (2) the subsequent formation of excessive cAMP by the activated adenylate cyclase.

(1)   *Inactivation of adenylate cyclase regulating protein*

NAD + pertussis toxin $\longrightarrow$ ADP-ribose-toxin + nicotinamide

ADP-ribose-toxin + regulating protein $\longrightarrow$ regulating protein-ADP-ribose + toxin

(2)   *Formation of cAMP in absence of regulating protein*

ATP $\xrightarrow{\text{adenylate cyclase}}$ cAMP + PPi

in an accumulation in the lungs of mucus, bacteria, and inflammatory debris.

Tracheal cytotoxin appears to be a breakdown product of the peptidoglycan in the gram-negative cell wall. The mechanism whereby this toxin destroys only the ciliated epithelial cells is unknown. A similar toxin has been shown to be produced by the gonococcus and is probably involved in damage to the ciliated cells of the fallopian tubes in cases of pelvic inflammatory disease.

***Phase Changes in* Bordetella pertussis**   It has been known for years that freshly isolated, virulent organisms could undergo a series of changes when cultivated in the laboratory. These modifications were manifested by changes in colonial morphology and a loss of one or more virulence factors. These phases were designated as phases 1, 2, 3, and 4, with phase 1 being highly virulent and phase 4 avirulent. These changes result from single mutations that are readily reversible.

Clinical Management   Cultures for pertussis are collected by using a swab of calcium alginate on a fine, flexible wire (cotton is inhibitory for growth of the organisms). The swab is inserted into one nostril and gently pushed in until it reaches the posterior nares, where it is left for 15 to 30 s. The swab is then plated on Bordet-Gengou medium and incubated for several days. The organisms die off rapidly, and best results are obtained if the plate is inoculated immediately after the nasal swab. Because *B. pertussis* is resistant to penicillin, the swab may be dampened with a drop of penicillin to limit the growth of throat contaminants.

Organisms growing on the culture plates may be identified by employing specific antiserum for an agglutination test or by staining with fluorescently labeled specific antibody. This latter technique can also be used to stain smears taken from the nasopharynx.

Treatment is not entirely satisfactory, but several antibiotics are of value. Chloramphenicol, erythromycin, or the tetracyclines may be used. Antibiotic therapy causes a reduction in the number of secondary infections, such as bronchitis and pneumonia, caused by other organisms.

The introduction of an effective vaccine has markedly reduced the prevalence of whooping cough. Formerly, almost every child had this disease, as shown by the fact that there were 265,000 reported cases of pertussis in the United States during 1934 with more than 7500 deaths. In contrast, there are presently fewer than 4000 cases with three to five deaths each year. It is obvious, therefore, that the vaccine is very effective but, as discussed in Chapter 21, the vaccine itself causes 5 to 20 deaths and approximately 50 cases

of permanent brain damage annually in the United States. It is normally given together with toxoids for diphtheria and tetanus (collectively termed the DPT vaccine), and there is currently a group of parents of children who were injured by the vaccine who call themselves DPT, for Dissatisfied Parents Together. In 1985 alone, there were 219 lawsuits filed against the producers of the vaccine and, as a result, two of the three manufacturers in the United States have ceased making the vaccine. Moreover, because of the necessity for an insurance reserve, the cost of the vaccine has gone from 11 cents in 1982 to $11.40 in 1986.

Frightening as this may be, statistically it would be unwise not to use the vaccine. The ultimate solution is to eliminate the toxicity of this preparation by using purified components of *B. pertussis* as an immunogen. A seemingly effective and considerably less toxic vaccine, consisting of formalin-treated pertussis toxin and filamentous hemagglutinin, is being used currently in Japan and several European countries. If proved to be as effective as the vaccine presently used in the United States, similar component vaccines will undoubtedly be used worldwide.

It is also worth noting that inactive toxins have been constructed by genetic manipulation of the pertussis toxin genes. Such strains of *B. pertussis* produce mutant pertussis toxin molecules that are nontoxic and immunogenic. Their use as a vaccine protects mice from an intracerebral challenge with virulent *B. pertussis.*

Because of the high mortality rate of the disease in infants younger than 1 year, current vaccine is administered first at 2 months of age, with booster injections at 4, 6, and 18 months of age. Another booster is given at the time the child enters school. Immunity acquired from immunization is not permanent, and data from epidemics have shown a 95 percent attack rate in individuals who had not received a booster within the previous 12 years.

Haemophilus influenzae: Meningitis

Haemophilus influenzae, a small gram-negative bacillus, is the major cause of bacterial meningitis and an infection of the epiglottis, causing an acute obstruction of the airway in children. Although these organisms are considered to be small bacilli, there is marked variation in their shape (pleomorphism), and many people prefer to call them coccobacilli, inasmuch as some forms look like bacilli and others like cocci. They are nonmotile and aerobic. These organisms have capsules, and on the basis of serological differences in the capsule, it is possible to recognize six antigenic types. Of these six types, only one, type b, is of real importance in medical microbiology (Figure 24.15).

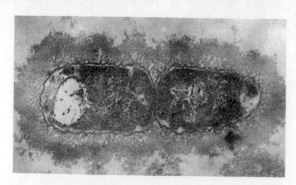

Figure 24.15 *Haemophilus influenzae* type b. Note the presence of a large capsule surrounding these organisms; antibodies directed against the capsule are protective.

The name *H. influenzae*, which is frequently misleading, stems from confusion about the role of this organism in producing influenza in the worldwide epidemics in the late 1800s and during World War I. It has since been recognized that in addition to producing the primary diseases previously mentioned, it is frequently a secondary invader and that following viral infections—particularly influenza virus—it may produce a severe secondary bacterial pneumonia.

The generic name *Haemophilus* evolved from the observation that for growth, these organisms have an absolute requirement for blood-containing medium. This need for blood is based on two essential substances: (1) a heat-stable material originally designated **X factor** and (2) a heat-labile component that was named **V factor**. It is now known that X factor is hematin, which is required for the synthesis of the organism's heme-containing cytochrome system and the enzyme catalase. The V factor can be replaced by the coenzyme NAD.

The incidence of *H. influenzae* infections is most marked in infants from 6 months to 3 years of age. These age limits occur because at 6 months of age the passively transferred antibody received from the mother has disappeared and the children have not yet synthesized their own bactericidal antibodies.

Pathogenicity *H. influenzae* is an obligate human parasite that is passed from person to person by way of the respiratory route. It is reported that 30 to 50 percent of children may carry the bacillus asymptomatically in their nasopharynx, generally as avirulent, nonencapsulated organisms, whereas only 1 to 2 percent will carry encapsulated type b organisms.

Meningitis is the most severe type of infection caused by this organism, resulting in 20,000 to 30,000 cases annually in the United States with a mortality rate of 3 to 7 percent. It follows the same pattern as does meningococcal meningitis in that the organism gets into the blood from the nasopharynx and from

there is able to invade the meninges. In view of the age group that is usually involved, that is, children from 6 months to 3 years of age, it is frequently difficult to obtain a history of headache or early symptoms. The spinal fluid of these patients is essentially identical to that seen in meningococcal meningitis so far as polymorphonuclear neutrophils and sugar are concerned.

One major complication of *H. influenzae* meningitis is the organic brain damage that can result if the thick exudate interferes with the flow of spinal fluid.

Acute epiglottitis is another serious disorder produced by *H. influenzae* type b organisms. In this infection an apparently healthy child or young adult suddenly has acute respiratory distress and may need immediate hospitalization for relief of the obstruction of the air passages. Usually the epiglottis is markedly swollen, red, and edematous. If it is not treated promptly, respiratory insufficiency may cause death.

H. influenzae is also one of the major causes of a middle ear infection termed *otitis media with effusion* (OME). Such infections affect 60 to 70 percent of children during the first 3 years of life and at least 40 percent experience more than one episode. Surprisingly, however, strains causing OME are routinely unencapsulated and, hence, nontypable. As a result, immunity is strain specific, explaining why multiple infections are common.

Similar strains are also being recognized as a common cause of septic arthritis in children younger than 2 years of age. Infections in adults over 50 years of age are also being seen, particularly in those who have some underlying disease. Such infections are manifested as pneumonia, meningitis, epiglottitis, or as an invasion of the female genital tract.

The virulence factors possessed by *H. influenzae* are not well characterized. The ability to resist phagocytosis appears to be the most obvious one, but there must be additional intrinsic factors inasmuch as essentially all serious systemic disease in otherwise healthy persons is caused by *H. influenzae* type b.

Clinical Management Suspected cases of *H. influenzae* meningitis must have immediate treatment; untreated cases have a high mortality rate, and survivors suffer about a 30 percent incidence of neurological disorders, including mental retardation, blindness, hydrocephalus, and convulsions. Gram stains of sedimented spinal fluid that show pleomorphic gram-negative rods are assumed to be *H. influenzae*, but they should be mixed with specific type b antiserum and observed for a positive quellung reaction.

Both blood and spinal fluid should be cultured on chocolate blood agar and incubated at 35°C in a candle jar or carbon dioxide incubator. Isolated organisms can be assayed for X and V factor requirements; a trypticase soy agar plate is streaked, and sterile disks (available commercially) containing X factor, V factor, and a combination of X and V factors are placed on the surface of the agar. Growth that is limited to the vicinity of the paper disks will make obvious any X or V factor requirement.

Because of the prevalence of *H. influenzae,* most people acquire protective antibodies between the ages of 3 and 8 years, either by becoming asymptomatic carriers or from an undiagnosed respiratory illness. Vaccines consisting of purified type b capsular material have been shown to be nontoxic and to stimulate the formation of protective antibodies in children over the age of 2 years. But, because the capsular material is not antigenic in very young children, numerous attempts have been made to make it more effective for this age group.

The newest, seemingly effective, vaccine consists of the type b capsular material conjugated to diphtheria toxoid. This converts the polysaccharide vaccine from a T-independent antigen (capable of inducing only an IgM response) to one that is T-cell dependent. Trials in Finland indicated that although the response to this vaccine was still age associated, it was far superior to that seen when the capsular material was used alone. Based on these and other studies, it has been estimated that the administration of three doses of this conjugate to infants could prevent up to 15,000 cases of *H. influenzae* type b infection annually in the United States. This conjugate vaccine, as well as one with a meningococcal protein conjugate, has now been licensed in the United States for use in children 2 months or older.

Rifampin has been used successfully as a prophylaxis for children who are exposed at home or in day care centers.

At one time, ampicillin was the therapy of choice for *H. influenzae* infections, but with the increase in reported ampicillin resistant organisms, one can no longer rely on ampicillin susceptibility. Chloramphenicol is usually effective and is used when ampicillin resistance is suspected. Recent data suggest that some of the cephalosporins may be as effective as chloramphenicol.

Haemophilus aegyptius: Infectious Conjunctivitis

H. aegyptius (considered by many to be a biotype of *H. influenzae*) is the etiological agent of a common conjunctivitis frequently called pinkeye.

Pathogenicity The illness may be mild or severe. Symptoms vary from vascular infection of the conjunctiva with slight discharge to severe irritation with lacrimation, swelling of the eyelids, photophobia, and a mucopurulent discharge. The incubation period is usually 1 to 3 days. Humans are the reservoir for this organism. Discharges from the conjunctiva and the upper respiratory tract contain the infectious agent.

Clinical Management Microscopic examination of smears, as well as cultures from the exudate, are useful diagnostic aids. Control measures require general cleanliness and the disinfection of all discharges from an infected person. Topical application of the tetracyclines provides effective treatment for this infection.

Recently, a series of severe and frequently fatal infections due to a strain of *H. aegyptius* was described in Brazil. Termed Brazilian purpuric fever, this invasive disease characteristically begins with a purulent conjunctivitis, which in a small percentage of patients progresses to an overwhelming endotoxemia resulting in irreversible shock.

Corynebacterium diphtheriae: Diphtheria

Although diphtheria is no longer common in the United States, it still occurs as sporadic cases or small outbreaks. Table 24.4 provides an annual case rate for the period 1976 to 1990. Although the morbidity rate is low, few areas of the country are completely free of the disease.

The organism was isolated in 1883 by Klebs and was shown to be the etiological agent of diphtheria in 1884 by Loeffler. The scientific name *Corynebacterium diphtheriae* was chosen because of the tendency of these organisms to be club-shaped. They are gram-

TABLE 24.4 Incidence of Diphtheria in the United States

Year	Cases Reported
1976	128
1977	81
1978	75
1979	60
1980	4
1981	4
1982	3
1983	5
1984	1
1985	1
1986	0
1987	3
1988	1
1989	2
1990	4

positive non-spore-forming nonmotile rods that can certainly be described as being pleomorphic. Some may be straight, others curved or club-shaped; when they divide by binary fission, the newly formed bacteria have a tendency to snap apart. As a result, stained smears show many cells forming sharp angles with each other (see Figure 24.16). Another outstanding characteristic of the diphtheria bacillus is its granular and uneven staining. When stained with methylene blue or toluidine blue, the cells show reddish granules. These refractile granules have been given a variety of names, such as Babes-Ernst bodies, metachromatic granules, and volutin. It is now known that they are composed of long chains of inorganic polyphosphate. The granules are not seen during active growth but start to appear toward the end of the logarithmic growth period. It appears that they represent storage depots of high-energy phosphate bonds, because it has been shown that the corynebacteria have enzymes that can directly phosphorylate glucose to form glucose-6-phosphate or can phosphorylate ADP to form ATP using the phosphate present in these storage granules.

C. diphtheriae grows well on a blood agar medium. However, because most other bacteria in the throat also grow under these conditions, media have been devised that favor the growth of the diphtheria bacillus and restrict the growth of some of the normal flora of the throat. Two media have been useful for this purpose: Loeffler's coagulated blood serum and blood agar or chocolate blood agar to which tellurite has been added. In the latter case, colonies of C. diphtheriae become dark grey to black (see Figure 24.17). This occurs because the tellurite or tellurous ions are able to diffuse through the cell wall and are reduced to tellurium metal, which is precipitated inside

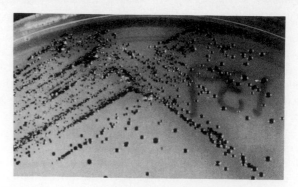

Figure 24.17 Gray-black colonies of *C. diphtheriae* from a throat swab streaked on cystine tellurite medium.

the cell. The organisms in this species have been divided into three types, based primarily on the appearance of their colonies on tellurite media. Originally it was thought that there was a correlation between the severity of the disease and the colony type, and although this is now known to be untrue, the names *gravis, intermedius,* and *mitis* are still used to refer to the three types of *C. diphtheriae.*

C. diphtheriae organisms are readily killed by heat, but they can remain alive for fairly long periods outside of the host.

Pathogenicity Basic understanding of the disease came in 1888, when Roux and Yersin discovered that the cell-free filtrates—free from diphtheria bacilli—were lethal for animals. Furthermore, the symptoms and pathological changes that occurred as a result of the injection of the toxic filtrate were the same as those of the disease itself. In brief, diphtheria is an acute infection in which the causative organisms remain in the respiratory tract, even though on some occasions, especially in the tropics, the organisms have produced wound infections. The site of the invasion, usually the throat, becomes inflamed as the bacteria grow and excrete their powerful exotoxin. Dead tissue cells—along with the host's leukocytes, red blood cells, and the bacteria—form a dull gray exudate called a *pseudomembrane.* If this forms in or extends into the trachea, the air passages may be blocked. In these cases, it is sometimes necessary to make an incision directly into the trachea (tracheotomy) in order to prevent suffocation. It is possible for non-toxin-producing strains of *C. diphtheriae* to be involved in a local infection with pseudomembrane formation, but these cases are mild. The symptoms and frequent deaths from diphtheria are the direct result of the action of the diphtheria toxin. The incubation period is 2 to 5 days.

Toxigenicity The ability to produce toxin is the only difference between avirulent and virulent

Figure 24.16 Gram-stained culture smear of *C. diphtheriae* (×4200). Note club-shaped cells and V-shapes of two attached bacteria.

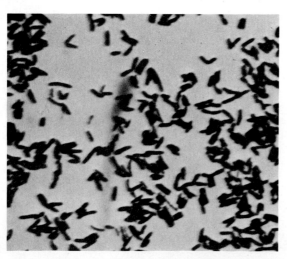

diphtheria organisms. This ability is possessed only by organisms that are lysogenic for phage-β or a closely related bacteriophage (see Chapter 14). All strains of diphtheria organisms that produce toxin are lysogenic, and avirulent strains of diphtheria bacilli can be made virulent by infecting them with the bacteriophage. This change in the properties of a bacterial cell as a result of becoming lysogenic is called **lysogenic conversion.** It occurs as a result of the incorporation of DNA from the phage into the bacterial DNA and the coding for a new property not previously possessed by the bacterium.

In the commercial production of the toxin for vaccine, the amount of iron present in the growth medium is critical. Good toxin production is obtained only at very low concentrations of iron (1.7 μM); at concentrations as low as 9 μM, toxin production becomes negligible. This occurs because normally the bacterium forms a repressor that prevents the expression of the phage tox$^+$ gene, and this repressor is an iron-containing protein. Thus, when the concentration of iron is abnormally low, the repressor is not synthesized and the tox$^+$ gene is transcribed, ultimately yielding toxin.

It has been known for some time that the toxin inhibits protein synthesis in mammalian cells and that this inhibition occurs in vitro only if NAD is present. It is now known that the toxin inhibits the translocase that moves the ribosome to the next codon on the mRNA after the peptide bond to the last amino acid is formed. The ribosome is therefore frozen, and protein synthesis stops. The toxin actually acts as an enzyme and catalyzes both the hydrolysis of NAD and the transfer of the ADP-ribose portion of NAD to the translocase (also called amino acyl transferase II). This causes the inactivation of the translocase (see Figure 24.18).

Sources of diphtheria include both the healthy carrier and the infected person. In the past there has been some spread of this disease by way of contaminated milk.

Clinical Management Diagnosis is based on isolating the organism from the infected area and demonstrating its toxin-producing ability. Specimens for direct smears and for cultures are collected with a cotton swab. It is always necessary to differentiate the avirulent non-toxin-producing strains from the virulent toxin-producing organisms, and the ability of the diphtheria organisms to produce toxin is the only distinguishing mark of the virulent strains.

In order to determine whether a particular strain produces toxin, a virulence test is performed. This is accomplished by injecting a suspension of the organisms intracutaneously into two guinea pigs, one of which has been protected by specific diphtheria an-

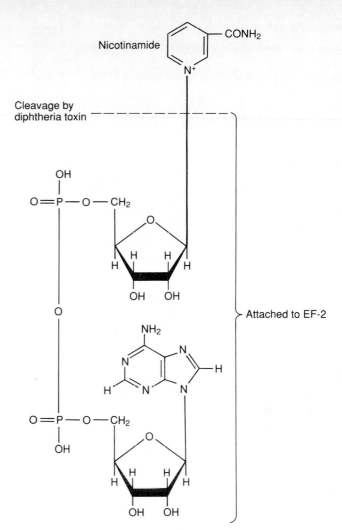

Figure 24.18 Structure of NAD, showing bond cleaved by diphtheria toxin.

titoxin. A toxin-producing organism will produce a severe inflammation at the site of the injection in the unprotected animal but little or no inflammation in the protected animal. There is also a virulence test in vitro in which the unknown organism is compared with a known toxin-producing diphtheria bacillus. In this case antiserum is placed on a piece of filter paper that is then laid vertically across the streaks of the organisms on a petri dish. If toxin is produced, a line of precipitate will form at the optimal concentration of the toxin and the antitoxin. These lines of precipitate can then be compared with those occurring from a known toxin-producing organism (see Figure 24.19).

In treating diphtheria, antitoxin should be administered promptly in order to neutralize the toxin being produced. Thus, the initial diagnosis must be made from the clinical picture, since it would not be safe to wait for a bacteriological confirmation before starting treatment. Most physicians agree that it is safer to err by occasionally administering antitoxin to

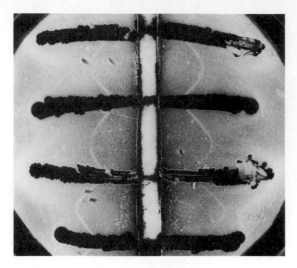

Figure 24.19 Curved lines of identity (precipitate) are visible between the dark streaks of organisms that cross a paper strip saturated with antitoxin. In this case, then, all four streaks are *C. diphtheriae*.

someone who was clinically misdiagnosed than to wait for positive bacteriological confirmation.

The organism is sensitive to penicillin and other antibiotics, but the antibiotics do not neutralize circulating toxin and, therefore, are of value only when used concurrently with antitoxin.

Control is based entirely on the mass immunization of children with a nontoxic toxoid prepared by treating purified toxin with formalin. This usually is administered along with tetanus toxoid and pertussis vaccine in three monthly injections beginning at 6 weeks of age. Because immunization does not provide lifelong immunity, booster shots should be given after 5 years, and every 10 years thereafter.

The Schick test can be used to determine whether an individual is susceptible to diphtheria. A small amount of diphtheria toxin is injected intradermally. A person without antibodies to the toxin (and hence susceptible to diphtheria) will develop an inflamed area at the site of the injection, which will reach a maximum level in about 48 h. In contrast, those who have antibodies will neutralize the toxin and therefore will not develop an area of inflammation at the site of injection. In actual practice, a control substance in which the toxin has been inactivated by heating is injected intradermally into the other arm in order to make certain that any inflammation is due to the toxin and not the result of a hypersensitivity reaction to some component of the toxin preparation.

Mycobacterium tuberculosis: Tuberculosis

Mycobacterium tuberculosis (more commonly called the tubercle bacillus) was shown by Robert Koch to be the causative agent of tuberculosis. This is a disease of humans that can be a brief, completely inapparent incident in the life of one individual whereas in another it may produce a chronic, progressive pulmonary disease resulting in the loss of almost all pulmonary tissue.

The tubercle bacilli are thin rods, usually straight but sometimes bent or club-shaped. They are non-motile and non-spore-forming, do not form capsules, and often appear beaded or granular when stained. One notable characteristic of the tubercle bacilli is their waxy appearance. This makes them quite difficult to stain; once stained, however, they are very resistant to decolorization. One can actually wash the stained cells with 95 percent alcohol that contains 3 percent hydrochloric acid without washing out the stain. Organisms with the ability to retain the original stain even after this washing are referred to as **acid-fast.** Only the members of the genus *Mycobacterium* and a few species of the genus *Nocardia* possess this property—an immense help in the study of body fluids such as sputum or gastric washings for the presence of the tubercle bacilli. Routinely, tubercle bacilli are stained using the Ziehl-Neelsen technique, in which a smear is covered with the red stain carbol-fuchsin and the dye-covered smear is heated to steaming for a few minutes to aid the penetration of the dye into the bacterium. A cold-stain modification of this procedure, in which the carbol-fuchsin is first dissolved in a detergent, is also used to stain these organisms. In either case the stained smear is then washed with acid alcohol and counterstained with methylene blue. Mycobacteria can also be stained with a fluorescent dye mixture of rhodamine and auramine. When examined with an ultraviolet light microscope, the stained mycobacteria will appear as bright rods (see Figure 24.20).

This property of acid-fastness is related to the very high lipid and wax content of the mycobacterial organisms. There has been a great deal of research to determine the structure and possible pathophysiological significance of these lipid and wax components, but much remains unknown. One large class of complex fatty acids found in mycobacteria are the mycolic acids. These substances make up a class of β-hydroxylated fatty acids that are found associated with both the waxes and glycolipids. It is believed that some of these substances may account for the acid-fastness of the mycobacteria.

One additional factor found in virulent mycobacteria deserves mention here. This substance—called **cord factor**—is also a very complex fatty acid linked to a carbohydrate component. The important thing about cord factor is that it is found only in virulent mycobacteria, and organisms possessing it will characteristically grow in large serpentine cords. Thus an acid-fast stain of *M. tuberculosis* may reveal millions of individual organisms all lined up parallel to

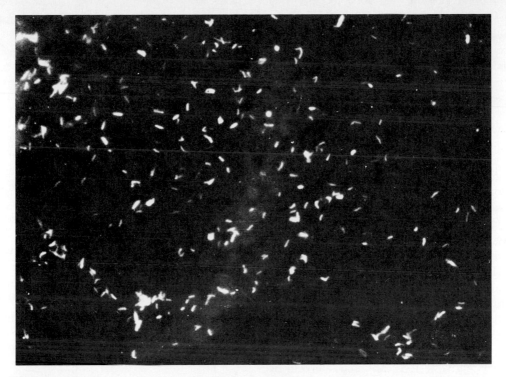

Figure 24.20 *M. tuberculosis.* This illustrates fluorescence microscopy of *M. tuberculosis* in lung tissue stained with auramine. The light rods are the tubercle bacilli. This method was found to detect the organisms in greater numbers in clinical specimens than did the Ziehl-Neelsen technique.

each other to form a cordlike structure. Although the precise action of cord factor is unknown, a recent report clearly shows that it induces the synthesis of cachectin (also called tumor necrosis factor alpha) in mice. When injected with cord factor, mice became severely wasted (cachectic), losing up to 25 percent of their weight within 48 h. The observation that antibodies to cachectin would prevent this effect supports the conclusion that cachectin was responsible for the wasting induced by cord factor. It also strongly suggests that cord factor is responsible for the cachexia observed in patients with tuberculosis.

Nonpathogenic mycobacteria are usually not as strongly acid-fast as are the virulent species. As a result, prolonged treatment of these species with acid alcohol may well result in their complete decolorization.

There are other characteristics of the tubercle bacilli that differentiate them from the other organisms discussed in this chapter. The growth of these organisms is considerably slower than that of most other bacteria, and it takes approximately 20 h for a new generation to appear. Therefore, it may take as long as 6 weeks before growth can be seen in a medium. These organisms are obligate aerobes and have the ability to grow on a relatively simple medium. For example, they grow well in a medium composed only of inorganic salts, asparagine, and glycerol. The organ-

isms are readily killed by heat, and the conditions of pasteurization practiced in the United States are sufficient to kill any tubercle bacilli that might be present in milk. On the other hand, they are quite resistant to chemical agents and particularly resistant to drying, which makes it possible for them to remain alive for long periods in rooms, bedding, sputum, and similar environments.

Pathogenicity of *M. tuberculosis*

Humans become infected with *M. tuberculosis* most frequently by inhaling droplet nuclei (small drops that remain suspended in air) that contain tubercle bacilli. Droplet nuclei are expelled by infected individuals and, because of their small size (1 μm to 10 μm in diameter), they remain airborne for long periods.

Tuberculosis appears to be a highly infectious disease, as manifested by the minor epidemics initiated by infected school teachers, students, bus drivers, or others who come in contact with large numbers of people.

Humans tend to vary considerably in their response to infection by tubercle bacilli, and active disease can, in general, be thought of as resulting either from a primary infection or from a subsequent reactivation of a quiescent infection.

Primary Infection Following inhalation, tubercle bacilli initiate small lesions in the lower respiratory tract and drain from these to the regional lymph nodes. This primary complex, as these lesions are called, frequently heals to form tiny "tubercles" too small to be seen by x-rays, but they may continue to harbor the viable tubercle bacilli indefinitely. In other cases, multiplication of the bacilli continues and the lesion expands, destroying the normal tissue and leaving the necrotic tissue in a semisolid, cheesy state. This process, called caseation necrosis, may eventually heal and become infiltrated with fibrous tissue and calcium deposits, or the lesion may continue to expand, leaving cavities in the lung after the clearance of necrotic tissue. The lesion can also involve a pulmonary vein, allowing the organisms to spread via the bloodstream. This development causes miliary tuberculosis, with lesions in other organs of the body. When a lesion ruptures into the meninges, a tuberculous meningitis will occur.

During the early part of the infection the organisms encounter little, if any, host resistance, and the lesions, called exudative lesions, are characterized by the presence of polymorphonuclear leukocytes, fluid, and inflammation. At this time most tubercle bacilli are found growing intracellularly in macrophages. Later, as the host develops a cellular immunity to the tubercle bacilli, the lesions are called productive lesions, or tubercles, and the organisms, which are few in number, are now mostly extracellular, surrounded by necrotic tissue and large mononuclear macrophages known as epithelioid cells. The necrotic tissue and epithelioid cells are, in turn, surrounded by lymphocytes and fibrous tissue, which wall off the tubercle from the normal tissues of the lung.

Reactivation Infection Whether the primary infection heals early or late, the walled-off tubercle is believed to contain viable organisms, which may persist for the remaining life of the host. It is estimated that two-thirds of all new active cases of tuberculosis represent a reactivation of a healed primary infection. This postulation is difficult to prove, but it is strongly supported by the observation that over 80 percent of all new cases reported occur in people over 25 years of age. Also, it has been shown that the majority of new cases of active tuberculosis among Navy recruits occur in people who exhibited positive skin reactions to tuberculin at the time of their enlistment.

One interesting exception to the concept of reactivation infection came from a study of 223 Arkansas nursing homes. Their data showed that the prevalence of tuberculin positive reactors was considerably lower than expected from their age group, but the infection rate for tuberculosis was higher than expected. The report concluded that many of these persons had outlived their initial infecting organism and, as a result, were reacquiring a primary infection.

Immunity and Host Response to *M. tuberculosis*

Immunity to tuberculosis is an elusive concept, and, although no one doubts its existence, there is no easy way that it can be quantitated. The earliest demonstration of immunity occurred when Robert Koch showed that a second injection of tubercle bacilli into a guinea pig produced a host response different from that of the initial injection. In essence, Koch's initial injection resulted in an ulcer that failed to heal at the injection site, followed by spread to the regional lymph nodes. A subsequent injection several weeks later was characterized by a more localized ulcer that eventually healed because of the guinea pig's ability to kill the tubercle bacilli before they reached the lymph nodes. This observation, called the Koch phenomenon, certainly demonstrates some immunity, because the lymph nodes draining the site of the second injection do not routinely become infected. The guinea pigs in such experiments, however, die from the initial infection.

Subsequently, Koch observed that the injection of a culture filtrate of *M. tuberculosis* into an animal that had been previously infected with tubercle bacilli evoked an allergic response characterized by induration (a hardened area that becomes maximal in 48 h), by erythema, and (in severe reactions) by an ulcerative necrosis at the site of the injection. Koch named this culture filtrate tuberculin, and, because the skin reaction was slower than the antibody-mediated Arthus reaction, similar reactions have been referred to as delayed-type hypersensitivity or tuberculin-type hypersensitivity. The proteins in the culture filtrate that evoke this allergic response have been partially purified and are available commercially as a purified protein derivative (PPD); a preparation of Koch's original type is called old tuberculin or O.T.

Clinical Management of Tuberculosis

Treatment Before the antibiotic era, complete bed rest was the primary treatment for tuberculosis, and every state had a number of sanitariums used exclusively for tuberculosis patients. The discovery of streptomycin in the mid-1940s marked the beginning of the chemotherapeutic treatment of tuberculosis. But a cure was neither quick nor easy, and it was soon found that long-term therapy with streptomycin resulted in damage to the eighth cranial nerve (involved in hearing) and in the occurrence of mutants of *M. tuberculosis* that were resistant to streptomycin.

Currently, there are a number of drugs available in the United States that have been approved for the

From the beginning of recorded history until the germ theory of disease was established in the late nineteenth century, sickness was believed to be caused by supernatural powers. Primitive man's attempts to frighten away the demons of disease employed tactics such as loud noises and dressing in the skins of large animals. This led to the establishment of the "medicine" man and eventually to the priest, who established the rules for worshipping and appeasing the supernatural powers who controlled one's fate.

Of these early diseases, perhaps none can rival tuberculosis as a consistent killer over the past sixty-three centuries of recorded history. This disease has been referred to by many names, but in the English language the most common of these was "consumption," a term meaning to consume or waste away. Consumption was particularly prevalent in large cities, causing approximately 20 percent of all deaths in England and Wales during the seventeenth century. The cause of tuberculosis remained unknown, and as late as 1881 the famous physician and Dean of Medicine at Johns Hopkins University, William H. Welch, wrote concerning consumption that "the doctrine of the contagiousness of the disease has its advocates, but the general belief is in its noncommunicability."

As one might guess, over the centuries a myriad of different treatments were tried, of which bleeding became one of the most widely used. By the beginning of the eighteenth century, however, the concept of rest and diet for the control of tuberculosis became popular. Many physicians recommended a diet high in milk, and it was generally believed that human milk was superior to cow's or goat's milk. Moreover, it was believed to be more efficacious if sucked directly from the animal or human. Thus, the "wet nurse" for consumption came into being and many patients attributed their recovery to the wet nurse method. Unfortunately, a number of wet nurses contracted tuberculosis because of the close association they had with their patients.

Considerable progress has been made since *Mycobacterium tuberculosis* was isolated in 1882 by Robert Koch. Gone are the sanitariums of the Western World, where tuberculosis patients were sent for rest and treatment. Although antibiotic therapy is long and difficult, it is moderately effective. In spite of all this, however, the World Health Organization estimates that there are 10 million new cases of active tuberculosis each year, resulting in an annual mortality rate of 3 million individuals.

One can readily see, therefore, that the "white plague," as the disease was termed in 1861 by Oliver Wendell Holmes, remains a major killer as we approach the end of the twentieth century.

---

treatment of tuberculosis. The primary drugs, isoniazid (INH) and rifampin, are the most effective and least toxic and, as a result, are the drugs most used for the effective treatment of tuberculosis. Secondary drugs include ethambutol, paraaminosalicylic acid (largely replaced by ethambutol), pyrazinamide, and streptomycin. At present, the most common treatment for active tuberculosis uses both INH and rifampin for at least 9 months. This may be supplemented with pyrazinamide and/or streptomycin. Reasons for such long treatments probably include the facts that the rate of metabolism of the mycobacteria is very slow and the chemotherapeutic drug does not easily penetrate the fibrotic or caseous lesions.

Control of Tuberculosis
The control of tuberculosis in a population requires the identification and treatment of infected individuals who spread tubercle bacilli via pulmonary secretions. However, even though there are annually over 20,000 new cases and approximately 2500 deaths reported in the United States, tuberculosis is usually a slow, chronic disease, and it is exceedingly difficult to find infected people until after months or years of active infection. For early detection, therefore, one must rely on the tuberculin skin test; a positive reaction is interpreted as denoting an infected person, whether the disease is quiescent or active. For this reason control now relies heavily on the preventive therapy of tuberculosis, and the Tuberculosis Advisory Committee to the Centers for Disease Control has recommended that the following people be considered potential candidates for active disease (in the order listed) and that they be treated with daily oral INH for one year: (1) household members and other close associates of those with recently diagnosed tuberculosis, (2) positive tuberculin reactors with findings on a chest roentgenogram consistent with nonprogressive tuberculosis even in the absence of bacteriological findings; (3) all those who have converted from tuberculin negative to tuberculin positive within the past 2 years; (4) positive tuberculin reactors undergoing prolonged therapy with adrenocorticoids, receiving immunosuppressive therapy, having leukemia or Hodgkin's disease, having diabetes mellitus, having silicosis, or who have had a gastrectomy; (5) all those under 35 years of age who are positive tuberculin skin reactors. INH therapy is not recommended for positive tuberculin reactors 35 years of age or older because prolonged treatment with INH sometimes causes progressive liver disease in older people. These procedures have been effective over the past couple of decades in decreasing the annual num-

ber of cases of active tuberculosis. This changed, however, in 1989 when some areas reported a 35 percent increase in tuberculosis cases and a 5 percent increase was reported on a national level. These increases are attributed primarily to increases in cases of the acquired immunodeficiency syndrome (AIDS), demonstrating how essential a functioning immune system is in our ability to handle an infection by *M. tuberculosis.*

A living, avirulent bovine strain named BCG (bacille Calmette–Guérin) has been widely used throughout the world to immunize negative tuberculin reactors. In England, a murine mycobacterium called the Vole bacillus has been used. As one might surmise, the resulting immunity is not solid, but statistics on immunized and nonimmunized people indicate a fourfold to fivefold reduction of active tuberculosis in the immunized group. The vaccine is not used routinely in the United States, primarily because a successful immunization converts an individual to a positive tuberculin reactor, eliminating the only good method for detecting early infections.

Interestingly, a number of genes (including some from the AIDS virus) have now been cloned in BCG, and immunization of animals with this recombinant vaccine induces both a humoral and a cellular response to the cloned antigens.

Laboratory Diagnosis of Tuberculosis

A tentative diagnosis of tuberculosis can be made by observing acid-fast rods in smears of sputum, gastric washings, or urine, but a definite diagnosis is established only after the isolation and identification of tubercle bacilli from the patient.

Cultural methods are considerably more reliable than staining direct smears, because only a few organisms are necessary for growth. Most cultures are grown on a medium containing egg yolk or oleic acid and albumin. Penicillin or dyes such as malachite green may be added to inhibit the growth of contaminating organisms.

Guinea pig inoculation is also a sensitive (but expensive) procedure for the demonstration of small numbers of tubercle bacilli. Samples of concentrated sputum or other body fluids are injected subcutaneously into a guinea pig, and after several weeks the animal is skin tested for tuberculin hypersensitivity. At that time the resulting lesion can also be cultured and examined for acid-fast organisms.

Mycobacterium bovis

M. bovis is closely related to *M. tuberculosis* in growth characteristics, chemical composition, and potential for virulence. Because it is normally a pathogen of cattle, human infections ordinarily result from the ingestion of contaminated milk. The organisms do not usually infect the lungs but, rather, produce lesions primarily in the bone marrow. If inhaled, however, *M. bovis* produces a pulmonary disease indistinguishable from that of *M. tuberculosis.*

Bovine tuberculosis has been essentially eradicated in many countries (including the United States) by a strict program for the destruction of tuberculin-positive cattle and by the widespread use of pasteurized milk.

Atypical Mycobacteria

During the past several decades it has become evident that there is an extremely large group of mycobacteria that are apparently normal inhabitants of soil and water. In the United States such organisms are found predominantly in the South, where, as judged by specific tuberculin reactions (using tuberculin prepared from these organisms), between one-third and one-half of the population has been infected with them. The pulmonary disease in diagnosed cases is milder than that caused by *M. tuberculosis* and, strangely, does not appear to be communicable from person to person.

This overall group of organisms has had several names, such as "the anonymous mycobacteria" (because no one knew enough about them to name them) or "the atypical mycobacteria" (because, unlike *M. tuberculosis* or *M. bovis,* they are completely avirulent for guinea pigs).

All are acid-fast bacilli. Some, such as *M. kansasii* and *M. intracellulare,* are most often associated with a mild pulmonary infection, although some may produce severe pulmonary disease that is similar to classic tuberculosis, whereas others may cause skin infections (*M. marinum*), lymphadenitis (*M. scrofulaceum*), or small abscesses (*M. fortuitum*).

Two of these organisms, *M. avium* and *M. intracellulare* have acquired a new significance in individuals with AIDS. These organisms are so closely related that many refer to them as the *M. avium–intracellulare* complex (MAC). The organisms are found worldwide and infect a variety of birds and animals. Both cause a pulmonary infection in humans similar to that caused by the tubercule bacillus, but such infections are seen most often in elderly persons with preexisting pulmonary disease.

In patients with AIDS, however, MAC is the most common cause of a systemic bacterial infection, usually seen as a late opportunistic disease occurring after one or more episodes of *Pneumocystis carinii* infection. Such individuals also often suffer from an intestinal infection and, in such cases, these organisms can be seen in acid-fast stains of stools.

Mycobacterium leprae: Leprosy

Leprosy is one of the most feared of known chronic infections. Approximately 15 million people are infected worldwide ($3\frac{1}{2}$ million cases in India alone), and over 225 new cases are diagnosed each year in the United States. Even though it is an ancient disease of humans, the etiological agent has yet to be grown in an artificial medium.

The causative agent of leprosy (known as Hansen's bacillus) is morphologically similar to *M. tuberculosis*. The organisms appear in lesions as acid-fast rods 3 to 5 μm long and 0.2 to 0.4 μm in diameter. Many attempts to infect human volunteers have been unsuccessful. Leprosy will, however, grow in the foot pad of the mouse and has recently been shown to produce a generalized, progressive, systemic infection in the armadillo, as well as certain species of monkeys.

Pathogenicity

Very little is known concerning the epidemiology of leprosy, but it is an infectious disease whose transfer may depend in large part on the susceptibility of a specific individual. In general, it seems that infection requires only relatively brief contact for "susceptible" people but that most individuals probably cannot be infected by any means. It has been generally believed that humans constituted the only source of *M. leprae* and that the disease was acquired by direct contact with an infected person. Recent evidence, however, suggests that exposure to wild, infected armadillos may be the source of infection for some Mexican patients. Clinically, the disease may occur in either of two major forms, lepromatous leprosy or tuberculoid leprosy.

Lepromatous Leprosy

Lepromatous leprosy is a progressive, malignant form of the disease, which, if untreated, routinely ends in death. The organisms are found in essentially every organ of the body, although the major pathological changes occur in the skin, nerves, and testes. In advanced cases, the eye is usually infected and eventual blindness is common.

Lepromatous leprosy occurs only in individuals who have a defective cellular immune system. Thus, lepromatous patients reject skin allografts very slowly, and their lymphocytes are unable to release macrophage inhibition factor when exposed to lepromin. It is still not known whether the defect in the cellular immune system results from lepromatous leprosy or precedes it.

Tuberculoid Leprosy

Tuberculoid leprosy is frequently a self-limiting disease that may even regress spontaneously. Skin lesions occur and nerve damage appears to result from the inflammation that occurs during a cellular immune response to the bacilli in the nerves. Thus, the major characteristic accounting for this form of leprosy is a normal cellular immune response mounted against the leprosy bacillus.

Clinical Management

Patients with leprosy can be treated with several chemotherapeutic agents. The most common drug is known as dapsone, and its use has revolutionized the care of lepers. The cost per patient is only about $5 per year, and India alone uses about 50 tons of dapsone per year. Patients with lepromatous leprosy must continue to take the drug for the rest of their lives, but this does allow them to remain at home rather than in a leprosarium. Unfortunately, the incidence of strains resistant to dapsone is increasing rapidly, and it is possible that dapsone will become ineffective in the future.

New drugs are being screened for the treatment of leprosy. Of these, rifampin has shown the greatest promise, but it, too, must be administered for years and at a cost that is several hundred times greater than that of dapsone. With the knowledge that even now less than 25 percent of leprosy cases receive any therapy because of its cost and availability, the widespread use of rifampin seems unlikely.

A newer approach for the control of leprosy involves the use of a vaccine in areas where the disease is endemic. Experimental vaccines using killed organisms isolated from the spleens and livers of infected armadillos are currently being evaluated. A second and promising vaccine combines killed *M. leprae* with viable BCG, which induces a specific immunity to *M. leprae* as well as a relevant immunity to BCG. Also, the cloning of a complete library of *M. leprae* genes in *Escherichia coli* will permit the extensive production of *M. leprae* proteins for serological analysis and diagnostic skin tests. In addition, the disease has now been transmitted to mangabey, rhesus, and African green monkeys, permitting primate investigations to be carried out with this organism.

Mycoplasma pneumoniae: Primary Atypical Pneumonia

During World War II, a pneumonia was described (notably in military camps, schools, and colleges) that did not respond to treatment with sulfonamides or penicillin. Although very debilitating, it was only rarely fatal. Researchers were unable to grow the agent on artificial media, and when Eaton and his coworkers transmitted the infection to hamsters and chick embryos using filtered bronchial washings, it was generally assumed that the agent of primary atypical pneumonia was a virus. This concept prevailed until 1962, when the agent was grown on an artificial medium and was shown to be a mycoplasma. The causative agent has been named *M. pneumoniae*.

Pathogenicity Primary atypical pneumonia is an acute respiratory infection characterized by chills, fever, malaise, headache, cough, and patchy areas of infiltration in the lungs that are visible on roentgenographic examination. In one study in which recruits were followed during their 3 months of basic training, it was shown that 44 percent developed antibodies to this agent but only about 3 percent had clinically diagnosed pneumonia. It would, therefore, appear that many individuals can have inapparent, or at least undiagnosed, infections with *M. pneumoniae*. The incubation period seems quite variable but in general ranges from 14 to 21 days.

It is interesting to note that the more serious infections appear to occur in persons over 20 years of age. Because recovery does not result in solid immunity to reinfection, it has been proposed that the more serious manifestations of this disease are the result of an autoimmune reaction occurring during the second infection with *M. pneumoniae*. This model suggests that after the attachment of the mycoplasma to a host cell, their membranes fuse and the mycoplasma membrane proteins are integrated into the host cell membrane. A subsequent infection induces an anamnestic response against these foreign antigens that is both humoral and cell mediated. This concept is supported by two lines of evidence: (1) repeated infections in animals produce an accelerated and exaggerated pulmonary histopathology, and (2) immunosuppressed animals and patients with immunodeficiencies may be infected without the occurrence of pulmonary pathology.

Clinical Management Diagnosis of *M. pneumoniae* is mainly clinical, based on physical findings and roentgenographic evidence. Late in the infection, during the second or third week, cold agglutinins usually appear in the patient's serum. The significance of cold agglutinins is not known, but they will agglutinate human type O red blood cells at 0 to 4°C but not at 37°C.

Because the source of the agent is the respiratory discharges of an infected person, control measures require the isolation of infected persons. Thus far, experimental vaccines have provided variable protection, and the development of an effective vaccine is still under study.

The organism is sensitive to both erythromycin and the tetracyclines. Many physicians prefer to use erythromycin because it is also effective for the treatment of legionellosis.

Legionella pneumophila: Legionnaires' Disease (also Legionellosis)

The discovery of legionnaires' disease was discussed in Chapter 22, and we shall briefly outline here the few facts that are known about legionnaires' disease and the bacterium responsible for it.

The disease is a pneumonia characterized by a dry cough and chest pain as well as abdominal pain and gastrointestinal symptoms in many patients. Of the 182 patients from the American Legion convention in Philadelphia, 29 died, indicating that this can be a fatal disease. On the basis of findings from an indirect fluorescent-antibody test, however, it now appears that the disease occurs worldwide and that cases may be mild or clinically asymptomatic. This did not seem to be true in Stafford, England, during the spring of 1985. This outbreak of legionellosis, which originated in a poorly designed cooling system, claimed between 39 and 46 lives, with a total of 163 cases.

L. pneumophila has been isolated and can be grown on Mueller-Hinton agar containing 1 percent hemoglobin and 1 percent Isovitalex (a vitamin supplement prepared by Baltimore Biological Laboratories, Baltimore, MD) in 5 percent carbon dioxide. This bacterium possesses a gram-negative-type cell wall (see Figure 24.21), but it contains considerably more branched fatty acids and phosphocholine than other bacteria. Thus, it is a truly different organism, one that had not been described before the 1976 epidemic in Philadelphia.

Twelve distinct serogroups of *L. pneumophila* have been isolated. Antiserum to the serogroup antigen appears to be protective, and each serogroup shows little cross-reactivity with any of the other serogroups. The serogroup-specific antigen appears to exist as a polysaccharide capsule.

L. pneumophila has been isolated from a number of air-conditioning cooling towers, and such contaminated water is believed to be a source of infection for those inhaling mist from these towers. Strains have also been isolated from creek water in areas where epidemics have occurred. Outbreaks have occurred also in areas where excavation has taken place, and it is thought that the organisms may reside in the soil, being transmitted to humans on contaminated dust. Most surprising was the report that *Legionella* had been isolated from 9 of 16 shower heads examined in a Chicago hospital where three people had contracted legionellosis while they were patients there. Moreover, a survey of 95 apartments and houses in one area of Chicago showed that 32 percent of the hot water systems were contaminated with *L. pneumophila* at concentrations of from 1 to 104 organisms per liter. In all of these systems, the hot water was maintained at under 60°C. These organisms appear to be both hardy and widespread in our environment.

Paradoxically, even though legionellosis is a respiratory disease and the lung is the only organ consistently involved in fatal cases, person-to-person spread does not seem to occur.

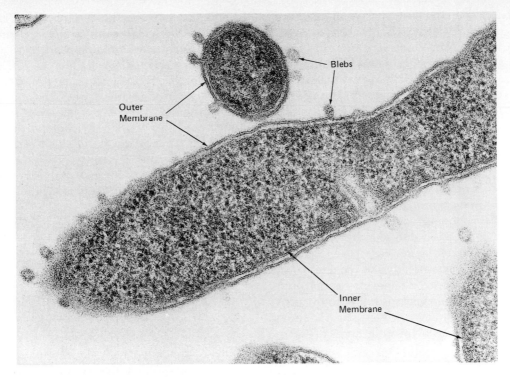

Figure 24.21 Electron micrograph of a legionnaires' disease bacterium (*L. pneumophila*) prefixed with 1.25 percent glutaraldehyde, 0.75 percent formalin, and 0.002 percent creosol. Both inner (cytoplasmic) and outer membranes can be seen along with several evaginations (blebs) at the outer membrane. (Original magnification, ×105,000.)

A definitive diagnosis of legionellosis requires the isolation and identification of the organism. But the organism is difficult to isolate, and the diagnosis is frequently based on serological evidence. Such tests have included both a direct immunofluorescence assay to detect organisms in tissues and an indirect immunofluorescence assay for determining the occurrence of antibodies to *L. pneumophila*.

Erythromycin is the drug of choice, but isolates are also sensitive to rifampin.

Other *Legionella* Species

The isolation of *L. pneumophila* in 1976 spurred investigators to look for other similar etiological agents of pneumonia. Twenty-two such species have been isolated and placed in the family Legionellaceae. Some species have been isolated from soil but not from humans. Other species have been isolated from cases of fatal pneumonia in humans and some from moist environments, such as respiratory therapy equipment and cooling towers. Surprisingly, these latter species appear to be only weakly virulent, causing overt disease only in individuals who have been immunocompromised (because of organ transplants, leukemia, or some infection for which corticosteroids are given) or

in people who may be subjected to an overwhelming initial contact with the organisms.

So it appears that there is a large widespread family of these organisms in nature. Their ubiquity is attested by reports that about half the air samples (containing airborne water droplets) near the erupting Mount St. Helens showed abundant levels of various *Legionella* species.

CHLAMYDIAE

Chlamydia psittaci: Psittacosis or Ornithosis

Psittacosis (also called parrot fever and ornithosis) is caused by *C. psittaci*, a member of the Chlamydiae. Original isolations of *C. psittaci* were from parrots and other psittacine birds (those resembling parrots), but it is now known that many other birds are also infected. The disease as it occurs in nonpsittacine birds is called ornithosis.

One of the characteristics of the avian form of ornithosis is its propensity to exist as an asymptomatic latent infection that can become overt following stresses such as crowding, unsanitary conditions, and bacterial infections—all of which are likely to occur during shipment of birds. Once activated, the chlamydiae cause a generalized infection in which the organisms can be found in essentially every organ of the bird. Fecal excretions contain many organisms and provide the major source of infection for humans and other birds.

Pathogenicity Humans usually acquire ornithosis by the inhalation of dried infected feces, and the disease in humans, as in birds, may be asymptomatic or mild. However, in many cases a severe and frequently fatal pneumonia develops after an incubation period of 1 to 3 weeks. About 100 such cases are reported in the United States each year.

Diagnosis requires the isolation of the etiological agent. Injection of acute-stage blood or sputum into cell cultures or embryonated eggs is followed by identification with fluorescently labeled specific antibody, complement-fixation tests, or neutralization of infectivity by the use of specific antibody.

Clinical Management Tetracycline is the drug of choice for the treatment of ornithosis, and adequate therapy with this antibiotic seems to produce a rapid and permanent cure.

Control is difficult because the disease prevails so widely as a latent infection of birds. Infection is an occupational hazard for workers in poultry slaughterhouses, particularly turkey abattoirs, although ducks and pigeons also provide a reservoir for ornithosis. Many human infections, however, are acquired from pet birds that have been imported into this country, in spite of the fact that all birds except those from Canada entering the United States are required to undergo a 30-day quarantine. All quarantined psittacine birds are given food containing chlorotetracycline, but as judged from the number of human cases of psittacosis acquired from newly purchased, imported birds, this treatment has not proved to be effective for curing infected birds. It has therefore been proposed that the 30-day quarantine be increased to 45 days.

Non-avian-Associated Psittacosis

Prior to 1986, it was assumed that essentially all chlamydial pneumonias were caused by *C. psittaci* and resulted from exposure to infected birds. Now, however, J. Thomas Grayston and his associates at the University of Washington have reported the isolation of a new species, which has been designated *C. pneumoniae* strain TWAR (first isolated in Taiwan, causing an acute respiratory disease) and which is believed to be spread from person to person, without the inter-

vention of an avian host. This organism causes an acute lower respiratory tract infection similar to that of the avian strains, and it appears to be an important etiological agent of pneumonia in college students. Using specific serological tests for antibodies, they demonstrated that almost half the adults from five different parts of the world had been infected at one time or another with *C. pneumoniae* strain TWAR, making it the most prevalent chlamydial infection occurring in humans. It seems probable that the reason this pathogen had not been previously recognized was because of difficulty in cultivating the organism.

RICKETTSIAE

Coxiella burnetii: Q Fever

C. burnetii is the causative agent of a disease called **Q fever.** As described in Chapter 13, the rickettsiae are minute rods and cocci. They are nonmotile and, if they are stained with the Gram stain, are gram-negative. Like other rickettsiae, *C. burnetii* requires living tissue for growth; however, an arthropod vector is not essential for transmission of the disease to humans. *Coxiella* is resistant to a temperature of 60°C for 2 h and may not be destroyed by ordinary pasteurization temperatures. For this reason, pasteurization temperatures have been raised to 62.9°C to inactivate the organism. This stability presented a paradox to the supporters of the leaky membrane explanation for the obligate intracellular growth of rickettsiae. All of it, however, began to make sense when Thomas McCaul and Jim Williams, from the Rocky Mountain Laboratories in Hamilton, Montana, discovered that suspensions of *C. burnetii* contained two morphological variants that they termed large-cell variant (LCV) and small-cell variant (SCV). Electron micrographs of these organisms strongly suggest that the SCV is a type of endospore that is formed at the polar end of the LCV, much as described for the members of the family Bacillaceae. Thin sections of *C. burnetii* demonstrating endospores in these organisms are shown in Figure 24.22. McCaul and Williams also have evidence suggesting that, after entry into a cell, the acid pH of the phagolysosome activates the general metabolism of the endospore, permitting it to undergo binary fission or to differentiate into the vegetative cell variant (LCV).

Pathogenicity Although ticks serve as both a vector and a reservoir, they are not usually the direct source of human Q fever infections. Many human infections originate from animal carcasses, and it is thought that the inhalation of dried tick feces from cattle hides is one mechanism of infection for slaugh-

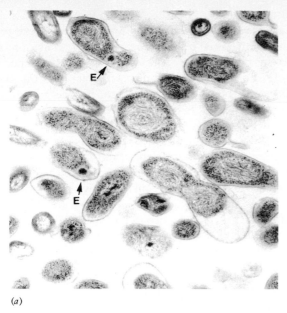

(a)

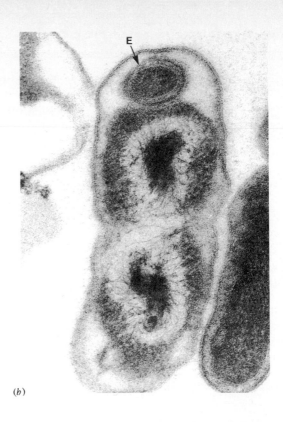

(b)

Figure 24.22 (a) An electron micrograph of thin sections of *C. burnetii* shows two large-cell variants with endospores (E). (b) Complete formation of the endospore (E) in a large-cell variant of *C. burnetii* undergoing unequal cell division. Note the nuclear regions of the dividing cell and the separation of the spore from the cytoplasmic contents by the membranes of the endospore.

terhouse workers. Moreover, infected cattle and sheep do not appear ill but shed large numbers of infectious organisms, particularly in their placental products. Thus, when such cattle are giving birth, aerosols containing billions of organisms contaminate the surrounding area, where they remain viable in dust and hides for extended periods. Many dairy herds are infected with *C. burnetii*; it is estimated that 10 percent of such herds in the Los Angeles area shed these organisms in their milk. However, even though the organisms seem to be highly virulent when inhaled, the ingestion of contaminated milk does not often cause disease.

Among the symptoms of Q fever in humans are chills, headache, malaise, weakness, and severe sweats. Pneumonia with a mild cough occurs in most cases, and the condition is in many ways clinically similar to primary atypical pneumonia. The incubation period is usually 2 to 3 weeks.

Clinical Management Because the clinical symptoms are similar to those of many other infectious diseases, an unequivocal diagnosis requires the isolation and identification of the causative agent or a demonstration of a rise in specific complement-fixing antibodies during the patient's convalescence. The organisms can be grown in the yolk sac of embryonated eggs or in cell cultures of mouse fibroblasts, and they can be identified by direct immunofluorescent antibody staining. Inoculation of blood or sputum into hamsters is also an effective method for isolation of the organisms.

Broad-spectrum antibiotics such as the tetracyclines are effective in treating Q fever, and relapses following adequate therapy are rare. However, it is becoming more apparent that Q fever can exist asymptomatically for periods of months or years and that such latent infections can become activated by the use of x-irradiation or multiple cortisone injections. Valvular heart disease is the most common manifestation of Q fever, although hepatitis is not an infrequent finding. Because of the seriousness of these infections, it has been suggested that patients with Q fever be treated with a combination of lincomycin and tetracycline for at least 12 months, with continued careful monitoring after withdrawal of antibiotics.

Apparently all strains of *C. burnetii* harbor a plasmid whose functions are unknown. Comparison of the plasmid sequences from a number of strains revealed that strains isolated from cases of endocarditis (heart involvement) were unlike those isolated from cases of pneumonia, suggesting that *C. burnetii* variants that cause endocarditis are unique and different from those that cause acute Q fever.

A vaccine consisting of killed phase I organisms has been used to immunize high-risk persons. Adverse reactions obtained during vaccine trials, however, have deterred the widespread use of this vaccine.

THE SYSTEMIC MYCOSES

With the exception of cryptococcosis, all the fungi causing systemic diseases (also called the deep mycoses) are **dimorphic**, existing as filamentous organisms in nature and as yeast cells in infected tissue. Moreover, their distribution is extremely limited within specific geographic areas. The diseases *all begin as lower respiratory infections* resulting from the inhalation of the fungal spores. It appears that the majority of such cases are asymptomatic or undiagnosed, and only a small percentage progress to serious systemic manifestations.

Blastomyces dermatitidis: Blastomycosis

Blastomycosis is caused by *B. dermatitidis*. The disease appears to be endemic in the eastern part of the United States, but cases have been reported in Mexico and Africa. A sexual stage for the organism, described in 1967, technically classifies this fungus in the Ascomycetes as *Ajellomyces dermatitidis*.

Pathogenicity   Although all evidence indicates that the human disease is initiated by inhalation of the fungal conidia, routine attempts to isolate the etiological agent from its natural habitat have usually failed.

Some insight was gained, however, following an epidemic of 48 cases of blastomycosis in Wisconsin in June 1984. All the infected individuals had attended a camp and had walked on an abandoned beaver lodge. Cultures from the lodge and the decomposed wood near the beaver dam grew *B. dermatitidis*. It is noteworthy that the isolations were made from moist soil with a high organic content, which had been exposed to animal excreta. *B. dermatitidis* has also been isolated from five sites near Augusta, Georgia, all of which were associated with animal excreta.

Blastomycosis may be seen as pulmonary, cutaneous, or systemic infections, but it is generally believed that all three types originate in the lung from a primary pulmonary blastomycosis. Unresolved pulmonary infections progress to lobar pneumonia, accompanied by spread of the organisms via the bloodstream to other internal organs, bone, and skin.

Skin lesions, which eventually evolve into ulcerated granulomas, are the most common symptom of the disease. Such lesions may progress over years, eventually involving large areas of the body. Multiple internal organs may become infected, and bone invasion, causing arthritis and bone destruction, occurs in 25 to 50 percent of cases.

Clinical Management   Blastomycin, a filtrate extracted from *B. dermatitidis* grown on a liquid medium, will induce a delayed-type skin reaction in infected persons. Individuals with histoplasmosis or coccidioidomycosis may also react to blastomycin, thereby reducing the value of this skin test. A specific fluorescently labeled antibody will react with the yeast cells in histological tissue section, but definitive laboratory diagnosis is best accomplished by the growth and identification of the etiological agent.

Conidia are the infectious elements of this disease; when grown at 25°C on Sabouraud's glucose agar, the organism bears spherical conidia from the sides of the hyphae. At 37°C the organism grows as a spherical yeast cell that can be readily identified by the single buds arising on a very broad base (see Figure 24.23).

Intravenous amphotericin B is the treatment of choice for all forms of blastomycosis.

Histoplasma capsulatum: Histoplasmosis

H. capsulatum is distributed worldwide, although certain areas in the central and midwestern United States appear to be the most heavily contaminated. It is estimated that 40 million people in the United States have been infected and that 200,000 new cases of histoplasmosis occur annually, making it one of the most common human mycoses. This dimorphic organism has recently been shown to have a sexual stage and is now classified with the Ascomycetes as *Ajellomyces capsulata*.

Pathogenicity   *H. capsulatum* is widely distributed in soil and preferentially grows in association with bird and bat feces. Thus, highly infectious areas may be found in chicken coops, bat caves, bird roosts, or any environment extensively inhabited by birds. In soil, *H. capsulatum* grows as branched hyphal filaments that form small single conidia and larger spiny spores referred to as tuberculated macroconidia (see Figure 24.24). Infection follows inhalation of the small

Figure 24.23   Yeast cells, many with buds, of *Blastomyces dermatitidis* in a culture of sputum.

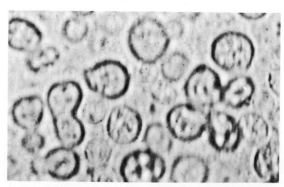

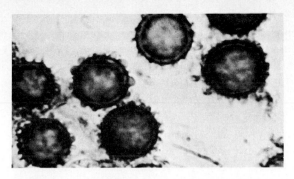

Figure 24.24 Tuberculated macroconidia of *H. capsulatum.*

microconidia into the lung, where they germinate and are transformed into budding yeast cells.

The yeast cells are distributed from the lungs via the bloodstream, and in the vast majority of cases the organisms are destroyed or sequestered by the host's immune responses. The lung lesions heal, followed by fibrosis and calcification similar to that found in healed calcified tuberculous lesions.

Recently, reactivation histoplasmosis has been commonly recognized in areas where the organism is endemic. The highest disease rates are seen in infancy and during the fifth and sixth decades of life. Such endogenous reactivation is identical to that seen for tuberculosis and, not infrequently, may coexist with tuberculosis. It thus appears that the calcified, healed lesions of histoplasmosis contain dormant but viable yeast cells.

Clinical Management Examination of white blood cells from centrifuged blood samples may reveal many yeast cells within the macrophages (see Figure 24.25). Such material, as well as sputum, can be cultured on Sabouraud's and blood agar. Colonies growing at 25°C produce characteristic tuberculated macroconidia and at 37°C will grow as ovoid budding yeast cells.

Figure 24.25 Yeast cells of *H. capsulatum* in a macrophage are shown in this example from a Giemsa-stained tissue smear (human patient).

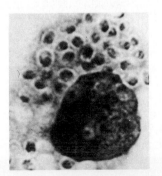

Histoplasmin (analogous to tuberculin) will give a delayed-type skin response in persons who have been infected with *H. capsulatum,* but cross-reactions with coccidioidin and blastomycin do occur. Serological tests include complement fixation, immunodiffusion using histoplasmin as the antigen, and latex particle agglutination, in which the latex particles are coated with histoplasmin and mixed with dilutions of serum. Specific fluorescently labeled antibody can also be used as a diagnostic aid.

The vast majority of *Histoplasma* infections are undiagnosed, and recovery is spontaneous. Even moderately severe cases of primary histoplasmosis can be adequately treated with bed rest. However, the disseminated disease requires antimicrobial therapy; amphotericin B is effective and recovery rapid. However, amphotericin B may cause kidney damage. As a result, ketoconazole, one of the newer imidazoles, may become the treatment of choice, since it appears to be equally effective with less serious side effects.

Coccidioides immitis: Coccidioidomycosis

C. immitis is found in soil only in arid regions and is highly endemic in the southwestern United States and northern Mexico. Endemic areas also exist in Central and South America.

Pathogenicity *C. immitis* grows in soil as a septated filamentous fungus that characteristically produces thick-walled arthrospores in alternating cells of a hyphal filament (see Figure 24.26). The mature hyphae easily fragment, and the liberated arthrospores are readily carried on surrounding dust particles. Human infection follows inhalation of the arthrospores and, as in histoplasmosis, may result in an asymptomatic or mild-to-severe respiratory infection.

Following inhalation, the arthrospore differentiates into a large, round, thick-walled structure 30 to 60 μm in diameter, which becomes multinucleated and eventually forms several hundred uninucleated endospores that are 2 to 5 μm in diameter (see Figure 24.27). This round structure, called a spherule, breaks open when mature, liberating the enclosed endospores. The morphological events of spherule and endospore formation are now recognized to be analogous to sporangium and sporangiospore production as found in the Zygomycetes, but no sexual stage has been observed. In the majority of cases, the liberated endospores are phagocytosed and killed, leaving no outward evidence that an infection has occurred. Symptomatic pulmonary infections are characterized by fever, chest pain, mild cough, headache, and loss of appetite. X-rays show small nodules in the lungs, which may form cavities that become fibrotic and later calcify.

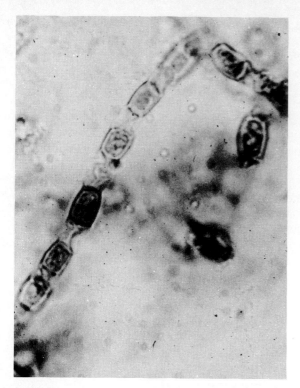

Figure 24.26 *C. immitis,* causative agent of coccidioidomycosis. This illustrates the typical arthroconidia formed by this species.

Figure 24.27 Endospore-filled spherule from a case of coccidioidomycosis.

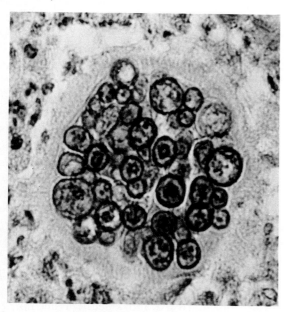

Disseminated coccidioidomycosis may produce an acute or chronic meningitis or a generalized disease characterized by lesions in many internal organs. Cutaneous lesions appear that may eventually heal or, in advanced disease, form draining ulcers.

Clinical Management Direct examination of sputum, pus, or gastric washings will frequently reveal the presence of mature spherules. The organism is readily grown on Sabouraud's medium, and the presence of arthrospores (in a formalized preparation) is indicative of *C. immitis.*

Coccidioidin is the best of the skin test materials. Immunodiffusion tests and latex particle agglutination assays have proved useful as serological acids in the diagnosis of the disease.

A primary infection, even though asymptomatic, induces a solid immunity to reinfection. As in other mycotic infections, amphotericin B is the therapy of choice for chronic pulmonary or disseminated disease; however, surgical removal of lung cavities may be required in cases of advanced pulmonary disease.

Cryptococcus neoformans: Cryptococcosis

Cryptococcosis is caused by *C. neoformans.* This species routinely grows in the yeast form and is distributed worldwide in soil. Most human infections, however, are thought to be acquired by the inhalation of dried, infected pigeon feces.

Pathogenicity Because the organisms are acquired by inhalation, primary infection occurs in the lungs. Most infections are either asymptomatic or undiagnosed. Occasionally they may be spread via the bloodstream, causing infections in many areas of the body, but the most frequent complication is involvement of the brain and meninges.

Clinical Management *C. neoformans* grows primarily as a yeast that produces a large capsule, but several strains now known can form a mycelium. An India-ink wet mount provides an easy method for visualizing this encapsulated yeast (see Figure 24.28), and the presence of characteristic organisms in such a preparation provides a tentative diagnosis. Specific identification of *C. neoformans* requires growth at 37°C and observed pathogenicity for mice.

Untreated meningitis due to *C. neoformans* is invariably fatal; however, amphotericin B is usually effective and is currently the therapy of choice.

Table 24.5 provides a summary of the bacteria and fungi that enter the body via the respiratory tract.

MISCELLANEOUS FUNGAL INFECTIONS

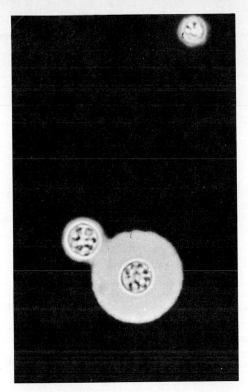

Figure 24.28 *C. neoformans,* the causative agent of cryptococcosis. The very large capsule is clearly evident around the yeast cell.

Paracoccidioidomycosis, caused by *Paracoccidioides brasiliensis,* is seen only in Central and South America. Spores are inhaled and most commonly the disease is subclinical; however, dissemination to the oral mucosa, causing ulcerative lesions in the mouth and nose, also occurs. Amphotericin B is the only effective chemotherapeutic agent for this disease.

Geotrichosis, caused by *Geotrichum candidum,* is found worldwide in decaying plant material, dairy products, normal skin, and in the stool, urine, and vaginal secretions of about one-third of all normal individuals. Only rarely does it cause pulmonary disease (which may be a secondary infection to tuberculosis) or an oral infection similar to thrush.

Aspergillus, Rhizopus, Mucor, and *Absidia* are opportunistic fungi that rarely produce progressive disease. However, they may cause serious pulmonary infections as well as disseminated disease in debilitated persons, diabetics, patients with leukemia or lymphoma, immunosuppressed individuals, or persons who are heavily exposed to large numbers of spores.

TABLE 24.5   Summary of Bacteria and Fungi That Enter the Body via the Respiratory Tract

Organism	Disease	Diagnostic Procedures	Control

Group A streptococci	Septic sore throat
	Scarlet fever
	Rheumatic fever
	Glomerulonephritis
		Isolation on blood agar
		Bacitracin-sensitive
		  fluorescently
		  labeled antibody
			Penicillin treatment of
			  infections

Group B streptococci	Neonatal septicemia
	  and meningitis
		Isolation on blood agar	?Possibly immunize
			  expectant mothers
			  to type III

S. pneumoniae	Lobar pneumonia	Isolation sensitivity to
		  optochin disks
			Polysaccharide
			  vaccine for most
			  common 23 types

N. meningitidis	Epidemic meningitis	Isolation from
		  nasopharynx,
		  blood, or spinal
		  fluid on chocolate
		  blood agar; quellung
		  test for groups A
		  and C
			Vaccine for groups A,
			  C, Y, and W135

B. pertussis	Whooping cough	Isolation on Bordet-
		  Gengou medium
			Vaccine of heat-killed
			  phase I organisms

H. influenzae	Meningitis and
	  epiglottitis
		Isolation: X and V
		  factor requirement;
		  quellung test for
		  type b
			Vaccine available

TABLE 24.5 (Continued)

Organism	Disease	Diagnostic Procedures	Control
C. diphtheriae	Diphtheria	Isolation from throat on Loeffler's or tellurite medium	Toxoid vaccine
M. tuberculosis	Tuberculosis	Isolation and positive tuberculin reaction	Vaccine (not used in United States). Follow tuberculin reactions
M. leprae	Leprosy	Clinical plus observation of acid-fast rods in lesions	Treat active cases
M. pneumoniae	Primary atypical pneumonia	Isolation plus roentgenographic examination	Vaccines effective, but not widely used
L. pneumophila	Legionnaires' disease	Isolation on Mueller-Hinton medium or injection into guinea pigs	None available
C. psittici	Ornithosis	Isolation from sputum	None available
C. pneumoniae	Pneumonia	Isolation from sputum	None available
B. dermatitidis	Blastomycosis	Fluorescently labeled antibody; appearance of yeast cells grown at 37°C	Amphotericin B for treatment
H. capsulatum	Histoplasmosis	Complement fixation Latex particle agglutination Fluorescently labeled antibody	Amphotericin B for treatment
C. immitis	Coccidioidomycosis	Coccidioidin Immunodiffusion Latex particle agglutination	Amphotericin B for treatment
C. neoformans	Cryptococcosis	India-ink mount Pathogenicity for mice	Amphotericin B for treatment

SUMMARY

Streptococci are gram-positive cocci that have a tendency to grow in chains. They can be classified both biochemically and antigenically. In Lancefield's antigenic classification, the streptococci are placed into groups based on the presence of a common group-specific C carbohydrate. Group A contains most human pathogens; it can be further subdivided into types on the basis of an M protein present in the cell walls. The streptococci produce many toxins, one of which—the erythrogenic toxin—is responsible for scarlet fever. The most serious aspect of group A streptococcal infections is the possibility of a subsequent late nonsuppurative complication—rheumatic fever or acute glomerulonephritis. Both of these sequelae apparently arise as a result of autoimmune reactions. Group B streptococci frequently cause a fatal neonatal infection. Early-onset group B infections, resulting in neonatal pneumonia, are acquired from the mother in utero or during delivery. Late-onset group B infections, resulting in neonatal meningitis, are acquired 1 to 3 weeks after birth.

Dental caries result from the dissolution of the tooth enamel by bacterially produced organic acids, formed primarily by *Streptococcus mutans*. Periodontal disease is caused by a bacterial infection characterized by the loss of attachment of the epithelium to the tooth and the resorption of the alveolar bone structure surrounding the roots of the tooth.

S. pneumoniae is the primary cause of lobar pneumonia in humans. These organisms can be subdivided into approximately 100 types based on antigenic differences in their capsules. They can be differentiated from the α-hemolytic streptococci by the use of optochin disks and by mouse lethality; they may be typed using the Neufeld quellung test.

Neisseria meningitidis is the cause of epidemic meningitis. It will not grow at room temperature and requires an enriched medium. The organism can be subdivided into eight groups on the basis of the antigenic differences in their polysaccharide capsules. Humans are the only hosts, and the nasopharynx is the common site for the isolation of *N. meningitidis.* Like all species of *Neisseria,* colonies of meningococci are oxidase positive.

Bordetella pertussis, the causative agent of whooping cough, is a small gram-negative coccobacillus. An effective vaccine against this disease employs heat-killed phase I organisms.

Haemophilus influenzae is the most common cause of meningitis in infants. The organism requires both hemin and NAD for growth in the laboratory. *H. aegyptius* is the etiological agent of a common conjunctivitis that is frequently called pinkeye.

Corynebacterium diphtheriae produces disease as a result of its ability to produce a very potent exotoxin. All exotoxin-producing strains are lysogenic, and their ability to produce exotoxin is a property of the prophage genome. The exotoxin acts on the host cells by preventing protein synthesis—by inhibiting the movement of the ribosome along the mRNA. Immunization is carried out using formalin-inactivated toxin (toxoid).

Mycobacterium tuberculosis, more commonly called the tubercle bacillus, is the causative agent of tuberculosis. The virulent organisms are acid-fast and grow very slowly. In infected individuals a delayed hypersensitivity against the organism develops that can aid in the diagnosis. *M. leprae,* the causative agent of leprosy, is similar in morphology and staining characteristics to the tubercle bacillus, but this organism has never been grown on artificial media. Leprosy may occur in either of two forms, lepromatous leprosy or tuberculoid leprosy.

M. pneumoniae is the etiological agent for primary atypical pneumonia. This disease is an acute respiratory infection that has been seen frequently in military recruits and college students.

Legionnaires' disease is a recently described respiratory infection. The etiological agent of legionnaires' disease possesses a gram-negative-type cell wall, and it was first isolated after the epidemic that occurred during the American Legion convention in Philadelphia in 1976. It is believed to be spread via an airborne route.

Chlamydia psittaci is a normal pathogen of birds, and it is spread to humans by the inhalation of dried infected bird feces, causing ornithosis (psittacosis). *C. pneumoniae* strain TWAR is the etiological agent for a chlamydial pneumonia that is spread directly from person to person.

Coxiella burnetii, the causative agent of Q fever, is a small rickettsial organism. Humans usually become infected by the inhalation of dried, infected tick feces or aerosols arising from placental products.

With the exception of *Cryptococcus neoformans,* the systemic fungi are all dimorphic. All diseases begin with a lower respiratory tract infection, but may be spread via the bloodstream to produce a systemic infection. These diseases include: (1) blastomycosis, caused by *B. dermatitidis;* (2) histoplasmosis, caused by *H. capsulatum;* (3) coccidioidomycosis, caused by *C. immitis;* and (4) cryptococcosis, a yeast infection caused by *C. neo-*

formans. Many of the infections caused by these organisms are asymptomatic.

Geotrichum and other opportunistic fungi such as *Aspergillus* may cause rare pulmonary infections.

QUESTIONS FOR REVIEW

1. On what medium are streptococci routinely grown in the laboratory?
2. How are streptococci classified?
3. Why is the spread of streptococci difficult to control? What control measures may be followed?
4. How could one prevent the late nonsuppurative complications of streptococcal infections?
5. What diagnostic tests may be performed to identify *S. pneumoniae?*
6. What diagnostic tests may be performed for epidemic meningitis, whooping cough, and secondary infections caused by *H. influenzae?*
7. What is meningitis? What organisms can cause this disease?
8. What disease is caused by *C. diphtheriae?* What are the characteristics of the organism?
9. What is the relationship of bacteriophage to toxin production of *C. diphtheriae?* Of iron concentration to toxin production?
10. How are active and passive immunity accomplished against *C. diphtheriae?*
11. Explain the procedure and use of the Schick test.
12. What procedures would you employ to prevent the spread of diphtheria?
13. Explain the mechanism of action of diphtheria toxin at the molecular level.
14. Describe the tubercle bacillus.
15. How may the tuberculosis organism be recognized?
16. What measures would a person assigned to the care of a patient with tuberculosis carry out for self-protection? How would the spread of the organisms to other patients be prevented?
17. How can you determine whether a person has been infected with the tuberculosis organism?
18. What is BCG and for what is it used?
19. What is meant by *reactivation infection?*
20. Describe what you know about legionnaires' disease (see also Chapter 22).
21. What is the reservoir for psittacosis and ornithosis and how do humans acquire the disease?
22. What disease is caused by *C. burnetii?* How can the disease be controlled?
23. Outline the epidemiology of Q fever.
24. What are three types of infections caused by *B. dermatitidis?* How can the disease be diagnosed?
25. What type of infection is caused by *H. capsulatum?* What is meant by reactivation histoplasmosis? How can this infection be diagnosed?
26. What are the endemic regions for coccidioidomycosis? How do humans become infected?
27. What is a spherule?
28. What are some opportunistic fungi? Under what conditions can they cause disease?

29. How does one acquire an infection caused by *C. neoformans*? What is the most common complication of such infections?

SUPPLEMENTARY READING

Bhakdi S, et al: Mechanism of membrane damage by streptolysin-O. *Infect Immun* 47:52, 1985.

Boyer KM, Gotoff SP: Prevention of early-onset neonatal group B streptococcal disease with selective intrapartum chemoprophylaxis. *N Engl J Med* 314:1665, 1986.

Centers for Disease Control: Streptococcal foodborne outbreaks. *MMWR* 33:669, 1984.

Centers for Disease Control: Diphtheria tetanus and pertussis: Guidelines for vaccine prophylaxis and other preventive measures. *MMWR* 34:405, 1985.

Davidson PT: Tuberculosis: New views of an old disease. *N Engl J Med* 312:1514, 1985.

Devoe IW: The meningococcus and mechanisms of pathogenicity. *Microbiol Rev* 46:162, 1982.

Draper P: Vaccines: Leprosy bacillus outwitted. *Nature* 316:388, 1985.

du Moulin GC, Stottmeier KD: Waterborne mycobacteria: An increasing threat to health. *ASM News* 52:525, 1986.

Edelstein PH: Environmental aspects of Legionella. *ASM News* 51:460, 1985.

Friedman RL, et al: *Bordetella pertussis* adenylate cyclase: Effects of affinity-purified adenylate cyclase on human polymorphonuclear leukocyte functions. *Infect Immun* 55:135, 1987.

Grayston JT, et al: Countrywide epidemics of *Chlamydia pneumoniae* strain TWAR in Scandinavia, 1981–1983. *J Infect Dis* 159:1111, 1989.

Kleemola M, et al: Epidemics of pneumonia caused by TWAR, a new chlamydial organism in military trainees in Finland. *J Infect Dis* 157:230, 1988.

Klein BS, et al: Serologic tests for blastomycosis: Assessments during a large point-source outbreak in Wisconsin. *J Infect Dis* 155:262, 1987.

Loosmore SM, et al: Engineering of a genetically detoxified pertussis toxin analog for development of a recombinant whooping cough vaccine. *Infect Immun* 58:3653, 1990.

MayoSmith MF, et al: Acute epiglottitis in adults. *N Engl J Med* 314:1133, 1986.

McCaul TF, Williams JC: Developmental cycle of *Coxiella burnetii*: Structure and morphogenesis of vegetative and sporogenic differentiations. *J Bacteriol* 147:1063, 1981.

Musher DM, et al: Natural and vaccine-related immunity to *Streptococcus pneumoniae*. *J Infect Dis* 154:245, 1986.

Newbrun E: Sugar and dental caries: A review of human studies. *Science* 217:418, 1982.

Pittman M, et al: *Bordetella pertussis:* Pathogenesis and prevention of whooping cough. In Leive L and Schlessinger D (eds): *Microbiology—1984*, Washington, DC, American Society for Microbiology, 1984, p 157.

Schlech WF, et al: Epidemic listeriosis: Evidence for transmission by food. *N Engl J Med* 308:203, 1983.

Schuchat A, et al: Population-based risk factors for neonatal group B streptococcal disease: Results of a cohort study in metropolitan Atlanta. *J Infect Dis* 162:672, 1990.

Stead WW, et al: Tuberculosis as an endemic and nosocomial infection among the elderly in nursing homes. *N Engl J Med* 312:1483, 1985.

Stephens DS, et al: Analysis of damage to human ciliated nasopharyngeal epithelium by *Neisseria meningitidis*. *Infect Immun* 51:579, 1986.

Stevens DL, et al: Severe group A streptococcal infections associated with a toxic shock-like syndrome and scarlet fever toxin A. *New Engl J Med* 321:1, 1989.

Weiss A, Hewlett EL: Pertussis toxin and extracytoplasmic adenylate cyclase as virulence factors of *Bordetella pertussis*. *J Infect Dis* 150:219, 1984.

Chapter 25

Viruses That Enter the Body via the Respiratory Tract

OBJECTIVES Completion of this chapter should familiarize you with

1. The kinds of viruses involved in the common cold syndrome.

2. What is meant by a "latent" virus infection.

3. How influenza virus makes drastic changes in its surface antigens.

4. The types of vaccines available for viruses causing respiratory infections.

5. Complications that may follow measles and mumps.

6. Why rubella infections are so serious in pregnant women.

7. The relationship between chickenpox and zoster.

Humans can be infected by any of a large number of viruses that gain entry to the body by way of the respiratory tract. Of these, some viruses eventually spread throughout the body to cause a generalized infection, while others remain localized in the respiratory tract. In the following sections of this chapter we shall discuss infectious organisms that enter the host mainly by way of the respiratory tract.

VIRUSES

Viruses Involved in the Common Cold Syndrome

The very commonness of the common cold was a stumbling block for many years in understanding the etiology of this disease. We probably acquire this infection more frequently than any other, and it is not unusual for a person to suffer several attacks annually. Although common colds are not serious in terms of morbidity or mortality, they are troublesome in that they are the cause of many lost work days. This makes the common cold an economically important disease.

Rhinoviruses

Although initially a single viral agent was sought, recent advances in virological techniques have demonstrated that a large variety of viruses cause the mild respiratory infections we refer to as the common cold. The largest number of these common cold viruses have been classified into a group known as **rhinoviruses.** Only a few of the viruses in this group have been extensively studied, but they are all small RNA viruses of the class **picornavirus** (see Figure 25.1).

Rhinoviruses can be grown in cell cultures, but success in propagating them occurs only under conditions similar to those in the normal nasal mucosa. Thus, rhinoviruses can be grown in human embryonic kidney tissue cells at 33°C (instead of 37°C), at neutral pH, and with good oxygenation. Using cell culture techniques, over 100 serological types of rhinoviruses have been isolated and grown. Because each type is immunologically distinct, an individual can have several infections over a fairly short period. Humans appear to be the only hosts for rhinoviruses, although many strains can be grown on monkey as well as human cells.

Much of the work concerning the disease caused by rhinoviruses has been done using human volunteers. A good bit of early research was done by Sir Christopher Andrews at the Common Cold Research Establishment in England. Since this type of research requires complete isolation of human volunteers, many

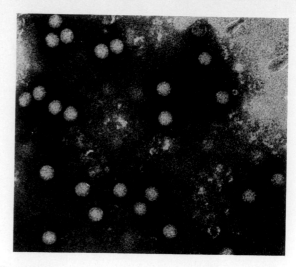

Figure 25.1 Virions of human rhinovirus type 14 (×120,000).

of the subjects were newlywed couples who were offered free accommodations in isolated honeymoon cottages in return for serving as guinea pigs for the common cold. Interestingly, this type of research indicates that exposure to inclement weather does not produce a cold. It appears that the major effect of cold weather is to place people closer together in poorly ventilated places and thus to enhance the possibility of transmitting the virus from an infected person to an uninfected one.

These studies have also demonstrated that immunity to rhinovirus infections is type specific and that it probably is effective for at least 2 years. As one might expect, it can be correlated better with the amount of IgA present in the nasal mucous secretions than with the serum IgG. After recovering from a rhinovirus infection, an individual is usually quite resistant to reinfection by any other rhinovirus for about a month. This nonspecific resistance is believed by some to be the result of the interferon production, although it seems unlikely that interferon levels would remain effective for more than a few days. Because over 100 different rhinoviruses have been described, an effective vaccine against the common cold does not appear feasible, even though a vaccine might well be effective for a homologous virus. Transmission studies have shown that the virus can be spread by way of aerosols under intimate conditions (Figure 25.2), but a more efficient mode of transmission appears to involve hand contact and self-inoculation of mucous membranes of the eye and nasal mucosa. Recent studies have shown that the use of virucidal paper handkerchiefs are highly effective in preventing the transmission of rhinoviruses from one person to another.

The diagnosis of the common cold is hardly a

Figure 25.2 Each droplet in this unstifled sneeze is laden with microorganisms.

laboratory procedure; however, diagnosing a rhinovirus infection requires the growth, isolation, and identification of the virus. Because this requires specialized techniques, it is done only by research laboratories working with these viruses.

Parainfluenza Viruses

A second group of viruses that can be considered common cold viruses are called the **parainfluenza viruses.** These are also RNA viruses, but, owing to their structure and their ability to agglutinate red blood cells, they are classified as paramyxoviruses. Like the influenza virus, the parainfluenza viruses have both a hemagglutinin and a neuraminidase on their surface.

There are four distinct immunological types of human parainfluenza virus. All can be grown in cell cultures, and they have been adapted to grow in the amniotic sac of chick embryos 7 to 9 days old.

The disease caused by the parainfluenza viruses in adults is very mild, and it is in all likelihood restricted to the upper respiratory tract. Thus, infection in adults produces a mild respiratory disease that would be diagnosed clinically as a common cold. In infants and children the parainfluenza virus may invade the lower respiratory tract, causing a severe pneumonia. Types 1 and 2 seem to be particularly frequent invaders of the larynx in infants, causing a syndrome called **laryngotracheobronchitis,** or **croup.**

The definitive diagnosis of a parainfluenza infection requires the isolation and growth of the actual virus. The virus can be grown on cell cultures of monkey kidney or embryonic human kidney, and it can be recognized by the ability of infected tissue cells to adsorb guinea pig red blood cells (hemadsorption). A formalin-inactivated vaccine is not protective, and living attenuated vaccines are now being developed with the hope that they will induce IgA-type antibody synthesis.

Respiratory Syncytial Virus

A third virus associated with the common cold syndrome is **respiratory syncytial virus (RSV).** Closely related to both measles and parainfluenza viruses, it is classified with the paramyxoviruses, where it is placed in the genus *Pneumovirus*. It differs, however, from other viruses in this group in that it does not possess a hemagglutinin and, therefore, cannot hemagglutinate red blood cells.

This virus was first described in 1956 and is now known to be the most important cause of serious lower respiratory tract infections in infants. It is estimated that RSV is responsible for one-half of all cases of bronchiolitis and one-fourth of all pneumonias occurring during the first 6 months of life. The disease can have a 20 percent mortality in newborns and, as was confirmed in Italy during 1979, epidemics may occur in which there is a high infant mortality. In adults, however, infection results in only a mild upper respiratory infection that would normally be diagnosed as a common cold. Since the newborn infant is so highly susceptible and RSV is highly transmissible, each winter sees major epidemics of this disease in infants. Maternal antibody is not effective; only IgA appears to be protective, and this type of immunoglobulin does not cross the placenta. Killed vaccine has been tried but was found to be ineffective in infants. It would seem that control of this disease may require an attenuated virus vaccine that will induce IgA-type antibodies.

Coronaviruses

Coronaviruses were so named because of a series of petallike projections that cover the surface of their envelopes, resembling a solar corona (see Figure 25.3). They are difficult to grow in cell culture, but it is known that they are RNA viruses multiplying in the cytoplasm of infected cells.

Coronaviruses cause gastroenteritis in swine and calves, and hepatitis in mice; however, it seems that the two human serotypes of these viruses are associated predominantly with upper respiratory tract infections indistinguishable from the common cold caused by rhinoviruses.

Coronavirus infections occur mainly in the winter and spring and are undoubtedly spread by way of respiratory secretions. It is estimated that coronaviruses cause about 15 percent of common colds. Studies using volunteers have shown an average incubation period of 3 days and an illness lasting about a week. Symptoms include a sore throat plus other effects normally associated with an acute upper respiratory infection.

Reoviruses

The last group of viruses considered part of the common cold syndrome are designated **reoviruses** (respiratory enteric orphan viruses). These viruses occur both in feces and in respiratory secretions. They have frequently been isolated from normal healthy adults as well as from people suffering from upper respiratory disease. As a result, virologists are not quite certain of the role of these viruses in the production of disease.

The reoviruses are unique both because their nucleic acid exists as double-stranded RNA and because it exists in each virion as 10 separate molecules within an icosahedral capsid (see Figure 25.4). They can be grown in cell cultures, and are usually identified using serological techniques (neutralization, complement fixation) with known antiserum. Three immunological types of reovirus have been described.

Because little is known concerning the disease entities caused by the reoviruses, there is little to be said concerning their control. Studies have shown that the majority of young adults possess antibodies to the reoviruses, so one may assume that they have a wide distribution.

Adenoviruses

Pathogenicity

The first of the viruses that now make up the large group called adenoviruses was isolated in 1953 from

Figure 25.3 Human coronavirus. Surface projections create a corona effect around each virion. (Original magnification ×230,000.)

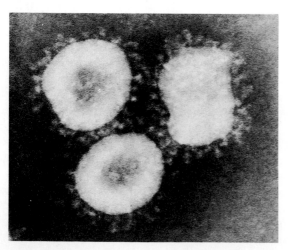

Figure 25.4 Reovirus capsids (×272,000).

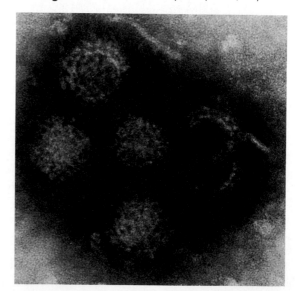

cell cultures containing normal adenoids. Subsequently, it has been found that there are at least 41 types of human adenovirus that produce a variety of symptoms. Each adenovirus type is specific for its own neutralizing antibodies, which are induced by one of the type-specific capsid proteins and one of the fiber proteins (see Figure 25.5). One of the really interesting observations is that these viruses may be isolated from both normal and infected tonsils and adenoids as well as from respiratory secretions and secretions of the eye. In fact, the majority of adenoids removed by surgery will grow out adenovirus in tissue culture even though the adenoids did not appear to be infected. The mechanism of this apparent "latency" is still obscure, but some investigators suggest that the viruses are not truly latent, for viral DNA is not integrated into host DNA in a manner analogous to that observed in the latent herpesviruses. Instead, the viruses seem to be replicating at a very slow or uneven rate. Using surgically removed tissue, these investigations showed that fewer than 1 out of 10 million cells were actually infected with adenovirus, even though the minced adenoids eventually grew out virus. An explanation for this low rate of infection is yet to be discerned.

The structure of adenoviruses has been studied more extensively than that of any other human virus (see Figure 25.5). The nucleic acid of adenoviruses is a double-stranded DNA. The virus produces a cytopathic effect on human cells in which the infected cells become swollen and rounded and aggregate into grapelike clusters. Adenoviruses may be the cause of acute respiratory disease and ocular infections.

At least 12 of the 41 types of human adenoviruses produce malignant carcinomas if injected into newborn hamsters, mice, or rats. Infection of cultured rat or hamster cells in vitro also results in transformation. Attempts to obtain the infectious virus from the transformed cells have generally failed, but a molecular analysis of such cells has revealed that they possess both viral DNA and viral RNA. Moreover, such studies have demonstrated that cells need to have only the leftmost 11 percent of the adenovirus genome integrated into their DNA in order to be transformed, explaining why it is not possible to obtain infectious virus from such cells.

The transforming region of adenovirus DNA encodes for a couple of proteins known as E1A and E1B, which are required for stable transformation. Recent data indicate that these adenovirus transforming proteins cause cell malignancies by virtue of their ability to bind to the gene products of two cellular antioncogenes known as the retinoblastoma gene and the p53 gene (see Chapter 14). The normal role of these antioncogene products is unknown but it is known that they are involved in the regulation of the cell cycle and that cell transformation occurs if they are not present.

The oncogenicity of adenoviruses in rodents and

Figure 25.5 (a) Negatively stained virion of adenovirus (about ×500,000); note fibers extending from vertices. (b) Crystalline aggregates of adenovirus in the nucleus of a HeLa cell 24 h after infection (×8870).

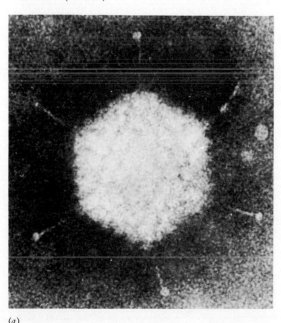

(a)

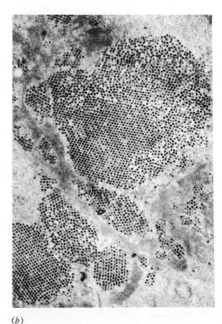

(b)

rodent cells in culture has given rise to the concern that human adenoviruses may cause cancer in humans. Several careful studies using DNA hybridization techniques capable of detecting less than one copy of adenovirus genome per cell failed to detect adenovirus DNA in any human tumors.

The incidence in adults of overt respiratory infection caused by adenoviruses appears to be quite low. However, the observation that adenoviruses can be grown from 50 to 80 percent of surgically removed tonsils and adenoids certainly suggests that the vast majority of adenovirus infections occur early in life. This is borne out by several studies showing that adenoviruses are responsible for a significant percentage of both upper and lower respiratory tract infections in children. For example, one study in Sweden concluded that adenoviruses are responsible for up to 19 percent of all febrile upper respiratory infections in children, and another study involving 18,000 infants and children estimated that at least 10 percent of all childhood respiratory infections are caused by adenoviruses. It is also interesting that different subgroups of adenoviruses may cause various manifestations of disease (as shown in Table 25.1). Subgroups of adenoviruses are determined by the percent guanosine and cytosine in their DNA, the types of erythrocytes they will agglutinate, and the length of the fibers extending from their capsid.

It has been noted that new army recruits frequently have a high incidence of adenovirus infections that are most commonly caused by types 3, 4, and 7. The Armed Forces have prepared vaccines containing killed adenoviruses of these types and have shown that they provide immunity to such infections. However, it seems that the vaccine would not be as valuable for the general population, since only a small number of respiratory infections in the civilian population seem to be caused by these serotypes. The reasons for this are not known.

Adenoviruses have been reported to be present in the stools of young children with diarrhea, and it appears probable that some serogroups are responsible for this syndrome. These viruses have also been implicated in fatal disseminated infections occurring in immunosuppressed renal transplant patients.

Other Respiratory Viruses

Influenza Virus

Influenza viruses are orthomyxoviruses that appear as spherical particles (although filamentous forms are produced by some strains) and whose ribonucleoprotein core possesses helical symmetry. Their nucleic acid exists as eight distinct pieces of RNA within the viral capsid. The capsid is enclosed in a lipid envelope covered with closely packed spikes about 10 to 14 nm in length. Electron microscopy of isolated spikes shows that they exist as two different morphological entities, of which one is a hemagglutinin and the other the enzyme neuraminidase (see Figure 25.6). The hemagglutinin spike allows the virus to attach to a glycoprotein on the surface of red blood cells, causing them to agglutinate. The neuraminidase aids in the penetration of the virus and, in time, it will also remove neuraminic acid from host cell glycoproteins. Antibodies directed against the hemagglutinin will prevent infection; thus, it is these antibodies that are protective.

Based on the antigenicity of the internal nucleoprotein, there are three completely unrelated serological types of influenza virus. Type A has been isolated from humans, animals, and birds; types B and C appear to infect only humans. In addition, type A and, to a lesser extent, type B can be subdivided into strains based on antigenic differences in their hemagglutinin and neuraminidase spikes. There is apparently only one strain of type C virus.

TABLE 25.1   Properties of the Human Adenovirus Subgroups

Subgroup	Type	Pathogenicity	Oncogenicity
A	12, 18, 31	Not defined	Rapidly form tumors in most animals
B	3, 7, 11, 14, 16, 21, 34, 35	Acute respiratory infections, conjunctivitis, pharyngitis, gastroenteritis	Slowly induce tumors in some animals
C	1, 2, 5, 6	Respiratory infections in infants	None
D	8–10, 13, 15, 17, 19, 20, 22–30, 32, 33, 36, 37, 38, 39	8, 19, and 37 keratoconjunctivitis	None
E	4	Acute respiratory infections	None
F	40	Diarrhea in children	None
G	41	Diarrhea in children	None

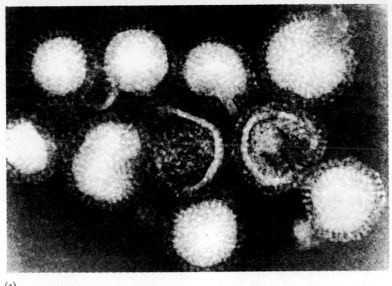

(a)

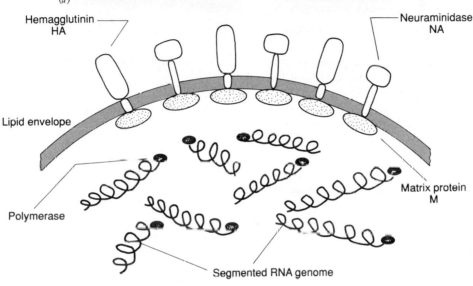

(b)

Figure 25.6 (a) Negatively stained influenza virus type A/Hong Kong (strain A HON2). Note the surface spikes on the viral envelopes (×214,200). (b) Diagram of a partial section through an influenza virion, illustrating the components making up a virus particle. Neuraminidase and hemagglutinin exist on separate, morphologically distinct spikes.

Influenza in Humans Infections in humans are normally characterized by fever, chills, headache, generalized muscular aching, and loss of appetite. The virus is generally restricted to the upper respiratory tract. Although a patient may be very ill during the acute infection, an uneventful recovery after 3 to 7 days usually occurs. Deaths from influenza are most frequently caused by an invasion of the lower respiratory tract by virus or bacteria, resulting in severe pneumonia. Though cases occur in which the influenza virus is the sole etiological agent of the pneumonia, secondary bacterial pneumonias are far more frequent

causes of death. The most common secondary bacterial invaders are *Staphylococcus aureus* (causing toxic shock syndrome), followed by the pneumococcus, *Haemophilus influenzae*, and somewhat less frequently by a β-hemolytic streptococcus.

Reye's Syndrome

Reye's syndrome is characterized by a severe encephalomyelitis and serious liver involvement. It occurs exclusively in children (2 to 16 years of age) and has a mortality rate of 10 to 40 percent. The etiology of this syndrome is unknown, but it occurs following infection by a number of different viruses such as chickenpox, rubella, measles, poliovirus, and adenovirus. Such occurrences are usually sporadic single cases; it is only during influenza epidemics that clustered cases of Reye's syndrome are seen.

After an extensive study of many cases, the Centers for Disease Control concluded that there is a possible association between the occurrence of Reye's syndrome and the use of aspirin to relieve the symptoms and fever of the viral infection. As a result, the Food and Drug Administration has suggested that aspirin labels carry a warning against the use of aspirin for children with influenza, chickenpox, or other viral disease.

Epidemiology of Influenza Epidemics of influenza A occur every 2 or 3 years, whereas those caused by influenza B virus are usually seen at 4- to 6-year intervals. To comprehend how these epidemics can occur, you must keep in mind that only antibodies directed against the hemagglutinin are protective, although antibodies to the neuraminidase do lessen the severity of the infection. The recurring epidemics are possible because of antigenic changes in the hemagglutinin, which permit the virus to be spread in a nonimmune population. It is proposed that these antigenic changes occur in the following two ways: (1) by spontaneous mutations within the virus, resulting in minor antigenic changes in the hemagglutinin (**antigenic drift**), and (2) by recombination, in which, for example, one virion from a bird and one virion from an animal (perhaps a pig) simultaneously infect a susceptible cell, producing progeny virus with an entirely new antigenic hemagglutinin (**antigenic shift**).

Two major factors contribute to the exchange of genetic information. First, influenza A viruses are found naturally in birds, animals, and humans. Second, the segmented nature of the influenza virus genome contributes to the observed high recombination frequency. One can readily produce antigenic hybrids between avian and animal influenza viruses by the multiple infection of cell cultures with different viruses (see Figure 25.7). Because most major pandemics begin in China, and because birds migrating from Australia to China, as well as swine in China, carry type A virus, it has been proposed that these two reservoirs are the progenitors for the major antigenic shifts responsible for worldwide influenza epidemics.

As one might surmise, the minor changes occurring during antigenic drift can still be partially or totally neutralized by antibodies present in a population that has recovered from previous influenza epidemics. Antigenic shift, however, may find a totally nonimmune population, and it is these strains that cause the major influenza pandemics. Type B influenza virus undergoes antigenic drift but not antigenic shift.

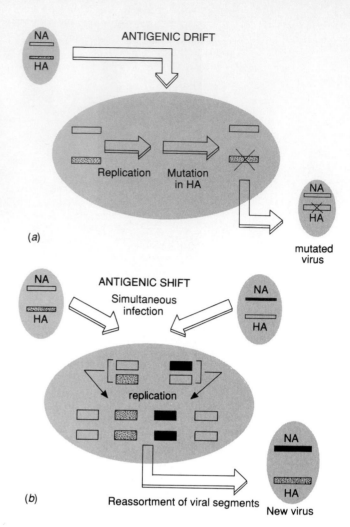

Figure 25.7 Antigenic drift and antigenic shift in influenza virus infections. (*a*) Antigenic drift due to mutations in the HA gene. Open boxes denote the RNA segments encoding the NA protein; hatched boxes denote the RNA segments encoding the HA protein. The X denotes a mutation in the HA coding sequence that gives rise to an amino acid alteration in the HA protein. (*b*) Antigenic shift due to the simultaneous replication of two genetically distinct influenza viruses. The individual virus segments encoding HA and NA from two different influenza viruses are denoted by the open, filled, and hatched boxes, respectively. The formation of a "new" influenza virus is depicted by the presence of a new complement of HA and NA gene, neither of which was present in the original infecting viruses.

Type A influenza viruses are designated for the area from which they were first isolated, the year they were isolated, and by the antigenic properties of their hemagglutinin (H) and neuraminidase (N), as shown in Table 25.2. Thus, the type A Asian influenza virus that caused the 1957 pandemic is designated as $A/Jap/305/57(H_2N_2)$, whereas the Hong Kong virus causing the 1968 pandemic (representing an antigenic shift) is called $A/Hong Kong/1/68/(H_3N_2)$. A 1977 epidemic by $A/USSR/90/77/(H_1N_1)$ was identical to the 1947 pandemic of influenza (see Table 25.2).

The isolation of a swine-influenza-like virus from a human infection in 1976 evoked much concern be-

TABLE 25.2 Hemagglutination and Neuraminidase Composition of Human Influenza A Viruses

Year Isolated	Composition
1934	H_0N_1
1947	H_1N_1
1957	H_2N_2
1968	H_3N_2
1977	H_1N_1
1989	H_2N_2

cause it is believed that a swine strain caused the 1918 pandemic. That virus was never isolated, but many people who lived through this period possess antibodies to the swine influenza virus. The magnitude of that pandemic can be appreciated by reviewing some mortality statistics from 1918. The U.S. death toll for a period of a few weeks is listed as 548,452 people, over 10 times the number of Americans killed during World War I (53,513). India put its toll at 12.5 million and the Dutch East Indies at 800,000. Many villages throughout the world were entirely wiped out by this virus, and worldwide mortality was finally placed at 20 million.

Control of Influenza The injection of a formalin-inactivated, chick-embryo-grown influenza virus is currently our major control mechanism. The efficacy of such vaccines has been the subject of much debate, because they do not induce an IgA response; however, the consensus is that even though such immunization does not always provide absolute protection, it will at least modify the disease. Maximum protection requires an annual immunization with concentrated vaccines; in the event of an antigenic shift of the virus, even this would be ineffective. Protection against new influenza strains would require that the virus be isolated, adapted to give high yields in chick embryos, and then grown in tremendous quantities that could be concentrated and dispensed for use in vaccines.

One very disturbing complication of influenza immunization came to light during the mass immunizations carried out by the U.S. government in 1977 against a possible swine influenza epidemic. A disproportionate number of individuals experienced an ascending paralysis known as Guillain-Barré syndrome, which began a few days after the swine influenza vaccine was received. The relationship of this paralysis to the vaccine is still unknown, but the statistical evidence that they were related is strong.

A newer approach to a more successful influenza vaccine is the use of live attenuated mutants of influenza virus to induce a more effective immune response. The most promising vaccines of this type have been developed from a master cold-adapted strain that was grown in chick embryo cells until a mutant was selected whose growth was optimal at 25°C. Using this master strain, living vaccines are prepared by appropriate recombination techniques to contain any desired hemagglutinin or neuraminidase. Such vaccines are administered intranasally, and clinical trials have indicated that they are more protective than the formalin-inactivated whole-virus preparations, prob-

MICROBIOLOGY MILESTONES

Influenza killed millions of people in less than 1 year during the pandemic of 1918. No other infection or war or famine has killed so many in so short a time. The worst of the pandemic occurred during the last months of World War I when many thousands of army and navy troops were also killed by this disease. The following facts describing the situation in the United States gives an insight into the seriousness of this disease, particularly in large cities around the world.

By December 1918, five hundred thousand Americans had died in the influenza epidemic and approximately twenty million had been infected. Philadelphia, one of the hardest hit cities in the United States, recorded as many as 300 deaths daily during the height of the epidemic. There the superintendent of the morgue noted that there were 61 bodies awaiting coffins and one casket manufacturer stated that he could use five thousand caskets in two hours if only he possessed them.

In Cincinnati, forty thousand cases of influenza were reported and on one day 21 men were each fined a dollar for spitting on the sidewalk. Also, in an effort to control the disease, officials in Chicago decreed that ten mourners were a healthy and respectful maximum.

And so it went. No one knew the cause or how to treat the disease. Letters to the editor of every newspaper contained numerous suggestions of how the epidemic could be aborted. These varied from sprinkling sulphur in one's shoes, inhaling smoke from wood or damp straw, to one that proposed that influenza was caused by excessive clothing and that nakedness might bring the epidemic under control.

Can it happen again? Possibly. Even though we have, for the most part, brought under control diseases such as bubonic plague, cholera, yellow fever, typhus, and smallpox, safety from influenza is not so well assured. Every few years minor epidemics of influenza occur in the United States, resulting in many thousands of deaths. We still do not fully understand why the virus causing the 1918 pandemic was so incredibly virulent, so it is a little difficult to be fully prepared for another onslaught.

ably because they induce an IgA response. Moreover, they do not cause illness when given to children because their optimal growth temperature is 10 to 12°C below that of the nasal mucosa. It would seem probable that such vaccines will be widely used in the not-too-distant future.

Another type of vaccine is the split-virus preparation, in which the hemagglutinin and the neuraminidase spikes are separated from the rest of the viral proteins. Such vaccines have no side effects and appear to be safer for children than are the formalin-inactivated whole-virus preparations. Unfortunately, the split-virus vaccine is not as immunogenic as the whole-virus preparation; two doses of the split-virus vaccine given about a month apart are required to induce protective levels of serum antibodies in children.

In spite of the inadequacies of current vaccines, annual immunization is recommended for high-risk groups such as those with diabetes, chronic pulmonary disorders, and all persons older than 65 years of age, as well as for health care workers and vital community personnel.

Mumps Virus

Mumps virus is classified with the large group of viruses called the paramyxoviruses. It is an RNA virus, and, like influenza virus, contains both a hemagglutinin and a neuraminidase on its surface. There is, however, only one antigenic type of mumps virus. The virus can be grown in chick embryos and will produce disease in monkeys as well as humans. In the human it can grow in the tissues of the brain, the meninges, the pancreas, the testes, the parotid gland, and even the heart. The presence of the virus in these various sites in the body may serve to explain the varied clinical signs of acute mumps infections.

Pathogenicity The familiar clinical picture of mumps hardly needs description. Infection of the parotid glands adjacent to the ears produces inflammation with marked swelling behind the ears and difficulty in swallowing 18 to 21 days after exposure. There is a viremia, and infection may develop in the brain, the meninges, the pancreas, the ovaries, the testes, and the heart. Meningitis may be the most severe complication, but the most feared is infection of the testes (orchitis) in the male who has reached puberty. Orchitis is not only extremely painful but may in rare cases result in sterility, particularly if it is bilateral.

This disease is not nearly as contagious as many of the other childhood diseases; therefore it is not unusual for people to reach adulthood and still be completely susceptible to infection by mumps virus. The virus is spread by infected individuals by way of discharges from the mouth and the nose.

Clinical Management The diagnosis is most frequently made from the clinical picture. However, in cases of atypical mumps, the diagnosis can be confirmed by isolation of the virus from saliva or spinal fluid after growth in the amniotic sac of chick embryos. Final identification can be made with the use of known antiserum, utilizing either the hemagglutination inhibition test or the neutralization test.

Control measures usually include isolation of the infected person to prevent further spread of the disease. The actual value of this procedure is questionable, because an individual may be spreading the virus up to 6 days before symptoms occur.

A live attenuated mumps vaccine grown in chick embryos has been widely used since 1967. It appears that at least 95 percent of susceptible individuals develop adequate antibody titers as a result of vaccine administration. Although the duration of the vaccine-induced immunity is not known, observations made over a 6-year period demonstrate continuing immunity to natural infection. The vaccine is of particular value in males who have no history of having had the natural disease. Because inapparent (or at least undiagnosed) infections account for approximately 30 percent of all infections by mumps virus, many people who are really immune may give no history of having had the disease. In these cases a skin test involving inactivated mumps virus may be used. Usually, those who are immune develop a delayed skin reaction that reaches maximum inflammation in about 24 to 48 h. However, the U.S. Public Health Service claims that the test is not a reliable indicator of immunity and recommends that persons who are unsure of their mumps disease history or mumps vaccination history should be immunized. There is no evidence that persons who have had mumps or who had previously received the vaccine suffer any adverse reactions to the live mumps virus vaccine.

Measles Virus

Measles virus is also classified as a paramyxovirus, although its position in this group is a little less solid than that of the other paramyxoviruses. It is an RNA virus morphologically similar to other paramyxoviruses. Furthermore, it contains a hemagglutinin that will weakly hemagglutinate monkey red blood cells. Measles virus does not contain neuraminidase.

Pathogenicity Measles (rubeola or morbilli) is a severe, acute, highly contagious disease that frequently occurs in epidemic form, particularly in the spring of the year. The severity of the clinical infection is frequently not well appreciated. High fever, delirium, cough, and eye pain from light (photophobia) are accompanied by severe conjunctivitis and a rash over the entire body.

The virus multiplies in the upper respiratory tract and conjunctiva during the early phase of incubation. Late in the incubation period, the virus gets into the blood (viremia) and is transported to all parts of the body, particularly the skin, resulting in the characteristic measles rash.

As a result of the spread of the virus throughout the body, complications are not uncommon. As in influenza, these may frequently be the result of secondary bacterial invaders, which cause pneumonia and ear infections. By far the most feared complication is encephalitis, which can cause permanent neurological injury and even death. Although encephalitis is relatively rare, it does occur in approximately 1 of 10,000 cases of measles. When it does occur, symptoms of encephalitis appear about 5 to 7 days after appearance of the rash.

The incubation period is quite uniform; the first symptoms of cough and fever begin 11 days after exposure and the rash occurs 3 days later. Infected individuals are usually contagious from approximately 3 days before to about 5 days after the appearance of the rash. Measles is one of the most contagious of all infectious diseases.

Clinical Management The diagnosis is almost always made on clinical grounds. Early diagnosis, before the rash occurs, frequently can be made by observing **Koplik spots** —small bluish-yellow spots that occur in the mouth on the buccal mucosa 2 to 3 days before the rash appears. In addition, the virus may be grown in tissue cultures from specimens taken from the nasopharynx or blood. Serological tests may be carried out for neutralizing and complement-fixation antibodies. Fortunately (for us) there is only one antigenic type of measles virus, so that recovery from the disease imparts lifelong immunity.

Isolation has not been an effective control measure because individuals are infective prior to the appearance of a rash. Antibodies to this virus pass through the placenta and protect infants for about the first 6 months.

Vaccines It has been estimated that during the first 20 years of vaccine use, immunizations prevented 52 million cases of measles, 52,000 deaths, and 17,400 cases of mental retardation. Because measles is a severe disease with frequently serious complications, much effort has been spent in developing an effective vaccine. Attenuated strains, given usually at 12 to 18 months of age, have been isolated. However, because of the ubiquity of measles in underdeveloped countries, the World Health Organization has recommended that measles vaccine be administered as soon as possible after 9 months of age. The use of a single injection results in antibody production in over 95

percent of the recipients. Mass immunization has reduced the incidence of measles in the United States by about 90 percent (see Figure 25.8) and if used has the potential of essentially eliminating the disease.

Gamma globulin may be given to exposed nonimmune individuals to prevent the disease completely and then may be followed 3 to 6 months later with the normal vaccination.

Rubella: German Measles

The virus that causes rubella is a small RNA virus that exists as a single antigenic type. The virus can be grown in a variety of human and primate cell cultures; in some, no noticeable cytopathic effect occurs but, in others, effects are readily detectable.

Pathogenicity Rubella is a rather mild disease spread via respiratory secretions. After replication in the cervical lymph nodes, the virus is disseminated via the bloodstream throughout the body, the first overt signs of disease being moderate catarrhal symptoms, mild fever, and a rash that tends to be variable. The incubation period varies from 2 to 3 weeks, but virus can be isolated from the nasopharyngeal secretions as early as a week before recognizable illness. The illness is of short duration and recovery is usually complete within 3 to 4 days after the appearance of the rash. Transient arthritis is a fairly frequent symptom in adult women; other complications are rare.

The tragic aspect of rubella may come to the fore if infection occurs during pregnancy. The virus crosses the placental wall, infecting the fetus, where it disseminates and grows in every fetal organ. Infection may result in the death of the fetus or in a large variety of congenital defects collectively referred to as the **rubella syndrome.** Possible defects include mental retardation, cerebral palsy, cataracts, microcephaly, and heart abnormalities as well as other congenital anomalies. Infected fetuses that survive may continue to shed rubella virus for 1 to 2 years after birth in spite of the presence of circulating antibodies to the virus; it is estimated that 10 to 20 percent of such babies die during the first year following birth.

The prognosis of an infected fetus born with the rubella syndrome is, in large part, dependent on its stage of development at the time the mother becomes infected. Congenital defects occur in as many as 80 percent of the fetuses of mothers infected during the first month of pregnancy, but this will drop to about 15 percent by the third month, and by the end of the first trimester the percentage of fetuses with congenital defects is low. It appears that the most serious congenital abnormalities occur when the embryo is infected during the period of maximum cell differentiation. The observation that rubella-infected human

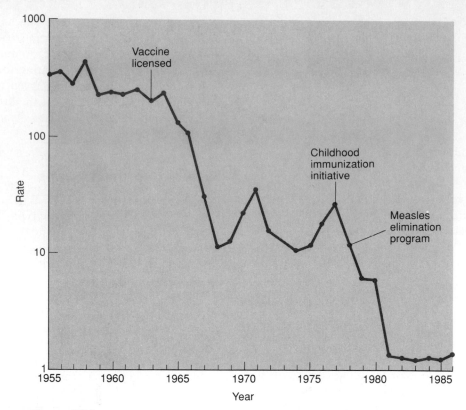

Figure 25.8 Measles (rubeola)—reported cases per 100,000 population by year, United States, 1912–1985.

embryo cells show significant chromosomal breakage and inhibition of normal mitosis sheds some light on the reasons for the congenital defects. Thus, infection of the embryo before cell differentiation is complete could easily interfere with development, causing a variety of defects.

Clinical Management Diagnosis is usually based on clinical observations; laboratory diagnosis requires the isolation of the virus, usually from nasopharyngeal secretions.

There is only one antigenic strain of rubella virus, and immunity appears to be lifelong, although second infections may occur occasionally as unapparent disease. Rash does not occur in some cases of rubella, even though such individuals are excreting virus.

Several live attenuated vaccines are now available that appear to induce some degree of immunity (see Figure 25.9). The vaccines may cause fever, mild rash, and, in adult women, transient arthritis. The major disadvantage of the available vaccines is that they can infect the fetus, although as yet the vaccines have not been associated with congenital abnormalities. However, the risk of this possibility makes it imperative for women who may become pregnant within 3 months to avoid being vaccinated. Rubella vaccine is routinely given at 15 months of age along with live measles and mumps vaccines.

Varicella-Zoster Virus: Chickenpox and Zoster

Chickenpox (varicella) and zoster (shingles) are two considerably different diseases caused by the same virus. The etiological agent for both diseases is called varicella-zoster virus. It is a double-stranded DNA herpesvirus that can be grown in human diploid cell cultures. There is a single antigenic strain of the virus, so recovery from chickenpox results in a solid immunity to chickenpox.

Pathogenicity Chickenpox in children is rather mild, but in adults it can be a very severe disease. The incubation period for chickenpox is 14 to 16 days. The disease is characterized by fever and a maculopapular rash that develops into vesicular lesions. After 3 or 4 days, a granular crust forms and in the absence of a secondary bacterial infection, the lesion heals without leaving a scar. Because a child with varicella is strongly tempted to scratch the encrusted lesions, the danger of secondary infection is great. One of the prominent characteristics of the lesions is that they occur in crops—that is, new lesions begin to develop while older lesions are crusting over and healing.

The virus is present in respiratory secretions, and the respiratory tract is the focus from which the virus enters the blood and is carried to the rest of the body, particularly the skin.

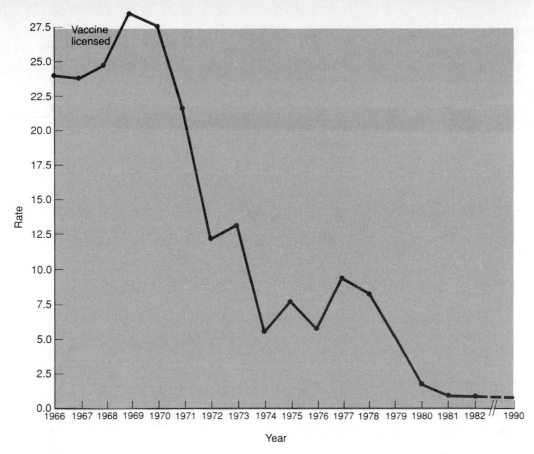

Figure 25.9 Rubella—reported cases per 100,000 population by year, United States, 1966–1990.

Zoster (also called shingles) is an infection of a sensory nerve by the varicella-zoster virus. It occurs in adults who had chickenpox in childhood and are partially immune to the virus. It is an extremely painful infection in which the lesions occur only along the sensory nerve that is infected (see Figure 25.10). Paralysis can result from infection of the spinal cord, but the more usual prognosis is for recovery in 2 to 4 weeks. Children who are exposed to an adult with zoster will develop a typical case of chickenpox.

There is good evidence that after recovery from chickenpox, the virus persists in nerve tissue. For most of us this seems to present no problem, but in a very small percentage of adults the latent virus is activated and causes an infection of a sensory nerve. We do not know exactly what triggers this activation, but it is known that zoster may occur following trauma and certain febrile illnesses and is particularly common in patients suffering from leukemia or other malignancies. The incidence of zoster increases rapidly with age, so that more than half the cases occur in individuals over 50. Because the origin of the virus is internal, exposure to either chickenpox or zoster is not a necessary part of the pathogenesis of zoster.

Clinical Management The diagnosis of either disease entity caused by this virus is almost always

Figure 25.10 Zoster of the ophthalmic branch of the trigeminal nerve. Note sharp midline demarcation on forehead.

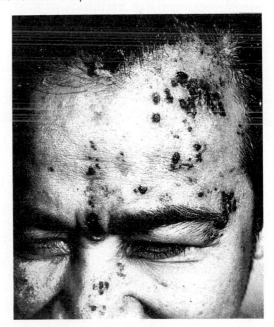

made on the basis of the clinical picture. Smallpox is the only major disease that has a clinical picture similar to that of chickenpox, and the differentiation between these two diseases can be made by histological observation of the cells occurring in the vesicles. Also, since the virus can be grown in cell cultures, a positive identification can be made using serological techniques with known antiserum.

Preventive measures for chickenpox are not particularly effective, because the infected individual spreads the virus before clinical symptoms are apparent. Since the source of infective virus is internal for zoster, there is no known way to prevent this manifestation of the disease.

There is no satisfactory treatment for either chickenpox or zoster, although both acyclovir and vidarabine have been used to alleviate symptoms in immunocompromised patients. Cortisone or steroids should not be used because they can greatly increase the severity of the disease. Children with leukemia or immunodeficiency syndromes or people who are taking immunosuppressive drugs may be given zoster-immune globulin to prevent or modify the disease.

An attenuated vaccine has been developed in Japan, and it is expected that it will be licensed for use in the United States in the near future. The vaccine is highly immunogenic, and trial tests have shown it to be safe for healthy children, immunocompromised children, and healthy adults. It will probably be offered to children at 15 months of age along with the measles, mumps, and rubella vaccine.

Variola Virus: Smallpox

The smallpox (variola) virus can be grown on the chorioallantoic membrane of 10- to 13-day-old chick

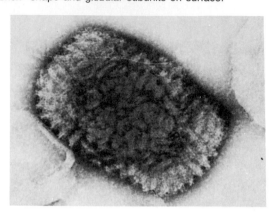

Figure 25.11 Smallpox (variola) virus (×147,000). Note "brick" shape and globular subunits on surface.

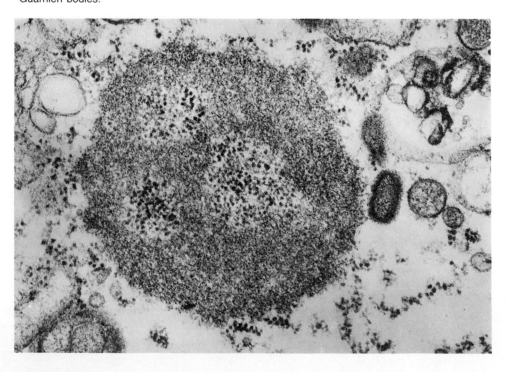

Figure 25.12 Electron micrograph of inclusion body in an L cell 3 h after infection with vaccinia virus. (×49,200). The appearance of inclusions resulting from smallpox virus infections is identical. As seen under the light microscope with appropriate staining such bodies are termed Guarnieri bodies.

embryos or in cell cultures of human, monkey, or chick embryos. The virus is a brick-shaped particle enclosed in a complex double membrane (see Figure 25.11). The core of the virus contains double-stranded DNA. It is a very hardy virus that remains infective for a considerable time outside host cells. Within an infected cell, the virus forms a cytoplasmic inclusion called a Guarnieri body (Figure 25.12).

Vaccination The conquest of smallpox began almost two centuries ago with a publication by an English physician. Edward Jenner's paper, published in 1798, long before anyone had any concept of the nature of a virus, pointed out that many of the milkmaids in England were immune to smallpox. He established that their immunity stemmed from an infection with cowpox virus, a disease not uncommon in cows. On the basis of this observation, a vaccination program using cowpox virus was initiated. There are a number of attenuated virus strains called vaccinia virus, and, although their true origin is not known, they probably represent attenuated cowpox virus. The use of these attenuated vaccines has resulted in the complete eradication of smallpox from the world. This was possible for the following reasons: (1) successful vaccination provided solid immunity; (2) the virus infects only humans; and (3) there are no asymptomatic carriers of the virus. Thus, since 1979 smallpox vaccination has no longer been recommended.

Although vaccinia virus is no longer used to immunize against smallpox, it has now acquired a new importance. DNA that encodes for surface proteins from other viruses has been incorporated into the DNA of vaccinia virus. Such strains of vaccinia are now being used in field studies to induce a long-term immunity to hepatitis B, influenza, and herpesvirus infections. If successful, undoubtedly other viral DNAs can be added to this novel immunogen.

Table 25.3 provides a summary of the viruses that enter the body via the respiratory tract.

TABLE 25.3   Summary of Viruses That Enter the Body via the Respiratory Tract

Virus	Disease	Incubation Period	Diagnotic procedures	Treatment	Control
Common cold	Common cold, acute coryza, or rhinitis	Usually 1 to 3 days	Clinical symptoms Isolation of virus and serological test for antibody rise (some cases)	No specific treatment	None satisfactory
Adeno viruses	Acute respiratory disease—pneumonia, pharyngitis, pharyngoconjunctivitis	Approximately 5 days	Clinical symptoms Isolation of virus Serological test for rise in antibody titer	No specific treatment	Virus vaccine in Armed Forces, not practical for civilians
Influenza	Influenza	Usually 1 to 3 days	Clinical symptoms Culture of virus in chick embryo Serological test for antibody rise	No specific treatment; sulfonamides and antibiotics for secondary invaders	Virus vaccine
Mumps	Mumps or epidemic parotitis	Usually 18 to 21 days	Clinical symptoms Isolation of virus in chick embryo Serological test for antibody rise	No specific treatment	Isolation of infected individual (virus vaccine available)
Measles	Measles (rubeola, morbilli)	11 days	Clinical symptoms Cell culture Serological test	No specific treatment Gamma globulin may be used	Isolation not effective Gamma globulin Viral vaccine
Rubella	Rubella (German measles	Usually 2 to 3 weeks	Clinical symptoms	No specific treatment	Isolation of infected individual Viral vaccine
H. varicellae	Chickenpox (varicella) Zoster (shingles)	2 to 3 weeks	Clinical symptoms	No satisfactory treatment	Control measures usually not effective

SUMMARY

Symptoms attributed to the common cold can result from infection by any one of 100 different serotypes of rhinoviruses, 4 different serotypes of parainfluenza viruses, respiratory syncytial virus, and 3 serotypes of reoviruses. Thus, the effectiveness of a vaccine seems highly unlikely. Adenoviruses produce a somewhat more severe upper respiratory infection than that we think of as a common cold. In addition, certain serotypes of adenovirus have been shown to induce cancer in newborn hamsters. There are three serological types of influenza virus, but epidemics are most frequently caused by type A. This virus undergoes periodic mutations resulting in antigenic changes that allow it to spread throughout the world in an essentially nonimmune population. There is only one antigenic type of mumps or measles virus. A living attenuated vaccine is routinely used to prevent both measles (rubeola) and mumps. Rubella is a very mild infection that is serious only in pregnant women. In these cases it can cross the placental wall, infect the fetus, and result in many congenital defects in the child. A living attenuated vaccine is now available that appears to stimulate immunity. Chickenpox (varicella) is a moderately mild childhood disease caused by a herpesvirus. Apparently, after recovery from chickenpox, the virus continues to exist for many years as a latent infection. This latent infection is sometimes activated in adults to produce a sensory nerve infection called zoster or shingles. Smallpox appears to have been eliminated from the world.

QUESTIONS FOR REVIEW

1. What viruses can cause the symptoms we call the common cold syndrome?
2. Why is it difficult to evaluate immunity to the common cold?
3. What is meant by a latent virus infection? Name one virus that can cause such infections.
4. Discuss the antigenic variation of influenza virus. How is this responsible for the frequent pandemics of influenza?
5. What measures may be carried out for the control of influenza?
6. What is mumps? What is one important complication that may follow infection?
7. What is the most serious complication that may follow infection with measles (rubeola)?
8. What are Koplik spots?
9. Why is rubella of particular significance in the first 1 to 3 months of pregnancy?
10. What is used as a vaccine for measles? Rubella?
11. What is zoster? How is it related to chickenpox?

SUPPLEMENTARY READING

Andrew ME, et al: Cell-mediated immune responses to influenza virus antigens expressed by vaccinia virus recombinants. *Microbial Pathogenesis* 1:443, 1986.

Centers for Disease Control: Rubella vaccination during pregnancy: United States 1971–1985. *MMWR* 35:275, 1986.

Centers for Disease Control: Toxic shock syndrome associated with influenza. *MMWR* 35:143, 1986.

Centers for Disease Control: Prevention and control of influenza. *MMWR* 35:317, 1986.

Douglas RG, et al: Respiratory virus infections and antiviral agents. In Leive L and Schlessinger D (eds): *Microbiology—1984*, Washington, DC, American Society for Microbiology, 1984, p 405.

Johnson PR, et al: Immunity to influenza A virus infection in young children: A comparison of natural infection live cold-adapted vaccine, and inactivated vaccine. *J Infect Dis* 154:121, 1986.

Sager R: Tumor suppressor genes: The puzzle and the promise. *Science* 246:1406, 1989.

Webster RG, et al: Molecular mechanisms of influenza virus (a review). *Nature* 296:115, 1982.

Chapter 26

Pathogens that Enter the Body via the Digestive Tract

OBJECTIVES   Study of this chapter should acquaint you with

1 The bacteria that cause a food intoxication by growing and secreting toxins in prepared foods.

2 Bacteria that cause a noninvasive food infection through their ability to secrete toxins in the intestine.

3 The role of mycotoxins in human disease.

4 Organisms that result in a localized invasion of intestinal epithelial cells followed by the destruction of such cells through their ability to secrete toxins.

5 The known molecular mechanism of action of all of the above toxins.

6 The vibrios that cause cholera as well as a number of other intestinal infections.

7 The mechanism of *Clostridium perfringens* food intoxication.

8 The pathogenesis of *Salmonella* gastroenteritis, *Salmonella* septicemia, and *Salmonella* enteric fevers.

9 The mechanism of disease production by the enteric yersiniae.

10 The epidemiology and pathogenesis of listeriosis and brucellosis.

11 The role of the obligately anaerobic enterics in disease production.

12 The protozoan responsible for amoebic dysentery.

13 The role of *Naegleria fowleri* as the etiological agent of amoebic meningoencephalitis.

14 The epidemiology of balantidiases, cryptosporidioisis, and giardiasis.

15 The role of toxoplasmosis in congenital disease.

16 The various types of tapeworms that can infect humans.

17 The epidemiology of a hydatid cyst.

18 The diseases caused by intestinal, liver, and lung flukes.

19 The various types of roundworms that can infect humans.

20 The causative agents of viral gastroenteritis and their role in the production of diarrheal disease.

21 The epidemiology, pathogenesis, and prevention of polio.

22 The role of other enteric viruses, such as coxsackieviruses and echoviruses as disease agents.

23 The epidemiology, pathogenesis, and disease characteristics of hepatitis A, hepatitis B, non-A, non-B hepatitis, and delta virus hepatitis.

24 Oral infections caused by *Candida albicans*.

25 Oral infections resulting in Vincent's angina and actinomycosis.

26 Infections caused by herpes simplex and cytomegaloviruses.

27 The role of Epstein-Barr virus in Burkitt's lymphoma and infectious mononucleosis.

This chapter will characterize the various groups of bacteria, viruses, protozoa, and helminths that enter the body via the mouth.

We shall divide these agents into three groups: (1) those that cause primarily a gastroenteritis resulting in a loss of fluid and electrolytes via diarrhea; (2) those that result in a systemic or neurological disease in which diarrhea is not a normal symptom; and (3) those in which the primary lesions are confined to the oral cavity. The reader will note, however, that these categories are not mutually exclusive, in that some strains of organisms may cause a devastating diarrhea while other strains of the same species may cause a systemic disease. We shall include here also infections caused by the obligately anaerobic normal flora of the intestinal tract. These obligately anaerobic organisms comprise about 99 percent of the normal flora of the large intestine and, as you will learn, they cause disease only when they are able to leave the large intestine and grow in other parts of the body. In such situations the resulting infections are extremely severe and frequently fatal.

BACTERIA ASSOCIATED WITH GASTROINTESTINAL DISEASE

Diseases of the gastrointestinal tract are most often characterized by diarrhea and vomiting, and probably no one old enough to read this book has not experienced multiple episodes of such gastroenteritis. In Asia, Africa, and Latin America, it is estimated that over 1 billion episodes of acute diarrhea occur annually in children under 5 years of age, resulting in about 5 million deaths. The bacterial causes of such illnesses can be grouped into three general categories: (1) food intoxications resulting from the ingestion of a preformed bacterial toxin; (2) food intoxications caused by noninvasive bacteria that secrete toxins while adhering to the intestinal wall; and (3) food intoxications that follow an intracellular invasion of the intestinal epithelial cells. The following sections provide examples of each of these categories.

Food Intoxication

A characteristic property shared by the bacteria that cause food intoxication (also termed food poisoning) is their ubiquitous occurrence. As a result, certain foods must always be considered to be contaminated with toxin-producing bacteria and, as a result, must be given special handling to keep them safe for consumption.

Staphylococcus

Staphylococcus aureus food intoxication is by far the most common type. It is caused by a small gram-positive coccus, the same staphylococcus responsible for so many problem infections in the hospital. The organisms grow readily on the usual nutrient media, and although many strains require several amino acids and one or more of the B vitamins, they cannot be considered highly fastidious bacteria. In stained preparations the cells appear as irregular groups of cocci (see Figure 26.1). This irregularity prompted their name, which comes from a Greek word meaning "bunch of grapes."

Figure 26.1 Clusters of gram-stained *S. aureus* (about ×4500).

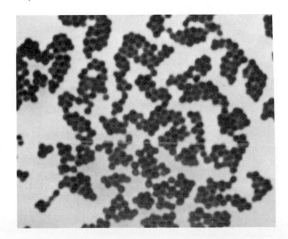

We find that quite a few people carry staphylococci—usually with no signs of illness. The nose appears to be the part of the body where they live and grow, but individuals who are carriers also have the organisms on their clothes, hands, and other parts of the body as well as in boils or skin infections. Thus, since at least 25 percent of adults carry staphylococci in their noses and on their bodies, one can assume that any food that is handled directly may become contaminated with staphylococci.

If we assume that a freshly prepared potato salad contains staphylococci, what will be the sequence of events that will render this salad toxic to those who eat it? First of all, the staphylococci must grow in the salad and secrete their toxins. Staphylococci grow optimally at body temperature of 37°C, but they will also grow, though more slowly, at temperatures below 10°C. However, for our discussion here, we choose a potato salad that will sicken everyone who eats it—so let us take it on a picnic and place it in a warm, sunny place for 3 or 4 h before we eat it.

While the staphylococci grow, the strains causing food poisoning release a toxin into the food; it is this toxin that causes the symptoms of food poisoning. The toxin has been given the general name **enterotoxin** because of the severe reaction it causes in the gastrointestinal tract. Not all staphylococci produce enterotoxin, but unfortunately we have no good method of determining which strains do and which strains do not, except to see whether the contaminated food will cause sickness. Humans are the best experimental animals for this purpose, but it is difficult to find willing volunteers. Other animals, such as rhesus monkeys and kittens, have been fed the suspected food to see if it would make them sick. The suspected food also can be extracted and, after gel electrophoresis, reacted with antibodies specific for the five different staphylococcal enterotoxins. Such a procedure is not, however, available in commercial laboratories and, therefore, could be done only in a few specialized laboratories. In most cases, however, the poisonous food can be identified by determining what food was eaten, by whom, and when.

One very important property of staphylococcal enterotoxin is its heat stability. The enterotoxin, once formed, may not be destroyed even if the food is heated sufficiently to kill all the viable staphylococci.

Having eaten a small amount of our theoretical potato salad and having ingested the staphylococcal enterotoxin, what can we expect as a clinical result? The outstanding characteristic of staphylococcal food poisoning is severe diarrhea, vomiting, and abdominal pain, while the outstanding diagnostic aid is the short incubation period—the time between eating the food and the illness itself. It is not unusual for people to become violently ill while still at the banquet table listening to the after-dinner speaker (such an incident occurred in the early 1950s at a banquet given during a national meeting of the American Society for Microbiology). The usual incubation period ranges from 2 to 4 h, and the ingestion of the contaminated potato salad at our picnic may well result in an extremely busy emergency room at the local hospital. The duration of the acute symptoms is usually less than 24 h, but the individuals may feel considerably debilitated for several days. There are usually no fatalities, but many people are hospitalized to receive intravenous fluids to replace the fluids lost through diarrhea and vomiting.

Bacillus cereus

This organism is a large gram-positive, spore-forming rod and is one of the most ubiquitous saprophytic members of the family Bacillaceae. *B. cereus* excretes a number of enzymes—such as penicillinase, phospholipases, proteolytic enzymes—and a hemolysin, but only relatively recently has this organism been recognized as a causative agent of food poisoning.

Human volunteer studies have shown that the ingestion of washed *B. cereus* cultures did not produce illness. However, when food was eaten in which *B. cereus* had grown for 24 h, severe abdominal pain and diarrhea occurred a few hours after ingestion. It is now known that *B. cereus* causes two distinct forms of food poisoning: (1) an illness with an incubation period of 10 to 12 h characterized by profuse diarrhea and occasional vomiting, lasting 12 to 24 h, and (2) an illness with an incubation period of 1 to 6 h characterized by vomiting with or without a mild diarrhea, lasting 6 to 24 h.

These two clinical entities result from the production of two different enterotoxins by *B. cereus*. The first, like cholera enterotoxin and the heat-labile toxin from *Escherichia coli*, stimulates the adenyl cyclase-cyclic adenosine monophosphate (cAMP) system in intestinal epithelial cells. When fed to rhesus monkeys, this heat-labile toxin causes primarily diarrhea. The second enterotoxin does not stimulate the synthesis of cAMP, and this heat-stable toxin causes mainly vomiting when fed to rhesus monkeys. Either of these enterotoxins, or both, may be produced by a single strain of *B. cereus*.

B. cereus is readily found in soil and on raw dry foods, including uncooked rice, a major source of *B. cereus* food poisoning. The spores may not be killed during cooking, and will germinate when the boiled rice is left unrefrigerated (to avoid clumping of the grains). Brief warming or flash frying does not always destroy the elaborated enterotoxins, particularly the heat-stable toxins. A diagnosis is usually based on finding 10^5 organisms per gram of incriminated food.

Prevention is best accomplished by the prompt refrigeration of boiled rice and other dried foods that have been cooked. Because the symptoms are mediated by preformed enterotoxins, antibiotic therapy is of no value.

Clostridium botulinum

C. botulinum, a large gram-positive rod in the family Bacillaceae (Figure 26.2), is the etiologic agent of a highly fatal food poisoning that usually follows the ingestion of a preformed exotoxin produced by the organisms while growing in the food. A total of 77 cases of botulism (infant plus foodborne) occurred in the United States during 1990 (see Figure 26.3).

Epidemiology of Botulism   *C. botulinum* is distributed in soil, on lake bottoms, and in decaying vegetation, so many foods, both vegetables and meats, become contaminated with these organisms. Numerous animals die each year after ingesting fermented grains. This is especially true of ducks (the disease is called "limber neck") and cattle (particularly in South Africa).

The endospores of *C. botulinum* are very resistant to heat and may withstand boiling water temperatures for several hours. Thus, botulism in humans usually occurs in food that has been inadequately sterilized and placed in an anaerobic environment where the surviving spores can germinate and produce toxin.

Most cases of botulism in Europe have resulted from eating smoked or spiced meats and fish, whereas botulism in the United States frequently follows the ingestion of home-canned vegetables. Among the

Figure 26.2 *C. botulinum.* The toxin secreted by this organism in food can cause a very severe type of food poisoning. Strict anaerobic conditions are necessary for growth and toxin production. Note the subterminal spores that cause swelling of the cell.

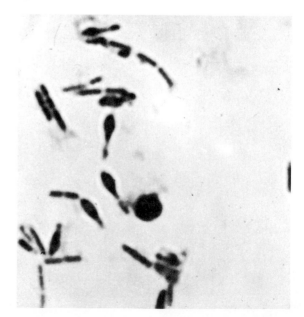

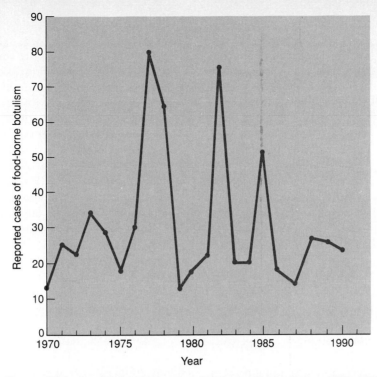

Figure 26.3 Cases of foodborne botulism in the United States, 1970–1990.

home-canned foods capable of supporting the growth of *C. botulinum* and permitting toxin production are asparagus, corn, peas, peppers, snap beans, and spinach. The toxin is not formed in 8 percent sodium chloride or 50 percent sugar syrups, so foods or fruits packed in high salt or syrup present no problem. Also, the toxin is not produced at a pH of 4.5 or below, so acid foods (such as tomatoes) are not usually a source of botulism. Botulism in commercially prepared foods is rare; when it does occur, it is the result of human error. During the past decades, botulism has resulted from eating commercial cheese, canned tuna, canned vichysoise, canned mushrooms, and smoked whitefish.

Pathogenesis of Botulism The toxins produced by *C. botulinum* are among the most toxic compounds known. It is estimated that 1 mL of culture fluid is sufficient to kill 2 million mice and that the lethal dose for humans may be in the range of 1 μg of toxin.

Symptoms in humans usually begin after an incubation period of 18 to 36 h and may include nausea and vomiting in addition to double vision, difficulty in swallowing, and some muscle paralysis. This may be followed by blurred vision and flaccid paralysis of the respiratory muscles, causing death.

Diagnosis of Botulism Even after a person develops symptoms of botulism, there may be free toxin circulating in the bloodstream. Mice are incredibly sensitive to the toxin; the intraperitoneal injection of a patient's serum may result in the death of the mouse. Usually the implicated food is no longer available. If it is, however, extracts should also be injected into mice and aliquots cultured anaerobically in an attempt to grow the organisms.

Prevention and Control of Botulism Anyone suspected of having botulism should be given antitoxin to types A, B, and E toxin. The antiserum cannot neutralize fixed toxin but can react with free residual toxin. All others who possibly could have eaten the same food should also be given antitoxin.

Unlike the endospores of the organism, botulism toxin is very heat-labile. Thus, home-canned vegetables should be boiled about 15 min before serving. Such treatment would inactivate any toxin that might be present.

Toxoids stimulate solid immunity but, because of the rarity of the disease in humans, their use would be unwarranted. Toxoids have, however, been used successfully for the prevention of botulism in cattle, particularly in South Africa. Such toxoids are prepared by incubating crude toxin preparation with dilute formalin. This treatment destroys the toxicity, but not the antigenicity, of the toxin.

Wound Botulism Wound botulism is a rare manifestation of this intoxication. It occurs when spores of *C. botulinum* (which are common in soil) are able to germinate and grow in an infected wound.

Toxin is produced, resulting in the typical symptoms of botulism. Only about three or four cases of wound botulism are reported annually in the United States. Usually the wounds are not serious, but wound botulism should be suspected for persons with even minor wounds who develop blurred vision, weakness, and difficulty in swallowing.

Infant Botulism A new variety of botulism was recognized in 1976 with the report of five cases of infant botulism. These occurred in babies as young as 5 weeks, some of whom were breast-fed, although all had had some exposure to other foods. Since then, hundreds of additional cases of infant botulism have been diagnosed, and it has become a significant pediatric clinical entity (see Figure 26.4).

Infant botulism has been diagnosed in infants ranging from 3 to 35 weeks of age. It is now well established that the disease is acquired by the ingestion of *C. botulinum* spores that subsequently germinate in the intestine and produce botulism toxin. Such spores are ubiquitous in soil; soil and dust samples from many homes have been shown to contain them. In addition, 9 out of 90 specimens of honey were also shown to contain such spores. Thus, even breast-fed infants are susceptible either through contaminated dust or possibly from supplemental feeding.

The major initial symptom of infant botulism is 2 to 3 days of constipation followed by a flaccid paralysis, resulting in difficulty in nursing and a generalized weakness that has been described as "overtly floppy."

There has been approximately a 3 percent mortality rate in such infants admitted to the hospital; some patients have required mechanical respirators because of respiratory distress. Death, however, may occur more frequently in undiagnosed cases, and there are some data linking infant botulism to at least some cases of the sudden infant death syndrome.

A tentative clinical diagnosis of botulism can be made for an infant with several days of constipation, an unexplained weakness, difficulty in swallowing, or respiratory distress. A laboratory diagnosis, however, requires the demonstration of botulism toxin in the feces, which is done by injecting fecal extracts intraperitoneally into a mouse. Death of the mouse within 96 h (which does not occur in controls in which the fecal extracts are first neutralized with botulism antitoxin) is taken as positive evidence for the presence of the toxin.

Infants themselves are not usually treated with antitoxin, primarily because it is a horse product and may induce lifelong hypersensitivity. Treatment has thus far been mostly symptomatic, requiring an average of about a month of hospitalization.

Clostridium perfringens In addition to being a major etiological agent in wound infections, *C. perfringens* is also an important cause of food poisoning. Such outbreaks are caused by strains that produce a

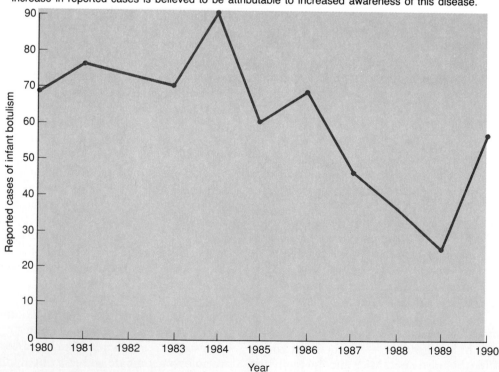

Figure 26.4 Reported cases of infant botulism in the United States, 1980–1990. The marked increase in reported cases is believed to be attributable to increased awareness of this disease.

heat-labile enterotoxin only when the vegetative cells form endospores. A typical sequence of events leading to *C. perfringens* poisoning includes the preparation of a meat dish that is eaten 1 or 2 days later. Because *C. perfringens* forms very heat-resistant endospores, the normal cooking frequently does not always destroy them. After the food has cooled, if it is not refrigerated the spores germinate and the resulting vegetative cells reproduce. When food containing large numbers of vegetative cells of *C. perfringens* is ingested, the vegetative cells enter the intestine and sporulate. It is only at the time of endospore formation that the food poisoning toxin is produced.

The release of this toxin by the sporulating cells results in acute abdominal pain and diarrhea after an incubation period of 8 to 24 h. Vomiting is not a normal symptom, and the duration of symptoms is usually less than 24 h.

Table 26.1 summarizes the characteristics of food intoxication.

MYCOTOXICOSIS

Mycotoxicosis is any illness that is induced by the consumption of food that has been made toxic by fungal toxins. These toxins, called mycotoxins, are found in a wide variety of foodstuffs consumed by humans and animals; in all likelihood, the diseases resulting from the ingestion of mycotoxin-contaminated food are considerably more serious than are the fungal infections discussed elsewhere in this text.

It is interesting to note that mycotoxins are low-molecular-weight secondary metabolites, and as a result, effective vaccines cannot be made. They occur frequently in moldy grain and for that reason, grains should be stored under low moisture conditions to preclude fungal growth. Moreover, grains should be assayed for the presence of fungi before being used for food.

Aflatoxin

Aflatoxin is produced by the saprophytic mold *Aspergillus flavus*. The ability of aflatoxins to cause liver damage has been demonstrated in many mammals, fish, and birds; initiation of liver carcinoma by aflatoxin is known to occur in ducklings, trout, rats, and ferrets. Its role in human disease is mostly circumstantial, but the fact that many foods consumed by humans are contaminated with *A. flavus* is cause for grave concern. Peanuts and peanut butter are one of the main sources of aflatoxins for humans, but these toxins have also been found in rice, cereal grains, beans, dried sweet potatoes, African beers, and cow's milk. The fact that half of all cancers occurring in Africa south of the Sahara are liver tumors may be correlated with a report that 40 percent of the foods screened in Uganda contained measurable amounts of aflatoxin. The considerably higher incidence of childhood liver cirrhosis in tropical countries (where the warm, moist climate provides ideal conditions for the growth of *Aspergillus*) can also be correlated with the presence of aflatoxins in foods such as breast milk. Moreover, of 50 urine samples taken from children with liver cirrhosis, 18 were shown to contain aflatoxins. Aflatoxin has also been implicated in the deaths of many farm animals that have eaten moldy hay or corn.

TABLE 26.1   Characteristics of Food Intoxication

Bacterial Agent	Usual Incubation Period (Hours)	Cause of Illness	Usual Symptoms	Common Source of Organism
S. aureus	2–6	Preformed enterotoxin	Vomiting and diarrhea	Human carrier, contaminated creamed foods, soups, and salads
B. cereus	1–12	Two different preformed enterotoxins	Vomiting and diarrhea	Cooked rice and dried foods
C. botulinum	18–36	Preformed neurotoxin	Muscle paralysis, double vision, difficulty swallowing	Undercooked canned foods, ingestion of spores for infant botulism
C. perfringens	8–24	Enterotoxin formed during sporulation in the intestine	Acute abdominal pain and diarrhea	Cooked meat dishes that have not been refrigerated

Because many molds grow readily on grains that have wintered under snow or that have been stored under moist conditions, it is not surprising that mycotoxins have had a dramatic role in the history of mankind. Of these, the most common form of fungal poisoning is ergotism, a disease resulting from the ingestion of one or more of a group of alkaloids known as ergot.

Ergot is formed by fungi in the genus *Claviceps*, an organism that grows on various grains but grows best on rye. When ingested, ergot may cause gangrene of the extremities, resulting in the loss of fingers, toes, and limbs, or it may cause a convulsive disorder that is characterized by writhing, tremors, or "fits." It is also known to produce strong hallucinations in which people believe they left their bodies and visited Heaven or Hell. Ergotism is also characterized by burning pains and sensations of intense heat, and affected individuals may feel as if they have been pinched, pricked, or bitten. There is about a 40 percent fatality in persons having overt symptoms of ergotism.

During the sixteenth and seventeenth centuries, hundreds of persons throughout Europe were put on trial and killed for witchcraft. Those who claimed to have been bewitched were generally children who suffered many of the nervous symptoms associated with ergotism. The fact that rye was widely grown and used for bread, supports the concept that such persons may well have ingested bread containing ergot.

Many also believe that ergotism was responsible for our own Salem witchcraft affair, which occurred in 1692. Court records show that the children who were "bewitched" suffered from "fits" and had sensations of being pinched or bitten. Three such children, as well as several cows, died from "bewitchment."

We may never know for sure, but the symptoms of bewitchment and the widespread use of rye for making bread have provided circumstantial evidence that the ingestion of a fungal toxin may have been responsible for the many accusations of witchcraft both in Europe and in the United States.

Miscellaneous Mycotoxins

There are many mycotoxins produced by other genera of fungi such as *Penicillium* and *Fusarium*, causing a variety of manifestations such as liver necrosis, paralysis, blindness, and death. One such mycotoxin causes a disease (alimentary toxic aleukia) that wiped out whole villages in Russia and Siberia when famine forced many people to collect and eat moldy grain.

The most frustrating aspect of the mycotoxicoses is that there is no ready solution. The ingestion of moldy foods in areas where poverty, starvation, and malnutrition are routine will undoubtedly continue.

NONINVASIVE FOOD INFECTIONS

There are many infectious disease agents that cause gastrointestinal disease (vomiting and diarrhea) through their ability to produce toxins in the intestine. Such organisms are taxonomically diverse, but the largest number are found in the families Enterobacteriaceae and Vibrionaceae.

Enterobacteriaceae

Some members of the Enterobacteriaceae are always considered to be pathogens, whereas others are routinely found as part of the normal flora of the intestinal tract or as saprophytes living on decaying plant matter. All, however, have the potential to produce disease under appropriate conditions and must be considered opportunists.

The family Enterobacteriaceae contains gram-negative rods that, if motile, are peritrichously flagellated. Because members of this family are morphologically and metabolically similar, much effort has been expended to develop techniques for their rapid identification. In general, biochemical properties are used to define a genus, and further subdivision is frequently based on sugar fermentations and antigenic differences. There are, however, many paradoxes that, for example, have resulted in the naming of over 1000 species of *Salmonella*, whereas the equally complex genus *Escherichia* contains a single species divided into about 1000 serotypes. It is obvious that over the years many taxonomists with different ideas have been involved in the classification of these bacteria, and it is not surprising that there is still disagreement concerning family and generic names. Table 26.2 gives an outline of the taxonomic scheme proposed by Ewing and Martin for the Enterobacteriaceae, compared with that published in *Bergey's Manual of Systematic Bacteriology*. Both schemes are used in diagnostic laboratories.

Biochemical Properties Used for the Classification of the Enterobacteriaceae

Early taxonomic schemes relied heavily on the organism's ability to ferment lactose, and numerous differential and selective media have been devised to per-

TABLE 26.2   Classification of the Enterobacteriaceae

| Ewing and Martin | | Bergey's Manual |
Tribe	Genus	Genus
Escherichieae	Escherichia	Escherichia
	Shigella	Shigella
		Edwardsiella
Edwardsielleae	Edwardsiella	Salmonella
		Citrobacter
Salmonelleae	Salmonella	
	Arizona	Klebsiella
	Citrobacter	Enterobacter
		Hafnia
Klebsielleae	Klebsiella	Serratia
	Serratia	
		Proteus
Proteeae	Proteus	Providencia
	Providencia	Morganella
		Yersinia
Erwineae	Erwinia	Erwinia
	Pectobacterium	

mit the recognition of a lactose-fermenting or a non-lactose-fermenting colony on a solid medium. When these organisms ferment lactose, acids are formed, whereas non-lactose-fermenters utilize the peptones present and do not form acids. The incorporation of an acid-base indicator into the agar medium results in a color change around a lactose-fermenting colony (see Figure 26.5). This has been a valuable technique for selecting the major non-lactose-fermenting pathogens that cause salmonellosis or shigellosis; we shall

Figure 26.5   Lactose-positive (left) and lactose-negative (right) enteric organisms grown on Hektoen enteric agar plates. Lactose-positive colonies are larger and become salmon to orange in color, whereas lactose-negative colonies remain colorless on the green agar medium.

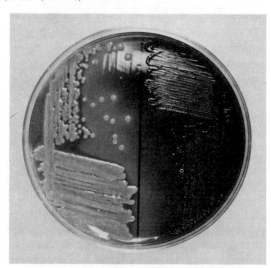

see, however, that under some circumstances, lactose-fermenters also cause a variety of infectious diseases.

Some enterics ferment lactose only slowly, requiring several days before sufficient acid is formed to change the color of the indicator. These organisms are a heterogeneous group, all of which synthesize beta-galactosidase (the enzyme that splits lactose into glucose and galactose) but lack the specific permease necessary for the transport of lactose into the cell. One can quickly determine whether an organism is a slow or non-lactose-fermenter by mixing a loopful of bacteria with ortho-nitrophenyl-beta-galactoside (ONPG) dissolved in a detergent. The linkage of galactose in ONPG is the same as its linkage in lactose; inasmuch as the ONPG can enter the cell in the absence of a permease, an organism possessing beta-galactosidase will hydrolyze ONPG to yield galactose and the very bright yellow compound ortho-nitrophenol. A slow lactose fermenter is an ONPG-positive organism that does not possess a specific lactose permease but does possess beta-galactosidase.

In addition, a number of selective media have been devised, containing bile salts, dyes such as brilliant green and methylene blue, and chemicals such as selenite and bismuth. The incorporation of such compounds into the growth medium has allowed for the selective growth of certain enterics while inhibiting the growth of gram-positive organisms.

Some other biochemical properties used to classify members of the Enterobacteriaceae include the ability to (1) form hydrogen sulfide; (2) decarboxylate the amino acids lysine, ornithine, or phenylalanine; (3) hydrolyze urea into carbon dioxide and ammonia; (4) form indol from tryptophan; (5) grow with citrate as a sole carbon source; (6) liquefy gelatin; and (7) ferment a large variety of different sugars.

No other group of organisms has been so extensively classified on the basis of cell surface antigens as the Enterobacteriaceae. The antigens used include the O antigen (the outer membrane lipopolysaccharide), the K antigens (capsular antigens), and the H antigens (flagellar antigens).

The scope of this book does not permit the listing of specific properties for each of the enterics. Rather, the reader is referred to the supplementary reading at the end of this chapter.

Pathogenicity of the Enterobacteriaceae

Many of the members of this family possess two major properties that contribute to their ability to produce disease: (1) the formation of **colonization factors,** which bind the organisms tightly to the epithelial cell lining the intestine, and (2) the secretion of one or more toxins that cause fluid loss. The fact that these properties are encoded in transmissible plasmids explains why these properties are occasionally found in many of the genera of this family.

Escherichia coli   Although *E. coli* is part of the normal flora of the intestinal tract, it was suspected for years of being able to cause a moderate to severe diarrhea in humans and animals. The disease in adults, known by many names such as **traveler's diarrhea** or **Montezuma's revenge,** may vary from a mild disease with several days of loose stools to a severe and fatal choleralike disease. It is now established that these organisms may cause gastroenteritis by several seemingly different mechanisms.

Enterotoxin-producing *E. coli,* called **enterotoxigenic *E. coli*** (ETEC), produce one or both of two different toxins. One is a heat-stable toxin called ST and the other is a heat-labile toxin called LT. Both toxins cause diarrhea.

LT, which is destroyed by heating at 65°C for 30 min, has been purified, and its mode of action is identical to that of cholera toxin. LT stimulates the activity of a membrane-bound adenyl cyclase. This results in the conversion of adenosine triphosphate to cAMP plus inorganic pyrophosphate (PPi), as shown below:

$$ATP \xrightarrow{\text{adenyl cyclase}} cAMP + PPi$$

Extremely minute amounts of cAMP will induce the active excretion of Cl^- and inhibit the absorption of Na^+, creating an electrolyte imbalance across the intestinal mucosa that causes the loss of copious quantities of fluid from the intestine.

ST, a small protein, will retain its toxicity after being heated to 100°C for 30 minutes. It has been shown to stimulate the activity of guanylate cyclase in intestinal epithelial cells. This enzyme forms cyclic guanosine monophosphate (cGMP) from guanosine triphosphate in a reaction analogous to that described for the formation of cAMP. ST does not alter cAMP levels, and it appears that the cGMP acts as an intracellular mediator for a change in ion transport by inhibiting Cl^- absorption, resulting in the loss of fluid from the intestine.

Interestingly, the genetic ability to produce both ST and LT resides on a single transmissible plasmid. Because of the ready transmission of this plasmid, it is not surprising that many serotypes of *E. coli,* as well as some other genera within the Enterobacteriaceae, can become enterotoxigenic organisms. Enterotoxigenic *E. coli* also possess fimbriae that provide for a close association between bacterium and epithelial cell (see Figure 26.6). Such fimbriae are essential for pathogenesis and antibodies that are specific for either the fimbriae or the enterotoxins are protective.

A second, but entirely different, group of *E. coli* are termed the enterohemorrhagic *E. coli* (EHEC).

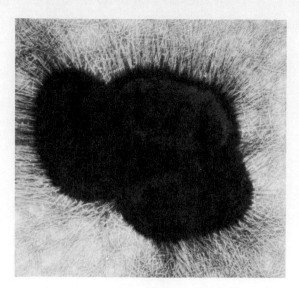

Figure 26.6   Cells of enteropathogenic *E. coli* isolated from a patient with diarrhea and bearing a large number of fimbriae (×8000).

These organisms cause a hemolytic inflammation of the intestine, a disease characterized by severe abdominal cramps and a copious, bloody diarrhea. The EHEC possess a fimbrial adhesin that permits them to adhere to intestinal epithelial cells where they produce one or both of two antigenically distinct toxins (Shigalike toxins I and II). Both toxins kill the epithelial cells by altering the 60S ribosomal subunit, resulting in a cessation of protein synthesis, the sloughing off of dead cells, and a bloody diarrhea.

A third group, called the enteropathogenic *E. coli* (EPEC), has been described. These organisms are characterized by the possession of an adhesin factor that causes a tight adherence of the bacterium to enterocytes of the small bowel. They may or may not form a Shigalike toxin but by a poorly understood mechanism they do cause an inflammation that results in cell destruction and fluid loss.

Klebsiella pneumoniae   This organism can be isolated from the respiratory or intestinal tract of about 5 percent of all healthy individuals. The organism is nonmotile and can be subdivided into many types based on the antigenicity of its capsule. *K. pneumoniae* may cause a severe and destructive pneumonia, and because *Klebsiella* can acquire from *E. coli* the plasmid that codes for LT and ST, it is not surprising that there has been a report linking *Klebsiella* to an epidemic of diarrhea in a newborn nursery.

Serratia   In the past, this genus was considered to be innocuous, but it has since been found to cause serious hospital-acquired (nosocomial) infections—particularly in the newborn, the debilitated, or the patient receiving immunosuppressive drugs. These organisms must be considered opportunistic pathogens.

Edwardsiella *and* Citrobacter *Edwardsiella* is a recently established genus of motile, hydrogen sulfide producing, non-lactose-fermenting enterics. Occasionally they are isolated from the stools of individuals with diarrhea, but they are also found in healthy humans.

Citrobacter contains citrate-utilizing bacteria that may be slow or fast lactose fermenters. They are not true pathogens, even though they have been isolated from individuals with diarrhea.

Protcus, Providencia, *and* Morganella These genera consist of very motile, non-lactose fermenters that possess the ability to deaminate phenylalaninc to phenylpyruvic acid.

All are found in feces, sewage, and soil, and all cause a number of opportunistic infections, particularly urinary tract infections. They have occasionally been implicated as a cause of diarrhea, but proof of their role in this disorder is equivocal.

Vibrionaceae

Vibrio cholerae: Asiatic Cholera

Vibrio cholerae is a small, slightly curved gram-negative organism possessing a single polar flagellum (see Figure 26.7). The organisms show many similarities to the members of the Enterobacteriaceae but can be readily differentiated by their positive oxidase reaction and their ability to grow at a pH between 9.0 and 9.5.

A number of serological types have been reported, based on antigenic differences in their O antigen. Of these, three serotypes have been given specific names: Inaba, Ogawa, and Hikojima. Some serotypes produce a soluble hemolysin, and these have been designated as El Tor biotypes of *V. cholerae.* Thus, an isolated strain might be designated as *V. cholerae* serotype Inaba, biotype El Tor.

Pathogenicity Cholera is spread as a fecal-oral disease and is acquired by the ingestion of fecally contaminated water and food. The organisms do not spread beyond the gastrointestinal tract, where they

Figure 26.7 *V. cholerae.* (*a*) The vibrios, which are curved rods are shown with rods and cocci in a fecal smear. (*b*) Fecal smear prepared from a cholera patient. The arrangement of the organisms is probably due to a mucus shred. (*c*) Smear prepared from organisms shown in *b* above, after culture on an agar medium. The vibrios appear as short plump rods and cocci.

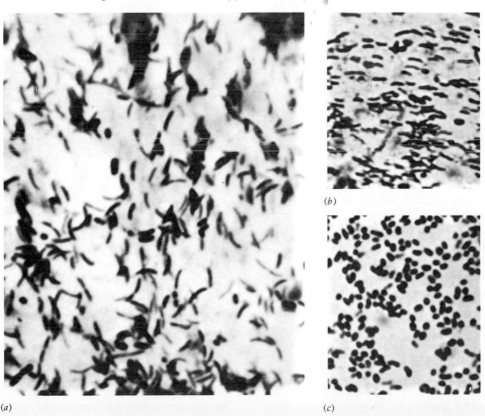

(*a*)

(*b*)

(*c*)

Figure 26.8 *V. cholerae* as it adheres to mucus and intestinal cells.

multiply to very high concentrations in the small and large intestines. They do not penetrate the epithelial layer but remain tightly adhered to the intestinal mucosa (see Figure 26.8).

The foremost symptom of cholera is a severe diarrhea in which a patient may lose as much as 10 to 15 L of fluid per day. The feces, containing mucus, epithelial cells, and large numbers of vibrios, have been referred to as "rice water stools." Death, which may occur in as many as 60 percent of untreated cases, results from severe dehydration and loss of electrolytes.

Diarrhea from *V. cholerae* is the result of the secretion of an enterotoxin called **choleragen** that stimulates the activity of the enzyme adenyl cyclase, which, in turn, converts ATP to cAMP. This activity is identical to that described for the LT enterotoxin produced by the enterotoxigenic *E. coli*. The cAMP stimulates the secretion of Cl^- and inhibits the absorption of Na^+, resulting in a copious fluid loss and an electrolyte imbalance.

Clinical Management The organisms can be viewed in stools, particularly with a darkfield microscope. Fluorescently labeled antiserum can be used to confirm the identification of the observed organisms.

The mortality of cholera can be reduced to less than 1 percent by the replacement of fluids and electrolytes. The observation that the inclusion of glucose in a salt solution permits oral replacement of electrolytes has made treatment of this disease (particularly in rural areas) much more effective. A 1974 epidemic in Portugal recorded 2241 cases but only 38 deaths. A major epidemic of cholera began in Peru in January 1991 resulting in over 160,000 cases and hundreds of deaths during a period of several months. The epidemic also spread to the neighboring countries of Ec-

uador, Chile, Columbia, and Brazil and is expected to eventually spread all over Latin and South America. Health officials in Brazil have predicted that there will be as many as 6 million cases and 40 thousand deaths from this epidemic over the next 3 years. The disease has also spread to the United States where 4 cases of cholera occurred in New Jersey in April 1991 from eating contaminated crabmeat that had been illegally imported from Ecuador. Antibiotics, especially tetracycline, reduce the number of intestinal vibrios and should be used in conjunction with fluid replacement.

Control of cholera requires proper sewage disposal and adequate water sanitation. Immunization with killed *V. cholera* or formalin-treated enterotoxin appears to give some protection, and recovery from the disease imparts a fairly solid immunity. Current research efforts are directed toward the preparation of a vaccine consisting of degraded enterotoxin, which will stimulate IgA antibody production against the toxin. This has now in large part been accomplished experimentally. Choleragen is composed of two subunits: (1) the B subunit, which binds to specific receptors on the epithelial cell; and (2) the A subunit, which enters the cell and catalyzes the increased production of cAMP. Using human volunteers, it has been shown that the oral administration of the B subunit of choleragen (which is nontoxic) along with killed whole cells of *V. cholerae* stimulates an immunity as strong as that acquired by recovery from the actual disease. These two components of the oral vaccine appear to induce a local IgA immunity that is both antitoxic (thus preventing the attachment of the toxin to its receptor) and antibacterial, which prevents the attachment of the bacterial cells to the host cells (see Figure 26.8). The exciting aspect of this observation is that many other toxins are also composed of an A and B subunit (such as *E. coli* LT), and similar vaccines have the potential of greatly diminishing the devastation caused by these diseases.

Vibrio parahaemolyticus This organism is a marine bacterium that requires a high sodium chloride concentration for growth. It has attained major importance as the etiological agent of food poisoning following the ingestion of uncooked or partially cooked seafood, particularly shellfish.

The organisms appear to be distributed worldwide and to have caused a multitude of cases of acute gastroenteritis in the United States. In countries such as Japan, where seafoods constitute a high percentage of the normal diet, *V. parahaemolyticus* is estimated to be the etiological agent of about half of all cases of bacterial food poisoning.

The nature of the *V. parahaemolyticus* enterotoxin is unknown, but almost all strains isolated from cases of gastroenteritis produce a heat-stable hemolysin capable of lysing human and rabbit erythrocytes.

Campylobacter Members of the genus *Campylobacter* are gram-negative curved rods, belonging to the family Vibrionaceae. All species appear to be inhabitants of the intestinal tract of wild and domestic animals, including household pets. Transmission to humans occurs via a fecal-oral route, originating from farm animals, birds, dogs, and processed poultry by way of contaminated food and water.

C. jejuni ranks along with the enterotoxigenic *E. coli* as a major cause of diarrheal disease in the world, particularly in developing countries. Clinical isolates of this organism have been shown to produce a heat-labile enterotoxin that raises intracellular levels of cAMP. Furthermore, the activity of this enterotoxin is partially neutralized by antiserum against *E. coli* LT and cholera toxin, demonstrating that the *Campylobacter* enterotoxin belongs to this same group of adenylate cyclase-activating toxins.

C. fetus also causes human diarrheal disease, but this species is more likely to progress to a systemic infection.

The incubation period for the diarrheal disease is 2 to 4 days. Gentamicin, erythromycin, and a number of other antibiotics may be used successfully for the treatment of *Campylobacter* infections.

A new species, named *Helicobacter* (formerly *Campylobacter*) *pylori*, was first described in 1983. This organism was found growing in gastric epithelium, and it is now established that *H. pylori* is the etiological agent of chronic gastritis and duodenal ulcers in humans. Interestingly, by age 60, 50 to 60 percent of individuals in the industrialized countries are infected, while in developing countries, infection begins earlier in life and reaches higher levels.

Symptoms of chronic gastritis include abdominal pain, burping, gastric distention, and halitosis. The disease can be reproduced in gnotobiotic (germ-free) piglets and in human volunteers following the ingestion of *H. pylori*. The observation that their eradication by antibacterial treatment results in normalization of the gastric histology also adds support to the role of a bacterial agent for chronic gastritis. It is noteworthy that a different subspecies of *H. pylori* can be routinely isolated from both normal and inflamed gastric mucosa of ferrets, and an analogous spiral-shaped bacterium routinely colonizes the gastric mucosa of cats.

It is interesting to note that over-the-counter medications containing bismuth salts have been used for years to treat gastritis (Pepto-Bismol) and the fact that *H. pylori* is quite sensitive to bismuth helps to explain its efficacy for the relief of gastric symptoms.

Other Pathogenic Vibrios *V. fluvialis* is a halophile (salt-loving) that has been isolated from the diarrheal stools of many patients in Bangladesh. It has also been found in the coastal waters and in shellfish on both the east and west coasts of the United States. This organism has been reported to produce both enterotoxinlike substances and an extracellular cytotoxin that kills tissue cells.

V. mimicus is similar to certain strains of *V. cholerae*. It produces a choleralike disease, presumably by the secretion of an adenylate cyclase-stimulating toxin.

V. vulnificus characteristically causes a systemic disease, frequently following the infection of a preexisting wound. The systemic disease also appears to follow the ingestion of undercooked or raw seafood, particularly raw oysters.

Table 26.3 summarizes the major properties of the noninvasive bacterial food infections.

TABLE 26.3 Food Infection by Noninvasive Bacteria

Bacterial Agent	Toxin Produced	Toxin Action	Source of Organism
ETEC	LT	Stimulate cAMP	Contaminated food or water
	ST	Stimulate cGMP	
EHEC	Shigalike toxins	Inhibition of protein synthesis	Contaminated food or water
0EPEC	?Shigalike toxins	?	Contaminated food or water
V. cholerae	Choleragen	Stimulate cAMP	Contaminated food or water
V. parahaemolyticus	?	?	Uncooked shellfish
V. fluviales	Enterotoxin	?	Uncooked shellfish
	Cytotoxin	?	
Campylobacter	LT	Stimulate cAMP	Animals
Clostridium perfringens	Enterotoxin	?	Rice, dried foods

LOCALLY INVASIVE FOOD INFECTIONS

The diseases we shall consider in this category are caused by bacteria that invade and grow intracellularly in the epithelial cells lining the intestine. They do not, however, ordinarily leave the intestine to cause systemic disease.

Shigella Members of the genus *Shigella* are nonmotile gram-negative rods belonging to the family Enterobacteriaceae. The genus is divided into four species, and (as shown in Table 26.4) each species may be additionally divided into serotypes. *S. sonnei* is a slow lactose fermenter, but no other species of *Shigella* can ferment lactose.

The shigellae cause a disease known as **bacillary dysentery,** and humans appear to be the only natural hosts for these organisms. Infection follows the ingestion of food or water that, frequently, has been contaminated by asymptomatic carriers.

To cause intestinal disease, shigellae must first invade the epithelial cells of the intestine. There, they secrete a toxin (Shiga toxin) that inactivates the 60S ribosomal unit, stopping all protein synthesis and killing the cell.

The most prevalent species of *Shigella* in the United States is *S. sonnei,* and the majority of persons infected by the organism are pre-school-age children, particularly those in day care centers.

Efforts to control the disease are usually directed toward sanitary measures designed to prevent the spread of the organism. The injection of killed vaccines is worthless because humoral IgG does not appear to be involved in immunity to the localized intestinal infection. Live vaccines have been used in clinical trials, but success has been equivocal. It appears that the organism must invade and colonize the intestinal cells in order to produce immunity so, in all likelihood, any effective vaccine will produce unwanted intestinal symptoms.

Interestingly, the ability to invade the intestinal epithelium is encoded in a large plasmid, and the conjugal transfer of this plasmid to avirulent strains of *E. coli* confers this invasive ability to the recipient. It is not surprising, therefore, that strains of *E. coli* are also able to cause a shigellalike gastrointestinal infection.

Enteroinvasive *E. coli* *E. coli* that produce diarrhea by the direct invasion of the intestinal epithelial cells are called **enteroinvasive *E. coli*** (EIEC). These organisms are very closely related to the shigellae, and it appears that their invasive potential results from the possession of a similar or identical virulent plasmid. Moreover, it appears that the EIEC also produce a Shigalike toxin that prevents protein synthesis and kills the cell, resulting in fluid loss in the intestine.

Salmonella

The term *salmonellosis* is used to describe any infection caused by members of the genus *Salmonella*. This is an extremely large group of gram-negative rods that can be distinguished from the normal flora of the intestine by means of biochemical and antigenic criteria.

The Kauffmann-White scheme is a complex antigenic classification in which each *Salmonella* is assigned to a group based on the O antigens present in its cell wall lipopolysaccharide. For example, each organism possessing O antigen 2 is placed in group A. All those possessing O antigen 4 are in group B, and so on; groups are lettered A to Z, and the remaining groups are numbered 51 through 65. Table 26.5 lists examples of some of the more common human pathogens, showing their group placement and O antigen designation. As you can see from this table, the group designation is based on the presence of one dominant antigen, even though other O antigens are present in the organism. Also, notice the subgroups that depend on the overall complement of O antigens possessed by a species.

The salmonellae in any one group can be further divided into serotypes on the basis of the H antigens that occur in either phase 1 or phase 2. Also, some salmonellae form a polysaccharide antigen on the outer surface of the cell that covers up the O antigen layer of the organism. Because this antigen is found most frequently in recently isolated virulent organisms, it is called the Vi antigen, indicating virulence. Because the Vi antigen surrounds the O antigen, organisms possessing a Vi antigen will not agglutinate in specific O antiserum unless the Vi antigen is first destroyed by placing the bacteria in a boiling water bath.

Salmonellae do not ferment lactose, but most form hydrogen sulfide and gases from carbohydrates and will decarboxylate lysine. *S. typhi* is an exception in that it produces very little or no hydrogen sulfide or gas from carbohydrate fermentation. *S. arizonae*, which consists of over 300 serotypes, is a second exception in that these organisms are slow lactose fer-

TABLE 26.4 Classification of the *Shigella*

Subgroup Species	Serological Subgroup	Number of Serotypes
S. dysenteriae	A	10
S. flexneri	B	8
S. boydii	C	15
S. sonnei	D	1

TABLE 26.5　Kauffmann-White Classification of Selected Salmonellae

Species or Serotype	O antigens[a]	H Antigens Phase 1	Phase 2
Group A			
S. paratyphi	1, **2**, 12	a	—
Group B			
S. schottmülleri	1, **4**, 12	b	1, 2
Group C₁			
S. choleraesuis	**6**, 7	c	1, 5
S. montevideo	**6**, 7	g, m, s	—
Group C₂			
S. manhattan	**6**, 8	d	1, 5
Group D₁			
S. typhi	**9**, 12, (Vi)	d	—
S. panama	1, **9**, 12	l, v	1, 5
Group D₂			
S. strasbourg	**9**, 46	d	1, 7
Group E₁			
S. anatum	**3**, 10	e, h	1, 6
Group E₂			
S. new-brunswick	**3**, 15	l, v	1, 7
Group E₃			
S. minneapolis	**3**, 15, 34	e, h	1, 6
Group H			
S. florida	1, 6, **14**, 25	d	1, 7

[a] O antigen in bold type is common to all members of the group.

menters. Some taxonomists prefer to place the organisms in this group in a separate genus called *Arizona*.

Originally, the salmonellae were given species names that were descriptive of the disease they caused. Later, as more antigenic types were described, a system of nomenclature was established by which each new antigenic type was named according to the geographical area where it was isolated. Thus we have names such as *S. typhi, S. choleraesuis, S. minneapolis,* and *S. arizonae*. Current practice in many laboratories, however, tends to use only three species: *S. typhi, S. choleraesuis,* and *S. enteritidis*. More than 1000 additional antigenic types are listed as serotypes of *S. enteritidis*, e.g., *S. enteritidis* serotype alabama or *S. enteritidis* serotype miami.

Epidemiology of Salmonellosis　The primary reservoir for the salmonellae is the intestinal tract of many animals, including birds, farm animals, and reptiles. Humans become infected through the ingestion of contaminated water or food. Water is polluted by the introduction of feces from any animal excreting salmonellae. Infection via food results from the ingestion of contaminated meat or via the hands, which act as intermediates in the transfer of salmonellae from an infected source. Thus, the handling of an infected, though apparently healthy, dog or cat can result in contamination with salmonellae. Another major source of *Salmonella* infections is pet turtles. In fact,

it is estimated that in the early 1970s there were almost 300,000 cases of turtle-associated salmonellosis annually in the United States and, as a result, it is now illegal to import turtles or turtle eggs, or even to ship across state lines domestic turtles with shells less than 4 in. in diameter.

In the United States, poultry and eggs comprise the most common source of salmonellae for humans. This occurs because a very large percentage of chickens are routinely infected with salmonellae. Thus, humans can acquire these organisms through direct contact with uncooked chicken or by the ingestion of undercooked chicken. And, because the organisms may occur both on the outer shell and in the yolk, consuming anything containing raw eggs (caesar salad, hollandaise sauce, mayonnaise, homemade ice cream) could result in a *Salmonella* infection. The CDC even cautions against eating eggs sunny side up and recommends that eggs be boiled for 6 to 7 minutes before being served.

On an industrial scale, slaughterhouse workers are faced with salmonellosis as an occupational hazard, primarily from poultry and pigs. Because humans can become asymptomatic carriers of *Salmonella*, infected food handlers are also responsible for the spread of these organisms.

Salmonella *Gastroenteritis*　Gastroenteritis is, without doubt, the most common type of infection caused by the salmonellae. It may be caused by any one of the thousands of serotypes of *Salmonella,* and it is characterized by the fact that the organisms remain in the gut. In the average case, symptoms occur 10 to 28 h after ingesting contaminated food, and the headache, abdominal pain, nausea, vomiting, and diarrhea may continue for 2 to 7 days. In all likelihood, most cases of gastroenteritis are acquired by the ingestion of foods or milk that have been contaminated with animal excreta. This infection initiated one of the nation's largest epidemics of *Salmonella*-related illnesses during the summer of 1985. This episode, involving over 17,000 *Salmonella* infections and six deaths, was traced to a large Chicago dairy where raw milk had been inadvertently mixed with pasteurized milk.

In spite of the prevalence of *Salmonella*-induced gastroenteritis, we still do not know the exact details of how these organisms produce disease. We do know that they must first invade the epithelial lining of the intestine and that this invasive ability resides on a plasmid. Strains that have been cured of their plasmids show little or no virulence in experimental animals.

The search for *Salmonella* toxins has not been as productive as one might wish, but there are multiple reports that *Salmonella* species may secrete a choleralike enterotoxin that induces increased levels of cAMP. In addition, a cytotoxin that inhibits protein synthesis in intestinal epithelial cells has been described. It is proposed that this cytotoxin may be responsible for the structural damage to the intestinal mucosa.

Interestingly, most species of *Salmonella* are readily killed if they leave their intestinal site and, as a result, gastroenteritis with local invasion of intestinal epithelial cells is the most common *Salmonella* infection. However, as we shall see in the next section, certain species do cause serious systemic disease.

One promising vaccine uses living organisms that have been cured of their plasmids. Such strains are unable to invade the intestinal epithelial cells and, hence, are unable to cause disease in experimental animals.

Table 26.6 summarizes the food infections caused by locally invasive bacteria.

Systemic Bacterial Infections Originating in the Gastrointestinal Tract

Salmonella typhi: Typhoid Fever   The classic enteric systemic disease is typhoid fever, infecting from 200 to 400 people annually in the United States. Following an incubation period of 1 to 3 weeks after ingestion of the organisms, there is inflammation in the small intestine, followed by invasion of the regional lymph nodes. From the lymphatic system the organisms enter the blood and infect various organs and tissues, including the liver, kidneys, spleen, bone marrow, gallbladder, and sometimes the heart. Enlargement of the spleen is characteristic, and multiplication of the organisms in the skin may result in the presence of rose spots, particularly on the abdomen. Symptoms may also include headache, loss of appetite, abdominal pain, weakness, stupor, and a continued fever.

Other species of *Salmonella* also cause enteric fevers, but usually these "paratyphoid" fevers are milder than that caused by *S. typhi.*

The organisms can be isolated from the blood during the first 2 weeks of the illness and from the urine usually between the first and fourth week. Stool specimens may remain positive for indefinite periods. Preincubation of fecal specimens in selenite or tetrathionate broth is occasionally used to advance the growth of the *Salmonella* beyond that of the coliforms prior to streaking the organisms on various selective and differential media.

A retrospective diagnosis can be made by demonstrating a rise in agglutinating antibodies to *S. typhi.* To be valid, this test (known as the Widal test) must show at least a fourfold increase in titer between the acute- and convalescent-phase serum.

Ampicillin and chloramphenicol are the antibiotics most used to treat typhoid fever. However, many strains are appearing that have acquired a plasmid encoding for resistance to chloramphenicol, ampicillin, and trimethoprim-sulfamethoxazole. Such strains are susceptible to third-generation cephalosporins such as cefoperazone. Convalescent patients may remain carriers for long periods, and ridding them of the organ-

TABLE 26.6   Food Infection by Locally Invasive Bacteria

Bacterial Agent	Toxin Action	Source
Shigella sp.	Inhibit protein synthesis	Human carriers
EIEC	Inhibit protein synthesis	Contaminated food
Salmonella		Animals

isms is sometimes extremely difficult. In 1971 British officials banned from their schools two children who were persistent carriers of *S. typhi*. Health officials stated that they had tried every known drug treatment on the children, and one official predicted that they would remain carriers for the rest of their lives. Because the organisms tend to grow in the bile ducts, some carriers have been cured by the surgical removal of their gallbladders.

The role of the carrier is exemplified by the 72 cases of typhoid fever that occurred in San Antonio, Texas, between August and October 1981. Seeking the source of this epidemic, epidemiologists from the Centers for Disease Control eventually traced it to a restaurant specializing in tortillas. Subsequent isolation of *S. typhi* from the intestinal tract of two asymptomatic employees clearly demonstrated that the epidemic was spread by food handlers who were carriers of the typhoid bacillus.

The most famous carrier was a Swiss immigrant named Mary Mallon who arrived in the United States in 1868. This woman, who later became known as Typhoid Mary, served as a cook and domestic primarily in the New York area. It is estimated that either directly or indirectly, she was responsible for hundreds of cases of typhoid fever.

Active immunization against *S. typhi* has been practiced for years, particularly in the military, using a killed vaccine containing *S. typhi* and two additional paratyphoid organisms. However, this vaccine is only moderately effective and is not widely used. A new and seemingly effective vaccine using living attenuated organisms that are given orally has been licensed for use in the United States. This vaccine is made up of mutant organisms that are unable to multiply in the body. It does, however, appear to induce both a humoral and a cell-mediated immune response, and trial immunizations in children in Middle East countries indicate that it is highly effective.

Genetic engineering has created still another strain that appears to possess the potential to serve as a vaccine against both typhoid fever and shigellosis. This strain was constructed by transferring to an avirulent strain of *S. typhi,* the plasmid from *Shigella* that is responsible for the invasion of the intestinal epithelial cell. This live vaccine was shown to protect experimental animals from challenge with either *S. typhi* or *S. sonnei.*

Salmonella *Septicemia* Septicemia caused by *Salmonella* is a fulminating blood infection that originates in, but does not involve, the gastrointestinal tract. Most cases are caused by *S. choleraesuis* and are characterized by lesions throughout the body. Pneumonia, osteomyelitis, or meningitis also may re-sult from such an infection. *Salmonella* osteomyelitis is especially prevalent in people who have sickle cell anemia.

Yersinia Members of the genus *Yersinia* are gram-negative rods that are taxonomically placed in the family Enterobacteriaceae. All species of *Yersinia* are primarily pathogens of wild and domestic animals, but all can infect humans as well.

Y. pseudotuberculosis is spread from animals or birds to humans via a fecal-oral route. Human infections frequently involve the mesenteric lymph nodes, resulting in symptoms that are easily confused with acute appendicitis. These organisms may also invade the bloodstream, causing a severe septicemia.

Y. enterocolitica has been isolated from humans with acute abdominal disease characterized by diarrhea and pain that mimics the symptoms of acute appendicitis. The organisms have frequently been isolated from swine and cattle, and it is likely that these animals provide a major reservoir for human infections. Moreover, it has now been shown that *Y. enterocolitica* produces a heat-stable enterotoxin that is identical to the ST toxin produced by enterotoxigenic *E. coli.*

Listeria monocytogenes: Listeriosis

L. monocytogenes is a small, gram-positive, motile bacillus resembling the corynebacteria in appearance. The organisms are facultative anaerobes that produce a narrow band of β hemolysis when growing on blood agar plates. Once isolated, they are easily subcultured, but initial isolation may be difficult. One unusual enrichment technique is to store specimens at 4°C for extended periods. This may be done by inoculating blood or spinal fluid into a 10-fold excess of blood culture medium that is then stored at 4°C and subcultured weekly for 3 to 6 months by streaking onto a blood agar plate. Tissues and fecal suspensions also may be stored in screw-cap tubes at 4°C and subcultured at weekly intervals. The organisms exhibit a tumbling type of motility if grown at 18 to 20°C.

At least 11 serotypes have been described, but 90 percent of clinical listeriosis infections are caused by types 1a, 1b, and 4b.

Pathogenicity It appears that *L. monocytogenes* is primarily an animal pathogen, and it is likely that many human infections are acquired through contact with domestic animals and birds. *L. monocytogenes,* however, has been isolated also from soil, water, sewage, and from the feces of a number of asymptomatic humans, rodents, swine, and poultry. Large numbers of *L. monocytogenes* have also been found in raw milk obtained from infected cows, and, when they are present in such high numbers, normal pasteurization procedures will not eliminate all of them. This became apparent in 1985 following an epidemic in which 49 adults appeared to have contracted listeriosis after the ingestion of pasteurized milk, resulting in a 29 percent fatality rate. It is probable that the intracellular existence of these organisms in leukocytes present in the milk contributes to their higher heat resistance, permitting some to escape the normal pasteurization process.

Another epidemic in 1985, which caused about 40 deaths, resulted from the ingestion of a Mexican-style cheese which was contaminated with *L. monocytogenes*. In this case, it is not known whether the milk used to make the cheeses (queso fresco and cotija) was properly pasteurized. In 1986, many people in four states reported suffering flulike symptoms after eating Kraft ice cream bars, and *L. monocytogenes* was subsequently isolated from both the ice cream and a plant in Richmond, Virginia, where the bars were made. In still another epidemic involving over 40 persons, coleslaw was identified as the vehicle of transmission. Investigators concluded that listeriosis can be spread readily by vegetables that are consumed uncooked, particularly if raw manure had been used to fertilize the growing fields.

Listeriosis can mimic a number of infectious diseases, but meningitis is the most common manifestation, occurring in more than 75 percent of infections. Moreover, *L. monocytogenes* is now the most common cause of bacterial meningitis in renal transplant recipients. The source of such infections is unknown but it appears that many persons must carry the organisms, or come into contact with them, without suffering an overt illness. The fact that immunocompromised individuals, such as recipients of renal transplants, are unable to mount an effective immune response may well explain their susceptibility to such infections. *Listeria* may also cause endocarditis, urethritis, conjunctivitis, and abortions, but the vast majority of infections occur in the newborn.

Neonatal listeriosis occurs in two forms, early-onset and late-onset. In the early-onset disease, the infant is critically ill at birth or becomes so during the first day or two of life. The most common manifestation is pneumonia, and it seems that the infection is acquired either in utero or from the vagina at the time of birth.

Late-onset listeriosis occurs between 1 and 4 weeks after birth and is seen as meningitis. In such cases, the infection was probably acquired after birth.

It is interesting to note the parallel between neonatal listeriosis and neonatal group B streptococcal infections. Either organism may cause an early-onset pneumonia or a late-onset meningitis, but neither is particularly virulent for the normal, healthy adult. In both cases, early-onset infection is acquired from the mother in utero or during delivery, whereas the late form of each disease is probably acquired through person-to-person contact. Both infections also have a serotype specificity in that most early-form listeriosis is caused by serotypes 1a or 1b, whereas the late-onset form more often results from serotype 4b. It is worth noting that *L. monocytogenes* grows as an intracellular parasite; as a result, immunity is cell-mediated rather than humoral. It is interesting to note that cell-mediated immunity is dampened during the last trimester of pregnancy, presumably to avoid fetal rejection. The lowering of this immune component permits the growth of *Listeria* in placental tissues.

Clinical Management *L. monocytogenes* is sensitive to penicillin, ampicillin, and the tetracyclines, and many investigators report that the tetracyclines are the drugs of choice.

Brucella: Brucellosis

Brucellosis is primarily a disease of domestic animals, from which humans occasionally become infected with the *Brucella* organisms. The disease is undoubtedly an ancient one, but its modern history began in the late 1890s and early 1900s. During this period, the causative organism was isolated by David Bruce, who was stationed with the British army on the island of Malta. Also during this period, a young Maltese physician named Zammitt showed that goats on the island of Malta were excreting the disease organisms in their milk and the people were contracting the disease from drinking it. This organism was named *Brucella melitensis*. Later, in Denmark, a related organism that caused abortions in infected cows was isolated from cow's milk. This organism was named *B. abortus*. Still later a third organism that normally infects swine was isolated, and this species was named *B. suis*. Recently, three additional species of *Brucella* have been isolated—*B. canis* from dogs, *B. ovis* from sheep, and *B. neotamae* from the wood rat. Because of the geographical areas in which this disease was frequently seen, some of the older names for the disease are Malta fever, Mediterranean fever, Gibraltar fever, and Cyprus fever. Also, because of the characteristic undulating fever, the disease has often been called undulant fever. It is now common to refer to all infections by *Brucella* species as **brucellosis**.

The organisms are small, gram-negative rods that are somewhat pleomorphic. Most strains of *B. abortus* grow on initial isolation only if given an atmosphere containing about 10 percent carbon dioxide. The various species can be differentiated from each other by sugar fermentations, rate of hydrogen sulfide production, rate of urea hydrolysis, and the ability to grow in the presence of the dyes thionine, basic fuchsin, and crystal violet.

Many of the biotypes of the various species are antigenically similar in that they contain varying quantities of two antigens designated A and M. *B. ovis* and *B. canis*, however, contain neither antigen A nor M.

Pathogenicity This acute systemic disease is characterized by intermittent or continued fever. The temperature usually falls during the night; this is followed by profuse sweating. Other symptoms include malaise, weakness, and aching. The disease may become chronic and persist for weeks or months.

Once in the body, the organisms invade the lymph nodes and then the blood. The granulomatous nodules that may develop in the liver, spleen, or bone marrow form abscesses. In lesions, the organisms are characteristically found to occur intracellularly (see Figure 26.9).

The incubation period may be from 1 to 6 weeks. Sources of the infection are found in the animal reservoirs of cattle, goats, swine, sheep, and horses. The organisms are found in tissues, secretions, and discharges from infected animals. In infected cattle, placentas, aborted fetuses, and vaginal discharges are infectious, as are blood and urine. Insofar as the general populace is concerned, the ingestion of raw milk from an infected animal is a primary source of infection. Because skin contact with infected animals or their secretions is also a frequent source, the disease is an occupational hazard for such groups as farmers, butchers, and veterinarians.

Clinical Management Diagnosis requires the isolation of the organisms from the blood or from a biopsy of infected tissue. Agglutination and opsonocytophagic (enhanced phagocytosis in the presence of specific antibody) tests may also show the presence of antibodies in the patient's serum. The brucellergen skin test may be performed to determine whether or not the patient has a delayed type of hypersensitivity to the organism, but it is limited in value. A negative test does not necessarily indicate absence of infection, even though the test is usually positive in the infected person.

Control of the disease requires the elimination of all infected domestic animals. The U.S. Department of Agriculture carries out an extensive program of testing and immunizing cattle in an attempt to stamp out the disease, and a number of states are now declared to be *Brucella*-free. A living, attenuated strain of *Brucella* is used to immunize cattle. Pasteurization of milk and milk products destroys the organisms.

Treatment with sulfonamides and antibiotics has been used, but results to date have been erratic and disappointing. A summary of systemic infections originating in the gastrointestinal tract is given in Table 26.7.

Obligately Anaerobic Enterics

It may seem paradoxical that the largest part of our normal flora is so obligately anaerobic that short periods of exposure to air may result in their death. It might appear that the coliforms make up the majority of our normal intestinal flora but, in truth, they are

Figure 26.9 (a) Phagocyte infected with *B. abortus* for 48 h. The brucellae are in a large vacuole (×21,170). (b) Phagocytosis of *B. abortus* by a phagocyte in a 48 h infection mix. Pseudopodia engulf the brucella (×75,300).

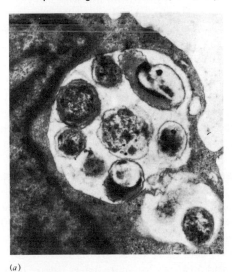

(a)

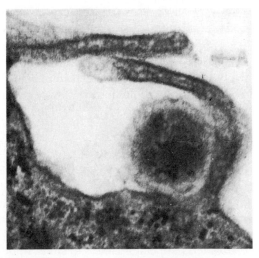

(b)

TABLE 26.7 Infections Originating in the Gastrointestinal Tract by Systemically Invasive Bacteria

Bacterial Agent	Disease	Major Source
S. typhi	Typhoid fever	Human carrriers, contaminated food
Yersinia sp.	Appendicitislike symptoms	Swine, cattle
L. monocytogenes	Flulike symptoms, frequent meningitis	Infected animals; contaminated milk, cheese, or vegetables
Brucella sp.	Brucellosis	Infected animals; milk products

outnumbered more than 100 to 1 by the obligately anaerobic gram-negative rods in the family Bacteroidaceae. Two genera of this family, *Bacteroides* and *Fusobacterium,* comprise the major opportunistic pathogens of this family.

Bacteroides

The major lesion produced by members of the genus *Bacteroides* is the abscess and, although these organisms can be considered as opportunists with little invasive ability, they can be extremely virulent, causing widespread tissue destruction. In most lesions, they exist as mixed flora, usually along with facultative organisms.

Intraabdominal abscesses follow contamination of the peritoneal cavity by fecal contents, resulting either from trauma or surgical procedures. *B. fragilis* is the major pathogen, being found in at least two-thirds of such infections. *B. fragilis* may also cause pelvic abscesses following trauma during delivery, induced abortion, malignancy, or the use of intrauterine contraceptive devices.

B. melaninogenicus is found in pulmonary abscesses, where it arrives apparently as a result of aspiration of mouth flora. *B. melaninogenicus,* along with spirochetes and other anaerobes, is also a major etiological agent of periodontal disease. Such infections, together with intraabdominal abscesses, provide a source for *Bacteroides* bacteremia, resulting in the spread of the organisms throughout the body, with a mortality in excess of 30 percent.

Fusobacterium

A number of species of *Fusobacterium* are found as normal flora of the mouth, large intestine, and female genital tract. They are characteristically obligately anaerobic, gram-negative organisms that appear as long, slender rods with tapered ends.

The fusobacteria are major causes of sinus, middle ear, and dental infections. They are also involved in

brain abscesses and are a major anaerobic organism involved in lung abscesses. *F. necrophorum,* the most common pathogen of this genus, is frequently associated with *Bacteroides* species in liver abscesses and intraabdominal abscesses.

PROTOZOA THAT ENTER THE BODY VIA THE DIGESTIVE TRACT

Sarcodina

Entamoeba histolytica: Amebiasis

Human infections with *E. histolytica* occur with or without clinical symptoms, but overt intestinal disease is termed amebic dysentery. The active amebas, called trophozoites, move by extruding pseudopodia; the remaining cytoplasm flows into the pseudopodium. During a usual asymptomatic infection, the amebas colonize on the intestinal wall and feed on bacteria and other amebas present in the intestine. In cases of amebic dysentery, however, the trophozoites invade the intestinal mucosa, and examination of stool specimens may reveal ingested red blood cells. Because *E. histolytica* is essentially the only parasitic ameba to engulf red blood cells, their presence within the ameba provides strong evidence for its identification. Figure 26.10 shows a drawing of a trophozoite as well as an immature and a mature cyst, which is a resting stage of this organism.

Pathogenicity Approximately 90 percent of all infections by *E. histolytica* are asymptomatic and undiagnosed. In a typical case of amebic dysentery, the organisms first penetrate the intestinal mucosa. The resulting lesions may be minor and the symptoms confined to a few daily loose stools containing flecks of blood and mucus. In acute cases there may be nu-

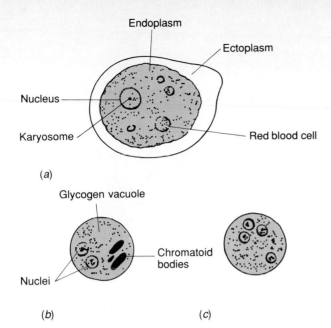

Endoplasm

Ectoplasm

Nucleus

Karyosome

Red blood cell

(a)

Glycogen vacuole

Chromatoid bodies

Nuclei

(b)   (c)

Figure 26.10 *E. histolytica.* (a) Trophozoite containing red blood cells. (b) Binucleated cyst containing chromatoid bars. (c) Mature tetranucleated cyst.

merous intestinal ulcers, and these may coalesce and become secondarily infected with bacteria. When this happens, an individual may experience severe diarrhea characterized by the presence of considerable blood and mucus.

In occasional severe cases the intestinal ulcers may erode into adjoining blood vessels, causing intraluminal bleeding and permitting the spread of the amebas to the liver. Painful hepatic abscesses can develop and may erode through the diaphragm into the lung, causing bronchial abscesses.

Clinical Management The presence and frequency of amebic dysentery can be correlated with the sanitary conditions prevailing within a given locality. The infection is spread primarily by the chronic, asymptomatic carrier who excretes mainly the cyst form of the organism, which is much more resistant to external conditions than are the active trophozoites.

A definitive diagnosis requires that the parasite be identified in feces or infected tissues. Trophozoites containing ingested red blood cells are indicative of active disease. Stained cysts over 10 μm in diameter that possess four nuclei (see Figure 26.10) also indicate an *E. histolytica* infection.

A number of effective drugs are available for the treatment of amebic dysentery. Most of these drugs, such as dehydroemetine, diiodohydroxyquin, chloroquin, or metronidazole exhibit varying degrees of toxicity and should be taken only under the supervision of an experienced physician.

Primary Amebic Meningoencephalitis

In recent years, strains of the free-living amoeboflagellates *Naegleria fowleri* and *Acanthamoeba* species have been isolated from patients with meningoencephalitis. *N. fowleri* is widespread in fresh water, moist soils, decaying vegetation, and fecal wastes. Cases of human meningoencephalitis occur during the summer months and are usually manifested within a week after a person has swum in contaminated water.

The dramatic clinical course is characterized by a fulminating meningoencephalitis and almost invariably ends with the death of the patient within 3 to 6 days after initial symptoms.

Little is known about the pathogenesis of this infection, but it is believed that the nasal mucosa may provide the portal of entry. Diagnosis is based on the observation of motile amebas in spinal fluid and at autopsy by finding stained amebas in sections of the brain.

Acanthamoeba is also known to cause a severe keratitis that may lead to blindness. Infection frequently follows ocular trauma or exposure to contaminated water, but more and more cases are now occurring in persons wearing contact lenses. One survey of such infected persons indicated they were significantly less likely than controls to disinfect their lenses as frequently or as thoroughly as recommended by lens manufacturers.

Treatment is exceedingly disappointing, but initial reports suggest that amphotericin B may possess some clinical potential for treating these diseases.

Ciliophora

Balantidium coli: Balantidiasis

B. coli may attain a length of 200 μm and is the largest of the parasitic protozoa found in the human intestine. Infections are frequently asymptomatic, but occasionally *B. coli* may cause a bloody diarrhea not unlike the acute episodes of amebic dysentery. Trophozoites of *B. coli* are oval and covered with cilia, whereas the cysts are rounded and covered with a thin refractile covering (see Figure 26.11).

B. coli appears to be routinely present in swine; it seems possible that humans become infected by the ingestion of water or food contaminated with cysts present in swine feces. Treatment with oxytetracycline, carbarsone, or diiodoquin appears to be effective for the elimination of this potential pathogen.

Mastigophora

The parasitic flagellated protozoa fall into two categories with respect to the type of disease produced in

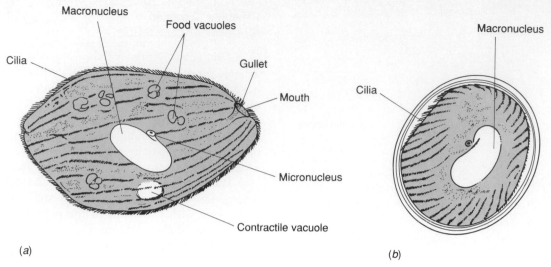

Figure 26.11 *B. coli.* (*a*) Trophozoite. (*b*) Cyst.

humans. One group, called the intestinal flagellates, is found only in the digestive or genital tract, producing infections varying in severity from asymptomatic to serious disease. Members of the second group, the hemoflagellates, are transmitted by bloodsucking insects and produce severe and frequently fatal infections.

Intestinal Flagellates

Giardia lamblia: *Giardiasis* G. *lamblia* is the only flagellated protozoan that produces frank intestinal disease; it is easy to recognize because the trophozoites are bilaterally symmetrical. Each of the organelles and structures is paired, as is shown in Figure 26.12.

Infections in adults are acquired via a fecal-oral route, and waterborne outbreaks are occurring more

---

Figure 26.12 *G. lamblia* trophozoite. Flagellates infecting the human genital and intestinal tract.

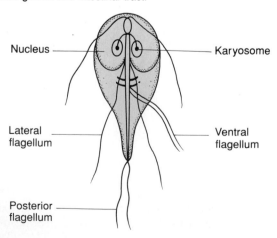

and more frequently, particularly among hikers and campers. Infections are worldwide, occurring most often in children. It is estimated that at least half of all infections are asymptomatic and that 4 to 7 percent of adults in the United States are chronic carriers. It is, as a result, the most frequent cause of waterborne diarrhea in the United States, as well as being the most frequently identified intestinal parasite found in stool specimens submitted to the U.S. Public Health Service laboratories.

Diagnosis can be made by observing stool specimens for ova and parasites, but this is more laborious and less sensitive than available enzyme immunoassay procedures. Infections can be effectively treated with quinacrine hydrochloride (Atabrine).

Sporozoa

Toxoplasma gondii: Toxoplasmosis

T. gondii is one of the most widespread of the parasites that infect vertebrate hosts. Serological tests indicate that over 50 percent of adults in the United States have been infected, and in some countries the infection rate exceeds 90 percent. The disease in humans is mild or asymptomatic and is rarely diagnosed. However, in the immunocompromised host, it may cause encephalitis, myocarditis, or pneumonia, and, when acquired congenitally by the human embryo, the infection may cause mental retardation, severe visual impairment and blindness, convulsions, and an enlarged liver.

Pathogenicity A common source of infection for adult humans is the ingestion of undercooked or raw meat containing trophozoites or cysts of *T. gondii*.

Interestingly, as is illustrated in Figure 26.13, the parasite will undergo its sexual reproductive cycle only in the intestinal cells of members of the cat family. From the cat, the oocysts are passed in the feces, and ingestion of the oocysts releases the enclosed sporozoites, which then multiply asexually in the new host.

Because 30 to 80 percent of cats in the United States have been infected with *Toxoplasma*, pregnant women should not change litter pans if they own a cat and should stay away from cats altogether if possible. Reinfection apparently does not occur, as shown in a prospective study of 25,000 women. In this series no case of congenital toxoplasmosis was found when maternal infection had occurred prior to pregnancy (as judged by the possession of serum antibodies to *T. gondii*), but there were 119 congenital infections in the infants of women who became infected during pregnancy.

Clinical Management Mice are susceptible to infection, but the usual diagnosis is based on a serological test (called the Sabin-Feldman dye exclusion test), in which antibodies to the parasite will prevent the dye methylene blue from staining viable cells of *T. gondii*. In the absence of antibodies, the parasites take up the dye and are colored blue.

Treatment of toxoplasmosis with pyrimethamine in combination with trisulfapyrimidines or sulfadiazine is the only effective therapy.

Cryptosporidium: Cryptosporidiosis

Cryptosporidium is a small protozoan that has been known for years to cause gastroenteritis in many different animal species. In 1976 the first case of cryptosporidiosis in humans was described, but because cases in humans appeared to be confined to immunosuppressed individuals, it was assumed that *Cryptosporidium* possessed little virulence for normal humans.

Recent reports have dispelled that conclusion, and it now appears that *Cryptosporidium* is a frequent cause of diarrhea in otherwise healthy persons, particularly children. One report from Liverpool, England, concluded that *Cryptosporidium* is at least as important as salmonellae, shigellae, campylobacter, or *E. coli* as a cause of childhood gastroenteritis.

The organisms appear to be spread through food and water that has been contaminated with oocysts excreted by animals or infected humans. The disease in immunocompetent persons is characterized by a self-limiting diarrhea and cramps that last 1 to 22 days. The disease in immunocompromised individuals, such as those with the acquired immunodeficiency syndrome, however, is a far different story. Such persons will have 6 to 25 bowel movements per day and, because there is no effective drug treatment, cryptosporidiosis becomes an important contributory factor in the death of the patient.

Table 26.8 summarizes the protozoa that enter the body via the gastrointestinal tract.

PLATYHELMINTHES

The platyhelminthes, or flatworms, are the most primitive of the helminths in that they characteristically have either no digestive tract or only a rudimentary one. They are flat, in rough terms shaped like a leaf or a measuring tape, and are in most examples hermaphroditic (both male and female organs occur in the same animal). Many of the parasitic species have complicated life cycles that require an alternation of hosts. Humans frequently are the definitive host for the adult worm, whereas other animals are the intermediate hosts for larval stages.

Cestodes

Intestinal Cestodes of Humans

The habitat of the adult cestode (tapeworm) is the intestinal tract, and the animal in which the larval stage develops into an adult worm is termed the **definitive host.** Animals in which the eggs develop into the larval stage are called **intermediate hosts.**

Adult tapeworms are usually long and ribbonlike and divide into segments called proglottids (Figure 26.14). The **scolex** (head) at the anterior end is provided with suckers and in some cases hooks that attach

Figure 26.13 Life cycle of *T. gondii.*

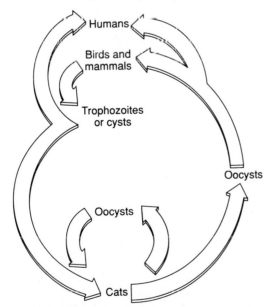

TABLE 26.8   **Protozoa Entering the Body via the Gastrointestinal Tract**

Parasite (Disease)	Definitive Host	Intermediate Host	Reservoir
E. histolytica (amebic dysentery)	Humans	None	Humans
N. fowleri (amebic meningoencephalitis)	Humans	None	?
B. coli (balantidiases)	mHogs, humans	None	Hogs, humans
G. lamblia (giardiasis)	Humans	None	Humans
T. gondii (toxoplasmosis)	Cats	Many domestic animals	Domestic animals; cats
Cryptosporidium	Animals	None	Animals, humans

the worm to the intestinal wall. Posterior to the scolex is a neck region, followed by immature proglottids, mature proglottids containing male and female sex organs, and fertilized segments, called gravid proglottids, containing fertilized eggs.

We shall not attempt to divide the class Cestoda taxonomically. Rather, we shall characterize selected important tapeworms and the ways in which humans become infected.

Taenia saginata   Humans are the only definitive host for *T. saginata,* or beef tapeworm. The infection is acquired by the ingestion of raw or rare beef infected with the larval stage of the parasite. After reaching the small intestine, the worm head emerges from the larva and attaches itself to the intestinal mucosa. The mature worm may reach a length of 8 to 12 m and contain as many as 2000 proglottids. The posterior gravid proglottids are filled with fertilized eggs, and as these break off they are passed in the feces.

Cattle and allied animals become infected from grazing on contaminated soil. After the fertilized eggs hatch in the animal's intestine, the embryos disseminate throughout the body of the intermediate host by way of the lymphatics and the bloodstream, terminating in the muscles. There they develop into a larval stage, the cysticercus, which is a fluid-filled bladder containing an invaginated scolex. If not eaten by the definitive host (humans), the cysticercus will die in

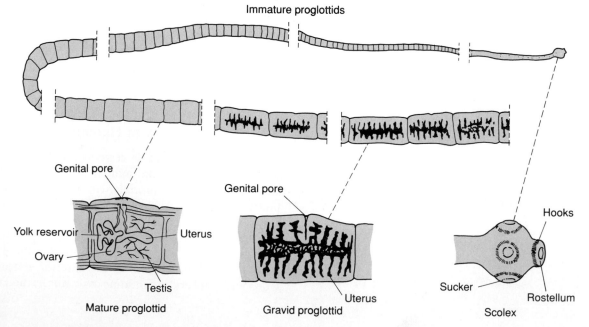

Figure 26.14   Major morphological parts of an adult tapeworm.

Immature proglottids

Genital pore
Yolk reservoir
Ovary
Uterus
Testis
Mature proglottid

Genital pore
Uterus
Gravid proglottid

Hooks
Sucker
Rostellum
Scolex

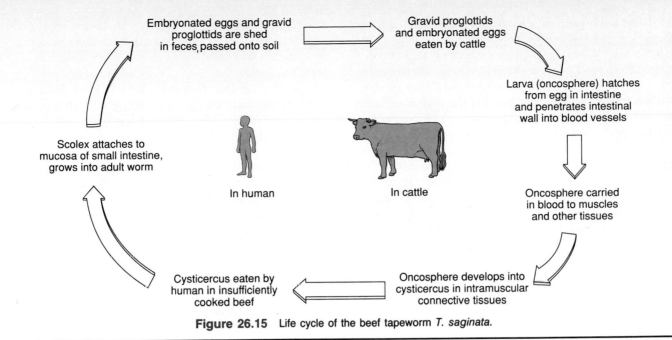

Embryonated eggs and gravid proglottids are shed in feces, passed onto soil

Gravid proglottids and embryonated eggs eaten by cattle

Larva (oncosphere) hatches from egg in intestine and penetrates intestinal wall into blood vessels

Scolex attaches to mucosa of small intestine, grows into adult worm

In human

In cattle

Oncosphere carried in blood to muscles and other tissues

Cysticercus eaten by human in insufficiently cooked beef

Oncosphere develops into cysticercus in intramuscular connective tissues

Figure 26.15 Life cycle of the beef tapeworm *T. saginata*.

about 9 months. Thus, as depicted in Figure 26.15, *T. saginata* would cease to exist if humans did not eat undercooked infected beef and cows did not eat food contaminated with human feces.

In humans, beef tapeworms commonly occur singly, and their presence is ascertained by the occurrence of gravid proglottids and embryonated eggs in the feces. The proglottids average 6 to 20 mm in size and can easily be mistaken for small adult worms. Surprisingly, most infected persons show no symptoms, although anemia and malnutrition, weight loss, ab-

dominal pain, and loss of appetite can occur. Many concoctions have been used to treat this infection, but the current drug of choice is niclosamide.

Taenia solium Many similarities exist in the life cycles and epidemiologies of beef and pork tapeworms. Humans are the only known definitive hosts for the pork tapeworm, *T. solium*; however, the cysticerci of pork tapeworms can also develop in humans as they do in the normal intermediate host, the pig. A life cycle is presented in Figure 26.16.

Figure 26.16 Life cycle of the pork tapeworm *T. solium*.

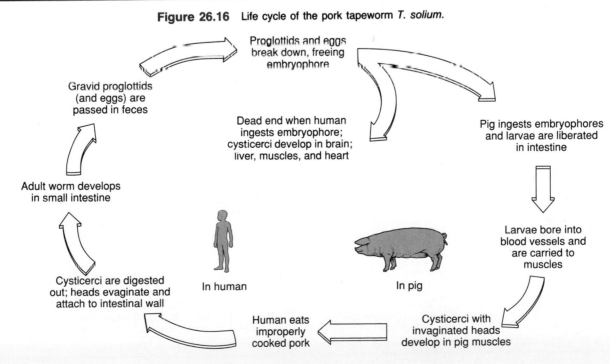

Proglottids and eggs break down, freeing embryophore

Gravid proglottids (and eggs) are passed in feces

Dead end when human ingests embryophore; cysticerci develop in brain; liver, muscles, and heart

Pig ingests embryophores and larvae are liberated in intestine

Adult worm develops in small intestine

In human

In pig

Larvae bore into blood vessels and are carried to muscles

Cysticerci are digested out; heads evaginate and attach to intestinal wall

Human eats improperly cooked pork

Cysticerci with invaginated heads develop in pig muscles

The adult pork tapeworm may reach a length of 2 to 3 m and normally consists of less than 1000 proglottids. When the pig ingests eggs on fecally contaminated food, larvae develop and disseminate throughout the musculature, where they may survive for several years. Humans become the definitive host for the adult worm by eating infected raw or undercooked pork. In general, symptoms are few or absent.

The situation is much different if a person ingests eggs instead of larvae through contaminated food or drink. In such cases, the eggs hatch in the small intestine, and the liberated larvae (also called oncospheres) are disseminated throughout the body, where they form cysticerci. This condition is called cysticercosis, and symptoms vary according to the number and location of the cysticerci. Thus, in muscle or subcutaneous tissue, a light infection may be asymptomatic, whereas involvement of the central nervous system can cause epilepsy and death.

The diagnosis of an intestinal pork tapeworm infection requires the identification of the eggs or egg-filled gravid proglottids passed in the feces. In human cysticercosis, however, diagnosis is usually based on biopsies or roentgenograms of characteristic calcified cysticerci. Niclosamide is used for the chemotherapy of the intestinal infection, but there is no satisfactory treatment for cysticercosis.

Diphyllobothrium latum *D. latum* enjoys a wide variety of definitive hosts, but in endemic areas the disposal of human sewage into fresh water maintains the cycle.

In the infected definitive host (e.g., a human), fertilized eggs are discharged from gravid proglottids into the intestine. After these reach fresh water, the life cycle of the worm requires two different intermediate hosts before it can infect the definitive host. The sequence of events, as diagrammed in Figure 26.17, occurs as follows: (1) the egg must first hatch in fresh water to form a motile embryo (**coracidium**) before it becomes infective; (2) this form is ingested by a crustacean, in which it penetrates the intestine and forms an elongated stage (called a **procercoid larva**) within the crustacean body; (3) when the crustacean is ingested by a suitable fish (pike, salmon, trout, whitefish, turbot), the larvae are disseminated throughout the body of the fish, in which they are transformed into a spindle-shaped larval stage (**plerocercoid**); (4) the definitive host becomes infected from eating raw or inadequately cooked infected fish. In the definitive host, the adult worm attaches to the intestinal wall, where it may reach a length of 10 m and contain 3000 to 4000 proglottids.

Symptoms may be mild or absent, but *D. latum* appears to possess an unusually high affinity for vi-

Figure 26.17 Life cycle of *D. latum*.

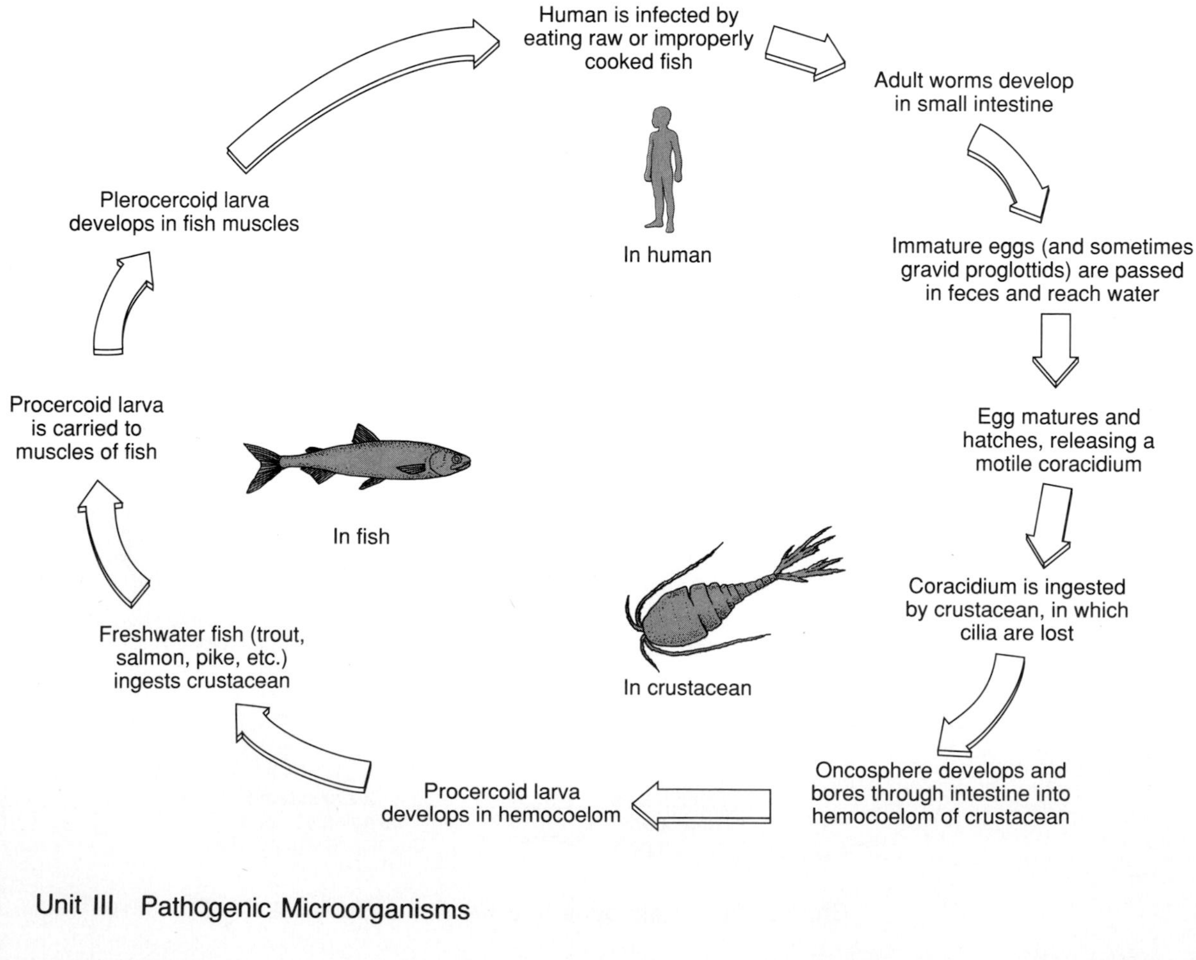

Human is infected by eating raw or improperly cooked fish

Adult worms develop in small intestine

Plerocercoid larva develops in fish muscles

In human

Immature eggs (and sometimes gravid proglottids) are passed in feces and reach water

Procercoid larva is carried to muscles of fish

Egg matures and hatches, releasing a motile coracidium

In fish

Freshwater fish (trout, salmon, pike, etc.) ingests crustacean

In crustacean

Coracidium is ingested by crustacean, in which cilia are lost

Procercoid larva develops in hemocoelom

Oncosphere develops and bores through intestine into hemocoelom of crustacean

tamin B_{12}. If the worm is attached high in the small intestine, it will absorb essentially all of the vitamin B_{12} ingested by the host, producing pernicious anemia; this does not occur when the worm is farther down in the intestinal tract.

Diagnosis is usually based on case histories plus the finding of typical eggs in the feces. Niclosamide is considered the therapy of choice.

Hymenolepis nana   The definitive hosts for *H. nana* are humans, mice, or rats, and no intermediate host is required. Humans are the major reservoir for this cestode, known also as the dwarf tapeworm. Infection occurs under conditions of poor sanitation, usually by the direct transfer of fertilized eggs from hand to mouth. After reaching the intestine, the embryos hatch from the eggs, enter the mucosa, and develop into a larval stage, the cysticercoid. These larvae subsequently enter the intestinal lumen and develop into adult worms.

Symptoms routinely are limited to mild abdominal discomfort unless the infection is heavy, in which case they may include abdominal pain, nausea, diarrhea, and headaches. Diagnosis is made by observing typical eggs in the feces. Niclosamide is used for therapy.

Extraintestinal Infections of Humans by Cestodes

Echinococcus granulosus   As described for *T. solium*, a tapeworm infection in which the larvae are distributed throughout the body of the intermediate host causes symptoms very different from those that occur in the definitive host in whom the adult worm is confined to the intestinal tract.

Humans may serve as hosts for several other larval cestodes in addition to *T. solium*, but our discussion includes only one common species, namely *E. granulosus*. The usual intermediate hosts of this worm are sheep, cattle, and other herbivores. Humans almost always acquire the infection from dogs, although wolves, coyotes, and foxes also serve as definitive hosts.

After the ingested eggs hatch, the liberated larvae (oncospheres) penetrate the intestinal wall and are disseminated throughout the body by way of the lymphatics and the bloodstream. The liver is the organ most frequently infected, but the larvae may also settle in the lungs, kidneys, bone, or brain. Each larva forms a fluid-filled bladder—a **hydatid cyst**—which continues to increase in size and to form secondary interior cysts known as **brood capsules**. The primary and secondary cysts contain countless scoleces, and each scolex may develop into an adult worm when ingested

by a dog. Rupture of the hydatid cyst may release thousands of scoleces into the surrounding tissues. The life cycle is summarized in Figure 26.18.

In humans, a single cyst usually develops, and it frequently impairs organ function by pressure on the surrounding tissues. An inflammation develops, mediated in part by a hypersensitivity response to the scoleces, and adjacent tissues either atrophy or undergo pressure necrosis.

The diagnosis of hydatid disease is commonly based on finding a tumor mass, usually in the liver, and a history of association with dogs in an endemic area. X-rays may reveal a calcified cyst wall. The diagnosis can be confirmed by the demonstration of brood capsules and scoleces in the fluid of the surgically removed cyst.

Most human cases of hydatid disease occur in grazing countries, where an intimate association between humans and dogs is common, and the disease is occasionally seen in the United States, particularly in California. Infections follow a hand-to-mouth transfer of the eggs from contaminated soil or dog fur. Most drugs are ineffective for the chemotherapy of hydatid disease; treatment must usually rely on the surgical removal of the hydatid cyst. Mebendazole has, however, been successfully used early in the disease. Control is directed at the prevention of the infection in dogs; dogs should not be fed uncooked viscera from slaughtered animals. Dogs in endemic areas should be treated periodically with quinacrine (Atabrine) or niclosamide. Table 26.9 summarizes the cestode infections for which humans may be the definitive host.

Trematodes

Trematodes are also referred to as **flukes,** with the adult worms varying in size from 1 mm to several centimeters in length. The reader is referred to Chapter 15 for a general review of the life cycles of trematodes.

Based on the major organ that is parasitized, flukes are divided into intestinal flukes, liver flukes, lung flukes, and blood flukes. All but the blood flukes are acquired by the ingestion of a larval stage termed a **metacercaria.** Because the blood flukes gain entrance to the body by burrowing through the skin, they will be discussed in Chapter 28.

Intestinal Flukes

Intestinal flukes are basically parasites of other animals, and humans become accidental hosts during the development of these flatworms.

Fasciolopsis buski   *F. buski* normally parasitizes pigs in Southeast Asia, although the incidence of infection in humans may reach 100 percent in some areas. After leaving the snail, the cercariae attach

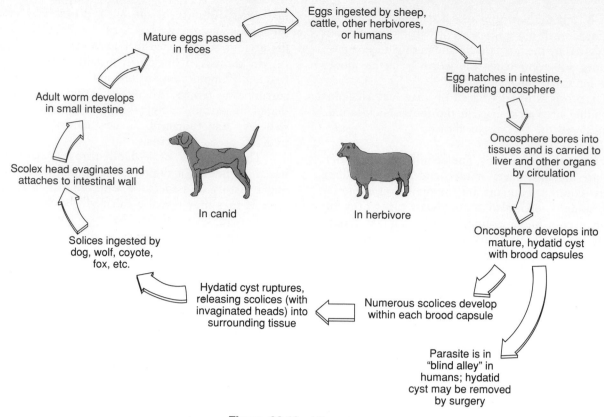

Figure 26.18 Life cycle of *E. granulosus*.

themselves to various types of vegetation, where they develop into metacercariae. Humans become infected by eating contaminated plants such as water chestnuts, bamboo, and water hyacinths (Figure 26.19).

The metacercariae evolve into adult worms that attach to the mucosa of the small intestine. Symptoms will vary according to the degree of infection. In small children, heavy infections may be particularly severe, occasionally resulting in death. Diagnosis is confirmed by the presence of characteristic eggs in the feces. Both hexylresorcinol and tetrachloroethylene are effective for treatment.

Metagonimus yokogawai *and* Heterophyes heterophyes The metacercariae of both *M. yokogawai* and *H. heterophyes* are formed under the scales of freshwater fish. Dogs and cats serve as the usual definitive hosts in Asia, and humans become infected by eating raw fish.

The adult worms attach to the mucosa of the small intestine, and symptoms are usually mild. Eggs are occasionally carried by the lymphatics to the heart or central nervous system, which can cause severe reactions. Hexylresorcinol is an effective treatment, as is tetrachloroethylene.

TABLE 26.9   Summary of Cestode (Tapeworm) Infections

Scientific Name	Common Name	Definitive Host	Intermediate Host
T. saginata	Beef tapeworm	Humans	Cattle
T. solium	Pork tapeworm	Humans	Swine
D. latum	Fish tapeworm	Humans, many animals	Crustaceans and fish
H. nana	Dwarf tapeworm	Humans, mice, rats	None required
E. granulosus	Hydatid disease	Dogs, foxes, humans	Sheep, cattle

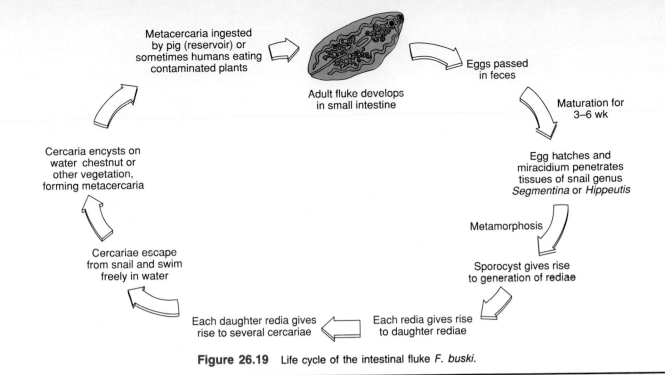

Figure 26.19   Life cycle of the intestinal fluke *F. buski.*

Liver Flukes

Flatworms that mature into adults in the bile ducts of the definitive host are commonly referred to as liver flukes.

Fasciola hepatica   Sheep and, to a lesser extent, cattle constitute the usual hosts for *F. hepatica.* Human infections occur in many parts of the world; cases have been reported from southern France, Algeria, Cuba, and the Latin American countries.

Fertilized eggs pass by way of the common bile duct into the intestinal tract. If they reach water that contains the appropriate snails, the miracidia will infect this intermediate host, and motile cercariae will be released at the end of larval development. The cercariae encyst on local vegetation (such as grass or watercress) to form metacercariae. Sheep, cattle, and occasionally humans become the definitive hosts following the ingestion of metacercariae on contaminated vegetation.

In the definitive host, the metacercariae penetrate the intestine and migrate from the peritoneal cavity to infect the liver parenchyma and eventually the bile ducts. There they produce mechanical and toxic injuries that, in severe infections, terminate with cirrhosis. Eggs entering the intestines from the common bile duct are passed and reinitiate the larval cycle.

Bithionol is currently the drug of choice for the treatment of infections caused by these liver flukes.

Clonorchis sinensis   *C. sinensis,* also known as the Chinese liver fluke, exists primarily in the Far East. The intermediary larval stages in the snail are as described for trematodes in general; the definitive hosts (usually dogs and cats but also humans) become infected from eating raw fish infected with the metacercariae (Figure 26.20).

Symptoms vary considerably, depending on the degree of infection. Light infections may be asymptomatic, whereas heavy and repeated ingestion of the larvae may cause abscesses and liver impairment.

Two other liver flukes that occasionally infect humans are *Opisthorchis felineus* and *O. viverrini.* Cats and, to a lesser extent, dogs comprise the primary definitive hosts for the adult stage of these parasites.

Experimental drugs for treating infections by *Clonorchis* or *Opisthorchis* species are under trial in the Far East, but there is no effective therapy available in the United States.

Lung Flukes

Paragonimus   westermani   *P.   westermani* causes a major disease, paragonimiasis, in which the adult worm parasitizes the lungs. It is restricted to the Far East. Other species of *Paragonimus* are responsible for human infections in Africa and South America.

After leaving the snail, the cercariae infect crabs or crayfish (in which they become encysted) as second

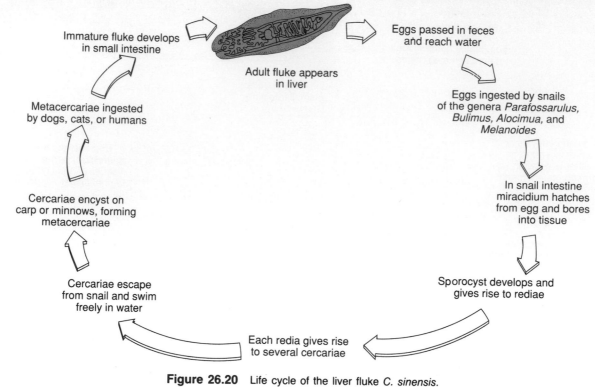

Figure 26.20 Life cycle of the liver fluke *C. sinensis*.

intermediary hosts. Humans become infected by eating raw or improperly cooked crabs or crayfish containing metacercariae (Figure 26.21).

Following ingestion, the cercariae leave the small intestine and migrate from the peritoneal cavity through the diaphragm into the bronchioles, where a fibrous capsule forms around the larva. After the development of the adult worm, the capsule ruptures

Figure 26.21 Life cycle of the lung fluke *P. westermani*.

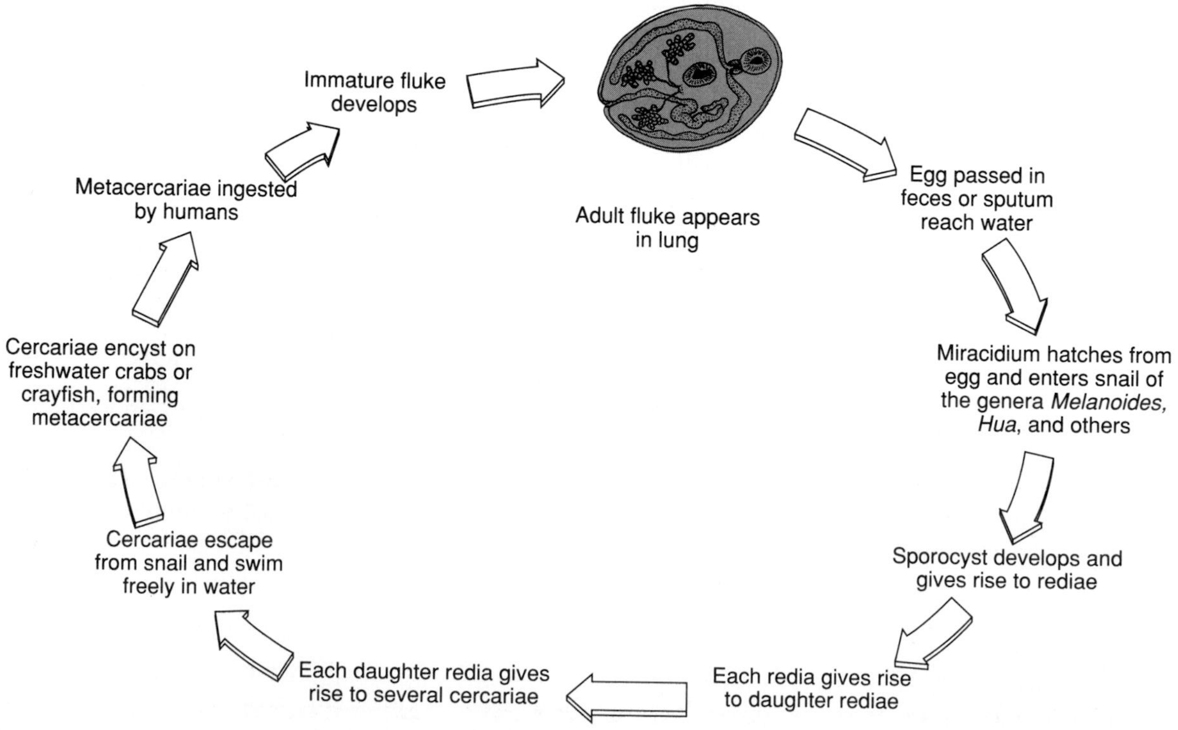

into the bronchioles, releasing eggs, which are coughed up and expectorated or swallowed. In fresh water, the eggs hatch into miracidia, and infection of the appropriate snails starts the cycle over again.

Symptoms of paragonimiasis are absent or restricted to occasional coughing up of rusty sputum, depending on the number of parasites in the lungs. The disease responds to the oral administration of bithionol. This drug, however, may cause side effects such as skin rashes, nausea, diarrhea, headaches, and dizziness. It can also induce a contact dermatitis. Because bithionol was at one time incorporated into a number of medicated soaps and shampoos, patients should be checked for hypersensitivity reactions before using this drug.

Human trematode infections are summarized in Table 26.10.

NEMATODA

Nematodes are round, elongated worms; the sexes are separate, and, unlike the flatworms, adult nematodes possess complete digestive systems that include both a mouth and an anus. Human infections caused by these roundworms are usually divided into two large categories, the intestinal roundworms and the blood and tissue roundworms.

Intestinal Nematode Infections in Humans

Roundworms that exist in the intestine during their adult stage are generally categorized as intestinal nematodes. We shall see, however, that the larval stages of some intestinal species may be widely distributed throughout the body of an infected person.

Trichinella spiralis *T. spiralis* is the etiological agent of **trichinosis**, a disease disseminated in carnivorous animals. Human infections occur worldwide, and as recently as the 1930s, incidence of the disease in the United States was estimated to include 20 percent of the population (control measures have now reduced this to approximately 4 percent). Human infections almost invariably result from the ingestion of improperly cooked pork containing the encysted larvae; however, bear meat has also been implicated in the northwestern United States. Following ingestion, the cysts reach the intestine, where larvae are liberated and develop into adult worms. After copulation, the male worms are passed in the feces and the female penetrates the intestinal mucosa, where she will produce as many as 1500 larvae over a period of 1 to 3 months. The larvae enter the bloodstream, primarily by way of the lymphatics, after which they penetrate the skeletal muscles and become encysted (Figure 26.22).

As with many helminthic infections, the symptoms are in large part dependent on the magnitude of the initial infection. Light infections are normally asymptomatic, but the ingestion of large numbers of encysted larvae may cause a severe disease culminating in the death of the patient. Initial symptoms, occurring within 1 to 3 days after infection and resulting from the intestinal activities of the adult worms, characteristically include diarrhea, malaise, and fever. This stage may be mild to severe but persists for only about a week. The second stage of the infection, involving the migration of the larvae into skeletal muscle, begins about 1 week after the initial infection; in cases of moderate to heavy infection, muscular pain throughout the body may be significant. Moreover, although encystment occurs only in muscle, the larvae may infect other organs of the body, including the lungs, heart, eyelids, meninges, and brain. Massive invasion of these latter organs may result in death during the

TABLE 26.10 Summary of Trematode Infections

Name	Intermediate Host	Usual Source of Human Infection
Intestinal flukes		
F. buski	Snail	Ingestion of contaminated water chestnuts
M. yokogawai	Snail	Ingestion of infected fish
H. heterophyes	Snail	Ingestion of infected fish
Liver flukes		
F. hepatica	Snail	Ingestion of contaminated vegetation
C. sinensis	Snail	Ingestion of infected fish
Lung flukes		
P. westermani	Snail	Ingestion of crabs or crayfish

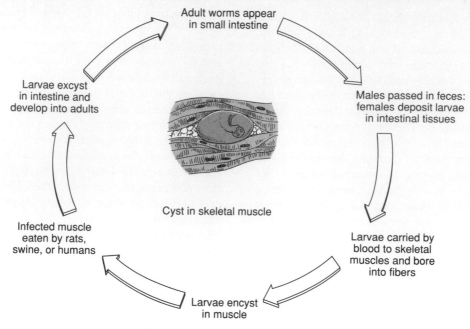

Adult worms appear
in small intestine

Larvae excyst
in intestine and
develop into adults

Males passed in feces:
females deposit larvae
in intestinal tissues

Cyst in skeletal muscle

Infected muscle
eaten by rats,
swine, or humans

Larvae carried by
blood to skeletal
muscles and bore
into fibers

Larvae encyst
in muscle

Figure 26.22   Life cycle of *T. spiralis*.

early weeks of the infection. The usual pathogenic effects, however, result from the destruction of striated muscle fibers and occur during the encystment and subsequent calcification of the larval cysts.

Barring reinfection, major symptoms subside in several months, but weakness, rheumatic pain, and loss of dexterity may persist for long periods.

There are several laboratory tests for the diagnosis of trichinosis, but the only definitive one is the observation of the larvae or larval cysts in biopsies of infected muscles. Complement fixation and a skin reaction using an antigen prepared from *Trichinella* are used to support a tentative diagnosis.

Thiabendazole appears to be effective against the intestinal phase of this parasite, but its efficacy against larvae in muscles is not clearly established. In severe cases, steroids are given to lessen the extent of inflammation and to provide symptomatic relief. Since ingestion of undercooked infected meat provides the only mechanism of infection, it might seem that it would be possible to eliminate this disease from domestic animals. However, the major reservoir for *T. spiralis* occurs in cannibalistic brown and black rats, and it is essentially impossible to completely control this source of infection for pigs. Laws requiring sterilization of garbage to be used for feeding hogs are in force throughout the United States, and such measures, coupled with the thorough cooking of pork (or freezing, which also destroys the encysted larvae), have done much to decrease the incidence of trichinosis.

Trichuris trichiura   Trichuriasis, or **whipworm disease,** is caused by the nematode *T. trichiura* (Figure

26.23). This disease occurs extensively in the tropics and occasionally in lower socioeconomic areas in the southern United States. It is a disease associated with filth and is seen primarily in children. Its prevalence is indicated by the estimate that about 500 million people are infected with this parasite.

The adult worm lives mainly in the human cecum attached to the intestinal mucosa, where it will continue to produce eggs for a period of 6 to 8 years. The eggs are passed in the feces, and in a moist, warm environment infective larvae will develop within the eggs in 3 to 6 weeks. Ingestion of the mature eggs, either directly from the soil or through contaminated food and water, initiates a new cycle during which the larvae will develop into mature adults in the cecum (Figure 26.24).

Heavy infections are usually accompanied by chronic diarrhea. Abdominal pain, vomiting, constipation, headache, and anemia may also accompany whipworm infections.

The observation of the barrel-shaped eggs in the feces provides a definitive diagnosis for trichuriasis. Adult worms are rarely seen in the feces, for they normally remain attached to the intestinal mucosa. By proctoscopy, heavy infections of worms can be observed attached to the rectal mucosa.

The drug mebendazole is an effective treatment when given orally for several days. Enemas containing hexylresorcinol were previously used to treat heavy infections. The major effective control measures are those directed toward the sanitary disposal of human feces. Secondary measures involve thorough sanitary procedures such as the washing of hands before meals

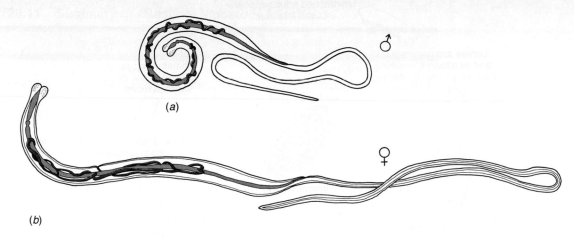

Figure 26.23 The nematode *T. trichiura*. (a) Male. (b) Female.

live and
male mi₁
in the p₁
sions, su
but infe₁
fertilized
an inten₁
inates th
hand-to-
such ma
where th
are asyn
spontane
matic cy

Beca
diagnosis
skin. If a
slide is fi
attach an
microsco₁
anus dur
curring sl
quently r

A nı
treat pin
pamoate
mebendaz
effective,
Treatmen
can prev₁
spontane₁
months.

Tabl₁
acquired
water.

and the complete cleansing of uncooked vegetables before consumption.

Ascaris lumbricoides   *A. lumbricoides* may attain a length of 20 to 30 cm and is the largest nematode that parasitizes humans. It occurs most frequently in young children in the tropics and, to a lesser extent, in certain areas of the southern United States. It is estimated that over 900 million persons are infected and that about a million of these live in the mountainous regions of the southern United States.

Second-stage larvae develop within the eggs while in the soil, and humans are infected by the ingestion of the infective eggs (Figure 26.25). After reaching the intestine, the larvae hatch, penetrate the intestinal mucosa, and (after being picked up in the portal circulation and passing through the heart and liver) eventually reach the lungs. There they undergo additional differentiation and are coughed up, swallowed, and returned to the intestine.

Adult worms remain unattached in the small intestine and, in the absence of a heavy infection, symptoms are mild or absent. Heavier infections cause abdominal pain, and complications may occasionally occur as a result of invasion of the liver, bile ducts, gallbladder, and appendix by adult worms. Hypersensitivity reactions occurring during the pulmonary migration of larvae in subsequent infections may result

VIRUS
BODY
TRACT

Viral Ga

It has bee₁
gastroente

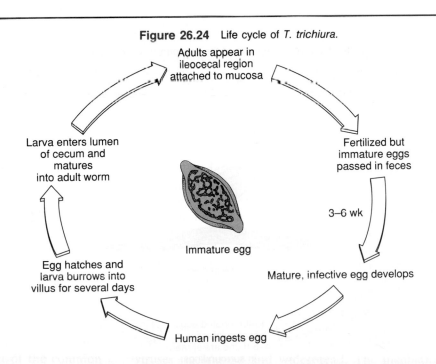

Figure 26.24 Life cycle of *T. trichiura*.

Adults appear in ileocecal region attached to mucosa

Fertilized but immature eggs passed in feces

3–6 wk

Mature, infective egg develops

Human ingests egg

Egg hatches and larva burrows into villus for several days

Immature egg

Larva enters lumen of cecum and matures into adult worm

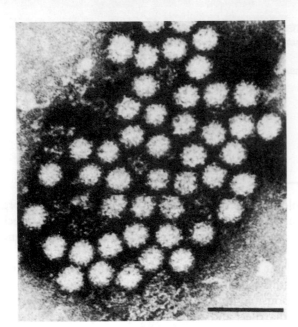

Figure 26.27 Norwalk agent virions. Diameter = 27 nm; bar = 100 nm.

Figure 26.27), but such a procedure would probably only be used to identify the cause of an epidemic of gastrointestinal illness.

Rotaviruses

Rotaviruses were named because of the morphological similarity of their double-walled capsid to a wheel (Figure 26.28). These viruses are related to the reoviruses, and each contains 11 segments of double-stranded RNA within its capsid.

The ubiquity of rotaviruses throughout the animal kingdom is supported by the observation that piglets and calves deprived of colostrum (the first milk, containing high concentrations of the mother's antibodies) will invariably die from a severe diarrhea caused by a virus related to human rotavirus. Approximately 75 percent of newborn humans possess maternally acquired antibodies to rotavirus. The antibody titer declines during the first 6 months of life, but by 5 or 6 years of age, approximately 90 percent of all children will have acquired an active immunity to rotavirus infections.

The incubation period of the gastroenteritis is 1 to 3 days; in infants, infection may result in death due to dehydration and loss of electrolytes from the diarrhea and vomiting. In one study involving 21 fatal cases of rotavirus gastroenteritis, the infants' ages varied from 4 to 30 months. Death occurred within 1 to 3 days following the onset of symptoms. Only cholera leads to dehydration with a frequency equal to or greater than rotaviral gastroenteritis.

Early hospitalization with fluid and electrolyte replacement all but eliminates the mortality, but such

period is 24 to 48 hours, and the infection is characterized by a sudden onset of nausea and vomiting, often severe. Other symptoms include fever and diarrhea.

Laboratory diagnosis of Norwalk and Norwalk-like virus infections has been hindered by the inability to grow these viruses in cell culture. The viruses can be visualized in the stools of infected persons using immune electron microscopy on stool specimens (see

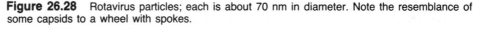

Figure 26.28 Rotavirus particles; each is about 70 nm in diameter. Note the resemblance of some capsids to a wheel with spokes.

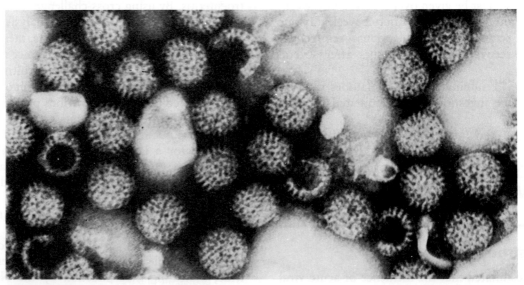

treatment is often unavailable, particularly in underdeveloped countries. Successful rehydration has been accomplished with the oral administration of electrolyte solutions containing sugar. Because of the severe vomiting, however, this can be difficult to achieve, especially with the very young infant.

Using neutralizing antibodies, five serotypes of human rotaviruses have been defined. Thus, it may be possible to produce an oral vaccine. Such a vaccine could also be used to immunize expectant mothers, because mother's milk is high in immunoglobulins and would readily protect the newborn from rotavirus infections.

ENTERIC VIRUSES THAT CAUSE SYSTEMIC ILLNESSES

Many enteric viruses have been isolated from the intestinal tract and may cause mild to severe intestinal symptoms as well as produce serious disease in other parts of the body. Any of the enteric viruses can cause a variety of symptoms: intestinal illness, respiratory illness with or without a rash, and, frequently, meningitis. They are responsible as well for a number of asymptomatic illnesses, and many of these viruses have been isolated from apparently healthy persons.

All enteric viruses are classified in the family Picornaviridae (pico [small] RNA viruses). The picornaviruses include two large subgroups, the rhinoviruses (see Chapter 25) and the enteroviruses. The enteroviruses include polioviruses, coxsackieviruses, echoviruses, and hepatitis A virus.

Enteroviruses are very small—approximately 20 to 30 nm in diameter. They contain single-stranded RNA. The enteroviruses are very stable and can survive for long periods in sewage. Human strains are easily grown on human or monkey cell cultures, in which they produce cytopathic areas of killed cells.

Poliovirus

Poliovirus is a very small picornavirus (see Figure 26.29). There are three immunologically distinct types of poliovirus: type 1, the Brunhilde or Mahoney strain, type 2, the Lansing strain; and type 3, the Leon strain.

Pathogenicity Poliomyelitis is an infectious disease that usually occurs in three phases. The first is an infection of the gastrointestinal tract. During this phase the virus is found in the throat and in feces. It is thought that the initial multiplication of the virus takes place in the oropharynx and the intestines. After the disease becomes symptomatic, the virus may be difficult to isolate from the throat, but it remains in the feces. The second phase is characterized by an invasion of the bloodstream, and in the third phase the virus migrates from the bloodstream into the meninges.

The abortive form of the disease is the most common type. It is characterized by fever, headache, malaise, drowsiness, sore throat, nausea, and vomiting. Recovery is usually rapid. Since the illness is mild, diagnosis can be confirmed only by isolating the virus or by serological tests. In some cases stiffness of the

Figure 26.29 (a) Crystalline array of poliovirus particles in a HeLa cell ($\times$ 50,400). (b) Highly purified stained type 1 Mahoney poliovirus ($\times$ 290,000).

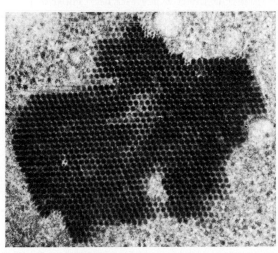

(a)

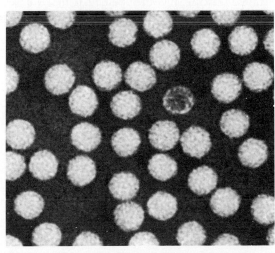

(b)

neck and the back, accompanied by pain, are found in this nonparalytic form of the disease.

Paralytic poliomyelitis occurs in about 0.1 percent of poliovirus infections. It develops as a result of invasion of the meninges and is characterized by destruction of the large nerve cells in the anterior horn of the brain. Because these cells give rise to motor fibers of peripheral nerves, their destruction is accompanied by paralysis. In bulbar poliomyelitis there is involvement of neurons in the medulla, and the respiratory or vasomotor center may be affected. The incubation period may be from 3 to 35 days, but it is more commonly 7 to 14 days.

Sources of the virus are mainly the intestinal tracts of infected people. Thus, the disease is spread primarily as a fecal-oral infection. It is possible that the virus could be spread as a respiratory infection early in the illness, but since the virus remains in the oropharynx for such a short time, this is probably a very minor route of transmission.

Clinical Management In most infections caused by poliomyelitis virus there are no distinctive symptoms. In the serious forms of the disease, the diagnosis can usually be made on the basis of the clinical picture. In abortive poliomyelitis, symptoms are usually not distinctive enough to make a definitive clinical diagnosis; undoubtedly the vast majority of such cases are not diagnosed. A diagnosis can be made in such persons either by isolating the virus or by demonstrating a rise in antibody titer to one of the strains of poliovirus.

Vaccines Poliomyelitis was for many years one of our most feared diseases because it threatens its older victims with paralysis. (Infection in an infant who still possesses maternal antibodies is much less likely to lead to paralysis.) As recently as 1950, 20,000 cases of paralytic polio were reported annually in the United States, but this has dropped to 0 to 10 cases per year. It is estimated, however, that there may be as many as 1 million such cases occurring annually in developing and Third World countries, and the World Health Organization is currently working toward the global eradication of paralytic polio by the year 2000.

Effective vaccines are available. The original vaccine, developed by Jonas Salk, contained formalin-killed viruses of each of the three serotypes. The injection of this vaccine stimulated the production of antibodies in the serum. Following infection, any virulent virus was neutralized as it entered the bloodstream, preventing involvement of the central nervous system.

A live vaccine, the Sabin vaccine, has now replaced the Salk vaccine in the United States. The Sabin vaccine is composed of attenuated polioviruses of each of the three types. The vaccine comprising all three types is administered as a liquid on the tongue in three successive doses. The live vaccine multiplies in the cells of the gastrointestinal tract and oropharynx, inducing the same solid immunity that the natural infection would provide. Supposedly, however, it is unable to multiply in the cells of the central nervous system. This, unfortunately, may not be always true. In the period 1969 to 1980, over 291 million doses of live polio vaccine were distributed in the United States. During this time, 93 cases of vaccine-associated poliomyelitis were reported. Of these, 36 occurred in vaccine recipients and 57 were household or community contacts of the vaccines. This represents 1 case per 5.1 million doses distributed. After considering the benefits and risks, an advisory committee to the Public Health Service has recommended that the use of live oral polio vaccine be continued for primary immunization of children in the United States.

The efficacy of the Salk and Sabin vaccines in preventing paralytic polio is well established, as is shown in Figure 26.30. There were, however, 5 confirmed cases of vaccine-associated paralytic polio in 1989 and cases of suspected polio were reported during 1990.

It should be kept in mind that, although the polio vaccine has essentially eliminated the paralytic aspects of this disease, the virus is still widespread. Public complacency with regard to immunization could easily result in a large population with no immunity, and the return of paralytic poliomyelitis is possible.

Coxsackieviruses

Coxsackieviruses were named for Coxsackie, New York, where they were first isolated from stools in 1948. They were originally thought to be a new type of poliovirus, but when it was shown that they would infect newborn mice, they were designated as a separate group, the coxsackieviruses. The group is divided into two subgroups based on their effects on newborn mice. There are 24 serotypes of Coxsackie A viruses, which produce a characteristic fulminating lethal infection in newborn mice, and six serotypes of Coxsackie B viruses, which produce much less serious disease in newborn mice (see Figure 26.31.)

Pathogenicity The coxsackieviruses have been implicated in a variety of diseases, from the common cold to lethal myocarditis of the newborn. The ordinary manifestations of coxsackievirus infections are herpangina (severe sore throat associated with patches of exudate), pleurodynia (pain in the chest—also called Bornholm disease), summer grippe (an influenzalike disease), and aseptic meningitis (see Table 26.12).

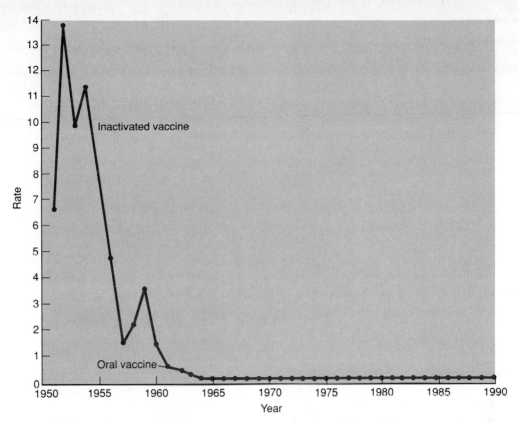

Figure 26.30 Poliomyelitis (paralytic)—reported cases per 100,000 population by year, United States, 1953–1990.

Figure 26.31 An array of coxsackievirus, Group B-3, in a thin section of mouse muscle (×70,000). Compare with the poliovirus array in Figure 26.29.

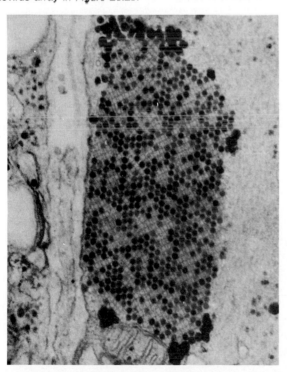

TABLE 26.12 Clinical Syndromes Associated with Coxsackieviruses

Syndrome	Coxsackievirus Group
Aseptic meningitis	A and B
Herpangina	A
Pleurodynia	B
Summer grippe	A and B
Myocarditis	B
Pericarditis	B

The virus grows in the pharynx and intestines and spread can probably occur via oral secretions or a fecal-oral route. Because most cases occur in the summer or fall swimming season, it seems probable that the fecal-oral spread is most responsible.

Clinical Management The virus can be isolated from throat washings, spinal fluid, or feces. Injection into newborn mice or the use of cell cultures showing cytopathic effects will allow recovery of the virus. Identification requires the use of specific neutralizing antiserum.

A retrospective diagnosis can also be made by showing a rise in neutralizing antibodies between the acute and convalescent phases of the disease.

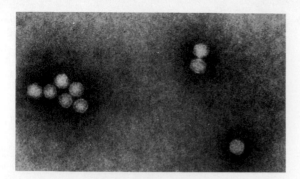

Figure 26.32 Virions of echovirus type 19 (× 120,000).

There are no effective control measures. Although vaccines would likely be effective, the multiplicity of virus types would make their development and production highly impractical.

Echoviruses

The echoviruses illustrate the dilemma virologists faced as they discovered many new viruses in the feces of people who had no clinical illness. At one time these agents were facetiously referred to as viruses in search of disease. Later they were given the rather long name of **enteric cytopathic human orphan**, indicating that they were isolated from human feces, caused a cytopathic effect in cell cultures, and could not be associated with any disease. From the acronym came the name **echovirus**.

Echoviruses are picornaviruses (see Figure 26.32), and they have now been shown to cause a number of clinical manifestations in humans. Thirty-one different serological types have been isolated and characterized. Although humans are the only hosts for echoviruses, other animals harbor similar enteroviruses. Monkeys

have their ecmoviruses, swine their ecsoviruses, and cattle their ecboviruses.

Pathogenicity The illnesses caused by the echoviruses are similar to coxsackievirus infections and, like other enterovirus infections, they are frequently unapparent and undiagnosed. The specific illness depends on the type of echovirus involved. Some of the echoviruses are still "orphans" because they have yet to be associated with any disease entity.

In general, echovirus illnesses can be subdivided into four categories: (1) aseptic meningitis, from which recovery is usually complete; (2) respiratory infections, which are clinically similar to the common cold; (3) gastroenteritis, resulting in a diarrhea that can be particularly severe in the newborn; and (4) disseminated infections, which are characterized by skin rashes, fever, and malaise.

Clinical Management Laboratory diagnosis of echovirus infections requires the isolation and serological identification of the virus. The viruses can be readily grown on human or monkey cell cultures in a laboratory that is equipped to grow echoviruses.

Control of these infections seems to be extremely difficult. Although it is likely that they are occasionally spread by way of respiratory infections, it appears that the fecal-oral route is more common. Immunization with a viral vaccine might well be effective, but the myriad of serological types makes vaccination impractical.

Hepatitis Viruses

Over 50,000 cases of viral hepatitis (inflammation of the liver) are reported annually in the United States (see Table 26.13). Early knowledge about the epidemiology and pathogenesis of viral hepatitis was dis-

TABLE 26.13   Cases of Viral Hepatitis Reported in the United States 1980–1990

Calendar Year	Acute Hepatitis A and Unspecified	Hepatitis B	Non-A, Non-B[a]
1980	27,181	17,550	
1981	23,700	19,521	
1982	28,670	16,131	2629
1983	28,017	23,654	3470
1984	26,567	25,862	3749
1985	27,533	25,808	3857
1986	23,117	25,867	3482
1987	27,558	25,170	2882
1988	28,553	22,223	2455
1989	36,780	22,512	2272
1990	30,108	19,651	2695

[a] Not reported until 1982.

covered by studying human volunteers and other primates infected with extracts of blood, feces, urine, and duodenal contents obtained from patients with hepatitis.

Such studies have shown that human hepatitis is caused by at least five distinct viruses that possess different antigens and can, therefore, be distinguished from each other by serological techniques. The diseases caused by these viruses are also characterized in part by the length of their incubation periods. These diseases are currently designated as hepatitis A, hepatitis B, at least two different viruses causing non-A, non-B hepatitis (now referred to as hepatitis C and hepatitis E), and the delta agent.

Hepatitis A

Hepatitis A virus appears to be one of the most stable viruses infecting humans. It can withstand heating at 56°C for 30 min and is remarkably resistant to many disinfectants. Electron microscopy of fecal extracts that have been mixed with antibody to hepatitis A virus reveals clumps of viruslike particles appearing to possess an icosahedral symmetry (see Figure 26.33); it has been shown that this virus is a picornavirus.

Pathogenicity Hepatitis A infections have an asymptomatic incubation period of 15 to 40 days; symptoms usually begin abruptly with fever, nausea, and vomiting. After several days, jaundice may become apparent, and at that time the acute symptoms usually subside. Complete recovery may require 8 to 12 weeks. Virus is present in the blood and feces for about 2 weeks before symptoms begin and for 1 to 2 weeks after the disappearance of jaundice. Asymptomatic carriers of the virus do not occur. The severity of the disease varies considerably with age; most cases occurring in young children are mild and undiagnosed.

Clinical Management The spread of hepatitis A is most often from person to person by way of a fecal-oral route. As shown in Table 26.12, an average of about 25,000 cases of hepatitis A are reported each year in the United States, but these cases represent only a small percentage of actual infections, because most hepatitis A infections are undiagnosed. This is particularly true for children in whom jaundice is not a common manifestion of the disease.

Outbreaks of hepatitis A have been reported in day care centers and in institutions for the mentally retarded. In some cities of the United States, 9 to 12 percent of reported cases of hepatitis A occur in parents, children, or staff of day care centers. Epidemics have resulted from drinking fecally contaminated water, but such waterborne epidemics are rare. Eating food prepared by an infected person or the ingestion of raw oysters, clams, or mussels harvested from fecally contaminated water has also resulted in many cases of hepatitis A infections. Because there is no persistent infection with continuous viremia (as in hepatitis B infections), hepatitis A transmission by blood products is rare.

Diagnosis of hepatitis A is difficult; it is usually made on the basis of clinical symptoms, liver function test, and by the presence of IgM antibody to hepatitis A virus in the serum of an infected person.

Because most adults have recovered from hepatitis A (even though undiagnosed), pooled gamma globulin contains antihepatitis A antibodies and is effective for the prevention of hepatitis A infections if it is administered before or shortly after exposure.

Hepatitis A has been successfully transmitted to marmoset monkeys and will grow in human embryo fibroblast cells. A vaccine is not yet commercially available but clinical trials using inactivated whole virus are currently being tested for immunogenicity and safety. Preliminary studies indicate that multiple doses are required to induce sufficient levels of antibodies.

Hepatitis B

It is estimated that over 300,000 Americans become infected with hepatitis B virus each year, resulting in about 5000 deaths from hepatitis B-related chronic liver disease and/or cancer of the liver. In countries such as China, a large percent of the population becomes infected during birth from infected mothers.

Figure 26.33 Immune electron microscopy of hepatitis A virus purified from a chimpanzee stool (×256,300). The coating of IgM antibody produces the "halo" around individual particles.

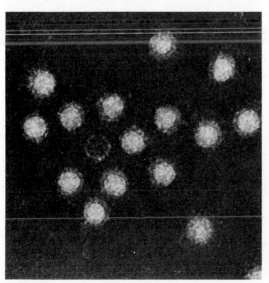

The largest known epidemic of hepatitis B occurred in 1942, before anyone even knew that viruses could cause hepatitis. This happened during World War II when hundreds of thousands of U.S. military personnel were immunized against yellow fever by injecting them with a vaccine containing a living attenuated strain of the yellow fever virus. Each lot of vaccine was stabilized by the addition of pooled human serum that had been obtained from 25 to 30 persons and that provided 25,000 to 86,000 immunizing doses. Unfortunately, some of these serum donors were hepatitis B carriers, and an estimated 427,000 military personnel received vaccine that was contaminated with hepatitis B virus. Of these, 51,000 vaccine recipients were hospitalized for acute hepatitis and, based on subsequent antibody studies, it is estimated that as many as 330,000 vaccinees probably had an undiagnosed or an asymptomatic infection with hepatitis B.

Because hepatitis B was not characterized until many years after this event, it is not now possible to assess the eventual mortality that may have occurred as a result of this catastrophe.

Hepatitis B virus has never been grown in cell culture, but the availability of viruslike particles in the blood of infected individuals has contributed to our knowledge of this virus. These particles, first discovered in the serum of an Australian aborigine, were originally termed Australia antigen or hepatitis-associated antigen. They do not, however, contain nucleic acid and are assumed to represent excess viral capsids; the current terminology, HB_sAg, specifies that they are the surface antigen of hepatitis B virus.

In 1970, larger particles (named Dane particles after their discoverer) were also found in the serum of hepatitis B patients. Treatment of the Dane particles with nonionic detergents causes them to dissociate into HB_sAg and an inner core. This inner core, called HB_cAg, contains double-stranded DNA and a DNA polymerase. The Dane particle represents the intact hepatitis B virion, composed of an inner double-stranded DNA-protein core (HB_cAg) and an outer coat (HB_sAg) (see Figure 26.34).

Pathogenicity Early studies failed to show a normal portal of exit for hepatitis B, and for years it was believed that a person could become infected only by the injection of blood or serum from an infected person or through the use of contaminated needles or syringes—hence, the older name of *serum hepatitis* for this disease. It has now been shown that this supposition is untrue. Using serological techniques, HB_s Ag has been found in feces, urine, saliva, vaginal secretions, semen, and breast milk. The fact that the disease is more prevalent among people living in crowded conditions—for example, institutionalized children—supports the oral route of transmission.

Figure 26.34 Fraction of the blood serum from a severe case of human hepatitis; the patient's immune system failed to counteract the infection. The larger spherical particles, or Dane particles, are 42 nm in diameter and are thought to be the complete hepatitis B virus. The smaller, more numerous spherical particles may be the cores (HBcAg) of the larger ones. Also seen are filaments of capsid protein (HBsAg).

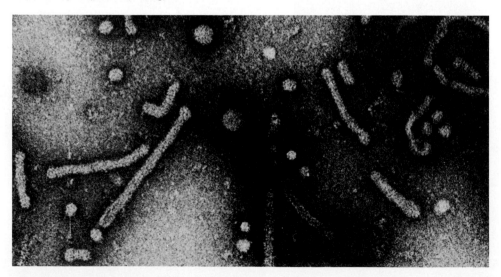

COLOR
PLATES
ON THE
IDENTIFICATION
OF THE
FAMILY
ENTEROBACTERIACEAE

Plate 1
Presumptive Identification of the Enterobacteriaceae

A bacterial isolate can be presumptively grouped in the family Enterobacteriaceae by assessing both colonial and biochemical characteristics. On blood agar the colonies appear gray, opaque, mucoid, or butyrous and may or may not produce hemolysis. *Proteus vulgaris* and *Proteus mirabilis* swarm on blood agar.

As a family, all members of the Enterobacteriaceae ferment glucose, reduce nitrates to nitrites, and are negative for cyrochrome oxidase. These characteristics are illustrated in this plate.

A

Blood agar plate of a 24-h culture showing gray, opaque, somewhat mucoid colonies characteristic of one of the members of the Enterobacteriaceae.

B

Higher-power view comparing the large, gray, mucoid colonies of a Klebsiella species with the smaller, convex, smooth, pale yellow colonies of *Staphylococcus aureus.*

C

Blood agar plate illustrating the swarming pattern of a motile species of Proteus. The growth from the primary colony may appear in sequential waves as shown here; or, the growth may cover the agar surface with a thin, almost invisible veil.

D–G

All members of the Enterobacteriaceae ferment glucose with the production of acid. The acid production is detected in the medium by the color conversion of a pH indicator.

D

Hektoen enteric (HE) agar with a 24-h growth of *Escherichia coli.* The yellow color of the medium indicates acid production through the fermentation of lactose or sucrose. Colonies not producing acid remain clear and do not form a yellow color in the surrounding medium.

E

Kligler Iron Agar (KIA) slant showing an acid (yellow) slant and acid deep. This reaction indicates fermentation of both lactose and glucose.

F

Purple broth medium showing fermentation of glucose in the right tube (yellow color from acid production) compared to the purple, uninoculated control tube on the left.

G

Purple broth medium showing glucose fermentation in the two tubes on the left and the production of a small amount of gas that has collected under the inverted Durham tubes.

H

Cytochrome oxidase paper test strips (Pathotec) revealing a positive purple reaction (top) compared to the colorless control (bottom). Any organism giving a positive cytochrome oxidase reaction is not a member of the Enterobacteriaceae.

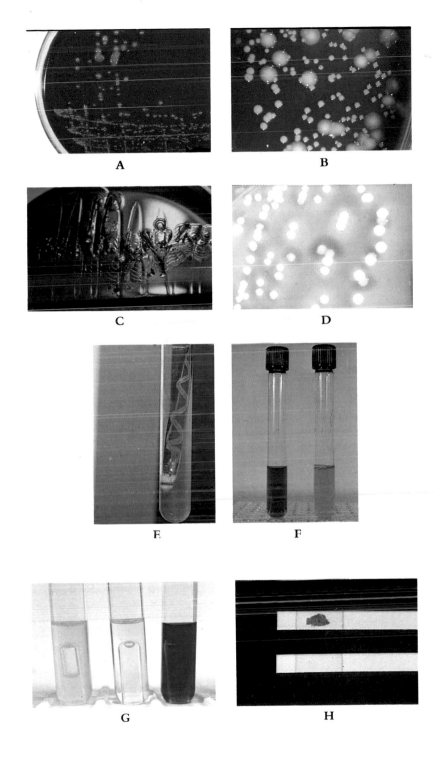

A

B

C

D

E

F

G

H

Neonatal transmission also appears to occur by an oral route during delivery, since gastric aspirates taken from the newborn immediately after birth from infected mothers, as well as vaginal secretions, almost always contain HB_sAg. The presence of HB_sAg in breast milk also provides an additional vehicle for the transmission of hepatitis B to the newborn. In addition, the demonstration of infectious virus in semen indicates that the virus can be sexually transmitted.

In the United States, current data indicate that approximately 9 percent of hepatitis B infections are attributable to homosexual men, 15 to 20 percent to IV drug users, 3 to 4 percent to health care workers, and 35 percent to heterosexual transmission. In the remaining 35 percent there is no known or readily identifiable source of infection. It is therefore noteworthy that of the estimated 300,000 new cases occurring in the United States each year (even though only about 25,000 are reported to the CDC), over 100,000 result from heterosexual contacts according to Dr. Harold S. Margolis, Chief, Hepatitis Branch, CDC. It is also noteworthy that each year over 4000 newborns in the United States acquire hepatitis B from an infected mother.

Clinical Management   Hepatitis B infections cannot be clinically distinguished from those caused by hepatitis A. There are, however, a number of characteristics that differentiate the infections caused by these viruses, as shown in Table 26.14. Hepatitis B infections are characterized by a very long incubation period, ranging from 50 to 180 days. The usual case is characterized by headache, mild fever, loss of appetite, and jaundice; however, the duration and severity of the disease may vary from a clinically inapparent to a fatal fulminating hepatitis.

The unusual aspect of this disease is that between 5 and 10 percent of clinically diagnosed patients become persistently infected and may continue to have HB_sAg in their blood for life; many of them will also have high concentrations of Dane particles, which may persist for life. Worldwide, it is estimated that there are over 200 million persistent hepatitis B carriers.

There is also a strong association between persistent hepatitis B infection and liver carcinoma. It has been noted, for example, that a large percentage of patients with liver carcinoma also have HB_sAg in their blood. Moreover, although this does not prove its oncogenic ability, it has been demonstrated with probes of labeled, cloned hepatitis B DNA that the cellular tissue of liver carcinomas contains an integrated genome of hepatitis B virus.

In view of the large percentage of hepatitis B carriers, the examination of all blood donors for the presence of HB_sAg is now routine. There are numerous techniques for measuring HB_sAg; the current ones include (1) red cell agglutination, in which red cells are coated with anti-HB_sAg prior to the addition of the serum to be tested; and (2) radioimmunoassay in which the HB_sAg in the patient's serum competes with known, radioactively labeled HB_sAg for reaction with the specific antibody. The radioimmunoassay is the more sensitive. A vaccine consisting of HB_sAg has been prepared from several sources. Originally HB_sAg was isolated and purified from the serum of persistent hepatitis B carriers. Now, using recombinant DNA technology, the DNA from Dane particles has been cloned in both *E. coli* and in yeast. Both are able to synthesize the HB_sAg and the yeast product has been licensed as a vaccine for hepatitis B. This is given to high-risk individuals such as blood bank workers, doctors, dentists, nurses, and people who might be expected to

TABLE 26.14   Differential Characteristics of Hepatitis A and Hepatitis B

Characteristic	Hepatitis A	Hepatitis B
Length of incubation period	15–40 days	50–180 days
Source of infection	Mostly fecal-oral	Transfer of blood products, sexual, and perinatal
Host range	Humans and some nonhuman primates	Humans and some nonhuman primates
Seasonal occurrence	Higher in fall and winter	Year round
Age incidence	Much higher in children	All ages
Occurrence of jaundice	Much higher in adults	Higher in adults
Virus in blood	2–3 weeks before illness to 1–2 weeks after recovery	Several weeks before illness to months or years after recovery
Virus in feces	2–3 weeks before illness to 1–2 weeks after recovery	Present in very small amounts
Probable size of virus	27 nm	42 nm
Diagnosis based on	Liver function tests, clinical symptoms, IgM against virus	Liver function tests, clinical symptoms, history, and presence of HB_sAg in blood
Value of gamma globulin for prophylaxis	Good	Ineffective

require multiple blood transfusions. Pooled gamma globulin is of little or no value for the prevention or treatment of hepatitis B, however, immune serum globulin (ISG) known to contain antibodies to hepatitis B virus will prevent or at least lessen the severity of the disease if administered early in the incubation period.

Non-A, Non-B Hepatitis

It has become apparent during recent years that the preponderance of hepatitis occurring after blood transfusions is caused by infectious agents not related to hepatitis A or B virus. Such disease is currently referred to as non-A, non-B hepatitis (NANB hepatitis) or hepatitis C. The CDC estimates that hepatitis C accounts for 90 percent of cases of posttransfusion hepatitis in the United States. However, these cases of posttransfusion hepatitis only account for about 15,000 of the 150,000 NANB cases occurring annually in the United States. A large number also occur in IV drug users and it is probable that there are other routes of transmission that are associated with sexual activity and/or occupational exposure.

Moreover, epidemiological data has shown that there is a second NANB virus transmitted via a fecal-oral route. This enterically transmitted non-A, non-B hepatitis (now called hepatitis E) has not been a problem in the United States, but large epidemics have been reported in India, Burma, Africa, and Mexico. Clinical disease from hepatitis E has a high mortality rate in pregnant women, perhaps because of the depressed cellular immunity that is associated with pregnancy.

Between 50 and 75 percent of posttransfusion hepatitis C infections are followed by a chronic, persistent infection analogous to, but developing slower than that described for hepatitis B infections. Such infections account for a substantial proportion of the chronic liver disease in the United States.

None of the viruses causing NANB hepatitis has been grown in cell culture, but a part of the genome of the virus involved in posttransfusion hepatitis C has been cloned, and an assay has been developed to detect antibodies to this epitope. Preliminary testing of a small number of individuals known to have had posttransfusion hepatitis revealed that all had antibodies to the cloned hepatitis C virus fragment although, in some cases, it was 6 to 12 months after the onset of clinical hepatitis before the antibodies could be detected. Other surveys have shown that about 80 percent of chronic posttransfusion NANB hepatitis patients in Italy and Japan had antibodies to hepatitis C virus, indicating that it is a major contributor to NANB hepatitis throughout the world. Undoubtedly, the use of this test to screen blood donors will help to curtail the spread of posttransfusion hepatitis.

The genome of hepatitis E virus has also been sequenced and shown to contain a positive, single-stranded molecule of RNA. It does not, however, appear to be related to the picornaviruses or hepatitis C.

Delta Virus Hepatitis

The delta (δ) virus (HDV) was first described in 1977 as a new antigen detected by immunofluorescence in liver cells of a patient with a chronic hepatitis B virus infection. It is now known that HDV is yet another virus that infects the cells of the liver, causing the inflammation known as hepatitis. This, however, is a defective virus, which does not encode for the synthesis of its own capsid (coat protein) and, as a result, is completely unable to replicate unless a cell is infected simultaneously with both HDV and hepatitis B virus. In such cases, it uses the hepatitis B capsid to enclose the virus RNA. Interestingly, HDV RNA exists as a single-stranded circular molecule that is similar in structure to certain pathogenic RNAs of plants known as viroids.

As you can see, therefore, HDV hepatitis always occurs together with hepatitis B virus infections and, as you may suspect, HDV and hepatitis B virus infections are usually more severe than infections caused by hepatitis B virus alone. In fact, a serious relapse in a chronic carrier of hepatitis B virus is frequently due to a superinfection with the HDV, and persons infected with HDV frequently become chronic, persistent carriers of both viruses.

Viruses causing systemic illnesses, which gain entry to the body via the gastrointestinal tract, are summarized in Table 26.15.

AGENTS ASSOCIATED WITH ORAL INFECTIONS

In this section it becomes more difficult to dissociate infections that are restricted to the oral cavity from closely related agents that cause systemic diseases. Thus, with the exception of genital herpesvirus and varicella-zoster infections, we shall include all other herpesvirus infections here, even though it is obvious that some are involved in a systemic reaction.

Bacteria

Treponema vincentii and *Bacteroides melaninogenicus:* Vincent's Angina

T. vincentii and *B. melaninogenicus* have been thought to be involved in a fusospirochetal disease (caused by

TABLE 26.15   Viruses Originating in the Gastrointestinal Tract that Cause Systemic Illnesses

Virus	Disease	Vaccine
Polioviruses	Polio	Salk—killed virus Sabin—attenuated virus
Coxsackieviruses	Respiratory, meningitis	None
Echoviruses	Respiratory, meningitis, gastroenteritis	None
Hepatitis A	Hepatitis	Killed virus
Hepatitis B	Hepatitis	HB$_s$Ag; recombinant, yeast vaccine
Non-A, non-B (Hepatitis C)	Hepatitis	None
Non-A, non-B (Hepatitis E)	Hepatitis	None
Delta virus	Hepatitis	None

both fusiform bacteria and spirochetes). This is commonly referred to as Vincent's angina, Vincent's infection, or trench mouth (it was prevalent among the infantry during World War I). *T. vincentii* is an active, motile spirochete, and *B. melaninogenicus* is a nonmotile, gram-negative, straight or slightly curved rod that frequently occurs in pairs with the outer ends pointed and the blunt ends together (fusiform). Both organisms are obligate anaerobes.

Vincent's angina is characterized by ulcerative lesions of the mouth or tonsillar area. There is a possibility that the real etiological agent is a herpesvirus and, for reasons unknown, the two bacteria, which are normal flora of the mouth, multiply during the active disease.

Diagnosis is based on the clinical picture and smears of the lesions. Penicillin will help to control the bacterial population, but it is not particularly effective in eliminating the ulcerative lesions, perhaps because the bacteria are not the true etiological agent.

Actinomyces: Actinomycosis

Both *A. bovis*, the causative agent of lumpy jaw in cattle, and *A. israelii*, the etiological agent of actinomycosis in humans, cause similar types of infections.

A. israelii occurs as part of the normal flora in the crypts of the tonsils, in dental caries, and occasionally in the intestinal tract or the lungs. Infections caused by this organism originate from an endogenous source; the initial lesions, which usually contain a mixture of actinomycetes and other endogenous bacteria, occur in cervicofacial, abdominal, or lung tissue. The lesions frequently progress to draining sinuses that tend to become chronic.

It is not known why the *Actinomyces* cause disease only in certain people; the infection apparently does not spread from person to person. It was believed at one time that there was some relationship between this disease in humans and the infection as seen in cattle, but studies show that is no more frequent in farmers than in nonfarmers.

Prolonged treatment with sulfonamides and certain antibiotics—including penicillin, tetracycline, and chloramphenicol—has been recommended.

Yeast

Candida albicans: Thrush

C. albicans usually appears as oval, yeastlike cells that reproduce by budding. However, in infected areas, filamentous hyphae plus pseudohyphae (which consist of elongated yeast cells that remain attached to each other) may also be seen. The yeast is easily grown at 25 to 37°C on Sabouraud's glucose agar, and, if grown on corn meal agar at 25°C, *C. albicans* can produce many characteristic thick-walled chlamydoconidia (Figure 26.35).

Under normal conditions, *C. albicans* occurs in small numbers in the alimentary tract, mouth, and vaginal area, and disease results only when a major change in normal flora or a disturbance of the immune response occurs. Such infections also occur in individuals who have physiological defects, such as endocrine disorders.

Oral Candidiasis   Oral candidiasis (also called thrush) occurs most frequently in the newborn and is probably acquired during passage through an infected or colonized vagina. The yeast appears as a creamy, gray membrane covering the tongue and seems able to produce disease only because of the absence of other resident flora. If thrush has not occurred by the third day of life, it is unlikely that it will appear. Treatment with oral nystatin is effective.

Oral thrush in older children or adults may occur as a result of endocrine disturbances or avitaminosis

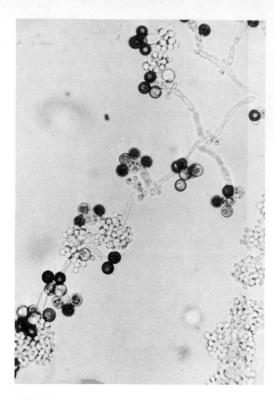

Figure 26.35 *C. albicans,* showing the pseudohyphae, clusters of budding cells, and chlamydoconidia.

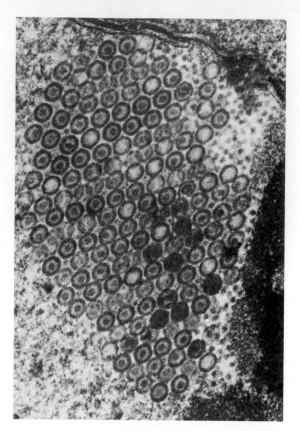

Figure 26.36 Electron micrograph of an intracellular inclusion body of herpesvirus; note the presence of both "empty" and DNA-containing particles (×32,400).

(particularly a deficiency of riboflavin), as a complication of diabetes, as a result of poor oral hygiene, or following the administration of corticosteroids or antibiotics. It is also reported to occur as an early manifestation of the acquired immunodeficiency syndrome (AIDS).

Herpesviruses

Herpesvirus infections occur in many forms, and the portal of entry can vary. With the exception of the varicella-zoster herpesvirus discussed in Chapter 25, most herpesviruses appear to be acquired by close personal contact, and all appear able to enter the body through the mucous membranes of the mouth.

Herpesviruses contain double-stranded DNA and are replicated in the nucleus of infected cells (see Figure 26.36). Probably their most unusual characteristic is their ability to exist as latent infections in people who have recovered from the overt disease. In Chapter 25, we discussed one such infection in which the herpesvirus causing chickenpox can be reactivated after many years to infect a sensory nerve, resulting in a disease called zoster or shingles. In the following sections, we shall discuss some other herpesvirus infections.

Herpes Simplex Type 1

The most frequent manifestation of a herpes simplex type 1 infection is the common fever blister (see Figure 26.37). The area most often infected is the mucous membranes of the mouth; however, herpes simplex type 1 can also infect the mucous membranes of the genitalia, the eye, the skin, or the central nervous system, causing a severe encephalitis.

Pathogenicity The initial infection usually occurs in children during their first 2 years of life and in the large majority of cases goes undiagnosed. In approximately 10 percent of initial infections of the mouth, however, the individual may develop quite severe multiple vesicles on the mucous membranes. Occasionally, a baby with eczema will contract herpes in passing through the genital tract during birth; the resulting infection is overwhelming and frequently fatal.

In people who have recovered from a primary infection, the virus remains in a latent state for life. Subsequently an individual may be subjected to re-

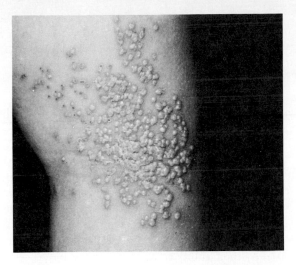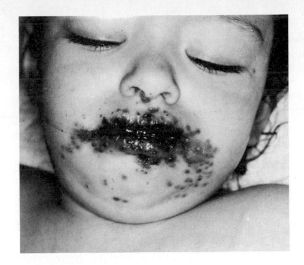

Figure 26.37 Herpes fever blisters on leg, lips, and face.

current attacks of fever blisters in the same area as the initial infection. We do not know specifically what stimulates the latent virus to multiply and cause disease, but we do know that many external factors such as heat, cold, sunlight, menstruation, or even emotional stress sometimes result in recurrent attacks of fever blisters.

Clinical Management Control is difficult because at least 80 percent of adults have antibodies to the virus. In fact, even individuals with frequent recurrent herpes attacks may have high antibody titers against the virus. The virus seems to be spread by close personal contact, particularly by people with inapparent infections or with minor recurrent fever blisters.

Herpes infection of the eye is one of the few virus infections that respond to chemotherapy. The effective drugs are analogues of DNA, so they inhibit DNA synthesis in the herpesvirus.

Cytomegaloviruses

Cytomegaloviruses (CMV) are typical herpesviruses, both in their structure and in the manner in which they replicate. Infected cells become distinctly enlarged (i.e., cytomegaly). The unusual appearance of salivary gland cells of children who had died from congenital infections prompted the name *salivary gland virus disease,* although the more common name is *cytomegalic inclusion disease.*

Pathogenicity Cytomegaloviruses are exceedingly widespread; surveys indicate that the majority of adults possess antibodies to them. They are prob-

ably acquired orally, most often through close personal contact, but spread may also occur through sexual contact. Adult infections appear to be asymptomatic but, as in other herpesvirus infections, the virus remains in a latent state following the primary infection, and recurrent exacerbations of viral multiplication may occur. The importance of these viruses lies primarily in their ability to infect the fetus before birth. Such infection may cause the death of the fetus, give rise to a wide variety of congenital defects, or produce no immediate obvious effects in the newborn.

The congenital defects of CMV are extremely varied, but it is postulated that CMV is the major viral cause of mental retardation, surpassing even rubella virus in this regard. Other congenital abnormalities caused by CMV include microcephaly (abnormally small head), central nervous system damage resulting in seizures and deafness, psychomotor retardation, ocular abnormalities, chronic gastroenteritis, jaundice, and pneumonia. One study revealed that the average IQ score of infants born with a CMV infection was significantly lower than that of a control group, even though these infants appeared normal in other respects.

Clinical Management Urine or saliva can be cultured on human fibroblasts, which are then observed for the cytopathic characteristics of swollen, rounded cells possessing large intranuclear inclusion bodies.

The ubiquity of these agents is attested by the observation that approximately 1 percent of babies born in the United States are infected with cytomegalovirus. Data are also available showing that fetal infection can occur both during a primary infection of the mother or during an exacerbation of a latent infection. Such data indicate, however, that infection of the fetus is much more likely to occur during a primary infection than in a recurrent bout with this virus. This would suggest that a vaccine might be of value, but inasmuch as the nonattenuated virus usually produces no symptoms in adults, one cannot be certain that a vaccine is avirulent.

Cytomegalovirus is also the most common infectious cause of death following bone marrow transplantation. This is frequently manifested as a cytomegalovirus pneumonia associated with a mortality of about 90 percent. This virus is also a major cause of morbidity and mortality in patients with AIDS as well as patients who are immunosuppressed after organ transplantation. A number of different antiviral drugs have been used to treat such infections but, as yet, none has been more than marginally effective.

Epstein-Barr Virus

Epstein-Barr virus (EB virus), named for its codiscoverers M. A. Epstein and Y. M. Barr, is a herpesvirus that has been isolated from children with Burkitt's lymphoma (BL), a malignancy originating in the cells of lymph nodes. In this disease, the jaw is the most common site of the tumor, although tumors may occur in many organs. BL is seen most frequently in Africa.

The specific role of EB virus as the etiological agent of BL remains elusive. In cell cultures, for example, EB virus can infect only human lymphoid cells and, in doing so, transforms some of these cells into permanent EB-virus-carrying cells that have all the characteristics of a malignant cell. Several inconsistencies exist, such as why this widespread virus produces so few actual malignancies and why (as described in the following section) EB virus routinely causes a communicable disease (infectious mononucleosis) from which recovery is complete.

Current data suggest that malignancy may occur when an individual is simultaneously infected with EB virus and *Plasmodium falciparum*, an etiological agent of malaria. It is believed that the malaria infection suppresses T-cell-mediated immunity, permitting the EB-virus-infected B cell to undergo rapid replication. During this replication, a specific chromosomal translocation occurs that results in the activation of the c-*myc* protooncogene (see Chapter 14). In addition, it has also been reported that a cloned gene of EB virus causes tumors when introduced into susceptible cells. It appears possible, therefore, that tumor production by EB virus is the end result of a chain of events.

EB virus has also been associated with nasopharyngeal carcinoma, a malignancy that occurs more frequently in China than in the Western Hemisphere.

The reason for this geographical distribution and the role of the virus in initiating this malignancy are unknown.

Infectious Mononucleosis

Infectious mononucleosis (IM) is primarily a disease of young adults in which the usual symptoms are high fever, headache, chills and sweats, fatigue, and a severe sore throat. The duration of the illness may vary from several days to several weeks.

Clinically, the disease is diagnosed by the above symptoms and by the presence of abnormal, large lymphocytes in the blood. In addition, most patients develop high titers of heterophile antibodies, characterized by their ability to agglutinate red blood cells from sheep.

Interestingly, lymphocytes from patients with IM can be grown in continuous cell culture (normal lymphocytes cannot), and EB virus can be isolated from the cultured lymphocytes. After recovery, however, it is no longer possible to cultivate a patient's lymphocytes as a continuous cell line.

It seems, therefore, that IM is an excellent example of a self-limiting malignancy, and that the disappearance of these transformed lymphocytes following an individual's recovery from IM may result from the host's immune responses to new membrane antigens occurring in the transformed lymphocytes.

IM has been referred to as "kissing disease" because of the belief that the disease is transmitted through close oral contact. The observation that throat washings from individuals with IM could transform EB-negative lymphocytes into EB-positive lymphocytes supports this assertion. Table 26.16 summarizes the herpesvirus infections originating in the mouth or gastrointestinal tract.

In 1985 a new syndrome termed **chronic Epstein-Barr virus disease** was described. This infection is seen most often in females who are 25 to 40 years of age, many of whom can trace their symptoms to the onset of acute infectious mononucleosis. Chronic EB virus disease is characterized by chronic fatigue, a low-grade fever, recurrent sore throat, swollen glands, headaches, difficulty in concentrating, and a highly variable antibody titer to EB virus. Many also have mild defects in their cellular immune system. There is no current effective treatment, but the use of intravenous gammaglobulin has shown promising results.

New Human Herpesviruses

A new human herpesvirus (termed human lymphotropic herpesvirus) has been identified from mitogen-stimulated peripheral blood mononuclear cells. Initial studies indicated that this virus replicated efficiently

TABLE 26.16 Herpesvirus Infections Originating in the Mouth or Gastrointestinal Tract

Virus	Disease	Diagnosis
Herpes simplex	Fever blisters	Clinical symptoms, isolation of virus
Cytomegalovirus	Birth defects	Isolation of virus
EB virus	Infectious mononucleosis, Burkitt's lymphoma	Heterophile antibody, clinical symptoms, isolation of virus

in cord blood B lymphocytes. However, recent studies have shown that the virus will also infect T cells and other human cell lines; therefore, it has been renamed human herpesvirus 6 (HHV 6). Virus can be recovered from patients with lymphoproliferative disorders and antibodies to HHV 6 are found in the serum of between 10 and 40 percent of normal American adults.

HHV 6 differs from EB virus in that an HHV 6 infection of B lymphocytes leads to cell lysis and not immortalization. Recent evidence indicates that HHV 6 is typically acquired in early infancy and causes exanthem subitum (Roseola infantum) in a proportion of infected children.

SUMMARY

Bacteria may cause gastrointestinal disease by any of three general methods: (1) elaboration of a toxin while growing in food; (2) the secretion of toxins while adhering to the intestinal wall; or (3) the release of toxins after the intracellular invasion of the intestinal epithelial cells. Those causing a food intoxication resulting from the ingestion of preformed toxins include the following organisms: *S. aureus, B. cereus,* and *C. botulinum.* Noninvasive food infections may be caused by *C. perfringens* but are most often the result of infections by members of the Enterobacteriaceae or Vibrionaceae.

The family Enterobacteriaceae contains organisms that may exist as normal flora, as saprophytes living on decaying plant matter, or as overt pathogens. Under appropriate conditions all may cause disease and hence be considered opportunists. *E. coli,* a normal inhabitant of the large intestine, may cause a moderate to severe diarrhea by adhering to the intestinal wall and producing one or two enterotoxins called ST and LT. LT stimulates adenylate cyclase, producing cAMP, while ST causes an increase in cellular cGMP. These strains of *E. coli* are termed ETEC. *K. pneumoniae* may cause a severe pneumonia, whereas members of the genera *Serratia, Edwardsiella, Citrobacter, Proteus,* and *Providencia* are all opportunists, causing infections in the newborn, the debilitated, or persons receiving immunosuppressive drugs. Other strains of *E. coli* (EHEC) also adhere to the intestinal epithelium and produce a Shigalike toxin that inhibits protein synthesis and, as a result, kills the cell.

V. cholerae, the cause of cholera, adheres to the intestinal epithelium and produces disease by the secretion of an enterotoxin that, like *E. coli* LT, stimulates the enzyme adenylate cyclase, increasing cAMP concentrations. It is the presence of excess cAMP in the gut that

causes the tremendous loss of fluids during the infection. There are a number of halophilic vibrios that also produce gastrointestinal infections. Most common among these are *V. parahaemolyticus, V. fluvialis,* and *V. vulnificus.* Members of the genus *Campylobacter* are inhabitants of the intestinal tract of wild and domestic animals and rank along with *E. coli* as a major cause of diarrheal disease. *H. pylori* causes a chronic gastritis by infection of the gastric mucosa.

Organisms producing a locally invasive food infection include members of the *Shigella,* the enteroinvasive *E. coli* (EIEC), and the *Salmonella.* All possess one or more plasmids that encode for their invasive abilities, and all produce toxins that kill local cells.

Systemic bacterial infections originating in the intestine include the enteric fevers, which are best characterized by typhoid fever. Other systemic infections are caused by members of the genus *Yersinia.* Listeriosis, an infection by *L. monocytogenes,* is most often manifested as a meningitis. It is acquired by adults by the ingestion of contaminated food but it may be acquired by the newborn from an infected mother. Brucellosis is spread via contaminated dairy products or contact with infected animals.

Bacteroides and *Fusobacterium* comprise the major genera in the obligately anaerobic enteric normal flora. These organisms possess little invasive ability, but they may cause widespread tissue destruction following trauma or surgical procedures. Protozoa entering the body by way of the gastrointestinal tract include *E. histolytica, B. coli, G. lamblia, T. gondii,* and *Cryptosporidium.*

The habitat of the adult cestode is the intestinal tract, and the animal in which the larval stage develops into an adult is termed the definitive host. Animals in which eggs develop into the larval stage are called intermediate hosts. Adult forms of cestodes are commonly called tapeworms. Beef, pork, and fish tapeworms are acquired by eating raw or undercooked infected meat, whereas dwarf tapeworm and hydatid disease are initiated by the direct hand-to-mouth transfer of fertilized eggs. Trematodes, also called flukes, must have a specific species of snail as an intermediate host. Intestinal, liver, and lung flukes are acquired by the ingestion of infected fish or contaminated vegetation, such as water chestnuts. Intestinal nematodes, or roundworms, are usually acquired by the ingestion of contaminated food or water.

Viral gastroenteritis may be caused by either the Norwalk-like agents or a rotavirus.

The enteric viruses include poliovirus, coxsackievirus, and echoviruses. All are spread via a fecal-oral route. Hepatitis A is also spread via a fecal-oral route. Hepatitis B is commonly spread via sexual intercourse or perinatally, and because so many individuals remain persistent carriers of this virus, it can readily be transmitted via transfused blood. This is largely prevented by the routine assay for the presence of hepatitis B surface antigen in blood. A third type of hepatitis commonly called non-A, non-B hepatitis, also results in persistent carriers; this virus appears to be responsible for about 90 percent of all transfusion-associated hepatitis. There are at least two different viruses that cause non-A, non-B hepatitis that have been designated as hepatitis C and E. Delta virus is a defective virus that can replicate only if a cell is simultaneously infected with hepatitis B virus. It also causes a severe hepatitis and is found in persistent chronic carriers of hepatitis B virus.

Microorganisms causing oral infections include a combination of *T. vincentii* and *B. melaninogenicus,* as possible agents of trench mouth, and *A. israelii,* the etiological agent of human actinomycosis. Herpes simplex type 1 is a cause of fever blisters on the oral mucous membranes. Other herpesviruses acquired via the oral route include cytomegalovirus, a major cause of birth defects, and EB virus, the etiological agent of infectious mononucleosis and a possible cause of a human malignancy called Burkitt's lymphoma. Human herpesvirus is also widespread and appears to be the cause of Roseola infantum. The protozoan, *N. fowleri,* also uses the oral or nasal mucosa as a portal of entry.

QUESTIONS FOR REVIEW

1. Describe three general ways in which bacteria can cause a gastroenteritis resulting in diarrhea and vomiting.
2. What two organisms are most frequently involved in food intoxication?
3. A diagnosis of food intoxication has been made following the ingestion of home-canned beans. (a) What is the most probable causative agent? (b) What are the characteristics of the organism that enabled it to survive the canning process and multiply in the sealed container? (c) What could have been done to prevent the disease before the ingestion of the food?

4. What are some sources from which infants contact botulism?
5. Discuss the major properties that are used to divide the Enterobacteriaceae into genera.
6. How does an ONPG-positive organism differ from a fast lactose fermenter?
7. What are O antigens? H antigens?
8. What special properties are possessed by enterotoxigenic *E. coli?*
9. How does the *E. coli* LT toxin cause diarrhea?
10. Discuss the Kauffmann-White scheme for the classification of the *Salmonella.* How are *Salmonella* given species names?
11. What three types of infections are caused by *Salmonella?*
12. What is the primary reservoir for *Salmonella?*
13. What disease is caused by the shigellae? What species of *Shigella* would most likely be found in the United States?
14. How does *V. cholerae* cause diarrhea?
15. How and from what source do humans usually become infected with *E. histolytica?*
16. What extraintestinal organs may be infected with *E. histolytica?*
17. How would you diagnose amebic dysentery?
18. What is the causative agent for primary amebic meningoencephalitis, and how do humans usually acquire this parasite?
19. Do any ciliates cause human disease? If so, what disease and what organism?
20. Name diseases caused by intestinal flagellates and describe how humans become infected.
21. Define *definitive host. Intermediate host.*
22. What precautions can one take to avoid infection by pork, beef, or fish tapeworms?
23. List three types of flukes. What is the intermediate host for each type?
24. List the larval form for each fluke that is infective for humans.
25. What is the best control measure for the prevention of trichinosis?
26. List the agents that cause viral gastroenteritis.
27. What is brucellosis? How may the infective agent be transmitted? Of what diagnostic use is the brucellergen test?
28. What are the two major genera that comprise our obligately anaerobic enteric normal flora? What type of infection do they cause?
29. Describe the etiological agent of poliomyelitis. How is the disease spread?
30. What control measures are recommended for poliomyelitis?
31. What are coxsackieviruses, and what types of illness do they cause in humans?
32. How did echoviruses acquire their name, and what types of illness do they cause in humans?
33. What do you know about the etiological agent of hepatitis A? How is it spread?
34. How is hepatitis B spread? What is used for a hepatitis B vaccine and where is it obtained?
35. How is non-A, non-B hepatitis usually acquired?
36. What techniques are being used to control the spread of hepatitis B?

37. What is delta virus? Why is it always associated with hepatitis B infection?

38. What is Vincent's angina, and what are the possible causes of it?

39. What is the cause of human actinomycosis, and what is the source of the infectious agent?

40. Describe the conditions leading to thrush in a newborn. In an adult.

41. What is meant by latency in herpesvirus infections? Name two herpesvirus infections in which latency is an established fact.

42. Under what conditions are cytomegalovirus infections extremely serious? Why?

43. What diseases appear to be caused by EB virus?

44. Why is infectious mononucleosis referred to as a "self-limiting" malignancy?

45. What infection is caused by rotavirus? What are possible methods for the control of this virus?

SUPPLEMENTARY READING

Arnon SS: Infant botulism: Anticipating the second decade. *J Infect Dis* 154:201, 1986.

Binder HJ: The pathophysiology of diarrhea. *Hosp Pract* 19:107, 1984.

Blaser MJ: *Helicobacter pylori* and the pathogenesis of gastroduodenal inflammation. *J Infect Dis* 161:626, 1990.

Bodey GP, Fainstein V (eds): *Candidiasis*. New York, Raven, 1985.

Brown JE, et al: Shiga toxin from *Shigella dysenteriae* 1 inhibits protein synthesis in reticulocyte lysates by inactivation of aminoacyl-tRNA binding. *Microb Pathog* 1:325, 1986.

Buck GE, et al: Relationship of *Campylobacter pyloris* to gastric and peptic ulcer. *J Infect Dis* 153:664, 1986.

Centers for Disease Control: Chronic fatigue possibly related to Epstein-Barr virus. *MMWR* 35:350, 1986.

Centers for Disease Control: Hepatitis B associated with a jet gun injection. *MMWR* 35:373, 1986.

Centers for Disease Control: Common-source outbreak of giardiasis—New Mexico. *MMWR* 38:405, 1989.

Centers for Disease Control: Community outbreaks of shigellosis—United States. *MMWR* 39:509, 1990.

Centers for Disease Control: *Salmonella enteritidis* infections and shell eggs—United States 1990. *MMWR* 39:909, 1990.

Chia JK, et al: Botulism in a adult associated with foodborne intestinal infection with *Clostridium botulinum*. *N Engl J Med* 315:239, 1986.

Cleary TG, et al: Shiga-like cytotoxin production by enteropathogenic *Escherichia coli* serogroups. *Infect Immun* 47:335, 1985.

Drasar BS, Barrow PA: *Intestinal Microbiology*, Aspects of Microbiology Series, Washington, DC, American Society for Microbiology, 1985.

Dreyfus LA, et al: Characterization of the mechanism of action of *Escherichia coli* heat-stable enterotoxin. *Infect Immun* 44:493, 1984.

Evered D, Whelan J (eds): *Microbiol Toxins and Diarrhoeal Disease*. London, Pitman (Ciba Foundation Symposium 112), 1985.

Flehmig B, et al: Simultaneous vaccination for hepatitis A and B. *J Infect Dis* 161:865, 1990.

Flemin DW, et al: Pasteurized milk as a vehicle of infection in an outbreak of listeriosis. *N Engl J Med* 312:404, 1985.

Gitler C, Merelman D: Factors contributing to the pathogenic behavior of *Entamoeba histolytica*. *Annu Rev Microbiol* 40:237, 1986.

Gotuzzo E, et al: An evaluation of diagnostic methods for brucellosis: The value of bone marrow culture. *J Infect Dis* 153:122, 1986.

Hatheway CL: Toxigenic clostridia. *Clin Microbiol Rev* 3:66, 1990.

Holley HP Jr, Dover C: *Cryptosporidium*: A common cause of parasitic diarrhea in otherwise healthy individuals. *J Infect Dis* 153:365, 1986.

Kapikian AZ, et al: Rotavirus: The major etiologic agent of severe infantile diarrhea may be controllable by a "Jennerian" approach to vaccination. *J Infect Dis* 153:815, 1986.

Knutton S, et al: Adhesion of enterotoxigenic *Escherichia coli* to human small intestinal enterocytes. *Infect Immun* 48:824, 1985.

Korlath JA, et al: A point-source outbreak of campylobacteriosis associated with consumption of raw milk. *J Infect Dis* 152:592, 1985

Lupski JR, Feigin RD: Molecular evolution of pathogenic *Escherichia coli*. *J Infect Dis* 157:1120, 1988.

Marques LRM, et al: Production of Shiga-like toxin by *Escherichia coli*. *J Infect Dis* 154:388, 1986.

Martinez-Palomo A (ed): *Amebiasis*, New York, Elsevier, 1986.

Matson DO, Estes MK: Impact of rotavirus infection at a large pediatric hospital. *J Infect Dis* 162:598, 1990.

Morse DL, et al: Widespread outbreaks of clam- and oyster-associated gastroenteritis: role of Norwalk virus. *N Engl J Med* 314:678, 1986.

Schlech WF III, et al: Epidemic listeriosis: Evidence for transmission by food. *N Engl J Med* 308:203, 1983.

Schuchat A, et al: Epidemiology of human listeriosis. *Clin Microbiol Rev* 4:169, 1991.

Snydman DS: Hepatitis in pregnancy. *N Engl J Med* 313:1398, 1985.

Taylor JP: Typhoid fever in San Antonio Texas: An outbreak traced to a continuing source. *J Infect Dis* 149:553, 1984.

Telzak EE, et al: A nosocomial outbreak of *Salmonella enteritidis* infection due to the consumption of raw eggs. *N Engl J Med* 323:394, 1990.

Thompson NE, et al: Isolation and some properties of an enterotoxin produced by *Bacillus cereus*. *Infect Immun* 43:887, 1984.

Wang K-S, et al: Structure sequence and expression of the hepatitis delta viral genome. *Nature* 323:508, 1986.

Yokosuka O, et al: Hepatitis B virus RNA transcripts and DNA in chronic liver disease. *N Engl J Med* 315:1187, 1986.

Chapter 27

Pathogens that Enter the Body via the Genitourinary Tract

In this chapter we shall be concerned with infections of the urinary tract as well as those of the reproductive system. We shall see that urinary tract infections are almost always caused by opportunistic, normal flora organisms, whereas infections of the reproductive system are usually the result of sexually transmitted diseases.

URINARY TRACT INFECTIONS

The urinary tract consists of two kidneys, each with a ureter carrying the urine to a single bladder. Urine is voided from the bladder through a tubular structure called the urethra. With the exception of the external portion of the urethra, this entire system is normally sterile. Infections of the urinary tract that are confined to the bladder are referred to as **cystitis,** whereas those occurring in the kidneys are termed **pyelonephritis.**

Cystitis

Most instances of cystitis originate from organisms within the urethra that have ascended into the bladder. Considering that the female urethra is only about 2.5 cm long whereas the male urethra extends to the end of the penis, it is understandable that cystitis is much more common in the female than in the male. It is also worth noting that catheterization (inserting a tube through the urethra into the bladder) probably always mechanically pushes a few organisms from the external urethra into the bladder. Fortunately, this does not always cause an infection, but many cases of cystitis are the result of this procedure.

It is not surprising, therefore, to find that most cases of cystitis are caused by opportunistic, normal flora organisms. *Escherichia coli* can unequivocally be awarded the number one position as the etiological agent of cystitis. Second place as a cause of cystitis must be shared by such organisms as *Klebsiella pneumoniae, Proteus mirabilis, Pseudomonas aeruginosa,* and *Enterococcus* sp. Others that are seen less frequently include *Serratia marcescens, Staphylococcus epidermidis,* and *S. saprophyticus.* In addition, there are a few systemic infections in which the urinary tract becomes infected via a hematogenous route. Such infections are secondary to the major disease and they will not be discussed here.

Pyelonephritis

An infection of one or both kidneys can be considerably more serious than one confined to the bladder. Such infections, termed pyelonephritis, may occur as a result of an ascending infection from the bladder via the ureter. In such cases we would find essentially the same organisms involved as were listed as etiological agents of cystitis. Infection by bloodborne organisms also occurs, but such cases rarely involve the gram-negative enterics. These infections are more likely to be caused by *S. aureus,* the group A streptococci, or *Mycobacterium tuberculosis.*

Laboratory Diagnosis Urethral catheterization may yield samples with minimum contamination, but there is always a danger of introducing organisms from the urethra into the bladder. As a result, catheterization is not performed routinely for the collection of urine samples. Instead, voided midstream samples are obtained after careful cleansing of the external genitalia.

One must remember, however, that all voided urine samples will contain some bacteria; therefore, the diagnosis of cystitis is based on a quantitative assay for the number of bacteria present. Considerable experimental data have led to the formulation of the following rules: (1) 10^5 bacteria or more per milliliter from a clean, voided specimen indicate a urinary tract infection; (2) when 10^3 to 10^4 bacteria per milliliter are found, the test should be repeated; and (3) 10^3 or less bacteria per milliliter are not considered significant in a voided sample.

Because of the necessity for quantitating the bacteria present in a sample of urine, a number of different techniques have been devised to accomplish enumeration as rapidly as possible. An old and still used procedure is to add 0.1 mL of urine to 9.9 mL of sterile water and then to prepare a pour plate by mixing 10 to 20 mL of melted nutrient agar with 0.1 mL of the urine dilution. Because this represents an overall dilution of 10^{-3}, each colony appearing on the pour plate after incubation would represent 1000 organisms in the original urine sample; thus 100 colonies would represent 10^5 organisms per milliliter, indicating an active cystitis.

A standard platinum dilution loop (commercially available) will hold approximately 0.001 mL of liquid. Such a loop can be used to streak a urine specimen directly on a nutrient agar plate. If the loop is calibrated monthly by comparing with counts obtained by the pour-plate method, the overall accuracy of the calibrated-loop technique is equivalent to that of a pour plate.

Clinical Management There is no one specific treatment for all urinary tract infections. The observation that most cases of cystitis result from infection by normal flora—gram-negative enterics—limits the routine choice of antimicrobial agents. Sulfonamides, with or without trimethoprim, are probably among the most common treatments. Other effective chemo-therapeutic agents include oral ampicillin, cephalexin, and nitrofurantoin. Tetracyclines are usually effective, but they should not be given to children or pregnant women because of dental staining that may occur in the child or newborn.

Vaginitis

Organisms causing an inflammation of the vagina (vaginitis) may be acquired as sexually transmitted diseases or from endogenous sources. Endogenous agents are normally infective only under rather restrictive conditions. For example, yeasts such as *Candida albicans* become symptomatic after prolonged antibiotic therapy, or in individuals whose hormone balance is abnormal, such as in diabetics or during pregnancy. Toxic shock syndrome vaginitis (Chapter 28) occurs only in women using superabsorbent tampons during menstruation.

Several other organisms that seem to be transmitted sexually may cause a vaginitis but may exist as an asymptomatic infection. For instance, large numbers of *Gardnerella vaginalis,* a small gram-negative rod, are found in exudates from some cases of vaginitis, but small numbers of these same organisms may be normally present in asymptomatic women. Similarly, the protozoan *Trichomonas vaginalis* undoubtedly has a role as a causative agent of vaginitis in spite of the fact that many such infections are asymptomatic.

T. vaginalis appears to be a sexually transmitted disease, infecting both men and women. In the latter, a thin, watery, vaginal discharge is the most prominent symptom, although many cases are accompanied also by burning and itching. Infection in males is usually asymptomatic except in cases involving the prostate and seminal vesicles.

Metronidazole is normally effective for the treatment of this parasite. Because the infection is transmitted sexually and frequently is asymptomatic in the male, both male and female partners should receive treatment.

The following section characterizes the infectious agents whose transmission is almost always through sexual contact but that may progress to cause a systemic infection.

SEXUALLY TRANSMITTED DISEASES

The list of infectious agents that can be transmitted sexually has lengthened during the past couple of decades. This may be the consequence of several factors.

First, chlamydiae might have been the most common sexually transmitted disease (STD) in 1950 but the organism was unknown at that time, ruling out a definitive diagnosis. Second, the so-called sexual revolution that began in the 1960s exposed far more individuals to STDs than previously. This is exemplified particularly by the fact that reported cases of gonorrhea among teenage girls rose from 56,000 in 1960 to over 275,000 in 1975. Moreover, this change in sexual mores also provoked an explosion in the incidence of genital herpesvirus infections. Finally, it appears that diseases such as the acquired immuno-deficiency syndrome (AIDS) represent a new STD that has been described only within the past decade.

The remainder of this chapter is devoted to a description of STDs and their etiological agents. We will begin with the historically older STDs and progress through the more recently described infections of the reproductive system.

Treponema pallidum: Syphilis

Syphilis does not appear to be as ancient as many diseases we have discussed. Early descriptions of the disease show that its initial spread throughout Europe began at about the time Columbus's sailors returned from the New World. This had led to the theory that the disease was brought to Europe by Columbus's sailors. Others, however, believe that this was mere coincidence, and that it evolved as a non-STD in Africa and became an STD in Europe about the time Columbus returned to Europe. In any event, the disease spread throughout Europe in epidemic proportions during the following century.

T. pallidum, the causative agent of syphilis, was first described by Schaudinn and Hoffman in 1905. The tightly coiled organisms vary from 5 to 15 μm in length and about 0.09 to 0.5 μm in diameter. Because this is about the limit of resolution of the light microscope, the treponemes are best observed using a darkfield microscope (see Figure 27.1).

Pathogenicity With the exception of congenitally acquired syphilis, the disease is usually acquired during sexual contact with an infected individual. Following initial contact, the organisms penetrate the mucous membranes and, within 1 to 4 weeks, produce a localized ulcer called a chancre. Because the lesion teems with treponemes, a darkfield examination of the fluid from the chancre is the fundamental laboratory method for an immediate diagnosis.

After several weeks, the chancre spontaneously heals and, outwardly, the disease appears quiescent. But during this period the organisms are transported throughout the body via the bloodstream, eventually resulting in the widespread lesions of secondary syph-

In all likelihood, there has never been a "new" disease that spread throughout Europe as rapidly as syphilis. There is evidence that suggests that it was brought to Spain from Haiti by Columbus's crew but this is not accepted by everyone. It does appear, however, that syphilis in Europe first occurred in Spain in 1493 and that its spread from Palos (where the *Nina* and the *Pinta* returned to Spain from the New World) via Seville to Barcelona followed the route taken by the crews of Columbus when they came to Barcelona to meet King Ferdinand and Queen Isabella. Many of the sailors then traveled by land to visit Rome and Naples, leaving a trail of syphilis as they traveled.

In 1494, Charles VIII of France invaded Italy to claim the throne of Naples. Italy offered no effective resistance, and the invasion became more a march of debauchery than a military campaign. By 1495, however, the "new" disease had permeated Charles's mercenary troops (who had come from practically every country in Europe), and Naples was evacuated. As the troops returned to their own countries, they carried syphilis throughout Europe. Records show that it became rampant in France, Germany, and Switzerland in 1495, in Greece and Holland in 1496, in England and Scotland in 1497, and in Hungary and Russia in 1499.

Interestingly, there was no name for this new disease until 1530, when it was described in a poem by Hieronymus Fracastorius entitled, "The Sinister Shepard." Prior to that the Italians called it the Spanish or the French disease; the French called it the Italian or Neapolitan disease; the English called it the French disease; the Russians called it the Polish disease; and the Turks called it the French disease. In Spain, where the disease was first recognized, it was called the disease of Haiti.

Syphilis appears to have evolved into a far less severe disease than that which spread through Europe in the fifteenth and sixteenth centuries. The 25 percent mortality prompted governments to take extreme measures to slow its spread. Many towns banned syphilitics from churches and inns. France ordered all foreign syphilitics to leave the country and herded their own into Saint Germain (a suburb of Paris), threatening death by drowning to anyone hiding the disease. Scotland ordered all inhabitants of Edinburgh who were afflicted with syphilis into banishment to the Island of Inchkeith.

A partial list of a Who's Who of Syphilis would include Charles VIII of France, Frederick the Great, Henry III of France, and Peter the Great. Included also were artists Benvenuto Cellini and Paul Gaugin, writers Guy de Maupassant, Christopher Marlowe, and Heinrich Heine, and musicians Robert Schuman, Nicola Paganini, and Ludwig von Beethoven.

We shall probably never know the true origin of syphilis, but a study of its spread throughout the world reveals considerable information about life during the fifteenth and sixteenth centuries.

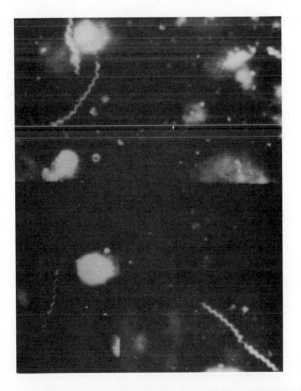

Figure 27.1 Darkfield micrograph of *T. pallidum* in fluid from a chancre (about ×2000).

ilis. The lesions occur on the skin and may also be present on the mucous membranes, eyes, bones, and central nervous system; like the primary chancre, they swarm with treponemes. Direct contact with these secondary lesions can result in nonsexually acquired syphilis.

After a period of months to a few years, the lesions disappear spontaneously. Approximately one-fourth of these cases appear to be true cures, based on the observation that such individuals will lose their antibodies to the treponema. Another one-fourth apparently retain a latent infection for life, since they maintain antibodies to the organisms but remain asymptomatic. In the remaining half, the infection will be reactivated as tertiary syphilis.

Tertiary syphilis may occur 5 to 40 years after the initial infection. Lesions, called gummata, may occur in the central nervous system, causing paresis (insanity), or in the cardiovascular system, resulting in aortic aneurysm (ballooning out of the aortic artery). They may also arise in the eyes, skin, bones, or viscera.

Congenital Syphilis

T. pallidum also passes through the placenta to infect the fetus. If such an infection does not kill the fetus, the newborn will have congenital syphilis. In the latter case, the disease may manifest itself immediately at birth and cause early death, or the organism may go undetected and cause death at a later date (latent syphilis). It has been found that congenital syphilis can often be avoided if the infected mother is treated early in pregnancy.

Laboratory Diagnosis Diagnosis can be made either by direct observation of the spirochetes from a lesion or by serological techniques. Direct observation is possible only during the primary or secondary stage of the disease. Inasmuch as this is a chronic infection with long periods of latency, serological techniques are the major diagnostic tool. These tests may be specific or nonspecific, depending on the antigen used.

An early serological test for syphilis was designed by Wassermann. Extracts from normal beef heart were mixed with the patient's serum and a positive reaction was accompanied by the destruction of complement. It is now known that the actual Wassermann antigen in beef heart is a normal constituent of tissues called cardiolipin, which is diphosphatidylglycerol (see Figure 27.2). Over the years a number of modifications of this test using the same antigen were devised. These modifications employed flocculation rather than a complement-fixation test and go by the names of their originators. Some of the more common ones are called the Kolmer, Kline, Hinton, and Kahn tests. One test commonly used today is called the **VDRL** (Venereal Disease Research Laboratory) test. This is a slide flocculation test that employs a lipid antigen (cardiolipin) extracted from normal beef heart. In actual practice, a buffered saline suspension of cardiolipin, lecithin, and cholesterol is mixed with the patient's serum on a slide. The slide is then agitated on a mechanical rotor for several minutes. A positive test results in a clumping of the antigen mixture.

Another widely used nonspecific test is the rapid plasma reagin (RPR) card test. The RPR test is performed by adsorbing the VDRL antigen on carbon particles and mixing this modified antigen with the patient's serum on a card. In a positive test, the flocculation of the carbon particles is visible to the naked eye. None of the tests employing cardiolipin as an antigen is completely specific, and other disorders occasionally give rise to antibodies resulting in false positive returns. What was needed was a more specific test to check out the positive results.

A number of specific tests are now available for the definite diagnosis of syphilis. All require either live *T. pallidum* grown in the testes of a rabbit or the Reiter strain of *Treponema*, which is thought to be an avirulent form of *T. pallidum* and can be grown on artificial media. By far, the most widely used serological test for specific treponemal antibodies is an indirect fluorescent antibody test, abbreviated FTA-ABS (fluorescent treponemal antibody-absorption). The absorption step in this test is necessary to eliminate nonspecific reactions that occur as a result of antibodies to common antigens shared by both pathogenic and saprophytic treponemes. The absorption is accomplished by mixing the patient's serum with a standardized extract from nonpathogenic Reiter treponemes. (This sorbent can be purchased commercially.) The absorbed serum is then used to cover a smear of pathogenic testes-grown treponemes (Nichols strain, which can be purchased commercially as lyophilized organisms). After 30 min at 37°C, the slide is thoroughly rinsed to remove unreacted serum proteins. It is then covered with a fluorescently labeled antihuman gamma globulin. This again is allowed to react for 30 min before it is rinsed; the slide is then examined microscopically (using an ultraviolet light source) for the presence of fluorescently stained trep-

Figure 27.2 Structure of cardiolipin. This molecule consists of three molecules of glycerol joined by phosphodiester bonds; the terminal two glycerol molecules are each esterified with two long-chain fatty acids.

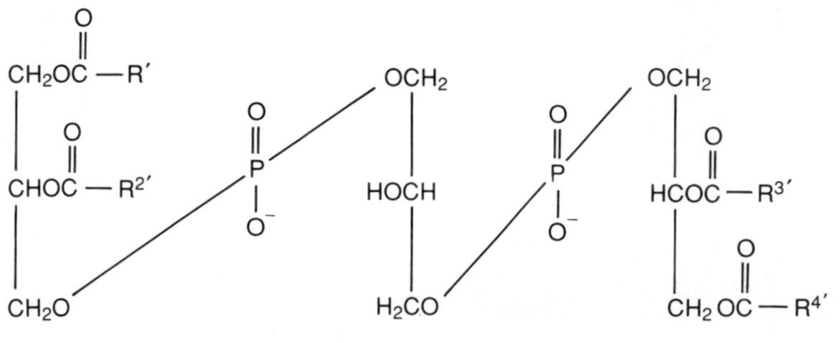

onemes. Because the fluorescently labeled antihuman gamma globulin can react with the treponemes only if the organisms are coated with human antibody, the presence of fluorescently labeled treponemes signifies the presence of specific antibodies to *T. pallidum*. On occasions when lesions are present, expressed material can be stained directly with specific antibodies to *T. pallidum* that have been conjugated to a fluorescent dye (DFATP—direct fluorescent antibody test of *T. pallidum*).

The microhemagglutination test for *T. pallidum* (MHA-TP) also assays for specific treponemal antibodies. This method employs specially treated sheep's erythrocytes that have been coated with antigen from *T. pallidum*. This test has been automated and adapted to a microvolume procedure. A positive result is signified by the agglutination of the red cells.

There are a number of other procedures for determining specific treponemal antibodies, such as the TPI (*T. pallidum* immobilization) test, TPA (*T. pallidum* agglutination), TPIA (*T. pallidum* immune adherence), and whole-body *T. pallidum* complement-fixation tests. None of these, however, is used routinely in diagnostic laboratories.

The choice of which test to use may be dictated in part by personal preference. The VDRL or RPR are less expensive than procedures for determining specific antibodies and should, therefore, be used in screening low-risk populations, such as individuals having premarital serological testing. Positive results can be confirmed with one of the specific tests. When it becomes possible to grow pathogenic treponemes on artificial media, the availability of an inexpensive antigen will undoubtedly influence the choice of tests used to diagnose syphilis serologically.

Control Measures Before the twentieth century, there was no treatment for syphilis. This problem led the brilliant scientist Paul Ehrlich (1854–1915) to carry out his systematic search for a "magic bullet" that would selectively attack microorganisms without undue toxicity to the host. Although Ehrlich never found his universal "magic bullet," he did develop an arsphenamine that was effective for the treatment of certain protozoan diseases and syphilis. The treatment of syphilis with arsenicals was a long and painful experience, but until penicillin became available in the middle 1940s, it remained the only treatment available.

This disease could now be eradicated, as the organism is sensitive to penicillin. Even neurosyphilis, the tertiary stage of the disease, may be arrested with adequate penicillin therapy. However, in spite of the ease with which the disease may be cured, public health officials report a rising incidence of syphilis since the 1960s (see Figure 27.3).

Neisseria gonorrhoea: Gonorrhea

Few diseases in the United States have grown to the epidemic proportions of gonorrhea during the past decade. Approximately 1 million cases are reported annually in this country, and these undoubtedly represent only a portion of the actual number occurring each year.

N. gonorrhoeae, frequently called the gonococcus, is the causative agent of gonorrhea. The gonococcus is morphologically similar to the meningococcus and other members of the genus *Neisseria* (see Figure 27.4). It is a gram-negative diplococcus and, like the meningococcus, requires an enriched medium for growth. Although the gonococcus is difficult to grow, the most common method is to plate the specimen on an enriched infusion agar medium to which 5 percent heated, defibrinated whole blood has been added (chocolate blood agar). The gonococcus grows best in an atmosphere of 10 percent carbon dioxide at 35°C. It is differentiated from the meningococcus and the other *Neisseria* on the basis of its fermentative abilities; it ferments glucose but not maltose, sucrose, or lactose. Frequently, colonies of *Neisseria* on a plate can be differentiated from other bacterial colonies on the basis of the oxidase test, in which the colonies are sprayed with tetramethylparaphenylene diamine hydrochloride. Colonies of *Neisseria* will turn dark within 30 to 60 s.

Pathogenicity Gonorrhea is a disease of the mucous membranes of the genitourinary tract and eye. In the male there is usually a urethral inflammation with pus and painful voiding 3 to 9 days after infection. The prostate and the epididymis may also be involved. Fibrosis sometimes follows the acute stage, causing urethral stricture. In the female, the infection is much more likely to be asymptomatic or accompanied by a minor discharge, which can easily go unnoticed. The organisms, however, may infect the urethra, vagina, cervix, and fallopian tubes, causing a chronic disease that results in sterility. Bacteremia may also occur, leading to lesions in the skin, heart, eyes, meninges, or joints, in which case a gonococcal arthritis results. Pregnant women are especially susceptible to the disseminated form of gonorrhea.

Gonorrheal ophthalmia neonatorum is acquired by the infant in passing through the birth canal of an infected mother. The first manifestation is conjunctivitis; later, ocular structures may be involved, and blindness can be the final result. The instillation of 1 percent silver nitrate (Credé's method) in the eyes of the newborn was required by law in most states to prevent this infection. This has now been largely superseded by the use of erythromycin or tetracycline.

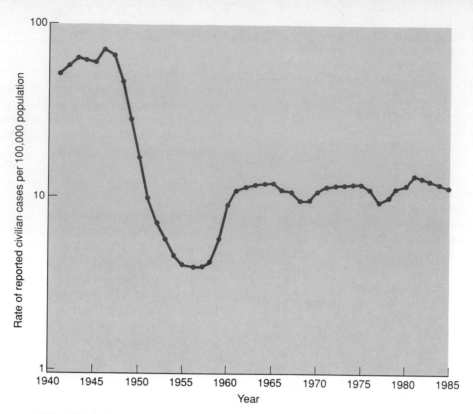

Figure 27.3 Reported cases of syphilis (primary and secondary) in civilians per 100,000 population by year, United States, calendar years, 1941–1985.

1941–1948 fiscal years: twelve-month period ending June 30 of the specified years
1947–1981 calendar years

Figure 27.4 (*a*) Drawing of "doughnut-shaped" diplococci of *N. gonorrhoeae* as they sometimes appear under the microscope. (*b*) Electron micrograph of a negatively stained pair of *N. gonorrhoeae* cells (×49,140).

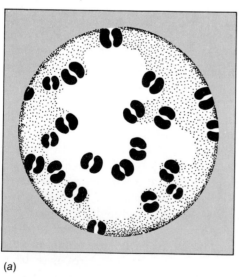

(*a*)

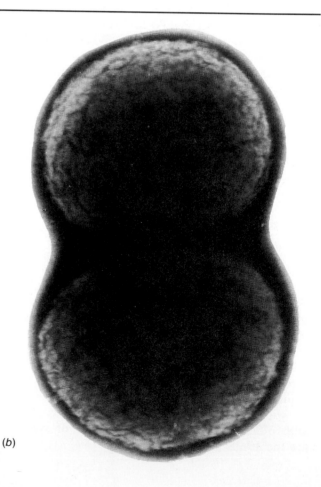

(*b*)

Clinical Management The diagnosis of acute gonorrhea in the male is usually not a difficult problem. The patient complains of pain on urination, and stained smears of urethral exudate will show intracellular gram-negative cocci. The disease in the female is somewhat more difficult to diagnose, since it may be asymptomatic. The magnitude of this problem is demonstrated by the fact that of almost 9 million women examined for gonorrhea during 1976 by culture of the organisms, 4.2 percent were found to be infected with the gonococcus. In any case, a positive diagnosis requires that the organism be grown from the exudate and identified biochemically. Chronic gonorrhea is much more difficult to diagnose, since the organisms may be present in very low numbers. Fluorescent antibody techniques unlike those described for syphilis, are not yet completely reliable.

As in syphilis, early diagnosis, treatment, and locating sexual contacts are most important. Recovery from gonorrhea does not confer immunity to reinfection.

Control Measures In the mid-1930s, when the sulfonamides were first introduced, essentially all cases of gonorrhea responded well to sulfonamide therapy. During the course of the next decade, however, most of the sulfonamide-sensitive strains were killed off; as a result, the strains of the gonococcus resistant to the sulfonamides became prevalent. Fortunately, by the time sulfonamide-resistant strains of gonorrhea had become a major problem, penicillin became available.

Penicillin has been the treatment of choice for over 30 years. But, alas, in 1976 penicillinase-producing gonococci were isolated from widely separated areas in the United States and England. Such strains have now been reported to occur worldwide. Depending on the locality, penicillinase-producing strains of gonococci may account for over 40 percent of clinical isolates. It is thus imperative to test all gonococcal isolates for penicillinase production. The Centers for Disease Control (CDC) still recommends that uncomplicated gonorrhea caused by penicillinase-negative strains be treated with an injection of aqueous procaine penicillin accompanied by oral probenecid or oral ampicillin together with probenecid. Probenecid is not antibacterial; it merely retards the urinary excretion of the penicillin. Penicillinase-producing strains have been treated successfully with 2 g of intramuscular spectinomycin. But here again, trouble looms on the horizon, because several cases of penicillinase-producing, spectinomycin-resistant strains have now been found. Such strains have, however, responded well to treatment with cefoxitin.

Great effort has been directed toward producing an effective vaccine for gonorrhea, but success is still "around the corner." The dilemma is not that antibodies do not initiate a bactericidal effect on the gonococcus, but rather that the gonococcus has developed an ingenious procedure for changing its antigenic structure. For example, antibodies directed against the organism's fimbriae are protective by preventing attachment of the gonococci to epithelial cells. The gonococcus, however, possesses multiple genes that encode for many different fimbriae sequences and, through a series of recombination events, may cease to produce fimbriae or may switch to an entirely different antigenic type of fimbriae. Thus, within a single culture, there are multiple antigenic types of fimbriae.

The gonococcus also possesses outer membrane proteins, one of which is involved in adherence of the bacterium to various types of host cells. Because antibodies to this protein (termed protein II) act as opsonins (enhance phagocytosis) it would seem that this protein might serve as an effective vaccine. But, alas, the gonococcus has also evolved a mechanism to circumvent this possibility by gene rearrangements in which a single strain may switch to form antigenically distinct types of protein II. It thus appears that the gonococcus has evolved some very complex genetic mechanisms to ensure its survival, making the advent of a successful vaccine remote.

Haemophilus ducreyi: Chancroid

H. ducreyi is the causative agent of chancroid, or soft chancre. There is no relation between these lesions and chancre, the initial lesion of syphilis. The organism is a small, gram-negative rod that is nonmotile and non-spore-forming. It can be grown on a rich blood agar medium, preferably under a reduced oxygen tension.

Pathogenicity Chancroid is characterized by a ragged ulcer on the genitalia, with marked swelling and pain. Regional lymph nodes increase in size and may suppurate. Autoinoculation may occur, and lesions are often multiple. There is no permanent immunity.

Chancroid is more commonly found in tropical climates, but there have been recent minor epidemics in the United States, particularly among immigrants.

The incubation period is 3 to 5 days. Sources of the organism are the infectious lesions of patients. Sexual contact is the chief means of transmission, although the disease may occasionally be acquired accidentally by medical personnel; then the lesion is found on the hands.

Clinical Management The diagnosis is usually confirmed by smears and cultures. Control of this disease depends on early diagnosis and the treatment and finding of contacts. Erythromycin appears to be the drug of choice, although sulfamethoxazole-trimethoprim is usually also effective.

Chlamydiae

Our discussion of the diseases caused by *Chlamydia trachomatis* will deviate slightly from our "point of entry" format. *C. trachomatis* can be subdivided into 15 different serotypes. Types A, B, Ba, and C are responsible for trachoma, a serious infection of the eye frequently resulting in blindness; types D through K are usually sexually transmitted, but they may also cause a serious conjunctivitis in the newborn and pneumonia during the first year of life; types L-1, L-2, and L-3 are the etiological agents of an STD known as lymphogranuloma venereum.

Trachoma

This clinical disease has been recognized for over 3000 years and was at one time listed as one of the three most important diseases of humans. The World Health Organization estimates that approximately 400 million people are infected with trachoma and that 6 to 10 million of these are totally blind as a result of the disease.

Pathogenicity Trachoma occurs only in humans and grows exclusively in the conjunctival cells. It is transmitted by direct contact with fingers or contaminated towels or clothing; in many areas of the world (for example, Egypt and the Middle East), it is so prevalent that children commonly become infected in early childhood. The infections occur primarily in areas where poor hygienic practices exist. In the United States, American Indians have probably been the most frequent victims.

The overt disease may begin suddenly with an inflammation of the conjunctiva. Leukocytes enter the area and form follicles under the conjunctiva. As vascularization and infiltration of the cornea continue, the resultant scarring of the conjunctiva may cause partial or complete blindness. Simultaneous bacterial infection is common and also contributes to the inflammation and scarring.

Clinical Management Treatment of trachoma with both systemic and topical sulfonamides and tetracycline is usually effective, but relapses are common and because chlamydiae characteristically produce latent infections, it seems possible that such relapses may result from a reactivation of the latent state.

Trachoma is usually diagnosed on the basis of the pathological findings associated with the disease. The agent can be grown by inoculating conjunctival scrapings into cell cultures or the yolk sacs of chick embryos. It may also be identified by staining the char-

acteristic inclusion bodies with fluorescently labeled antibody.

Inclusion Conjunctivitis

The etiological agent of this disease is so similar to that of trachoma that the two organisms are frequently called the TRIC (trachoma-inclusion conjunctivitis) agents. However, the pathogenesis of infection and the pathology of the conjunctivitis are considerably different from those of trachoma.

Pathogenicity The agent of inclusion conjunctivitis is most commonly found in the human genitourinary tract, whence it is passed from human to human by sexual contact. As an STD, the organisms grow in the epithelial cells of the female cervix or in the lining of the urethra of both sexes. The genital symptoms may be very mild to absent, and the majority of cases of sexually transmitted inclusion conjunctivitis are undiagnosed.

However, as the name implies, infections of the conjunctiva are not so benign. They are seen most frequently in the newborn 5 to 12 days after birth. In such cases, the infection is acquired from the mother by the infant as it moves through the birth canal. The symptoms usually begin with an acute purulent conjunctivitis, which starts to subside after several weeks and spontaneously disappears in a few months. Inclusion conjunctivitis does not result in blindness.

Adult infections of the conjunctiva may also occur; prior to the use of chlorine in swimming pools, such infections were often called "swimming pool conjunctivitis." Now most conjunctival infections of adults occur from contact with contaminated fingers or towels.

Clinical Management Both the conjunctivitis and the genital infection respond to treatment with tetracycline and sulfonamides. Reinfections can occur, and it is not known whether specific immunity is ineffective or whether subsequent infections are caused by a different one of the six serotypes of inclusion conjunctivitis.

Diagnosis relies heavily on the history and the clinical findings. Laboratory confirmation requires the isolation of the agent or the demonstration of cellular inclusions that will stain with fluorescently labeled specific antibody.

Infant Pneumonitis

It is estimated that the same serotypes of *C. trachomatis* responsible for inclusion conjunctivitis cause 30 percent of all pneumonias in infants under the age of 6 months. The newborn acquires the organism from an infected mother during birth, although its lower respiratory infection may have originated from a conjunctival infection.

Chlamydial infant pneumonia is not usually a severe disease, but it may last several weeks. Infants with this disease may also have chlamydial infections of the middle ear.

Erythromycin or the sulfonamides are the drugs of choice, inasmuch as the tetracyclines are contraindicated for this pediatric group.

Genital Tract Infections

As is obvious from the preceding sections, *C. trachomatis* (serotypes D through K) is acquired by adults primarily as an STD. Chlamydiae have been isolated from about one-third of all women attending venereal disease clinics and from two-thirds of those with gonorrhea; as for men, approximately 20 percent of those with gonorrhea also have a chlamydial infection.

Nongonococcal Urethritis   *C. trachomatis* appears to be the most common STD occurring in the United States. Some 3 to 10 million new cases are estimated to occur each year, and the rapid spread of this infection is probably due in part to the minimal nature or absence of symptoms, particularly in women. The disease is especially prevalent among sexually active, adolescent females. In spite of the paucity of symptoms, the CDC has estimated that as many as 50,000 American women may be rendered infertile each year from this infection, primarily from a scarring of the fallopian tubes. This organism is also the most common cause of nongonococcal urethritis in men, and because many such men are dually infected with the gonococcus, it is also responsible for the majority of postgonococcal urethritis infections. Such infections are usually recognized by the failure of penicillin therapy to relieve the symptoms of urethritis.

C. trachomatis has also been found to be the cause of most infections of the epididymis in sexually active men under the age of 35.

Lymphogranuloma Venereum

Lymphogranuloma venereum is caused by the L-1, L-2, or L-3 serotype of *C. trachomatis*. This disease is strictly a sexually transmitted disease; it is spread only by sexual contact. It is most prevalent in the tropics but is seen also in the United States—particularly in the South.

Pathogenicity   Approximately 7 to 10 days after infection, a small erosion or painless papule appears in the genital area. The organisms migrate to the regional lymph nodes, and most symptoms result from this lymph node involvement. The original papule soon heals, but after 1 to 2 months the regional lymph nodes become enlarged and tender and may break open and drain. These enlarged lymph nodes are called buboes; as they enlarge, the draining lymph channel may be completely obstructed. Such restriction can cause tremendous enlargement of the genitalia, and rectal strictures may occur as an effect of perirectal scarring.

Clinical Management   Control measures are the same as for other sexually transmitted diseases. Both tetracycline and sulfonamides are effective for the treatment of lymphogranuloma venereum.

The diagnosis is made from the clinical picture, history, complement-fixation test for antibodies, biopsy of infected nodes for isolation of the organisms, and the Frei skin test. The Frei test is conducted by injecting killed organisms into the skin and observing for a delayed-type skin reaction similar to the tuberculin reaction. The Frei test, however, is not specific and will give positive results in people who have had psittacosis.

Mycoplasmas

The mycoplasmas (see Chapter 13) have been isolated from a variety of human sources in numerous pathological conditions. However, at this time the only human disease known to be caused by a member of this group is primary atypical pneumonia, caused by *M. pneumoniae* (see Chapter 24).

T strains of mycoplasma (*Ureaplasma urealyticum*), however, are believed by many investigators to cause urethritis in humans. The difficulty in proving that they cause disease lies in the fact that they can be isolated from the genital tract of up to 50 percent of asymptomatic people and from 50 to 80 percent of individuals with nongonococcal urethritis. It seems probable that they can exist as parasites in the genital tract without causing illness, but in some cases they are probably the etiological agents of a urethritis.

Viruses

In this section, we discuss viruses that cause important sexually transmitted diseases. As such, only three groups of viruses will be considered: (1) herpes simplex type 2; (2) human immunodeficiency virus; and (3) human papillomaviruses. All three may gain entrance to the body via other means, but in each case, it is the STD that has assumed major consequences.

Herpes Simplex Type 2

Herpes simplex type 2 (HSV-2) is transmitted during sexual intercourse. Primary infections may be asymptomatic, or, in the female, painful lesions may occur on the vulva, vagina, cervix, or perineum. In the male, lesions may occur on the glans, prepuce, or shaft of the penis, producing symptoms of burning, itching, and painful urination. The primary lesions usually appear 6 to 8 days after contact and disappear after 2 or 3 weeks, but, as in the oral infections of type 1, the virus becomes latent and may cause sporadic recurrences of overt disease. And therein lies the problem with this disease. It is estimated that there may be as many as 20 million infected people in the United States with perhaps 500,000 new infections occurring annually. This undoubtedly accounts for the observation that antibodies to HSV-2 have been detected in 80 percent of female prostitutes. The unfortunate aspect of a herpes infection is that there is, at present, no cure available. Thus, an infected individual may be subject to painful, recurrent attacks for life. Moreover, it has been reported that the disease can be transmitted sexually even during periods of apparent quiescence.

Although less common, it should be noted that herpes simplex type 1 can cause genital lesions that are clinically indistinguishable from those caused by the type 2 virus.

But a dim light has appeared on the horizon. A new antiviral drug, commonly called acyclovir, appears to shorten the time of viral shedding and accelerate the healing of some genital herpes simplex infections, particularly in men. However, this does not appear to result in a cure, and later exacerbations of this infection may continue to occur.

Neonatal Infections Childbirth during an episode of herpes simplex type 2 infection may result in an unusually severe infection in the newborn during the first month of life. Such infections are frequently fatal, and survivors may have residual damage of the central nervous system or the eye, although current evidence indicates type 2 infections of the newborn may, in some instances, be asymptomatic. As a result, it is general practice to deliver a child by cesarean section if the mother has an active herpes infection at the time of birth.

Neonatal infections can also be acquired in utero, but such situations appear to be confined to women who acquire a primary infection during the first 20 weeks of pregnancy. Such cases may result in spontaneous abortion or congenital malformations in the newborn. The changing incidence of this type of infection was illustrated by a study in Seattle that reported that neonatal herpesvirus infections increased from 2 cases per 100,000 during 1966 to 1969 to 13.8 cases per 100,000 during 1978 to 1982.

Oncogenic Potential There is currently evidence linking herpes simplex type 2 infections to cervical cancer. This postulation is based in part on the fact that in several large, worldwide studies antibodies

to the type 2 virus were found in a much higher percentage of women with cervical carcinoma than in appropriate control groups. The virus has also been isolated from degenerating cervical tumor cells, and hybridization studies have suggested that such tumor cells do contain herpes simplex type 2 DNA. Furthermore, if one infects hamster embryo fibroblasts with type 2 virus that has been partially inactivated with ultraviolet light to prevent it from completing its lytic cycle, such fibroblasts are transformed into cancer cells that produce highly malignant tumors after inoculation into hamsters. Thus, at least in this in vitro situation, the potential oncogenicity of the virus has been demonstrated.

Human Immunodeficiency Virus: AIDS

Human immunodeficiency virus (HIV) is the etiological agent of **acquired immune deficiency syndrome**, more commonly known as AIDS. This disease, first described in 1981, is characterized by the occurrence of multiple opportunistic infections and malignant diseases in patients without an otherwise recognized cause for immunodeficiency. A number of opportunistic infections have been observed, of which the principal ones are caused by *Pneumocystis carinii*, cytomegalovirus, atypical mycobacteria, *Toxoplasma gondii, Candida albicans,* herpes simplex, *Cryptococcus neoformans*, and *Cryptosporidium*. Another highly distinctive feature of AIDS is the frequent occurrence of Kaposi's sarcoma, a malignancy of pigmented cells, frequently seen in the skin.

HIV is a retrovirus (Figure 27.5) which infects a number of different cells in the body, including cells in the central nervous system, but the primary symptoms (from which the disease gets its name) result from the destruction of T-helper (Th) cells. Because Th cells are required for both antibody synthesis and delayed-type hypersensitivity reactions, many of those who are infected eventually lose all capability to mount an immune response.

HIV is present in essentially all body fluids and its transmission to uninfected persons can occur via blood, semen, vaginal fluids, or breast milk. In the United States it occurs most frequently in homosexual males (75 percent) and intravenous drug users (17 percent). The remaining cases are found in persons who received blood transfusions, women who were sexual partners of bisexual men, and infants born to infected mothers. Transmission of HIV from women to men appears to occur infrequently. There are no proven cases in which the virus has been acquired through casual contact with an infected person, al-

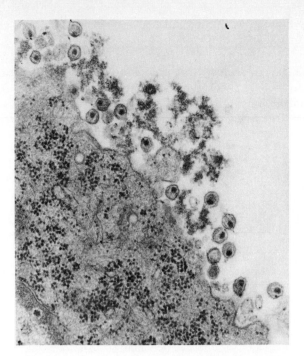

Figure 27.5 Human immunodeficiency virus (×114,800). Note dark, slightly off center, nucleoid structure in each virion.

though in 1991 it was reported that a number of patients were infected by their dentist.

There appears to be a very long and **extremely** variable incubation period between infection and the occurrence of symptoms. Swollen lymph nodes (indicative of a chronic infection) are most commonly one of the earliest symptoms of this disease. These may occasionally regress or, after years, develop into full-blown AIDS. In the United States, the number of infected persons is believed to be between 1 and 2.5 million, and it is predicted by some scientists that 75 to 90 percent of these will eventually develop fatal AIDS.

HIV may be present in the blood for years in the absence of overt symptoms, and such individuals provide the source for the spread of the disease through transfusions. As a result, all blood is now screened for the presence of antibodies to HIV using an enzyme-linked immunosorbent assay (ELISA). Because this test may give an occasional false positive reaction, all sera testing positive for HIV antibodies are further screened using a Western blot analysis in which the virus antigens are first subjected to electrophoresis and the resulting proteins then transferred to a nitrocellulose paper that is reacted with the serum specimen in question. Antibody binding to the separated viral polypeptides is detected by using a second antihuman antibody that is conjugated to an indicator system.

Because the vast majority of cases of AIDS occur in homosexual men, intravenous drug users, and their contacts, the avoidance of intimate contact with these high-risk groups can greatly increase one's safety.

There is no cure for AIDS, but a drug introduced in 1986 appeared to alleviate many of the symptoms. This drug, azidothymidine, (also called AZT or zidovudine) inhibits the reverse transcriptase of HIV and it is now routinely given to individuals who are antibody-positive for HIV even in the absence of overt symptoms of AIDS. Clinical trials have shown that the use of AZT does prolong survival but, unfortunately, long-term use has resulted in the appearance of drug-resistant HIV.

Two other nucleoside analogs, dideoxycytidine and dideoxyinosine, also have been shown to have anti-HIV activity and both are currently being clinically evaluated.

There will undoubtedly be more such agents introduced to combat this infection.

At present, there are few bright lights on the horizon signalling a successful vaccine. Current possibilities include mutants derived from HIV that no longer kill T cells; Simian-monkey-derived AIDS virus (SIV), which is distantly related to HIV but does not kill T cells; human viruses that are related to HIV but do not kill T cells; and recombinant cell lines that secrete certain HIV glycoproteins. None of these is beyond the laboratory stage and it may be some years before a successful challenge to this disease is available.

Meanwhile, the experts can give us only the grim projection that by 1992, some 500,000 people in the United States will have AIDS or will have died from it; that 1 to 2 percent of the entire African population carries the AIDS virus; and a major catastrophe seems likely.

Human Papillomaviruses

There are 40 different types of human papillomaviruses (HPV), which for years have been known to cause abnormal host cell growth. Most appear to be specific for certain areas of the body, and the majority induce benign growths, such as warts on the hands or soles of the feet.

HPV is also recognized as causing anogenital and cervical warts, which are spread by sexual contact. The increasing incidence of such infections can be seen in the observation that 5 to 10 percent of young women are now infected as opposed to 1 to 2 percent of all women.

The frightening aspect of this incidence is the increasing evidence linking HPV with cervical and other genital cancers. The conclusion that HPV is involved in these cancers comes from the fact that essentially all such tumor cells contain HPV DNA integrated into their chromosomes. In fact, Harold zur Hausen states that the majority, if not all, of cervical, penile, and vulval cancers carry HPV DNA.

Other data linking genital cancers to an infectious agent are as follows: (1) neoplasia of the cervix occurs most frequently in women who have had multiple partners; (2) male partners of females with benign or premalignant cancers are at high risk for penile cancer; and (3) cell culture studies with bovine papillomavirus show that the product of one early gene could induce a malignant transformation in infected cells.

Of the many HPVs, DNA from HPV 16 and HPV 18 are the ones most commonly found in cervical carcinoma cells. Other types include HPV 11, 31, 33, and 35.

There are no vaccines, and prevention requires avoidance of promiscuous sexual relations.

Table 27.1 summarizes the pathogens that enter the body via the genitourinary tract.

TABLE 27.1   Summary of Pathogens that Enter the Body via the Genitourinary Tract

Organism	Description	Disease	Incubation Period	Diagnosis	Treatment	Control
Urinary tract						
Escherichia	Gram-negative rod	Urethritis	Variable	Culture and identify	Ampicillin Sulfonamides	Medical asepsis
Proteus	Gram-negative rod	Urethritis		Culture and identify	Ampicillin Broad-spectrum antibiotics	Aseptic catheter-ization techniques
Enterobacter	Gram-negative rod	Urethritis		Culture and identify	Ampicillin	
Pseudo-monas sp.	Gram-negative rod			Culture and identify	Variety of antibiotics	
Klebsiella	Gram-negative rod	Urethritis		Culture and identify	Ampicillin	
Genital tract						
T. pallidum	Slender spiro-chete (not stained by Gram stain)	Syphilis	About 3 weeks	Darkfield examination for spiro-chetes Specific sero-logical tests	Penicillin	Diagnosis and prompt treat-ment Finding con-tacts Treatment of pregnant syphilitics
N. gonor-rhoeae	Diplococcus Gram-negative bacterium	Gonorrhea	3–5 days	Smears Cultures and biochemical tests	Penicillin or spectinomycin	Early diagnosis and treatment Finding contacts
H. ducreyi	Small bacillus Gram-negative bacterium	Chancroid or soft chancre	3–5 days	Smears Cultures	Sulfonamides Tetracycline Streptomycin	Early diagnosis and treatment Finding contacts
C. trachomatis	Chlamydia	Trachoma	2–9 days	Clinical picture Conjunctival scrapings Growth in cell culture or chick embryo	Sulfonamides Broad-spectrum antibiotics	Cleanliness
C. trachomatis	Chlamydia	Inclusion conjunc-tivitis	5–12 days	Clinical picture History	Sulfonamides Broad-spectrum antibiotics	

TABLE 27.1 *(Continued)*

Organism	Description	Disease	Incubation Period	Diagnosis	Treatment	Control
C. trachomatis	Chlamydia	Lympho-granu-loma vener-eum	7–12 days, usually	History and clinical picture Complement-fixation test Biopsy of in-fected nodes Frei skin test	Sulfonamides Tetracyclines Chloramphenicol Penicillin	Same as for other STDs
T-strain myco-plasma (*Ureaplasma*)	Highly pleo-morphic Filterable	Arthritis (?) Conjunc-tivitis (?) Urethritis (?)	Uncertain	Cultures in stained agar blocks	Tetracyclines Streptomycin Neomycin Kanamycin, etc.	Uncertain
Herpes simplex type 2	Virus	Lesions on mucous mem-branes, cervical carci-noma (?)	6–8 days	Clinical picture; culture	Acyclovir?	
HIV		AIDS	?	Multiple infections and malig-nancies	AZT	Same as for other STDs
Human papillo-maviruses		Warts, genital cancers (?)	?	Clinical picture; culture	None	

SUMMARY

Members of the genera *Escherichia, Proteus, Enterobacter, Klebsiella,* and *Pseudomonas* are the most common causes of urinary tract infections. The infections may remain localized in the urethra or bladder to cause a urethritis or cystitis, or they may move up the ureters to infect the kidney, causing a pyelonephritis. *C. albicans* and *G. vaginalis* occur in small numbers and may cause a vaginitis in some persons. *T. vaginalis* is a protozoan that is sexually transmitted and may or may not cause a symptomatic vaginitis.

Many of the other genital infections occur as a result of sexual contact with an infected person. The spirochete *T. pallidum* is the causative agent of syphilis. This disease is characterized by three stages: a primary stage in which a chancre develops at the initial site of infection, a secondary stage in which the organisms are spread throughout the body, and a tertiary stage that may occur many years after apparent recovery from the secondary stage. Syphilis is usually diagnosed by serological techniques. These may use a nonspecific antigen, such as the VDRL or RPR tests and its many modifications, or they may

use *Treponema* organisms grown in rabbit testes. These latter tests are much more specific.

N. gonorrhoeae, the causative agent of gonorrhea, is a gram-negative diplococcus morphologically identical to the meningococcus. In addition to an infection of the genital mucous membranes, the gonococcus can also infect the eyes of a newborn passing through the birth canal. Erythromycin or tetracycline placed in the eyes of a newborn will prevent this eye infection and possible blindness. *H. ducreyi* is the etiological agent of a venereal disease called chancroid.

Various strains of *C. trachomatis* cause trachoma (a serious infection of the eye often resulting in blindness), inclusion conjunctivitis (a mild to asymptomatic sexually transmitted disease that also causes a moderately severe but self-limiting conjunctivitis), and lymphogranuloma venereum (a sexually transmitted disease that can result in severe and permanent pathological changes in the genitalia and rectum).

T-strain mycoplasmas probably cause occasional urinary tract infections, but they are also found in the genital tract of up to 50 percent of asymptomatic persons.

Herpes simplex type 2 is transmitted primarily as a venereal disease that can exist as a latent infection with sporadic recurrences of the genital lesions. Neonatal infections with this virus can be very severe. There is circumstantial evidence that herpes simplex type 2 may be linked to human cervical cancer.

Human immunodeficiency virus (HIV), the cause of AIDS, infects and destroys T-helper cells, resulting in an inability to make antibody or to mount a delayed-type hypersensitivity reaction. The virus is spread primarily by sexual contact, by an exchange of blood through transfusions, by the use of contaminated needles, or transplacentally to the developing fetus. Major symptoms are the occurrence of multiple opportunistic infections and malignant diseases.

Human papillomaviruses (HPV) cause a variety of abnormal growths and some types are spread sexually, resulting in the occurrence of genital warts. The majority of cervical, penile, and vulval cancers contain HPV DNA, and many investigators believe that these viruses are the cause of such cancers.

QUESTIONS FOR REVIEW

1. Differentiate between cystitis and pyelonephritis. What simple laboratory procedure can be used to make a diagnosis of cystitis?
2. What are some of the possible reasons for the increase in numbers of STDs during the past 25 years?
3. What symptoms would one expect to see from disseminated gonorrhea?
4. Why does it seem unlikely that a successful vaccine for gonorrhea will be developed?
5. What is characteristic of each stage of syphilis? What laboratory tests may be used as diagnostic aids for each stage of syphilis?
6. What is the significance of gonorrheal ophthalmia neonatorum? How may this be prevented?
7. What infections are caused by *C. trachomatis*? List the means of transmission for each.
8. What are the major pathological changes that result from untreated lymphogranuloma venereum?
9. What are T strains of mycoplasma and what is their role in human disease? (Refer to Chapter 9 for a more complete description of T-strain mycoplasmas.)
10. What types of infections are caused by herpes simplex type 2?
11. What is the evidence that links herpes simplex type 2 with human cervical cancer?
12. What is the most common cause of nongonococcal urethritis?
13. What host cell type is destroyed after infection with HIV, and what disease symptoms result from this destruction?
14. How can one test to determine if an individual has ever been infected with HIV?
15. What type of sexually transmitted disease is caused by human papillomaviruses?
16. What data link HPV to genital cancers?

SUPPLEMENTARY READING

Anand A, et al: Human parvovirus infection in pregnancy and hydrops fetalis. *N Engl J Med* 316:183, 1987.

Beaudenon S, et al: A novel type of human papillomavirus associated with genital neoplasia. *Nature* 231:246, 1986.

Britigan BE, Cohen, MS, Sparling PF: Gonococcal infection: a model of molecular pathogenesis. *N Engl J Med* 312:1683, 1985.

Dillon HC, Jr: Prevention of gonococcal ophthalmia neonatorum. *N Engl J Med* 315:1414, 1986.

Fransen L, et al: Ophthalmia neonatorium in Nairobi Kenya: The roles of *Neisseria gonorrhoeae* and *Chlamydia trachomatis*. *J Infect Dis* 153:862, 1986.

Hirsch MS: Chemotherapy of human immunodeficiency virus infections: Current practice and future prospects. *J Infect Dis* 161:845, 1990.

Laurence L: The immune system in AIDS. *Sci Am* 253(6):84, 1985.

Lifson AR, et al: Detection of human immunodeficiency virus DNA using the polymerase chain reaction in a well-characterized group of homosexual and bisexual men. *J Infect Dis* 161:436, 1990.

Luftig RB, et al: Update on viral pathogenesis. *ASM News* 56:366, 1990.

Jorizzo JL, et al: Role of circulating immune complexes in human secondary syphilis. *J Infect Dis* 153:1014, 1986.

Norgard MV, et al: Cloning and expression of the major 47-kilodalton surface immunogen of *Treponema pallidum* in *Escherichia coli*. *Infect Immun* 54:500, 1986.

Ostrav RS, et al: Detection of papillomavirus DNA in human semen. *Science* 231:731, 1986.

Rosenblum LS, et al: Heterosexual transmission of hepatitis B virus in Belle Glade, Florida. *J Infect Dis* 161:407, 1990.

San Joaquin VH, Rettig PJ: Role of *Chlamydia trachomatis* in upper-respiratory infections in children. *J Infect Dis* 154:193, 1986.

Schoolnik GK: How *Escherichia coli* infects the urinary tract. *N Engl J Med* 320:804, 1989.

Schwalbe RS, Sparling PF, Cannon JG: Variation of *Neisseria gonorrhoeae* protein II among isolates from an outbreak caused by a single gonococcal strain. *Infect Immun* 49:250, 1985.

Segal E, et al: Role of chromosomal rearrangement in *N. gonorrhoeae* pilus phase variation. *Cell* 40:293, 1985.

Spruance SL, et al: Treatment of recurrent herpes simplex labialis with oral acyclovir. *J Infect Dis* 161:185, 1990.

Stamm WE: Urinary tract infections from pathogenesis to treatment. *J Infect Dis* 159:400, 1989.

Werness BA, et al: Association of human papillomavirus types 16 and 18 E6 proteins with p53. *Science* 248:76, 1990.

Chapter 28

Pathogens that Enter the Body via the Skin or by Animal Bites

OBJECTIVES Study of this chapter will acquaint you with

1. The role of the staphylococci as etiological agents of human infections.

2. The proposed mechanism of action of staphylococcal toxic shock syndrome toxin.

3. The problem of *Pseudomonas* infections in burn victims.

4. The epidemiology of leptospirosis.

5. The pathogenesis of tetanus and gas gangrene.

6. The types of infections caused by the dermatophytes.

7. The organisms causing subcutaneous mycoses.

8. The virulence determinants of *B. anthracis.*

9. The etiological agents of rat-bite fever.

10. The epidemiology of rabies and the type of vaccine being used in humans.

11. The epidemiology of the blood flukes.

The intact skin is an effective protective barrier that prevents many infectious agents from gaining entrance to the body. However, during the course of normal living, the skin does not always remain intact. Skin breaks so minor that they go unnoticed may allow bacteria to enter and multiply.

Some organisms enter the body via direct contact with the skin; others gain entrance through insect bites. This chapter is concerned primarily with infectious agents that gain entrance to the body through the skin without the help of an insect vector.

BACTERIA THAT ENTER VIA SKIN ABRASIONS

Staphylococcus

Members of the genus *Staphylococcus* are spherical cells that occur singly or occasionally in pairs but most frequently as irregular clusters. They are gram-positive, nonmotile, non-spore-forming facultative anaerobes (see Figure 28.1).

Classification The classification of the staphylococci has changed frequently as a result of attempts to differentiate the large number of avirulent micrococci from morphologically similar organisms. The current system of classification separates the staphylococci from the micrococci on the basis of the ability to ferment glucose with the production of acid. Thus, the members of the genus *Staphylococcus* will ferment glucose and produce acid, whereas the members of the genus *Micrococcus* (which are for the most part morphologically identical to the staphylococci) are unable to ferment glucose to produce acid. The micrococci can, however, oxidize glucose aerobically, but this does not result in the formation of acid.

The appearance of a Gram-stained smear is usually sufficient to distinguish the staphylococci from the streptococci, since the streptococci are much more prone to form chains of cells. Where there is doubt, these two genera can be separated on the basis of the presence of the enzyme catalase. Catalase will break down hydrogen peroxide to form water and oxygen. If one mixes a loopful of staphylococci on a slide with 3 percent hydrogen peroxide, bubbles of oxygen will be visible to the naked eye. Streptococci do not form catalase—hence, no bubbles.

Under the present classification there are three recognized species of staphylococci: *S. aureus*, *S. saprophyticus*, and *S. epidermidis*. Essentially all of the serious diseases involving the staphylococci are caused

Figure 28.1 (a) Clusters of gram-stained *S. aureus* (about ×4500). (b) Electron micrograph of cells of *S. aureus* (×49,000).

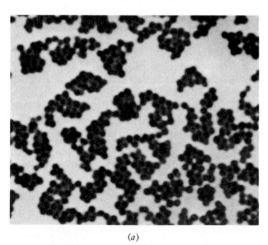

(a)

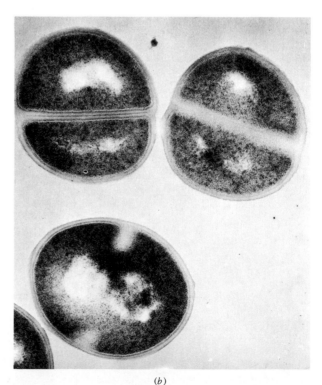

(b)

by *S. aureus*. The ability of this organism to produce disease depends both on its ability to resist phagocytosis and the effect of some of the toxins and enzymes secreted by the cell.

Toxins and Enzymes of *S. aureus*

Coagulase is an enzyme that is secreted by the bacterium into the environment. It activates a coagulase-reacting factor (CRF) normally present in plasma, causing the plasma to clot by the conversion of fibrinogen to fibrin. All coagulase-producing staphylococci are by definition *S. aureus*. The role of coagulase in the production of disease is unclear, but it has been postulated that it may coat the organisms with fibrin to inhibit their phagocytosis. *S. aureus* also produces a bound coagulase that causes the organisms to clump when mixed with plasma. Bound coagulase can convert fibrinogen directly to fibrin and does not require the presence of CRF for activity.

All hemolysins produced by *S. aureus* give a β-type hemolysis. However, because of their varied effect on many types of cells, they are more frequently referred to as toxins. There are four such toxins: alpha (α), beta (β), gamma (γ), and delta (δ). These can be differentiated from each other by antigenic distinctions and by the type of erythrocytes they preferentially lyse. All four toxins possess properties that result in varying degrees of toxicity for leukocytes and tissue cells. Some (particularly α toxin) kill skin cells and are lethal if injected into mice and rabbits.

Leukocidin activity, which is separable from these toxins, can be demonstrated in many pathogenic staphylococci. The leukocidin appears to combine with membrane phospholipid, resulting in increased permeability and death of the cell. It may also result in a degranulation of the polymorphonuclear leukocytes, similar to that described for streptolysin O.

Exfoliatin is an exotoxin that causes a sloughing of the epidermis. The infection is characterized by wrinkling and peeling of the epidermis, resulting in a considerable loss of fluid from the denuded skin (see Figure 28.2). The disease, called staphylococcal scalded-skin syndrome (SSSS), is caused by the diffusable exfoliatin; the infecting staphylococci may or may not be present at the denuded site. SSSS is seen most commonly in newborn infants, but it has occurred in adults receiving immunosuppressive therapy.

S. aureus also secretes a series of enterotoxins (see food poisoning, Chapter 26), penicillinase (which destroys penicillin), hyaluronidase (spreading factor), lipases, and a staphylokinase that lyses fibrin clots in a manner analogous to the action of streptokinase (see Chapter 24). In addition, *S. aureus* possesses a surface component, protein A, that inhibits its phagocytosis by binding to the Fc portion of IgG; protein A thereby

Figure 28.2 Neonatal mouse injected 24 h earlier with 1 × 10⁹ staphylococci from phage group II. The staphylococci were obtained from a patient with scalded skin syndrome.

competes with leukocytes for the Fc portion of specific opsonins.

A new class of staphylococcal toxins, termed pyrogenic toxins, has recently been described. There are three such toxins, designated A, B, and C, that can be separated by gel electrophoresis. They are characterized by their ability to cause fever in rabbits (and, presumably, humans) and to greatly enhance susceptibility to lethal shock by endotoxins. These toxins are very similar to the streptococcal erythrogenic toxins in that they also can cause a scarlet-fever-like rash. Pyrogenic toxins types A and B are secreted by most strains of *S. aureus*, whereas pyrogenic toxin C is found in a more limited distribution (see toxic shock syndrome).

Pathogenicity Throughout life we are surrounded by virulent or potentially virulent staphylococci. It is estimated that 15 to 20 percent of the general population carry coagulase-positive *S. aureus* in their noses and throats, and this incidence increases to 40 to 70 percent for hospital personnel. However, overt disease in a healthy individual is not a frequent event. People most susceptible to serious staphylococcal infections include those whose ability to phagocytose and destroy the staphylococci is not completely developed or is significantly inhibited. Such individuals include newborns, surgical or burn patients, people receiving immunosuppressive drugs, or those with immunodeficiency diseases such as chronic granulomatous disease (see Chapter 17). Individuals with lower respiratory viral infections such as influenza or measles and diabetic patients are also more susceptible to staphylococcal infections.

Infection of a hair follicle resulting in a localized superficial abscess or boil is undoubtedly the most frequent manifestation of staphylococcal disease (see Figure 28.3). These lesions usually heal spontaneously but may spread to the subcutaneous layers of the skin to produce a furuncle. They may then continue to

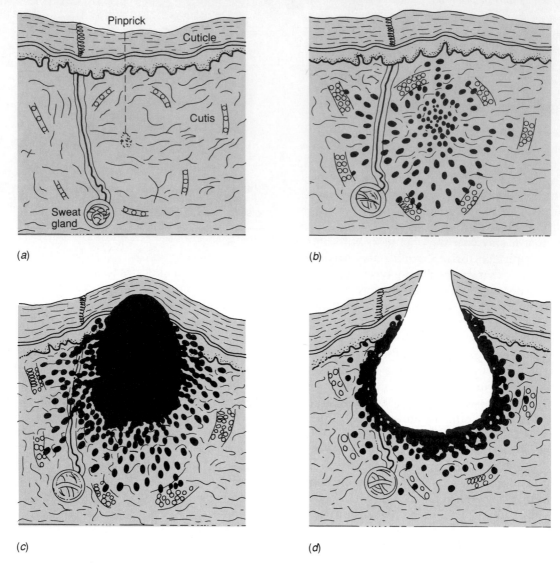

Figure 28.3 The history of a boil. (*a*) A cross section of normal skin. Note the surface layer, or cuticle, and the "true skin," or cutis. In the cutis the blood capillaries are just wide enough for the blood cells to pass through in single file. The skin has just been pricked by a dirty pin. On the point of this pin were several virulent organisms that were deposited subcutaneously. (*b*) The toxin from these organisms diffuses through the cutis. The capillaries dilate. The leukocytes force their way through the walls of the capillaries and travel toward these bacteria. The skin in this region is now swollen, red, hot, and painful. (*c*) The migration of leukocytes has continued until they now form a dense mass surrounding the organisms. The poison has killed all the leukocytes and also all the cutis immediately around them, and now digestive fluids from the dead leukocytes are turning the whole dead mass into liquid pus. The boil has "come to a head." There is a little lump on the skin, and through its thin covering of cuticle can be seen the yellow exudate. (*d*) The boil has finally ruptured. The liquid pus has escaped, carrying with it the organisms and most of their toxins; the migration of leukocytes has stopped. The capillaries are returning to their normal size, and now new tissue will grow and fill up the hole.

spread to include multiple contiguous lesions, which are then referred to as carbuncles.

An infection of the bones, staphylococcal osteomyelitis, occurs most frequently in boys under the age of 12 and, if untreated, may have a mortality rate of 25 percent. The source of such infections may be adjacent tissue infections, or the infection may be spread through the bloodstream.

Staphylococcal wound infections are less common now than they were in the preantibiotic era, but they are still the single most common hospital-acquired surgical or wound infection. Such infections most often appear to arise from endogenous strains carried by the patient, but they can also be transmitted by medical personnel or by contaminated bedding or equipment.

Staphylococcal enteritis is a severe infection of the small or large intestine that may follow gastrointestinal surgery or occur after intensive antibiotic therapy, which destroys much of the normal intestinal flora. The necrotic lesions in the intestinal wall are believed to result from the secretion of enterotoxin by the invading staphylococcus.

Toxic shock syndrome (TSS), a seemingly new syndrome, was first described in 1978, with almost all confirmed cases occurring in teen-aged and young women. Of these cases, the vast majority had their onset during a menstrual period. Subsequent epidemiological and microbiological studies have established that the disease results from the vaginal colonization with *S. aureus* associated with the use of tampons during the menstrual period. Nonmenstrual toxic shock syndrome has been seen following childbirth, therapeutic abortions, and the appearance of infected surgical wounds, deep abscesses, and infected lesions such as burns, abrasions, lacerations, furuncles, and insect bites.

Symptoms are variable but include fever, rash, peeling of the skin on the palms and soles, diarrhea, and vomiting. Since bacteremia rarely occurs, these symptoms are suggestive of a staphylococcal toxin. Such a toxin has been independently isolated and characterized by Merlin Bergdoll and his associates at the University of Wisconsin and by Patrick Schlievert and his associates at the University of Minnesota.

The precise nature of toxic shock syndrome toxin (TSST-1) is still under investigation, but it is generally accepted that TSST-1 is synonymous with pyrogenic toxin C. Moreover, there have been a number of investigations reporting that TSST-1 can enhance the susceptibility to endotoxic shock by several thousand-fold, which appears to explain the hypotension and fatal shock associated with this disease. It is also noteworthy that TSST-1 is very similar to the streptococcal pyrogenic toxins, suggesting that TSS may actually be a staphylococcal scarlet fever. One of the mysteries of this disease is that although TSST-1 seems to be produced by all vaginal strains causing TSS, many non-vaginal cases of TSS have yielded strains that do not appear to produce TSST-1. This has led to the postulation that more than one toxin may be able to cause the symptoms of TSS. This has been borne out by reports in which staphylococcal enterotoxins A, B, and C, as well as streptococcal pyrogenic toxins A, B, and C have been implicated as causes of TSS.

The ubiquity of TSST-1-producing strains was demonstrated by a report in which 23 strains of *S. aureus* were isolated from 44 asymptomatic males. Of these, 9 produced TSST-1 and, surprisingly, all persons still carried the same strains 3 months later.

Some insight into the mechanism of action of TSST-1 was gained when it was found that TSST-1 is a potent inducer of both interleukin-1 and cachectin (tumor necrosis factor). Moreover, TSST-1 and any endotoxin present will act synergistically to induce the synthesis of still more cachectin. These observations explain how TSST-1 causes the fever and irreversible shock seen in TSS.

One of the curious aspects of this disease is the recent occurrence of large numbers of tampon-associated TSS, in spite of the fact that tampons have been in use for more than 40 years. One possible explanation is that the newer, highly absorbent tampons provide a more aerobic environment for the staphylococci than did the tampons used a decade or two ago. Under such conditions, it appears that toxin production is stimulated. Another explanation came from the work of Edward Kass and his associates at Harvard University. They reported that low Mg^{2+} concentrations stimulated the synthesis of TSST-1. Furthermore, they provided data demonstrating that fibers within the newer, superabsorbent tampons bound Mg^{2+} from vaginal fluids, thus stimulating TSST-1 production. Undoubtedly, future studies will provide a more definitive explanation for the emergence of this syndrome. It seems very probable, however, that the sequence of events leading to this syndrome begins with the contamination of the tampon by the fingers.

Women can reduce their risk of TSS by not using tampons or by not using tampons continuously during the menstrual period.

Staphylococcal pneumonia is rarely a primary event in an otherwise healthy person. Infection with influenza virus, however, predisposes one to a serious, and frequently fatal, pulmonary infection by staphylococci. There is now considerable evidence that the high fatality rate of such infections may result from TSST-1 producing strains of staphylococci.

Clinical Management Since staphylococci are so frequently found on the skin and in the nasopharynx, one cannot always be certain that one is dealing with the true etiological agent of the disease or with a contaminating *S. epidermidis*.

S. aureus normally produces a light golden pigment and possesses a ribitol teichoic acid in its cell wall that is antigenically distinct from the glycerol teichoic acid of *S. epidermidis*. However, it is the ability of *S. aureus* to produce coagulase that is used to differentiate these two species (see Figure 28.4).

Further division of *S. aureus* employs phage typing to assign an unknown strain to one of four phage groups (see Table 28.1). In practice, one places a small drop of each phage group onto a plate previously seeded with the unknown strain of *S. aureus*. After overnight incubation, clear plaques of lysis permit one to rank the unknown strain in one of the four phage

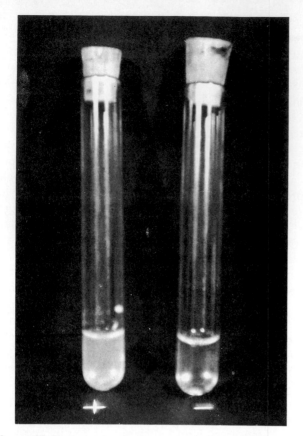

Figure 28.4 Coagulase test for staphylococcus pathogenicity. If the staphylococcus secretes the enzyme coagulase, it will cause blood plasma to clot. The tube to the left shows a positive test. The tube to the right (no coagulation) shows a negative test.

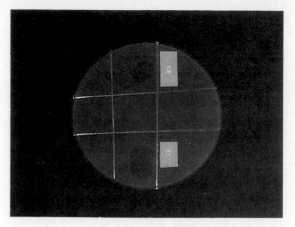

Figure 28.5 Bacteriophage typing of *Staphylococcus aureus*. Note that this strain is lysed by phages 53 and 84, thus placing it in phage group III.

groups (see Figure 28.5). Additional breakdown can be accomplished through the use of individual phages within the assigned group.

S. aureus appears to be consistently adaptable in becoming resistant to drug therapy; as a result, many antibiotics soon become ineffective for the treatment of staphylococcal infections. Many strains of staphylococci that produce penicillinase can be successfully treated with the penicillinase-resistant semisynthetic penicillin called methicillin, but strains resistant to this

TABLE 28.1 Lytic Phage Groups of *S. aureus*

Group	Phage Numbers
I	29, 52, 52A, 79, 80
II	3A, 3C, 55, 71
III	6, 42E, 47, 53, 54, 75, 77, 83A, 84, 85
IV	42D
Not grouped	81, 187

antibiotic are now being found more frequently. Hence, any isolated staphylococcus must be assayed for antibiotic susceptibility as soon as possible, even though therapy should usually commence before such results are available. Once therapy has begun, it should be intensive, so all organisms are killed before mutation to a slightly higher level of drug resistance can occur.

Control is directed primarily at individuals who are most susceptible to staphylococcal infections. The newborn nurseries and surgical operating rooms of hospitals are particularly dangerous areas, where rigid aseptic procedures must be followed. In the event of a hospital outbreak, the etiological staphylococcus should be phage-typed and all attending personnel cultured to locate the possible source of the organism.

S. epidermidis exists as normal flora on the human skin and, in general, is not a problem for the normal, healthy individual. However, it has now become an opportunistic pathogen causing nosocomial infections in joint and vascular prostheses. It is also a cause of urinary tract infections, particularly in children and in elderly male patients who have undergone urethral instrumentation.

S. saprophyticus is a frequent cause of urinary tract infections in women, particularly in the 16-to-30-year age group. It is seldom observed in males. The source for these infections is obscure, since this organism is only rarely isolated from the urine, rectum, or skin of uninfected humans. It can, however, be isolated from the skin of a variety of animals and from lesions on the hands of people handling animals. *S. saprophyticus* can be differentiated from *S. epidermidis* only by its poor growth under anaerobic conditions and by its resistance to novobiocin. Therefore some believe it should only be considered as a variant of *S. epidermidis*.

Streptococcus pyogenes

In addition to producing scarlet fever and streptococcal sore throat, *S. pyogenes* (group A β-hemolytic streptococci) can produce serious infections by invading the host through the skin. This organism is a gram-positive coccus that typically occurs in chains and, when grown on blood agar, produces a clear zone of β hemolysis around the colony.

Pathogenicity   Among the variety of infections that can be caused by the β-hemolytic streptococci are cellulitis, erysipelas, puerperal fever, and wound infections. **Cellulitis** is a rapidly spreading infection involving the skin, the subcutaneous tissues, and the lymphatic tissues. Symptoms may be marked due to the liberation of an exotoxin. **Erysipelas** is an acute infection of the skin resulting in characteristic red, edematous lesions found most commonly on the face and the legs. **Puerperal fever** is an acute infection caused by the introduction of the infectious agent into the uterus following childbirth. In addition, skin abrasions or surgical wounds contaminated with the organisms may result in infections of varying severity. Streptococci may also be secondary invaders in impetigo.

The incubation period varies. In puerperal fever it is usually 1 to 3 days. Humans are the reservoirs of infection. A person may be self-infected or may acquire the organisms from an external source. The organisms are found in the respiratory tract, in the intestinal tract, and on the skin.

Clinical Management   The diagnosis is confirmed by isolating and identifying the causative organism based on colonial appearance, β-hemolysis, sensitivity to bacitracin, and appearance of Gram stain.

Control measures call for strict asepsis, especially around open wounds. All dressings and discharges from an infected person should be burned or disinfected. Penicillin is by far the drug of choice, but the tetracyclines and the sulfonamides are bacteriostatic against these organisms.

Pseudomonas aeruginosa

P. aeruginosa is a gram-negative, nonsporulating motile rod. It is a strict aerobe and grows readily in ordinary culture media. The organisms are found as part of the normal flora of the intestinal tract as well as on human skin. As they grow, they produce pyocyanin and fluorescein; these two compounds color the culture blue-green.

Pathogenicity   *P. aeruginosa* does not, as a rule, act as a primary invader, but it does cause infections and severe disease under the following circumstances:

1. It can cause infections when it is mechanically placed into the urinary tract during catheterization or into the meninges during a lumbar puncture.
2. It is able to contaminate respiratory ventilators and deliver large numbers of organisms directly into the lungs of an already debilitated person.
3. It may cause a fatal sepsis in people with leukemia or those who are receiving immunosuppressive drugs.
4. Because of its resistance to many antibiotics, it can cause severe infections in people receiving antibiotic therapy for burns or wounds.

Table 28.2 provides data for the incidence of *P. aeruginosa* infections in a large number of hospitalized individuals. As can be seen, burn patients are by far the most susceptible in that approximately 25 percent become infected.

In spite of its limited invasive ability, *P. aeruginosa* produces several exotoxins, a leukocidin, and several enzymes that enhance its virulence. Of these, the product that has been most extensively studied is

TABLE 28.2   **Risk of Infection Due to *Pseudomonas aeruginosa* in 90,000 Patients Under Surveillance in Community Hospitals**

Hospital Service or Area	Site of Infection	Patients at Risk per 1000
Burn unit	Burn	246
Burn unit	Urinary tract	16
Burn unit	Surgical wound	11
Urology	Urinary tract	6
Burn unit	Lower respiratory tract	5
General surgery	Surgical wound	4
Medicine	Urinary tract	4
General surgery	Urinary tract	4
General surgery	Lower respiratory tract	3
Gynecology	Urinary tract	2

exotoxin A. The potency of this toxin can be attested by the fact that it has an LD_{50} (lethal dose for 50 percent of the population) of less than 1 μg for the mouse, and on a weight basis shows a similar LD_{50} for monkeys, dogs, and cats. Surprisingly, the mechanism of action of exotoxin A is identical to that of diphtheria toxin (see Chapter 24), in which the toxin acts enzymatically to cleave the nicotinamide moiety from NAD and then to catalyze the transfer of the resulting ADP-ribose to form a covalent bond with elongation factor 2, thus freezing the ribosome and blocking protein synthesis. It is also interesting to note that these two toxins are antigenically distinct and that the diphtheria toxin is encoded in a temperate phage genome, whereas the *Pseudomonas* toxin is encoded in the bacterial chromosome. We can only speculate as to why *P. aeruginosa* rarely infects a healthy person whereas literally millions of individuals have died from diphtheria. The answer would appear to lie in that elusive concept of invasive ability: *Corynebacterium diphtheriae* has it, *P. aeruginosa* does not.

P. aeruginosa is also responsible for skin infections that are most frequently acquired in hot tubs that are used by numerous persons. Such tubs are usually not chlorinated and provide a suitable medium for the growth of the organisms. *P. aeruginosa* also may be the cause of a chronic otitis media known as "swimmers ear."

P. aeruginosa also elaborates two proteolytic enzymes that may be involved in disease production. One (an elastase) digests elastin, a component of arterial walls. Elastase also inactivates some of the components of the complement system. The other enzyme is a collagenase that may be involved in the spread of the organisms within the body.

Clinical Management Control measures lie in the aseptic treatment of wounds, extreme care with burns, proper disposal of dressings, and care in the sterilization of urinary catheters. Current therapy uses tobramycin, carbenicillin, colistin, and gentamicin. However, because of the frequency with which these organisms become resistant to antibiotics, new antimicrobials will undoubtedly become the therapy of choice.

Leptospira: Leptospirosis

Leptospira are morphologically characterized as spirochetes exhibiting a fine, tightly coiled spiral and usually possessing a hook on one or both ends. They are very thin (0.1 μm), usually quite long (10 to 20 μm), and readily visible with darkfield microscopy (see Figure 28.6). Their extremely active motility appears to be due to the rhythmic contractions of their flagella (axial fibrils), particularly at the hooked ends.

Figure 28.6   (a) Darkfield micrograph of *Leptospira interrogans* serotype *illini* ($\times$1000). (b) Electron micrograph of one end of a negatively stained leptospire, showing clearly the axial fibril ($\times$51,000).

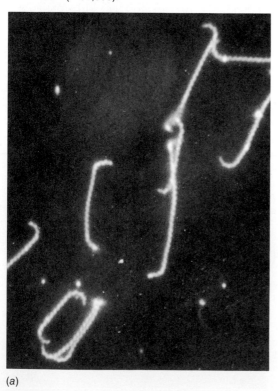

(a)

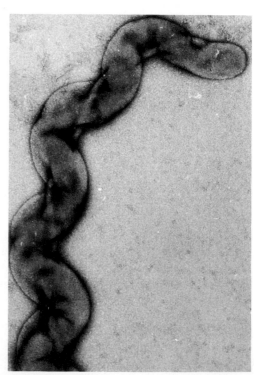

(b)

Leptospira can be grown in a number of artificial media supplemented with bovine albumin or sterile rabbit serum. All are obligately aerobic and appear to grow best in a neutral medium at 30°C.

The taxonomy of the *Leptospira* is in a fluid state and is currently being studied by the subcommittee on *Leptospira* of the International Committee on Systematic Bacteriology. Most taxonomists, however, divide the genus into two species—the pathogenic *L. interrogans* and the saprophytic *L. biflexa*. *L. interrogans* is additionally divided into serogroups and many serovarieties (termed **serovars**). Each serovar is defined by a microscopic agglutination test using serovar-specific rabbit antiserum. Thus, current terminology changes the older designation of *L. icterohemorrhagiae* to *L. interrogans* serovar *icterohemorrhagiae*.

Pathogenicity *Leptospira* are primarily parasites of vertebrates other than humans—such as rodents, dogs, pigs, cattle, and racoons—in which the bacteria appear to persist in a lifelong asymptomatic infection of the kidney, and they are continuously shed in the urine by some infected animals. Certain serovars are routinely associated with specific hosts (for example, *icterohemorrhagiae* with rodents, *canicola* with dogs, *pomona* with pigs, *hardio* with cattle, and *autumnalis* and *grippotyphosa* with mice), but cross-infections do occur. Interestingly, leptospirosis has become a major cause of bovine abortion in certain parts of the world, notably the United Kingdom. Moreover, human infections by serovar *hardio*, originating from such bovine sources, are not unusual. The same serovar has been reported also to be transmitted from mother to child via human breast milk.

Disease in humans may be caused by any one of the serovars of *Leptospira*. The usual sequence of events begins when a person becomes infected through skin abrasions or intact mucous membranes following contact with urine from an infected animal or by exposure to urine-contaminated water or soil. The organisms enter the blood, invading various tissues and organs, particularly the kidney, liver, meninges, and conjunctiva. Physical symptoms frequently include muscular pain, headache, photophobia, fever, and chills. Infections may or may not result in jaundice or meningitis. One serovar, *icterohemorrhagiae*, causes a more severe illness known as Weil's disease, in which the fatality may run as high as 25 percent. Another unusual leptospiral infection, pretibial (or Fort Bragg) fever, characterized by a rash on the shins, is caused by serotype *fort-bragg*.

Clinical Management Leptospirosis is an occupational hazard for sewage workers, slaughterhouse employees, or individuals who come into contact with rat-infested areas. Interestingly, the prevalence of leptospirosis during the past several decades has been increasing, and it is now apparent that the vast majority of cases are not related to occupation but, rather, to contact with pets, particularly dogs.

The observation of *Leptospira* in a darkfield examination of blood would provide strong support, but an unequivocal diagnosis requires that the organisms be grown and serologically identified. Organisms from the patient's blood can be cultured in artificial media, or the blood can be inoculated into young hamsters or guinea pigs to enrich the number of organisms present.

Treponema

The diseases known as bejel, yaws, and pinta are closely related to syphilis in that they are all caused by species of the genus *Treponema*, and the spirochetes present in their lesions are indistinguishable morphologically from *T. pallidum*, the causative agent of syphilis. They are also similar in that they cause severe tissue destruction.

They all differ from syphilis in that they are not sexually transmitted diseases. For the most part, each disease is fairly well limited to a specific geographic area, and none is found in the United States. Yaws is found exclusively in the tropics, bejel is restricted for the most part to Arabia, and pinta is found in Central and South America, primarily in Mexico and Cuba.

Bejel and yaws are primarily diseases of children, while pinta may infect any age group. It is believed that all three are transmitted either by direct contact or by the bite of flies that mechanically carry the infectious agent from an infected individual.

These diseases may result in extensive disfigurement. All respond to treatment with penicillin, and a major effort is being made through the World Health Organization to control them where they are endemic.

Clostridium tetani: Tetanus

C. tetani is the etiological agent of tetanus, or lockjaw. The organisms in the genus *Clostridium* all have the unique property of being obligate anaerobes; they cannot grow in the presence of air. In addition, the clostridia are large, gram-positive, motile rods that, like the members of the genus *Bacillus*, produce extremely resistant endospores. The position and size of the endospore as it occurs inside the vegetative cell will vary from one species to another, but in the case of *C. tetani* the endospores occur as terminal swollen spores that make the organisms look like miniature tennis rackets (see Figure 28.7).

The direct cause of disease is the very potent exotoxin liberated from the organisms growing in the tissue. However, tetanus is a little unusual in two respects: the causative organisms have essentially no invasive ability and they are able to grow only under

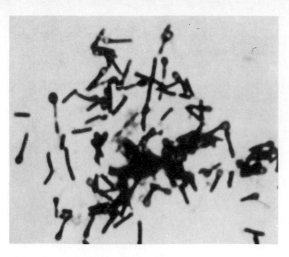

Figure 28.7 Cells of *C. tetani* after 24 h on a cooked-meat glucose medium (×4500). Note spherical terminal endospores.

anaerobic conditions. One might wonder how such organisms could ever cause disease. In brief, they accomplish this by contaminating a wound in which a number of the host cells have been killed and are no longer receiving an adequate amount of oxygenated blood.

The species can be divided into a number of serological types, but the exotoxin released by all of them is identical. This is extremely important, since immunity is directed only against the exotoxin. The exotoxin produced by *C. tetani* is an extremely potent and lethal neurotoxin. It is believed to bind to receptors on nerve cells in the spinal cord and, by doing so, interfere with the regulation of neurotransmitters that control the relaxation of muscles. The major manifestation of this disease is a convulsive contraction of voluntary muscles. Since the spasms frequently involve the neck and jaws, the disease is sometimes referred to as "lockjaw." Death ordinarily results from muscular spasms affecting the mechanics of respiration.

The incubation period is usually between 5 and 10 days but in some cases may be as long as several weeks.

The organism is widely disseminated in soil, especially in fertilized farmlands, and is found as a normal inhabitant in the intestinal tract of humans and animals. Thus, soil, dust, and feces are the common sources of infection. Although the tetanus spores are usually introduced through deep wounds such as gunshot or puncture wounds, minor abrasions and burns may also be infected.

Clinical Management A clinical picture of tetanus with a history of injury is usually sufficient to provide a clinical diagnosis. A more specific diagnosis requires the isolation and identification of the organism as well as the demonstration of tetanus toxin pro-

duction by the isolated bacterium. However, this is rarely done.

Good active immunity can be acquired from tetanus toxoid. The initial process requires three injections (usually given at 2, 4, and 6 months of age along with diphtheria toxoid and pertussis vaccine). Booster doses are not necessary for 10 years.

Following an injury, tetanus antitoxin is usually administered to individuals who have not been immunized previously with the tetanus toxoid. Individuals who have been actively immunized receive only a booster injection of the toxoid.

The organism is sensitive to penicillin, but the antibiotic has no effect on the neutralization of the toxin. Thus, the use of antibiotics for the treatment of tetanus is far less important than is the treatment with antitoxin or the use of booster injections of toxoid.

Clostridium perfringens and Other Members of the Gas Gangrene Group

There are a number of different species of the genus *Clostridium* that can infect wounds and produce gas gangrene. All of these species are morphologically similar—they are gram-positive, motile, spore-forming rods, and all are obligate anaerobes. Furthermore, they are pervasive in soil, so that contamination of a wound with dirt will very likely mean contamination with one of the gas gangrene clostridia. Another common property of these organisms is their ability to produce a large variety of very powerful exotoxins that can diffuse out of the contaminated wound into the host, causing death.

C. perfringens is the most common causative agent of gas gangrene. Other species of *Clostridium* frequently associated with this disease are *C. novyi*, *C. septicum*, and *C. histolyticum*, as well as a number of less common species. It is not uncommon for a wound to be infected with more than one species.

Pathogenicity Gas gangrene results from the contamination of wounds with *Clostridium* spores that are universally present in the soil. Because all of these organisms are obligate anaerobes, they are unable to grow in the presence of an oxidation-reduction potential like that found in an area freely exposed to atmospheric oxygen. However, deep wounds that become necrotic as a result of diminished blood supply provide an excellent environment for the growth of the anaerobic clostridia. Thus, in necrotic tissue, the spores are able to germinate, reproduce, and secrete their exotoxins into the surrounding environment. Liberation of the toxins destroys more tissue, resulting in a rapid and fulminating spread of the organism in the necrotic environment. In addition, carbohydrates are fermented with the production of large quantities of gas in the tissues. The pressures in the tissues resulting from the gas formation may cause still more restriction of blood supply to adjoining healthy tissues and still more necrosis. In the absence of surgical and antitoxic treatment, severe toxemia and death frequently ensue.

The toxins produced by the various species of the gas gangrene group are quite similar but not identical from one species to another. Actually, a number of them have not been purified and characterized; they are referred to by the general name of **lethal toxins.** The toxins produced by *C. perfringens* have received the most study. At least 12 different toxins have been identified. Of these, one of the most extensively studied is a lecithinase that is frequently called the alpha toxin. Lecithinase hydrolyzes the phospholipid lecithin to a diglyceride and phosphorylcholine (Figure 28.8). Since lecithin is a component of cell membranes, its hydrolysis can result in cell destruction throughout the body. Another toxin produced by members of this group is the θ (theta) toxin. This lethal, hemolytic substance is characterized by its effect on the heart (i.e., its cardiotoxic properties). Other toxic enzymes produced by the gas gangrene group of organisms include a collagenase that hydrolyzes the body's collagen, a hyaluronidase that hydrolyzes the body's hyaluronic acid (the intercellular cementing ground substance of tissue), a fibrinolysin that breaks down blood clots, a DNase, and a neuraminidase that can remove the neuraminic acid from a large number of glycoproteins. With such a large array of toxic substances,

it is no wonder that gas gangrene was one of the major causes of death in the American Civil War—and undoubtedly in many other wars.

The organisms are very widely disseminated. The spores of *C. perfringens,* which account for 70 to 80 percent of all gas gangrene, may be found in soil, dust, manure, human feces, rivers, and various foods. *C. perfringens* forms a part of the normal flora of the intestinal tract of both humans and animals. Of the six serological types of the organism, type A is the most prevalent.

The incubation period is usually about 8 to 48 h, depending on the extent of contamination and the severity of the wound.

Clinical Management Diagnosis is based on the clinical picture. However, final proof as to the causative organism requires its isolation and identification.

Control depends mostly on the surgical cleansing of wounds to eliminate any extraneous material and necrotic tissue. Because the various toxins are antigenic, antitoxin is commercially available. Antitoxin may be administered to patients whose wounds are deep and dirty. Anyone who suffers a severe wound can develop gas gangrene, but the condition is particularly a concomitant of war wounds.

Sulfonamides, penicillin, or the tetracyclines should be used with the antitoxin. However, it must be remembered that the antibiotics will not penetrate to the necrotic areas where the organisms grow and that these areas must be cleaned surgically.

Figure 28.8 Hydrolysis of lecithin by a toxin (lecithinase) to a diglyceride and phosphorylcholine.

Lecithin
(phosphatidylcholine)

Diglyceride

Phosphorylcholine

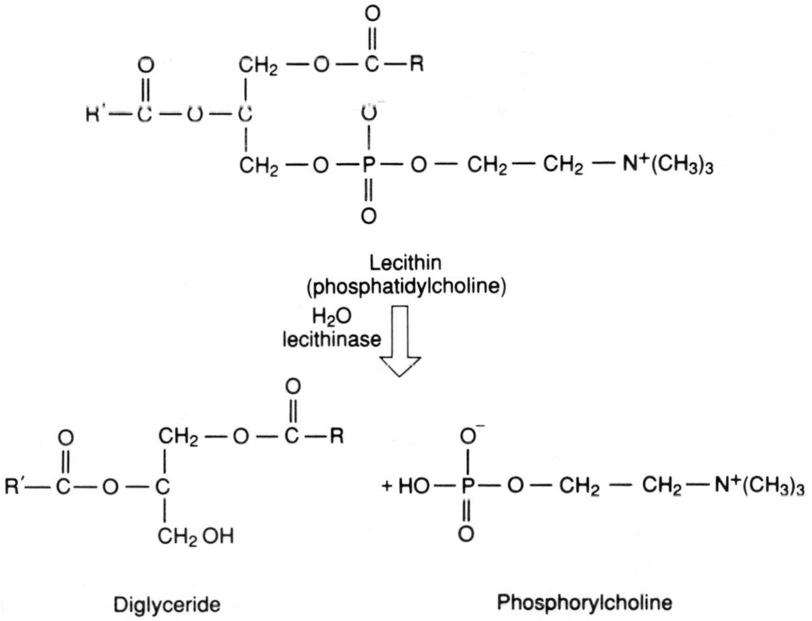

DERMATOPHYTOSES

The etiological agents of the dermatophytoses are closely related fungi that utilize keratin for growth. (Keratin is present in the skin, hair, and nails.) These organisms share antigenic and physiological properties and, for the most part, cause similar types of infections. The major differentiation between the three genera of dermatophytes (*Trichophyton*, *Microsporum*, and *Epidermophyton*) is based on the kinds and the appearance of spores and hyphae (see Figure 28.9). In spite of the fact that the dermatophytes are usually considered to be members of the Deuteromycetes, sexual spores have been observed in about half of the known species of *Trichophyton* and *Microsporum*.

Dermatophytes invade only keratinized tissues— such as hair, nails, and skin—causing an infection that does not extend into the subcutaneous areas of the body. In general, dermatophyte infections begin in the horny layer of the skin and spread in a centrifugal pattern with a "ringworm" appearance. Early concepts attributed these infections to insects. The Romans named these diseases *tinea*, meaning small insect larva, and this prefix is still used as part of the clinical terminology for them. All genera of dermatophytes may cause any of these clinical entities. However, species of *Microsporum* usually invade the hair and skin but not the nails; *Epidermophyton* invades the skin and nails but usually not the hair; and species of *Trichophyton* infect hair, skin, and nails.

Pathogenicity The major diseases caused by the dermatophytes are the following:

1. Tinea pedis, or athlete's foot, is characterized by itching between the toes and the formation of small blisters. The infective agents are species of *Trichophyton* or *Epidermophyton floccosum*.

2. Tinea corporis, or ringworm of the smooth skin (see Figure 28.10), is characterized by annular lesions with a border containing vesicles. *Trichophyton rubrum* and *T. mentagrophytes* are the most common etiological agents.

3. Tinea capitis, or ringworm of the scalp, appears as expanding rings on the scalp, with organisms

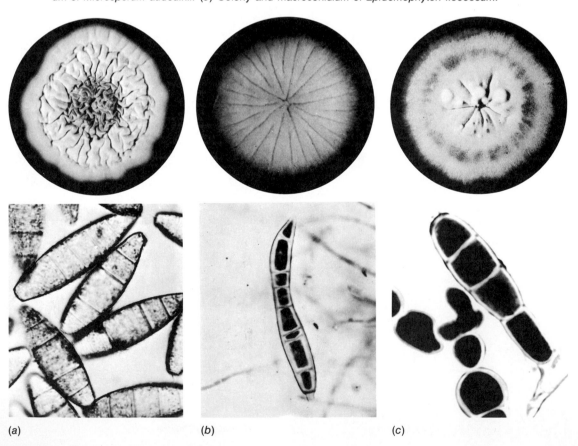

Figure 28.9 (a) Colony and macroconidia of *Trichophyton gallinae*. (b) Colony and macroconidium of *Microsporum audouinii*. (c) Colony and macroconidium of *Epidemophyton floccosum*.

(a)　　　　　(b)　　　　　(c)

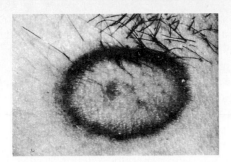

Figure 28.10 *Tinea corporis* (ringworm of the body), showing characteristic ring of inflammation.

growing in and on the hair. Inflammatory reactions may cause deep ulcers, which heal with scarring and permanent loss of hair. The most common etiological agents are *Microsporum canis, M. audouinii,* and *T. tonsurans.*

4. Tinea unguium, or ringworm of the nails, is characterized by thickened, discolored, brittle nails (see Figure 28.11). All species of dermatophytes have been involved as etiological agents, but *T. rubrum* is found most commonly.

The incubation period for tinea capitis, corporis, and pedis is about 10 to 14 days; it is unknown for tinea unguium.

Humans and infected animals, such as dogs and cats, serve as reservoirs of the infective agents. The organisms are spread from human to human indirectly by contaminated floors, towels, combs, theater seats, bed linens, and similar objects that come into contact with infected areas of the body.

Clinical Management Topical applications of long-chain fatty acids, salicylic acid, selenium sulfide, sulfur in ointment, and many other compounds have for years been used to treat the dermatophyte fungal infections. Currently, the treatment of choice is a top-

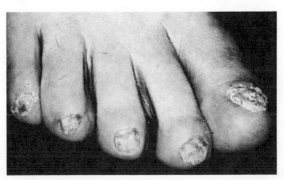

Figure 28.11 *Tinea unguium* (ringworm of the nails, or onychomycosis) in the toenails of a 61-year-old male.

ical preparation containing one of the newer imidazole drugs. Griseofulvin is often used in conjunction with a topical agent for the treatment of tinea capitis, or it may be used alone for severe nail infections or extensive skin lesions. Given orally, it becomes incorporated into newly synthesized keratin layers, rendering them resistant to fungal infection, but has little effect on the keratin structures already infected. Thus, depending on the site of the body involved and the causal organism, therapy may vary from a daily treatment for several weeks with griseofulvin and/or topical agents (for tinea capitis caused by *M. audouinii*) to many months of treatment for nail infections caused by *T. rubrum.*

Diagnosis is based on a microscopic examination of infected material and cultures on Sabouraud's dextrose agar. Chronically infected persons often develop a hypersensitivity to the fungus, which may cause an allergic response. This is called a dermatid (usually abbreviated "id") reaction and is manifested by the appearance of vesicles, usually on the hands. Such individuals will also react strongly to an extract of the fungus (called trichophytin), which will produce a delayed-type skin reaction.

SUBCUTANEOUS MYCOSES

Subcutaneous mycoses are fungal infections that gain entrance to the body as a result of trauma to the skin. Once established, these mycoses generally remain localized in the traumatized area but may extend via draining lymph channels to regional lymph nodes. Fungi involved in subcutaneous types of mycoses are normal soil inhabitants. Since the infections they cause are localized and progress slowly, they are considered organisms of low virulence.

Sporothrix schenckii: Sporotrichosis

S. schenckii is found in soil and plants in a worldwide but sporadic distribution. It also occurs in wood and moss, and infections frequently result from inoculation by splinters, thorn pricks, sphagnum moss, grasses, or garden soil.

Pathogenicity Sporotrichosis is usually a localized infection of the skin, subcutaneous tissues, and regional lymphatics. The organism gains entrance to the body following trauma or through open wounds. After an incubation period varying from 1 week to 6 months, a subcutaneous nodule appears and develops into a necrotic ulcer. The initial lesions heal as new ulcers appear in adjacent areas. Lymphatics in the area develop nodules and ulcers along the lymph channels and these may persist for months or years. In areas where *S. schenckii* is highly endemic, many persons develop a cellular immunity to the organism, which can be demonstrated as a delayed-type skin reaction to sporotrichin. Rarely, *S. schenckii* may invade the blood and spread to various organs of the body; occasionally, the fungi may enter the host via the respiratory tract, resulting in a primary pulmonary sporotrichosis.

Clinical Management *S. schenckii* is a dimorphic fungus that grows in culture, soil, or plant material as septate branching hyphae with clusters of pear-shaped conidia at the ends of the branches. In contrast, the organisms appear in infected tissue as cigar-shaped yeast cells that reproduce by budding (see Figure 28.12).

A definitive diagnosis requires the growth and identification of the infecting organism, even though sporotrichosis with lymph channel ulceration can be diagnosed clinically with reasonable confidence.

Oral potassium iodide administered over a period of weeks is the most common treatment for localized sporotrichosis; however, amphotericin B is used for relapsing cases as well as for pulmonary and disseminated sporotrichosis.

Chromomycosis

Chromomycosis (also known as chromoblastomycosis and verrucous dermatitis) is a clinical infection that may be caused by any one of several different fungi. The taxonomy of this group of agents is not yet firm, but the principal genera involved are *Fonsecaea*, *Phialophora*, and *Cladosporium*.

Pathogenicity The causative agents of chromomycosis are soil saprophytes found in decaying vegetative matter and rotting wood. Most infections appear to originate from puncture wounds; the great majority of these are seen in the tropics, particularly in Mexico and South America.

Most infections occur on the feet and legs, in rural localities where shoes are seldom worn. The original lesion appears as a small, raised, violet papule, and over a period ranging from months to years additional lesions appear in adjacent areas.

The infection generally remains localized, but lesions may become secondarily infected with bacteria, resulting in a purulent exudate. Rarely, spreading may occur via the bloodstream to involve other areas of the body, including the lungs and brain.

Clinical Management Skin scrapings mounted in potassium hydroxide show brown branching hyphae and brown chlamydospores that can be cultured on Sabouraud's glucose agar.

Potassium iodide given orally over a period of years has been a standard treatment for chromomycosis. More recently, topical and intravenous amphotericin B has been successfully employed. Surgical excision of the lesion, particularly during the early stages of the disease, has also proved effective.

BACTERIA THAT ENTER VIA ANIMAL BITES OR CONTACT WITH HIDES

Bacillus anthracis: Anthrax

B. anthracis is a large, aerobic, gram-positive spore-forming rod. The isolation of this organism by Robert Koch in 1877 marked the beginning of medical microbiology. It was in large part Koch's work with the anthrax bacillus that was responsible for the well-known Koch postulates. It was also with this organism that the first systematic vaccination was accomplished. Pasteur demonstrated that organisms grown at 42°C were no longer capable of causing anthrax but did stimulate immunity to this infection.

The virulent organisms are very large rods with square ends (see Figure 28.13). They possess a well-defined capsule composed solely of a polymer of D-glutamic acid. Although the organism forms spores readily in cultures, soil, and dead animals, it does not sporulate in the blood or the tissues of a living animal. The ability of the organism to produce disease depends both on the antiphagocytic action of its capsule and on the formation of a potent exotoxin.

Pathogenicity Anthrax, the disease caused by *B. anthracis,* is primarily a disease of sheep and cattle

---

Figure 28.12 Gram stain of yeast cells in pus from a mouse infected with *Sporothrix schenckii*.

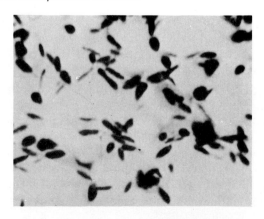

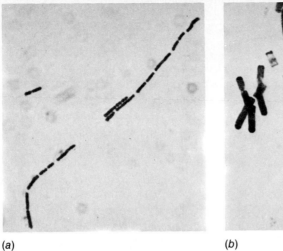

(a)   (b)

Figure 28.13 Giemsa-stained cells of *B. anthracis* (about ×2000). (a) Chains of cells after 18 hours of nutrient broth. (b) B. anthracis as it appears when growing in animals (media was supplied with all constituents necessary for antigen production). Note broader cells, square corners, and irregularly stained cytoplasm.

and, to a lesser extent, other herbivorous animals. Although of worldwide distribution, it is relatively rare in the United States. It is a serious problem in certain parts of Europe, particularly in Germany and France. Once the disease is established in an area, bacterial endospores from a dead infected animal are able to contaminate the soil, and, since the spores are very resistant, the pasture area may remain infectious for other animals for many years.

Humans become infected through the skin by contact with hides of infected animals or by inhaling the spores from infected hides. Because the pulmonary form of the disease has occurred frequently in people who sort wool, it has been commonly referred to as wool-sorters' disease. The lesion resulting from the skin inoculation (cutaneous infection) is sometimes called a malignant pustule.

Cutaneous infection is frequently an occupational hazard for people who handle livestock. The organisms get into the body through small cuts or abrasions—usually on the hands. An initial lesion occurs at the site of entry, which soon develops into a black necrotic area (see Figure 28.14). If the lesion is not treated, the organisms will invade the regional lymph nodes as well as the blood, frequently causing death within 5 to 6 days after infection. In such cases the causative agents are found in all body excretions except the urine.

A seemingly rare source of anthrax infections was reported in 1979, when an epidemic of anthrax occurred in Sverdlovsk, U.S.S.R. This epidemic, which is believed to have killed hundreds of people, supposedly originated as gastric anthrax caused by eating black-market anthrax-infected meat. In such cases, it is assumed that the organisms enter the bloodstream from the intestine and eventually reach the lungs to cause a fatal pneumonia. Although gastric anthrax is rare in the Western world, Russian textbooks have described previous epidemics of gastric anthrax that resulted in 100 percent mortality.

The pulmonary form is more rare than the cutaneous form, but when it does occur, it has a high mortality rate.

Until 1955 it was believed that anthrax caused death in infected animals by its ability to grow in the blood to tremendous numbers, resulting in the eventual blockage of the capillaries. It is now known, however, that virulent *B. anthracis* contains a plasmid that encodes for a powerful exotoxin. Anthrax toxin con-

Figure 28.14 Lesion of cutaneous anthrax (eighth day of illness) on the arm of a woman who had been a carder in a wool factory.

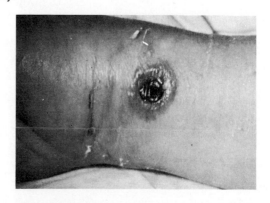

sists of a complex of several proteins whose primary known function is to increase intracellular levels of cAMP, primarily in polymorphonuclear leukocytes. This, in turn, activates various protein kinases, which phosphorylate a number of cellular proteins, inhibiting the ability of the phagocyte to carry out normal phagocytosis. Virulent organisms also contain a second plasmid that encodes for the synthesis of the poly-D-glutamic acid capsule. Both plasmids are required for virulence.

The primary source of infection for humans is direct contact with the hides of infected animals. However, a number of cases of cutaneous anthrax have occurred as a result of using shaving brushes made from contaminated animal hair. This is no longer a problem in this country, since all such commercially imported items are now sterilized before being sold. Souvenir items purchased in foreign countries are not always safe, as was exemplified by the individual who contracted anthrax from contaminated goat-hair fringe on a souvenir drum purchased in Haiti in 1974. Other Haitian products using goat hair also have been contaminated with anthrax spores (see Table 28.3); hence the Centers for Disease Control in Atlanta has recommended that all such items be considered potentially contaminated and placed in the hands of local health authorities for disposal.

Clinical Management Direct smears from lesions or sputum are helpful in making a tentative diagnosis. However, a positive diagnosis requires the isolation and identification of the causative agent.

Treatment is quite effective if started early. Antianthrax serum, sulfonamides, penicillin, and the broad-spectrum antibiotics have been effective.

Control measures are directed toward prevention, both of the disease in animals and the spread of the virulent spores to humans. Any animals dying from anthrax should be burned or disposed of in such a way that the surrounding area is not contaminated.

A vaccine composed of living attenuated strains of *B. anthracis* is used in animals—particularly in areas where pasture land is already contaminated with spores. All effective living vaccines, however, possess some toxicity, and they have not been used in the United States for humans. An experimental vaccine that is currently being evaluated consists of a strain of *B. subtilis* into which the exotoxin moiety responsible for binding to mammalian cells has been cloned. This engineered vaccine is not toxic, but it does induce the synthesis of antibodies that prevent binding of the complex virulent toxin. The vaccine used for humans is a preparation of alum-precipitated toxoid that induces a short-lived immunity and requires annual boosters.

Control of the disease is difficult for people who are occupationally exposed. However, sterilization of wool, hair, and other "risk" products does prevent the spread of the disease to people not normally exposed directly to infected hides.

Spirillum minor: Rat-Bite Fever

S. minor is the causative agent of one of the two diseases known as rat-bite fever. The organism is a short spiral rod possessing two or three rigid coils. It is a gram-negative, actively motile organism. Although there have been some reports that this organism has been grown, there are many who believe that it has never been cultured.

Pathogenicity Rat-bite fever (also called Sodoku) is characterized by a local lesion at the site of entrance of the infective agent. Fever, swelling of regional lymph nodes, and a skin rash are also promi-

TABLE 28.3 *B. anthracis* **Culture Results for Haitian Goat-Skin Products Imported into the United States, 1974**

Item	Number Cultured	Number Positive	Percent Positive
Drums	219	22	10
Rugs	58	45	78
Mosaic pictures	55	20	36
Voodoo dolls	13	3	23
Goat skins	10	4	40
Purses	10	1	40
Stool	1	1	100
Hat	1	0	0
Bottle holder	1	0	0
Total	368	96	26

Source: MMWR, June 22, 1974.

nent symptoms. This disease is far more common in Asia than in the United States.

The incubation period is usually 5 to 14 days but may be 6 weeks or longer.

Humans become infected following the bite of an infected rat or other rodent.

Clinical Management Because the organism has not been grown with certainty in artificial media, it is necessary to see the infecting agent in the blood in order to make a definitive diagnosis. Blood from a patient may be examined directly, or it may be injected into a mouse or guinea pig. The occurrence of gram-negative spirilla in the blood of a patient with a history of a rat bite is sufficient evidence for a definitive diagnosis of this disease.

Control measures are directed toward exterminating rodents. Penicillin is effective for the treatment of the disease.

Streptobacillus moniliformis: Haverhill Fever

S. moniliformis is a common parasite of both wild and white laboratory rats and mice. The disease it causes is sometimes referred to as rat-bite fever but is also called Haverhill fever.

The organism is a gram-negative, highly pleomorphic branching organism. Since the branching filaments fragment, a stained smear of the organism will contain both bacillary and coccobacillary forms. Another very interesting property of this organism is its ability to form a minute, pinhead-sized colony which is known as an L form.

Pathogenicity This disease is clinically similar to that caused by *S. minor*. Probably all sporadic cases caused by *S. moniliformis* occur as a result of bites of infected rats.

Several large epidemics of the disease have occurred through the ingestion of contaminated milk. The first occurred in Haverhill, Massachusetts, in 1925; hence the name Haverhill fever. It is not known whether the milk in this and subsequent milk-borne epidemics became infected from rat contamination or from an infected cow.

The incubation period is usually 7 to 10 days, but it may be as short as 2 days.

Clinical Management The diagnosis is usually based on a history of rat bite and is confirmed by the isolation and the identification of the causative organism. This disease is much more common in the United States than Sodoku caused by *S. minor*.

Because the organism is so widespread in the rodent population, it is impossible to control the disease

except by the elimination of rats and mice. Any person who develops fever following a rat bite should be treated promptly with penicillin without waiting for a definitive diagnosis.

TREMATODES THAT ENTER THROUGH THE SKIN

Blood Flukes

The human blood flukes belong to the genus *Schistosoma*. The life cycles of these parasites differ from that of the trematodes previously described in the following ways: (1) The adult worms are not hermaphroditic and exist as two separate sexes. The sexes can easily be differentiated because the male worm is much larger than the female (Figure 28.15). (2) A specific snail intermediate host is involved, but schistosomes do not require a secondary intermediate host and, hence, do not form metacercariae; animals, including humans, become infected by the direct penetration of the cercariae through the skin.

In general, after penetrating the skin, the cercariae migrate through the heart and the lungs, eventually reaching the intrahepatic part of the portal system. Here they mature and mate, after which (depending on the species) they migrate together to the mesenteric venules, where the fertilized eggs are deposited.

It is estimated that over 200 million people worldwide are infected with schistosomes, and perhaps as many as 400,000 cases of schistosomiasis exist in immigrants to the United States. The disease cannot,

Figure 28.15 Adult schistosomes. The larger male is coupled with the slender female worm.

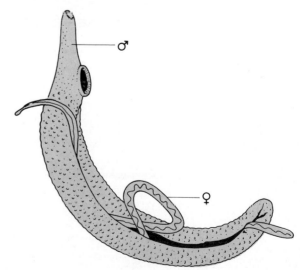

however, be spread in the United States because of the absence of the specific snail intermediate host.

Human schistosomiasis occurs in three fairly distinct stages: (1) The invasion stage, unless overwhelming, is usually asymptomatic. This stage includes the migration of the cercariae to the hepatic veins. (The cercariae are referred to as schistosomula after losing their tails following penetration.) (2) The acute stage occurs with the start of egg laying, after the mature worms have migrated to the mesenteric veins. This stage is characterized by diarrhea, fever, malaise, and general discomfort. Most of the symptoms of this second stage appear to be due to a hypersensitivity reaction to the eggs. (3) The final stage is chronic and occurs as a tissue reaction to the eggs. Eggs are passed in the urine or feces and may be deposited in various organs of the body, particularly the liver. In these cases of deposition, a tissue reaction is manifested by walling off of the foreign material to form granulomas not unlike the tubercles that occur in tuberculosis.

Three species of *Schistosoma* constitute the major causes of human schistosomiasis, namely, *S. mansoni*, *S. japonicum*, and *S. haematobium*. Each of these is tightly restricted to the genus of snail that it can use for its intermediate host. In addition, during the adult stage (a period that may last as long as 30 years), each species is located in a different area of the human body.

Schistosoma mansoni

S. mansoni is found in Africa, as well as in South America, the West Indies, and Puerto Rico. In human infections, the parasites characteristically leave the liver and the adult worms take up permanent residence in the inferior mesenteric veins of the large intestine. In this environment, they deposit eggs in the wall of the intestine; the eggs eventually break through the mucosa into the lumen and are passed with feces. After reaching fresh water the eggs hatch, and, if the appropriate species of snail is present, the miracidia enter the intermediate host to become cercariae that will initiate a new cycle of human infection (Figure 28.16).

Schistosoma japonicum

S. japonicum occurs exclusively in the Far East, and its definitive hosts include horses, cattle, pigs, dogs, cats, rodents, and water buffalo, in addition to humans.

The life cycle and pathogenesis of *S. japonicum* is as described for *S. mansoni*. The final residence for the adult worms is found in the superior mesenteric veins of the small intestine. The eggs break through the wall into the lumen and are passed with the feces. *S. japonicum* is the most prolific egg layer of the parasites causing human schistosomiasis, and it is common for the eggs to be carried to the liver and other organs of the body, including the central nervous system. As a result, both cirrhosis and brain lesions are more commonly associated with this species than with the other schistosomes.

Schistosoma haematobium

Large areas of Africa, the Nile River Valley, and the Near East are endemic for *S. haematobium*, a pre-

Figure 28.16 Life cycle of the blood flukes in the genus *Schistosoma*. Refer to text for minor differences among species.

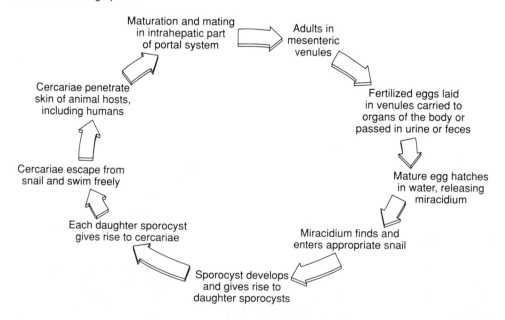

dominantly human parasite. The adult worms characteristically migrate into the rectal vessels and eventually reach veins surrounding the bladder. Unlike the other schistosomes, the eggs of *S. haematobium* penetrate the bladder mucosa and are passed in the urine. Ulceration of the bladder causes numerous small hemorrhages, and the passage of blood in urine is not rare.

Diagnosis, Therapy, and Control of Schistosomal Infections

The diagnosis of all schistosomal infections is based on a history of exposure to an endemic area and on the finding of eggs in the urine or stools of the infected person.

Success of therapy varies among the species, with *S. haematobium* the most readily cured and *S. japonicum* the most difficult to eradicate. The drugs used successfully in treatment contain trivalent antimony; all of these are quite toxic, and considerable effort has been expended to produce less toxic chemotherapeutic agents. Oral drugs used with some success include agents such as lucanthone hydrochloride, 1-diethyl-amino-ethylamino-4-methylthiaxanthone hydrochloride (Miracil-D), and niridazole (Ambilhar). Currently, stibophen is the drug of choice for treatment of infections by *S. mansoni*, and antimony potassium tartrate is used for *S. japonicum* infections.

Efforts to control schistosomal infections (estimated by the World Health Organization to be second only to malaria as a cause of morbidity and mortality in the tropics) are directed toward the elimination of the snails that act as intermediate hosts and the protection of water from human fecal contamination. A number of chemicals, such as copper sulfate, can be used to kill snails, but the migration of snails from untreated areas frequently neutralizes this method of control. In the Far East, the task is further complicated by the common use of human feces as a primary source of fertilizer. It would seem that the only hope for controlling schistosomiasis lies in a program of education designed to teach methods for the sanitary disposal of feces and urine; however, such programs alone have not been effective to date.

Table 28.4 provides a summary of pathogens that enter the body via skin abrasions.

TABLE 28.4 Summary of Pathogens that Enter the Body via Skin

Organism	Description	Disease	Incubation Period	Diagnostic Procedures	Treatment	Control
Bacteria						
C. perfringens and gas gangrene group	Bacillus Gram-positive	Gas gangrene	8–48 hours	Clinical picture Isolate and identify organisms	Sulfonamides, penicillin, or tetracyclines with antitoxin	Surgical cleansing of wounds Antimicrobial drugs Antitoxin
C. tetani	Bacillus Gram-positive	Tetanus	5–10 days; but may be several weeks	History and clinical picture (isolation and identification of organisms usually not necessary)	Antitoxin (more satisfactory for prevention than treatment)	Tetanus toxoid Tetanus antitoxin for immediate protection following injury
Leptospira sp.	Spirochetes	Leptospirosis	Varies, may be 6–12 days	Cultures Serological tests	Streptomycin Chloramphenicol	Rodent control
P. aeruginosa	Bacillus Gram-negative	Infections of wounds, burns, and urinary tract	Varies	Cultures Smears	Carbenicillin Colistin Tobramycin Gentamicin	Aseptic treatment of wounds, burns, etc. Proper disposal of contaminated materials

TABLE 28.4   *(Continued)*

Organism	Description	Disease	Incubation Period	Diagnostic Procedures	Treatment	Control
S. aureus	Spheres—single, pairs and clusters Gram-positive	Boils or furuncles Carbuncles Wound infections Impetigo (pneumonia, meningitis, etc.)	Variable	Cultures Coagulase test Phage typing	Antibiotics, determined by sensitivity testing of organisms	Strict aseptic technique Isolation of infected persons Disinfection of all discharges
S. pyogenes	Spheres in chains Gram-positive	Cellulitis Erysipelas Puerperal fever Wound infections	Variable (usually 1–3 days in puerperal fever)	Isolate and identify organism	Penicillin Tetracyclines	Strict asepsis around wounds Careful disposal of all discharges
Treponema sp.	Spirochetes (indistinguishable morphologically from agent of syphilis)	Bejel, yaws, and pinta	Bejel and yaws may be 2 weeks to 3 months; pinta is 7–20 days	Clinical symptoms Darkfield examination Serological test	Penicillin	Avoid contact with infected persons Fly control
Schistosomes	Blood fluke worms	Schistosomiasis	1–3 months to acute stage	Finding eggs in urine and stool	Stibophen	Sanitary disposal of urine and feces

VIRUSES THAT ENTER VIA ANIMAL BITES

Rabies Virus

Rabies is a fatal viral infection involving the central nervous system; as far as is known, all warm-blooded animals are susceptible, and, although human rabies is rare in the United States, the World Health Organization estimates that more than 20,000 humans die each year from this disease. Other estimates claim that 40,000 to 50,000 people in India die of rabies yearly and that the incidence in other developing countries in Asia, Africa, and Latin America appears to be at the same general level.

The virus is bullet-shaped, about 70×210 nm (see Figure 28.17). It can be grown in laboratory animals and cell cultures as well as in chick and duck embryos.

Pathogenicity   Prominent symptoms of the disease are headache, nervousness, fever, malaise, and paralysis. Delirium, convulsions, and coma may fol-

low. Painful spasms of the swallowing muscles cause the patient to fear both food and water, the basis for the common name of hydrophobia.

Characteristic inclusion bodies are usually seen in the brain cells of infected animals. These inclusions, called **Negri bodies,** are oval or globular and approximately the size of a red blood cell (see Figure 28.18).

The source of infection is almost always a bite by an infected animal. However, the disease can also be contracted by inoculation of a skin wound or abrasion with infectious saliva from an animal or by spelunkers who inhale sprayed urine of infected bats. There have been a few laboratory infections in which the virus was inhaled. Essentially any animal may serve as a reservoir, but most commonly a bat, dog, raccoon, fox, wolf, skunk, coyote, or cat is the reservoir. Until a few years ago, it was believed that the only kind of bat that carried rabies was the vampire bat found in Central and South America. However, it is now known that the insectivorous bats common in the northern hemisphere may also carry rabies and can infect humans. It is interesting that only the vampire bat seems to be able to carry the rabies virus yet remain asymptomatic. All other animals succumb to the infection.

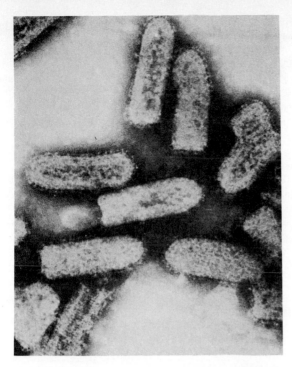

Figure 28.17 Bullet-shaped virions of vesicular stomatitis virus, a typical rhabdovirus that infects animals. Note the external spikes protruding from the virion membrane and striations within the core resulting from the spirally wound RNA within the nucleoprotein.

The incubation period is usually quite long, that is, from 1 to 3 months, although in a recent human case it was reported to be 14 to 32 months. The location of the bite seems to influence the length of the incubation period. Infections resulting from bites around the head and the face characteristically will have a much shorter incubation period—in some cases as short as 10 days.

Figure 28.18 An oval Negri body is seen to the right of the nucleus in this brain cell from a human rabies patient. Lendrum's stain was used in this instance, although most manuals call for Seller's stain.

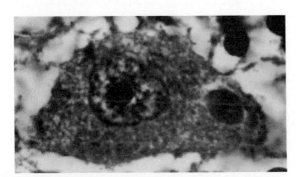

Clinical Management There is no satisfactory treatment once symptoms have appeared. However, because of the long incubation period, an individual can frequently be successfully immunized to the disease during this time. This was first done by Pasteur in 1885, when he used ground-up spinal cords of rabbits that had been infected with rabies virus. Although the viral vaccine used today is somewhat different from that used in 1885, the principle is the same, and the process of injecting rabies vaccine into an individual bitten by a rabid animal is still spoken of as the **Pasteur treatment.** Present-day prophylactic treatment of exposed people also requires the injection of hyperimmune rabies antiserum. This should be administered within 24 h, if possible, after exposure to the disease. It is fairly well accepted that the use of hyperimmune antiserum plus the vaccine is more effective than either antiserum or vaccine alone.

Control of rabies is based on the elimination of the disease in the wild animal and dog population. Most cities and counties require that all dogs be vaccinated with rabies virus. A live attenuated virus (Flury LEP) is used for dog immunization. In addition, any reports of rabies in the wild animal population must be followed by hunting and trapping of these animals to prevent the occurrence of a widespread epidemic of rabies in the animal population. The widespread occurrence of rabies in the wild animal population can be appreciated by knowing that approximately 6000 cases of animal rabies were reported during 1985 in the United States. Most of these cases were in racoons, skunks, and bats.

A novel way to control rabies in the wild animal population has been proposed by scientists at the Wistar Institute in Philadelphia. They have genetically engineered a living rabies vaccine by inserting the gene that encodes for rabies virus capsid glycoprotein into vaccinia virus (see Chapter 21). Animals that are fed this living hybrid virus show no effects from the vaccine but they do develop neutralizing antibodies to rabies virus. The proposal is to incorporate this vaccine virus into a bait that would be spread by airplanes in areas that are known to have a significant rabies problem in their wild animal population. Ingestion of the bait would immunize animals, resulting in a rabies-free area. Large-scale field trials have met with considerable resistance because of what seem to be unjustified fears of releasing a recombinant virus. Trials have, however, been carried out in parts of Europe with considerable success, and it is likely that this vaccine will soon be used on a worldwide basis for the control of rabies in the animal population.

There is no cure once the symptoms have appeared. Thus, any unprovoked bites by wild animals, bats, or dogs must be considered a potential exposure to rabies. If at all possible, the animal should be ap-

MICROBIOLOGY MILESTONES

Louis Pasteur had already made many contributions to the budding field of microbiology when he began his studies on rabies in 1880, but it was this work that brought him his greatest worldwide acclaim. The concept of a virus as we know it today was still many years away, and all that was really known was that rabies was caused by an infectious agent.

By infecting laboratory animals with tissues from rabid animals, Pasteur found that the lowermost portion of the brain (the medulla oblongata) was the most virulent source of infective material. He also found that if this infective material was allowed to dry at room temperature for 14 days, it became totally avirulent. Because Pasteur had already had experience making a vaccine for anthrax, he commenced a series of laboratory experiments in which he first injected dogs with an infective brain emulsion that had been dried for 14 days. Two days later the dogs were injected with an emulsion that had been dried only 12 days, continuing on until he was finally injecting them with fresh, highly infective rabid brain cells. As he may have expected from the beginning, all such treated dogs were com-

pletely protected from a rabies infection.

Although Pasteur was not ready to use this material on humans, he was faced with a difficult decision when a young boy named Joseph Meister was brought to him. Meister had been badly bitten 14 times by an obviously rabid dog, and Pasteur felt certain that he would die of rabies if nothing was done and so he began the injections on July 7, 1885. Joseph Meister lived!

Pasteur immediately became world famous. Funds were raised to establish an institute for research and the treatment of rabies and on November 14, 1888, the now internationally renowned Pasteur Institute was dedicated in Paris.

People came from all over the world to receive the "Pasteur treatment" after being bitten by rabid animals. Of the first 350 people so treated, only one person died of rabies, and she was a girl who was brought to Pasteur 37 days after being infected. And, a nice postscript to this story is that Joseph Meister, Pasteur's first patient, became a scientist and later joined the staff of the Pasteur Institute in Paris.

prehended and taken to public health authorities for 10 days of observation, during which the symptoms will appear if rabies is present. If the animal is well after 10 days, there is no danger from the bite. However, if the animal is destroyed immediately after the bite, microscopic examination of the brain may or may not show Negri bodies (which provide a positive diagnosis), since they sometimes do not form early in the disease. When in doubt, material from the brain of the suspected animal is injected into the brain of a mouse, rabbit, or guinea pig, which will develop symptoms as well as Negri bodies in the brain within approximately a week if the original animal was rabid. The absence of either of these after 21 days indicates that the animal did not have rabies. In some areas

there may be a slight variation in the procedures just described, but the principles are much the same for all modern public health agencies.

The current rabies vaccine licensed in the U.S. consists of killed rabies virus that has been grown in human diploid cell cultures. This new vaccine provides a good antibody response after only five injections. Because the vaccine contains essentially no extraneous foreign proteins, allergic reactions do not occur. Another promising experimental rabies vaccine contains only the glycoprotein spikes isolated from rabies virus that has been grown in human diploid cells.

The disease-producing organisms that enter the body following animal bites or contact with hides are summarized in Table 28.5.

TABLE 28.5   Pathogens that Enter the Body via Animal Bites or Contact with Hides

Organism	Description	Disease	Incubation Period	Diagnostic Procedures	Treatment	Control
Bacteria *Bacillus anthracis*	Gram-positive bacillus	Anthrax	Usually less than 4 days	Direct smear Isolate and identify organisms	Antiserum Sulfonamides Penicillin Broad-spectrum antibiotics	Control disease in animals; prevent soil contamination from infected animals by burning; vaccines for animals; sterilization of wool, hair, etc.

TABLE 28.5 *(Continued)*

Organism	Description	Disease	Incubation Period	Diagnostic Procedures	Treatment	Control
Spirillum minor	Rigid spiral, gram-negative	Rat-bite fever	Usually 5–14 days; may be 6 weeks or longer	History and clinical picture Examine blood directly or inject mouse	Penicillin	Rodent control
Streptobacillus moniliformis	Pleomorphic, branching, gram-negative	Rat-bite or Haverhill fever	Usually 7–10 days	History of rat bite Isolate and identify organism	Penicillin	Rodent control
Dermatophytes						
Trichophyton	Fungus	*Tinea pedis Tinea corporis Tinea capitis*	10–14 days	Clinical picture Observe microscopically	Fatty acids Griseofulvin	
Epidermophyton	Fungus	*Tinea pedis Tinea corporis Tinea capitis*	10–14 days	Clinical picture Observe microscopically	Fatty acids Griseofulvin	
Microsporum	Fungus	*Tinea pedis Tinea corporis Tinea capitis*	10–14 days	Clinical picture Observe microscopically	Fatty acids Griseofulvin	
Subcutaneous mycoses						
Sporothrix schenckii	Fungus	Sporotrichosis	1 week–6 months	Clinical picture Cultures	Oral potassium iodide; amphotericin B	
Cladosporium and *Phialophora*	Fungus	Chromoblastomycosis	Months	Clinical picture	Potassium iodide; amphotericin B	
Viruses						
Rabies virus		Rabies	Usually 1–3 months	History and clinical symptoms Negri bodies in nerve cells Inoculate lab animal with brain tissue	No specific treatment after symptoms appear; hyperimmune rabies antiserum can be used	Viral vaccine Antiserum Eradication of disease in animals

SUMMARY

Staphylococci are probably the most common cause of skin infections. They also cause generalized infections that may involve any organ of the body. The characteristic lesion of the staphylococci is the abscess. One of the major problems in the treatment of staphylococcal infections results from the tendency of these organisms to mutate to antibiotic resistance. *P. aeruginosa* is part of our normal flora but also a frequent cause of urethritis, wound infections, and burn infections. This organism excretes an exotoxin that is similar to diphtheria toxin. Humans usually acquire leptospirosis by coming in contact with urine from infected rats and dogs. Although these organisms gain entrance to the body through the skin, they mainly infect the kidney, liver, and meninges.

Bejel, yaws, and pinta are nonvenereal diseases caused by spirochetes in the genus *Treponema*. These diseases are not found in the United States.

Tetanus and gas gangrene are both diseases resulting from the contamination of wounds with clostridial spores present in soil. In the presence of a diminished blood flow to the wounded area, necrosis of the tissue provides a sufficiently low oxidation-reduction potential so that these obligately anaerobic organisms may grow and secrete their powerful exotoxins. Immunity is directed against the exotoxins.

Dermatophytes are fungi that invade only keratinized tissue such as hair, nails, and skin. Genera making up the dermatophytes include *Trichophyton, Microsporum,* and *Epidermophyton.*

Subcutaneous mycoses, such as sporotrichosis and chromoblastomycosis, are localized infections caused by soil fungi.

Humans most frequently acquire anthrax by contact with hides of infected animals. The organism is able to produce disease by its ability to resist phagocytosis and by the elaboration of a powerful exotoxin. *Spirillum minus* and *Streptobacillus moniliformis* both cause infections as a result of rat bites, although the latter organism has caused epidemics through contaminated milk.

Schistosoma, or the blood flukes, use snails as intermediate hosts. Humans become infected when the cercariae penetrate the skin.

Rabies is usually caused by the bite of a rabid animal. There is no satisfactory treatment once symptoms have appeared, but individuals can frequently be successfully immunized with a killed virus vaccine during the long incubation period of the disease.

QUESTIONS FOR REVIEW

1. What are the major sources of staphylococci?
2. What is the coagulase test and why is it necessary to run this test on isolated staphylococci?
3. How can the various strains of *S. aureus* be typed?
4. List the various skin infections caused by the streptococci.
5. What is exotoxin A of *P. aeruginosa*?
6. What is the reservoir for leptospirosis? How do humans usually acquire this disease?
7. What do bejel, yaws, and pinta have in common?
8. What are the characteristics of the genus *Clostridium*?

9. Name all of the species of *Clostridium* and the disease with which each is associated.
10. How may tetanus be controlled?
11. What is the basis of the name *gas gangrene*?
12. Name the organisms that may cause rat-bite fever. How do they differ morphologically?
13. List the specific infections caused by the dermatophytes.
14. What is an "id" reaction?
15. What is sporotrichosis? How is the disease transmitted?
16. Under what conditions is one most likely to become infected with the organisms causing chromoblastomycosis?
17. What are the three stages in the pathogenesis of schistosomiasis?
18. What is the reservoir for rabies? How is it transmitted to humans? What should you do if you are bitten by a wild animal or a stray dog?

SUPPLEMENTARY READING

Baer GM, Fishbein DB: Rabies postexposure prophylaxis. *N Engl J Med* 316:1270, 1987.

Centers for Disease Control: Toxic shock syndrome associated with influenza. *MMWR* 35:143, 1986.

Centers for Disease Control: Compendium of animal rabies: 1986. *MMWR* 34:770, 1986.

Centers for Disease Control: Acute schistosomiasis in U.S. travelers returning from Africa. *MMWR* 39:141, 1990.

Crass BA, Bergdoll MS: Toxin involvement in toxic shock syndrome. *J Infect Dis* 153:918, 1986.

Iglewski B: Probing *Pseudomonas aeruginosa,* an opportunistic pathogen. *ASM News* 55:303, 1989.

Ivins BE, et al: Immunization studies with attenuated strains of *Bacillus anthracis. Infect Immun* 52:454, 1986.

Ivins BE, Welkos, SL: Cloning and expression of the *Bacillus anthracis* protective antigen gene in *Bacillus subtilis. Infect Immun* 54:537, 1986.

Kass EH et al: Effect of magnesium on production of toxic-shock-syndrome toxin-1: A collaborative study. *J Infect Dis* 158:44, 1988.

Sheagren JN: *Staphylococcus aureus:* The persistent pathogen. *N Engl J Med* 310:1368, 1984.

Wickboldt LG, Fenske NA: Streptococcal and staphylococcal infections of the skin. *Hosp Pract* 21(3A):41, 1986.

Chapter 29

Pathogens that Enter the Body via Arthropod Bites

OBJECTIVES   This chapter should familiarize you with

1. The epidemiology of bubonic plague.

2. The spread and management of tularemia.

3. The role of ticks and lice in relapsing fever

4. The characteristics and epidemiology of Lyme disease.

5. The role of arthropod vectors in the spread of rickettsial diseases such as typhus, Rocky Mountain spotted fever, rickettsialpox, scrub typhus, and Q fever.

6. The mechanisms whereby protozoa such as *Trypanosoma, Leishmania, Plasmodium,* and *Babesia* are acquired by humans.

7. The epidemiology of the blood and tissue nematode infections.

8. The characteristics and epidemiology of the arthropod-borne virus diseases, including yellow fever, dengue fever, and the viral encephalitides.

There are many pathogenic microbes found primarily in wild animals and birds. In this chapter we are concerned with diseases that are carried to humans from the infected animal by an arthropod vector, frequently a mosquito or a tick. With the exception of tularemia, the diseases included in this chapter are normally spread to humans only by an arthropod vector and, as such, can usually be controlled or prevented by eliminating the vector, even though the disease continues to be present in its animal reservoir.

One should note also that many diseases that require an arthropod vector are seasonal in temperate climates and, for the most part, are usually spread during summer months.

BACTERIA

Yersinia pestis: Bubonic Plague

The genus *Yersinia* was created in 1970 by a subcommittee studying the taxonomic position of members of the genus *Pasteurella*. It was noted that some species in the genus *Pasteurella*, namely, *P. pestis* and *P. pseudotuberculosis*, possessed many properties that made them more closely related to the enteric bacteria than to the other species in the genus. As a result, the new genus *Yersinia* was created in commemoration of Yersin, the discoverer of the plague bacillus. It has been shown that yersiniae can conjugate with *Escherichia coli* and accept various plasmids, such as resistance transfer factors. In addition, *Yersinia* has several antigens in common with *E. coli*. Thus, it appears that the creation of the genus *Yersinia* as part of the family Enterobacteriaceae places these organisms in a better taxonomic position.

Y. pestis is the etiological agent of **bubonic plague.** This disease is an ancient one that has killed untold millions of people over the centuries. For example, it is believed to have killed more than 100 million in an epidemic in the sixth century. Another epidemic in the fourteenth century killed one fourth of the European population. A more recent epidemic started in 1893 in Hong Kong. From there it spread to India, where over a 20-year period many millions of people died of bubonic plague. The disease reached the United

---

MICROBIOLOGY MILESTONES

If one were to examine the fall of many great civilizations during the past 3000 years, it would be evident that epidemics of infectious diseases were often involved in their decline. Of such diseases, none can rival the pandemics of bubonic plague that swept through the world.

In the Old Testament, there are descriptions of innumerable deaths as early as 1325 B.C. that are interpreted as epidemics of bubonic plague, but the earliest documented epidemic began in 430 B.C. during the Peloponnesian Wars. During this epidemic, the Athenian empire was destroyed and it is thought that at least one third, and possibly two thirds, of the population died.

The first real pandemic (worldwide epidemic) reached Europe in 542 A.D., causing millions of deaths. It is claimed that as many as 10,000 persons died each day in Constantinople. The second pandemic (also called the medieval plague) struck Europe in 1347, and by the time it waned, 25 million persons were dead. Because it was generally assumed (and supported by the Church) that this was due to the wrath of God, monuments were erected all over Europe to appease His anger. The citizens of a German village named Oberammergau vowed to present a passion play each decade if they were spared. The first such play was presented in 1634 and it is still being performed.

There were also many individual tragedies that were only indirectly attributed to the plague. One occurred in Milan in 1630 when Guglielmo Piazza, a commissioner of health, was making an inspection through the city. As he made notes, he periodically wiped his ink-stained fingers on the walls of houses as he passed. Observers accused him of smearing the walls with plague poison. After a third round of torture, he confessed and, with another turn of the screw, he implicated the barber, Mora, as his source of the plague poison. Both were sentenced to death. Their flesh was torn from their bodies with red-hot pincers. Their right hands were cut off and their bones were broken with the wheel. Then, after being stretched on the wheel for six hours, they were burned alive. Such was the fear of plague.

There was also a somewhat lighter side. Many wore garlands around their necks to ward off "the evil vapors," but as indicated in the last line of a nursery rhyme attributed to that period, it was to no avail:

> Ring a ring of rosies
> Pocket full of posies
> Achoo, Achoo
> All fall down

The final pandemic began in China in 1892, killing millions before it too dissipated. It is estimated that during this pandemic, 6 million people died in India alone.

States during that epidemic and was first seen in San Francisco around 1900.

Fortunately, we have not had a major epidemic for some years, although there are occasional reports of possible epidemics in China. The disease is, however, endemic in many areas of the world.

Y. pestis is a small, plump bacillus (see Figure 29.1). It is gram-negative and nonmotile, becoming very pleomorphic if grown under suboptimal conditions such as a high salt concentration. *Y. pestis* is not a particularly fastidious organism and can easily be grown on routine laboratory media. The organism produces a variety of different antigens, but only a few have been characterized sufficiently well to postulate their role in the production of disease: (1) a capsular antigen, designated Fraction 1 (F1), is antiphagocytic; (2) V/W antigens consisting of a protein (V) and a lipoprotein (W) are produced together and act to prevent phagocytosis; and (3) an intracellular toxin that is lethal for the mouse ($LD_{50} < 1$ μg) and rat, apparently acting on the vascular system and causing irreversible shock and death.

Pathogenicity Normally a disease of rodents, plague (termed **sylvatic plague** when it occurs in animals) exists in two kinds of epidemic centers, namely, the permanent but relatively resistant wild rat population, where the organisms reside during interepidemic periods, and the temporary but susceptible rodents, particularly the domestic rat population.

The spread of plague to humans is a function of the relative balance between resistant and susceptible species of rats. When the domestic rat population overlaps into the wild rat population, the domestic rats become infected by the bite of an infected rat flea.

The domestic rat fleas then become infected and, when biting other rats, regurgitate the plague bacillus into the new host, instituting a new epidemic. As the domestic rats continue to die, the rat fleas will bite humans—intruders in the normal wild rat-domestic rat flea cycle.

The endemic focus in the United States is confined to the southwestern area, where the organisms infect ground squirrels, prairie dogs, wood rats, mice, and other rodents. Between 1960 and 1987, 281 cases of human plague were reported from this area, mostly in people who camped frequently in woods and who often slept in the open in endemic areas. Of the 40 cases of human plague reported in New Mexico and Arizona during 1983, 6 were fatal. Most cases came from handling infected animals.

Human plague in Africa has occasionally been acquired by people who had handled meat from goats or camels presumably infected by fleas or ticks that, in turn, had recently fed on plague-afflicted rodents.

Following the bite of the flea, the microbes are transported to the regional lymph nodes. Because most bites occur on the legs, the nodes in the groin are most frequently involved. The microbes grow in the lymph nodes, causing enlarged tender areas called buboes—hence the name of the disease, bubonic plague. From the lymph nodes, the microbes get into the blood, whence they are disseminated to the spleen, liver, lungs, and skin. This results in frequent subcutaneous hemorrhages that give the skin a black appearance; hence the common name "Black Death." In some cases, an individual may acquire a pneumonic plague that can be spread directly from person to person via the respiratory route. Untreated pneumonic plague is almost always fatal, whereas bubonic plague has a mortality rate of 60 to 90 percent. The incubation period of bubonic plague varies from 2 to 6 days; in pneumonic plague, it varies from 3 to 4 days.

Clinical Management The diagnosis is confirmed by cultures from the blood and the local lesions (buboes) and by the characteristic bipolar staining of the organism (see Figure 29.1). Animal inoculation and serological tests are also used as diagnostic aids. A specifically fluorescently labeled antibody is available for a definitive identification.

Control of plague may be accomplished in two ways. The most effective method is the elimination of the rat population. However, during the course of an epidemic, it is considered unwise to kill off large numbers of rodents because this dispossesses even larger numbers of hungry, infected rat fleas who may then prey on humans. In such cases the fleas are attacked directly with large amounts of an insecticide. A procedure that was very effective in Africa after World War II used the now-banned insecticide DDT.

Figure 29.1 Wayson-stained smear of spleen from an experimentally infected mouse showing bipolar staining of cells of *Y. pestis* (×1000).

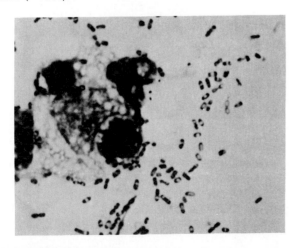

Vaccine　There are a number of plague vaccines. They are either heat-killed virulent organisms or living attenuated plague bacilli. None is 100 percent effective, but they do help to reduce both the attack rate and the mortality of the disease. An attack of the disease usually confers permanent immunity.

Sulfonamides, streptomycin, or the tetracyclines are of value in the specific treatment of plague. Practical considerations are essentially the same as for any infectious disease. All dressings should be burned or disinfected. With pneumonic plague in particular, the patient should be isolated to prevent the respiratory spread of the organisms.

Francisella tularensis: Tularemia

Although tularemia occurs throughout the world, it was first described and the organism isolated and characterized in the United States in the early 1900s. Much of the early work with this organism was carried out by Edward Francis.

Tularemia, like plague, is primarily a disease of wild animals. It is transmitted from animal to animal via various arthropod vectors, of which flies, fleas, lice, and ticks are the most important.

The tularemia organism *Francisella tularensis* is a short, gram-negative, very pleomorphic rod (see Figure 29.2). Nutritionally, *F. tularensis* requires a rich medium to which the amino acid cysteine has been added; although it is a facultative anaerobe, better growth is obtained when it is grown aerobically.

In 1919 in Utah, Dr. Edward Francis isolated *F. tularensis* from a fatal case of deer fly fever and, during this same period, *F. tularensis* was shown to be the etiological agent of "market fever," a severe illness occasionally contracted by butchers after cleaning wild rabbits. Also, the organism was shown to exist in the wood tick, where it is passed transovarially from generation to generation, providing both a reservoir and a vector for the transmission of the disease. Thus, there is a varied epidemiology in which the disease may be transmitted by the deer fly in the Southwestern United States, by the wood tick in the Northwest and Midwest, and all over the world by direct contact with infected animals, especially rabbits. *F. tularensis* has been isolated from about 100 species of mammals, birds, fish, and invertebrates. The majority of infections in the United States are transmitted by the bite of an infected tick. In one tick-associated epidemic of 28 cases of tularemia in North Dakota, it was shown that ticks collected from 8 of 46 dogs were infected with *F. tularensis.* Direct contact with infected animals, particularly rabbits, also provides a frequent source of infection. Domestic cats that have eaten in-

Figure 29.2　Pleomorphic cells of *F. tularensis* (×42,000).

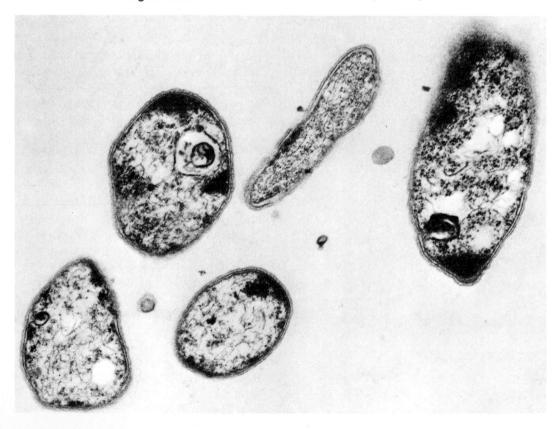

fected animals have provided a source of human infections. It is also noteworthy that dead infected animals on the banks of streams can contaminate the water, which, if drunk, also serves as a source of infection. There were 133 cases of tularemia reported in the United States during 1990.

Pathogenicity When humans contract tularemia, either through skin contact or insect bites, a local area of inflammation develops at the site of entrance of the organisms. This primary lesion eventually becomes an open necrotic ulcer. The organisms then move up the lymphatics, causing an extensive infection in the regional lymph nodes. Occasionally they enter the bloodstream from the nodes, causing septicemia or meningitis.

Another very serious type of infection caused by *F. tularensis* develops if the organisms get directly into the eye. This may result from the spattering of infected material or from wiping the eye with infected fingers. Such a condition leads to severe ulceration of the eye as well as the regional lymph nodes and may continue to spread, causing septicemia or meningitis. Pulmonary infections from this organism occur very rarely.

It is interesting to note that although *F. tularensis* is considered to be among the most virulent organisms that infect humans, it is still not possible to pinpoint the factors that contribute to this virulence. Certainly, its ability to grow and survive in monocytes and polymorphonuclear leukocytes protects the organisms from lysis by humoral antibody and complement. However, no toxic factors have been described that can explain the extreme virulence of these organisms.

The incubation period averages about 3 days, but it may vary from 1 to 10 days.

Clinical Management The diagnosis is usually made on the basis of the history presented by the patient and the clinical picture. Because of the extreme virulence of this organism, most diagnostic laboratories are unwilling to grow it for the purpose of identification. Because the organism requires a medium supplemented with cysteine, the deliberate avoidance of such a medium assures the staff of a diagnostic laboratory that they will not inadvertently culture this organism from a blood specimen sent into the laboratory. However, an increase in antibody agglutinins, as well as the use of skin test material, is used to support a clinical diagnosis. Fluorescently labeled antibody is also used for identification.

Treatment and Control Treatment with any antibiotic is difficult, probably because of the intracellular existence of the organisms. However, streptomycin seems to be the antibiotic of choice, although the tetracyclines, gentamicin, or chloramphenicol are also efficacious. Relapses are not uncommon, and prolonged treatment may be necessary.

Both phenol-killed and attenuated vaccines have been used but, because immunity is primarily cell mediated, the inactivated vaccine appears to be of little value. An attenuated vaccine is available from the Centers for Disease Control and can be given to persons whose risk of exposure is high.

Control is virtually impossible in the wild animal population but, because many human cases result from direct contact with infected rabbits, one should either use rubber gloves when cleaning wild rabbits or, for absolute protection, follow the advice given in Leviticus 11:6–8, which, in part, states, "and of the hare . . . of their flesh shall ye not eat, and their carcass shall ye not touch; they are unclean to you."

Isolation of the patient is not required, since the disease is not transmitted from person to person. However, draining discharges may be infectious and should be sterilized promptly.

Borrelia: Relapsing Fever

Members of the genus *Borrelia* can be distinguished morphologically from other spirochetes by their coarse, irregular coils. Organisms may be up to 0.5 μm wide and 20 μm long and are, therefore, much easier to see than are other spirochetes (see Figure 29.3).

In humans, many different species of *Borrelia* may cause relapsing fever. *B. recurrentis* is the only species that is transmitted from human to human, by the body louse; all other species are transmitted by various species of the tick *Ornithodoros*. The body louse feeds only on humans, so humans are, then, the sole vertebrate reservoir for *B. recurrentis*. However, because other species of *Borrelia* are passed transovarially in the tick, this vector can also serve as a reservoir for all species of *Borrelia* except *B. recurrentis*. Other primary reservoirs include rodents, ground squirrels, armadillos, and monkeys.

Species of *Ornithodoros* ticks prefer warm, humid climates and are found primarily at altitudes between 1000 and 2000 m. They feed mainly at night, completing their blood meal in less than 30 min; thus, a sleeping person is frequently unaware of being bitten. Relapsing fever occurs in the United States most frequently among campers, particularly those sleeping in rodent-infested cabins.

A few strains of *Borrelia* have been grown on artificial media, and these strains require a microaerophilic atmosphere and an enriched medium.

Pathogenicity Humans acquire the infection from the bite of an infected vector. The microorganisms enter the blood and cause multiple lesions in the

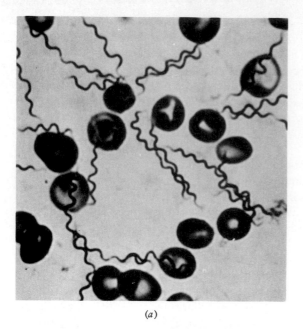

(a)

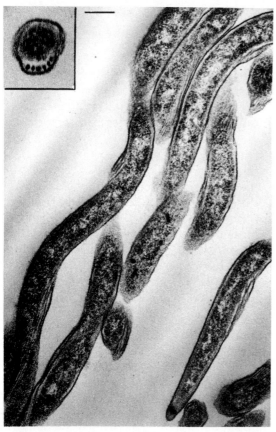

(b)

Figure 29.3 (a) Giemsa stain of *B. hermsii* in the blood of an experimentally infected pine squirrel. *B. hermsii* is one of the causes of relapsing fever in the United States. (×2700). (b) The Lyme disease spirochete *B. burgdorferi* (transmission electron microscopy—thin section profile of five-day-old BSK-II culture of prototype strain B31). Bar = 0.2 μm. Insert: Thin section profile showing the endoflagella, located in the area of asymmetry between the outer membrane and protoplasmic cylinder (×2 enlarged).

spleen, liver, kidneys, and gastrointestinal tract. After 4 to 5 days of fever, the microorganisms seem to disappear and the patient becomes afebrile for a week or 10 days; then a relapse occurs and the organisms can be found in blood and internal lesions. After 4 or 5 more days, the organisms again disappear, only to recur 3 to 10 more times before complete recovery.

The mechanism of the relapse appears to result from the spontaneous appearance of new antigenic types. Accordingly, when each succeeding humoral immune response clears the organisms, a new antigenic type becomes prevalent, causing the ensuing relapse. These antigenic changes also occur when the organisms are grown in the laboratory. One investigation showed that *B. hermsii* gave rise to progeny that represented 26 separate serotypes.

The molecular mechanism whereby new antigenic variants occur is somewhat like that described for the pilin genes of the gonococcus. In the case of the borreliae, however, the antigenic serotype is based on a surface protein associated with the outer membrane. Antibodies directed to this protein result in a clearance of the organisms either through complement-induced lysis or Fc C3b-induced phagocytosis. Because of the ability of the borreliae to change from one surface protein to an antigenically distinct protein, this structure has been termed the "variable major protein" (VMP). It is now known that the different VMPs are encoded in linear plasmids and that each VMP exists as a separate gene. Moreover, only one such gene is actively expressed, while the remaining VMPs occupy silent loci. Antigenic variation, occurring in *B. hermsii* at a rate of 10^{-4} to 10^{-3} per generation, results from the transposition of a silent VMP gene to the expression site. The borreliae possessing the new VMP are then able to induce a relapse because they do not react with antibodies to the old VMP.

The diagnosis of relapsing fever is based on clinical symptoms in addition to the observation of *Borrelia* organisms in stained blood smears or darkfield microscopy. Also, blood can be injected into mice and the mouse blood examined for *Borrelia* after 2 or 3 days.

Penicillin is the drug of choice, although other antibiotics such as tetracycline and chloramphenicol are effective. Because the organisms are transmitted only by vectors, tick and louse control are effective measures for the control of relapsing fever.

Borrelia burgdorferi: Lyme Disease

A seemingly new syndrome was described in 1975 by Allen Steere and his colleagues during their studies of a cluster of arthritis cases in the Lyme, Connecticut, area. This disease occurs during the summer and usually begins with a skin lesion accompanied with headache, fever, stiff neck, malaise, and swollen lymph glands. Weeks or months later, many patients develop encephalitis, myocarditis, or migratory musculoskeletal pain. Still later, an arthritis may develop, which may recur intermittently over a period of several years, or it may become chronic. This disease, which was originally described as juvenile arthritis, has now been reported to occur in at least 24 states, as well as in 19 different countries.

The seasonal occurrence of Lyme disease, plus the observation that penicillin or tetracycline given early in the illness prevented or attenuated the subsequent arthritis, suggested a bacterial disease transmitted by an arthropod vector. This was found to be the case when a previously undescribed spirochete was isolated from both *Ixodes dammini* ticks and from blood, skin, and spinal fluid of patients suffering from Lyme disease.

This organism, named *B. burgdorferi,* causes the most frequently reported tick-borne illness in the United States. The disease is transmitted in the northwestern and midwestern United States by the *I. dammini* tick and in the west by *I. pacificus.* Moreover, *B. burgdorferi* has also been isolated from deer flies, horse flies, and mosquitoes, indicating that these hematophagous (blood-eating) insects may also be involved in the transmission of these organisms to humans from the wild animal population. Major animal reservoirs include white-tailed deer, mice, voles, raccoons, dogs, chipmunks, and birds. Also, high levels of serum antibodies against *B. burgdorferi* have been shown to occur in 24 to 53 percent of healthy dogs from enzootic areas. The most common clinical finding associated with seropositive dogs was lameness. It did not appear, however, that dog ownership was associated with human Lyme disease.

Interestingly, the *Borrelia* appear to remain alive in an infected host for years, in some cases. This conclusion is based on the observation that high doses of penicillin result in about a 50 percent cure rate for Lyme arthritis even when treatment was initiated several years after the onset of the arthritis.

Ideally, a definitive diagnosis would be based on the isolation and identification of the Lyme spirochete from the skin lesions or the blood of an infected patient. At present, however, this has proved to be a low-yield procedure, and the most reliable laboratory diagnosis is currently based on the presence of antibodies to the spirochete as measured in an enzyme-linked immunosorbent assay using *B. burgdorferi* as an antigen.

Table 29.1 summarizes the diseases caused by the arthropod-borne bacteria.

RICKETTSIAE

A review of Chapter 13 will point out the general characteristics of the rickettsia. All rickettsial diseases (with the exception of Q fever) are transmitted to humans via the bites of infected arthropod vectors. Furthermore, these diseases are primarily diseases of wild animals and ticks; humans are infected only when they become the animal in the normal animal-to-arthropod-to-animal cycle of the organism. The single exception to this is seen in epidemic typhus, in which humans may be the major reservoir, and the organism is transmitted from person to person by way of the body louse or the head louse.

In general, rickettsial diseases are diagnosed on the basis of the clinical illness and with the use of a serological test called the **Weil-Felix test.** In this test, the antibodies are measured in the patient (preferably during the acute illness and after convalescence, so as to demonstrate a rise in antibody titer); these will agglutinate certain strains of a gram-negative bacterial rod in the genus *Proteus. Proteus* organisms have nothing whatsoever to do with causing the disease, but certain strains of this bacterium have an antigen in their cell walls that is also present in the rickettsiae. Three different strains of *Proteus* are used in this test; they have been designated *Proteus* OX-2, *Proteus* OX-19, and *Proteus* OX-K. As shown in Table 29.2, a positive Weil-Felix test is not specific and must be interpreted on the basis of the reaction to all three *Proteus* strains as well as the clinical illness and history.

The major rickettsial diseases that may infect humans are discussed in the following sections.

Spotted Fever Group

Diseases of this group have for the most part acquired common names based on the area of the world in which they are found. Thus, a few examples of rickettsial diseases included in the spotted fever group are Rocky Mountain spotted fever, Queensland tick spotted fever, North Asian tick-borne spotted fever, Boutonneuse spotted fever, and rickettsialpox. This group of rickettsiae all share a common group antigen, but they also possess a type-specific complement-fixing antigen. They are also characterized by the fact that they grow in both the nucleus and the cytoplasm of infected cells. Of these diseases, only Rocky Mountain spotted

TABLE 29.1   Summary of Arthropod-borne Bacteria

Organism	Description	Disease	Incubation Period	Diagnosis	Treatment	Control
Bacteria						
F. tularensis	Short gram-negative bacillus	Tularemia	1–10 days, average 3 days	Agglutination test Skin sensitivity test	Streptomycin Chloramphenicol Tetracyclines	Use of rubber gloves in cleaning rabbits Thorough cooking of rabbits Disinfection of discharges (Vaccine, experimental)
Y. pestis	Short gram-negative bacillus	Plague: bubonic, septicemic, pneumonic	Bubonic, 2–6 days Pneumonic, 3–4 days	Cultures Smears Animal inoculation Serological tests	Sulfonamides Streptomycin Tetracyclines	Rodent and flea control Proper disposal of all contaminated materials Isolate pneumonic cases Bacterial vaccine
B. recurrentis	Spirochete	Relapsing fever	3–10 days	Blood smears Animal inoculation Serological tests	Tetracyclines Erythromycin Penicillin	Avoid exposure to ticks
B. burgdorferi	Spirochete	Lyme disease	10–30 days	Serological tests	Penicillin	Avoid exposure to ticks

fever and rickettsialpox occur in the United States. (See Figure 29.4.)

Rickettsia rickettsii: Rocky Mountain Spotted Fever

Rocky Mountain spotted fever (RMSF) has been reported to occur in almost every state in the United States. There are approximately 1000 cases each year, with the majority occurring in children living in the "tick-belt" of Maryland, Virginia, North Carolina, South Carolina, and Georgia. The causative organism, *R. rickettsii,* is named after Howard Taylor Ricketts,

TABLE 29.2   Weil-Felix Reactions for Several Rickettsial Diseases

Disease	*Proteus* Strain Giving a Positive Agglutination Reaction with Convalescent Serum
Epidemic typhus	OX-19
Endemic typhus	OX-19
Rocky Mountain spotted fever	OX-19 and OX-2
Scrub typhus	OX-K
Trench fever	None
Rickettsialpox	None
Q fever	None

an early pioneer in the study of rickettsial diseases, who died as a result of his investigations of typhus.

A person invariably contracts this disease as a result of being bitten by an infected tick. In the western part of the United States the wood tick *Dermacentor andersoni* carries the disease, and the dog tick *D. variabilis* serves as the vector in the eastern United States.

Pathogenicity   The disease is characterized by fever, severe headache, and a maculopapular rash that usually appears first on the palms and soles. If untreated, the patient may die, usually during the second week of illness, or recovery will occur after an illness of about 3 weeks.

The organisms are widespread in the animal population as well as in the ticks themselves. Thus the tick may become infected either as a result of a blood meal from an infected animal or by the transovarian passage of the rickettsia from mother to offspring. Reports have shown that dogs are readily infected and that rickettsiae can be isolated from their blood for at least 2 weeks after infection, indicating that dogs may serve as an important reservoir for RMSF. The disease does not harm the tick. It is estimated by some that approximately 3 percent of ticks are infected, although this figure varies greatly from one geographic area to another. It is interesting to note that the average mortality from untreated cases of RMSF varies from 90 percent in areas of Montana to as low as 5

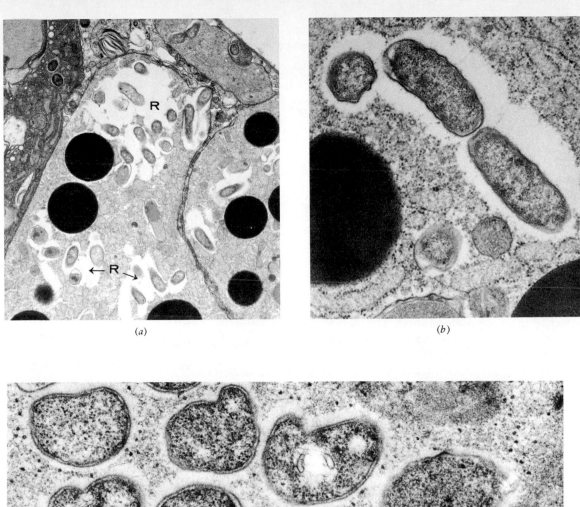

(a)

(b)

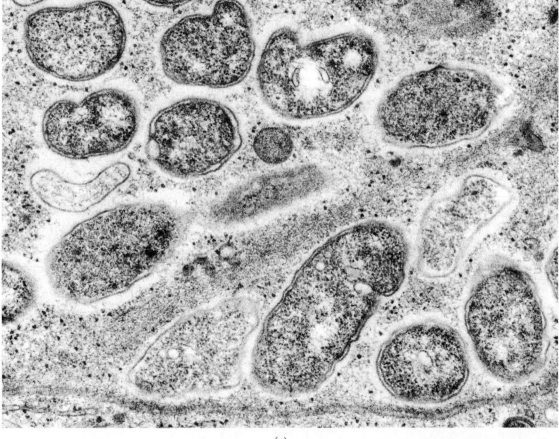

(c)

Figure 29.4 Morphology of rickettsia. (a) *R. rickettsii* growing in the salivary gland of the Rocky Mountain wood tick, *D. andersonii.* Note how the rickettsiae grow within the acinus of the salivary gland. (b) *R. rickettsii* growing within the acinus of the salivary gland of the acinar tissue of the wood tick, *D. andersonii.* This magnification clearly shows the gram-negative characteristic cell wall. (c) *R. prowazekii* in experimentally infected tick tissue. Note the double-layered cell wall similar to that of gram-negative bacteria. (×45,000).

percent in other areas of the United States. It seems that this variation in severity of the disease must be a result of different strains of the etiological agent in different geographic areas.

The incubation period varies from 3 to 12 days.

Clinical Management The diagnosis is usually made on the basis of the clinical picture and a history of a tick bite. This is confirmed by use of the Weil-Felix test demonstrating a rise in titer to the *Proteus* OX-19 organisms. The rickettsiae can be grown in the yolk sac of embryonated hen's eggs, and specific complement-fixation tests can be carried out using washed rickettsiae that have been so grown.

Control of this disease is possible in part by frequent examinations of the body for the presence of ticks whenever a person is in an environment in which ticks are likely to be found. Fortunately, a tick is unable to infect a person until approximately 4 h after it has attached itself for a blood meal. This period is sometimes referred to as the **rejuvenation period**. Thus, a periodic removal of ticks, even though they may have been infected with rickettsiae, is very effective in preventing the disease.

A vaccine composed of yolk-sac-grown, formalin-inactivated *R. rickettsiae* organisms has in the past been the only means of inducing immunity to this disease. This vaccine, however, is poorly antigenic and, in addition, contains egg yolk contaminants that cause severe allergic reactions in some people and, as a result, is no longer commercially available.

Specific treatment with chloramphenicol or the tetracyclines is quite effective, particularly if started early in the illness. Immunity for people who recover from the disease appears to be permanent.

Rickettsia akari: Rickettsialpox

Rickettsialpox was first described after an epidemic in some apartment houses in New York City in 1946. Subsequently the causative organism, *R. akari*, has been shown to be of virtually worldwide distribution. In the United States, there are about 200 cases reported annually, but this may represent only a small fraction of cases actually occurring.

Pathogenicity The reservoir of this organism is the common house mouse, and humans become infected after being bitten by infected mouse mites. The infection varies considerably in its severity from one person to another but for the most part is a mild disease. A primary lesion similar to that seen in scrub typhus develops at the site of the bite. Fever and chills occur early in the illness, and after about 3 to 4 days a rash appears. The incubation period is about 10 to 24 days.

Clinical Management The diagnosis may be based primarily on the clinical picture but can be confirmed by the use of a specific complement-fixation test.

Chloramphenicol and the tetracyclines are used for the specific treatment. Control of this infection is directed toward the elimination of rodents.

Typhus Group

There are two major types of disease in this category. The etiological agents for both are morphologically similar and share a common group antigen, but they can be differentiated on the basis of a specific complement-fixation antigen and the epidemiology of the respective diseases.

Rickettsia prowazekii: Epidemic Typhus

R. prowazekii (Figure 29.4) is the causative agent of epidemic louse-borne typhus fever. Epidemic typhus has probably been more important than any other disease in shaping world history. The disease is associated with filth, and no major war has ever escaped its ravages. Lice have defeated the most powerful armies of Europe and Asia, and it is frequently stated that Napoleon's retreat from Russia was started by a louse. Until very recently it was believed that humans served as the sole reservoir for these organisms, and that the human body louse was the only arthropod vector (in spite of the fact that Marilyn Bozemann and her collaborators had isolated *R. prowazekii* in 1963 from the eastern flying squirrel in Virginia). Now it appears that the flying squirrel is truly the first known nonhuman reservoir for this organism. In the late 1970s, there were at least eight cases of serologically proven epidemic typhus in Georgia, Tennessee, Pennsylvania, and Massachusetts. Several of the victims had known exposure to flying squirrels, and it is now presumed that the fleas from the animals may be an additional vector for this disease.

The infected human body louse carries the rickettsiae in its alimentary canal; when it bites, the organisms can be introduced either by the insect's bite or by its feces, vomitus, or entire carcass crushed and scratched into the skin.

The incubation period is usually about 12 days, but it may vary from 6 to 15 days.

Pathogenicity The manifestations of the disease result from an overwhelming bacteremia with growth of the rickettsiae in the endothelial cells of the blood vessels. After the fifth or sixth day of illness, a macular rash begins on the trunk of the body and spreads to the extremities. Neurological changes characterized by delirium and stupor may also occur. The major symptoms are headache, chills, fever, malaise, and general aches and pains.

Clinical Management The diagnosis is usually not difficult because the disease occurs in epidemics. Both the Weil-Felix test and a specific complement-fixation test may be used as diagnostic aids.

Control is directed toward sanitation and the elimination of human lice. Very impressive results were obtained in World War II, when epidemics were controlled immediately by spraying a susceptible population with DDT to kill any human lice present. Vaccines were originally prepared from rickettsiae obtained by washing the intestines of infected lice. However, they can now be prepared from organisms grown in chick embryos. The vaccine should be administered to anyone entering an area where the disease may be found. Chloramphenicol or the tetracyclines are used for the specific treatment of epidemic typhus.

Brill's disease (sometimes referred to as Brill-Zinsser disease) occurs only rarely, but it demonstrates an important property of the rickettsiae that cause epidemic typhus. It was first observed in early immigrants to the United States who acquired an illness resembling typhus fever. Because these people were not infested with lice, the disease was originally thought to be the endemic (flea-borne) type of disease. Subsequent work by Brill established that these infections were identical with the louse-borne epidemic typhus. Zinsser correctly postulated that these cases of typhus represented individuals who had recovered from epidemic typhus many years before their current illness but had carried the virulent rickettsiae in their bodies in a latent state. Then, under conditions that we still do not understand, the latent organisms started to multiply and produced overt disease. However, as one might expect, cases of Brill's disease are usually quite mild, because the individual already possesses a partial immunity and responds with an anamnestic immunological response. It is interesting, though, that these people undoubtedly provide an additional reservoir for this disease during interepidemic periods.

Rickettsia typhi: Endemic Flea-borne Typhus

Rickettsia typhi is the etiological agent of endemic flea-borne typhus, which is also frequently referred to as murine typhus. This disease occurs sporadically rather than in large epidemics as seen with epidemic typhus.

Pathogenicity *R. typhi* is closely related to the etiological agent of epidemic typhus, for either disease confers immunity to both. The disease in humans is

considerably less severe than the epidemic form but in all other details is clinically indistinguishable from epidemic typhus.

The rat serves as the primary reservoir, although other wild rodents such as ground squirrels may also carry the disease, as is the case in the southwestern United States. In the normal sequence of events, the disease is transmitted from rat to rat by the rat flea. Only when humans accidentally interrupt this cycle by being bitten by the rat flea do they acquire the disease. Because neither the rat nor the rat flea becomes ill as a result of this infection (in contrast to epidemic typhus caused by *R. prowazekii*), they constitute the reservoir and vector, respectively, for murine typhus. The incubation period is 6 to 14 days.

Clinical Management The diagnosis makes use of the Weil-Felix test as well as the more specific complement-fixation reaction. Clinically, this disease can be confused with Rocky Mountain spotted fever, but it can be differentiated in that the rash of typhus begins on the trunk and later moves out to the periphery of the body.

Control measures are directed against both the rodent and the flea population in an endemic area. In the past, the use of DDT has been very effective in reducing the flea population in rat-infested endemic areas; in all likelihood DDT would be used to bring an epidemic under control. It is important to use the DDT before attempting to exterminate the rat population so as to prevent the infected fleas from leaving dead rats and biting people. Broad-spectrum antibiotics are effective in the treatment of this disease.

Rickettsia tsutsugamushi: Scrub Typhus

R. tsutsugamushi is the causative agent of scrub typhus, a disease found in Asia and the southwest Pacific. It achieved particular importance in the United States during World War II and in Vietnam, when large numbers of troops in the southwest Pacific were infected by this organism. The death rate varied from 0.6 to 35 percent.

Pathogenicity Scrub typhus (also called tsutsugamushi fever) normally occurs in rodents and is transmitted from rodent to rodent by a mite. Mites lay their eggs in the soil and the larvae (chiggers) hatch later and feed on animals, including humans. After feeding, the larvae drop off and eventually develop into adult mites. Infected mites transmit the rickettsiae to their eggs, resulting in infected larvae. In such ways, the mite can act both as a vector and as a reservoir for scrub typhus.

Humans acquire the disease from the bite of an infected mite. Initial symptoms, which occur 7 to 14 days after the victim is bitten, include severe headache, chills, and fever. Recovery begins about 3 weeks after initial symptoms, but convalescence may last for several months.

Clinical Management The diagnosis employs the Weil-Felix test as well as a specific complement-fixation test.

Control is directed against the mite. During World War II a miticide was used quite effectively both as a repellent on clothing and to kill the mites within a given geographic area.

The use of chloramphenicol prophylactically was also effective, but because of the known toxicity of this drug over long periods, it would not be recommended except under unusual circumstances.

Vaccines have not been too successful owing to the large number of antigenically different strains. There is no evidence that the disease is spread directly from human to human. Treatment with chloramphenicol over long periods, in the range of 4 weeks, seems to be quite effective.

Rochalimea quintana: Trench Fever

Pathogenicity R. quintana is the causative agent of trench fever. Prominent symptoms are fever, headache, exhaustion, and a roseolar rash. An interesting aspect of this infection is that it has been found only among armies during periods of fighting, and then only in Central Europe. It is a disease tied to filth and poor sanitation, in that the reservoirs and the vectors appear to be humans and lice, respectively. The organism is the only rickettsia that has been grown in cell-free media.

Treatment with chloramphenicol seems to be very effective in eliminating the symptoms but acts only as a rickettsiostatic drug, since it does not eliminate the carrier state.

Ehrlichia canis: Ehrlichiosis

Rickettsia that characteristically grow intracellularly in leukocytes are placed in the genus *Ehrlichia* and, until recently, it was believed that each species was host specific. Thus, *E. risticii* is the etiological agent of equine ehrlichial colitis or, as it is sometimes called, Potomac horse fever, a disease with a mortality rate for horses of about 30 percent. Similarly, *E. canis*, the cause of canine ehrlichiosis, has been responsible for a high mortality in dogs. In this case the organisms are spread from dog to dog by the brown dog tick, *Rhipicephalus sanguineus*. The only human agent was

E. sennetsu, the cause of a human mononucleosislike syndrome that occurred only in Japan and Malaysia.

Now, however, there is strong evidence that *E. canis* may be spread via tick bites to humans, mimicking Rocky Mountain spotted fever. Major symptoms include headache, exhaustion, and a rash in about one half of infected patients. Essentially all cases have been diagnosed by demonstrating a rise in antibodies specific for *E. canis*, and it is suggested that ehrlichiosis be considered for any febrile illness following a history of a tick bite and displaying negative serology for Rocky Mountain spotted fever, Lyme disease, or tularemia. A retrospective study in Oklahoma showed that 11 percent of 144 patients who had been suspected of having Rocky Mountain spotted fever actually demonstrated at least a fourfold rise in antibodies to *E. canis* in paired serum samples. Treatment with tetracycline has been shown to reduce the severity of the illness in both humans and dogs.

A summary of rickettsial diseases is presented in Table 29.3.

PROTOZOA

Hemoflagellates

Flagellated protozoa transmitted to humans by the bite of infected bloodsucking insects are referred to as hemoflagellates and are classified in either the genus *Trypanosoma* or *Leishmania*. Cells of *Leishmania* and *Trypanosoma* pass through similar morphological stages during their life cycles in vertebrate and invertebrate hosts (see Figure 29.5); they differ, however, in that *T. gambiense* grows only as a trypomastigote form in a vertebrate host but is seen in both the trypomastigote and epimastigote forms in the invertebrate host. On the other hand, all four stages are seen when *T. cruzi* grows in the vertebrate host, but only the trypomastigote and epimastigote forms are in the invertebrate host. All species of *Leishmania* grow only in the amastigote form in the vertebrate host and in the promastigote form in the invertebrate host. It is therefore obvious that there is no sharp anatomical dividing line between these two genera.

Trypanosoma gambiense: Gambian Trypanosomiasis

Gambian trypanosomiasis, also called West African sleeping sickness, is transmitted to humans by the bite of an infected tsetse fly.

Pathogenicity During the 2 to 3 weeks of the incubation period, trypanosomes can be seen in small

TABLE 29.3   Summary of Arthropod-borne Rickettsiae

Organism	Arthropod Host	Disease	Incubation Period	Diagnostic Procedures	Treatment	Control
Rickettsiae						
R. rickettsii	Tick	Rocky Mountain spotted fever	3–12 days	Weil-Felix agglutination test Specific complement-fixation test	Chloramphenicol Tetracyclines	Tick control Rickettsial vaccine for those in infested areas
R. prowa-zekii	Body louse	Epidemic typhus fever	6–15 days, usually 12 days	Weil-Felix test Specific complement-fixation test	Chloramphenicol Tetracyclines	Sanitation Eliminate body louse Vaccine for those in areas where disease may be prevalent
R. typhi	Flea	Endemic or murine typhus	6–14 days	Weil-Felix test Specific complement-fixation test	Broad-spectrum antibiotics	Rodent and flea control
R. tsutsu-gamushi	Mite	Scrub typhus	Usually 10–12 days; may be 6 days to 3 weeks	Weil-Felix test Specific complement-fixation test	Chloramphenicol	Mite control Chloramphenicol prophylaxis
R. akari	Mouse mites	Rickettsialpox	10–24 days	Specific complement-fixation test	Chloramphenicol Tetracyclines	Elimination of rodents
R. quintana	Louse	Trench fever	14–30 days	Specific complement-fixation test	Chloramphenicol	Sanitation Eliminate lice

Figure 29.5 Morphological forms occurring in the hemoflagellates.

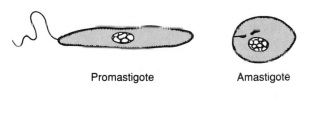

Promastigote   Amastigote

Epimastigote

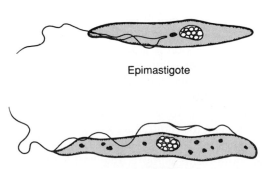

Trypomastigote

numbers in blood smears, but the attacks of fever do not commence until the organisms invade the lymph nodes. At this time the number of trypanosomes in the bloodstream and lymph nodes increases dramatically. After several days to a week, the fever subsides, and the patient is asymptomatic for several weeks before the occurrence of a subsequent, similar episode. These intermittent attacks may continue for several months before the trypanosomes invade the central nervous system, causing a meningoencephalitis manifested by slurred speech and difficulty in walking. Later stages are characterized by convulsions, paralysis, and mental deterioration. The central nervous system symptoms may last for months before the individual finally becomes comatose and dies.

Clinical Management T. gambiense does infect domestic animals such as cattle, pigs, and goats, but the cycle of infection for humans is from human to tsetse fly and back to human. Note the change in morphological form of the trypanosome occurring in tsetse fly and human, as shown in Figure 29.6.

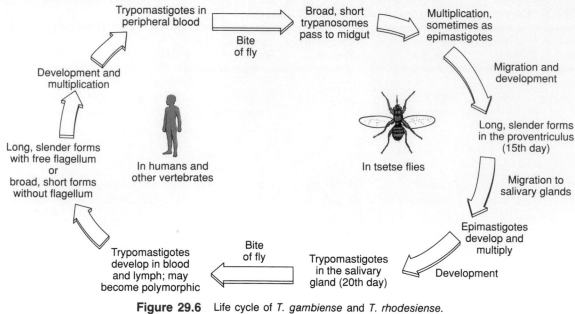

Figure 29.6 Life cycle of *T. gambiense* and *T. rhodesiense*.

For many years, the most puzzling aspect of this disease was the unexplained, recurrent febrile attacks and the apparent lack of an effective immune response to the trypanosomes, in spite of high titers of IgM antibodies to trypanosomal antigens. Using recombinant DNA techniques, it has now been shown that a single trypanosome may possess several thousand different genes that encode for a surface glycoprotein. Only one of these genes is transcribed at any one time, but the ability of the parasite to switch from one surface glycoprotein to another effectively counteracts any immune response by the host.

The diagnosis of trypanosomiasis is dependent on observation of the parasites in blood, lymph nodes, or spinal fluid, or by the demonstration of IgM antibody to the trypanosome. Control is directed toward the elimination of the tsetse flies.

A variety of drugs can be used for the successful treatment of trypanosomiasis, such as pentamidine isethionate and melarsoprol.

Trypanosoma rhodesiense: Rhodesian Trypanosomiasis

Rhodesian trypanosomiasis, also called East African sleeping sickness, is similar to Gambian trypanosomiasis but progresses more rapidly, and death frequently occurs prior to the development of meningoencephalitis.

The vector of Rhodesian trypanosomiasis is also a tsetse fly, but this vector becomes infected from feeding on wild animals, making control more difficult than it is for the West African variety.

Treatment uses a toxic drug called suramin as well as the drugs normally used for the treatment of Gambian trypanosomiasis.

Trypanosoma cruzi: American Trypanosomiasis

American trypanosomiasis, also called Chagas' disease, has an epidemiology and pathogenesis different from those of the African trypanosomiasis, as is shown in Figure 29.7. This disease occurs in the southern part of the United States and in Central and South America.

Pathogenicity The reservoir for the trypanosomes includes a wide variety of wild animals, particularly rodents, opossums, and armadillos. The vector may be any of a number of species of infected reduviid bugs (also called triatomids). These vectors characteristically inhabit houses. Trypanosomes grow in the gut of the bug, and humans are infected because the bug defecates while feeding, introducing the organisms into the site of the bite. Note the changes in morphological forms that occur during the life cycle of *T. cruzi*, shown in Figure 29.7.

The organisms infect the regional lymph nodes and are spread via the bloodstream to other organs of the body. The liver and macrophages of the spleen are commonly infected, but infection of the heart by the amastigote forms is more frequent, leading to inflammatory responses and an enlarged heart. Central

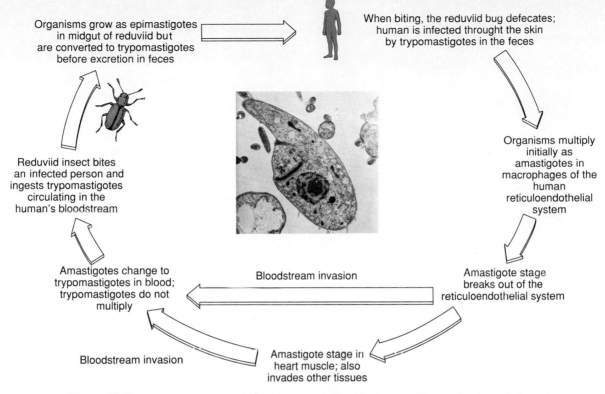

Figure 29.7 Life cycle of *T. cruzi*. The electron micrograph shows a thin section through the epimastigote form.

The arrows in the figure, read clockwise starting at top, contain:

Organisms grow as epimastigotes in midgut of reduviid but are converted to trypomastigotes before excretion in feces

When biting, the reduviid bug defecates; human is infected throught the skin by trypomastigotes in the feces

Organisms multiply initially as amastigotes in macrophages of the human reticuloendothelial system

Amastigote stage breaks out of the reticuloendothelial system

Bloodstream invasion

Amastigote stage in heart muscle; also invades other tissues

Bloodstream invasion

Amastigotes change to trypomastigotes in blood; trypomastigotes do not multiply

Reduviid insect bites an infected person and ingests trypomastigotes circulating in the human's bloodstream

---

nervous system involvement is rare in adults but not unusual in infants.

    Clinical Management Most fatal infections occur in children under the age of 5; the disease may be chronic in older children and adults. A diagnosis can be made by observing the trypanosomes in blood smears, growing the trypanosomes from blood cultures, or using fluorescently labeled antibody on blood smears.

    Chagas' disease is difficult to control, and its elimination will require improved housing and the elimination of reduviid bugs. The disease can be treated with nifurtimox (Bayer 2502).

Leishmaniasis

All species of *Leishmania* are transmitted from one animal to another by the bite of infected sandflies belonging to the genus *Phlebotomus*. There are a large number of serological types of *Leishmania* but no agreement on the taxonomy of this genus. In all cases the organisms grow in the sandfly in the promastigote form and in the vertebrate host in the amastigote form (see Figure 29.7). Leishmaniasis occurs in humans as two major types, cutaneous leishmaniasis and visceral leishmaniasis.

Leishmania tropica: Cutaneous Leishmaniasis

Numerous serological variants of *L. tropica* serve as the etiologic agents of cutaneous leishmaniasis.

    Pathogenicity The prototype of the cutaneous ulcer is called an Oriental sore (see Figure 29.8); the disease occurs primarily in the Near East, the Mediterranean region, Africa, southern Russia, and southern Asia. Several weeks after the bite of an infected sandfly, a papule evolves into an ulcer, which usually heals in about a year, leaving a disfiguring and depigmented scar. Immunity is solid, and since usually only one serological type of *L. tropica* is endemic in any one area, it is not uncommon for parents to infect their children intentionally in a part of the body where the resulting scar will not normally be visible.

    Mucocutaneous leishmaniasis is a variant of the cutaneous variety, but this disease is considered to be sufficiently distinct to warrant the naming of a new species, *L. braziliensis*, as the causative agent. Lesions occur in the nasal septum, lips, and soft palate; if the disease is untreated, death usually results from secondary bacterial infections.

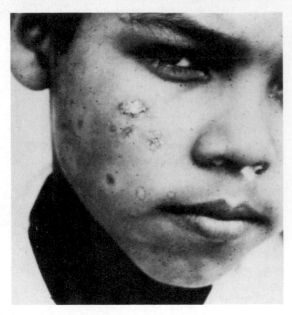

Figure 29.8 Oriental sores resulting from cutaneous leishmaniasis.

Clinical Management The normal reservoir for *L. tropica* varies from one endemic locale to another. For example, in India the dog seems to be the only reservoir of infection for humans. Elsewhere, rodents, gerbils, and ground squirrels provide the source of infection for the sandflies.

Diagnosis is usually based on the observation of the parasites in stained skin scrapings from infected areas. Treatment with arsenicals or amphotericin B is reported to yield good results.

Leishmania donovani: Visceral Leishmaniasis

Visceral leishmaniasis is more commonly known by its Indian name of *kala-azar*. It is seen in the Near and Far East, southern Russia, the Mediterranean area, parts of Africa, and in Central and South America.

Pathogenicity Several weeks to a year after the bite of an infected sandfly an individual will experience abdominal swelling due to the gross enlargement of the liver and spleen. *L. donovani* is also found in the skin and subcutaneous areas, and it is estimated that 90 percent of untreated cases of visceral leishmaniasis terminate in death about 2 years after the initial symptoms.

Clinical Management Humans and dogs apparently serve as the vertebrate reservoirs for *L. donovani* in urban districts; in rural sections, the reservoirs are rodents and other wild animals.

The diagnosis is dependent on the demonstration of the parasites, either by culture of the blood or bone marrow or by biopsies of the liver and spleen. Treatment is similar to that described for cutaneous leishmaniasis.

Sporozoa

Members of the Sporozoa are obligate parasites of animal hosts. The major human diseases include malaria and toxoplasmosis (see Chapter 26 for discussion of toxoplasmosis). In all cases the life cycles of the sporozoa are considerably more varied than those described for the other pathogenic protozoa; indeed, a new vocabulary is necessary to describe the stages in the incredibly complex life cycles of these parasites.

Plasmodium Species: Malaria

Malaria has been for centuries the greatest killer of all the infectious diseases. It is estimated that about 300 million cases occur each year, with the majority of these occurring in Africa. Of these, approximately 1 million terminate in the death of the host.

Four species of the Sporozoa are recognized as etiological agents of human malaria: *Plasmodium vivax*, *P. ovale*, *P. malariae*, and *P. falciparum*. The clinical picture in humans varies somewhat with each species, but the usual symptoms are chills and fever at more or less regular intervals, followed by profuse sweating. The source of human infections is the bite of an infected female *Anopheles* mosquito, and humans constitute the usual reservoir for infecting the mosquito.

The life cycle of the malarial parasites varies slightly among the different species, but we shall describe a general life cycle for all species of *Plasmodium*. Keep in mind that in humans the parasites can undergo only asexual reproduction and that sexual reproduction takes place only in the mosquito.

Pathogenicity An individual becomes infected when the **sporozoite** form of the parasite is injected into the bloodstream by the bite of an infected mosquito. The sporozoites immediately invade the parenchymal cells of the liver, where, depending on the species, they undergo one or more cycles of asexual reproduction before being liberated back into the bloodstream (see Figure 29.9). After entering a liver cell, the filamentous sporozoite rounds up to form a **trophozoite,** which continues to enlarge for approximately a week. When mature, the nucleus of the trophozoite divides to form thousands of nuclear masses. During this stage the parasite is called a **schizont.** A cytoplasm and membrane surround each nucleus to form a **merozoite** after which the infected cell ruptures,

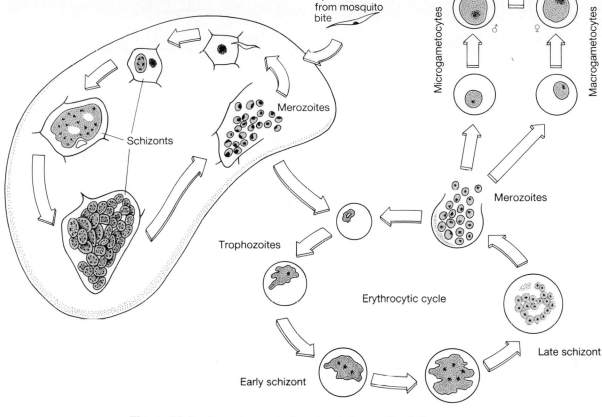

Figure 29.9 Asexual reproduction of malaria parasites in humans.

releasing the merozoites, which, depending on the species, will either infect red blood cells or begin a new round of reproduction in liver cells.

The erythrocytic cycle begins when a merozoite infects a red blood cell. The general sequence of events during the erythrocytic cycle is similar to that described for the preerythrocytic cycle. The merozoite differentiates into a trophozoite, which appears as a round or crescent-shaped structure called a ring stage. This structure becomes a schizont, which eventually liberates 12 to 28 merozoites. It is at this point that an individual experiences chills and fever.

After one or more cycles of asexual reproduction in the red cells, some of the merozoites will not divide but will instead form microgametes (male) and macrogametes (female). In humans these forms are the end of the line; unless ingested by a mosquito, they will degenerate within 6 to 12 h or be destroyed by the host's immune system. When the gametocytes are ingested during a blood meal by a mosquito, they are converted to male and female gametes and sexual fusion occurs within the stomach of the mosquito (see Figure 29.10). The resulting zygote undergoes a developmental stage to become an **ookinete;** this rounds

up outside the stomach to become an **oocyst,** within which are formed hundreds of spindle-shaped sporozoites. When the sporozoites are liberated from the oocyst, they become dispersed throughout the mosquito's body and many reach the salivary glands. This chain of events requires 10 to 12 days, and once the sporozoites are liberated, the mosquito is capable of transmitting the infection during its next blood meal.

Clinical Management If untreated, attacks may continue for 3 to 6 weeks, after which the parasites can no longer be found in the blood. Relapses, however, are not uncommon for a year or more after the initial infection, suggesting that the parasites can exist in an exoerythrocytic state during these periods of latency.

The diagnosis of malaria requires the visualization of the parasites in a stained blood smear, as shown in Figure 29.11. Fluorescently labeled antibody is also used to detect the presence of the parasites in the blood smear.

Control is directed toward finding and treating infected persons so as to render them noninfective and toward the elimination of adult mosquitoes and their

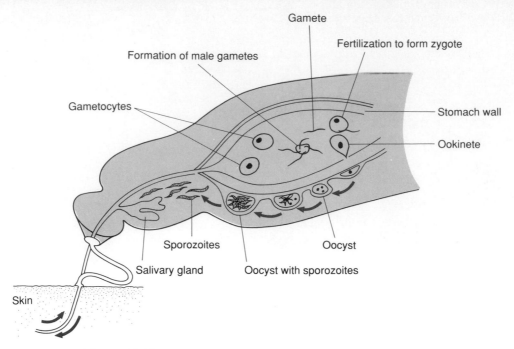

Figure 29.10 Sexual reproduction of malaria parasites in mosquito.

breeding places. Treatment utilizes several synthetic drugs such as quinacrine (Atabrine), chloroquine (Aralen), amodiaquine (Camoquin), pyrimethamine, and primaquine. The occurrence of drug resistance and the fact that a single drug is not effective against all of the stages of a particular malarial parasite necessitates the use of a combination of drugs for successful treatment. The combination depends on the species of *Plasmodium* involved and the properties of the current local strains.

An encouraging hope for the future centers about the production of an effective vaccine for the prevention of malaria. The feasibility of such a vaccine is based on the observation that antibodies directed against a sporozoite surface protein (termed the **circumsporozoite protein**) will neutralize the infectivity of the sporozoites, thus stopping the infection at its first stage. The gene encoding the circumsporozoite protein of *P. falciparum* has been cloned in *E. coli*, and the bacterially produced antigen has been shown in animals to induce the synthesis of protective antibodies, making such proteins a prime candidate for a human malaria vaccine.

Babesia microti: **Babesiosis**

Human occurrence of this disease was first reported in 1976. It is now apparent that babesiosis in the United States is limited to a very specific geographic area which includes New York, Massachusetts, and its offshore island, Nantucket.

Babesia species are sporozoan parasites found in many animals, particularly dogs, cats, rodents, and cattle, but *B. microti* is the only species known to infect humans. Their life cycle is similar to that of the malaria parasites except that the vector is a tick in the genus *Ixodes*. Clinical symptoms mirror those of malaria, including chills, fever, nausea, and vomiting. Such symptoms may persist for several months.

Interestingly, there are a number of cases in which it appears that an individual acquired both Lyme disease and babesiosis from a single tick bite. Because *Ixodes* is the vector for both infections, such situations seem highly possible.

Diagnosis requires either the visualization of the parasite in blood smears or the use of an indirect

Figure 29.11 *P. vivax* in red blood cells. One cell has two rings; two other cells contain growing trophozoites.

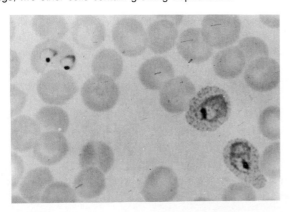

immunofluorescence test. Treatment with chloroquine gives symptomatic relief but clindamycin and quinine are more effective as curative agents.

Table 29.4 summarizes the protozoan diseases spread via arthropod vectors.

Blood and Tissue Nematode Infections in Humans

Unlike intestinal roundworm infections, nematodes infecting blood and tissue are not spread through fecal contamination. Most of these latter parasites are carried from human to human by the bite of an arthropod vector.

The worms to be discussed in this section belong to the superfamily Filarioidea, and a resulting human infection is called **filariasis**. Adult worms generally range from 2 to 30 cm in length, with the female about twice the size of the male. One property distinguishing the filariae from other nematodes is that the female does not lay eggs, but instead gives birth to prelarval forms called **microfilariae**. As we shall see, the ingestion of the microfilariae by blood-sucking insect vectors provides for the transmission of filariae from one person to another.

Wuchereria bancrofti

Bancroftian filariasis (a cause of elephantiasis) is seen extensively in the Pacific Islands as well as in much of Africa. Sporadic cases occur in European countries bordering the Mediterranean Sea, and the disease is also found scattered throughout the Near and Far East and in Central and South America.

W. bancrofti is transmitted to humans through the bite of an infected mosquito—a species of *Culex, Aedes,* or *Anopheles* (Figure 29.12). The mosquito, however, provides more than a mere mechanical vector for transmission; the ingested microfilariae must undergo transformation within the mosquito to form larvae infective for the definitive host, that is, humans. This development requires about 10 days, after which time infective larvae enter the proboscis of the mosquito and are transmitted to a human during the next blood meal.

Within the human, the larvae enter the lymphatic vessels and nodes, where they develop into adult worms during the ensuing 6 months. The worms appear to preferentially infect the lymphatics of the lower extremities, and pathological changes occur most frequently in the groin and genital areas. Once ensconced, mating occurs and the resulting microfilariae migrate through the walls of the lymphatics into the adjacent blood vessels.

Symptoms result from the presence of the adult worms in the lymphatics, but light infections are often unnoticed, with the only physical signs being slightly enlarged lymph nodes, particularly in the groin. However, in endemic areas where frequent exposure occurs, attacks are characterized by fever, chills, vomiting, malaise, and tender, swollen lymph nodes. Such attacks last several days and may result in the formation of draining ulcers in the lymph nodes or along the lymphatic vessels. Surprisingly, the major symptoms

TABLE 29.4 Major Arthropod-borne Protozoa Causing Specific Human Diseases

Parasites (Diseases)	Definitive Hosts	Intermediate Hosts	Important Reservoir Hosts	Transmission to Humans
T. gambiense, T. rhodesiense (African sleeping sickness)	Humans, animals	Tsetse flies (*Glossina* species)	Humans, animals	By inoculation (bite of fly)
T. cruzi (Chagas' disease)	Animals, humans	Reduviid bugs	Armadillos, opossums, humans	By contamination (infective feces of bug)
L. donovani (kala-azar) *L. tropica* (Oriental sore) *L. braziliensis** (espundia)	Humans, dogs	Sand flies (*Phlebotomus* species)	Dogs, humans	By inoculation (bite of fly; direct transmission possible)
P. vivax *P. falciparum* *P. malariae* *P. ovale* (malaria)	Anopheline mosquitoes	Humans	Humans	By inoculation (bite of mosquito; also by transfer of infected blood)
B. microti	*Ixodes* ticks	Humans, animals	Animals	By bite of *Ixodes* ticks

*Dogs and other animals have been implicated in the life cycles of *L. donovani* and *L. tropica,* but the part played by hosts other than humans in the case of *L. braziliensis* is questionable. Naturally infected dogs have been found in South America.

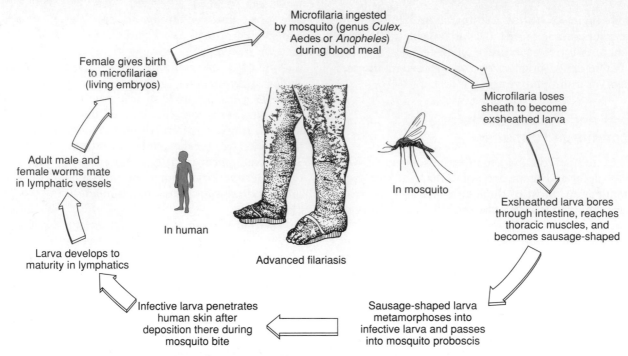

Figure 29.12 Life cycle of *W. bancrofti*. The photo demonstrates an advanced case of filariasis caused by this parasite's blocking the lymphatics in both legs of a child.

seem to be the result of a hypersensitivity reaction directed against antigens present in the adult worms. In rare cases, multiple exposure to the filariae results in the proliferation of fibrous tissue around the dead worms, leading to an obstruction of lymphatic flow and causing extensive edema (swelling owing to the accumulation of fluid) in the legs, scrotum, female genitalia, or breasts. This obstructive filariasis, more commonly called elephantiasis (shown in Figure 29.12), is fortunately rare even in endemic areas and occurs only after repeated infections and many years of chronic filariasis.

The physical signs of inflamed lymph vessels (lymphangitis); swollen, tender lymph nodes (lymphadenitis); and edema of the extremities, genitalia, or breasts (elephantiasis) occurring in an endemic area are sufficient to make a tentative diagnosis of chronic filariasis. An unequivocal diagnosis, however, requires the observation of the microfilariae in the blood. This may prove difficult for two reasons: (1) very few microfilariae are present during the chronic stage and (2) with the exception of the South Pacific strain, the microfilariae show a definite periodicity in which very few are present in the blood during daylight hours while the patient is active but large numbers may be found during the night. This nocturnal periodicity is not fully understood, but it could be influenced by an increased partial pressure of oxygen in the lungs during normal activity. This phenomenon necessitates collection of blood samples between 10 P.M. and 2 A.M.

Diethylcarbamazine (Hetrazan) kills the microfilariae of *W. bancrofti,* but it is not always effective in eliminating adult worms. Antihistamines and steroids are used to decrease hypersensitivity reactions. Advanced cases of elephantiasis, however, usually require surgery to control lymphatic obstructions. This may require removal of an enlarged scrotum or the anastomosis (joining together) of the deep and superficial lymphatics in the legs.

Efforts to control *W. bancrofti* can be directed toward two vulnerable stages in the life cycle of the filariae: (1) the elimination of the mosquito, as was achieved against yellow fever, and (2) the mass administration of diethylcarbamazine to all persons within an endemic region. This latter procedure destroys the microfilariae and has proved successful in several isolated island areas such as the Virgin Islands and Tahiti.

Brugia malayi

B. malayi, the etiological agent of Malayan filariasis, differs slightly in morphology from *W. bancrofti* but shares many life-cycle characteristics. In addition, the clinical and pathological features of Malayan filariasis are similar to those described for the bancroftian variety. The Malayan Peninsula comprises one of the major endemic areas of this disease, although the parasites are seen also in India, Vietnam, Indonesia, Thailand, and Ceylon. The microfilariae can develop in the same mosquito species as *W. bancrofti*, but the principal mosquito vector for *B. malayi* is a mosquito belonging to the genus *Mansonia*. Unlike *W. bancrofti*, for which humans comprise the only known reservoir, *B. malayi* has been found in monkeys, dogs, and cats, and it is assumed that there may be any number of mammalian hosts able to serve as reservoirs for this nematode.

Laboratory diagnosis requires the visualization of the microfilariae in a blood smear. As described for *W. bancrofti*, microfilariae from most strains also exhibit a nocturnal periodicity, but some strains have been reported to lack this periodicity.

Treatment and control of this parasite employ essentially the same procedures as for the bancroftian variety. The major vectors, mosquitoes of the genus *Mansonia*, breed in ponds heavily populated with the water plant *Pistia stratiotes*, an essential component of the life cycle of these mosquitoes. As a result, herbicides such as sodium methyl chlorphenoxyacetate provide effective control.

Loa loa

The scientific designation of *L. loa* for this species of nematode is derived from its African name. It is found only in Africa, and a common name is "the **African eye worm**."

The transmission of loiasis, the disease resulting from an infection of *L. loa*, is much the same as with diseases from other filarial parasites, except that in this case the vector is one of several species of mango or deer flies (genus *Chrysops*). The microfilariae undergo development within the fly and are transmitted to humans through the bite of an infected fly. Monkeys and humans appear to be the only definitive hosts.

Following infection, the larvae develop into adult worms, which migrate throughout the subcutaneous tissues of the definitive host. The infection is generally asymptomatic, although occasional inflammatory reactions known as Calabar swellings occur at irregular intervals. These swellings are thought to result from hypersensitivity reactions and normally disappear within a week. The disquieting and somewhat painful manifestation of a *Loa* infection occurs when adult worms migrate into the facial area. There they can frequently be seen (and removed) as they pass over the bridge of the nose or migrate through the subconjunctival tissue of the eye. Untreated, the infection may last for many years without major apparent damage to the host. Chronic loiasis sometimes results in an allergic dermatitis in which abscesses may form following secondary bacterial infections.

The occurrence of Calabar swellings in an endemic zone suggests a *Loa* infection; however, a definitive diagnosis of loiasis is based on finding the microfilariae in the blood or observing the adult worms beneath the conjunctiva. The oral administration of diethylcarbamazine is very effective for eliminating adult worms and microfilariae. Also, the worms can be removed surgically, but since chemotherapy is effective, this procedure is probably unwarranted.

Prevention of infection is directed at eliminating the carriers by mass treatment with diethylcarbamazine in endemic sections; nets, screens, and repellants are also used to ward off infected flies.

Onchocerca volvulus

Most infections by *O. volvulus* are found in Central Africa, but they also occur in restricted areas of Central America and northern South America. Larvae are transmitted to humans through the bites of infected black flies (genus *Simulium*). The adult worms occur in the subcutaneous tissue of the definitive host. There the male and female worms routinely become enclosed in fibrous capsules that can be seen grossly as small nodules under the skin. Microfilariae migrate from the capsules and move throughout the dermis; these microfilariae induce the major pathological lesions of onchocerciasis. The most serious lesions occur following the migration of the microfilariae into the eyes. Ocular lesions may be caused, in part, by toxic products, but it is generally thought that the chief destructive changes are the effect of allergic reactions to the invading microfilariae. The severity of the damage can be emphasized by statistics obtained from some endemic regions of Africa, which reveal that about 18 million people in the developing world are affected, and 500,000 of these become permanently blinded each year as a result of onchocerciasis. Rivers and streams provide the breeding areas for the black fly vectors and, as a result, the disease is also called "river blindness."

Skin inflammation is also a frequent manifestation of this disease. Onchodermatitis, which is the result of an allergic reaction to the microfilariae in the skin, may after long-established, chronic disease be seen as thick, wrinkled, hyperpigmented skin. In extreme cases, a sac of pelvic tissue may hang down to the knees owing to loss of elasticity in the skin.

It now appears possible to rid the world of this disease. Merck and Company has developed a drug termed "mectizan," a derivative of another drug called ivermectin. Two doses per year provide complete protection with minimal side effects. Mectizan kills the microfilariae in the body and prevents the adult worms from producing new microfilariae. Thus, if an entire population can be treated for a period of about 10 years, all adult worms should have died and the disease eliminated from the world.

To make this possible, Merck and Company has offered the drug without cost to developing countries. The World Health Organization, along with a review committee set up by Merck, will establish distribution and reporting systems. It now seems possible that if appropriate antiparasite drugs can be developed, any of the filarial diseases in which humans comprise the sole definitive host can be eliminated.

Prevention of infection is directed toward the elimination of the disease in humans (the only definitive host from which the vector becomes infected with the microfilariae) and the eradication of the vector itself. The latter procedure involves the addition of insecticides to local waters to destroy aquatic developmental stages of the black flies.

Table 29.5 summarizes the nematode diseases spread by arthropod vectors.

VIRUSES

The viruses discussed in this section belong to the families Togaviridae and Bunyaviridae and are referred to as the **arboviruses.** This designation stems from the fact that they are able to infect vertebrates only as a result of being transmitted from one host to another by an arthropod vector; hence, the infections they cause are commonly called **arthropod-borne diseases.** The most important vector of this group of viral diseases is the mosquito, although the tick is involved in several arbovirus diseases. Without an arthropod vector, these diseases would soon disappear from the earth.

There are at present between 150 and 200 different arboviruses, and they fall into many antigenic groups. Many of these viruses are found predominantly in one geographic area of the world, and they have been given names corresponding to these areas. Examples of this are seen in such names as Venezuelan equine encephalitis, Semliki forest (Africa) encephalitis, St. Louis encephalitis, Japanese B encephalitis, West Nile encephalitis, Colorado tick fever, and others. Togaviruses are all enveloped RNA viruses that cause the hemagglutination of chicken red blood cells. Based on the hemagglutination-inhibition test, the togaviruses can be divided into two large genera—*Alphavirus* and *Flavivirus*. All alphaviruses stimulate in a host the production of antibodies that inhibit the hemagglutination caused by other alphaviruses, whereas the flaviviruses stimulate antibody production that will inhibit hemagglutination by all members of flaviviruses. Figure 29.13 illustrates the reproduction of Sindbis virus (an alphavirus) as it buds from the membrane of an infected cell. We will confine our discussion of this large group to several examples that illustrate the types of diseases occurring in humans.

Yellow Fever Virus

The yellow fever virus is a very small spherical particle measuring approximately 38 nm in diameter. The agent belongs antigenically to the genus *Flavivirus*. It can be grown in chick embryos, cell cultures, and certain animals, particularly monkeys. There is only one major antigenic strain of yellow fever virus, so a single vaccine is effective.

Pathogenicity   Yellow fever is an acute infectious disease that is characterized by severe liver damage. Symptoms include headache, backache, fever, prostration, nausea, and vomiting. The disease may

TABLE 29.5   Blood and Tissue Nematode Infections in Humans

Disease	Etiological Agent	Reservoir	Vector
Bancroftian filariasis	*Wuchereria bancrofti*	Humans	Mosquito
Malayan filariasis	*Brugia malayi*	Humans	Mosquito
Loiasis (African eye worm)	*Loa loa*	Monkeys and humans	Mango and deer flies
Onchocerciasis (River blindness)	*Onchocerca volvulus*	Humans	Black flies

Figure 29.13 Numerous viral buds are seen here on the outer surface of the inner leaflet of the plasma membrane surrounding a cell infected with Sindbis virus. (Original magnification ×69,000.)

mogogus act as vectors to transmit the disease to humans.

Clinical Management Clinical grounds are usually sufficient for diagnosis during periods of epidemics, but mild cases may be difficult to diagnose. Various laboratory tests include the inoculation of blood into various animals, particularly mice, and the demonstration of neutralizing antibodies in the patient.

The eradication of *Aedes aegyptius* mosquitoes in urban areas is a very effective control measure. It is this kind of attack that has virtually eliminated the disease from the urban populations of Central and South America as well as a few of the southern cities of the United States. However, since monkeys constitute a major reservoir of the disease, such control in forest and jungle areas is not possible.

There is a living attenuated vaccine for yellow fever (17 D strain of virus), which is grown in chick embryos and has proved to be very effective in stimulating active immunity to the disease. There is no specific treatment.

Dengue Fever Virus

Dengue fever virus is also a flavivirus, about 25 nm in diameter (see Figure 29.14), which is transmitted from person to person by several species of *Aedes* mosquitoes.

Epidemics of dengue fever are indigenous to southeast Asia. In the western hemisphere the disease is found almost entirely in Central America and the Caribbean Islands where over 43,000 cases were reported in 1984. In the United States, the only reported cases during the past several decades occurred in Texas, near the Mexican border.

Pathogenicity Dengue fever is an acute infection usually manifested by headache, backache, fatigue, stiffness, loss of appetite, chilliness, and occasionally a rash. Probably the most characteristic symptom is emphasized by another name for this disease—break-bone fever. Many of these symptoms may precede the first rise in temperature. In some cases the onset may be sudden, with a sharp temperature rise, severe headache, backache, pain behind the eyes, and muscle and joint pains.

The virus of dengue fever is found in the blood of the infected person shortly before the onset of the febrile stage. Four types of virus have been described. They possess considerable cross-reactivity, but recovery from an infection by one type does not provide cross-immunity against infection by other types.

The incubation period is usually 5 to 8 days but seems to be influenced by the amount of infecting virus and may vary from 3 to 15 days.

be mild or severe. As a result of the liver damage, jaundice may be evident as early as the fourth or the fifth day of the illness. Following recovery, a person is permanently immune to reinfection. The incubation period is usually 3 to 6 days.

A great advance in medicine occurred when Walter Reed and his associates worked out the epidemiology of yellow fever. Their observations not only made it possible to complete the building of the Panama Canal but have also been responsible for the elimination of this dread disease in many urban areas of the world. It is now known that the disease is transmitted from person to person, monkey to human, or monkey to monkey by one or another species of mosquito. Humans are the reservoirs of infection in urban areas, and monkeys, marmosets, and perhaps marsupials serve as reservoirs in the jungle areas. Jungle yellow fever differs from yellow fever only in its vectors. Mosquitoes in the genera *Aedes* and *Hae-*

Although we seldom think today of yellow fever in the United States, our ancestors were not so fortunate. It is recorded that 1411 soldiers out of a total of 1500 died on the island of St. Lucia in the Caribbean in 1664, and in 1802 yellow fever killed 29,000 of 33,000 soldiers and sailors sent by Napoleon to conquer Santo Domingo in Haiti, perhaps changing the future of North America. New York, Boston, and Philadelphia had their first epidemics of yellow fever in 1668, 1691, and 1695, respectively. Epidemics of yellow fever also occurred in New Orleans, Memphis, Charleston, Norfolk, Galveston, Baltimore, and many other cities, killing well over 100,000 people during a 200-year period.

During that 200 years, scientists searched for clues that would tell them how to prevent this disease, but it took until the year 1900 before the first experiments were designed that provided answers to that question.

Who were the conquerors of yellow fever? Certainly, Dr. Carlos Finlay deserves the credit for proposing in a published paper in 1881 that mosquitoes appeared to be the carriers of yellow fever. But, it was 19 years later that Dr. Walter Reed came to Havana, Cuba, and designed the experiments that showed that yellow fever was transmitted only by the mosquito. His experiments conclusively showed that one could not acquire yellow fever by living in the same room with an infected person as long as mosquitoes were excluded.

What were these experiments? Human volunteers permitted themselves to be bitten by mosquitoes that had fed on patients with yellow fever. Other volunteers lived in isolation rooms for as long as 3 weeks using bedding that had been taken from yellow fever patients. Of these volunteers, one physician, Dr. Jesse W. Lazear, died of yellow fever after permitting himself to be bitten by an infected mosquito.

What was the final result of these experiments? Following the report of Major Walter Reed in 1901, Major W. C. Gorgas headed a campaign to eliminate mosquitoes from Havana. Every cistern and water container that could support mosquito breeding was covered. Spigots were installed free of charge at the bottom of all water barrels so water could be drawn off without removing the lid. All open, standing fresh water was eliminated, and people were fined if mosquito larvae were found on their premises.

What happened was astounding. Within the first year, deaths from yellow fever dropped from an average of 500 to 13 and during the next 3 years, Havana remained completely free of yellow fever. Similar campaigns against mosquitoes were started in Brazil, Mexico, Panama, and the United States, with the result that yellow fever has been essentially eliminated from urban areas. In fact, if the number of deaths from yellow fever had not been drastically reduced by mosquito control, it is possible that the Panama Canal would not have been completed. The disease still exists, however, in the monkey population in the jungles of South America, and in the absence of mosquito control, it could return to its old haunts.

Humans and the *Aedes* mosquito are the recognized reservoir and vector for this disease. However, as in yellow fever, the monkey may well be another reservoir.

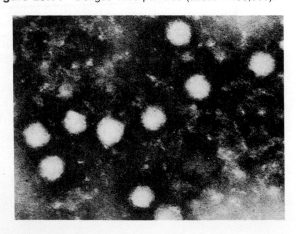

Figure 29.14 Dengue virus particles (about ×150,000).

Clinical Management Serological tests that are useful as diagnostic aids include complement fixation, hemagglutination, and neutralization. The major types of the virus are adapted for growth in mice, and it is this source that is used for the preparation of the specific antigen.

Control measures are directed toward the elimination of the *Aedes* mosquitoes. A vaccine has been effective in experimental work, but it is not yet available for general use. There is no specific treatment that is effective against the virus.

Epidemic Encephalitis Viruses

Pathogenicity A very large number of different mosquito-borne arboviruses throughout the world may infect humans and cause encephalitis. We shall discuss only four viruses in this group. Three of these occur in the United States; western equine encephalitis (WEE), eastern equine encephalitis (EEE), and St. Louis encephalitis (SLE). The other viral encephalitis of major importance occurs in Japan and the eastern part of Asia and is called Japanese B encephalitis (JBE).

All these encephalitides are transported to humans by one or another species of mosquito, the species being determined by the geographic area involved. There have been occasional epidemics of EEE in the United States, particularly in New Jersey. This appears to be the most severe of the encephalitides that occur in the United States. It has a high mortality rate, and a large percentage of those who recover are left with a severe neurologic disorder. WEE and SLE are considerably less severe; the majority of infected individuals recover completely from these two viral infections. However, a major epidemic of SLE in Texas in 1964 resulted in approximately a 10 percent mortality. Also, Southern Florida experienced an epidemic of SLE during the autumn of 1990, which forced the rescheduling of night football games and other activities after dark, when mosquitoes are most active.

JBE is also a severe disease in Japan. Like so many infections, viral encephalitis must cause a very large number of undiagnosed infections. This can be shown by demonstrating protective antibodies in a large percentage of adults and by the fact that a large percentage of patients afflicted are children.

The pathogenesis of all the encephalitides begins when an individual is bitten by an infected mosquito. The virus grows in the lymphatic tissue and internal organs and eventually reaches the central nervous system, causing neurological symptoms.

The major reservoir of the encephalitis viruses is the bird population. Thus, even though many animals may be infected by a mosquito, it seems apparent that the real reservoir is birds and that the infected mosquito acquires the virus while taking a blood meal from an infected bird. During epidemics, horses frequently become infected (hence *equine* encephalitis), but it is not believed that horses act as reservoirs for the continued spread of the disease (see Figure 29.15).

The incubation periods vary with each disease, but they fall within a range of 4 to 21 days.

Clinical Management The general symptoms of encephalitis are high fever, headache, vomiting, convulsions, and, in the most severe cases, drowsiness and coma. The common term for these infections is sleeping sickness, even though they are not related to the infamous African sleeping sickness. Although all these viruses can be grown in animals or cell cultures, they are very dangerous organisms with which to work. Thus, diagnosis usually is made by serological techniques such as complement-fixation tests on the patient's serum.

The most effective control measure is the eradication of the mosquito vectors. A virus vaccine may be used in individuals who are likely to be exposed,

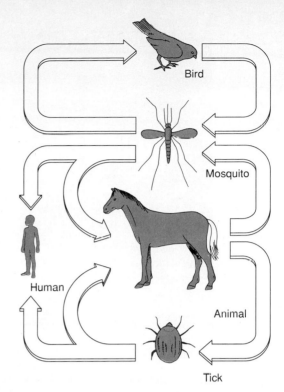

Figure 29.15 Epidemiology of togavirus infections. Although the bird-mosquito cycle is probably most important for the maintenance of these viruses, lower animals and ticks also participate in the epidemiology of the togaviruses. Humans become infected by either mosquitos or ticks (varying, of course, among the different togaviruses) when they come into contact with one of these cycles.

but it is not recommended for general use. Those accidentally exposed may be given immune serum to provide passive immunity. There is no specific treatment for these diseases.

Bunyaviridae

This group contains approximately 180 viruses and is divided into subgroups based on complement-fixing cross-reactions. Subgroups are further divided by the use of hemagglutination-inhibition and neutralizing antibodies.

All members of the Bunyviridae have been isolated from mosquitoes, but only a few subgroups have been shown to infect humans. We shall discuss only a couple of these diseases.

The California encephalitis group contains 10 serotypes found in many geographic areas of the world (see Figure 29.16). They have been isolated from rabbits and rodents as well as from mosquitoes of the genera *Aedes* and *Culex*. Human infections have been reported throughout the Midwest and Far West of the

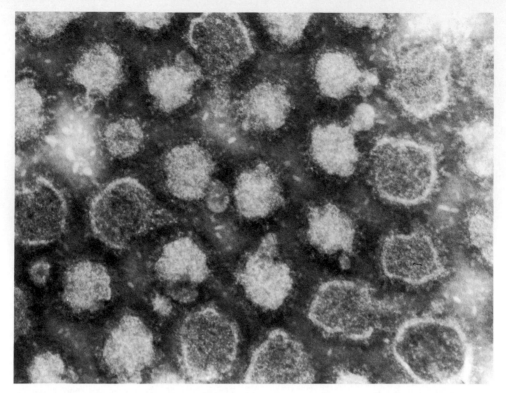

Figure 29.16 Nucleocapsids of California encephalitis virus, a representative bunyavirus (×212,500).

United States. Symptoms may be moderately severe, including central nervous system involvement; death is rare, and recovery is usually complete.

Sandfly fever virus is spread to humans via the bite of the female sandfly, *Phlebotomus papatasii*. Found primarily around the Mediterranean Sea, the disease causes headache, nausea, fever, photophobia, and abdominal pain. Complete recovery is usual.

TABLE 29.6 Summary of Arthropod-borne Virus Diseases

Virus	Disease	Incubation Period	Diagnostic Techniques	Control
Dengue fever virus	Dengue fever	Usually 5–8 days; may be 3–15 days	Complement-fixation, hemagglutination, and neutralization test	Elimination of *Aedes* mosquitoes (Viral vaccine, experimental)
Yellow fever virus	Yellow fever	3–6 days	Inoculate blood into mice Neutralization test	Eradication of mosquito vector Viral vaccine
Epidemic encephalitis viruses	Epidemic encephalitis	4–21 days	Complement-fixation test	Eradication of mosquito vector Viral vaccine for the highly exposed Immune serum for the accidentally exposed
California encephalitis virus	Encephalitis		Complement-fixation test	Mosquito control
Sandfly fever virus	Sandfly fever	3–4 days	Complement-fixation test	Control of sandflies
Colorado tick virus	Colorado tick fever	4–5 days	Complement-fixation test	No control

Colorado tick fever is a mild disease that occurs throughout much of the western United States. Major symptoms in humans include fever, headache, joint pain, nausea, and vomiting. Mortality is very low. Little is known concerning the etiological agent, but it contains double-stranded RNA and is now classified with the Orbiviruses. The wood tick constitutes both the vector and the reservoir for the disease, since the virus is passed via the transovarian route in the tick.

The arthropod-borne virus diseases are summarized in Table 29.6.

SUMMARY

Of the bacterial diseases transmitted to humans as a result of an arthropod bite, none is so well known as bubonic plague or, as it is frequently called, the Black Death. Its reservoir is the rat population, and it is spread to humans via the rat flea. Tularemia, on the other hand, is frequently contracted as the result of direct contact with infected animals—most frequently rabbits—even though the disease may be spread by deer fly or tick bites.

Rocky Mountain spotted fever, rickettsialpox, epidemic typhus, murine typhus, scrub typhus, and trench fever are all diseases caused by rickettsial organisms that are spread to humans by the bite of an arthropod vector. The vector varies from one disease to another; ticks, lice, fleas, and mites are the common ones.

The hemoflagellates include the genera *Trypanosoma* and *Leishmania*. Gambian and Rhodesian trypanosomiasis are acquired by humans from the bite of an infected tsetse fly. American trypanosomiasis, or Chagas' disease, is spread by infected reduviid bugs. Leishmaniasis may occur as a cutaneous variety, called Oriental sore, or as a frequently fatal visceral variety. Cutaneous leishmaniasis is caused by *L. tropica* and visceral leishmaniasis by *L. donovani*. Both are spread by bites of infected sandflies belonging to the genus *Phlebotomus*.

Malaria may be caused by any one of four species: *Plasmodium vivax*, *P. ovale*, *P. malariae*, or *P. falciparum*. The parasites undergo a very complex life cycle in which the asexual reproduction occurs in humans and the sexual reproduction occurs in an infected mosquito.

Babesiosis is caused by a protozoan that is spread to humans through the bite of an infected *Ixodes* tick.

Blood and tissue nematodes are most frequently acquired as a result of a bite by an infected mosquito or fly. All such worms belong to the superfamily Filarioidea and the resulting human infection is called filariasis.

One very large group of viruses that is spread to humans as a result of the bite of an infected mosquito is called the togaviruses.

Another group of viruses spread by arthropod vectors is the bunyaviruses, which includes the causative agents of California encephalitis and sandfly fever. Colorado tick fever is caused by a virus that is now classified with the orbiviruses.

QUESTIONS FOR REVIEW

1. Why is it considered dangerous to poison the rat population during a plague epidemic? What measures can be taken to control an epidemic?

2. What are the major ways in which tularemia may be transmitted to humans?
3. Name all of the characteristics the *Rickettsia* have in common. (Refer to Chapter 13 for growth characteristics.)
4. What diseases are caused by rickettsiae? How is each transmitted to humans? List the diagnostic and control measures for each.
5. List the different kinds of trypanosomiasis. Include vector, reservoir, and etiological agent for each.
6. What are the various manifestations of human leishmaniasis? List vector, reservoir, and etiological agents for each.
7. Outline the sequence of events in the asexual and sexual reproduction of the genus *Plasmodium*.
8. How is malaria diagnosed?
9. Where and how would you be most likely to become infected with *Babesia microti*?
10. Describe the life cycle of *Wuchereria bancrofti*.
11. Give the causative agent, the mode of spread, and the means of control for dengue fever, yellow fever, epidemic encephalitis viruses, California encephalitis, sandfly fever, and Colorado tick fever.

SUPPLEMENTARY READING

Anderson JF, et al: Involvement of birds in the epidemiology of the Lyme disease agent *Borrelia burgdorferi. Infect Immun* 51:394, 1986.

Badaro T, et al: New perspectives on a subclinical form of visceral leishmaniasis. *J Infect Dis* 154:1003, 1986.

Benach JL, et al: Serologic evidence for simultaneous occurrences of Lyme disease and babesiosis. *J Infect Dis* 152:473, 1985.

Donelson JE, Turner MJ: How the trypanosome changes its coat. *Sci Am* 252:44, 1985.

Fox JL: Interest in Lyme disease grows. *ASM News* 55:65, 1989.

Hadley TJ, Klotz FW, Miller LH: Invasion of erythrocytes by malaria parasites: A cellular and molecular overview. *Annu Rev Microbiol* 40:451, 1986.

MacDonald GA, et al: Cloned gene of *Rickettsia rickettsiae* surface antigen: Candidate vaccine for Rocky Mountain spotted fever. *Science* 235:83, 1987.

Magnarelli LA, Anderson JF, Barbour AG: The etiologic agent of Lyme disease in deer flies, horse flies and mosquitoes. *J Infect Dis* 154:355, 1986.

Marshall E: Malaria research—What next? *Science* 247:399, 1990.

McDade JE: Ehrlichiosis—A disease of animals and humans. *J Infect Dis* 161:609, 1989.

McDade JE, Newhouse VF: Natural history of *Rickettsia rickettsii. Annu Rev Microbiol* 40:287, 1986.

McEvedy C: The bubonic plague. *Sci Am* 258:118, 1988.

Nussenzweig V, Nussenzweig RS: Circumsporozoite proteins of malaria parasites: A minireview. *Cell* 42:401, 1985.

Walker DH: Rocky Mountain spotted fever: A disease in need of microbiological concern. *Clin Microbiol Rev* 2:227, 1989.

Walsh J: Human trials begin for malaria vaccine. *Science* 235:1319, 1987.

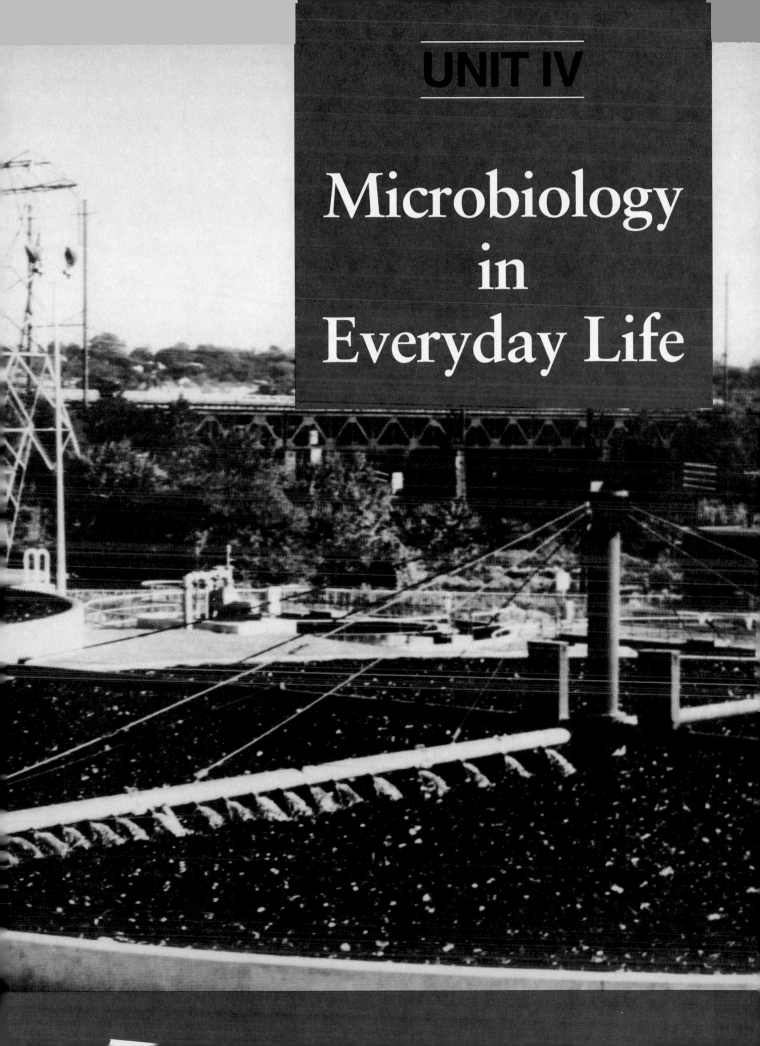

UNIT IV

Microbiology
in
Everyday Life

Chapter 30

Microbiology of Water and Sewage

Few, if any, substances are as crucial to our daily living as water. Unfortunately, no human requirement is so subject to contamination with injurious materials. The purpose of this chapter is to provide a brief introduction showing how our society safeguards this precious commodity from infectious disease organisms.

Water purification for disease prevention is so nearly universal in the western world today that it is difficult for most of us to realize how serious water contamination can be. Unfortunately, epidemics of waterborne diseases such as cholera, dysentery, and typhoid fever still occur in parts of the world, and these diseases could easily become epidemic in any country if it were not for rigid government control of drinking water and sewage disposal. Recreational waters are less easily controlled against waterborne infections.

Water that is fit for drinking, that is, free from harmful and unpleasant substances, is said to be **potable.**

It is important, however, to realize that water can contain either poisonous chemicals or pathogenic organisms, yet still be clear and sparkling. In such a case, we refer to it as **contaminated water. Polluted water,** in turn, may or may not be contaminated but has an undesirable appearance or taste.

This chapter discusses the techniques of determining whether water is microbiologically safe for use and control measures used to ensure the safety of drinking water and recreational water.

MICROBIAL FLORA OF WATER

Water provides a habitat for many different microorganisms. Among the bacteria most commonly found in natural waters are sulfur bacteria, iron bacteria, free-living spiral forms, certain pigmented and nonpigmented species, and some spore formers. Since water is in contact with soil, many common soil inhabitants, such as *Bacillus* species, are usually present. Pathogenic microbes do not usually make up a part of the normal flora in any area; therefore, water is dangerous only if it is contaminated from some external source.

TESTING THE POTABILITY OF WATER

Our primary concern in testing water for potability is the presence of organisms capable of causing disease.

In addition, we would like to know how many bacteria of all types are present in a given sample of water in order to determine the efficiency of a community's water purification system.

Standard Plate Count

Water being prepared for a standard plate count is usually diluted with sterile water. Measured quantities of each dilution are mixed with a nutrient medium in a petri dish and, after incubation, the standard plate count (the count per milliliter) is determined by multiplying the number of colonies on a plate by the dilution of the sample. One must keep in mind that this will not necessarily be a true count, since strict anaerobes, autotrophs, and some other species will not grow under the conditions provided in this test. In fact, numbers obtained with the standard plate count will represent only about 1 percent of the total number of bacteria detected by staining with acridine orange.

However, even though a low bacterial count is desirable, it is the kind of organisms present in a water supply that determines its sanitary quality and not the total number of bacteria. A few of the common diseases caused by such organisms are typhoid fever, bacillary dysentery, amebic dysentery, giardiasis, hepatitis, and poliomyelitis. Even if these were the only dangerous organisms, it would be a really major achievement if one could say, "I have tested the water for all of those pathogens and can find none present."

Luckily, there is an easier way to determine whether water might contain any of these enteric pathogens. It is based on one assumption: **Any water contaminated with feces contains enteric pathogens and is, therefore, unpotable.** Fecal contamination is established by the discovery of an organism that occurs only in feces, never free-living in nature. There are several such organisms, and in this country the indicator organism is *Escherichia coli.* The finding of *E. coli* is sufficient evidence that the water in question is not safe, since enteric pathogens may be presumed to be present where there is fecal contamination.

It is interesting to note that the decade 1980–1990 was designated by the World Health Organization as the International Drinking Water Supply and Sanitation Decade.

Laboratory Tests

Routine tests performed by health departments to determine fecal contamination of water include the following:

Presumptive Test Test tubes of a nutrient medium containing lactose are inoculated with measured quantities of water samples. These tubes also contain an inverted vial to trap any gas produced and an acid-base indicator to show whether acid is formed. Because *E. coli* can ferment lactose, the presence of acid and gas in the inoculated tubes after 48 h of incubation at 35°C is presumptive evidence for the presence of *E. coli* and, hence, fecal contamination. If the lactose is not fermented, it can safely be assumed that *E. coli* is not present and, as a result, the water is free from fecal contamination. The fermentation of lactose, however, may result from nonenteric organisms, and it is therefore necessary to identify *E. coli* specifically as being present in the fermented lactose broth.

One can obtain a semiquantitative measure of the number of lactose-fermenting organisms in the following way: for each water sample, 5 tubes of lactose broth are inoculated with 0.1 mL of sample, 5 others with 1.0 mL, and another 5 with 10.0 mL of the water samples. Based on how many of each of the lactose broths show acid and gas formation after 48 h, one can refer to a standard table to determine the **most probable number** (MPN) of coliform bacteria per 100 mL of the original water sample (see Table 30.1).

Confirmed Test All tubes showing gas in the lactose broth must be rechecked to ascertain that the gas resulted from the fermentation of the lactose by an enteric organism. This is done by transferring a loopful of medium from tubes in the presumptive test that showed gas into a fermentation tube containing brilliant green lactose bile broth. Both the brilliant green dye and the bile will inhibit the growth of gram-positive organisms and, as a result, select for the growth of enteric organisms. The tubes are incubated at 35°C for 48 h, and the formation of gas in any amount in the inverted vial of the brilliant green bile broth fermentation tube confirms the presence of coliforms.

Completed Test Each brilliant green lactose broth tube that showed gas formation is streaked on Endo or eosin methylene blue plates to provide isolated, discrete colonies. After incubation at 35°C for 24 h, coliform-appearing colonies are transferred to a nutrient agar slant and a lactose broth fermentation tube. Typical coliform colonies will have a green metallic sheen; however, atypical mucoid colonies that appear pink should also be transferred. The formation of gas in the lactose broth and the demonstration of gram-negative rods on the nutrient agar slant constitute a positive completed test for the presence of coliform bacteria.

In a variation of these tests, a loopful of broth from a positive presumptive test is streaked onto an eosin methylene blue-lactose agar plate. The appearance of "green-sheen" colonies provides a positive confirmatory test. The completed test consists of inoculating a green-sheen colony into a phenol-red lactose fermentation tube and observing for acid production resulting from the fermentation of lactose. An IMViC test (see below) is run simultaneously with the completed test.

Fecal Coliforms The term *coliform* is used to describe all gram-negative non-spore-forming, facultative organisms that produce acid and gas from the fermentation of lactose. However, even among the enteric bacteria that can ferment lactose (the coliforms), not all are obligate intestinal parasites like *E. coli*. For example, *Enterobacter aerogenes* would give a positive coliform test, but because it also grows free in nature, on decaying plant material, its presence does not necessarily indicate fecal contamination. Therefore, when testing any raw water source, sewage treatment systems, bathing waters, or seawaters, it is important to differentiate between fecal and nonfecal coliforms. This can be accomplished by transferring organisms from a tube showing a positive presumptive test into a sterile lactose broth and incubating the inoculated tubes in a water bath at 42.5°C for 24 h. Only coliforms from the intestines of warm-blooded animals can grow at this temperature; therefore, gas in these tubes indicates the presence of fecal coliforms.

Differentiation of *E. coli* and *Enterobacter*

To provide biochemical proof that the isolated organism is indeed *E. coli*, a series of tests, usually referred to as the IMViC tests, are used. Each letter (except the i, which is used for phonetic reasons) stands for a property or product that can be used both to characterize *E. coli* and to differentiate it from *E. aerogenes*. Colonies from the nutrient agar slant of the completed test, described above, are used to inoculate

TABLE 30.1   Some Examples of the Most Probable Numbers When Dilutions Are 10, 1, and 0.1 mL

Combination of Positives	Most Probable Number per 100 mL
1:0:0	2
2:3:0	12
4:1:0	17
4:3:1	33
5:0:2	43
5:3:1	110
5:4:3	280
5:5:1	350
5:5:4	1600

TABLE 30.2 IMViC Reactions

	Indole	Methyl Red	Voges-Proskauer	Citrate
E. coli	+	+	−	−
E. aerogenes	−	−	+	+

several different media. The media used and the interpretations are as follows:

I A complex medium rich in the amino acid tryptophan is inoculated and allowed to grow for 24 h. *E. coli* makes the enzyme tryptophanase, which forms indole, pyruvic acid, and ammonia from tryptophan. Because *E. aerogenes* cannot catabolize tryptophan, one need only test for the presence of indole to differentiate these two organisms.

CH₂—CH—COOH ⟶
NH₂

Tryptophan

+ CH₃CO COOH + NH₃

Indole   Pyruvic   Ammonia
acid

M Methyl red is an acid-base indicator. Thus, if methyl red is added to a culture medium containing glucose in which an organism has grown for 18 to 24 h, a red color would indicate that organic acids had been formed as a result of fermentation of the glucose. *E. coli* forms large amounts of acids and is methyl-red positive. *E. aerogenes*, on the other hand, carries out a 2,3-butylene glycol type of fermentation and thus produces only small amounts of organic acids, with the result that it is methyl-red negative.

V The Voges-Proskauer test detects the presence of acetoin (also called acetylmethylcarbinol), the immediate precursor of 2,3-butylene glycol. Thus its presence is indicative of a 2,3-butylene glycol fermentation, which is positive for *E. aerogenes* and negative for *E. coli*.

C The final letter of the IMViC tests represents citrate. This test merely determines whether the organism in question can grow using citrate as its sole source of carbon. Although both *Escherichia* and *Enterobacter* possess the necessary enzymes to metabolize citrate, only *Enterobacter* can use it as a carbon source, because citrate cannot enter the cells of *Escherichia*. This test, therefore, de-

termines the difference in the ability of these two organisms to transport citrate across their cell membranes. Table 30.2 gives a summary of IMViC test results for these enteric organisms.

Membrane Filter Technique

Membrane filters can be used to detect the presence of bacteria in water and other materials. The membrane of cellulose acetate permits water to pass easily, but it traps bacteria on its surface. This very useful technique, which can be used to determine the coliform count in water, had its origin in Germany during World War II, and it has received acceptance by the American Public Health Association.

The water sample is filtered through a special device, so that microorganisms collect on the membrane (see Figure 30.1). The membrane is removed aseptically and placed on a sterile filter pad (Figure 30.2), saturated with a special differential medium, such as Endo broth, and then incubated. Coliform colonies grow on the surface of the membrane and

Figure 30.1 A measured volume of water is filtered through this membrane filter, and any bacteria are retained on the membrane.

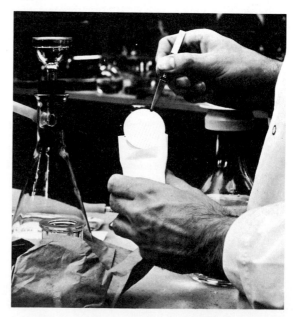

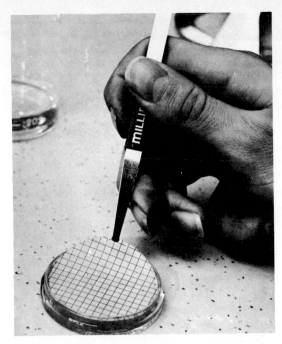

Figure 30.2 The membrane filter can then be placed in a petri dish, covered with a medium, and the growth of individual cells into visible colonies observed after 24 to 48 h.

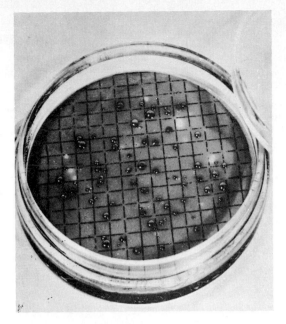

Figure 30.3 Typical coliform "sheen" colonies on Millipore filter indicate fecal contamination in the water sample. The plate was incubated for 18 h on MF-Endo broth.

can be counted and observed for the characteristic metallic sheen (see Figure 30.3). This method has definite advantages over the conventional ones in that it can test larger volumes of water.

ROLE OF BACTERIA IN THE PURIFICATION OF STREAMS AND RIVERS

Since the beginning of history, streams and rivers have been used for the disposal of waste. In nonindustrial countries, this system of disposal has been effective and harmless to the ecology, because the biological activities of the waterways resulted in a self-purification. Bacteria are by far the major factors involved in this self-purification.

It is now becoming quite obvious that the industrial nations of the world have overloaded many of our inland waters and even oceans to the point where they can no longer handle the volume of waste products. Some of the major reasons for the inability of bacteria to keep pace with the modern industrial waste output might be summarized as follows:

The bacterial decomposition of organic materials to inorganic matter (a process called mineralization) is carried out by both aerobes and anaerobes. How-

ever, it is primarily the aerobic metabolism that results in the complete oxidation of organic waste to carbon dioxide and water. When a stream or river is vastly overloaded with industrial waste, the dissolved oxygen in the water is quickly used up by the aerobic bacteria. Under these overloaded conditions, the diffusion of oxygen from the atmosphere is not nearly rapid enough to allow the aerobic metabolism of the waste products. Because the remaining anaerobic bacteria are much less efficient in the breakdown of waste material, the normal self-purification cannot occur. A secondary effect of the overloading of our streams and rivers with industrial waste is that, when the oxygen levels of the water are essentially depleted by the aerobic degradation of the waste, the fish will die. This in itself, of course, creates an additional waste load on an already overloaded waterway.

In addition, many of the industrial and domestic wastes that are dumped into our streams and rivers contain high concentrations of minerals essential for microbial growth. Perhaps the best example of this type of waste is high-phosphate detergents. These waters can nourish algae, and it is now not unusual for lakes to be covered with a green algae scum. Of course, as the algae grow and die, they too contribute to the overloading of our waters. Another form of pollution that is becoming more common is the spilling of oil and petroleum products at sea, either from tankers or directly from undersea oil wells. There are bac-

The three major waterborne diseases, dysentery, typhoid fever, and cholera, have been responsible for many millions of deaths during the past century. Typhoid and dysentery were particularly common among fighting troops, and in most wars these diseases accounted for more deaths than did battlefield casualties. They also had no respect for rank, as can be seen by looking at some of the English monarchs that succumbed to these infections. William the Conqueror died from a ruptured ulcer of the large bowel, a late result of typhoid fever. Edward I died of dysentery, as did the hero of Agincourt, Henry V. And, in more recent times, Albert, the husband of Queen Victoria, died of typhoid fever in 1861.

Let us look at some war casualties. During the American Civil War, the Union army lost 93,433 men on the battlefield and 186,216 men to disease, of which 81,360 deaths were attributed to typhoid fever. Figures are not available for the Confederate forces, but they were in all likelihood higher. During the Boer War in South Africa in 1899–1901, over 6000 died in battle and more than 11,000 died of disease—mostly typhoid fever and dysentery. And so it went for many centuries.

Cholera is undoubtedly the most important of the waterborne diseases. As far as is known, cholera was confined to India for the almost 2000 years between its first description by Hindu physicians in 400 B.C. and its spread to Arabia, Persia, Turkey, and Southern Russia in the early 1800s. There were six major pandemics of cholera during the 1800s covering the entire world, killing millions wherever it struck. During one such outbreak in London during 1849, the famous physician, John Snow, traced the spread of the disease to a Broad Street pump from which area residents obtained their water. The spread of cholera in this area was stopped when Snow recommended that the handle of the pump be removed. This is particularly remarkable when one remembers that the germ-theory of disease had not yet been formulated.

Improvements in living conditions have resulted in considerable diminution of waterborne disease for most of the developed world, but they still remain a serious problem in many Third World Countries.

---

teria that can degrade these petroleum products, but they act too slowly. The layer of oil prevents the diffusion of oxygen into the water and the suffocated fish compound the pollution.

It is generally assumed that there are in the world microorganisms that can degrade any biological product: wood, petroleum products, leather, and all of the organic compounds. The present generation has learned that the "principle of microbial infallibility"—the idea that for every material on earth there is a microbe or combination of them capable of enzymatically breaking it down—is no longer true. Many materials are produced that are absolutely impervious to microbes. The day might well come when the world is covered with nondegradable plastics, the waters are permanently polluted by nondegradable detergents, and the beaches are clogged with petroleum products.

DISEASE ORGANISMS IN WATER AND SEWAGE

The proper disposal of sewage in a community is one of the most important factors in keeping water clean. The wastewater from bathing, laundry, and so on, as well as body excreta, contains much organic material that is literally swarming with both pathogenic and nonpathogenic bacteria. Water that comes in contact with this sewage may easily be contaminated by disease agents. Those that are of special significance are usually excreted in feces and are chiefly associated with enteric diseases. Modern methods of sewage disposal are based on the principle of decay of organic matter by microbial enzymes. This decomposing process is so effective that the refuse can be safely used as a soil fertilizer. Waterborne diseases include chiefly typhoid fever, paratyphoid fever, bacillary and amebic dysentery, giardiasis, cholera, pathogenic *E. coli*, and a whole host of parasitic diseases. Viral agents excreted in feces include those that cause hepatitis, enteroviruses such as those causing poliomyelitis, and probably a number of agents responsible for infectious nonbacterial gastroenteritis. Spirochetes from the genus *Leptospira*, causing a hemorrhagic jaundice, are occasionally found in contaminated water.

Because pathogenic bacteria or viruses do not multiply in the water, their presence is only transitory. Therefore, to be a hazard, a water supply must have recent or continuous contamination.

CONTROL OF WATERBORNE DISEASES

Control of enteric diseases transmitted through water is accomplished through purification of water supplies

From the time that humans began to congregate in cities, sickness and death due to enteric fevers were commonplace. This is not surprising when one remembers that it was only about 100 years ago that the germ theory of disease was proposed. Prior to the late nineteenth century, water sanitation and sewage disposal were essentially unknown, and epidemics of dysentery, cholera, and typhoid fever were almost routine.

The one exception to this was Rome. A drainage system, the Cloaca Maxima, was built in the sixth century B.C., and this became a system for the disposal of sewage. By the third century B.C., an aqueduct was built that brought pure water into Rome, and by the time of Christ, there were six such aqueducts supplying water to Rome. About half this water was used for the public baths, but it still left about 50 gallons daily per person.

In the year 67 A.D., during the reign of Nero, most of Rome was destroyed by fire. In rebuilding the city, a master plan was followed providing straight, broad streets and large squares. Rome's water supply and sanitation were more analogous to twentieth century New York or London than to other European cities a thousand years later. Public roads were cleaned, and

regulations were enforced that ensured the freshness and quality of food. Burials were prohibited within the city walls, and cremation was routine until the general acceptance of Christianity implanted the idea of a physical resurrection.

So what caused the downfall of the Roman empire? One reads that the decadent luxury imported from the East was a contributing factor as well as attacks by their neighbors from the North and East. Infectious diseases, however, also had a major role. Beginning in the first century B.C., a severe form of malaria became frequent in the agricultural areas around Rome, and this remained a major problem for 500 years, causing many persons to leave their small farms and move to Rome, bringing the disease with them. In addition, an epidemic of plague began in 125 A.D., killing over 1 million people in North Africa and Italy. During the next three centuries, there were recurrent outbreaks of plague. Malaria remained rampant, and famine was widespread.

One can see, therefore, that in spite of the advanced state of the Roman empire, it was brought to its knees by the Three Horsemen of the Apocalypse: Pestilence, Famine, and War.

---

and proper sewage disposal. Frequent testing of water is necessary to determine and maintain its purity. Methods of purification aim at the destruction and removal of pathogens as well as of other undesirable materials.

However, in spite of the extensive treatment of municipal water supplies, outbreaks of waterborne diseases still occur in the United States. This is clearly substantiated by the data compiled by Edwin Geldreich as shown in Table 30.3. As can be seen, the majority of identified waterborne illness were due to microorganisms.

Water Purification Methods

Procedures commonly employed in artificial purification include coagulation, sedimentation, filtration, and the use of chemicals such as chlorine, ozone, and iodine. The extent of purification required will vary for clean, municipal reservoir waters containing very little particulate matter on the one hand and, on the other, situations in which water for drinking is taken from a muddy river, which may even have a sewage disposal plant some miles upstream.

In general, the first step is to remove suspended materials, and this usually is accomplished by the addition of alum (aluminum potassium sulfate). Alum forms a gelatinous floc that gradually settles out, car-

rying along particulate matter that includes a large number of microorganisms.

After the alum floc has sedimented, the water is pumped to filters to remove any remaining particles as well as much of the residual bacteria. Filters are made of sand and gravel, with fine sand particles nearer the surface. The filter bed may extend to a depth of 6 ft and cover a wide area. It is this filtration step that is necessary for the removal of *Giardia* cysts, because the subsequent chemical disinfection of the water frequently is ineffective for the destruction of these parasites. It is noteworthy that, as shown in Table 30.3, the two most common causes of waterborne diseases (*Giardia* and *Salmonella*) are widespread in the animal population, and it is most often from this source that water contamination occurs.

The final step in the purification of water for drinking is to treat it chemically to ensure that no potential enteric pathogens can enter the system. This is accomplished by adding chlorine to the water. Chlorine has several qualities that favor its use in water supplies. Its primary advantage lies in the fact that it is a very effective bactericidal compound even when used in a concentration of 1 or 2 parts per million. In addition, it is fairly stable (in the absence of excess organic matter) and reasonably inexpensive. It is not, however, particularly effective against enteric viruses.

TABLE 30.3 Outbreaks of Waterborne Diseases in the United States, 1961–1983

Type of Etiological Agent	Number of		
	Outbreaks	*Cases*	*Deaths*
Bacterium			
Shigella spp.	52	7,462	6
Salmonella spp.	37	19,286	3
Campylobacter spp.	5	4,773	0
Toxigenic *E. coli*	5	1,188	4
Vibrio spp.	1	17	0
Yersinia spp.	1	16	0
Virus			
Hepatitis A	51	1,626	1
Norwalk	16	3,973	0
Rotavirus	1	1,761	0
Protozoan			
Giardia spp.	84	22,897	0
Entamoeba spp.	3	39	2
Chemical			
Inorganic (metals, nitrate)	29	891	0
Organic (pesticides, herbicides)	21	2,725	7
Unidentified agents	266	86,740	0
TOTAL	572	153,394	23

Ozone, a powerful oxidizing agent, is also an effective water disinfectant, but it is expensive. It has an advantage over chlorine in that it eliminates undesirable tastes, but the cost limits its practical use at present. In addition, it lacks the continuous antimicrobial effect of chlorine. Figure 30.4 summarizes the steps used by large cities to purify their water.

Swimming Pools

The possibility of disease transmission in swimming pools is a necessary consideration. Pools should be drained and cleaned at frequent intervals, and the water should be treated chemically with chlorine at the required concentration for safety.

Infections that may be transmitted in swimming pools include those of the respiratory tract, the eye, the ear, and the skin as well as those of the intestinal tract. Probably the most common infection associated with bathing and swimming establishments is transmitted not through water but by contact with damp contaminated floors. This is epidermophytosis, a fungal infection commonly known as athlete's foot.

Figure 30.4 This schematic applies particularly to large cities that obtain their water supply from rivers and lakes. Final filtration bed serves to remove the last trace of floc material.

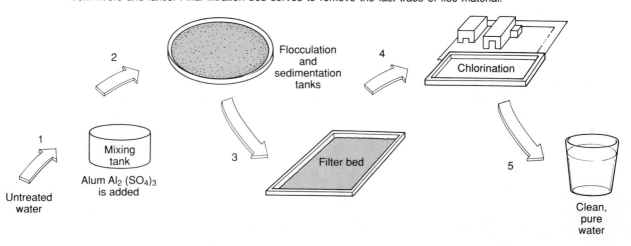

Drinking Fountains

Only fountains that prevent direct mouth contact should be used. Certain organisms deposited from the mouth may remain viable around the cold fountain for quite some time. Organisms associated with diseases of the respiratory area may be transmitted through contact with a contaminated fountain.

Ice

Since the freezing process does not destroy bacteria, only potable water should be used in the preparation of ice. Many bacteria can survive at this low temperature for relatively long periods and have been found to be the cause of some intestinal disease outbreaks. It is important, therefore, to remember that when traveling in a country where the water is unsafe to drink, it is equally unsafe to put ice into a beverage. Precautions should also be observed in handling ice that is to come in contact with food or drinking fluids. Scoops should be used and ice machines cleaned routinely.

Sewage Disposal

The primary goal in the disposal of sewage is to reduce its **biological oxygen demand (BOD)** as much as possible before dumping it into available rivers and streams. BOD is an indirect measure of the amount of organic material present because it represents the amount of oxygen necessary to oxidize all carbon compounds to carbon dioxide and water. Secondary goals of sewage disposal are to inactivate any pathogenic microorganisms that may be present and to remove excess amounts of phosphate and nitrates.

These goals are accomplished by the bacterial conversion of wastes into inoffensive and often usable substances. Fortunately, most pathogenic bacteria are unable to survive the following procedures used for the disposal of sewage.

Primary Treatment   The purpose of this step is to remove the greater part of the solids occurring in the incoming sewage. This is accomplished by allowing the sewage to flow into large sedimentation tanks. As in water purification, flocculating materials may be added to hasten the sedimentation process. The sedimented material, called sludge, can be disposed of in several ways, but it is frequently sent to a second tank where it will undergo anaerobic digestion.

This anaerobic digestion is not particularly efficient, but much of the residual organic material is fermented to form organic acids, and hydrogen and they, in turn, are further converted by bacteria primarily to methane and carbon dioxide as shown below:

$$CO_2 + 4H_2 \longrightarrow CH_4 + 2H_2O$$

and

$$CH_3COOH \longrightarrow CH_4 + CO_2$$

These gases are frequently collected, and the methane is burned to heat the disposal plant or to dry out the final residual sludge. Sludge is then pumped to open beds for air-drying or passed through filters to remove much of the water. The final residue can be used for fertilizer or a soil conditioner, but it is quite low in phosphorus. It is sometimes incinerated, or it can be used for land-fills. Occasionally, it is taken on barges out to sea and dumped.

A noteworthy problem of sludge disposal came to light in 1980, when Gabriel Bitton and Samuel Farrah looked for the presence of viruses in sludge after it had been used as a fertilizer. They found that the secondary sludge treatment did not inactivate the enteroviruses and that these viruses could be found for as long as a month after the sludge was spread, depending on the weather conditions. Because enteroviruses include polioviruses, coxsackieviruses, echoviruses, and hepatitis A virus, such sludge could be a major source of exposure to these agents.

Secondary Treatment   The effluent from the primary treatment contains all of the soluble organic materials as well as about one third of the total incoming solids. The purpose of the secondary treatment is to oxidize all organic material, thus removing any residual BOD when it is finally dumped into a river or stream. This is accomplished by an aerobic breakdown of these materials that is carried out by organisms within the flora of the sewage water. But herein lies the trick. There are not sufficient microorganisms in the incoming sewage to accomplish this in any reasonable time, nor would the rate of diffusion of oxygen be rapid enough to keep the process aerobic. Two different procedures are used to overcome these problems.

The **activated sludge process** pumps the effluent from the primary treatment tank into a large tank in which air is continually forced through the liquid to keep it fully aerated. And to hasten the metabolic breakdown, there is a continuous feedback of part of the sludge that settles out in a subsequent settling tank. This material, called **activated sludge,** consists of flocs whose surface is heavily colonized with bacteria, yeasts, and protozoa, thus providing a continuous heavy inoculation of the incoming sewage. Figure 30.5 provides a schematic illustration of this process.

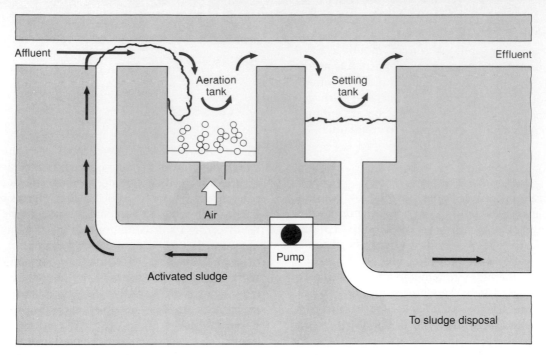

Figure 30.5 Secondary treatment of sewage using activated sludge process.

The **trickle filter** is a second method for the oxidation of the organic matter in sewage. In this process, the sewage is sprayed over a filter composed of gravel that is 4 to 6 ft deep. Such filters require about 6 months to mature after initial use, but during that time many aerobic organisms colonize in gelatinous masses on the rocks. Air circulates through the filter bed and, once mature, such filters are quite efficient for the oxidation of the organic matter trickling down over them (see Figure 30.6).

Figure 30.6 Both physical and biochemical processes are involved in sewage treatment with the trickling filter. The filter is a bed of stones, clinkers, and other coarse materials about 5 to 10 ft deep. Distributor pipes discharge the sewage over the stone beds; oxygen is absorbed as the spray passes through the air. The gelatinous layer that forms on the surface of the bed teems with bacteria and other microbes that work in the oxidation process to convert impurities to simpler compounds.

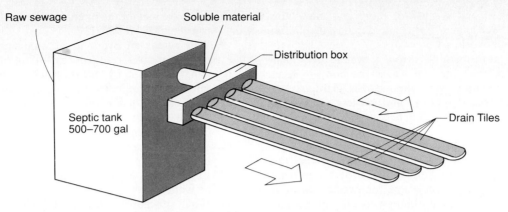

Figure 30.7 As raw sewage enters the septic tank, particulate material settles to the bottom. Here, anaerobic bacteria break down the particulate matter into soluble complex compounds. These soluble compounds then flow out near the top of the septic tank where they are distributed to a series of drain tiles, which are usually laid in coarse gravel just under the surface of the ground. As the solutes flow through the drain tiles, aerobic organisms complete the breakdown of the complex organic molecules. For a one-family unit, approximately 300 lin ft of drain tile would be used.

Tertiary Treatment Many sewage disposal treatment plants pump the effluent from the secondary treatment directly into local rivers or irrigation ditches. However, such material still contains some organic material (and hence, a BOD) as well as large amounts of nitrogen and phosphorus. As you can see, large amounts of secondary eluate could easily lead to the development of algae downstream. Tertiary treatment is not carried out widely because of the expense, but it is designed to remove chemically dissolved nitrogen and phosphorus. This is accomplished by converting all nitrogen to ammonia, which can be removed in scrubbing towers, and to precipitate all phosphorus as calcium phosphate.

Septic Tanks On a smaller scale, as in a home septic tank, the oxygen does not diffuse below the surface, and the major breakdown within the septic tank is anaerobic. Bacteria hydrolyze the complex organic compounds, bringing the solid matter into solution. In the absence of oxygen, however, such decomposition is not complete. Most of the organic material is still organic matter and cannot be com-

pletely mineralized unless the effluent is exposed to air. This is done by passing the material through long, tiled drain lines embedded in gravel. The material seeps into the gravel, where it can be aerobically oxidized (see Figure 30.7).

Disposal of Solid Wastes

One need only observe the streets and sidewalks of a large city following a strike by garbage collectors to appreciate the enormous amount of solid waste that must be disposed of each day. Much of this material consists of biodegradable materials such as paper and food scraps, which eventually will be degraded by microorganisms to simpler molecules. The rate at which this occurs will depend, in large part, on the environment in which the solid waste is disposed. That which remains in contact with the air is decomposed rapidly by aerobic organisms, whereas waste material that is buried deep in landfills may require many years before it is recycled to become again the paper and food scraps that constitute our garbage.

SUMMARY

Water provides a normal habitat for many species of bacteria, and the environmental conditions in a particular area will influence the microbial flora for that area. Soil organisms, including both heterotrophs and autotrophs, are commonly found.

Several tests are performed to determine the sanitary quality of water. If organisms that normally inhabit the intestinal tract are found in the water, there is a strong

possibility that pathogens are also present. In the United States, *E. coli* is the species searched for as an index for fecal pollution.

Tests for the differentiation of *E. coli* and *E. aerogenes* are called the IMViC tests.

Under normal conditions, bacterial degradation of waste products results in a self-purification of our rivers and streams. However, the industrial overload of our rivers and lakes, and the use of more and more non-biodegradable substances, has resulted in heavy pollution of many of our world's waterways.

Diseases transmissible through water are chiefly those in which the organism leaves the body in feces. These include typhoid fever, salmonellosis, giardiasis, bacillary and amebic dysentery, and cholera. Viruses of poliomyelitis, rotaviruses, and hepatitis A also are excreted in feces.

Waterborne diseases can be controlled by the purification of water for drinking taken from rivers and lakes. Purification of drinking water involves removing suspended material by flocculation, followed by filtration through sand beds and chlorination to kill any residual pathogens.

Treatment of sewage involves sedimenting out most of the solids, called sludge, and the anaerobic fermentation of organic matter present in the sludge. The effluent from the sedimented sludge is aerobically oxidized using either an activated sludge process or a trickle filter.

QUESTIONS FOR REVIEW

1. Differentiate between potable, polluted, and contaminated water.
2. What types of bacteria are commonly found in water? What bearing do these have on the sanitary quality of a water supply?
3. What is the usual means by which pathogens are introduced into drinking water?
4. What is the basis for determining that water is safe to drink?
5. What is the significance of a positive presumptive test? Confirmed test? Completed test?
6. What organism is sought as an index of fecal pollutions?
7. Why are individual pathogens generally not searched for?
8. How are membrane filters used in the bacteriological examination of water?
9. Why might an overload of waste products into a stream or river result in the killing of fish?
10. What is meant by the statement "The principle of microbial infallibility is no longer true"?
11. As a citizen of the world, how would you attempt to solve the problem of pollution of our waterways, lakes, and oceans?
12. What diseases may be transmitted through contaminated water?
13. Explain the procedures employed in the purification of water.
14. What procedures should be followed to maintain safe water in swimming pools?
15. Why are the methods of sewage disposal so important?
16. Outline the activated sludge process for sewage disposal. How does this differ from a trickle filter?

SUPPLEMENTARY READING

Bonde GJ: Bacterial indicators of water pollution. *Adv Aquatic Microbiol* 1:273, 1977.

Dart RK, Stretton RJ: *Microbiological Aspects of Pollution Control.* Amsterdam, Elsevier, 1977.

Geldreich EE: Potable water: New directions in microbial regulations. *ASM News* 52:530, 1986.

Greenberg AE, et al (eds): *Standard Methods for the Examination of Water and Wastewater*, 16th ed. Washington, D.C., American Public Health Association, 1985.

Moulin GC, Stottmeier KD: Waterborne mycobacteria: An increasing threat to health. *ASM News* 52:525, 1986.

Schroeder ED: *Water and Wastewater Treatment.* New York, McGraw-Hill, 1977.

Taber WA: Wastewater microbiology. *Annu Rev Microbiol* 30:263, 1976.

Chapter 31

Microbiology of Food and Milk

OBJECTIVES   Study of this chapter will familiarize you with

1. How foods can be preserved by a bacterial lactic acid fermentation.

2. The production and uses of single-cell protein.

3. The types of bacteria that are normally present in milk and their role in souring milk.

4. The source and types of diseases that may be transmitted in milk.

5. The techniques used for the control of milkborne diseases (including pasteurization).

6. Methods of testing milk for bacterial contamination.

7. The role of bacterial fermentations of milk to preserve and prepare milk products.

8. The immunological properties of milk.

Since time began, the major preoccupation of all animals has been with obtaining food. Prehistoric humans undoubtedly went through periods of feast and famine, since they had no knowledge of how to preserve food for future use. However, as the food-producing era evolved, which may have begun about 8000 years ago, the necessity to preserve food products became imperative. Early methods included heavy salting to prevent the spoilage of meat and drying of meats, vegetables, and fruit for their preservation. These methods were used for thousands of years, but it is unlikely that anyone had any real comprehension of the causes of food spoilage until about 100 years ago.

Of course, we now know that microorganisms use the same foods that we do so that, for the most part, food preservation involves either killing the microorganisms present on the food or preventing them from growing. Heating or cooking the food is the major technique for killing contaminating organisms, although the addition of high concentrations of salt and the use of radiation are also effective for some foods. Food preservation methods that do not kill the organisms but are effective because they prevent their growth include both freezing and drying.

What do we really mean by the term "spoiled food"? Quite obviously we have all experienced occasions when the food in question had acquired a bad odor or a bad taste as a result of microbial action. These cases do not present a problem because such food is not eaten. However, few of us are so fortunate that we do not at some time in our lives become ill from eating improperly handled food. Illnesses of this type may vary from only mild distress (indigestion) to extremely serious and even fatal conditions.

We discussed the various food intoxications and food infections in Chapter 26, and the reader is encouraged to review that material. In this chapter, we are concerned with the techniques of food preservation, including the use of microorganisms to prevent food spoilage. In addition, the control of milkborne diseases and the use of microorganisms to produce fermented milk products are discussed here.

FOOD PRESERVATION BY MICROORGANISMS

It may seem paradoxical that microorganisms can preserve as well as spoil food, but there is a large group of organisms, called the **lactic acid bacteria,** that are used to preserve a wide variety of foodstuffs. These organisms are effective in food preservation because they metabolize available carbohydrate in the food, producing lactic acid as a major end product of the fermentation. The lactic acid lowers the pH to about 3.5 to 4.0, which is far too acid for the growth of most pathogenic bacteria. The following sections describe a few such fermentations.

Fermented Vegetables

Almost any vegetable can undergo a lactic-acid-type fermentation, usually carried out by various species of *Streptococcus, Lactobacillus, Leuconostoc,* and *Pediococcus.* These organisms convert sugars present in the vegetable primarily to lactic acid, which restricts the growth of other organisms and imparts a unique flavor to the fermented vegetable. After fermentation, such vegetables are frequently referred to as "pickled," and it is not unusual to see jars of pickled beets, pickled green beans, or pickled carrots. Because the process is much the same for all fermented foods, we shall describe only a few of the most common ones.

Sauerkraut

Sauerkraut is the product of a lactic acid fermentation of shredded cabbage. Fresh cabbage invariably contains a number of species of *Leuconostoc* and *Lactobacillus,* so it is not necessary to add bacteria to start the fermentation. One merely shreds the cabbage and mixes it thoroughly with about 2.5 percent salt by weight. The salt draws out the plant juices containing the sugars and inhibits the growth of many non-lactic-acid-producing organisms. The salted, shredded cabbage is covered with a weight to keep the fermentation anaerobic. After about 7 to 10 days at about 21°C, the fermentation is completed and the pH of the final product should be approximately 3.5. Sauerkraut can then be kept in the cool for extended periods or, after pasteurizing, can be canned.

Pickles

If cucumbers are placed in water, they will become soft within 24 to 48 h as a result of microorganisms that hydrolyze pectin. If, however, a 5 to 9 percent salt solution is added, the pectin-hydrolyzing bacteria are inhibited and the lactic acid bacteria become the dominant species. Dill may be added to produce dill pickles, and the addition of garlic results in kosher dill pickles. The organisms responsible for fermented pickles are essentially all species of the genus *Lactobacillus,* and the final product has about the same acidity as sauerkraut.

Ensilage

Anyone who has traveled through farm country, particularly in the midwestern part of the United States, could not have missed seeing the many large silos present on almost every farm. Each fall these silos are filled with finely chopped green corn plants, a material termed **ensilage.** The sugars within this chopped plant material undergo a lactic acid fermentation that lowers the pH sufficiently to preserve the ensilage for an extended period. Moreover, because of the action of facultative organisms, much of the area within the silo becomes sufficiently anaerobic to prevent the growth of molds, thereby further prolonging the preservation time of this material for use as feed for cattle. It is also interesting to note that the lactic acid formed during this fermentation can be further metabolized in the rumen, providing additional nourishment for the cow.

Olives

Green olives are initially treated with a 1 to 2 percent solution of lye for about 24 h to remove part of the bitter taste. After thorough washing to remove the lye, the olives are placed in a barrel and covered with a 6 to 9 percent salt solution. The ensuing lactic acid fermentation lasts for 6 to 10 months, after which the green olives are sorted and packed into commercial containers.

Ripe olives are treated more extensively with lye to remove essentially all of the bitter taste. During this process the skin darkens to the color typical of a ripe olive. The lye is then washed out of the olives, and they are suspended in a 2 to 3 percent salt solution for several days. A slight lactic acid fermentation may occur during this time, but it is not essential for the final product. The ripe olives are then canned in glass or tin containers and processed at about 115°C for 60 min.

Fermented Meats

Dried or summer sausage is about the only fermented meat product used in the western hemisphere. This is prepared by placing the chopped meat in casings, along with spices and salt. The sausages are then held at about 8°C for 40 or more days, during which a lactic acid fermentation takes place along with considerable dehydration of the meat. This, of course, increases the salt content, which, along with the lactic acid, prevents the growth of any spoilage organisms.

Dry, or Italian-type sausages are prepared in a manner similar to summer sausage except that the ground meat is incubated in the casing at 25 to 35°C for varying lengths of time, depending on whether starter cultures are added to hasten the fermentation. Such sausages are referred to as Hungarian, Genoa, or Milano salami and they are generally eaten without cooking.

Lebanon bologna is an all-beef, heavily smoked and spiced product that is fermented by a culture of *Pediococcus cerevisiae.* Nitrates or nitrites are added as a preservative, and such sausages usually reach a temperature of about 68°C during smoking.

Oriental Fermented Foods

There are a large number of Oriental foods that owe their flavor to the action of microorganisms. Soy sauce, made from cooked soybeans, is probably the one best known to westerners; the fermentation occurs as follows: (1) Enzymes secreted by the mold *Aspergillus* hydrolyze the carbohydrates and proteins of the soybeans and undoubtedly are responsible for much of the characteristic flavor of the soy sauce; (2) *Lactobacillus delbrueckii* ferments the carbohydrate, forming sufficient acid to prevent spoilage. Other lactic acid bacteria as well as several genera of yeast contribute to the final flavor of the soy sauce.

There are other Oriental foods in which various molds contribute to their flavor. Most are products of either soybeans (tempeh or miso) or cooked rice (koji), although the Japanese prepare a mold-fermented fish in which the major fungus is a species of *Aspergillus.*

FOOD PRESERVATION WITH CHEMICALS

Many chemicals have been added to foods and beverages, but we shall consider here only those widely used in the food industry.

Benzoic acid is the first chemical preservative approved by the U.S. Food and Drug Administration (FDA) to be added to foods. It is effective only in its undissociated form and is therefore effective only at an acid pH. For example, at a pH of 4, 60 percent is undissociated, while at a pH of 6, only 1.5 percent is undissociated. Its use is thus restricted to high-acid products such as apple cider, tomato catsup, soft drinks, and salad dressings.

Sorbic acid is also most effective at an acid pH and is essentially ineffective above pH 6.5. Sorbic acid is mainly used in preventing yeast or mold growth, but it also inhibits the growth of a wide range of bacteria. It is employed primarily in cheeses, salad dressing, bakery products, beverages, and fruit juices.

Propionic acid is also most effective in an acidic environment, and it is incorporated into bakery products and cheeses as a mold inhibitor. It is interesting to note that the major species making up the normal flora of the skin produces propionic acid as a metabolite.

Sodium nitrite and sodium nitrate are used in a number of meat products. Although they do inhibit the growth of some organisms, the ability to stabilize red meat color is the main reason for their use. This occurs when nitrate (NO_3) is reduced to nitrite (NO_2) and nitrite is reduced to nitric oxide (NO), which then reacts with myoglobin in the meat to form a red pigment, nitrosomyoglobin. A disquieting effect of nitrite, however, is that it will react with other amines to form nitrosamines, many of which are known to be carcinogens. Whether this is a significant problem is unknown.

FOOD PRESERVATION BY DRYING

Because water is required for microbial growth, one of the oldest methods of preserving foods is by drying. A quantitative measurement of this water requirement is termed the **water activity** (a_w); this is defined as the ratio of the water vapor pressure of the food (P) to the vapor pressure of pure water at the same temperature (P_0). Thus, $a_w = P/P_0$, and one can see that pure water would have an a_w of 1.00. Most food-spoilage bacteria will not grow below an a_w of 0.91, although some molds will grow at an a_w as low as 0.80. Thus, any procedure that will lower the a_w to these values will prevent microbial growth.

The a_w of most fresh foods is above 0.99, and the most common methods of lowering the a_w are either to remove water by dehydration or to add solutes such as sodium chloride or sucrose. For example, a 22 percent solution of sodium chloride has an a_w of 0.86 and a saturated solution of sodium chloride has an a_w of 0.75. Similarly, the a_w for honey is 0.75 and for jams it ranges from 0.80 to 0.91. Thus, one can see that foods that are preserved in brine or have a high sugar content are quite resistant to bacterial growth but may support the growth of molds.

FOOD PRESERVATION BY CANNING

The primary purpose of the canning industry is to sterilize food using heat, followed by sealing it in the can to prevent access by environmental microorganisms. Because excess heating destroys the quality of many foods, one wants to use the minimum heat necessary to accomplish that goal. However, because botulism is frequently a fatal food poisoning, and because the endospores of *Clostridium botulinum* are quite heat resistant, a lot of research has gone into deter-mining just how much heat must be used to minimize quality changes and still ensure a safe product.

The industry has adopted what is termed a 12-D concept. This means that if a food contained 10^{12} botulism endospores, it must be heated sufficiently to reduce that number by 12 decimal places or, in other words, to one survivor. Because you would never have a food containing 10^{12} endospores, this ensures sterility at the end of the cooking period.

The temperatures used for canning vary for different foods, being as low as 100°C for fruits and high-acid vegetables (tomatoes), or as high as 121°C for low-acid vegetables and meats.

PRESERVATION OF FOOD BY LOW TEMPERATURES

The average refrigerator maintains a temperature between 0 and 6°C, and many organisms will grow slowly at these temperatures. Therefore, normal refrigeration only extends the storage time of most foods.

Foods kept at −18°C, a normal freezer temperature, can be kept for extended periods because no bacterial growth will occur at that temperature. Bacteria present on such foods do die slowly and some parasites, such as the nematode causing trichinosis, will die in less than a week at such temperatures. It should be kept in mind, however, that foods do undergo chemical changes even while frozen and they cannot, therefore, be preserved indefinitely in a freezer.

PRESERVATION OF FOOD BY RADIATION

Ionizing radiation such as gamma rays from either ^{60}Co or ^{137}Cs, are used in a number of European countries and in Japan for the preservation of food. In the United States, however, radiation is used primarily to kill insects living in spices and on certain fruits. Such food must be labeled to indicate that it was treated with gamma radiation to extend shelf life.

SINGLE-CELL PROTEIN

Single-cell protein (SCP) refers to microorganisms that are used as food for either humans or animals. It may consist of yeast, algae, or bacteria, although most SCP processors currently use yeast. The production of SCP provides a method of converting an inexpensive car-

It is undoubtedly true that mankind has come a long way during the past century in its ability to control the spread of pathogenic organisms. Progress, however, can sometimes fool us, resulting in an increased spread of infectious organisms, as shown by the following examples.

A large company in Norfolk, Virginia, bills itself as the nation's largest manufacturer of packaged sandwiches by turning out over 500,000 each week for sale in convenience stores and vending machines. In the summer of 1990, the FDA threatened to close them down for sanitary reasons because inspections repeatedly found contamination with *Listeria monocytogenes.* This organism causes about 1700 infections per year in the United States with about a 25 percent mortality. It is particularly virulent for pregnant women, newborns, the elderly, and those with immune system disorders. No specific infections were reported to occur from eating these sandwiches, but the potential was present for a serious epidemic.

The use of a microwave oven provides a rapid and convenient method for cooking a wide variety of foods.

One must keep in mind, however, that microwave ovens tend to heat food very unevenly, leaving some parts undercooked and others well done. These cold spots may well harbor live organisms. There has, for example, been a report of an individual who developed trichinosis after eating pork sausages prepared in a microwave oven. The Agriculture Department also recommends that stuffed poultry not be cooked in a microwave oven because the moist stuffing and the potential for uneven cooking provide excellent conditions for the growth of *Salmonella.* In another example, the *Journal of the American Medical Association* told of a woman who had cooked some haddock in her microwave. She and her son each ate some of the fish, leaving a third portion cooling on the kitchen table. After a brief time, a number of small, thin wormlike creatures were observed crawling out of the cooled fish. These were subsequently identified as anisakis larvae, a nematode frequently infecting fish and capable of invading the human stomach or intestine, producing a disease of varying severity.

bohydrate source into an edible food containing as much as 70 percent protein by dry weight, as well as many of the B vitamins. The carbohydrate sources used are usually byproducts that have little value and include such things as n-paraffin, whey, sulfite waste (from paper mills), methane, potato starch, and molasses.

In some parts of the world, cyanobacteria grow in shallow waters, and after harvesting and drying, they are used for food. This, however, is practical only in temperate to tropical areas, where sunlight is available for most of the year.

Thus far, there are only a few commercial plants in the world producing SCP, but many are in the pilot plant stage or under consideration. Almost all SCP is now used as a supplement to animal feeds, but a number of nutritionists believe that SCP may one day provide a major source of protein and vitamins for human use, especially in underdeveloped countries, where hunger and starvation are endemic.

MICROBIOLOGY OF MILK

Since milk contains proteins, carbohydrates, fats, vitamins, and minerals and has a pH of about 6.8, it is no wonder that, in addition to being an excellent food for humans, it provides an excellent growth medium for microorganisms. Nonpathogenic bacteria are found in milk even when the most rigid sanitary conditions prevail. However, our main concern is that milk provides potential means for the dissemination of pathogenic microbes. As we shall see in this chapter, milk may become contaminated with pathogens anywhere along the line from cow to home. Contamination may come from an infected cow or an infected milk handler, who may not actually be ill. In the following sections, some of the milkborne diseases and the precautions taken against them by the dairy industry are discussed.

NORMAL FLORA OF MILK

Unless a cow is ill, milk is sterile until it reaches the milk ducts in the cow's udder; but because the ducts contain bacteria, the first milk drawn always contains microorganisms. However, this is not the major source of bacteria in milk. Most bacteria come from milk pails, dairy equipment, barn dust, handlers, and similar sources. Disease-free dairy personnel and utilization of sanitary equipment help reduce the number of bacterial contaminants from external sources; yet in spite of all precautions, certain nonpathogenic microbes are always found in the milk.

By far the most abundant bacteria making up the normal flora of milk are the lactic acid bacteria belonging to the families Lactobacillaceae and Streptococcaceae. They are gram-positive, nonmotile microaerophilic or anaerobic rods or cocci. These organisms have rather complex nutritional requirements, and all require varying numbers of amino acids and vitamins for growth. In addition, all lactic acid bacteria require a fermentable carbohydrate as their source of energy. Conventionally these organisms are divided into two groups. The group that essentially produces only lactic acid ($CH_3CHOHCOOH$) from fermentable carbohydrates is known as the **homofermentative lactic acid bacteria.** The group that produces acetic acid (CH_3COOH), ethanol (CH_3CH_2OH), carbon dioxide, and lactic acid from a fermentable carbohydrate is called **heterofermentative lactic acid bacteria.**

Some members of the family Streptococcaceae produce disease in humans (for example, *S. pneumoniae* and some other members of the genus *Streptococcus*). The remaining members of this family are of great importance in applied microbiology and especially in dairy microbiology. The major organisms in this group that are found as part of the characteristic flora of milk are *S. lactis, S. cremoris,* and a number of species of the genus *Lactobacillus* such as *L. casei, L. acidophilus, L. plantarum,* and *L. brevis.* These organisms do not produce disease in humans, but they do ferment the carbohydrate in milk to form acids (mainly lactic acid), thus lowering the pH of the milk. When the pH is lowered to about 4.5, the casein in milk becomes curdled and forms a lumpy precipitate. Also, because of the acid formed, the milk tastes **sour.** It is evident, therefore, that the souring of milk is a result of the activities of the bacteria normally present in milk.

Other bacteria usually found in raw (unpasteurized) milk include members of the following genera: *Micrococcus, Pseudomonas, Staphylococcus,* and *Bacillus.* In addition, *Escherichia coli* is frequently present, but this organism is a completely undesirable one, and the extent of its presence is directly related to the sanitary conditions of the dairy.

Thus, one can summarize by saying that although milk is sterile within a healthy cow, it acquires a characteristic flora of nonpathogenic organisms as soon as it leaves the cow. The role of dairy sanitation is to ensure that disease-producing microorganisms do not get into the milk to spread disease to consumers.

DISEASES TRANSMITTED IN MILK

As a result of increased sanitation, pasteurization, and public health controls, the number of cases of diseases that occur from the consumption of infected milk and milk products has dwindled to a relatively small number. However, because milk is such a universal food for both young and old, it can easily be seen that any relaxation in the controls over the production and marketing of milk products could result in the transmission of disease-producing organisms.

The diseases spread by milk are numerous and may be divided into two main categories: (1) diseases caused by organisms that infect the cow and gain entrance to the milk and (2) diseases resulting from contamination of milk by human sources. The best-known examples of the first group are bovine tuberculosis (*Mycobacterium bovis*) and brucellosis (*Brucella abortus*). Also, streptococcal sore throat and scarlet fever have been transmitted to humans through milk from cows that had an udder infection from *S. pyogenes.* Similarly, *S. aureus* may infect the cow and cause food poisoning through contaminated milk. A rickettsial disease, Q fever, may also be transmitted to humans by way of milk from infected cows. This disease may be acquired either by consuming contaminated milk or by inhaling dust or dried feces from infected pastures or barns. And, as previously described, *Listeria monocytogenes* may be spread from an infected cow via contaminated milk, producing a severe, and frequently fatal, disease.

Direct or indirect contamination of milk and milk products by dairy personnel may occur via dirty hands, coughing, sneezing, or poorly cleaned equipment. Salmonellosis (infection with salmonellae), in the form of food infection or typhoid fever are major diseases that can be spread through milk. Other diseases in which milk has been implicated include shigellosis, or bacillary dysentery (*Shigella* species), diphtheria (*C. diphtheriae*), and scarlet fever or septic sore throat (*S. pyogenes*). *Campylobacter jejuni* has also caused numerous cases of gastroenteritis, particularly from the ingestion of raw milk. Poliomyelitis may also be milkborne, but it probably is rarely so. There are undoubtedly other diseases spread by way of this medium that have not caused large or severe epidemics and hence are not usually thought of as infections spread by milk.

CONTROL OF MILKBORNE DISEASES

Two general techniques are employed for the specific control of milkborne diseases. These are (1) the inspection of cattle and the elimination of infected animals and (2) the killing of disease agents by the process of pasteurization or sterilization.

The U.S. government has spent many millions of dollars for the inspection of cattle. The inspection is directed against two diseases—bovine tuberculosis and brucellosis. Skin tests are done on all cattle, using material derived from the organism causing tuberculosis or, in the case of brucellosis, material derived from the *Brucella* organisms. Any cow showing an allergic reaction to either of these materials is judged to have (or to have had) the corresponding disease and is destroyed. Through use of these techniques, bovine tuberculosis has essentially disappeared from the United States, although some countries in Europe are still plagued with this disease. In like manner, the incidence of brucellosis has been greatly reduced, but, unfortunately, the disease occurs occasionally in this country. Other diseases (such as streptococcal infections) transmitted from the cow into the milk are the result of a localized infection in the udder or the milk ducts.

The presence of an infection in a cow is most easily ascertained by determining the number of leukocytes (white blood cells) present in the milk. This can be done by several techniques. A quick screening method is to mix the milk sample with a standardized detergent (obtained commercially for this test). The detergent liberates the DNA (deoxyribonucleic acid) from the leukocyte nuclei, causing the milk sample to clump or gel. Even a slight precipitate means that the milk contains in excess of 500,000 leukocytes per milliliter, indicative of an infected udder. More precise leukocyte counts can be obtained by spreading 0.01 mL of milk on a slide, staining, and counting the number of leukocytes (cells possessing a nucleus) in a predetermined number of microscope fields. As a confirmatory test, the number of leukocytes can be determined by an electronic cell-counting instrument called a Coulter counter. This method provides quick, accurate leukocyte counts but requires more expensive equipment.

The use of antibiotics for treating infections in the cow results in the presence of some antibiotic in the milk. Such milk should not be consumed because some people are extremely allergic to antibiotics such as penicillin. To test for the presence of antimicrobials, a sterile filter paper disk is wet with the milk sample and is then placed on an agar medium in a petri dish that has been previously inoculated with a standardized spore suspension of *Bacillus subtilis*. After incubating overnight, a clear zone surrounding the paper disk is indicative of inhibitory substances in the milk.

Milk handlers (like all other food handlers) should be checked by public health officials, and any detected carrier of pathogenic organisms must be kept away from contact with milk until he or she is cured. In addition, the number of bacteria present in milk is considerably smaller if the milk is obtained under rigid sanitary conditions. Because it is the nonpathogens in milk that cause it to sour, it follows that with fewer

bacteria, the keeping quality of milk is improved, and disease agents, if present, are more scattered. Much contamination is prevented in the modern dairy barn, in which automatic machines milk the cows and deposit the milk directly into cans.

Even with all of the precautions we have already discussed, no doubt much disease still would be spread by milk if it were not for the essentially universal (at least in the United States) practice of pasteurization. Pasteurization is a process of heating the milk to destroy all pathogenic microbes; this is followed by rapid cooling. Today most of us invariably associate pasteurization with milk. However, the process was first devised by Louis Pasteur not for milk but to destroy the organisms that could spoil wine. Now it is important primarily as a means of ensuring safe milk, since it destroys pathogens. One of the most difficult disease agents to destroy in milk is the rickettsia *Coxiella burnetii*, the cause of Q fever. A drop of even 2°C in the pasteurization temperature will allow some of these organisms to survive. It is also believed that the virus causing hepatitis A can survive the routine pasteurization process and has occasionally spread the infection to humans. In addition to destroying the pathogens in milk, pasteurization causes a marked decrease in the numbers of nonpathogens such as *S. lactis* and various *Lactobacillus* species. This enhances the keeping quality of the milk, because it is the lactic acid formed by the nonpathogens that is responsible for souring. Therefore, with fewer organisms, less lactic acid is produced and the time required for the milk to sour is extended.

There are two processes whereby milk may be pasteurized, both of which are effective in eliminating pathogens:

1. Heating the milk (or other substance) to 62.9°C for 30 min (holding method).
2. Heating to 71.6°C for not less than 15 s (high-temperature method).

Both are followed immediately by rapid cooling.

The sanitary handling of milk with pasteurization is important in preventive medicine. Organisms killed by pasteurization include those of tuberculosis, brucellosis, Q fever, typhoid fever, paratyphoid fever, bacillary dysentery, diphtheria, scarlet fever, and foot-and-mouth disease virus, all of which can cause illness in humans. Of course, it is important that the milk not be contaminated after completion of the pasteurization process.

A third technique for preserving milk is sometimes referred to as the **ultrahigh temperature method.** In this procedure, the raw milk is heated to 148.9°C for 1 to 2 s before being rapidly cooled. This treatment eliminates the bacteria that cause milk to sour and, as a result, such milk can be stored for several months without refrigeration.

Methods of Testing Milk

As has been emphasized repeatedly in this chapter, sanitary methods of handling milk must be adhered to rigidly in order to provide safe milk for human consumption. Furthermore, since milk is a good growth medium, even a small number of nonpathogens can multiply considerably if the milk is not kept refrigerated. Because the consumer has no way of knowing whether or not the milk delivered to the home or purchased in the store is contaminated, most localities require that a number of standard tests be carried out periodically on milk sold in that area.

Tests commonly employed are the (1) phosphatase test, (2) standard plate count, (3) Breed count (direct microscopic count), (4) reductase test, (5) tests for coliform organisms, and (6) tests for specific pathogens.

Phosphatase Test

The only purpose of the phosphatase test is to determine whether the milk has been heated adequately during the pasteurization process. Phosphatase itself is an enzyme that will liberate inorganic phosphate from organic compounds containing phosphate in an ester linkage. The test merely involves mixing an aliquot of milk with an organic phosphate ester and, after an appropriate incubation period, measuring the release of inorganic phosphate by its ability to form a colored product when it reacts with ammonium molybdate. The enzyme is present in milk but is inactivated by the heat if the pasteurization was effective. This is important because improperly pasteurized milk could still transmit tuberculosis, brucellosis, and Q fever. The test itself, then, simply determines if the milk has been properly pasteurized by determining whether the phosphatase has been destroyed.

Standard Plate Count

The standard plate count gives an indication of the approximate number of bacteria in the milk. Obviously, very clean milk will have lower bacterial counts than milk collected or handled under unsanitary conditions or improperly refrigerated milk. The standard plate count is a basis for grading milk.

The count is carried out in a manner similar to that used in determining the number of bacteria in water. Diluted samples of the milk to be tested are mixed (in measured quantities) with a melted nutrient medium. After incubation for 48 h, the original counts per milliliter of milk are determined by multiplying

the colony count by the dilution. For example, 1 mL of a 1:100 dilution of milk is mixed with a nutrient medium and incubated. If 30 colonies appear on that plate, the standard plate count for that sample of milk is 3000 bacteria per milliliter.

Breed Count or Direct Microscopic Count

The Breed count is another way to determine the number of bacteria present in a sample of milk. The procedure is carried out by transferring 0.01 mL of the milk sample to a slide. The milk is spread over an area of 1 cm^2. Then the smear is stained with methylene blue and counts of stained bacteria are made from about 30 microscopic fields. From these results it is possible to calculate the number of bacteria present in the sample of milk. This count is usually somewhat higher than the plate count, since bacteria that do not grow under the conditions of the standard plate count will also be included.

Reductase Test

If methylene blue is added to a measured quantity of milk, the time required for the blue dye to be reduced to a colorless compound is a measure of the number of bacteria present in the milk. This color change occurs because microorganisms present ferment the sugar in the milk; as a result, the reduced coenzymes reduced nicotinamide adenine dinucleotide (NADH) and reduced nicotinamide adenine dinucleotide phosphate (NADPH) are formed. Through the mediation of enzymes called diaphorases (which are present in the microorganisms), the reduced coenzymes are able to pass their electrons directly to methylene blue, forming a reduced, colorless compound. Sterile milk could not reduce methylene blue, hence the rate of reduction for the methylene blue is directly proportional to the number (and therefore the metabolic activity) of microorganisms in the milk. The best grades of milk will not reduce methylene blue to a colorless compound in $5\frac{1}{2}$ h, while poor milk may accomplish this in less than 2 h.

Tests for Coliform Bacteria

Coliform bacteria include the organisms *E. coli* and *Enterobacter aerogenes*, both normal inhabitants of the large intestine. The milk test is the same as that for *E. coli* in water; the presence of these organisms indicates fecal contamination. The sanitary significance of this lies in the fact that the presence of coliform bacteria usually points to unsanitary handling after the completion of the pasteurization process (assuming that the phosphatase test gave evidence of effective pasteurization). In raw (unpasteurized) milk, the coliform test indicates the degree of contamination, whether from careless handling or unsanitary equipment.

Tests for Specific Pathogens

Unless there is some evidence that a particular disease is being transmitted through milk, tests for specific pathogens are not run. The procedure to be followed would depend on the specific organism in question. The most recent large-scale epidemics spread via milk involved *Listeria* and *Salmonella*.

FERMENTED MILK PRODUCTS

As was mentioned elsewhere, lactic acid is produced when the milk sugar lactose is fermented by the lactobacilli and streptococci that make up the usual flora of milk. Lactic acid increases the acidity of the milk, and since the beginning of history people have prepared various palatable forms of sour or fermented milk to prevent its spoiling. It has long been believed that fermented milks prolong life and promote good health; as a result, the preparation of such milks has become a major industry in many countries.

Butter

Although it is true that butter can be churned from sweet, unfermented milk, such products lack the typical flavors we ordinarily associate with butter. Commercially, butter is prepared from pasteurized cream that has been inoculated with *Streptococcus lactis* or *Streptococcus cremoris*. These organisms lower the pH of the cream by fermenting the sugar lactose to form lactic acid. The inoculum also contains *Lactobacillus cremoris* or *S. lactis* subsp. *diacetilactis*, which, at the lowered pH, carries out a 2,3-butylene glycol fermentation and converts residual lactose to acetoin and diacetyl. It is these two compounds that contribute most to the delicate flavor of butter. Cultured buttermilk, the fluid remaining after removal of the butter, also owes much of its flavor to these compounds.

Yogurt

Yogurt may be made from the milk of cows, goats, sheep, or buffalo. It is usually scalded to partially concentrate the milk and, after slight cooling, is inoculated with strains of *Lactobacillus bulgaricus* and *Streptococcus thermophilus*. The milk can then be poured into tubs and kept warm until it coagulates,

or it may be constantly stirred until it thickens. Various flavorings such as strawberry, raspberry, cherry, or other fruits may be added to this semisolid yogurt before it is poured into smaller containers.

Acidophilus Milk

This fermented milk is essentially the same as yogurt except that *L. acidophilus* is used for the fermentation instead of *L. bulgaricus*. It is a popular "health" drink in Europe and the United States.

Kumiss and Kefir

These beverages are prepared from mare's, camel's, or cow's milk. Both begin as a *Lactobacillus* fermentation, but a yeast fermentation follows that converts the lactic acid to alcohol. In some cases, sugar may be added to the milk to increase the alcoholic content.

Cottage Cheese

Cottage cheese can be made by allowing milk to sour and, after draining, washing and salting the remaining curd. Commercially, it is made by inoculation of milk with a mixture of *S. lactis* and *Leuconostoc citrovorum*. After a short period of fermentation, a preparation of rennet containing the enzyme rennin is added to curdle the milk. The curd is then collected, salted, and sold as cottage cheese.

Cured Cheeses

Literally hundreds of different cheeses are made throughout the world. The one thing they have in common is that all cheeses are the product of the fermentation of milk. They differ, however, in the temperature and time of fermentation and in the species of organisms added to the milk.

One of the best known is cheddar cheese, which comprises about 75 percent of all cheese produced in the United States. The production of cheddar cheese begins much like that described for cottage cheese except that the curd is pressed and then "cheddared," which consists of cutting the curd into large squares and stacking them into a pile, allowing more of the whey to be pressed out. The curd is then placed in cheesecloth or in a plastic-lined hoop, salt is added, and the curd is permitted to cure for about 6 months. During this period, many products, such as diacetyl, acetoin, lactic acid, and acetic acid are formed by the bacteria. It is these products that give cheddar cheese its characteristic flavor. This process yields about 1 lb of cheese from 10 lb of milk.

Swiss or Emmenthaler cheeses use three organisms as starter cultures, *Streptococcus thermophilus*,

Lactobacillus helveticus, and *Propionibacterium shermanii*. After an initial fermentation, rennet is added and the resulting curd is cooked at 50 to 55°C. This destroys most contaminating bacteria, but the starter cultures remain viable. The curd is then pressed and the wheels of cheese are immersed in a salt solution for several days, during which time a third rind is formed. The cheese is then cured for 3 to 6 months at 12 to 18°C.

During the curing, much of the lactic acid produced by the streptococci and lactobacilli is converted by the propionibacteria to propionic acid, acetic acid, and carbon dioxide (which produces the holes characteristic of Swiss cheese). It is, however, the propionic acid that gives Swiss cheese its characteristic sharp flavor.

Roquefort cheese is made in France from the milk of ewes that have been bred for their high milk-producing ability. After an initial streptococcal fermentation, the curd is drained in hoops and inoculated with finely ground dried bread on which has grown the mold *Penicillium roqueforti*. The ripening takes place in caves in France where the temperature remains at 8 to 9°C and the humidity is high. The pungent taste of Roquefort is due to the action of a fungal lipase that liberates caproic, caprylic, and capric acids from the milk fats. Blue cheeses are made in the United States by a similar process, except that cow's milk is used rather than sheep's milk. Other mold-ripened cheeses include Camembert, Stilton, and Gorgonzola cheeses.

Some cheeses, such as Brie, begin as a lactic acid fermentation of milk, but the final product is the result of a surface bacterial fermentation by *Brevibacterium linens*.

IMMUNOLOGICAL PROPERTIES OF MILK

Unit II of this text was concerned with host resistance to infection, and here we shall discuss briefly the role of milk in this regard. The milk of most mammals (including humans) is rich in immunoglobulins, and their importance to the newborn varies from one animal to another. For example, human antibodies of the IgG class readily pass from the mother's blood to the unborn fetus, so that at the time of birth the newborn has the same IgG antibodies as the mother. Antibodies of the IgA class cannot pass from mother to fetus but are normally found in body secretions such as saliva, mucus, and human milk. In the human, for example, secretory IgA makes up 97 percent of the total protein of the first milk (colostrum). Although

this value drops to 10 to 25 percent in later milk, it appears that such milk provides a major source of protection against gastrointestinal infections in the newborn.

Cows, on the other hand, are unable to pass their antibodies to the unborn fetus and at the time of birth the newborn calf is extremely susceptible to infection. Cow's milk, particularly that secreted for a few days after birth of the young (colostrum), is exceedingly rich in antibodies and, therefore, provides the newborn calf with its only source of specific protective antibodies.

Human milk also contains a large amount of lactoferrin as compared with cow's milk, thus providing the newborn with a much needed source of iron. Moreover, since lactoferrin is released unsaturated with respect to iron, it serves as an antimicrobial substance by sequestering any ingested iron.

PRACTICAL CONSIDERATIONS

Because milk is a good medium for bacterial growth and since pasteurized milk is not sterile, a cold temperature to inhibit bacterial growth is a necessity. Milk should be removed from the refrigerator only when actually ready for use. Refrigerated trucks (for the transport of milk) are essential for reducing microbial growth. The one exception to this rule is milk that has undergone the ultrahigh-temperature treatment. Such milk is stable at room temperature for several months as long as it has not been opened and exposed to environmental bacteria.

Since dust always carries bacteria, milk containers should be kept covered as well as cold. Ideally, only disposable containers should be used in areas such as hospital wards, where additional contamination is a danger.

Home pasteurization can be carried out successfully with simple equipment. The milk must be maintained at the required temperature for 30 min, care being taken to ensure that all parts of the sample are kept at the required temperature. This can be accomplished by placing the container of milk in a water bath. Rapid cooling, preferably in a container of ice water, should follow the heating period.

Coliform bacteria, anaerobes, and some yeasts may produce gas and undesirable flavors in milk. Milk of questionable odor or taste should not be consumed even after pasteurization. It should be remembered also that ice cream, though frozen, can serve as a medium for disease-producing organisms.

SUMMARY

Many foods may be preserved from spoilage as a result of a lactic acid fermentation. Examples include sauerkraut, ensilage, and fermented vegetables such as beets, green beans, and pickles, as well as green olives. Dried summer sausage also owes its stability and flavor to a lactic acid fermentation.

The characteristic flavor of a number of Oriental foods is the result of a combined fungal and lactic acid fermentation. Soy sauce is one of the more common examples in the western world.

Foods may be preserved by removing water or by adding sodium chloride or sucrose. These procedures are designed to lower the water activity below that which will support microbial growth. Chemicals, such as benzoic acid, sorbic acid, and propionic acid, as well as nitrates and nitrites are also added to food to prevent spoilage.

Milk is an excellent medium for the growth of many bacterial species, including some pathogens. It has no natural flora, but certain bacteria are always present in the cleanest raw milk.

Bacteria may be introduced into milk from a variety of sources such as workers, infected cows, the cow's udder, feces and dust in barns, and milk containers or other equipment.

A number of tests are performed to determine the sanitary quality of milk. These provide information about the conditions of collection, handling, pasteurization, and refrigeration.

Disease organisms in milk may come either from an infected animal or from an infected human or carrier. Disease agents that may enter milk from infected cattle are *Mycobacterium bovis*, *Brucella* species, streptococci (udder infections), *C. burnetii*, and the virus of foot-and-mouth disease. Agents from human sources are *Salmonella* species, *Shigella* species, *C. diphtheriae*, streptococci, *M. tuberculosis*, and the virus of poliomyelitis.

Milkborne diseases are controlled by general sanitary procedures and by pasteurization. All pathogens likely to gain access to the milk are destroyed by careful pasteurization. The keeping quality of milk is also enhanced by pasteurization and refrigeration.

All cheeses begin as a lactic acid fermentation of milk but differ in the time and temperature of curing and in the type of bacteria or molds that are used to produce the final product.

QUESTIONS FOR REVIEW

1. Name as many foods as you can that are preserved as a result of a bacterial fermentation.
2. What is single-cell protein? What are some of the carbon sources from which it is produced?
3. Why does milk provide a good medium for the growth of many bacterial species?
4. What are the common sources of bacteria found in milk?
5. Explain the tests used to determine the sanitary quality of milk.
6. What are two sources from which disease agents gain access to milk? What diseases may be associated with each source?
7. What measures may be carried out to reduce the spread of disease through milk?
8. Aside from disease prevention, why is it important to keep bacterial numbers in milk low?
9. Explain the significance of the pasteurization process.
10. A student usually spends her summer vacation on a farm. Raw milk is the only type of milk served at the table. How can the student safely handle the situation?
11. Why is butter that is made from fermented cream tastier than that made from sweet cream?
12. What is meant by water activity and how can it be lowered?
13. List several chemicals that are added to foods as preservatives.
14. How are the holes made in Swiss cheese?
15. Name three mold-ripened cheeses.

SUPPLEMENTARY READING

Edelson E: Milk. *Science* 4(6):66, 1983.

Knorr D, Sinskey AJ: Biotechnology in food production and processing. *Science* 229:1224, 1985.

Jay JM: *Modern Food Microbiology* New York, Van Nostrand Reinhold, 1986.

Kosikowski FV: Cheese. *Sci Am* 252(5):88, 1985.

Levine AS, Labuza TP, Morley JE: Food technology: A primer for physicians. *N Engl J Med* 312:628, 1985.

Plucknett DL, Smith NJH: Agricultural research and third world food production. *Science* 217:215, 1982.

Rao VC, Melnick JL: *Environmental Virology*. Aspects of Microbiology Series, Washington, D.C., American Society for Microbiology, 1986.

Rose AH: The microbiological production of food and drink. *Sci Am* 245(3):126, 1981.

Chapter 32

Agricultural and Industrial Microbiology

OBJECTIVES

After study of this chapter, you should comprehend

1. The recycling of elements through putrefaction, decay, and fermentation.

2. The steps involved in the carbon cycle, nitrogen cycle, and sulfur cycle.

3. Production of wine, beer, and vinegar.

4. Production of industrial chemicals, enzymes, amino acids, and pharmaceuticals.

5. The role of recombinant DNA techniques in plant engineering, vaccines, human products, probes for diagnostic procedures, and gene therapy.

There is a general tendency to associate microorganisms with disease. However, if we were to study all the microorganisms on earth, we would find that very few have the ability to produce disease in humans. Many of the nonpathogenic microbes are absolutely essential for the continuance of the organic cycles in nature. Others are of major importance in making our lives easier and more enjoyable.

PUTREFACTION, DECAY, AND FERMENTATION

The processes referred to as putrefaction, decay, and fermentation represent mechanisms through which microorganisms break down large organic molecules into simple substances that can be reused by other forms of life, both plant and animal. Without microbial enzymes, the bodies of dead plants and animals and their wastes would accumulate on the earth's surface. Imagine what it would be like if every plant and every animal—large and small—never decomposed after death. We would not only be faced with the gigantic problem of disposing of all this material, but eventually much of the usable matter in the world might exist in an unusable form, and then all life would end.

Fortunately for us, the myriad populations of microorganisms in the world prevent this. With their complex enzyme systems they perform the amazing task of converting organic material into inorganic material so that it can again be utilized for plant growth. Thus, the amino acids in the steak you eat will eventually be converted by bacteria into water, carbon dioxide, nitrates, and a few other inorganic substances. These, in turn, will nourish a blade of grass, then a steer—and, finally, another steak is ready.

It may be misleading to consider these processes as separate and distinct mechanisms, since all occur simultaneously. However, strictly speaking, they can be defined as follows.

Putrefaction is the anaerobic decomposition of proteins by bacterial enzymes. Some microorganisms secrete proteolytic enzymes that hydrolyze the large protein molecules into their component parts—that is, the amino acids. The amino acids can then be taken into the bacterial cell and further dissimilated to provide a carbon source, a nitrogen source, and an energy source for the bacterium. Not all the amino acids are completely broken down; some may be only deaminated (removal of amino group) or decarboxylated (removal of amino group) to yield a basic amine. Many of these basic amines have very foul odors—thus the use of the word *putrid,* meaning "bad-smelling." An excellent example of this is the odor of well-rotted meat. The end result of the process of putrefaction is the breaking down of the very large protein molecules (like those that occur in animals) and their conversion to smaller, soluble compounds that can be reused by other forms of life. Thus, putrefaction is one of the useful activities of certain microorganisms, and it is essential so that some of the elements can be used over and over.

Decay is a somewhat looser term that can apply to any aerobic breakdown of complex material. Like putrefaction, it begins with the excretion of extracellular enzymes that can hydrolyze the large, complex molecules into smaller usable compounds. A rotten stump or log is one example of decay. Another is the decaying leaves and grass in a compost pile; this material can be used later as a source of plant nutrients.

Fermentation is defined as the anaerobic breakdown of carbohydrates resulting in the formation of stable fermentation products. Examples of usable fermentation products produced by microorganisms include such things as ethyl alcohol, lactic acid, acetic acid, glycerol, butylene glycol, acetone, butanol, and butyric acid. In addition, many fungi are used for the commercial production of organic acids such as citric acid, fumaric acid, malic acid, and succinic acid.

Soil and the Cycles of the Elements

The earth's surface (which we usually speak of as soil or dirt) is actually composed of both inorganic and organic material. It is from this soil that plants obtain all their physical requirements for growth except carbon dioxide. In addition to the plants, which are easily seen, the soil contains an extensive microbial population of bacteria, yeasts, molds, algae, and protozoa.

The type of soil, the available nutrients, and the pH influence the number and the types of organisms in the microbial flora. Topsoil, in which there is an abundant supply of oxygen, has the greatest number. There is a considerable decrease in number of organisms below a depth of about 4 ft, and at a depth of 8 to 10 ft there are usually very few. Not only is oxygen not available, but other constituents required by anaerobic bacteria are usually lacking. In well-cultivated soils where fertilizer has been added, the bacterial count is much higher than it is in sandy or clay soils.

Perhaps in no place are microorganisms as important in these cycles of nature as they are on the farm. Here, soil microorganisms not only decompose complex organic materials but also transform them into compounds that can be used by plants for growth. Of the many inorganic elements needed for growth, the most important are nitrogen, sulfur, carbon, and phosphorus. Each of these elements must be in a form that green plants are able to use. In the case of phosphorus, this is not particularly complex, because the decay of any organic compound liberates phosphorus as a phosphate ion, which is ready to be assimilated directly by plants and microorganisms. Other cycles are more complex.

The Carbon Cycle

Because all organic molecules contain carbon, it is obvious that the cycling of this element is of utmost importance to all forms of life. The basic points to remember about the cycling of carbon compounds are: (1) the final oxidation product for all carbon compounds is carbon dioxide, and (2) photosynthesis is, by far, the main route whereby carbon dioxide is reduced to form the organic compounds that are essential for the survival of the animal kingdom. A simplified version of this cycle can therefore be represented by two major steps: (1) fixation of carbon dioxide into organic compounds by green plants, algae, and autotrophic bacteria, and (2) the oxidation of organic compounds to carbon dioxide to yield energy and heat. This latter step may be represented by a single reaction that occurs during the burning of wood or fossil fuels, or it may involve many stepwise reactions carried out by myriads of heterotrophic organisms. In either case, the carbon is returned to the atmosphere as carbon dioxide, where it is again available for recycling through plant, animal, fungal, or procaryotic cells before being again oxidized to carbon dioxide.

Carbon dioxide in the atmosphere acts to absorb some of the infrared radiation (heat) that the earth normally loses to space. Because of our increased use of fossil fuels, the level of carbon dioxide is increasing. Also, because of the millions of acres of forests being cut in South America, less carbon dioxide is being fixed in photosynthesis. The net result is more carbon dioxide in the atmosphere, and many scientists believe this excess carbon dioxide will produce a "greenhouse" effect, resulting in an overall warming of the earth's atmosphere. An overall warming of even a few degrees could have a devastating effect by melting large amounts of the polar ice caps and substantially raising the level of the oceans.

The Nitrogen Cycle

The nitrogen cycle is somewhat more complex than the carbon cycle and includes a number of microbial steps in the conversion of this element to a usable form.

The microbial degradation of proteins starts with the enzymatic hydrolysis of a protein into its individual amino acids; next, the released amino acids are metabolized further. During the course of this metabolism, the amino group is most frequently released as ammonia, a process known as **ammonification**:

$$R\text{—}\underset{\underset{\text{Amino acid}}{NH_2}}{\overset{|}{C}}HCOOH \xrightarrow{\text{deaminase}}$$

$$R\text{—}\underset{\underset{\alpha\text{-Keto acid}}{\overset{\|}{O}}}{C}\text{—}COOH + \underset{\text{Ammonia}}{NH_3}$$

Since plants can use this released ammonia as a nitrogen source, the cycle could stop there insofar as a balance in nature is concerned. However, there are a large number of autotrophic bacteria that derive their sole source of energy from the oxidation of ammonia to nitrite.

$$NH_4^+ + 2O_2 \longrightarrow NO_2^- + 2H_2O$$

This oxidation is carried out by a group of closely related gram-negative aerobic organisms. *Nitrosomonas* is the most thoroughly studied genus of this group.

At this point another group of autotrophic bacteria takes over, of which *Nitrobacter* is the most common genus; these bacteria obtain their energy by the oxidation of nitrite to nitrate:

$$2NO_2^+ + O_2 \longrightarrow 2NO_3^-$$

As a result, the major form of nitrogen in the soil is nitrate, which can also be used by plants as a nitrogen source. This process in which ammonia is converted to nitrate is termed **nitrification.**

A great many bacteria are able to use nitrates as final electron acceptors in place of oxygen (anaerobic respiration), and they reduce the nitrates back to nitrites. Much more critical to our ecology are the organisms that are capable of reducing nitrites to nitrogen gas, which then escapes into the atmosphere. Free nitrogen gas cannot be assimilated by plants; thus, the production of nitrogen gas from inorganic nitrogen sources constitutes a direct loss in fertility. This process, called **denitrification,** is carried out by many bacteria, particularly members of the genus *Pseudomonas*, and by the autotrophic bacterium *Thiobacillus denitrificans.* The latter organism oxidizes sulfur for its source of energy, and, when growing anaerobically, uses nitrate as an electron acceptor. The nitrate is eventually reduced to nitrogen gas.

The process of denitrification appears to consist of at least four steps for the reduction of nitrate to nitrogen gas. These are (1) reduction of NO_3^- to

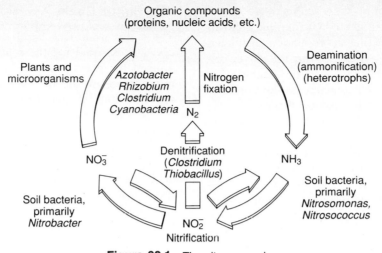

Figure 32.1 The nitrogen cycle.

NO$_2^-$ by the enzyme nitrate reductase, (2) reduction of NO$_2^-$ to NO by the enzyme nitrite reductase, (3) reduction of NO to N$_2$O by the enzyme nitric oxide reductase, and (4) reduction of N$_2$O to N$_2$, presumably by a nitrous oxide reductase. Figure 32.1 shows a schematic representation of the cycle involved in the conversion of organic nitrogen to inorganic nitrogen and back again to organic compounds.

Nitrogen Fixation

The nitrogen cycle that we have discussed so far presents a situation in which many bacteria are converting nitrogen compounds to nitrogen gas, an element that cannot be utilized by green plants. Thus, if there were no microbial mechanism to convert the nitrogen gas back to usable nitrogen compounds, the biological world would cease to exist when all the nitrogen was converted to nitrogen gas. Fortunately, many bacteria have the ability to fix atmospheric nitrogen and to render it again available to green plants as ammonia or nitrates. Bacteria that fix atmospheric nitrogen can be categorized either as **free-living nitrogen fixers** or as **symbiotic nitrogen fixers**.

The most important free-living nitrogen fixers are found among the cyanobacteria and in the bacteria classified in the genus *Azotobacter*. Many other bacteria, such as the clostridia and the photosynthetic bacteria, are also capable of fixing atmospheric nitrogen.

MICROBIOLOGY MILESTONES

Of all the examples of how life forms use the cycles of nature for survival, none appears stranger than that observed around the hydrothermal vents on the ocean's floor.

Such vents, occurring at depths of 1500 to 3000 m, are openings in the ocean floor from which sulfide-rich water spews into the ocean. Surrounding such vents are found 8-ft-long tube worms, possessing neither mouths nor intestines. Also found are giant white clams and jellyfishlike animals.

How do such creatures survive in such a hostile environment, where there is neither light nor appreciable organic matter? They exist because of a symbiotic relationship with bacteria that use the sulfide coming from the vents as their source of energy. The giant tube worms do not need a mouth or intestines because they contain internal colonies of such bacteria, which provide them with both food and energy. Similarly, the gi-

ant clams and other marine life existing in these places, feed on the mats of sulfur bacteria growing free on the ocean floor.

The story becomes even more strange when one realizes that such vents exist for only 10 to 15 years and that the distance between vents may be 600 miles or more. One must then ask how such animals can get from one vent to another.

A plausible explanation has been proposed by Stan Smith from Arizona State University. Based on observations of dead whales on the ocean floor that were colonized by the same type of animals, he has proposed that dead whale carcasses provide stepping stones that sustain the vent animals as ocean currents propel them on their journey to a new vent.

Thus we see another example of the interdependence of eucaryotic and procaryotic life.

The symbiotic nitrogen fixers are small gram-negative rods classified in the genus *Rhizobium*. The rhizobia are capable of infecting the roots of a large class of plants called legumes (peas, soybeans, clover, alfalfa, etc.). After infecting the root, the bacteria become irregularly shaped cells (bacteroids) and nodules enclosing the bacteria are formed on the root at the point of infection. Within this nodule they fix atmospheric nitrogen (thus helping the plant) and in turn receive nutrients from the plant that they can use in their own metabolism (hence it is a mutualistic relationship). Thus, soil that is nitrogen-poor can be replenished with ammonia and nitrates for plant growth by the planting of a legume, such as alfalfa, for 1 year. This is the reason farmers rotate their crops from a nitrogen-depleting crop (such as corn) to a nitrogen-replenishing crop (such as soybeans or alfalfa). It is estimated, for example, that an acre of alfalfa may fix up to 400 lb of nitrogen in one season.

Actinomycetes in the genus *Frankia* are also symbiotic nitrogen fixers. These organisms form characteristic nodules on the roots of alder trees.

Let us summarize our current knowledge. The system for the fixation of nitrogen appears to be essentially identical in all organisms (i.e., photosynthetic bacteria, *Azotobacter*, *Clostridium*, and *Rhizobium*). The major fixation process consists of two separate reactions: (1) the formation of a reductant and (2) the binding of nitrogen gas. Adenosine triphosphate (ATP) is essential for the first reaction, in which electrons are passed from reduced ferredoxin to an as yet unidentified reductant. In the second reaction nitrogen gas is fixed to an enzyme (nitrogenase), which contains both molybdenum and iron. The overall reduction—which consists of at least two steps—is as follows:

$$N_2 + 6 \text{ electrons} + n(ATP + H_2O) \longrightarrow$$
$$2NH_3 + n(ADP + Pi + H^+),$$

where ADP denotes adenosine diphosphate, and Pi inorganic phosphate. As the equation indicates, it is not known just how many molecules of ATP are required for this process.

Inasmuch as the availability and replenishment of nitrogen in the soil are essential for food production, it is interesting to note the direction of current research in this field. Using recombinant DNA techniques, the nitrogen fixation genes have been transferred into a variety of different microorganisms. The objective is to produce new nitrogen-fixing strains that will function under crop and soil conditions unfavorable to the growth of *Rhizobium*. Attempts are also being undertaken to transfer the nitrogen-fixing genes directly into plant cells. The success of this endeavor would permit the growth of cereal grains in nitrogen-poor soil.

The Sulfur Cycle

The sulfur cycle (see Figure 32.2), like the nitrogen cycle, is essential for the reutilization of sulfur compounds for plant growth. Many different groups of microorganisms are involved in carrying out the various reactions of this cycle.

The formation of hydrogen sulfide (H_2S) from the degradation of proteins can be accomplished by a wide variety of heterotrophic bacteria. Since essentially all proteins contain cysteine and methionine—amino acids containing sulfur—the complete degradation of these proteins releases the sulfur as sulfides.

Several large groups of microorganisms carry out the next two steps in the sulfur cycle: the oxidation of H_2S to sulfur (S) and the oxidation of S to SO_4^{2-}. The first reaction is carried out by several groups of organisms, of which the major ones are the green sulfur and purple sulfur photosynthetic bacteria. These organisms use H_2S as a reductant in photosynthesis, converting it to elemental sulfur. There is also a group of filamentous, gliding bacteria (*Beggiatoa, Thioploca, Vitreoscilla*) that obtain their energy from the oxidation of H_2S to S. It is interesting to note that many important sulfur deposits, such as those in Louisiana and Texas, were formed by the H_2S-oxidizing bacteria.

The oxidation of S to SO_4^{2-} is most often carried out by a group of chemoautotrophic gram-negative rods classified in a number of different genera, such as *Thiobacillus*, *Thiomicrospira*, and *Sulfolobus*. Many of these organisms are very acid-tolerant and, thus, grow best at a pH of 3 to 4.

Because of its impact on the environment, we shall discuss one species in a little more detail. *Thiobacillus ferrooxidans* is a small gram-negative rod that can obtain its energy by oxidizing either ferrous ions or sulfide. It has presented a serious problem in strip mines that contain large amounts of ferrous sulfide (FeS) in the coal. After such mines are opened to the

Figure 32.2 The sulfur cycle.

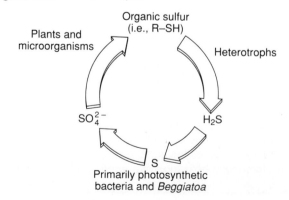

The requirement for vitamin B_{12} marks one of the most unusual dependencies of the animal kingdom for products of procaryotic organisms. This vitamin is necessary for red blood cell maturation and, if not present, an individual will develop pernicious anemia. It is also involved in the proper functioning of the nervous system and in protein and fat metabolism. All animals have an absolute requirement for B_{12}, yet it is not synthesized by any animal cell. Moreover, unlike the other vitamins required by humans, vitamin B_{12} cannot be obtained from eating fruits or vegetables because it is not synthesized by members of the plant kingdom either.

Where do animals normally get their vitamin B_{12}?

Ruminants, such as cows and sheep, obtain it from the cellulose-digesting bacteria in the rumen. Nonruminant herbivores obtain their B_{12} either by absorption from their intestinal tract or through coprophagy, a practice common among many species of animals. Carnivores (including humans) get their B_{12} from eating meat, a source rich in B_{12}. And some species, such as fruit-eating bats, appear to ingest sufficient contaminating bacteria on the surface of fruits to satisfy their B_{12} requirement.

One can see, therefore, that we are not only dependent on bacteria for recycling of the elements, but for the very existence of animal life as we know it.

---

air and water, *T. ferrooxidans* oxidizes ferrous sulfide as follows:

$$2Fe^{2+} + 3S^{2-} + 6O_2 \longrightarrow Fe_2(SO_4)_3$$

The ferric sulfate in the presence of water spontaneously reacts as follows:

$$Fe_2(SO_4)_3 + 6H_2O \longrightarrow 2Fe(OH)_3 + 3H_2SO_4$$

The resulting ferric hydroxide is insoluble, and the sulfuric acid (H_2SO_4) is a major source of pollution for many streams and rivers. The consequent low pH kills fish and inhibits sewage breakdown, producing a high **biological oxygen demand** until the water can leech enough limestone from the soil to raise the pH.

MICROORGANISMS IN INDUSTRIAL PROCESSES

Because of their putrefactive, fermentative, and synthesizing abilities, microorganisms have attained a useful place in many industrial processes, including the manufacture or processing of food, clothing, and drugs.

In the manufacture of leather, microbial enzymes decompose undesirable parts of the hides. Likewise, in the manufacture of linen, they decompose the carbohydrate that binds the fibers of flax together. Bacterial enzymes are at work in curing processes, improving the flavor of coffee, cocoa, vanilla pods, and tobacco. Ripening of various cheeses is a result of the action of several species of bacteria, molds, and yeasts. Swiss cheese, for example, owes much of its characteristic flavor to the propionic acid produced by the

propionibacteria; its holes are due to the carbon dioxide produced during fermentation. And, as described in Chapter 31, there are a number of other cheeses and fermented foods that acquire their properties from a microbial fermentation, but such edibles represent only a small part of the number of microbial products that benefit humankind. Table 32.1 provides a summary of just what industrial microbiology is all about. In the remainder of this chapter we will discuss a few of these processes.

Production of Wine and Beer

The production of alcoholic beverages for human consumption relies on the conversion of sugar to alcohol, as shown below:

$$C_6H_{12}O_6 \longrightarrow 2CO_2 + 2C_2H_5OH$$

Glucose   Carbon   Ethyl alcohol
dioxide

The pathway of this fermentation is described in Chapter 6, here we shall be concerned only with the industrial process itself.

Wine

Wine can be defined as the product resulting from the alcoholic fermentation of any fruit juice; by and large, however, most wine is made from grape juice. There are many different varieties of grapes and hence numerous different wines. We shall describe only the general procedures involved in wine making.

The grapes are usually harvested when the content of their juice is 19 to 22 percent sugar. The grapes are transported to the winery, where they are stemmed and crushed by machine. Because grapes routinely

TABLE 32.1 Major Microbial Industrial Products

Organism	Type	Product
Foods and beverages		
Saccharomyces cerevisiae	Yeast	Baker's yeast, wine, ale, sake
Saccharomyces carlsbergensis	Yeast	Lager beer
Saccharomyces rouxii	Yeast	Soy sauce
Candida milleri	Yeast	Sour French bread
Lactobacillus sanfrancisco	Bacterium	Sour French bread
Streptococcus thermophilus	Bacterium	Yogurt
Lactobacillus bulgaricus	Bacterium	Yogurt
Propionibacterium shermanii	Bacterium	Swiss cheese
Gluconobacter suboxidans	Bacterium	Vinegar
Penicillium roquefortii	Mold	Blue-veined cheeses
Penicillium camembertii	Mold	Camembert and Brie cheeses
Aspergillus oryzae	Mold	Sake (rice-starch hydrolysis)
Rhizopus	Mold	Tempeh
Mucor	Mold	Tofu (soybean curd)
Monascus purpurea	Mold	Ang-kak (red rice)
Industrial chemicals		
Saccharomyces cerevisiae	Yeast	Ethanol (from glucose)
Kluyveromyces fragilis	Yeast	Ethanol (from lactose)
Clostridium acetobutylicum	Bacterium	Acetone and butanol
Aspergillus niger	Mold	Citric acid
Xanthomonas campestris	Bacterium	Polysaccharides
Amino acids and flavor-enhancing nucleotides		
Corynebacterium glutamicum	Bacterium	L-lysine
		5′-Inosinic acid and 5′- guanylic acid
Single-cell proteins		
Candida utilis	Yeast	Microbial protein from paper-pulp waste
Saccharomycopsis lipolytica	Yeast	Microbial protein from petroleum alkanes
Methylophilus methylotrophus	Bacterium	Microbial protein from growth on methane or methanol
Vitamins		
Eremothecium ashbyi	Yeast	Riboflavin
Pseudomonas dentrificans	Bacterium	Vitamin B_{12}
Propionibacterium	Bacterium	Vitamin B_{12}

have a profusion of yeasts on their surface, a natural fermentation will begin immediately. However, some of this normal flora may impart undesirable flavors to the finished wine. Therefore sulfur dioxide (SO_2) is added to kill most of these organisms. After several hours, an inoculum of a cultured yeast (one of many strains of *Saccharomyces cerevisiae*) is added to the crushed grapes to begin the fermentation. Such strains are more resistant than wild yeasts and are able to grow in the presence of any residual SO_2.

The nature of the fermentation will depend in part on the type of wine desired. Wine color is derived entirely from the grape skins; thus, white wine can be made from white grapes or from red grapes if the skins are removed at the time the grapes are crushed. Rosé wines result when crushed red grapes are allowed to ferment for about a day before the juice is removed

from the skins and seeds with wine presses. In such wines only a part of the color is extracted from the skins. For the heavier red wines (such as the burgundies and clarets), the fermentation is allowed to proceed for 3 to 5 days before the juice is separated from the skins.

When the major fermentation is complete, the wine will usually contain 12 to 14 percent alcohol; it is then removed from the yeast and aged for various periods during which protein and tannins precipitate, esters are formed, and the wine loses much of its harsh flavor. The wine is then clarified and bottled. White and rosé wines are usually drunk within a few years after production, as are some red wines like Beaujolais. Other red wines, such as the burgundies and clarets, may be aged from 5 to 40 years before they reach their peak of flavor.

TABLE 32.1 *(Continued)*

Organism	Type	Product
Enzymes		
Aspergillus oryzae	Mold	Amylases
Aspergillus niger	Mold	Glucamylase
Trichoderma reesii	Mold	Cellulase
Saccharomyces cerevisiae	Yeast	Invertase
Kluyveromyces fragilis	Yeast	Lactase
Saccharomycopsis lipolytica	Yeast	Lipase
Aspergillus	Mold	Pectinases and proteases
Bacillus	Bacterium	Proteases
Endothia parasitica	Mold	Microbial rennet
Polysaccharides		
Leuconostoc mesenteroides	Bacterium	Dextran
Xanthomonas campestris	Bacterium	Xanthan gum
Pharmaceuticals		
Penicillium chrysogenum	Mold	Penicillins
Cephalosporium acremonium	Mold	Cephalosporins
Streptomyces	Bacterium	Amphotericin B, kanamycins, neomycins, streptomycin, tetracyclines and others
Bacillus brevis	Bacterium	Gramicidin S
Bacillus subtilis	Bacterium	Bacitracin
Bacillus polymyxa	Bacterium	Polymyxin B
Rhizopus nigricans	Mold	Steroid transformation
Arthrobacter simplex	Bacterium	Steroid transformation
Mycobacterium	Bacterium	Steroid transformation
Hybridomas		Immunoglobulins and monoclonal antibodies
Mammalian cell lines		Interferon
Escherichia coli (via recombinant-DNA technology)	Bacterium	Insulin, human growth hormone, somatostatin, interferon, blood factor VIII, tissue plasminogen activator
Carotenoids		
Blakeslea trispora	Mold	Beta-carotene
Phaffia rhodozyma	Yeast	Astaxanthin
Entomopathogenic bacteria		
Bacillus thuringiensis	Bacterium	Bioinsecticides
Bacillus popilliae	Bacterium	Bioinsecticides

Dessert wines such as port or muscatel are made by halting the fermentation at the end of the second day; brandy, which is made by distilling wine, is then added to bring the alcohol content to about 20 percent. Because the fermentation was interrupted early, such wines are very sweet, containing 12 to 15 percent sugar.

Beer

Beer is made from barley and other cereals, and here we have a situation much different from that described for the production of wine. These starting substances contain no fermentable material. In other words, the carbohydrate in barley exists as long polymers of starch that must be broken down into simple sugars before the yeast can convert them to alcohol. This is accomplished by using the enzymes formed in the barley itself when it germinates. The major steps in beer production can be summarized as follows: (1) Barley is allowed to germinate by soaking in water. The germinated barley is rich in enzymes that will hydrolyze starch to simple sugars; after drying, this is termed malt. (2) Malt is dissolved in water, and other ground cereals may be added at this time. The temperature is maintained at 65 to 70°C to facilitate the enzymatic hydrolysis of the starch. (3) After filtering, hops are added (this imparts the bitter taste characteristic of beer) and the liquid (called wort) is boiled for an hour or two to inactivate any enzymes, precipitate some proteins, and extract the flavor of the hops. (4) The wort is again filtered and pumped to fermentation tanks, where it is inoculated with the yeast *S. carlsbergensis*. (5) After fermenting for 8 to 14 days at 4 to 6°C, it is aged several weeks to permit proteins, yeast, and resins to precipitate before it is filtered and bottled.

The following paragraph is included because the author believes that any microbiologist should know the difference between regular beer and that which the advertising agencies call **light beer**. As shown schematically in Figure 32.3, the starch present in barley consists of polymers of glucose that are linked to each other by two different linkages (i.e., an $\alpha1-4$ linkage and an $\alpha1-6$ linkage). The hydrolytic enzymes in the malt can hydrolyze only the $\alpha1-4$ linkages, leaving short chains of glucose each time they reach an $\alpha1-6$ linkage. These residual glucose polymers are called **limit dextrans**; because the yeast cannot metabolize them either, they remain in the final beer, and therein lies the caloric problem. Limit dextrans account for approximately 22 percent of the total starch in regular beer; when metabolized by the body, they add to the total calories present in the beer. If, however, before fermenting the wort, a mold enzyme that will break these $\alpha1-6$ linkages is added, all the glucose will be fermented, reducing the number of calories to yield light beer. Interestingly, the techniques of recombinant DNA technology have now entered this field, and the DNA that encodes for the enzyme to break the $\alpha1-6$ linkage has been cloned directly into a yeast cell. It is likely that in the near future light beer will be made from this yeast without the use of the mold enzyme.

Production of Distilled Beverages

Because the natural fermentation of a carbohydrate source such as fruit juice or malt will yield only 12 to 14 percent alcohol, higher alcohol content is ob-tained by distilling the fermented material to concentrate the alcohol.

Throughout the world, there are numerous distilled beverages, some of the most common of which are whiskey, gin, vodka, rum, and a myriad of brandies. All begin with the fermentation of a carbohydrate source but each differs either in the origin of the carbohydrate source or in the method of distillation and aging. Table 32.2 provides a summary of the starting materials and treatment after distillation. Keep in mind, however, that all grains (corn, barley, wheat, rye) or potatoes must be first treated with malt to digest the complex starches into fermentable sugars.

Production of Vinegar

Because wine or beer provide the initial material for making vinegar, this seems an appropriate place to describe its production.

A gram-negative rod-shaped organism in the genus *Acetobacter* carries out the following oxidation of ethanol:

$$\underset{\text{Ethanol}}{C_2H_5OH} + \underset{\text{Oxygen}}{O_2} \longrightarrow \underset{\text{Acetic acid}}{CH_3COOH} + \underset{\text{Water}}{H_2O}$$

When *Acetobacter* oxidizes the ethanol present in wine or beer, the result is vinegar. The various types of vinegar produced are merely representative of the starting material. Thus, the use of apple wine yields cider vinegar, grape wine produces wine vinegar, and the use of beer results in malt vinegar.

The actual production may utilize a batch method whereby a barrel is partly filled with wine or beer before it is inoculated with *Acetobacter*. More frequently, however, a continuous generator consisting

---

Figure 32.3 Structural formula of amylopectin from starch showing both $\alpha1-4$ and $\alpha1-6$ linkage between glucose molecules. Because the barley enzymes cannot hydrolyze an $\alpha1-6$ linkage, nor, in most cases, the $\alpha1-4$ linkage adjacent to the $\alpha1-6$ linkage, the boxed-in portion of the molecule is a limit dextran that normally remains unhydrolyzed and unfermented.

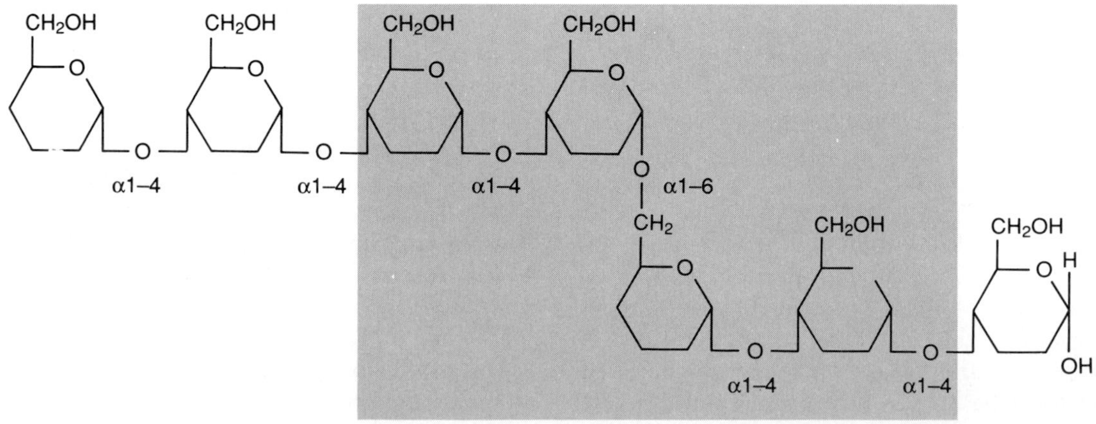

TABLE 32.2 Production of Distilled Spirits

	Primary Fermented Material	Special Treatment after Distillation
U.S. whiskey	Corn (bourbon) Rye	Aged in charred white oak barrels for 2–8 yr
Scotch whiskey	Barley malt dried with smoldering peat	Aged in casks; frequently blended with other grain distillates
Gin	Grains, principally barley	Distilled through a receptacle containing juniper berries and other herbs or barks; not aged
Vodka	Principally wheat but also rye, corn, barley, and potatoes	Consists almost solely of diluted alcohol; may be filtered through charcoal
Rum	Sugar cane syrup	Aged in casks
Brandy	Any fruit juice, mainly grape juice	Aged, usually in new plain white oak barrels

of a column filled with wood shavings is used. *Acetobacter* develops on the surface of the shavings and, as the liquid is percolated through the column, the alcohol is converted to acetic acid.

Production of Industrial Chemicals

Microorganisms can synthesize literally hundreds of different chemicals, although many can be produced more economically using chemical synthesis. Microorganisms are, however, still used in a part of the industrial production of ethanol, acetone, butanol, lactic acid, acetic acid, and citric acid. In addition, a number of different amino acids and enzymes are commercially produced using microbial processes. It is beyond the scope of this book to describe the methods of manufacture of these industrially important microbial syntheses, but the following sections will provide a summary of a few such products.

Organic Solvents

Ethanol, acetone, and butanol are all, in large part, made from petroleum. However, they have in the past been produced commercially through microbiological fermentation; if the price of petroleum should continue to rise, it is probable that microbial syntheses will again be the major source of these chemicals. Some countries, particularly Brazil, have established a huge industry to ferment molasses to ethanol with the objective of using the ethanol to replace gasoline as a fuel. In the United States, a number of such plants were developed in the late 1970s and early 1980s. Here, however, corn was the source of carbohydrate converted to ethanol. Keep in mind that the starch in

corn, like that in barley, must be broken down enzymatically into simple sugars before the yeast can ferment it. The industrial alcohol produced by this means in the Midwest was mixed 1:9 with gasoline and sold under the name of gasohol.

It seems unlikely that the conversion of food starches, such as corn, into ethanol for fuel will ever become a major industry in the United States. If, however, through genetic engineering, strains of microorganisms are developed that are capable of fermenting the cellulose from wood, this will undoubtedly spawn a new era in industrial microbiology.

Enzymes

Many hundreds of tons of bacterial enzymes are produced annually. Commercially important enzymes include glucoamylase, alpha-amylase, glucose isomerase, and proteases. The first three of these enzymes are all involved in the production of sweeteners from starch. In brief, the glucoamylase and the alpha-amylase are used to hydrolyze starch into its basic constitutent, glucose. Glucose by itself is not terribly sweet, but by converting it into fructose using the enzyme glucose isomerase, the soft drink industry obtains a sugar that is many times sweeter (and thus more economical) than either glucose or our normal cane or beet sugar, sucrose.

The bacterial proteases include several different enzymes (made primarily from *Bacillus licheniformis*) that are able to hydrolyze proteins into soluble peptides. They have been widely incorporated into a variety of detergents because of their ability to "solubilize" things such as the gravy spot on your shirt or blouse. Other proteases are used as meat tenderizers to convert that tough steak into a filet mignon.

Amino Acids

Of the 20 amino acids comprising a protein, 8 cannot be synthesized by most animals and must, therefore, be obtained through diet. Two of these, lysine and methionine, are present only in small amounts in many cereal grains and are, therefore, added as supplements to a number of animal foods. Methionine is made synthetically, but approximately 40,000 tons of lysine are produced annually as a bacterial product.

Glutamic acid, an amino acid, is widely used as a flavor enhancer in the form of monosodium glutamate. Approximately 300,000 tons of this amino acid are produced each year from the bacterial degradation of glucose.

All species of bacteria used to produce amino acids consist of mutants that have lost the ability to regulate the synthesis of these amino acids. As a consequence, such syntheses become the major metabolic activity, resulting in the formation of large amounts of the specific amino acid.

Production of Pharmaceuticals

The production of biologically active compounds may well represent the largest and most important aspect of industrial microbiology. However, because of its incredible diversity and complexity, we can here only summarize the highlights of this field.

Antibiotics are by far the largest and most important class of pharmaceuticals made by microorganisms. The reader is referred to Chapter 11 for a review of the major antibiotics and the mechanism by which they prevent microbial growth. Interestingly, however, the synthesis of antibiotics—unlike that of the industrial products described earlier in this chapter—appears to have no role in the growth and reproduction of the microorganism. Such products, therefore, are called **secondary metabolites,** to distinguish them from degradation products of energy-yielding reactions (such as ethanol, butanol, etc.) or from synthetic products essential for cell growth (such as amino acids or vitamins). The selection of various mutants by the microbial geneticist has resulted in strains of molds or bacteria that produce hundreds of times more antibiotic than did the original parent strain; it seems likely that microbiologists will continue to improve both strains and production methods for these important biological molecules.

A second area in which microorganisms are intimately involved in the synthesis of pharmaceutical products has been termed **bioconversion.** Here, microorganisms are used to catalyze only certain steps of a synthesis, the remaining process being accomplished chemically. Examples include the conversion of the plentiful six-carbon alcohol sorbitol into sorbose by an organism in the genus *Gluconobacter.* The sorbose is then easily converted chemically to vitamin C. Steroid synthesis also utilizes a number of bioconversions in which distinct reactions—such as adding a specific hydroxyl group or removing hydrogen and hydroxyl groups from specific carbons—are carried out by microorganisms. Such bioconversions, along with a number of chemical steps, are used to synthesize hormones such as cortisone and prednisone.

RECOMBINANT DNA TECHNOLOGY

The general techniques used to insert foreign or modified DNA into a plasmid and the subsequent transfection of a bacterium or yeast with the hybrid plasmid are discussed in Chapter 8. Here we are concerned primarily with the potential benefits of these procedures.

Agricultural Applications of Plant Engineering

It has been known for many years that the bacterium *Agrobacterium tumefaciens* could infect a number of different plants, causing a cancer known as crown gall disease. The surprising aspect of this infection is that, after the initial infection, the continued presence of the bacterium is not required to maintain tumor growth. It now is known that its tumor-inducing principle (TIP) resides in a plasmid (termed Ti) that integrates part of its DNA into the infected plant cell chromosome. The incredible value of such a plasmid resides in the possession of eucaryotic-type regulation signals that permit its transcription by plant enzymes.

Quite obviously, the production of a plant tumor would not be a desirable effect of gene transfer into plants. Considerable efforts have been directed toward the isolation of mutant Ti plasmids that will transfect and integrate into the DNA of plant cells but will not produce tumors. It is hoped that future research on plant recombinant DNA will result in plants with improved growth in their normal environment. The insertion of genes encoding for nitrogen fixation would be an achievement of this goal. Additional aims of such research are to learn more about plant gene regulation as well as to introduce genes that will make the plant more resistant to herbicides and pathogens.

Not all recombinant DNA work in plants has used the Ti plasmid as a vector for the insertion into the plant chromosome. Some procedures have been designed to produce bacterial hybrids that colonize the plant and protect it from other predators. For example, gene splicing has produced a strain of *Pseudomonas fluorescens* that colonizes the roots of corn plants and produces a toxin that kills the cutworms that feed on corn plant roots. Another engineered organism, *Pseudomonas syringae,* is being used in trial experiments to protect plants from freezing. Such organisms are effective by replacing normal flora bacteria that produce proteins that act as nuclei for the growth of ice crystals at 0°C. The recombinant strains cannot make this protein, and initial results indicate that plants colonized with these bacteria may be able briefly to sustain temperatures as low as $-3°C$.

Recombinant Vaccines and Human Products

A number of products resulting from recombinant DNA techniques are now available for use as vaccines. Some have already been approved for use, while others are currently undergoing trials to assess their safety and effectiveness. These vaccines fall into two main categories: (1) those that consist of a virus or microorganism that is altered so as no longer to cause disease, and (2) those that result from the insertion of a single gene into *E. coli* or *S. cerevisiae.* In the latter case, the hybrid microorganism synthesizes the foreign protein (such as the surface protein of hepatitis B virus), which can then be used as a vaccine to induce an immune response to the virulent agent. A few other examples of such vaccines that are currently being tried include surface proteins from influenza virus, foot-and-mouth-disease virus, and malaria parasites. Because of their safety and effectiveness, many other such vaccines seem destined to be produced from an isolated gene that has been inserted into another bacterium or yeast.

In addition to recombinant vaccines, a number of human genes have been cloned in *E. coli* and yeast. These techniques have succeeded in the synthesis of many human products, including insulin, growth hormones, and the interferons.

DNA Probes for the Diagnosis of Infectious Diseases

Cloned DNA from various bacteria or viruses is capable of recognizing and undergoing a very specific nucleic acid hybridization with complementary nucleic acid present in cells from which the cloned nucleic acid was obtained. Such DNA is now available from a number of commercial sources and can be used to detect infinitely small amounts of either DNA or specific ribosomal RNA in a target organism. As a result,

such probes are used for the very rapid identification of organisms present in sputum, serum, feces, urine, or tissue homogenate. Probes may be made radioactive or may be linked to an enzyme conjugate such as acid phosphatase, horseradish peroxidase, or alkaline phosphatase. In any case, the hybridized, double-stranded DNA is separated from any nonreacted probe, and the reaction is quantitated radioactively or enzymatically.

DNA probes are exquisitely sensitive and many hybridization reactions can be completed in 1 to 3 h. They are now available for a number of agents which are difficult to diagnose, such as *Mycoplasma pneumoniae, Legionella pneumophila,* enterotoxigenic *E. coli, Chlamydia trachomatis, Campylobacter jejuni, Mycobacterium tuberculosis,* and *Neisseria gonorrhoeae.* Viruses for which probes have been developed include hepatitis B, herpes simplex, Epstein-Barr, human papillomavirus, cytomegalovirus, adenoviruses, and human immunodeficiency virus. It seems safe to assume that DNA hybridization will be used increasingly often for the diagnosis of infectious diseases.

Gene Therapy

One of the ultimate goals of recombinant DNA techniques is to supply functional DNA to individuals suffering from a deficiency of a particular gene. The scientific and ethical hurdles to be cleared are momentous. Three such genetic deficiencies that have been considered for gene therapy are adenosine deaminase deficiency, purine nucleoside phosphorylase deficiency, and the Lesch-Nyhan syndrome, which is due to a deficiency of hypoxanthine-guanine phosphoribosyltransferase. All of these genetic diseases result in serious neurological malfunctions that could theoretically be prevented by the insertion of a normal gene that encodes for the nonfunctional enzyme.

About 25 percent of children suffering from severe combined immunodeficiency (SCID) have a defective gene for the synthesis of the enzyme, adenosine deaminase (ADA). The FDA has now given permission to carry out experiments to correct this deficiency by replacing the defective gene with a functional gene that has been cloned into a retrovirus. This will be attempted by isolating defective bone marrow cells (containing stem cells) from the patient, infecting such cells with the ADA retrovirus, and injecting the engineered cells back into the patient. If the ADA is correctly transcribed in these cells, infected stem cells should give rise to normal B and T cells.

The apparent initial success of these trials hopefully will lead to additional attempts of gene therapy to correct some of the thousands of known genetic defects.

SUMMARY

Microorganisms contribute to our existence through the processes of putrefaction, decay, and fermentation. Without these useful activities, dead plants and animals and their wastes would accumulate and make the earth unfit for life.

Organic substances are not only decomposed but also converted into substances that are used again. Many soil bacteria are involved in breaking down organic matter. Some of the results of these actions are represented by the cycles of elements such as nitrogen, carbon, and sulfur. Each group of species of bacteria has its own function to perform in making dead organic matter again available for human use.

Microorganisms also contribute to human welfare in many other ways, as in the processing of leather, linen, coffee, cocoa, vanilla pods, tobacco, alcoholic beverages, antibiotics, sauerkraut, vinegar, and a host of other important products.

Wine is the result of an alcoholic fermentation of any fruit juice such as grape juice. Beer is produced by the fermentation of cereal grains, primarily barley. Using molasses or hydrolyzed cornstarch, these same processes can be used for the industrial production of ethanol.

Bacteria are also employed to produce enzymes used to make sweeteners for the soft drink industry and proteases that are incorporated into detergents. Other bacterial products of industrial importance include the amino acids lysine and glutamic acid. Microbes are also put in action in carrying out specific steps in the synthesis of vitamin C and certain steroid hormones. In addition, antibiotics—microbial products as well—are of major importance in the control of infectious diseases. Recombinant DNA techniques have been used to modify plants by the insertion of new genes into the plant chromosome. Bacterial flora of plants have also been modified so as to kill predators by the secretion of toxins or to lower the temperature at which plants are killed. A number of vaccines are currently being produced from hybrid organisms into which genes from pathogenic viruses and parasites have been inserted.

A number of human genes have also been cloned in bacteria and yeast that encode for the production of human products such as insulin, growth hormone, blood factor VIII, tissue plasminogen activator, and interferons.

Recombinant DNA is used also as a probe to provide a rapid and sensitive identification of an unknown organism based on the ability of the cloned DNA to hybridize with complementary nuclei acid. Gene therapy, the insertion of a gene into an animal deficient in that gene, is projected for the future.

QUESTIONS FOR REVIEW

1. How do putrefaction, decay, and fermentation contribute to human welfare?
2. Why does soil serve as a good habitat for microbes?
3. How do microorganisms contribute to the farm industry?
4. What is meant by the cycles of the elements?
5. Explain the nitrogen cycle. What is its significance to humankind?
6. What is meant by free-living and symbiotic nitrogen fixers?
7. What are the steps in the conversion of organic sulfur to a form usable by green plants?
8. How does light beer differ from regular beer?
9. What starting material is used for the production of vinegar?
10. What is the major industrial use of bacterial proteases?
11. What human products are now being produced by yeast or *E. coli* as a result of recombinant DNA technology?

SUPPLEMENTARY READING

Alexander M: *Introduction to Soil Microbiology*. New York, Wiley, 1977.

Ausubel FM: Molecular genetics of symbiotic nitrogen fixation. *Cell* 29:1, 1982.

Bauer WD: Infection of legumes by rhizobia. *Annu Rev Plant Physiol* 32:407, 1981.

Day LE: Genetics and molecular biology of industrial microorganisms. In Schlessinger D (ed): *Microbiology—1981*. Washington, D.C., American Society for Microbiology, 1981.

Deacon, JW: *Microbial Control of Plant Pests and Diseases*. Aspects of Microbiology Series, Washington, D.C., American Society for Microbiology, 1983.

Friedmann T: Progress toward human gene therapy. *Science* 244:1275, 1989.

Gardner RC, Knauf VC: Transfer of *Agrobacterium* DNA to plants requires a t-DNA border but not the VirE locus. *Science* 231:725, 1986.

Gasser CS, Fraley RT: Genetically engineering plants for crop improvement. *Science* 244:1293, 1989.

Marx JL: How rhizobia and legumes get together. *Science* 230:157, 1985.

Payne WF: Centenary of the isolation of denitrifying bacteria. *ASM News* 52:627, 1986.

Pursel VG, et al: Genetic engineering of livestock. *Science* 244:1281, 1989.

Smith S: Afterlife of a whale. *Discover* 11:46, 1990.

Sun M: Engineering crops to resist weed killers. *Science* 231:1360, 1986.

Verma IM: Gene therapy. *Sci Am* 263(5):68, 1990.

A Brief Review of Microbiology

DEVELOPMENT OF MICROSCOPES; EARLY MICROSCOPIC OBSERVATIONS

1590 Johannes Janssen and his son Zacharias, of Holland, developed a compound microscope.

1624 Galileo Galilei, Italian astronomer, constructed both a microscope and a telescope.

1656 Athanasius Kircher, German mathematician, used a microscope to study plague. He reported seeing "small worms" in what was apparently the first application of the microscope to the study of disease. It is uncertain whether the "small worms" were actually bacteria, since the magnification possible at the time probably was inadequate.

1661 Marcello Malpighi, Italian physician, was the first person to use the microscope in the study of anatomy. He discovered capillary circulation and demonstrated the termination of the trachea in bronchial filaments.

1665 Robert Hooke published *Micrographia,* which contained the first reference in science to "little boxes or cells, distinct from one another," and which described his studies of fungi.

1674 Antony van Leeuwenhoek, Dutch merchant and microscopist, is believed to have made the first recorded observations of microscopic life, which he described in voluminous letters to the Royal Society of London. His descriptions of protozoa are considered to be highly accurate even by modern standards.

THEORY OF SPONTANEOUS GENERATION

c. 1650 Francesco Redi, Italian naturalist, was the first person to confute the notion that maggots and grubs develop spontaneously in decaying matter. In one of his early experiments he placed pieces of animal flesh in flasks, leaving some flasks open and closing some. Flies were able to enter and leave the open flasks freely. Before long, the flesh in the open flasks was covered with maggots, while none appeared in the covered flasks. In a later experiment, Redi placed pieces of meat and fish in a flask and covered it with wire gauze that permitted air but not flies to enter. No maggots ever appeared in the meat, whereas flies were seen to light on the outside net and there deposit eggs, from which maggots developed.

1711 Louis Joblot boiled hay in water, placed equal quantities of the infusion in two containers of the same size, closed one container with parchment, and left the other one uncovered. A few days later organisms appeared in the open vessel but none appeared in the closed vessel.

1748 John Turberville Needham, English biologist and priest, poured boiled mutton gravy into a glass vial and corked the vial. A few days later numerous organisms appeared in the gravy. Needham concluded that the organisms arose spontaneously. He conducted later experiments using seeds of corn and other grains, with similar results. His experiments lent support to the theory of spontaneous generation.

1776 Lazzaro Spallanzani, Italian biologist, performed experiments similar in some ways to Needham's. He placed his infusions into eight containers and boiled them. He closed four of the containers with corks and four with airtight seals. No organisms appeared in the airtight vessels, whereas the corked vessels showed organisms in abundance. His experiments did not disprove spontaneous generation, but they strengthened the case against it.

1858–1861 Louis Pasteur, French chemist, boiled his nutrient solution in a gooseneck flask that permitted untreated air to enter. The organisms settled in the gooseneck; none appeared in the solution. In later experiments the infusions were boiled in glass flasks; the necks were sealed with a flame while the infusions were hot. All the infusions remained sterile.

1877 John Tyndall, British physicist, constructed a special box in which he placed six culture tubes. A beam of light was passed through the tubes. When no dust was visible in the beam, he filled the tubes with nutrient broth and boiled them. Organisms failed to grow, but when the air in the box was stirred the nutrient broth quickly spoiled. The experiment demonstrated that bacteria spread through the air on dust particles and as spores.

FERMENTATION

1836 Theodore Schwann, German physiologist, discovered the organic nature of yeast. By demonstrating that the yeast plant causes fermentation, he was able to associate yeasts with alcohol production. He proved that putrefaction is due to living organisms.

1837 Charles Cagniard-Latour confirmed the organic nature of yeasts, including budding as a mechanism of reproduction. He also observed the relationship of fermentation to yeast growth.

1839 Justus von Liebig, German chemist, made important contributions to chemistry and to medicine, among them studies of uric acid compounds, the discovery of chloroform, and the concept of metabolism. Nevertheless, he held that fermentation was a purely chemical process in which living cells had no part. He believed that fermentation and putrefaction were merely physical disturbances of equilibrium.

1857 Louis Pasteur's studies on the optical isomers of tartaric acid provided the background for his work on fermentations. He considered fermentation to be due to living organisms, in contrast to Leibig's belief.

1866 Louis Pasteur's *Studies on Wines* was published. He postulated that souring of wine was due to extraneous organisms that produced acid; thus undesirable secondary fermentations were the cause of the so-called wine diseases. He introduced the method of gentle heating to kill the undesirable contaminants. This process is now known as pasteurization.

1878 Joseph Lister, English physician, conducted studies on lactic fermentations and related these to disease processes. He introduced a pure culture technique for lactic acid bacteria by a serial dilution method.

1897 Eduard Buchner, German chemist, showed that extracts of yeast cells could produce an alcoholic fermentation.

MICROORGANISMS AND DISEASE

Early Observations on the Relationship of Microorganisms to Disease

1530 Girolamo Fracastoro, physician and scholar in Venice, published a poem entitled *Syphilis sive Morbus Gallicus*. In it he described the shepherd Syphilus, who had the disease that has since come to be known as syphilus.

1546 Girolamo Fracastoro recognized the contagious nature of diseases. He considered contagions to develop from three sources: contact, fomites, and from a distance. For lack of proof, his ideas had little influence during this period.

1835 Agostino Bassi, Italian in Lodi, showed that the silkworm disease muscardine was caused by a parasitic fungus called Botrytris.

1843 Oliver Wendell Holmes, poet and physician in the United States, wrote the essay *On the Contagiousness of Puerperal Fever*, in which he pointed out that a woman who had just been delivered should not be attended by a physician who was conducting autopsies or attending patients with puerperal fever, and that puerperal fever could be thus conveyed from patient to patient. Holmes stated that washing the hands in calcium chloride and changing clothes after attending a patient with puerperal fever were preventive measures. These ideas met with fierce opposition from the established obstetricians of the day.

1847 Ignaz Semmelweis discovered the etiology of puerperal fever.

1850 Ignaz Semmelweis, Hungarian physician, was an assistant in the first obstetric ward of a Vienna hospital, where the puerperal fever rate was exceedingly high; students came to this ward directly from the dissecting room without washing their hands. In the second ward, where midwives received instructions, much greater attention was paid to cleanliness, and here the puerperal fever rate was far lower. This observation, as well as other pieces of evidence, convinced Semmelweis that puerperal fever was a septicemia. He enforced the washing of hands in calcium chloride and by this simple measure brought about a dramatic fall in the puerperal fever rate. Yet he too was subjected to much ridicule. This scorn was believed to have contributed to his subsequent insanity and early death.

1865 Louis Pasteur isolated the parasite causing the silkworm disease pébrine. He showed that the disease could be eliminated by use of only healthy, disease-free worms for breeding stock.

1867 Joseph Lister, British surgeon, introduced antiseptic surgery by the use of phenol to soak surgical dressings. The conquest of wound infection was his great contribution.

1876 Robert Koch, German physician, was the first person to demonstrate the relationship of a specific organism with one type of disease process by proving that *Bacillus anthracis* was the etiological agent of anthrax.

1879 Albert Neisser, German physician, discovered the gonococcus, the causative agent of gonorrhea.

1880 Kark Joseph Ebcrth, German pathologist, isolated the typhoid bacillus, which for many years was named *Eberthella typhosa*.

1880 Charles Louis Alphonse Laveran, French physiologist and bacteriologist, discovered the parasite of malaria.

1882 Robert Koch, German bacteriologist, isolated the tubercle bacillus (Koch's bacillus) and proved that it was the causative agent of tuberculosis. His four experimental steps have since become known as Koch's postulates.

1882 Friedrich Löffler, bacteriologist in Germany, demonstrated the bacillus of glanders.

1882 Carl Friedländer, German pathologist, discovered the bacillus of *Klebsiella pneumoniae,* often called Friedländer bacillus.

1882 Paul Ehrlich, German bacteriologist, experimented extensively on dyestuffs and tissue staining. He developed the fuchsin stain for tubercle bacilli on the basis of his discovery that this bacillus was acid-fast.

1894 Shibasaburo Kitasato and Alexandre Yersin independently discovered the plague bacillus.

1898 Kiyoshi Shiga, Japanese physician, reported the discovery of the dysentery bacillus.

1898 Ronald Ross, surgeon with the Indian Medical Service, demonstrated malarial parasites in the salivary glands of the anopheles mosquito.

1898 Giovanni Battista Grassi and Amico Bignami succeeded in transmitting malaria via infected mosquito bites. Their experiments were conducted in malarious country and thus were not as conclusive as those conducted in London by Patrick Manson.

1900 Patrick Manson, British physician, had demonstrated previously that *Filaria bancrofti* was transmitted by the culex mosquito and that it was the cause of elephantiasis. The experiments of Grassi were repeated by Manson in London. Mosquitoes that had bitten malarious patients were sent to London from Rome and permitted to bite Manson's son and another person. Both contracted the disease.

1900 Walter Reed, physician in the United States, was appointed Chief of the Army's Yellow Fever Commission to investigate the disease among American troops in Cuba. The Commission experimentally produced cases of yellow fever and proved that it was transmitted by mosquito bite.

1903–1906 David Bruce, British Army surgeon, demonstrated that African sleeping sickness was caused by *Trypanosoma gambiense* and that the tsetse fly was the vector.

1905 Fritz Schaudinn and Erich Hoffman, in Germany, discovered the etiological agent of syphilis, *Treponema pallidum*. Other contributions of Schaudinn were the discovery of Entamoeba, the cause of amebic dysentery, identification of the cause of tertian malaria, and identification of hookworm disease.

1977 First isolation of *Legionella pneumophila,* the causative agent of legionnaires disease.

1983 First isolation of *Borrelia burgdorferi,* the causative agent of Lyme disease.

Viruses

1892 Dmitrii Iwanowski, Russian scientist, reported that the agent of tobacco mosaic disease was filterable.

1898 M. W. Beijerinck independently recognized that the agent of tobacco mosaic disease was filterable.

1898 Friedrich Löffler and P. Frosch, in Germany, demonstrated that the agent of foot-and-mouth disease of cattle was filterable.

1911 Peyton Rous, pathologist in the United States, transferred the malignant sarcoma tumor to Plymouth Rock chickens by inoculating them with a cell-free filtrate.

1915 Frederick William Twort, in England, discovered transmissible lysis of bacteria (bacteriophage).

1917 Felix d'Herelle, in Canada, discovered the bacteriophage of the dysentery bacillus.

1935 Wendell M. Stanley, in the United States, published *Isolation of a Crystalline Protein Possessing the Properties of Tobacco Mosaic Virus.*

1979 World declared free of smallpox.

1979 First official case of the acquired immunodeficiency syndrome (AIDS) reported in the United States.

Rickettsiae

1906 Howard Taylor Ricketts, in the United States, was the first person to describe bacillary forms of rickettsiae in the blood of patients infected with Rocky Mountain spotted fever.

1910 Howard Taylor Ricketts, studying typhus fever in Mexico, discovered that it was transmitted via lice by an organism similar to that of Rocky Mountain spotted fever. He accidentally contracted typhus fever and died.

1913–1915 Stanislaus von Prowazek, of Bohemia, investigating the cause of typhus in Serbia and Turkey, discovered the organisms, as described by Ricketts, in lice that had fed on typhus patients. Like Ricketts, he died after contracting typhus during the course of his studies.

THE BEGINNING OF IMMUNOLOGY

1798 Edward Jenner, English physician, introduced vaccination for the prevention of smallpox via inoculations with cowpox.

1880–1881 Louis Pasteur accidentally left virulent cultures of chicken cholera virus in his laboratory during a vacation. On his return he found that these cultures had become sterile or inactive and, when injected, acted as a preventive vaccine against later injection of a virulent character. This attenuated virus, carried through several generations, retained its immunizing property. Applying the technique of attenuation, Pasteur developed a vaccine against anthrax.

1884 Elie Metchnikoff, Russian biologist, showed how the ameboid cells in the connective tissues and the blood engulf solid particles and bacteria, absorbing and thus destroying the bacteria. He called these cells "phagocytes."

1885 Louis Pasteur, again applying the technique of attenuation, developed a successful vaccination against rabies using attenuated rabies virus. The first patient to be treated was a 9-year-old boy who had been bitten by a rabid dog.

1886 Daniel Salmon and Theobald Smith, American pathologists working with inoculations of killed cultures of microorganisms, demonstrated that immunity to many infections could be thus produced.

1890 Emil von Behring and Shibasaburo Kitasato developed antitoxins against tetanus and diphtheria.

1898 Jules Bordet discovered bacterial hemolysis and, with Octave Gengou, discovered fixation of the complement, the principle of the Wassermann reaction.

1924 Albert Calmette and Alphonse Guérin introduced a preventive vaccine against tuberculosis prepared from living avirulent tubercle bacilli—the BCG vaccine.

1933 First use of a pertussis vaccine.
1937 First use of an influenza vaccine.
1938 First use of a tetanus vaccine.
1954 First use of Salk inactivated polio vaccine.
1956 First use of Sabin live attenuated polio vaccine.
1958 First use of a measles vaccine.
1962 First use of a rubella vaccine.

CHEMICAL AGENTS IN THE TREATMENT OF DISEASE

1908 Paul Ehrlich, German scientist, synthesized an arsenical compound, Salvarsan, which was used to treat syphilis. It marked the introduction of chemotherapy—the use of a specific drug capable of curing or alleviating disease without causing great damage to the patient.

1929 Alexander Fleming, English bacteriologist, noticed that an agar plate inoculated with *Staphylococcus aureus* in his laboratory had become contaminated with a mold. A clear zone surrounding the mold indicated that bacterial growth had been inhibited. Fleming and his colleagues isolated the mold and identified it as *Penicillium notatum*. The inhibitory substance, or antibiotic, was thus called penicillin.

1935 Gerhard Domagk, in Germany, investigating a group of dye compounds, reported that Prontosil was an effective bactericide. Further study revealed that this activity was due to its sulfonamide portion.

1939 Rene Dubos, in the United States, isolated a culture of *Bacillus brevis,* from which the active principles gramicidin and tyrocidine were isolated.

1944 Selman A. Waksman and coworkers isolated the antibiotic streptomycin, produced by a soil Actinomycete, *Streptomyces griseus.*

1947 John Ehrlich, Paul R. Burkholder, and David Gottlieb isolated the antibiotic chloramphenicol.

1948 B. M. Duggar and associates isolated chlortetracycline.

REFERENCES

Brock TD: *Milestones in Microbiology.* Englewood Cliffs, NJ, Prentice-Hall, 1961.

Bulloch W: *The History of Bacteriology.* Oxford University Press, London, 1938.

Dobell C: *Antony van Leeuwenhoek and His Little Animals.* New York, Staples Press, 1932.

Grainger TH, Jr: *A Guide to the History of Bacteriology.* New York, Ronald Press, 1958.

Lechevalier HH, Solotorovsky M: *Three Centuries of Microbiology.* New York, McGraw-Hill, 1965.

Major RH: *A History of Medicine.* Springfield, Ill., Charles C Thomas, 1954, vols 1 and 2.

Vallery-Radot R: *The Life of Pasteur.* New York, Garden City Publishing, 1926.

Glossary

Abscess a localized accumulation of pus in tissues.

Acid curd coagulation of casein in milk by an acid pH of 4.7 or lower.

Acid dye a stain in which the colored part of the dye is negatively charged.

Acid-fast not decolorized with alcohol which contains 3 percent hydrochloric acid after staining with carbol fuchsin.

Acid-fast stain staining procedure in which certain bacteria, primarily the mycobacteria, resist decolorization with alcohol containing 3 percent hydrochloric acid.

Acquired immunodeficiency syndrome (AIDS) an infection caused by a retrovirus resulting in the destruction of T cells, which are necessary to mount an immune response.

Acquired immunity resistance resulting from infection with a specific microorganism. It may be active or passive. It is acquired during life, in contrast to natural or innate immunity.

Acquired pellicle deposition of glycoproteins from saliva on a tooth.

Activated sludge flocs of sewage whose surface is heavily colonized with bacteria, yeast, and protozoa. Provides a continuous, heavy inoculation for the degradation of incoming sewage.

Activated sludge process the addition of partially degraded flocs of sedimented sewage to inoculate raw sewage, resulting in a more rapid breakdown of incoming material.

Activation energy the minimum energy required to initiate a chemical reaction.

Active immunity resistance built up in an individual as a result of experience with the disease agent or its products, either by having the disease or by receiving a vaccine.

Active transport the energy-dependent transport of a solute across a membrane against a concentration gradient.

Acute most severe stage of a disease; usually of short duration.

Adansonian analysis numerical type of analysis of the characteristics of an organism; each property of an organism (e.g., motility, sugar fermentations) is enumerated. Strains yielding the same or a similar total number are considered to be taxonomically closely related.

Adenine a purine constituent of nucleic acids.

Adenosine triphosphate (ATP) compound in which energy is stored in high-energy phosphate bonds; its components are the purine adenine, D-ribose, and three phosphoric acid groups.

Adenovirus an iscosahedral, DNA-containing virus causing respiratory infections in humans.

Adjuvants insoluble materials that act to keep antigens in tissues for longer periods and thus cause a longer stimulation of antibody production.

Aeration the process of introducing oxygen so that organic material can be degraded by aerobic bacteria. The procedure is used in sewage treatment.

Aerobic requiring oxygen for growth.

Aerosols vapors or particles suspended in air; chemical aerosols may be used to kill microorganisms.

Aerotolerant organisms organisms that can grow in the presence of air but do not use oxygen as a final electron acceptor.

Aflatoxin toxin produced by the mold *Aspergillus flavus*.

Agammaglobulinemia defect in antibody synthesis due to failure to develop antibody-forming cells.

Agar polysaccharide derivative of seaweed; used as a solidifying agent in bacteriological media.

Agglutination clumping of cells, as occurs with a particulate antigen and its antibody, agglutinin.

Agglutinin antibody that causes agglutination of its particulate antigen, as bacteria or other cells.

AIDS *see* Acquired immunodeficiency syndrome.

Algae the simplest plants that contain chlorophyll and require sunlight; they vary from microscopic forms to giant seaweeds.

Allergen substance that induces allergy.

Allergy hypersensitivity; the harmful reaction of antibody with its specific antigen.

Alloantibodies antibodies that exist in some, but not all, members of a species, i.e., anti-blood-group antibodies.

Alloantigens antigens that exist in some, but not all, members of a species, i.e., blood group antigens.

Allografts tissue grafts between different strains of the same species of animal.

Allosteric enzyme an enzyme whose activity is changed because of reaction with a compound at a nonactive site. Allosteric enzymes are subject to feedback inhibition, in which the end product of a series of reactions will react directly with the enzyme of the first reaction and inactivate it.

Ameba class of protozoans that move by extruding pseudopodia.

Ames test a procedure for measuring the ability of a chemical to induce bacterial mutations.

Amino acids building blocks of proteins: contain amino ($-NH_2$), and carboxyl ($-COOH$ groups).

Aminoglycoside group of broad-spectrum antibiotics containing multiple amino groups and sugar residues.

Ammonification breakdown of organic nitrogen compounds with the release of ammonia.

Amorphous without form; characteristic of plasmodium stage of slime molds.

Amphitrichous polar type of flagellation, with a single flagellum at each end of the cell.

Amylase enzyme that catalyzes the hydrolysis of starch.

Anabolism metabolic process consisting of the synthesis of protoplasm; requires energy.

Anaerobic growing in the absence of atmospheric oxygen.

Anamnestic response recall phenomenon, or secondary response to an antigen with very rapid production of antibodies.

Anaphylaxis extremely severe allergic reaction involving IgE type of antibody.

Anemia deficiency of red blood cells.

Animal kingdom eucaryotic organisms that obtain their energy by the oxidation of chemical compounds.

Antibiotic chemotherapeutic agent that occurs as a byproduct of the metabolic activity of bacteria or fungi.

Antibodies specific molecules formed in the animal body in response to the presence of an antigen; once formed, they react with that particular antigen.

Antibody-mediated immunity immune reactions between humoral antibodies and an infectious agent.

Anticodon the three nucleotides on a tRNA molecule that recognize the codon on the mRNA.

Antigenic drift minor antigenic changes occurring in the hemagglutinin or neuraminidase of influenza virus, which are the result of spontaneous mutations.

Antigenic shift major antigenic changes occurring in the hemagglutinin or neuraminidase of influenza virus, which are the result of a reassortment of RNA fragments following the infection of a single cell with two different influenza virions.

Antigens foreign substances that induce a specific immune response.

Antimetabolite an antibacterial analog of an essential metabolite. Examples are the sulfonamides and trimethoprim.

Antiseptic a chemical agent that prevents growth, by either inhibiting the growth of or destroying microorganisms; used for topical application to living tissue, primarily skin.

Antiserum blood serum containing specific antibodies against an antigen, such as a bacterial cell, a virus, or an exotoxin.

Antitoxin antibody formed against a specific exotoxin and capable of neutralizing that toxin.

Apoenzyme protein structure of an enzyme.

Arbovirus a virus that is spread from one animal to another via the bite of an arthropod vector such as a mosquito.

Arthropod an invertebrate animal with jointed legs, such as an insect or a crustacean.

Arthropodborne disease infections that are carried by and spread to mammals by arthropods such as fleas, mites, and ticks.

Arthrospores asexual spores formed by fragmentation of a filament in certain fungi.

Arthus reaction reaction resulting from injection of an antigen into the skin; inflammation results from activation of complement by IgG-antigen complexes.

Artificial active immunity resistance to disease that

develops after administration of a vaccine or inactivated toxin.

Artificial passive immunity immunity resulting from the injection of preformed antibodies.

Artificial medium a medium for microbial growth that contains complex substances such as beef extract, yeast extract, tryptones, and blood.

Ascomycetes filamentous fungi that form their sexual spores within a sac called an ascus.

Ascospore sexual spore, characteristic of fungi of the class Ascomycetes.

Ascus sac in which are developed the sexual spores, or ascospores, characteristic of fungi of the class Ascomycetes.

Asepsis a condition in which there are no infectious or contaminating microorganisms; the absence of sepsis.

Aseptic free of microorganisms that can cause contamination or infection.

Aseptic meningitis nonbacterial meningitis.

Asexual spores spores generated by one cell, without fusion of two cells.

Attenuation change; used in reference to certain vaccines in which the organism or its products are reduced in virulence.

Autoclave steam-pressure sterilizer.

Autografts transplant of tissue from one area to another area of a body.

Autoimmune disease due to antibodies produced against the individual's own tissues; failure to differentiate between self and nonself.

Autolysis self-lysis, or disintegration of cells due to the action of their enzymes.

Autotroph an organism capable of growth in a completely inorganic environment, using carbon dioxide as its sole carbon source and obtaining its energy from the oxidation of inorganic compounds such as ammonia, nitrite, or sulfur.

Autotrophic organisms that obtain energy by the oxidation of inorganic compounds.

Auxotroph mutant strains of bacteria that require growth factors that the parent strain, or prototroph, can synthesize.

Avirulent lacking disease-producing ability.

Axial filament fibrillar structure that spirals around the organism, is attached at each end of a spirochete.

B cell lymphocyte that differentiates into an antibody-producing plasma cell.

Babes-Ernst bodies metachromatic granules, or volutin; granules of long chains of inorganic phosphate; are characteristic of *Corynebacterium diphtheriae*.

Bacillus a cylindrical or rod-shaped bacterium; *Bacillus* is genus name for aerobic spore-forming rods.

Bacillus anthracis aerobic spore-forming rod; the causative agent of anthrax.

Bacteremia the presence of bacteria in the blood.

Bactericidal bacteria-killing.

Bactericide agent that kills bacteria but not necessarily endospores.

Bacteriology the science that deals with bacteria.

Bacteriophage bacterial virus; following maturation, the phage is released as a result of lysis of the bacterial cells.

Bacteriostatic inhibiting growth or multiplication of bacteria.

Balantidium coli protozoan ciliate; causative agent of balantidiasis, a type of dysentery.

Basidiomycetes filamentous fungi that form their external spores externally on a stalk called a basidium.

Basidiospore sexual spore characteristic of fungi of the class Basidiomycetes; spores are produced externally on a club-shaped cell, the basidium.

Basophils leukocytes that are involved in IgE-mediated hypersensitivities and are characterized by being stained with basic dyes.

BCG vaccine bacillus of Calmette and Guérin; vaccine containing the living attenuated bovine strain of *Mycobacterium tuberculosis*.

Bdellovibrio very small microscopic procaryotic cells that parasitize other bacteria.

Benign mild; not malignant.

Bile solubility the property of undergoing lysis in the presence of surface-active agents like bile; characteristic of the pneumococcus.

Binary fission asexual method of reproduction; splitting of a cell into two similar cells.

Binomial nomenclature scientific method for naming organisms, using a genus and a species name.

Bioconversion usually refers to the microbial conversion of waste products into usable compounds.

Biological oxidation any chemical reaction occurring within a cell that results in a release of energy. Oxidation results from the addition of oxygen, or the loss of hydrogen and electrons.

Biological oxygen demand (BOD) the amount of oxygen required to oxidize the organic matter in a given quantity of water. It is thus a measure of the amount of organic matter present in the water.

BOD *see* Biological oxygen demand.

Blastospore asexual spore, produced by budding.

Blocking antibodies antibodies that act to intercept the allergen from the IgE antibody. The allergen is neutralized by the IgG or IgA before it can react with the IgE antibody type. This term is also used to describe nonagglutinating antibodies that may be present in patients with brucellosis.

Blood group antigens genetically determined antigens found on the surface of red blood cells.

Boil furuncle; localized pyogenic infection; frequently caused by *Staphylococcus aureus*.

Booster dose additional injection of an antigen to maintain antibody production at its peak.

Bordetella pertussis Bordet-Gengou bacillus; gram-negative rod, the causative agent of whooping cough.

Borrelia vincentii a spirochete found in the mouth in cases of Vincent's angina. It occurs together with *Fusobacterium fusiforme*.

Botulism food poisoning, caused by the exotoxin of *Clostridium botulinum*.

Bradykinin a peptide released from most cells in an IgE-allergen reaction; causes contraction of smooth muscles, increased vascular permeability, and a decrease in blood pressure.

Breed count direct microscopic count of bacteria in milk.

Broad-spectrum antibiotic effective against wide variety of gram-positive and gram-negative bacteria.

Bromthymol blue a dye used as a pH indicator; pH range is 6.0 to 7.6.

Brood capsules secondary internal cysts occurring within a hydatid cyst. They occur during infections by *Echinococcus granulosus*.

Brownian movement random movement of bacterial cells and small particles due to bombardment by water molecules.

Brucella species small, gram-negative, coccoid to rod-shaped cells; causative agents of brucellosis in animals and humans.

Brucellergen concentrated culture medium after growth of *Brucella;* used to elicit skin reaction in diagnosis of brucellosis.

Brucellosis an infection caused by a bacterium in the genus *Brucella*.

Buboes enlarged lymph nodes seen in bubonic plague and lymphogranuloma venereum.

Bubonic plague an infection caused by *Yersinia pestis* and spread to humans from infected rat fleas.

Budding asexual method of reproduction in which a small bud (blastospore) develops on the parent cell; typical of yeast.

Buffer a substance containing the salt of a weak acid or a weak base that tends to resist pH change when either acid or alkali is added.

Bursa of Fabricius avian organ from which B cells originate.

Burst size the number of phage particles released from a bacterium after maturation of the phage. The burst size is usually in the range of 50 to 200 new bacteriophages.

Burst time the number of minutes required from attachment of the phage to its bacterial host until the liberation of mature phages. The average burst time is between 20 and 40 min.

Cachectin a cytokine secreted by macrophages that have been stimulated with endotoxin. Also called tumor necrosis factor alpha.

Calculus a rough, stony crust consisting of calcium deposited on existing tooth plaque.

Candida albicans yeastlike fungus; a member of the normal flora of humans, and associated with certain infections, e.g., thrush.

Capsid protein coat of a virus.

Capsomers repeating structural subunits of protein that make up the capsid in viruses.

Capsule a mucilaginous envelope surrounding certain microorganisms, external to the cell wall; frequently consists of polysaccharide.

Carboxide a gaseous mixture of 10 percent ethylene oxide and 90 percent carbon dioxide.

Carbuncle enlarged skin abscess.

Carcinogen chemical substance that induces cancer.

Cardiolipin a phospholipid extracted from beef heart that is used as an antigen in nonspecific tests for syphilis.

Caries tooth decay.

Carrier a person who harbors and spreads disease-producing microorganisms without having the overt symptoms of the disease.

Caseation the process by which a tubercular lesion develops into a coagulated, cheesy mass.

Catabolism metabolic process consisting of the ultimate oxidation of a substrate, accompanied by the release of energy.

Catalase an enzyme that catalyzes the breakdown of hydrogen peroxide to yield water and oxygen.

Catalyst a substance that alters the speed of a chemical reaction without itself being changed in the process.

Catheter tubular, flexible device for withdrawing fluid from body cavity, especially urine.

Cation a positively charged ion.

Cell fission asexual method of reproduction whereby a cell splits to form two cells.

Cell-mediated hypersensitivity allergic reactions carried out by sensitized T cells.

Cell-mediated immunity immune reactions between sensitized T cells and an infectious agent or a foreign antigen on the surface of a host cell.

Cell-mediated toxicity killing of cells by cytotoxic T cells.

Cell-specific a condition in which the effect produced is directed to cells of a particular species; e.g., interferon inhibits viral replication most effectively in the species in which it was produced—it is cell-specific.

Cell wall peptidoglycan a structural component of the bacterial cell wall that provides the rigidity necessary to maintain the integrity of the cell. It is a very large molecule made up of *N*-acetylglucosamine and *N*-acetylmuramic acid to which is attached a pentapeptide. Bridges of amino acids cross-connect the pentapeptides.

Cellular immunity immunity developed by certain cells that have acquired an increased ability to ingest or destroy specific foreign substances.

Cellulases hydrolytic enzymes that degrade cellulose to glucose; also called a β glucosidase because it hydrolyzes the β linkages of glucose that make up cellulose.

Cellulitis rapidly spreading infection, involving the skin and the subcutaneous and lymphatic tissues; can be caused by a β-hemolytic streptococcus.

Cercaria the final larval stage of a trematode, resembling the adult worm in appearance.

Chancre initial lesion in syphilis; primary stage.

Chemoautotroph an organism that obtains its energy by chemical oxidations of inorganic substances and that has the ability to fix carbon dioxide as its sole source of carbon.

Chemolithotroph synonym of autotroph.

Chemostat a growth chamber for maintaining a bacterial culture in the logarithmic phase of growth by the continual addition of fresh medium.

Chemosynthetic the ability to trap energy biologically from chemical oxidations.

Chemotaxis movement of organisms in response to a chemical substance; the attraction of leukocytes to an area of injury is due to a release of a chemical substance from the injured tissue.

Chemotherapy treatment of disease with chemical compounds.

Chitin polysaccharide containing glucosamine; characteristic of cell walls of some fungi; also found in insects.

Chlamydiae obligate intracellular parasites that appear to be related to gram-negative bacteria and that have a complex method of reproduction that sets them apart from Rickettsiae; they are not transmitted by an arthropod vector.

Chlamydoconidia a thick-walled, asexual fungal spore.

Chlamydospore resistant asexual spore characteristic of some fungi.

Chlorophyll a green pigment essential for photosynthesis.

Chloroplast internal membranous structure of photosynthetic plants that contains chlorophyll.

Choleragen the exotoxin secreted by *Vibrio cholerae*, causing an increase in cAMP levels.

Chromatin body nuclear material found in the procaryotic cell.

Chromatophores discrete vesicles or special membrane systems that contain photosynthetic pigments in procaryotic cells.

Chromosome strands of DNA that contain the major heritable properties of a cell.

Chronic of long duration; not acute.

-cide [suffix] killing effect.

Cilia (*sing.* **cilium**) short, hairlike structures that provide for movement; characteristic of protozoa of the class Ciliata.

Citric acid cycle a system of reactions that converts pyruvic acid to carbon dioxide (and water) with the energy captured in ATP molecules.

Clonal deletion a proposal that during neonatal life all clones of B or T cells that recognize self are eliminated.

Clone cells derived from a single cell.

Clostridium genus name, refers to anaerobic spore-forming rods.

Clostridium botulinum anaerobic spore-forming rod, gram-positive; causative agent of botulism.

Clostridium perfringens anaerobic spore-forming rod, gram-positive; the most common causative agent associated with cases of gas gangrene; frequent cause of food poisoning.

Clostridium tetani anaerobic spore-forming rod, with terminal spores, gram-positive; its exotoxin causes tetanus, or lockjaw.

Coagulase an enzyme that, in concert with certain serum factors, causes blood plasma to clot; produced by *Staphylococcus aureus*.

Coagulation the formation of a mass or clot.

Coccidioidin test skin sensitivity test used as an aid in the diagnosis of infections caused by *Coccidioides immitis*. The injected material consists of dilution of a culture filtrate of *Coccidioides immitis*.

Coccobacillus oval bacterium that is intermediate between the coccus and bacillus forms.

Coccus (*pl.* **cocci**) spherical bacterium.

Codon a series of three consecutive nucleotides in a nucleic acid molecule that codes for one specific amino acid in a polypeptide chain.

Coenocytic hypha a hypha lacking septations or cross walls.

Coenzymes small organic nonprotein carrier molecules; may carry a portion of a substrate or electrons to a next or final acceptor.

Colicins protein substances produced by some enteric bacteria; they are lethal to some other strains of enteric bacteria.

Coliform bacteria small, aerobic, gram-negative, non-spore-forming bacilli; ferment lactose, with acid and gas production, normal inhabitants of lower intestine.

Collagenase enzyme that causes breakdown of collagen, the ground substance of bone, skin, and cartilage; may aid in the spread of the pathogen from the initial site of entry.

Colonies clumps of microorganisms that developed from single cells or groups of cells; visible to the naked eye on solid or semisolid medium.

Colonization factor surface structures occurring on a gram-negative bacteria that are responsible for their adherence to epithelial cells.

Columella the rounded apex of the sporangiophore within the sporangium of fungi of the class Zygomycetes.

Commensal relationship of coexistence in which the parasite receives benefit and the host is unaffected.

Communicable disease a disease that is transmitted directly or indirectly from one person to another.

Competence transitory period during which the bacterium is capable of taking up transforming DNA.

Competitive inhibition interference with a synthetic process due to an inhibitor that combines reversibly with the active site on the enzyme; e.g., sulfonamides inhibit folic acid synthesis.

Complement a blood serum protein group involved in certain antibody-antigen reactions.

Complement system a complex series of normal serum proteins that, when activated by antibody-antigen reactions or by the presence of certain foreign proteins, can result in the lysis or phagocytosis of foreign cells or bacteria.

Complex medium a medium containing undefined macromolecules such as serum, yeast extracts, or other large proteins.

Compound microscope a microscope that has two sets of lenses, one in the objective next to the object to be studied, and the other in the ocular next to the eye.

Concurrent disinfection disinfection procedures carried out throughout the course of an illness with rigid aseptic disposal of contaminated materials.

Conditional lethal mutant an organism having a mutation that will permit growth under one set of conditions but not under other situations in which the parent strain could grow.

Congenital acquired while in utero.

Conidiophore the hyphae of certain fungi that support and bear the conidia.

Conidium (*pl. conidia*) an asexual spore, not enclosed, borne on certain hyphae.

Conjugation sexual method of reproduction involving contact between mating cells; in bacteria, genetic material is transferred from donor to recipient cell through actual contact of the two cells.

Constitutive genes genes that are continually expressed.

Constitutive enzyme enzyme that is always present in the cell regardless of growth conditions.

Contact inhibition the observation that tissue cells growing in a monolayer will stop growing when they come into contact with other cells.

Contagious transmitted directly from one person to another.

Contaminated water water that contains either poisonous chemicals or pathogenic organisms.

Contamination undesirable organisms in a culture or other material.

Continuous cell line animal tissue cells that are propagated in a manner analogous to that used for a bacterial culture and that can be used for virus replication.

Continuous disinfection concurrent disinfection; disinfection procedures carried out throughout the course of an illness with careful disposal of contaminated materials.

Convalescence the period of recovery from a disease.

Coordinated enzyme repression the process or condition whereby the synthesis of all of the enzymes involved in the biosynthesis of a particular molecule are simultaneously repressed.

Coracidium a motile, embryonic stage of *Diphyllobothrium latum*. It is the freshwater form ingested by a crustacean.

Cord factor a substance formed by virulent tubercle bacilli that causes the bacteria to grow in a serpentine, cordlike formation.

Corynebacterium diphtheriae gram-positive, pleomorphic rod that shows granular staining; exotoxin produced by lysogenic cultures causes the symptoms of diphtheria.

Covalent bond one in which electrons are shared by two atoms.

Coxiella burnetii a species of rickettsia; the causative agent of Q fever.

Creutzfeldt-Jakob disease a rare viral disease causing presenile dementia.

Cryptococcus neoformans yeastlike organism; causative agent of cryptococcosis.

Culture a medium containing living microorganisms.

Cyanobacteria photosynthetic procaryotic organisms containing chlorophyll.

Cyclic photophosphorylation the overall system in which an electron leaves chlorophyll and, after passing through a series of electron acceptors, returns to chlorophyll. During passage of the electrons from ferredoxin back to chlorophyll, two molecules of ATP are formed from ADP for each pair of electrons passing through the cycle.

Cyclosporin an immunosuppressive drug used to prevent organ transplant rejection.

Cyst encystment; the cell forms a heavy protective coating and becomes very resistant to adverse conditions; characteristic of many protozoa.

Cysticercus the larval tapeworm stage existing in the muscles of the intermediate host. Humans become infected by ingestion of this larval stage.

Cystitis infection of the bladder.

Cytopathic effect destruction of tissue culture cells by an infecting virus.

Cytoplasm the living cell substance within the cell membrane and surrounding the nucleus.

Cytoplasmic streaming continuous movement of cytoplasm within the cell that results in a constant distribution of intracellular contents; provides amoebic motility to some types of cells.

Cytotoxic reaction allergic reaction directed against antigens on the cell surface.

Cytotoxic T cells subgroup of T cells that bind to a specific antigen on a foreign cell and kill the cell.

Dalton weight of a single hydrogen atom.

Dane particle hepatitis B virion.

Deaminase an enzyme that catalyzes the removal of an amino group from a molecule with the liberation of ammonia; the process is deamination.

Death curve the graph that shows the rate at which a microbial population dies. It can be plotted as a logarithm of the number of survivors at any given time. The curve shows that death occurs as a first-order reaction.

Death phase phase in growth of a culture when reproduction usually will have stopped. As the rate of death exceeds the rate of reproduction, the actual number of viable microorganisms declines.

Debilitated weakened.

Decarboxylase an enzyme that catalyzes the removal of carbon dioxide from the carboxyl group of an acid; the process is decarboxylation.

Decay aerobic breakdown of complex materials by microbial enzymes.

Defective viruses viruses lacking one or more genes necessary for replication; they can replicate only when a cell is simultaneously infected with a helper virus that provides the missing functions.

Defined medium a medium containing only constituents whose specific chemical formulas are known.

Definitive host the animal in which the larval stage of a helminth develops into an adult worm.

Degranulation fusion of lysosomal granules in a polymorphonuclear leukocyte into a phagosome.

Dehydrogenases oxidoreductases, which act by removing hydrogen and electrons from a substrate.

Dehydrogenation the usual type of biological oxidation that occurs in microorganisms. The enzyme dehydrogenase causes the hydrogen donor to lose part of its hydrogen, and as hydrogen and electrons are transferred energy is released.

Delayed hypersensitivity an allergic or hypersensitive reaction to a foreign substance resulting from a cellular immunity to that substance.

Denatured the destruction of a protein's tertiary structure, usually by heat or acid.

Denitrification reduction of nitrogen compounds with the liberation of free nitrogen.

Dental caries dental cavities.

Dental plaque dense masses of bacteria and glycoproteins that adhere to the surface of a tooth.

Deoxyribonucleic acid (DNA) the nucleic acid primarily found in the nucleus of cells; it contains adenine, guanine, cytosine, thymine, deoxyribose, and phosphoric acid.

Deoxyribose a five-carbon sugar found in DNA; it contains one oxygen atom less than ribose.

Dermatophytes fungi that infect superficial tissues—the skin, hair, and nails.

Desiccation drying, or removal of water.

Detergent a surface-active agent used in cleaning.

Determinant group the portion of an antigen that specifically combines with the antibody.

Deuteromycetes Fungi Imperfecti; fungi in which no sexual stage has been recognized.

Dextran polymer of glucose.

Diabetes disease in which the pancreas fails to produce insulin.

Diatoms unicellular eucaryotic marine algae characterized by their rigid silica-containing cell walls.

Differential medium a medium that shows different color reactions due to the enzyme activity of different kinds of bacteria; e.g., Hektoen agar will be changed in color by lactose fermenters but unchanged by non-lactose fermenters.

Differential stains stains that reveal chemical differences in bacterial structure; more than one stain is employed.

Digest to break down enzymatically into smaller units.

Digestion the process whereby large, complex molecules of protein, carbohydrate, and lipid are hydrolyzed to simple molecules that go readily into solution and can enter the cell.

Dimorphic occurring in two forms; characteristic of certain fungi, growing in the yeast phase under one set of conditions and filamentous under other conditions.

Dipeptide two amino acids joined together by a peptide bond.

Dipicolinic acid component of bacterial endospores.

Diplococci cocci that occur in pairs.

Diploid possessing two sets of duplicate chromosomes.

Diploid cell strains cell cultures derived from human embryonic tissue capable of continued growth for 40 to 50 subcultures.

Disaccharide compound sugar in which two monosaccharides are linked via a glycosidic bond.

Disease any change from a stage of health; interruption in the normal functioning of a body structure.

Disinfectant a chemical agent that destroys disease organisms; used on inanimate objects.

Disinfection the process of destroying pathogenic agents by the use of disinfectants.

Disulfide bond bond between two sulfur atoms, usually within a protein.

DNA *see* Deoxyribonucleic acid.

DPT vaccine containing diphtheria and tetanus toxoids and whole killed cells of *Bordetella pertussis*.

Droplet infection infection acquired from recently contaminated air via suspended droplets.

Droplet nuclei small droplets that lose moisture and remain suspended in the air, and that contain living microorganisms.

Dysentery gastrointestinal infection characterized by bloody, mucoid diarrhea.

Echovirus an acronym for *e*nteric *c*ytopathic *h*uman *o*rphan viruses.

Eclipse period the time between a viral infection of a cell and the production of completed virions.

Ecology study of the relationship of the organism to its environment.

Edema swelling due to the accumulation of fluid.

Elementary body the small infectious chlamydial cell.

ELISA enzyme-linked immunosorbant assay.

Embden-Meyerhof pathway a series of reactions for the conversion of glucose to pyruvic acid. There is a net yield of two ATP molecules per molecule of glucose degraded.

Encephalitis inflammation of the brain.

Endemic constantly present in a locality.

Endocarditis inflammation of the heart valves or chambers.

Endocytosis uptake of material into a eucaryotic cell.

Endoenzymes intracellular enzymes, concerned with cell synthesis and energy production.

Endospore minute, highly durable body developed within certain bacterial cells and capable of developing into new vegetative cells; characteristic of the genera *Bacillus*, *Clostridium*, *Sporosarcina*, *Sporolactobacillus*, and *Desulfotomaculum*.

Endotoxins large molecules of lipopolysaccharide that are normal components of cell walls of gram-negative bacteria; may cause fever and shock during infection.

Energy force, power, or the capacity for work.

Ensilage ground-up corn and corn stalks that have undergone a lactic acid fermentation.

Enteric relating to the intestinal tract.

Enteric cytopathic human orphan human enteric RNA picornaviruses that were originally thought not to cause disease.

Enteritis inflammation of the intestine.

Enteroinvasive *Escherichia coli* strains of *Escherichia coli* that invade and grow intracellularly in the epithelial cells lining the small intestine, characterized by the secretion of a Shigalike toxin.

Enteropathogenic *Escherichia coli* strains of *Escherichia coli* that adhere to intestinal epithelium causing diarrhea by the secretion of a Shigalike toxin.

Enterotoxigenic *Escherichia coli* those strains of *Escherichia coli* that cause diarrhea by the secretion of one or two toxins designated as LT or ST.

Enterotoxin exotoxin released in food by certain microorganisms; causes a severe reaction in the gastrointestinal tract.

Enzyme repression inhibition of enzyme synthesis at the genetic level.

Enzymes organic catalysts, protein in nature; speed up chemical reactions.

Eosinophils leukocytes involved in allergic reactions as well as in immunity to helminth infections.

Epidemic disease found in a large number of individuals in a community within a short period.

Epidemiology systematic study of the various factors that determine the spread of pathogenic organisms within a human community.

Episomes genetic material within the cell that may replicate outside the chromosome but may also become integrated in the chromosome.

Epitope portion of an antigen that actually binds to its antibody.

Ergot a fungal toxin.

Erysipelas acute infection of the skin that results in characteristic red, edematous lesions; found most commonly on the face and legs; caused by a group A β-hemolytic streptococcus.

Erythema red area on skin.

Erythrogenic toxin streptococcal product that causes rash of scarlet fever.

Essential metabolite a substance required by a microorganism; must be supplied in a growth medium.

Ester bond the bond between the carboxyl carbon of a fatty acid and an oxygen molecule, usually on a sugar residue.

Etiology the cause of a disease.

Eucaryotic possessing a true nucleus with a well-defined nuclear membrane.

Eumycetes true fungi; yeasts and molds.

Excision repair an enzyme system that can repair DNA by the elimination of ultraviolet-light-induced thymine dimers.

Exoenzymes hydrolases, or digestive juices, excreted into the environment.

Exotoxins toxic substances excreted from the bacterial cell.

Exudate fluid, often from formed elements of the blood, discharged from tissues to a surface or cavity.

Fab fragment portion of immunoglobulin containing one light chain and one half heavy chain.

Facultative aerobes organisms that will use oxygen as a final electron acceptor if it is available but will grow fermentatively in its absence.

Fc fragment fragment crystallized; contains carboxy end of heavy chain.

Feedback inhibition temporary inactivation of an enzyme due to the end product of a series of reactions reacting directly with and inhibiting the activity of the enzyme of the first reaction.

Fermentation anaerobic breakdown of carbohydrates resulting in the formation of stable fermentation products; organic compounds are final electron acceptors in this energy-yielding process.

F factor fertility factor necessary for fusion between enteric organisms.

Filterable virus a virus that can pass through filters that do not allow bacteria to pass.

Filtration the removal of substances by filtering through any one of several types of filters. The pore size is usually small enough to prevent the passage of bacteria.

Fimbriae pili; filamentous appendages, shorter, straighter, and considerably smaller than flagella.

Flagellum (*pl.* **flagella**) slender, threadlike appendage, generally several times the length of the cell; provides for motility.

Flocculent masses of cells floating, or settled to the bottom, in a liquid medium.

Flora microorganisms typically found in a given environment.

Fluctuation test test to determine the development of mutants in a culture. The test is based on the concept that mutation is a completely random event, and that spontaneous mutation would result in a large fluctuation in the number of mutants in a series of identical cultures.

Flukes general name for trematodes.

Fluorescence light emitted by a substance after absorption of radiation of a shorter wavelength from another source.

Fluorescence microscopy the staining of microorganisms with a fluorescent dye for microscopic study, using ultraviolet light as the illumination source.

Fluorescent antibody antibody to which a fluorescent dye has been conjugated.

Fluorosis mottling of tooth enamel due to excessive amounts of fluoride in water.

Focal infection local infection that acts as a nucleus for the spread of organisms to other sites.

Fomites objects contaminated with pathogenic microorganisms.

Food infection a gastrointestinal disturbance that results from eating food contaminated with certain living organisms, frequently *Salmonella*.

Food poisoning a gastrointestinal disturbance due to ingestion of food containing toxins or poisonous substances that have been excreted into the food

by certain species of bacteria, frequently *Staphylococcus aureus*, less commonly *Clostridium botulinum* and *Clostridium perfringens*.

Forespore a structure that develops in spore-forming bacteria after the active growth period. It later becomes refractile, forms a thick wall, and becomes the resistant endospore.

Formalin formaldehyde in a 37 to 40 percent aqueous solution.

Fraction sterilization tyndallization. Sterilization of culture medium by exposure to steam for 30 min each day for 3 successive days to kill vegetative cells.

Frame-shift mutation loss of one or two nucleotide bases causing a shift of one or two bases in each triplet code in the mRNA.

Free-living nitrogen fixers organisms that do not require a host in order to grow and fix nitrogen.

Freeze-drying a technique in which microorganisms are frozen rapidly and then dehydrated in high vacuum directly from the frozen state.

Frei test a skin test to determine sensitivity to the chlamydiae that cause lymphogranuloma venereum.

Freund's adjuvant a mixture of mineral oil and killed tubercle bacilli that is mixed with an antigen, resulting in a higher stimulation of antibody response.

FTA-ABS test a serologic test for antibodies to *Treponema pallidum* employing whole cells of *T. pallidum* and fluorescently labeled antihuman immunoglobulin.

Fungi Eumycetes; eucaryotic cells, frequently filamentous, that lack both a vascular system and chlorophyll; includes molds and yeasts.

Fungicide chemical agent that destroys fungi.

Fusiform spindle-shaped, or tapering at the ends; characteristic of *Fusobacterium fusiforme*.

Gamete a sex cell; gametes fuse in sexual reproduction.

Gamma globulin globulin fraction from pooled human serum; contains preformed antibodies.

Gangrene areas of necrotic tissue resulting from loss of blood supply.

Gastroenteritis inflammation of the stomach and intestines.

Gelatinase a hydrolytic enzyme that degrades the protein gelatin.

Gene DNA sequence that encodes for a specific product.

Generalized transduction the packaging of any bacterial DNA during phage maturation.

Generation time time required for a population to double.

Genes carriers of hereditary information; located on chromosomes.

Genetic engineering the use of recombinant DNA techniques to place functional foreign DNA into a bacterium or yeast cell.

Genotype genetic makeup of an organism.

Genus a group of related species.

Germicide a chemical agent that kills bacteria but not necessarily endospores; generally considered synonymous with *bactericide*.

Globulin protein found in blood serum; antibodies are made up of globulins.

Glomerulonephritis disease of kidney, characterized by inflammatory changes in glomeruli; may occur as a result of an autoimmune disease in which the streptococci induce manufacture of antibodies against the individual's own kidney tissue.

Glucan a polymer of glucose molecules.

Glycocalyx a term describing a poorly bound capsule.

Glucose carbohydrate, six-carbon monosaccharide.

Glycogen polysaccharide which, on hydrolysis, yields glucose.

Glycolysis anaerobic dissimilation of glucose to lactic acid.

Glycoprotein a protein to which is attached chains of carbohydrate.

Glycosidic bond a bond joining two sugar residues together.

Golgi apparatus stacks of internal membranes whose function it is to add carbohydrate to proteins prior to their secretion from the cell.

Graft-versus-host reaction the immune response of transplanted lymphocytes toward cells in the recipient.

Gram-negative describing bacteria that possess a unimolecular peptidoglycan cell wall bounded on one side by the cytoplasmic membrane and on the other by the outer membrane; such cells are decolorized by 95 percent alcohol during the Gram-staining procedure.

Gram-positive describing bacteria that possess a thick peptidoglycan cell wall that will retain the initial Gram stain during washing with 95 percent alcohol.

Gram stain differential stain that provides for grouping of bacteria as either gram-positive or gram-negative. If the organism retains the initial stain after contact with a decolorizer, it is gram-positive; if the organism loses the stain after exposure to a decolorizer, it is gram-negative.

Granulocyte leukocytes (mostly polymorphonuclear) containing lysosomal granules.

Growth curve representation of population changes in the growth of a culture.

Growth factor a nutrient required by an organism which it is unable to synthesize.

Guillain-Barré syndrome characterized by muscle weakness and paralysis. Etiology is unknown, but a number of cases followed influenza immunizations.

Guinea worm common term for *Dracunculus medinensis*.

H antigen flagellar antigen of bacteria; (H from *hauch*, Ger., "breath").

Haemophilus "blood-loving"; genus name.

Halophilic "salt-loving"; growth is dependent on or enhanced by high salt concentration.

Hanging-drop preparation a method for observing bacterial motility in a liquid medium.

Haploid single set of chromosomes in nucleus; characteristic of a mature gamete, or sex cell.

Hapten incomplete or partial antigen.

Healthy carrier a person who harbors and spreads pathogenic microorganisms without giving evidence of disease.

Heat-labile inactivated or destroyed readily by heat, usually 56°C for 3 min.

Heat-stable heat resistant, or not readily inactivated by heat.

Helminth parasitic worms.

Helper T cells lymphocytes that develop in the thymus and help B cells differentiate into antibody-producing plasma cells.

HLA complex human major histocompatibility complex.

Hemagglutination agglutination or clumping of red blood cells.

Hemagglutination inhibition test used in identification of viruses that cause red blood cells to agglutinate. If specific antibodies to the virus are present, agglutination is inhibited.

Hemoglobin an iron protein pigment in red blood cells; serves as an oxygen carrier.

Hemolysins substances that cause lysis of red blood cells; e.g., exotoxins or antibodies.

Hemolysis lysis, or disintegration, of red blood cells.

Hepatitis inflammation of the liver; frequently due to a virus.

Hetero- prefix, meaning different.

Heterofermentative lactic acid bacteria gram-positive bacteria that ferment carbohydrates, forming end products consisting primarily of lactic acid, acetic acid, ethanol, and carbon dioxide.

Heterogamy unlike gametes.

Heterolactic acid fermentation a series of reactions for glucose dissimilation with end products of lactic acid, acetic acid, ethanol, and carbon dioxide.

Heterophile antibodies antibodies that combine with antigens from unrelated species; e.g., found in infectious mononucleosis.

Heterophile antigens antigens present in various organs or tissues of humans and animals that can react with antigens from unrelated species; e.g., Forssman antigens.

Heterotroph an organism capable of obtaining its energy from the chemical oxidation of organic substrates.

Heterotrophic "nourished by others"; requiring organic nutrient material.

Hexose monophosphate shunt a series of reactions for glucose dissimilation with intermediates varying from two to seven carbons.

Hfr strain high frequency mating strain; e.g., strains of *Escherichia coli* that produced a thousand times more recombinants were isolated from F+ cultures, and these new donor strains, in which the F factor had been integrated into the chromosome, were called Hfr strains.

Histamine a pharmacologically active compound released from mast cells when an allergen binds to surface-bound IgE.

Histoplasmin filtrate from cultures of *Histoplasma capsulatum;* used in hypersensitivity skin tests as an aid in diagnosis of histoplasmosis.

HLA complex human leukocyte antigens existing as cell surface antigens that are encoded in the major histocompatibility complex genes.

Holophytic nutrition typical of plants; nutrients are in solution.

Holozoic nutrition typical of animals; digesting solid food.

Homofermentative lactic acid bacteria gram-positive bacteria that ferment carbohydrates forming lactic acid as a sole end product.

Homograft transplant of tissues from one member to another of the same species; does not refer to identical twins.

Host an organism that supports a parasite.

Human T-cell lymphoma viruses retroviruses that are the etiological agents of a malignancy known as T-cell lymphoma.

Humoral immunity antibody-mediated immunity.

Humus organic material in soil after decomposition by microorganisms.

Hyaluronidase "spreading factor"; enzyme capable of degrading hyaluronic acid, the intracellular material of connective tissues; may help the spread of the invading organism in the tissues.

Hybridoma antibody-secreting cells that will grow in vitro; they are made by fusing spleen cells with malignant myeloma cells.

Hydatid cyst a fluid-filled bladder containing many scoleces of *Echinococcus granulosis.*

Hydrogen acceptor a substance that accepts the hydrogen and electrons from the appropriate coenzyme in the dehydrogenation process.

Hydrogen donor the cellular substance that donates the hydrogen and electrons when it is oxidized.

Hydrolases enzymes that break chemical bonds by the addition of water; hydrolyze large molecules into smaller usable components.

Hydrolysis splitting a molecule by the addition of water.

Hypersensitivity increased sensitivity, or allergy, to foreign material.

Hyphae filaments of a mold.

Icosahedron 20-sided figure—characteristic shape of many viruses.

Immediate hypersensitivity allergic reactions due to humoral antibodies.

Immune complex circulating aggregates of antibodies and antigens.

Immune hemolysis red blood cell hemolysis in the presence of specific antibody and complement.

Immune serum antiserum; blood serum containing antibodies.

Immune surveillance a theory that our cellular immune system continually surveys all cells in the body, destroying those acquiring new antigenic determinants, such as malignant cells.

Immunity state of being highly resistant to a specific pathogenic organism; the increased resistance to infection.

Immunogen antigen.

Immunoglobulins (Ig) general name for antibodies.

Immunology a study of the mechanisms of resistance to disease.

Immunosuppression suppression of the immune response.

IMViC differential tests used to separate *Escherichia* and *Enterobacter* in the analysis of water.

Inapparent infection an illness in which symptoms are so mild that it goes undetected and thus, undiagnosed.

Incineration the process of burning to ashes; e.g., sterilizing the inoculating loop in the Bunsen burner.

Inclusion bodies round or oval bodies found in the nucleus or cytoplasm of cells during virus infections; may contain viral particles in some cases.

Incubation maintenance of cultures under conditions favorable for growth.

Incubation period the time between infection and the appearance of symptoms.

Indicator a substance that can be used to reflect pH changes by changing color.

Induced enzymes enzymes synthesized only in the presence of a specific substrate or inducer.

Induced mutants mutants resulting from the use of a mutagenic agent.

Inducible enzymes enzymes that are synthesized only in response to a specific inducer.

Industrial microbiology the branch having to do with

the use of microorganisms for commercial and/or trade purposes.

Inert not active.

Infection invasion of the body by pathogenic microorganisms.

Infectious capacity to be transmitted and produce an infection.

Inflammation the body's response to injury resulting in a localized reaction.

Infusion the clear liquid obtained after boiling ground meat with water and filtering off the solid material.

Initial body reproductive chlamydial cell.

Inoculation the transfer of microbial cells from a culture to a sterile medium by means of a sterile inoculating needle or loop; also, the introduction of microorganisms into the animal body, as in a vaccine.

Inoculum microorganisms used for transfer into a medium.

Insertion sequences discrete sequences present in chromosomal or plasmid DNA, which have the ability to move as a unit.

Interferon antiviral soluble protein substance, produced by cells infected with almost any animal virus; it is cell specific but not virus specific.

Intermediate host animals in which the helminth eggs develop into a larval stage.

Invasiveness the property that enables an organism to overcome the normal body defenses and cause an overt infection.

In vitro "in glass," as compared with in vivo; performed in a test tube or other glassware.

In vivo in the living organism.

Involution forms abnormal forms that appear in old cultures of bacteria; they are likely to assume various odd shapes and may swell or show rudimentary branching.

Iodophor iodine complexed to a detergent.

Isoagglutination agglutination of blood cells following a blood transfusion by a different donor blood group. The result is agglutination of the person's own blood cells by serum of the donor or agglutination of the donor's cells by the person's own serum.

Isoagglutinins antibodies, found in the blood serum of all persons, against the major agglutinogens or antigens that the person does not have in his or her own red blood cells.

Isoantibodies antibodies formed in response to isoantigens.

Isoantigens genetically determined antigens that vary among individuals of the same species; common examples are the A and B antigens on human red blood cells.

Isografts grafts between genetically identical animals.

Isomerases enzymes that catalyze the transfer of two hydrogen atoms from an adjacent carbon to an aldehyde group. This results in the reduction of the aldehyde group to a primary alcohol group, and the formation of a keto group on the carbon atom that donated its hydrogen atoms.

Isomers molecules with the same chemical formula but different structures.

Isoniazid structural analog of pyridoxine; used in the treatment of tuberculosis.

Jaundice bile pigments in blood, resulting in a yellow skin color.

K antigen enteric capsular antigen.

Kahn test precipitin test for diagnosis of syphilis, using individual's blood serum.

Keratin insoluble protein found in hair, skin, and nails.

Killer T cells T cells that exert a cytotoxic activity toward antibody-coated target cells.

Kline test a slide precipitin test for the diagnosis of syphilis, using individual's blood serum.

Koch's postulates a series of rules to establish the etiology of an infection.

Koplik spots bluish-white specks on mouth lesions in early measles, before skin eruption.

Krebs cycle *see* Citric acid cycle.

Kuru a degenerative disease of the central nervous system caused by an as-yet unidentified infectious agent.

Labile easily destroyed, usually by heat.

Lactic acid bacteria gram-positive rods or cocci that ferment carbohydrates in which lactic acid is a sole or major end product.

Lactose a disaccharide that yields glucose and galactose on hydrolysis; milk sugar [β-galactosyl (1–4) glucose].

Lag phase the period of adjustment to a medium after the inoculation of a culture.

Latent inapparent; not evident.

LD$_{50}$ dose of a substance that will kill 50 percent of the recipients within a given time.

Lecithinase alpha toxin; toxin that hydrolyzes the phospholipid lecithin, a component of cell membranes; produced by *Clostridium perfringens*. Also a component of many snake venoms.

Lepromin extract from *Mycobacterium leprae* used in a skin test for the diagnosis of leprosy.

Lesion an area of injury, or a circumscribed pathological tissue change.

Lethal capable of causing death.

Lethal toxin a series of toxins, secreted by certain strains of *Clostridium perfringens,* that cause death if injected in a laboratory animal.

Leukemia an excessive increase in number of leukocytes due to malignancy.

Leukocidins substances that kill leukocytes.

Leukocyte cells occurring in the blood stream whose function is to remove debris, including bacteria, by engulfing and destroying the foreign material.

Leukocytosis explosive increase in the number of leukocytes in the blood.

Leukopenia decrease in number of circulating leukocytes.

Leukotrienes substances released during the degranulation of mast cells and basophils; they are believed to be a major mediator of human asthma.

L forms bacterial forms devoid of a cell wall; some can revert to normal cells.

Lichen a plant consisting of a symbiotic relationship between a fungus and a photosynthetic alga.

Ligases enzymes that catalyze the linking together of two molecules. They take part in many of the steps involved in the synthesis of macromolecules, such as proteins and nucleic acids. (They are also known as synthetases.)

Light beer beer made from starch that does not contain limit dextrans.

Limit dextrans $\alpha1-6$ linkages in cereal starch that are not hydrolyzed by either the malt enzymes or the yeast.

Lipases enzymes that catalyze the splitting of fats to glycerol and fatty acids.

Lipopolysaccharide outer leaflet of the outer membrane of gram-negative bacteria. Also called endotoxin.

Liquefaction change of gel to a liquid state; e.g., hydrolysis of gelatin due to the proteolytic enzyme gelatinase.

Litmus plant extract, used as a pH indicator, also an oxidation-reduction indicator; it turns blue when alkaline and red when acid in reaction.

Log phase the period of most rapid reproduction in the growth phases of a culture; the generation time is constant when plotted on a graph, and the log of the number of organisms appears as a straight line.

Lyases enzymes that (1) remove groups from substrates nonhydrolytically, usually leaving double bonds, or (2) add groups to double bonds. This most commonly involves the removal of water, ammonia, or a carboxyl group.

Lymphocyte a type of leukocyte arising in lymphoid tissues, and the most important cell in specific immunity.

Lymphokines soluble factors secreted by lymphocytes when they bind to the specific antigen to which they are sensitized.

Lymphoma malignancy of lymphoid cells resulting in a solid tumor.

Lymphotoxin a lymphokine that acts on the plasma membrane of target cells.

Lyophilization freeze-drying, or dehydration from the frozen state.

Lysis disruption, or breaking apart, of cells.

Lysogenic term applied to bacteria that are carrying prophage.

Lysogenic conversion change in the properties of a bacterial cell as a result of becoming lysogenic.

Lysogeny a condition in which the nucleic acid of the bacteriophage exists inside the host bacterium and replicates along with the bacterium's nucleic acid but does not produce mature phage particles.

Lysosomes intracellular membrane vesicles containing a variety of enzymes involved in intracellular digestion.

Lysozyme enzyme that hydrolyzes the bond between *N*-acetylglucosamine and *N*-acetylmuramic acid of the peptidoglycan of the bacterial cell wall; this causes the cell wall to break, and lysis of the cell usually follows.

Lytic cycle the sequence of events that occurs when a phage infects a bacterium. It includes adsorption to a receptor site on the bacterium's cell wall, injection of nucleic acid, replication of phage nucleic acid, the assemblage of complete phage particles, and the breaking open of the bacterium with the release of mature phage particles. The sequence of events terminates with the lysis and death of the bacterial cell.

Macro- prefix, meaning large, in contrast to micro-.

Macrophage large mononuclear phagocyte found in various tissues of the body.

Macrophage-activating factor secreted by delayed-type sensitivity T cells, causing macrophages to become more metabolically active.

Macrophage activation an enhanced bacteriostatic or bactericidal activity acquired through the antigen-mediated release of a lymphokine.

Macroscopic visible to the naked eye, without the use of a microscope.

Macule small, round, colored area on the skin.

Maculopapule raised, small, round, colored area on the skin.

Major histocompatibility complex a single complex locus involved in the regulation of the immune response and encoding for a large number of major histocompatibility antigens.

Malaise general feeling of illness.

Malt sprouted, dried barley containing starch and enzymes to hydrolyze starch to maltose.

Maltase an enzyme that will cleave maltose into two molecules of glucose.

Mantoux test tuberculin skin test. A dilution of purified material prepared from culture filtrates of tubercle bacilli is injected into the skin. An inflamed area at the site of injection that reaches a maximum in about 48 h provides a positive test indicating that the person has been previously infected with the tubercle bacillus.

Mast cell small cells that are widely distributed in connective tissue.

Medical microbiology the branch having to do with disease-producing microorganisms.

Medium a nutrient substance used to grow microorganisms; it may be a liquid medium or a solid medium to which agar has been added.

Meiosis reduction-division process of nuclear division; daughter nuclei have half the chromosome number of the parent nucleus; occurs in the formation of sex cells.

Membrane attack unit the end product of complement activation, C5b–9, which attaches to and forms holes in certain cell membranes.

Meningitis inflammation of the meninges, the membranes covering the brain and spinal cord.

Merozoite a stage in the preerythrocytic cycle of reproduction of the malaria parasite.

Mesophile an organism whose growth temperature is in the range of 20 to 40°C.

Mesosomes relatively large, irregular invaginations of the plasma membrane.

Messenger RNA (mRNA) nucleic acid that carries the message to make the specific protein that had been coded in the DNA segment.

Metabolism the total chemical activities of an organism; consists of anabolism and catabolism.

Metacercaria an encysted trematode stage that, when ingested, provides the only source of infection for humans.

Metachromatic granules see Babes-Ernst bodies.

Metastasis spread of cancer cells from a primary location to other parts of the body.

Michaelis-Menten constant constant used to characterize an enzyme; it expresses the concentration of the substrate that will give half of the maximum velocity and is thus a measure of the affinity of an enzyme for a substrate.

Micro- prefix, meaning small, in contrast to macro-.

Microaerophilic organisms that grow best at an oxygen tension less than that of air.

Microaerophile a microorganism that grows best with reduced oxygen tension.

Microbe a microscopic organism.

Microbiology study of living organisms individually too small to be seen without the aid of a microscope.

Microcephaly abnormally small head.

Micrometer unit of measurement for microorganisms; 1 micrometer (μm) equals 0.001 mm or 1/25,400 in. (*Micrometer* replaces the older term *micron*.)

Microorganisms microbes; minute forms of life, individually too small to be seen without the aid of a microscope.

Microscopic anything so small that it cannot be seen without the aid of a microscope.

Microtubules tubular proteins that provide a cytoskeleton for cell shape.

Migration inhibition factor (MIF) soluble substance released by sensitized lymphocyte in the presence of its specific antigen; this inhibits migration of macrophages; in vitro system for measuring cellular immunity.

Miracidium a small, ciliated trematode larva that is the infective form for the snail.

Missense mutation a change in the cell's DNA with the effect that a wrong amino acid has been put into an essential protein.

Mitochondria internal membranous organelles that carry out many metabolic reactions, providing the source of most of the cell's ATP.

Mitosis nuclear division that follows duplication of chromosomes and results in daughter nuclei with chromosomes identical to the parent nucleus.

Molds filamentous fungi.

Molecular genetics study of how genes function to control the activities of a cell.

Monera a kingdom that includes all procaryotic cells.

Monoclonal antibodies those produced by cells arising from a single clone of antibody-producing cells.

Monocyte actively phagocytic mononuclear white cell found in the bloodstream; as monocytes mature, they emigrate into tissues and differentiate into macrophages.

Monokines soluble factors secreted by monocytes that act on lymphocytes; the best-studied monokine is named interleukin-1.

Mononuclear phagocyte macrophages that engulf and destroy foreign debris; some wander throughout the body and some are fixed to the lining of certain blood vessels.

Monosaccharide simple sugar, usually consists of five or six carbons.

Monotrichous having a single flagellum at one end of the cell.

Montezuma's revenge a slang expression to describe a diarrheal illness usually caused by pathogenic strains of *Escherichia coli*.

Mordant a substance that fixes a stain; e.g., iodine

in the Gram stain. In flagella staining of bacteria, the mordant also increases the diameter of the flagella so that they can be observed microscopically.

Morphology systematic study of the form and structure of living organisms.

Most probable number a statistical value often used to describe the number of coliform bacteria in a water sample.

Motility the ability of a organism to move by itself.

M protein protein present in the cell walls of group A streptococci.

mRNA *see* Messenger RNA.

Multiple resistance factors (MRF) R factors; episomes responsible for multiple drug resistance.

Murein synonymous with peptidoglycan.

Mutagenic agents substances that increase the occurrence of the overall number of mutants.

Mutant a cell with new properties due to a change in the cellular DNA.

Mutation genetic change that is subsequently inherited.

Mycelium network or mass of filaments of a fungus.

Mycology systematic study of the true fungi, i.e., yeasts and molds.

Mycoplasma smallest microorganisms known that are capable of reproduction outside of living cells; lack rigid cell wall and are pleomorphic.

Mycosis (*pl.* mycoses) fungus disease.

Mycotoxin fungal toxin.

Myocarditis inflammation of the myocardium, a muscle in the heart.

Myxomycetes slime molds.

Nanometer unit of measurement; one nanometer (nm) equals 0.001 μm, or 10^{-9} m; formerly millimicron.

Natural active immunity immunity following recovery from disease.

Natural killer cells T cells whose major target appears to be tumor cells.

Natural passive immunity antibodies acquired by the fetus in utero or by the infant through maternal milk.

Necrotic tissue refers to pathological death of tissue.

Negative control prevention of operon transcription directly by the repressor. Transcription occurs only after an inducer binds to the repressor to prevent the repressor from binding to the operator site.

Negative stain stain that colors the background while the cells appear clear, or unstained.

Negative-strand viruses single-stranded RNA viruses whose viral RNA must be transcribed to form mRNA.

Negri bodies inclusion bodies found in brain cells of animals infected with rabies virus.

Neissera genus of gram-negative diplococci; genus name for bacteria of gonorrhea and meningococcal meningitis.

Nematodes roundworms.

Neufeld (quellung) test a capsular swelling that occurs when an encapsulated organism is mixed with antibodies that are directed against the capsule.

Neuraminidase an enzyme that will remove neuraminic acid from a larger molecule (usually a glycoprotein). It is present in certain bacteria (i.e., *Vibrio cholerae*) as well as certain viruses (i.e., orthomyxoviruses and paramyxoviruses).

Neurotoxin a toxin that acts on the nervous system.

Neutralizing antibodies antibodies that inactivate viruses or toxins.

New World hookworm an infection by *Necator americanus*.

Nitrate reduction reduction of nitrates to nitrites or ammonia.

Nitrification oxidation of ammonia to nitrites and nitrates.

Nitrogen fixer a microorganism that fixes atmospheric nitrogen; it converts nitrogen gas to ammonia.

Nocardia genus of moldlike bacteria, may be acid-fast and pathogenic.

Noncyclic phosphorylation formation of ATP from electrons released from photosystem II that enter photosystem I.

Nondefective viruses viruses able to replicate in the absence of a helper virus.

Nonself foreign substances that will induce an immune response.

Nonsense mutation a change in the cell's DNA with the effect that an amino acid has been left out of an essential protein.

Nonseptate coenocytic; lacking cross walls, thus multinucleated.

Normal flora microorganisms that have become established in a given area; e.g., the organisms that have found a permanent home in some area of the human body.

Nosocomial infection one acquired by a patient in a hospital.

Nuclease an enzyme that will hydrolyze nucleic acid.

Nucleocapsid viral nucleic acid surrounded by a protein capsid.

Nucleotide compound composed of a pentose, a phosphate, and a pyrimidine or purine base.

Nucleus structure within cells that contain DNA, the hereditary material, usually organized into chromosomes.

Nutrients substances, inorganic and organic, that actually pass through the cytoplasmic membrane and provide the direct requirements of a cell. Food is the raw material from which nutrients are derived.

Nutrition the means by which an organism assimilates its foods.

O antigen somatic antigen, or that associated with the cell body, in contrast to flagellar; (O from *ohne hauch*, Ger., "without breath").

Obligate necessary.

Obligate aerobes organisms that must use oxygen as their final electron acceptor.

Obligate anaerobes organisms unable to grow in the presence of oxygen.

Obligate parasite one that is unable to multiply apart from a living host.

Old World hookworm an infection by *Ancylostoma duodenale*.

Oncogene potentially cancer-producing gene present in normal chromosomal DNA.

Oncogenic virus one that causes infected cells to become malignant.

Oncology the systematic study of tumors (neoplasms).

Oocyst a resting stage formed during the sexual reproduction of *Toxoplasma* in the cat; it is excreted in the feces and is ingested by humans and other animals causing toxoplasmosis.

Ookinete a motile zygote formed by sporozoans such as the malaria parasite.

Oospore sexual spore that is the product of sexual fusion of gametes of unequal size; found in some Phycomycetes.

Operator gene a specific region of the chromosome located adjacent to the sequence of genes, making up the operon.

Operon the entire sequence of genes responsible for the synthesis of the enzymes involved in the biosynthesis of a given molecule. The operator gene controls the function of this group.

Opportunists potentially harmful microorganisms, in reference to the normal flora of the animal body.

Opsonin antibody that combines with bacterial cells and makes them more susceptible to phagocytosis.

Optimum condition most favorable.

Order taxonomic group made up of families.

Organelles specialized structures within a cell that serve to perform a specific function.

Organic compound a carbon compound.

Osmosis the passage of water through a semipermeable membrane from a region of higher concentration to a region of lower concentration of water.

Osmotic pressure a function of the rate at which water will pass from a solution of low concentration to a solution of high concentration.

Ouchterlony test double diffusion of antigen and antibody in agar.

Oxidase test a test to distinguish colonies of *Neisseria* from some other bacteria; it does not differentiate *Neisseria* species. The test may be carried out by flooding the colonies on a plate with either dimethyl- or tetramethylparaphenylene diamine hydrochloride and observing for color changes in the colonies. *Neisseria* are oxidase positive and the colonies first turn pink, then dark red, and finally black.

Oxidation the addition of oxygen, or the loss of hydrogen and electrons.

Oxidative phosphorylation production of high-energy phosphate bonds of ATP by the electron transport system.

Oxidoreductases enzymes that carry out the specific energy-releasing reactions of the cell.

Oxyfume a mixture of 20 percent ethylene oxide and 80 percent carbon dioxide.

Pandemic an epidemic that has become very widespread, or is worldwide.

Papule small raised area on the skin.

Parainfluenza viruses ssRNA viruses causing upper respiratory tract infections in humans.

Parasite an organism that lives on or within a living host from which it derives its nourishment.

Parasitism the act of living at the expense of a host organism.

Parenteral not through the oral route; usually by injection.

Passive immunity immunity acquired as a result of receiving preformed antibodies.

Pasteur treatment an attempt to immunize an individual to rabies virus after a potential infection.

Pasteurization heating to destroy pathogenic microbes. Methods: heating to 62.9°C for 30 min (holding method); or heating to 71.6°C for not less than 15 sec. The heating is followed by rapid cooling.

Patch test a tuberculin skin test in which the tuberculin preparation is applied to a clean skin area on a piece of gauze or tape. A positive reaction, indicating sensitivity to the tubercle bacillus, will show an inflamed area on the skin.

Pathogenic disease-producing.

Pathogenicity ability of a microorganism to cause disease.

Pébrine a disease of silkworms caused by a protozoan.

Pellicle a film due to bacterial growth on the surface of a liquid medium; also refers to the tough flexible covering of protein surrounding certain protozoa, e.g., class Mastigophora.

Penicillinase enzyme that destroys penicillin.

Pentose a five-carbon sugar, as ribose.

Peptidase an enzyme that will cleave peptide bonds.

Peptide bond the bond between the carboxyl carbon

of one amino acid and the α amino nitrogen of a second amino acid.

Peptidoglycan a polymer of *N*-acetylglucosamine and *N*-acetylmuramic acid plus a few amino acids that make up the rigid cell wall in procaryotic organisms.

Peptonization the digestion of milk casein by proteolytic enzymes.

Pericardium a sac that encloses the heart.

Peridontal ligament structures that attach the tooth to the alveolar bone.

Peritrichous having flagella around the entire cell.

Permanent cell line eucaryotic diploid cell lines that can be cultivated for an unlimited number of generations.

Permeability allowing the passage of molecules through cell membranes.

Permease system that transports substances against a concentration gradient across the plasma membrane into the cell.

Pertussin an exotoxin produced by *Bordetella pertussis* that is responsible in part for the symptoms of whooping cough.

pH the extent of acidity or alkalinity of a solution or medium; expresses the negative logarithm of the hydrogen ion concentration.

Phage bacteriophage; virus that infects a susceptible bacterial cell.

Phagocyte white blood cell, capable of ingesting foreign particles, including microorganisms.

Phagocytosis the engulfment of bacteria and other foreign particles by white blood cells, or phagocytes.

Phase difference the difference in brightness, that is, light and shade, of different cellular structures, as observed with the phase-contrast microscope.

Phenol coefficient a test for evaluating the effectiveness of a disinfectant by comparing it with phenol under identical conditions.

Phenotype characteristics that can be observed, i.e., the expression of the genes.

Phosphatase test test employed to determine the adequacy of the pasteurization process; the enzyme is destroyed in adequately pasteurized milk.

Photoreactivation recovery of cells from ultraviolet damage by exposure to visible light; an enzyme, activated by visible light, hydrolyzes the thymine dimers formed as a result of ultraviolet exposure.

Photosynthesis process whereby carbohydrates are synthesized from carbon dioxide and water in the presence of light energy and chlorophyll.

Photosynthetic the ability biologically to trap energy from light.

Phototroph an organism that utilizes the energy of light.

Phylum a large subdivision within a kingdom that consists of a number of organisms that have certain broad characteristics in common.

Physiology systematic study of functions of living organisms.

Phytoplankton a collective term to describe microscopic algae in an aquatic environment.

Picornaviruses small single-stranded RNA viruses.

Pili fimbriae; filamentous appendages, shorter, straighter, and considerably smaller than flagella; they appear to make bacteria more adhesive. Sex pilus is essential for genetic transfer in bacterial conjugation.

Pinworms common term for *Enterobius vermicularis*.

Plant kingdom eucaryotic cells that usually obtain their energy from light.

Plaque a clear area in the confluent growth of a bacterial or cell culture due to lysis by a phage or virus.

Plasma the liquid part of blood that still contains clotting factors.

Plasma cell lymphoid cell, producer of circulating antibody.

Plasmid extrachromosomal DNA that encodes for its own replication.

Plasmolysis shrinkage of plasma membrane and cell contents away from the cell wall as a result of osmosis.

Pleomorphic having many forms.

Plerocercoid spindle-shaped larval stage of *Diphyllobothrium latum*. It develops after an infected crustacean is ingested by a suitable fish.

Plus-strand viruses single-stranded RNA viruses whose viral RNA functions directly as mRNA.

PMN *see* Polymorphonuclear neutrophil.

Pneumococci a slang name for *Streptococcus pneumoniae*.

Pneumonia inflammation of the lungs.

Polar flagellation flagella arise at the ends of cells. Types: monotrichous (single flagellum), amphitrichous (single flagellum at each end), and lophotrichous (tuft of flagella at one or both poles).

Polluted water water that has an undesirable appearance or taste; it may or may not be contaminated.

Poly *see* Polymorphonuclear neutrophil.

Polycistronic mRNA messenger RNA encoding more than one protein.

Polymerases general name for enzymes concerned with the synthesis of nucleic acids.

Polymorphonuclear neutrophil a type of leukocyte in the circulating blood that contains granules that stain with neutral dyes. The nucleus is divided into large lobes. These cells function in phagocytosis and are the most prevalent type of leukocyte in the circulating blood.

Polypeptide a chain of amino acids linked together by peptide bonds.

Polysaccharide complex carbohydrate consisting of many monosaccharide molecules.

Polysome a series of many ribosomes simultaneously attached to a single strand of mRNA.

Porins protein structures in the outer membrane of gram-negative bacteria, which permit the passage of small molecules through the membrane.

Portal of entrance the means by which a pathogenic organism enters the body to produce disease.

Positive control the condition under which transcription of an operon occurs only after the inducer binds to the repressor changing its conformation so that it can bind to the operator site, permitting transcription.

Potable water water free from harmful and unpleasant substances; water fit for drinking.

Pour-plate method a procedure designed to obtain separate colonies on a nutrient agar plate. The process consists of inoculating a culture into a cooled melted nutrient agar medium, mixing, and then pouring into a petri dish to solidify. The culture can be diluted as required, by transferring an aliquot from one melted agar tube to another before pouring the plates.

PPD purified protein derivative, a skin test material for the diagnosis of tuberculosis.

Precipitin antibody that causes precipitation of its soluble antigen.

Precipitin reaction a reaction between a soluble antigen and its antibody resulting in the formation of an antigen-antibody complex too large to stay in solution.

Primary cell culture specific tissue, e.g., kidney, liver, taken directly from an animal, broken up into individual cells, and then infected with the virus to be cultivated.

Prion infectious particles of certain slow virus diseases that are thought to be composed solely of protein.

Procaryotic lacking a true nucleus in that no nuclear membrane separates the DNA from the cytoplasm.

Procercoid an elongated larval stage of *Diphyllobothrium latum* that develops in a crustacean after it ingests the motile coracidium.

Proglottids segments of an adult tapeworm.

Promoter site attachment site on DNA for the enzyme RNA polymerase.

Prophage phage nucleic acid that becomes incorporated into the bacterial chromosome or exists as an episome without causing cell lysis.

Prophylactic preventive measure for protection against infection.

Proteases proteinases; enzymes that hydrolyze proteins to simpler molecules.

Protein organic compound composed of many amino acids joined together by peptide bonds.

Protista a kingdom that includes the unicellular algae and protozoa.

Protooncogenes regulatory genes occurring in normal eucaryotic cells that may become incorporated into a retrovirus, resulting in the formation of an RNA tumor virus.

Protoplast a cell devoid of its cell wall.

Prototrophs bacterial strains that do not require added growth factors such as vitamins or amino acids; this is in contrast to auxotrophs, which require added factors for growth.

Protozoa single-celled eucaryotes.

Protozoology systematic study of protozoa.

Provirus inactive virus, due to repression of the viral nucleic acid, but may become active and replicate when the repressor is blocked.

Pseudo- prefix, meaning false.

Pseudomembrane false membrane; in diphtheria, it consists of dead tissue cells, leukocytes, red blood cells, and bacteria that form a dull gray exudate.

Pseudopodia false feet; fingerlike projections of protoplasm that provide for movement and securing food in protozoa, class Sarcodina; e.g., Amoeba.

Psychrophile cold-loving microorganism, may grow at near-freezing temperatures; optimum growth temperature below 15°C and may grow at temperatures as low as 2 to 3°C.

Puerperal fever acute infection following childbirth due to introduction of the infectious agent into the uterus; can be caused by β-hemolytic streptococci.

Pure culture culture containing only one species, or type, of microorganism.

Pus accumulation of dead leukocytes, along with bacteria and serum.

Putrefaction anaerobic decomposition of proteins by bacterial enzymes.

Pyelonephritis an infection of the kidney.

Pyemia pus in the blood, or the release of purulent material into the blood.

Pyogenic pus-forming.

Q fever an infection, usually pulmonary, caused by *Coxiella burnetii*.

Quellung reaction capsular swelling test; Neufeld's test; when a specific antibody in blood serum combines with its specific capsular antigen, it gives the appearance of capsular swelling; e.g., typing of *Streptococcus pneumoniae*.

Racemases enzymes in the group of isomerases that convert L isomers to D isomers and vice versa.

Recall phenomenon anamnestic response, or secondary response to an antigen with rapid production of antibodies.

Receptor site the site on a host cell to which the viral protein coat must first be adsorbed in order for the virus to infect the cell.

Recombinant DNA a fragment of DNA that has been inserted into a self-replicating unit and that can be expressed when introduced into an appropriate bacterium.

Recombinants a cell or clone of cells resulting from conjugation.

Redia differentiated larval stage of a fluke that develops in the liver of an infected snail.

Reductase test test employed to determine the approximate bacterial count in milk; the time required for the reduction of methylene blue to a colorless compound is proportional to the number of bacteria present.

Reduction the loss of oxygen, or the gain of hydrogen and electrons.

Rennet curd sweet curd; coagulation of casein in milk catalyzed by the enzyme rennin.

Reoviruses double-stranded RNA viruses causing mild respiratory infections.

Replica plating a technique designed to show that mutations are spontaneous and not specifically induced in bacteria. In the test, a large number of bacterial cells are streaked out on the nutrient agar plate and allowed to develop into microcolonies; then a sterile velvet pad is pressed over the plate; a few bacteria from each colony adhere to the velvet so that when the velvet is pressed down on a second plate, it produces an exact copy of the original plate. The copy plate can be used to test resistance to various chemical agents without previous contact with the agent; e.g., penicillin.

Repressor a gene product that, when active, will prevent the transcription of a specific operon.

Resistance the ability to prevent disease.

Resolving power the smallest detail that can be seen in a microscope as a separate image; depends on the wavelength of light used.

Respiration any chemical reaction occurring within a living cell that results in a release of energy; biological oxidation.

Respiratory syncytial virus single-stranded RNA paramyxovirus causing serious respiratory infections in young children.

Reticuloendothelial system (RES) a system of phagocytic cells that are fixed in certain organs or tissues, including blood vessels in the liver, spleen, bone marrow, lungs, and lymph nodes. They function in removing bacteria and other particulate matter from circulating fluids.

Retrovirus RNA viruses whose replication requires that their RNA be transcribed into DNA that then inserts into the host cell chromosome.

Reverse transcriptase RNA-dependent DNA polymerase.

Rh factor an antigen found on red blood cells. If the blood cells contain this factor, the individual is Rh-positive; those without it are Rh-negative.

Rhinoviruses single-stranded RNA picornaviruses causing the majority of cases of the common cold.

Rhizoids rootlike structures characteristic of some molds.

Ribonucleic acid (RNA) nucleic acid which contains D-ribose, adenine, guanine, cytosine, uracil, and phosphoric acid.

Ribosome a small oval body within the cell that attaches itself to an initiation site on the mRNA, resulting in the synthesis of a protein molecule. Ribosomes consist of protein and RNA.

Rickettsiae obligate intracellular parasites that are closely related to gram-negative bacteria; carried in arthropod vectors.

RNA polymerase an enzyme that synthesizes RNA.

Rotavirus double-stranded RNA viruses that cause acute infectious diarrhea, particularly in the very young.

Rough endoplasmic reticulum internal membranes to which ribosomes are attached; its major function is the synthesis of secretory proteins.

Rubella German measles.

Rubella syndrome collective term describing congenital abnormalities resulting from an infection of the fetus by rubella virus.

Rubeola measles.

Sabin vaccine living attenuated polio vaccine.

Salk vaccine killed polio vaccine.

Sanitize to reduce microbial numbers to safe levels as judged by public health standards.

Sanitizer agent that reduces microbial numbers to safe levels according to public health requirements.

Saprophyte organism that lives on dead organic matter.

Sarcina spherical cells that divide in three planes at right angles to each other to form cubical packets.

Sarcoma tumor, usually highly malignant; neoplasm.

Schick test susceptibility test for diphtheria; diphtheria toxin is injected intracutaneously.

Schizont a stage in the preerythrocytic reproductive cycle of the malaria parasite.

Schultz-Charlton reaction diagnostic test for scarlet fever; antitoxin to β-streptococcus erythrogenic toxin is injected intracutaneously and the skin rash is observed for blanching.

Scolex head of a tapeworm.

Scrapie a disease of sheep caused by an unidentified slow agent, possibly a prion.

Secondary metabolite a product formed by an organism that is not necessary for the continued life of the cell. Antibiotics are prime examples.

Sedimentation the settling out of materials so that they can be removed; useful in water purification.

Selective IgA deficiency a syndrome wherewith an individual is unable to synthesize the IgA class of immunoglobulins.

Selective medium a medium that will select out certain organisms while inhibiting others; e.g., a medium containing bile salts is selective for pathogenic enteric bacteria.

Selective toxicity toxicity directed toward specific cells.

Septa dividing or cross walls characteristic of the filaments of molds with septate hyphae. The septa divide each hypha into many cells with individual nuclei.

Septate septations or cross walls separate the individual cells; characteristic of fungi in classes Ascomycetes, Basidiomycetes, and usually Deuteromycetes.

Septate hypha hypha with cross walls, or septa, that divide each filament into individual cells with individual nuclei.

Septicemia condition in which bacteria are actively multiplying in the blood.

Sequela morbid condition that results as a consequence of another disease.

Serology systematic study of blood serum, i.e., reactions between antibodies and antigens.

Serotonin pharmacologically active metabolite of tryptophan released from mast cells of rats and mice by the interaction of antigen and bound IgE.

Serovar synonymous with *serotype*. Basic taxon for members of the genus *Leptospira*.

Serum sickness a reaction that sometimes follows the injection of a foreign serum. The patient forms antibodies to an antigen in the foreign serum and the antibodies react with residual circulating antigen; the resulting antigen-antibody complexes are toxic and may cause fever and joint pains.

Sex factor genetic factor occurring in the cytoplasm of some bacteria. During conjugation it is transferred to recipient cells. The sex factor, or F factor, is found in male strains, and after conjugation, recipient cells are converted to F+.

Sex-linked agammaglobulinemia an inability to make antibodies that is inherited on the Y chromosome; it is thus restricted to males.

Sexual spores spores produced by the fusion of two cells.

Slime molds a group of higher protists that resemble both protozoa and the true fungi; they are like protozoa in their ameboid plasmodial stage and similar to the true fungi in spore formation.

Slow-reacting substance of anaphylaxis *see* Leukotrienes.

Sludge insoluble material generated in sewage disposal plant.

Smear thin layer of material spread across a slide for microscopic study.

Smooth endoplasmic reticulum smooth-surface internal membranes containing membrane-bound enzymes within the cytoplasm of the cell; functions in the biosynthesis of steroid hormones, metabolism of lipid-soluble drugs, and the metabolism of glycogen.

Somatic associated with the cell body of the organism.

Species a taxonomic category subordinate to a genus and superior to a subspecies; one kind of organism.

Specific immunity resistance that is specifically directed against a particular microorganism or its products.

Spheroplast a gram-negative cell that lacks peptidoglycan components but possesses outer and inner membrane structures.

Spherules thick-walled structures within which endospores develop in tissues; characteristic of *Coccidioides immitis*.

Spirillum corkscrew-shaped cells, rigid.

Spirochete a flexible spiral or corkscrew-shaped bacterium.

Spontaneous generation theory of development of living forms from nonliving matter.

Spontaneous mutations mutations that occur during normal growth.

Sporadic when used in reference to a disease, of only occasional occurrence.

Sporangiophore aerial filament of a fungus that supports the sporangium.

Sporangium closed sac in which asexual spores develop; characteristic of Zygomycetes.

Sporicide an agent that will kill spores.

Sporocyst first larval trematode stage developing in an infected snail.

Sporozoite form of malaria parasite injected into the bloodstream by an infected mosquito.

Sporulation the process of forming endospores.

Spreading factor name given to the enzyme, hyaluronidase, because it is believed to accentuate the spread of streptococci within the tissues.

Sputum mucopurulent matter expectorated from air passages.

Stained bacteria bacteria that have been colored with a chemical stain to make them easier to see and study.

Staphylococci (*sing.* **staphylococcus**) spherical cells in irregular clusters.

-stasis suffix, refers to inhibition.

Stationary phase the growth phase of a culture when

the rate of reproduction equals the death rate and the number of viable bacteria remains constant.

STD sexually transmitted disease.

Stem cells undifferentiated cells that give rise to blood cells, leukocytes, B cells, and the various subsets of T cells.

Sterilization killing of all forms of life in a given area.

Stock cultures pure cultures of microorganisms held in stock for laboratory use.

Streak plate a method of separating bacterial cells on a solid surfaced so as to obtain a pure culture from an isolated colony.

Streak-plate method a culture is streaked on the surface of a nutrient agar medium in a manner that separates the organisms so that isolated colonies develop.

Streptococcal pyrogenic toxin one of several toxins secreted by group A streptococci which cause fever, increased susceptibility to endotoxic shock, cardiotoxicity, and/or a skin rash.

Streptococci (*sing.* **streptococcus**) spherical cells in chains.

Streptokinase substance that activates a proteolytic enzyme (plasminogen) normally present in the host's plasma; this causes dissolution of blood clots and may allow for the spread of the organism.

Streptolysins hemolysins produced by β-streptococci; streptolysin S, stable in the presence of atmospheric oxygen, causes hemolysis on a blood medium; streptolysin O, inactivated in the presence of oxygen, stimulates antibody production, which provides a useful serological test for a recent streptococcal infection.

Streptozyme test immunological assay for antibodies to group A streptococcal products.

Strict aerobes organisms that must use oxygen as their final electron acceptor.

Strict anaerobes organisms that cannot grow in the presence of oxygen.

Subacute bacterial endocarditis infection of a previously damaged heart valve, usually by normal flora bacteria such as viridans streptococci or enterococci.

Substrate substance acted on by an enzyme.

Substrate phosphorylation the oxidation of a phosphorylated compound resulting in the direct formation of a high-energy phosphate bond.

Subterminal situated between the center and the end of the cell, as subterminal endospores.

Suppurative pus-forming.

Suppressor T cells the T-lymphocyte subset that regulates the immune response through the negative control of other subsets of T cells.

Sylvatic plague infections of ground squirrels, prairie dogs, and rodents with *Yersinia pestis*.

Symbiont a helpful organism in reference to the normal flora of the animal body. In a symbiotic relationship, the microbe and the host receive benefit from each other.

Symbiotic a relationship in which two microorganisms or a microorganism and its host are mutually beneficial to each other.

Symbiotic nitrogen fixers organisms that can fix atmospheric nitrogen only when growing in a host plant.

Synchronous growth a pattern of growth when all cells of a culture are at the same stage of division. This may be accomplished by several techniques, e.g., a chilling period followed by a favorable growth temperature.

Synergistic effect the greater effect produced by two chemotherapeutic agents given simultaneously than would result from either alone; also, an effect produced by two species together that neither could produce alone.

Synthesize to build up.

Synthetic medium a medium in which the chemical formula for each ingredient can be written; it may be very complex and may vary widely for different species.

Syphilis a venereal disease caused by the spirochete *Treponema pallidum*.

Tartar deposits of calcium on tooth plaque, forming a rough, stony crust.

T cell lymphocyte originating in the thymus.

T-dependent antigens antigens that require T-cell interaction in order to induce an antibody response.

T-independent antigens antigens that can induce an IgM response in the absence of T cells.

Teichoic acids polymers of ribitol or of glycerol phosphate.

Temperate phage a phage that produces lysogeny, rather than lysis.

Temperature-sensitive mutants organisms able to grow at low temperatures but not at the higher temperatures in which the parent strain could grow.

Terminal disinfection disinfection procedures carried out at the end of an infectious period. This involves cleaning the entire area that may have been contaminated by the patient.

Tertiary refers to the final structure of a protein; it is governed by a number of factors including hydrogen bonding and disulfide bonds.

Tetanus a disease caused by the exotoxin produced by *Clostridium tetani*.

Tetrapeptide four amino acids joined together by peptide bonds.

Therapy treatment of disease.

Thermal death point the temperature required to sterilize a given substance in 10 min.

Thermal death time the time required for sterilization at a given temperature.

Thermoduric the capacity to endure high temperatures that are not conducive to growth.

Thermolabile heat-labile.

Thermophiles organisms whose optimum growth temperature is usually above 45 or 50°C; some may grow at temperatures above 85°C; heat-loving.

Thermostable heat-stable.

Thrush infection of the mucous membranes of the mouth by *Candida albicans*.

Tinea ringworm.

Torula (*pl. torulae*) false yeast, or yeast in which a sexual stage of spore formation does not occur or has not been recognized.

Toxemia the presence of toxins in the blood.

Toxin poisonous substance produced by certain microorganisms.

Toxoid an exotoxin with reduced toxicity due to treatment with formalin or some other agent.

Transcription the copying of DNA into its corresponding RNA by the enzyme RNA polymerase.

Transduction transfer of genetic material from a donor to a recipient bacterial cell with bacteriophage as a carrier.

Transfer RNA (tRNA) RNA that reacts with an amino acid; the latter activates it and carries it to the mRNA.

Transferases a very large group of enzymes that catalyze the transfer of chemical groups from one substrate to another; e.g., glucose kinase transfers a phosphate group from ATP to glucose.

Transformation passage of DNA from donor to recipient bacterial cell, with subsequent change in character of recipient.

Translation the process by which the message contained in the mRNA directs the synthesis of a specific protein.

Transpeptidization using the energy within one peptide bond to form a second peptide bond as in the final cross-linking of cell wall peptidoglycan.

Transplantation antigens antigens naturally present in the cells of one person that are foreign to another person.

Transposase an enzyme that excises a transposon from the cell's DNA.

Transposon a movable genetic element.

Traveler's diarrhea usually refers to diarrhea caused by a pathogenic strain of *Escherichia coli*.

Treponema pallidum spirochete, causative agent of syphilis.

Treponema pallidum immobilization (TPI) test for syphilis, using patient's blood serum with live spirochetes; in a positive test, the spirochetes lose motility because of specific antibodies in the blood serum.

Tricarboxylic acid cycle *see* Citric acid cycle.

Trichinosis a disseminated disease of carnivorous animals (including humans) caused by *Trichinella spiralis*.

Trickle filter a 4- to 6-ft-deep bed of gravel over which sewage is sprayed. Organisms growing on the gravel oxidize the organic material in the sewage.

Triglyceride three fatty acids linked to glycerol via ester bonds.

Tripeptide three amino acids joined together by peptide bonds.

tRNA *see* Transfer RNA.

Trophozoite active cells of amoebae.

Tube dilution technique a quantitative technique in which the chemical agent in question is diluted in the growth medium in a series of twofold dilutions. All tubes are then inoculated with the organism to be studied.

Tuberculin concentrate of *Mycobacterium tuberculosis* culture; used in a hypersensitivity skin test, for recognition of past infection, but not necessarily active disease.

Turbid cloudy.

Tyndallization fractional sterilization; exposure to live steam for 30 min each day for 3 successive days to kill vegetative cells.

Ultramicroscopic visible with the electron microscope; too small to be seen with the conventional light microscope.

Ultrasonic waves high-frequency sound waves, used for destruction of bacterial cells.

Undulant fever brucellosis; disease caused by *Brucella* species.

Urethritis inflammation of the urethra.

Vaccine a product that contains an antigen, consisting of killed pathogenic organisms, living avirulent (attenuated or weakened) organisms, or inactivated exotoxins (toxoids).

Varicella chickenpox.

Variola smallpox.

VDRL Venereal Disease Research Laboratory.

Vector any agent that carries a disease from one host to another; may be animate, as insects, or may be inanimate.

Vegetative phase active growth period.

Venereal spread by sexual contact.

V-factor nicotinamide adenine dinucleotide, nicotinamide adenine dinucleotide phosphate, or nicotinamide riboside, a growth factor requirement for *Haemophilus influenzae*.

Vi antigen polysaccharide antigen, found most frequently in recently isolated virulent *Salmonella typhi*.

Vibrio curved organism resembling a comma; characteristic of *Vibrio cholerae*, the causative agent of cholera.

Viremia the presence of viruses in the blood.

Viricide agent that destroys viruses.

Virion infectious virus particle.

Viroids small molecules of naked RNA that cause several plant diseases.

Virology systematic study of viruses.

Virulence power to infect; the extent of pathogenicity.

Virulent phage phage in which the nucleic acid is injected into a susceptible bacterial cell, followed by replication, maturation, and lysis of the host bacterium.

Virus an obligate intracellular parasite that lacks certain components absolutely essential for its own replication and must depend on a host cell to provide the missing factors; it lacks an ATP generating system and ribosomes for protein synthesis.

Voges-Proskauer reaction test for acetylmethylcarbinol (also called "acetoin") production, used to differentiate *Escherichia* and *Enterobacter* of the coliform group.

Volutin *see* Babes-Ernst bodies.

Von Pirquet test tuberculin skin test that makes use of a small amount of the tuberculin preparation that is rubbed into a scratch on the skin. An inflamed area provides a positive reaction and indicates sensitivity to the tubercle bacillus. This test is less sensitive than the Mantoux test.

Wassermann test a nonspecific complement-fixation test used as a diagnostic aid in syphilis.

Weil-Felix test agglutination test used as a diagnostic aid in certain rickettsial diseases; certain *Proteus* strains are used as the antigen with the patient's blood serum.

Whipworm disease a nematode disease caused by *Trichuris trichiura*.

Widal test agglutination test using killed *Salmonella typhi* organisms and the patient's blood serum; a diagnostic aid in typhoid fever.

Xenograft tissue graft between members of different species.

X-factor hematin, a growth factor requirement for *Haemophilus influenzae*.

Yeasts microscopic fungi that exist as simple independent cells; e.g., *Saccharomyces cerevisiae*.

Zoonosis disease of animals that may be transmitted to humans.

Zoospore asexual motile spore; characteristic of Zygomycetes.

Zygospore sexual spore that results from the fusion of like gametes; characteristic of Zygomycetes.

Credits

COLOR INSERT

Plates 1 through 8 Copyright © 1979 by J. B. Lippincott Company. Reprinted by permission of the publisher.

Unit One Opener, page 1, E. R. Degginger.

Unit Two Opener, page 249, Juneann W. Murphy, Michelle R. Hidore, University of Oklahoma Health Sciences Center, Department of Microbiology and Immunology, from *INFECTION AND IMMUNITY,* April, 1991, © American Society for Microbiology.

Unit Three Opener, page 329, Anthony T. Murelli, Department of Microbiology, Uniformed Services University of the Health Sciences, Bethesda, MD.

Unit Four Opener, page 521, E. R. Degginger.

FIGURES

1.9 A. P. Craig-Holmes/BPS.
1.10 T. J. Beveridge, Univ. of Guelph/BPS.
1.12, 28.15 N. Salmonsky, Dept. of Microbiology, Univ. of Virginia School of Medicine.
1.13 From J. T. Sinski, *Dermatophytes in Human Skin, Hair and Nails.* Courtesy of Charles C. Thomas, Springfield, IL, 1974.
1.14, 1.15 Dr. T. E. Jensen, Dept. of Biological Sciences, Herbert H. Lehman College of CUNY, New York.
2.3 Centers for Disease Control, Atlanta, GA.
2.4 Courtesy of Carl Zeiss, Inc., New York.
2.6, 3.2, 3.23 S. C. Holt, Dept. of Microbiology, Univ. of Massachusetts, Amherst.
2.7 T. J. Beveridge, Univ. of Guelph/BPS.
2.8 Courtesy of Jay Brown, Dept. of Microbiology, Univ. of Virginia School of Medicine, Charlottesville, VA.
2.9 N. S. Hayes, K. E. Muse, A. M. Collier, and J. B. Baseman, Dept. of Bacteriology and Immunology, School of Medicine, The Univ. of North Carolina at Chapel Hill.
3.4e W. L. Dentler, Dept. of Physiology and Cell Biology, Univ. of Kansas, Lawrence.

3.6 S. C. Holt, Univ. of Texas Health Science Center, San Antonio/BPS.
3.7 Courtesy of Dr. George Galasso.
3.10b and d, 3.19, 3.22, 24.9 Courtesy of Carl A. Schnaitman, Univ. of Virginia School of Medicine.
3.13 G. W. Brown, Jr., J. L. Pate, and H. Sugiyama, *J. Bacteriol.* 105:1207, 1971.
3.14 S. J. Singer and G. L. Nicolson, *Science* 175: 720–731, 1972. Copyright © 1972 by the American Association for the Advancement of Science.
3.20a S. C. Holt, Univ. of Texas Health Science Center, San Antonio/BPS.
3.20b S. C. Holt, Univ. of Texas Health Science Center, San Antonio/BPS.
3.21 P. Fitz-James and E. Young, in G. W. Gould and A. Hurst (eds), *The Bacterial Spore,* Academic Press, Inc. Ltd., London, 1969.
7.10 Courtesy of Yvonne Osheim, Dept. of Microbiology, Univ. of Virginia School of Medicine, Charlottesville, VA.
8.4 C. C. Brinton, Jr. and J. Chapman, Dept. of Life Sciences, Univ. of Pittsburgh.
9.6, 30.1, 30.2, 30.3 Courtesy of Millipore Corp., Bedford, Mass.
11.3, 11.14 Courtesy of Dieter Groschel, Dept. of Pathology, Univ. of Virginia School of Medicine, Charlottesville, VA.
13.1 H. Plotz, J. Smadel, T. Anderson, and L. Chambers, reproduced from *The Journal of Experimental Medicine,* 77:355, by copyright permission of the Rockefeller University Press.
13.2, 13.3 L. A. Page, National Animal Disease Center, Ames, Iowa.
13.4a E. S. Boatman and G. E. Kenny, *J. Bacteriol.* 106: 1005, 1971.
13.4b, 13.5 M. G. Gabridge, Dept. of Microbiology, Univ. of Illinois, Urbana.
13.6 M. C. Shepard and D. R. Howard, *Ann. N.Y. Acad. Sci.* 174:809–819. 1970.
13.7 T. J. Beveridge, Univ. of Guelph/BPS.
13.8a–d K. Stephens, Stanford Univ./BPS.
14.2a S. U. Emerson, "Vesicular stomatitis virus: Structure and function of virion components," *Curr. Topics Mi-*

crobiol. Immunol, Vol. 73, Springer, Wein-New York, 1976, pp. 1–34.

14.8 F. Fenner and D. O. White, *Medical Virology,* 2nd ed., Academic Press, New York, 1976.

14.9a Adapted from W. B. Wood and R. S. Edgar, "Building a bacterial virus," *Sci. Amer.* 217:61. 1967.

14.9b Courtesy of Professor Stefan Höglund.

15.2 Courtesy of Robert Huskey, Dept. of Biology, Univ. of Virginia, Charlottesville, VA.

15.3 M. Murayama, Murayama Research Lab./BPS.

15.4a L. E. Roth, The Univ. of Tennessee/BPS.

15.4b J. R. Waaland, Univ. of Washington/BPS.

15.12 D. D. Kunkel, Univ. of Washington/BPS.

15.13 H. H. Najarian, *Textbook of Medical Parasitology,* © 1967, Williams and Wilkins Co., Baltimore, MD.

17.2 D. F. Bainton, *J. Cell Biol.* 28:277–301, 1966.

17.4 D. F. Bainton, Dept. of Pathology, Univ. of California, San Francisco.

18.7, 18.9, 18.10, 18.13, 19.7, 19.8, 19.10, 19.13, 20.3 Adaptation Courtesy of David C. Benjamin, University of Virginia.

19.2a, b, c, 19.3 D. E. Normansell, Dept. of Pathology, Univ. of Virginia School of Medicine.

19.9 Dr. R. Dourmashkin, M.D., St. Bartholomew's Centre for Clinical Research, London.

24.4b, c P. P. Cleary, Dept. of Microbiology, Univ. of Minnesota Medical School.

24.6 E. Newbrun, *Science* 217:418–423, 1982. Copyright © 1982 by the American Association for the Advancement of Science.

24.13 K. E. Muse/BPS.

24.15 F. L. A. Buckmire, The Medical College of Wisconsin.

24.16 S. M. Gibson, General Bacteriology Lab., Texas Dept. of Health Resources, Austin.

24.20 O. W. Richards and D. K. Miller, *Am. J. Clin. Path.* 11:1.

24.21 A. R. Flesher, S. Ito, B. J. Mansheim, and D. L. Kasper, *Ann. Intern. Med.* 90:628–630, 1979.

24.23, 24.25, 24.27, 28.12, 28.14, 29.8, 29.11 Centers for Disease Control, Atlanta.

24.24 C. T. Dolan et al. *Atlases of Clinical Mycology II, Systemic Mycosis—Deep Seated,* American Society of Clinical Pathologists, Chicago, 1975.

24.26, 24.28, 30.6 Courtesy of U.S. Public Health Services.

25.1, 26.32 H. D. Mayor, Dept. of Microbiology, Baylor College of Medicine, Texas Medical Center, Houston.

25.2 M. W. Jennison.

25.6a W. G. Laver, *Advan. Virus Res.* 18:62, 1973.

25.6b, 25.7 Adaptation Courtesy of J. Thomas Parsons, University of Virginia.

25.8 From *Morbidity and Mortality Weekly Report Annual Summary,* 1976.

25.10 Courtesy of G. C. Sauer, *Manual of Skin Diseases,* 4th ed., J. B. Lippincott, Philadelphia, 1980.

26.1, 28.1 L. J. LeBeau, Depts. of Pathology and Microbiology, Univ. of Illinois Hospital at the Medical Center, Chicago.

26.2 Courtesy of Clay Adams, Inc.

26.6 D. G. Evans, *Program in Infectious Disease and Clinical Microbiology,* The Univ. of Texas Medical School.

26.7 H. A. Reimann et al. *Am. J. Trop.* Med. 26:631.

26.8 G. T. Cole, Univ. of Texas, Austin/BPS.

26.9 M. J. Tufte, Dept. of Biology, Univ. of Wisconsin, Plattville.

26.28 A. Z. Kapikian et al. Science 185:1049, 1974. Copyright © 1974 by the American Association for the Advancement of Science.

26.29 B. A. Phillips, Dept. of Microbiology, Univ. of Pittsburgh School of Medicine.

26.33 E. H. Cook, Jr., Phoenix Labs., Center for Disease Control, Phoenix.

26.36 B. Rolzman, Committee on Virology, Univ. of Chicago.

26.37 P. Weary, Univ. of Virginia School of Medicine.

27.5 Division of Viral Diseases, Centers for Disease Control.

28.2 American Society for Microbiology, Washington, D.C.

28.6 D. Bromley, Dept. of Microbiology, West Virginia Univ. Medical Center.

28.7 V. R. Dowell, Centers for Disease Control, Atlanta.

28.8, 28.9, 28.10 © 1978 by Carroll H. Weiss RPB.

28.13 R. J. Heckley and E. Goldwasser, reprinted from *J. Infect. Dis.* 84:94–97, © 1949 by permission of The University of Chicago Press.

28.18 F. A. Murphy, Centers for Disease Control, Atlanta.

29.1 D. J. Bibel and T. H. Chen, *Bacteriol. Rev.* 40:633–651, 1976.

29.2 U.S. Army Medical Research Institute of Infectious Diseases, Fort Detrick, Frederick, MD.

29.3b Courtesy of Stanley F. Hayes, Rocky Mountain Laboratories, National Institute of Allergy and Infectious Diseases, Hamilton, MT.

29.4a, b Courtesy of Gregory A. McDonald, Rocky Mountain Laboratories, National Institute of Allergy and Infectious Diseases, Hamilton, MT.

29.7 J. J. Paulin, Dept. of Zoology, Univ. of Georgia.

29.14 R. W. Schlesinger, "Dengue Virus," *Virology Monographs,* Vol. 16, Springer, Wein-New York, 1977, p. 24.

Traveler's diarrhea, 412
Trematodes, *See* Flukes
  characteristics of, 243–244
  infections in humans, 244T
  metacercaria of, 243–244, 244F
  miracidium of, 243, 244F
  redia of, 243, 244F
  replication of, 243
Trench fever, 504
Trench mouth, 446–447
Treponema pallidum, 456, 457F
Treponema vincentii, 446–447
Triatomids, 506
TRIC agents, 462
Tricarboxylic acid cycle, 90, 91F
Trichinella spiralis, 433–434, 434F
Trichinosis, 433–434, 434F
Trichomonas vaginalis, as a STD, 456
Trichophytin, 481
Trichophyton, 480–481, 480F
Trichuris trichiura, 434–435, 435F
Trickle filter, 532, 532F
Triglyceride, 62F
Trimethoprim, action of, 161, 161F
Trophozoites, 508–509, 509F
Trypanosoma cruzi, 506–507
Trypanosoma gambiense, 504–506
Trypanosoma rhodesiense, 506
Trypanosomiasis
  American, 506–507, 507F
  Gambian, 504–506, 506F
  Rhodesian, 506, 506F
Trypomastigote, 505F
Tryptophan operon, control of, 104,
    105F
Tsetse fly, as a vector in trypanosomiasis,
    504
Tsutsugamushi fever, 503–504
Tubercle bacilli, *See* Mycobacterium
    tuberculos
Tuberculin, *See* PPD
Tuberculosis
  bovine, 374, 541
  diagnosis of, 373–374
  history of, 373
  immunity to, 372
  management of, 372–374
  tubercles of, 372
  vaccine for, 374
Tularemia, 496–497
Tumor antigens, 317, 317T
Tumor necrosis factor alpha, *See* TNF
Tumor suppressor genes, 222–223
Tumors, virus etiology of, 317T
Tyndall, John, 5–6
Typhoid fever, 418–419
  casualties from, 528
  vaccines for, 419
Typhoid Mary, 419
  life of, 253
Typhus
  Brill's disease and, 503
  endemic, 503
  epidemic, 502–503
  reservoir of, 502–503
Typing, bacteriophage, 188, 188F

Ulcers, peptic, 345
Ultraviolet light
  mutations by, 108–109, 109F
  sterilization with, 134–135, 135F
Undulant fever, *See* Brucellosis
Ureaplasma, 199
Ureaplasma urealyticum, 463
Urethra, normal flora of, 346, 344T
Urethritis, 455, 463
Urinary tract infections, 455
Urine
  bacteria in, 455
  collection of, 334–335

V-factor, 366
Vaccines
  allergic reactions from, 324
  attenuated, 322
  bacterial, 321–322
  booster of, 323
  DPT, 325
  duration of immunity from, 323–324
  immunity induced by, 323
  polio, 323
  rabies
  recombinant, 322–323, 557
  rickettsial, 322
  schedule for use, 324T
  subunit, 322
  toxoids, 322
  types of, 321–322, 323T
  viral, 322–323
Vaccinia virus
  current uses of, 401
  use in recombinant vaccines, 322–323
Vagina, normal flora of, 346, 344T
Vaginitis, 456
Vancomycin, 168
Variable major protein, in Borrelia, 498
Varicella, 398–400
  vaccine for, 400
Varicella-zoster virus, 398–400
Variola, *See* Smallpox
Variolation, 321
VDRL test, 458
Vegetables, fermented, 536–537
Vi antigen, 416
Vibrio cholerae
  biotypes of, 413
  morphology of, 413, 413F
  serotypes of, 413
Vibrio fluvialis, 415
Vibrio mimicus, 415
Vibrio parahaemolyticus, 414
Vibrio vulnificus, 415
Vincent's angina, 446–447
Vinegar, production of, 554–555
Viroids, 225
Virulence factors, 256–259, 260T
Viruses
  assay of, 210
  capsids of, 207
  characteristics of, 206
  classification of, 213, 214T, 215T
  envelopes of, 208

growth of, 209–210
inclusion bodies of, 212, 213F
latent infections by, 212–213
mutations in, 224
negative strand of, 210
nucleic acids of, 207
nucleocapsids of, 208
oncogenic, 218–219
plant, *See* Plant viruses
plaques of, 210, 212F
plus strand of, 210
properties of, 16–17
reaction of host cell to, 212,
    212T
replication of, 210, 211F
shapes of, 206–207, 207F, 216F
sizes of, 206–207
vaccine types of, 322–323
Visceral leishmaniasis, 508
Vitamin C, production of, 556
Vitamins, function in bacteria, 71T
Vodka, production of, 554, 555T
Voges-Proskauer test, 526
Volutin, 50, 368, 55T
Volvox, 230–231, 231F
von Prowazek, Stanislous, 8

Waksman, Selman, 162
Warts, genital, 465
Wasserman test, 458
Water
  activity of, 538
  completed test of, 525
  confirmed test of, 525
  diseases spread in, 528, 530T
  fecal coliforms in, 525
  fecal indicator for, 524
  flora of, 524
  plate count of, 524
  polluted, 524
  potable, 524
  presumptive test of, 525
  purification of, 529–530, 530F
Weil's disease, 477
Weil-Felix test, 499, 500T
West African sleeping sickness,
    504–506
Western equine encephalitis, 516–517
Wheel and flare reaction, 305
Whipworm, 434–435, 435F
Whiskey, production of, 554, 555T
Whooping cough, *See* Pertussis
Widal test, 418
Wine, production of, 551–553
Winogradsky, Serge, 7
Woese, Carl, 184
Woolsorters disease, 483
Wound botulism, 407–408
Wuchereria bancrofti, 511–512

X-factor, 366
Xenografts, 316, 316T

Yaws, 477
Yeasts